HIV-1: MOLECULAR BIOLOGY AND PATHOGENESIS

CLINICAL APPLICATIONS

HIV-1: MOLECULAR BIOLOGY AND PATHOGENESIS

CLINICAL APPLICATIONS
The Second of a Two-Volume Set

Edited by

Kuan-Teh Jeang
Laboratory of Molecular Biology
NIAID/NIH
Bethesda, Maryland

ADVANCES IN
PHARMACOLOGY

VOLUME 49

ACADEMIC PRESS
A Harcourt Science and Technology Company
San Diego   San Francisco   New York   Boston   London   Sydney   Tokyo

Academic Press
A Harcourt Science and Technology Company
525 B Street, Suite 1900, San Diego, California 92101-4495, USA
http://www.academicpress.com

Academic Press
Harcourt Place, 32 Jamestown Road, London NW1 7BY, UK
http://www.academicpress.com

International Standard Book Number: 0-12-032950-6

PRINTED IN THE UNITED STATES OF AMERICA
00   01   02   03   04   05   EB   9   8   7   6   5   4   3   2   1

Contents

HIV Therapeutics: Past, Present, and Future

Osama Abu-Ata, Jihad Slim, George Perez, and Stephen M. Smith

HIV Drug Resistance and Viral Fitness

François Clavel and Fabrizio Mammano

Inhibitors of HIV-1 Reverse Transcriptase

Michael A. Parniak and Nicolas Sluis-Cremer

HIV-1 Protease: Maturation, Enzyme Specificity, and Drug Resistance

John M. Louis, Irene T. Weber, József Tözsér, G. Marius Clore, and Angela M. Gronenborn

HIV-1 Integrase Inhibitors: Past, Present, and Future

Nouri Neamati, Christophe Marchand, and Yves Pommier

Selection of HIV Replication Inhibitors: Chemistry and Biology

Seongwoo Hwang, Natarajan Tamilarasu, and Tariq M. Rana

Therapies Directed against the Rev Axis of HIV Autoregulation

Andrew I. Dayton and Ming Jie Zhang

HIV-I Gene Therapy: Promise for the Future

Ralph Dornburg and Roger J. Pomerantz

Assessment of HIV Vaccine Development: Past, Present, and Future

Michael W. Cho

HIV-1-Associated Central Nervous System Dysfunction

Fred C. Krebs, Heather Ross, John McAllister, and Brian Wigdahl

Molecular Mechanisms of Human Immunodeficiency Virus Type I Mother-Infant Transmission

Nafees Ahmad

Molecular Epidemiology of HIV-1: An Example of Asia

Mao-Yuan Chen and Chun-Nan Lee

Simian Immunodeficiency Virus Infection of Monkeys as a Model System for the Study of AIDS Pathogenesis, Treatment, and Prevention

Vanessa M. Hirsch and Jeffrey D. Lifson

Animal Models for AIDS Pathogenesis

John J. Trimble, Janelle R. Salkowitz, and Harry W. Kestler

Contributors

Numbers in parentheses indicate the pages on which the authors' contributions begin.

Osama Abu-ata (1) Department of Internal Medicine, Saint Michael's Medical Center, Newark, New Jersey 07102 and Seton Hall School of Graduate Medical Education, Seton Hall University, South Orange, New Jersey 07079

Nafees Ahmad (387) Department of Microbiology and Immunology, College of Medicine, The University of Arizona Health Sciences Center, Tucson, Arizona 85724

Mao-Yuan Chen (417) Department of Internal Medicine, National Taiwan University Hospital, Taipei, Taiwan 10016

Michael W. Cho (263) AIDS Vaccine Research and Development Unit, Laboratory of Molecular Microbiology, National Institute of Allergy and Infectious Diseases, National Institutes of Health, Bethesda, Maryland 20892

François Clavel (41) Laboratoire de Recherche Antivirale, IMEA/INSERM, Hôpital Bichat-Claude Bernard, 75108 Paris, France

G. Marius Clore (111) Laboratory of Chemical Physics, National Institute of Diabetes, Digestive and Kidney Diseases, National Institutes of Health, Bethesda, Maryland 20892-0580

Andrew I. Dayton (199) Laboratory of Molecular Virology, Division of Transfusion Transmitted Disease, Office of Blood Research and Review, Center for Biologics Evaluation and Research, Food and Drug Administration, Rockville, Maryland 20852-1448

Ralph Dornburg (229) The Dorrance H. Hamilton Laboratories, Center for Human Virology, Division of Infectious Diseases, Jefferson Medical College, Thomas Jefferson University, Philadelphia, Pennsylvania 19107

Angela M. Gronenborn (111) Laboratory of Chemical Physics, National Institute of Diabetes, Digestive and Kidney Diseases, National Institutes of Health, Bethesda, Maryland 20892-0580

Vanessa M. Hirsch (437) Laboratory of Molecular Microbiology, NIAID, National Institutes of Health, Twinbrook II Facility, Rockville, Maryland 20852 and Laboratory of Retroviral Pathogenesis, AIDS Vaccine Program, SAIC Frederick, NCI/FCRDC, Frederick, Maryland 21702

Seongwoo Hwang (167) Department of Pharmacology, Robert Wood Johnson Medical School, Piscataway, New Jersey 08854

Harry W. Kestler (479) Department of Science and Mathematics, Lorain County Community College, Elyria, Ohio 44035

Fred C. Krebs (315) The Pennsylvania State University, College of Medicine, Department of Microbiology and Immunology, Hershey, Pennsylvania 17033

Chun-Nan Lee (417) School and Graduate Institute of Medical Technology, College of Medicine, National Taiwan University, Taipei, Taiwan 10016

Jeffrey D. Lifson (437) Laboratory of Molecular Microbiology, NIAID, National Institutes of Health, Twinbrook II Facility, Rockville, Maryland 20852 and Laboratory of Retroviral Pathogenesis, AIDS Vaccine Program, SAIC Frederick, NCI/FCRDC, Frederick, Maryland 21702

John M. Louis (111) Laboratory of Chemical Physics, National Institute of Diabetes, Digestive and Kidney Diseases, National Institutes of Health, Bethesda, Maryland 20892-0580

Fabrizio Mammano (41) Laboratoire de Recherche Antivirale, IMEA/INSERM, Hôpital Bichat-Claude Bernard, 75108 Paris, France

Christophe Marchand (147) Laboratory of Molecular Pharmacology, Division of Basic Sciences, National Cancer Institute, Bethesda, Maryland 20892

John McAllister (315) The Pennsylvania State University, College of Medicine, Department of Microbiology and Immunology, Hershey, Pennsylvania 17033

Nouri Neamati (147) Laboratory of Molecular Pharmacology, Division of Basic Sciences, National Cancer Institute, Bethesda, Maryland 20892

Michael A. Parniak (67) McGill University AIDS Centre, Lady Davis Institute for Medical Research, Montreal, Quebec H3T, IE2, Canada

George Perez (1) Department of Internal Medicine, Saint Michael's Medical Center, Newark, New Jersey 07102 and Seton Hall School of Graduate Medical Education, Seton Hall University, South Orange, New Jersey 07079

Roger J. Pomerantz (229) The Dorrance H. Hamilton Laboratories, Center for Human Virology, Division of Infectious Diseases, Jefferson Medical College, Thomas Jefferson University, Philadelphia, Pennsylvania 19107

Yves Pommier (147) Laboratory of Molecular Pharmacology, Division of Basic Sciences, National Cancer Institute, Bethesda, Maryland 20892

Esther Race (41) Laboratoire de Recherche Antivirale, IMEA/INSERM, Hôpital Bichat-Claude Bernard, 75108 Paris, France

Tariq M. Rana (167) Department of Pharmacology, Robert Wood Johnson Medical School, Piscataway, New Jersey 08854

Heather Ross (111) The Pennsylvania State University, College of Medicine, Department of Microbiology and Immunology, Hershey, Pennsylvania 17033

Janelle R. Salkowitz (479) Case Western Reserve University, Department of Infectious Diseases, Cleveland, Ohio 44106

Jihad Slim (1) Department of Internal Medicine, Saint Michael's Medical Center, Newark, New Jersey 07102 and Seton Hall School of Graduate Medical Education, Seton Hall University, South Orange, New Jersey 07079

Nicolas Sluis-Cremer (67) McGill University AIDS Centre, Lady Davis Institute for Medical Research, Montreal, Quebec H3T, IE2, Canada

Stephen M. Smith (1) Department of Internal Medicine, Saint Michael's Medical Center, Newark, New Jersey 07102 and Seton Hall School of Graduate Medical Education, Seton Hall University, South Orange, New Jersey 07079

Natarajan Tamilarasu (167) Department of Pharmacology, Robert Wood Johnson Medical School, Piscataway, New Jersey 08854

József Tözsér (111) Department of Biochemistry and Molecular Biology, University Medical School of Debrecen, H-4012, Debrecen, Hungary

John J. Trimble (479) Biology Department, Saint Francis College, Loretto, Pennsylvania 15940

Irene T. Weber (111) Department of Microbiology and Immunology, Kimmel Cancer Center, Thomas Jefferson University, Philadelphia, Pennsylvania 19107

Brian Wigdahl (315) The Pennsylvania State University, College of Medicine, Department of Microbiology and Immunology, Hershey, Pennsylvania 17033

Ming Jie Zhang (199) Laboratory of Molecular Virology, Division of Transfusion Transmitted Disease, Office of Blood Research and Review, Center for Biologics Evaluation and Research, Food and Drug Administration, Rockville, Maryland 20852-1448

Preface

As we enter a new millennium, it is clear that the AIDS pandemic is not going away. United Nations AIDS statistics indicate that in 1999 alone 2.6 million individuals succumbed to this disease. Cumulatively, since 1980, 16 million have died from AIDS, and despite remarkable medical advances, HIV-1 infections remain on the increase. Today, more than 33 million persons globally are living with HIV-1; 5.6 million of these were newly infected in 1999. Regrettably, the disease burden is highest in nations that have the most limited medical resources. Hence, 25% of all adults in Botswana, Zimbabwe, Swaziland, and Namibia are AIDS carriers. Every minute, in sub-Saharan Africa, 10 new individuals become infected with HIV-1. Worldwide, 95% of HIV-1 infections are in developing countries.

In view of these rather daunting numbers, one might think that researchers have made little progress in the study of HIV. This, in fact, is far from the truth. Since its initial identification and isolation in 1983, HIV-1 has become one of the best-elucidated viruses. Arguably, today we understand more about the workings of HIV-1 than of any other virus. In parallel, chemotherapeutic advances in the treatment of AIDS have also been impressive. From the combined efforts of many investigators, we currently have a large armamentarium of specific anti-HIV-1 reverse transcriptase (RT) and protease inhibitors. These antivirals work. Mortality from AIDS in developed countries that use RT and protease inhibitors has been significantly reduced. However, it is equally evident that, as yet, no chemotherapeutic regimen is curative.

How then might one view the AIDS question in the coming years? Better chemical antivirals are unlikely to be the final answer. The cost of chemical antivirals (currently around US$ 20,000 per person per year) makes this route prohibitive for developing nations. Chemical antivirals also have serious side effects that render questionable the ability of patients to sustain lifelong therapy. Additionally, we do not yet fully understand the ultimate scope of multi-drug-resistant viruses that are emerging in RT- and protease-inhibitor-treated individuals. Considered thusly, it is understandable how one prevailing view is that a practical, globally applicable solution for AIDS rests with the development of effective mass vaccination. However, the possibility that there are other yet-thought-of means for resolving the AIDS pandemic cannot be excluded.

This two-volume set of *Advances in Pharmacology* brings together 26 teams of authors for the purpose of describing where we have been in HIV-1 research and to explore where we might want to go in the future. The goal was to combine expositions on fundamental mechanisms of viral expression and replication with findings on viral pathogenesis in animal models and applications of chemotherapeutics in human patients. The authors were asked to survey the structures and functions of all the open reading frames of the HIV-1 genome as well as the roles of several noncoding regulatory RNA sequences. More importantly, each was encouraged to propose new ways for therapeutic intervention against HIV-1. In this regard, many interesting and novel ideas on HIV vaccines, gene therapy, and small-molecule inhibitors for viral envelope–cell fusion, viral assembly, integration, and gene expression are presented.

It is no accident that most of the authors in this set are some of the younger (late-30s, mid-40s), albeit authoritative, researchers on HIV. Likely, AIDS will defy a quick solution. Considering this and the fact that a major goal of this set is to explore new ideas rather than simply review past progress, I particularly wanted to assemble colleagues who, after having proposed interesting solutions for HIV, will be around to test, refine, and execute those ideas even if such were to require 10, 20, or 30 years. With some measure of luck, I and most of my co-authors will be here when the AIDS pandemic is solved. I wait with anticipation and interest to see whose ideas raised in these two volumes will be the ones that withstand the test of time.

Putting together this set has been an interesting learning experience for me. I thank all the authors for their enthusiastic participation. At the outset, I had thought that it would be difficult to recruit a sufficient number of busy researchers to write chapters for this project. However, I was pleasantly surprised when 26 of the 30 invited colleagues promptly agreed to contribute. Along the way, another colleague remarked to me that he did not want to write a chapter because "nobody important reads HIV books anymore." That comment taken, it remains my hope that some "unimportant" readers of these two volumes might nevertheless be spurred by its content to do important work in the fight against AIDS.

In closing, I thank Tom August for inviting me to edit these volumes. I am grateful to the late George Khoury, who asked me 14 years ago to work on the HIV Tat protein, and to Malcolm Martin, with whom I have discussed and debated various aspects of HIV biology for the past 12 years. Tari Paschall, Judy Meyer, and Destiny Irons from Academic Press have been wonderfully helpful, and Michelle Van and Lan Lin have provided excellent secretarial assistance. Finally, I appreciate the endless patience and understanding of my wife (Diane) and children (David, Diana, and John), who have, year-upon-year, put up with the abnormal working hours of an HIV researcher.

Kuan-Teh Jeang
January 14, 2000

Osama Abu-ata
Jihad Slim
George Perez
Stephen M. Smith
Department of Internal Medicine
Saint Michael's Medical Center
Newark, New Jersey 07102 and
Seton Hall School of Graduate Medical Education
Seton Hall University
South Orange, New Jersey 07079

HIV Therapeutics: Past, Present, and Future

Since the early 1990s, the number of available drug therapies for human immunodeficiency virus type 1 (HIV-1) infection has increased from 1 to 14. The number of drug types or classes has simultaneously increased from one to three. With this explosion in pharmaceutical armamentarium, potent anti-HIV-1 combination therapies became possible. In 1994, there was debate over whether drug therapy conferred long-term benefit to HIV-1-infected patients (1994). By 1996, researchers were suggesting that with prolonged combination drug therapy, eradication of HIV-1 might be possible (Perelson *et al.*, 1997). While the initial hypothesis regarding eradication of HIV-1 has proved to be wrong (Finzi *et al.*, 1999), many patients can now achieve indefinite suppression of HIV-1 replication using combination therapy (Stanley and Kaplan, 1998).

This chapter is an overview of antiretroviral therapy in three parts. Under the first section we discuss the individual drugs and their pharmacology, including mechanism of action, pharmacokinetics, major adverse ef-

Advances in Pharmacology, Volume 49
1054-3589/00 $35.00

TABLE I   Nucleoside Reverse Transcriptase Names, Abbreviations, and Formulations

Compound	Abbreviation	Trade name	Analog of	Formulation(s)
Zidovudine[a]	ZDV[b] or AZT	Retrovir	Deoxythymidine	100-mg capsules, 300-mg tablet; 10-mg/ml IV solution; 10-mg/ml oral solution
Lamivudine[a]	3TC[b]	Epivir	Deoxycytidine	150 mg tablets; 10-mg/ml oral solution
Didanosine	ddI[b]	Videx	Deoxyadenosine	25-, 50-, 100-, 150-mg tablets, 167-, 250-mg sachets
Zalcitabine	ddC[b]	HIVID	Deoxycytidine	0.375-, 0.75-mg tablets
Stavudine	d4T[b]	Zerit	Deoxythymidine	15-, 20-, 30-, 40-mg tablets
Abacavir	ABC[b]	Ziagen	Deoxyguanosine	300-mg tablets; 20-mg/ml oral solution

[a] Zidovudine (300 mg) and lamivudine (150 mg) formulated together in Combivir tablets.
[b] Abbreviation used in text of article.

fects, and significant drug interactions. Under the second section we summarize the major clinical trials of anti-HIV-1 compounds in adults. Finally, under the third section we discuss the current role of these drugs in the management of HIV-1 infection of adults.

I. Individual Drugs and Their Pharmacology

HIV-1 has three viral enzymes, reverse transcriptase, protease, and integrase; each of these enzymes is essential in the virus life cycle (Watkins *et al.*, 1995). To date, anti-HIV-1 drugs have targeted reverse transcriptase and the viral protease. As mentioned above, there currently are three classes of anti-HIV-1 drugs:

1. Nucleoside reverse transcriptase inhibitors (NRTIs) (Table I).
2. Nonnucleoside reverse transcriptase inhibitors (NNRTIs) (Table II).
3. Protease inhibitors (PIs) (Table III).

TABLE II   Nonnucleoside Reverse Transcriptase Inhibitors Names and Abbreviations

Compound	Abbreviation	Trade name	Formulation
Nevirapine	NVP	Viramune	200-mg tablets
Delavirdine	DLV	Rescriptor	100-mg tablets
Efavirenz	EFV	Sustiva	50-, 100-, 200-mg capsules

TABLE III   Protease Inhibitors Names, Abbreviations, and Formulations

Compound	Abbreviation	Trade name	Formulation(s)
Ritonavir	RTV	Norvir	100-mg capsules; 600-mg/ 7.5-ml oral solution
Saquinavir-HGC[a]	SQV-HGC	Invirase	200-mg capsules
Saquinavir-SGC[a]	SQV-SGC	Fortovase	200-mg capsules
Indinavir	IDV	Crixivan	200-, 250-, 400-mg capsules
Nelfinavir	NFV	Viracept	250-mg tablets; 50-mg/ gm oral powder
Amprenavir	AMP	Agenerase	50-, 150-mg capsules

[a] Saquinavir is available in hard-gel capsules (HGC) and soft-gel capsules (SGC).

For convenience, dosing recommendations in patients with renal or liver dysfunction are summarized in Table IV. Since antiretrovirals are usually used in combination, we have included Table V, which summarizes drug interactions between these compounds.

A. Nucleoside Reverse Transcriptase Inhibitors (NRTIs)

NRTIs were the first anti-HIV-1 drugs developed. This class consists of six Food and Drug Administration (FDA)-approved drugs. As shown in Table I, each drug has a commonly used abbreviation, a compound name, and a trade name. For instance, the drug zidovudine is more frequently referred to as ZDV or AZT. Each NRTI is referred to by its abbreviation (see Table I); the other drugs are referred to by their generic names (Tables II and III). The NRTIs suppress HIV-1 replication via a similar mechanism of action. Each of these drugs is an analog of a cellular nucleoside. As nucleoside analogs, the drugs are incorporated into the proviral DNA by the viral enzyme reverse transcriptase (Furman *et al.*, 1986). However, nucleoside analog incorporation results in chain termination, since the analogs lack the 3′ -OH group and cannot be linked to additional nucleosides. In this way, nucleoside analogs terminate reverse transcription. Without reverse transcription, the virus life cycle is also terminated. The mechanism of action of ZDV is discussed in detail as an example of all nucleoside-analog inhibitors.

B. Zidovudine (ZDV)

ZDV was the first agent approved for treating HIV-1 infection (McLeod and Hammer, 1992). ZDV was synthesized in 1964 as a potential therapeutic (Horwitz *et al.*, 1964). In 1974, it was shown that ZDV inhibited a retrovirus

TABLE IV Oral Dosage Recommendations for Antiretroviral Agents in Patients with Organ Dysfunction[a]

Drug	Dosage in patients with renal dysfunction					Dosage in patients with hepatic dysfunction	References
	CLcr,[b] 50 ≥ ml/min	CLcr, 26-49 ml/min	CLcr, 10-25 ml/min	CLcr, < 10 ml/min	Hemodialysis		
Zidovudine	200 mg q 8 h	200 mg q 8 h	100 mg q 8 h	100 mg q 8 h	100 mg q 8 h	100 mg q 8 h	1998h; Bareggi et al. (1994); Dudley (1995); Fletcher et al. (1992); Gallicano et al. (1992); Kremer et al. (1992); Taburet et al. (1990)
Didanosine	200 mg q 12 h	200 mg q 24 h	100 mg q 24 h	100 mg q 24 h	100 mg q 24 h	Consider empirical dosage reduction	1998j; Singlas et al. (1992)
Zalcitabine	0.75 mg q 8 h	0.75 mg q 12 h	0.75 mg q 12 h	0.75 mg q 24 h	0.75 mg q 24 h	0.75 mg q 8 h	1998d
Stavudine	40 mg q 12 h	40 mg q 24 h	20 mg q 24 h	20 mg q 24 h	20 mg q 24 h	40 mg q 24 h	1998m
Lamivudine	150 mg q 12 h	150 mg q 24 h	100 mg q 24 h[c]	25–50 mg q 24 h[c]	25–50 mg q 24 h[c]	150 mg q 12 h	1998b
Abacavir	300 mg q 12 h	NR[d]	NR	NR	NR	NR	1999i
Nevirapine	200 mg q 12 h	200 mg q 12 h	200 mg q 12 h	200 mg q 12 h	NR	Consider empirical dosage reduction	1998l; Havlir et al. (1995)

Drug							Recommendation	Reference
Delavirdine	400 mg q 8 h	400 mg q 8 h	400 mg q 8 h	NR	400 mg q 8 h	NR	Consider empirical dosage reduction	1998g
Efavirenz	600 mg q 24 h	NR	NR	NR	NR	NR	NR	1999h
Saquinavir	600 mg q 8 h	600 mg q 8 h	600 mg q 8 h	NR	600 mg q 8 h	NR	Consider empirical dosage reduction	1998c
Ritonavir	600 mg q 12 h	600 mg q 12 h	600 mg q 12 h	NR	600 mg q 12 h	NR	Consider empirical dosage reduction	1998f
Indinavir	800 mg q 8 h	800 mg q 8 h	800 mg q 8 h	NR	800 mg q 8 h	NR	Mild-moderate dysfunction: 600 mg q 8 h; Severe dysfunction: Consider further dosage reduction	1998a; Balani et al. (1996); Lin et al. (1996)
Nelfinavir	750 mg q 8 h	750 mg q 8 h	750 mg q 8 h	NR	750 mg q 8 h	NR	Consider empirical dosage reduction	1998k
Amprenavir	1200 mg q 12 h	NR[e]	NR	NR	NR	NR	300–450 mg q 12 h for severe-moderate dysfunction	1999a

[a] Originally published in Hilts and Fish (1998)© American Society of Health-System Pharmacists, Inc. All rights reserved. Adapted with permission (R9934).
[b] CLcr, creatinine clearance.
[c] Initial dose of 150 mg and then followed by listed dosing regimen.
[d] Not reported.
[e] No reported recommendations, but <3% amprenavir is excreted renally.

TABLE V Effect of Anti-HIV Drugs on Other Anti-HIV Drugs[a,b]

	ZDV	ddI[c]	d4T	Nevirapine	Delavirdine	Efavirenz	Saquinavir	Ritonavir	Indinavir	Nelfinavir	Amprenavir	Reference
ZDV	—	—	A[d]	⇧	—	—	—	⇧	—	—	⇧	1998I; Cato et al. (1998); Miller et al. (1998); Sadler et al. (1998)
ddI	—	—	—	—	—	—	—	—	—	—	—	Cato et al. (1998)
d4T	A[d]	—	—	—	—	—	⇧	⇧	—	—	—	Miller et al. (1998)
Nevirapine	—	⬆	—	—	—	—	⇧	—	—	—	—	Sahai et al. (1997)
Delavirdine	—	—	—	—	—	—	⇧	—	—	⇧	—	1998g
Efavirenz	—	—	—	—	⇧	—	—	⇧	—	—	—	1999h; Fiske et al. (1998a)
Saquinavir (SQV)[e]	—	—	—	—	—	Not to be used together	—	Dose: SQV 400 mg BID ⬆	⬆	Dose: SQV 800-mg TID ⬆	—	1999h; Buss (1998); Cox et al. (1997); Merry et al. (1997); Sahai et al. (1997)
Ritonavir	—	—	—	⇧	⇧	⇧	⇧	—	—	—	—	Fiske et al. (1998a); Morse et al. (1998); Murphy et al. (1997)
Indinavir (IDV)	—	⬆	—	⇧	Dose: IDV 600 mg TID ⬆	Dose: IDV 1000-mg TID ⬆	—	Dose: IDV 400 800-mg BID ⬆	—	⬆	—	1998a; Cox et al. (1997); Hsu et al. (1998); Murphy et al. (1999); Riddler et al. (1997)
Nelfinavir	—	—	—	—	⇧	⇧	⇧	⇧	⬆	—	—	Cox et al. (1998); Fiske et al. (1998b); Murphy et al. (1997); Sahai et al. (1997)
Amprenavir	—	—	—	—	—	⇧	—	—	⬆	—	—	Piscitelli et al. (1998); Sadler et al. (1998)

[a] ⇧, Clinical significance not established or clinically insignificant changes in "row" drug level; ⬆, Clinically significant decrease or increase in "row" drug level.

[b] Drug interactions reviewed in Guidelines (1999d).

[c] ddI only interferes with absorption when administered within 2 h of "row" drug.

[d] ZDV and d4T antagonize each other and should not be used together.

[e] Saquinavir dosage refers to Fortovase formulation only.

in tissue culture (Ostertag *et al.*, 1974). In 1985, investigators showed that ZDV inhibited HIV-1 replication *in vitro* (Mitsuya *et al.*, 1985). This observation led to clinical trials and finally to FDA approval in 1987.

1. Mechanism of Action

ZDV is a deoxynucleoside that is structurally related to thymidine, but differs in having an azido (-N3) group in place of the $3'$ -OH group (1998h). ZDV enters the cell by passive diffusion and is phosphorylated via three cellular kinases into its biologically active triphosphate form (zidovudine triphosphate). The second phosphorylation step (by the enzyme thymidylate kinase) is the rate-limiting step. ZDV triphosphate is recognized by HIV-1 RT as a thymidine analog and is incorporated into proviral DNA in place of deoxythymidine triphosphate. As above, once ZDV is incorporated into proviral genome, the DNA cannot be extended. ZDV monophosphate may also decrease viral replication by inhibiting the RNase activity of reverse transcriptase. ZDV only inhibits cellular DNA polymerases at concentrations 50- to 100-fold greater than required to inhibit reverse transcriptase.

2. Pharmacokinetics

ZDV is well absorbed after oral administration with 60–65% bioavailability (1998h). The serum half-life is 1–1.5 h; however, the intracellular half-life is 3 h. The ratio of the cerebrospinal fluid (CSF) concentration of ZDV to plasma concentration ranges from 0.1 to 1.35 (average = 0.6). This wide range reflects differences in the blood–brain barrier permeability in HIV-1-infected patients. ZDV is primarily metabolized in the liver to $3'$-azido-$3'$deoxy $5'$-O-B-D-glucopyranuronosylthymidine (GZDV). GZDV has no antiviral activity and is excreted renally. ZDV (8–15%) is also excreted renally. Dosing regimens have varied over the years. Initially ZDV was given 250 mg every 4 h. However, with the recognition of the long intracellular half-life, ZDV is currently given as 200 mg every 8 h (TID) or 300 mg every 12 h (BID). ZDV (300 mg) is also formulated with lamivudine, also known as 3TC, (150 mg) in Combivir tablets, which are given one tablet BID. Dosage reduction is suggested for patients with severe liver disease and recommended for patients with end-stage renal disease (Table IV).

3. Adverse Effects

The toxicity of ZDV has lessened with lower dose regimens (Fischl *et al.*, 1990a). In the early studies, hematologic abnormalities, especially anemia and granulocytopenia, were common (30% of patients) (Fischl *et al.*, 1987). Myelosuppression is worsened by coadministration of ganciclovir or interferon-α (1998h). With the lower dose regimen, these abnormalities are rare (<2%). Macrocytosis occurs in more than 90% of patients on ZDV, but does not correlate with the development of anemia (McLeod and Hammer, 1992; Fischl *et al.*, 1990a). In fact, macrocytosis or its absence can be used to assess whether the patient is taking the ZDV. Myopathy occurs in 6–18% of patients who have been taking the drug for more than 6 months (Helbert *et al.*, 1988; Till and MacDonell, 1990). This process can progress insidiously. Discontinuation of the drug results in gradual resolution of symptoms over a 6- to 8-week period in >70% of patients. Rare occurrences of fatal lactic acidosis associated with hepatic steatosis have been reported (1998h).

4. Drug Interactions

ZDV has few drug–drug interactions. Ribavirin inhibits ZDV phosphorylation and these drugs should not be used together. ZDV should not be used with d4T because the two drugs may antagonize each other (Miller *et al.*, 1998).

C. Didanosine (ddI)

I. Pharmacokinetics

Didanosine has an oral bioavailability of 33% (1998j). Acid degrades ddI, which must be coadministered with an antacid. The commercial tablet is buffered with calcium carbonate and magnesium hydroxide. The CSF to plasma ratio is 0.21. The plasma half-life is 1.6 h and the intracellular half-life is 25–40 h. The drug is cleared by renal mechanisms and is excreted largely unchanged. ddI should be taken 0.5 h before or 1 h after a meal. The usual dose is 200 mg twice per day for individuals >60 kg and 125 mg twice per day for those <60 kg. However, ddI has been used effectively at twice the regular dose given once per day (Federici *et al.*, 1998; Hoetelmans *et al.*, 1998).

2. Adverse Effects

The most serious of ddI is pancreatitis, which has been fatal (1998j). Pancreatitis occurs in 6–7% of patients on ddI (Dolin *et al.*, 1995; Kahn *et al.*, 1992). As with ZDV, acute, fulminant liver failure associated with lactic acidosis has been reported (Kahn *et al.*, 1992). ddI also causes peripheral neuropathy commonly (14–20%) (Dolin *et al.*, 1995; Kahn *et al.*, 1992). Manifestations of this neuropathy include numbness and tingling sensations in hands and feet. Symptoms usually improve with drug discontinuation.

3. Drug Interactions

The buffer, which is given with ddI, can interfere with the absorption of certain drugs, such as ketoconazole, dapsone, some quinolones, and tetracyclines (1998j). These drugs should be administered 2 h before or after ddI. Methadone decreases ddI levels by 41% and ddI dose increase should be consider when the two drugs are used together.

D. Dideoxycytidine (ddC)

1. Pharmacokinetics

ddC has excellent oral bioavailability (>80%) (1998d). The CSF to plasma ratio is 0.09:0.37. ddC is excreted primarily unchanged through the kidneys. The serum half-life is 1.2 h and the intracellular half-life is 3 h. The normal dose is 0.75 mg (one tablet) TID.

2. Adverse Effects

The most frequent adverse reaction is peripheral neuropathy, which occurs in up to 30% (Fischl *et al.*, 1995). This neuropathy is usually a sensorimotor neuropathy manifested often by a burning sensation and pain. These symptoms may be irreversible, but usually resolve slowly after drug discontinuation. Pancreatitis and hepatic toxicity have also been reported (Fischl *et al.*, 1995).

3. Drug Interactions

ddC has minimal drug interactions (1998d).

E. Stavudine (d4T)

1. Pharmacokinetics

d4T has excellent oral bioavailability (>80%) (Dudley, 1995; Lea and Faulds, 1996). d4T is eliminated through renal (40%) and unknown mechanisms. The serum half-life is 1.0 h, but it has an intracellular half-life of 3.5 h, which allows twice-daily dosing. Dosage should be adjusted for renal failure. The usual dose is 40 mg (one capsule) BID for weight >60 kg and 30 mg BID for weight <60 kg.

2. Adverse Effects

The toxicities of d4T are similar to ddI and ddC (Skowron, 1995). d4T can cause peripheral neuropathy in 15% of patients. Pancreatitis occurs uncommonly.

3. Drug Interactions

As above, d4T should not be coadministered with ZDV because of potential antagonism (Miller *et al.*, 1998).

F. Lamivudine (3TC)

1. Pharmacokinetics

3TC is well absorbed orally (>80%) (1998b). Absorption is unaffected by food. It is excreted in the urine unchanged. The serum half-life of 3TC is 3–6 h and the intracellular half-life is 12 h. The usual dose is 150 mg (one tablet) BID.

2. Adverse Effects

3TC is well-tolerated and few adverse effects are seen. As with other nucleoside analogs, lactic acidosis and severe hepatomegaly has been reported. Compared with ZDV monotherapy, a 3TC and ZDV combination had no increased incidence of adverse effects (Eron *et al.*, 1995).

3. Drug Interactions

Trimethoprim (160 mg)/sulfamethoxazole (800 mg) increases the level of 3TC (Moore *et al.*, 1996). The clinical relevance of this change is not clear.

G. Abacavir (ABC; 1592U89)

1. Pharmacokinetics

ABC had an 83% bioavailability and a CSF:plasma ratio of 0.27 (Parker *et al.*, 1993), (1999i). ABC is metabolized by alcohol dehydrogenase and its metabolites are excreted in the urine. The serum half-life of ABC is 1.5 h and the intracellular half-life is 3.3 h. The standard dose is 300 mg (one tablet) BID.

2. Adverse Effects

The most common side effects of ABC are gastrointestinal and CNS related (1999i). Nausea, vomiting, and/or diarrhea occur in up to 40%. Headache and dizziness can occur in 30%. However the most serious side effect is a hypersensitivity syndrome, which can occur in up to 5% of patients (1999f). Typically, this syndrome develops within the first 6 weeks of therapy. Patients with this syndrome can develop fever, fatigue, nausea, myalgia, shortness of breath, lymphadenopathy, and oral ulcers. The symptoms begin mildly, but quickly worsen with continued ABC usage. Laboratory tests may show increased liver function tests and increased creatinine. Liver and renal failure have also occurred in association with this hypersensitivity reaction. Symptoms improve rapidly once ABC is discontinued. However, if patients with this syndrome are rechallenged with ABC, death can occur. Patients who develop these symptoms while on ABC should be educated to seek medical attention immediately and to stop taking ABC.

3. Drug Interactions

No significant drug interactions have been reported with ABC.

H. Nonnucleoside Reverse Transcriptase Inhibitors (NNRTIs)

NNRTIs have been demonstrated to be potent and selective inhibitors of HIV-1 replication (reviewed in De Clercq, 1999). These drugs noncompetitively inhibit the viral enzyme RT. Presumably, NNRTIs bind to HIV-1 reverse transcriptase at a site away from the catalytic domain. This drug–enzyme interaction produces a conformational change which results in the inactivation of HIV-1 RT. These drugs have great selectivity for HIV-1 RT and are not active against HIV-2 or other animal retroviruses. As a class of drugs, listed in Table II, they are highly active *in vivo,* for short periods, but are limited by the relatively rapid emergence of resistant HIV-1 strains. NNRTIs do not require enzymatic modification to become active. NNRTIs also share a common toxicity, an eryathematous skin eruption, which is usually self-limited.

I. Delavirdine

1. Pharmacokinetics

Delavirdine is rapidly and readily absorbed after oral intake (Cheng *et al.,* 1997; 1998g). It can be administered with or without food. The usual adult dose of delavirdine is 400 mg (four 100-mg tablets) orally TID. Delavirdine inhibits its own metabolism by cytochrome P450 3A and consequently its half-life increases with increasing doses. In individuals on the usual dose, the half-life is 5.8 h.

2. Adverse Effects

Rash is the most frequent side effect and is seen in 18–36% (1998g; Para *et al.,* 1996). The rash is usually self-limited, occurs within 1–4 weeks of initiation of therapy, and is often pruritic and maculopapular eruptions over the upper half of the body. Rarely, hepatitis and neutropenia have been associated with delavirdine therapy.

3. Drug Interactions

Delavirdine interacts with many compounds. Antacids and ddI lower gastric pH and reduce the absorption of delavirdine (1998g). When used with ddI, delavirdine should be taken 1 h before or after ddI. Inducers of the cytochrome P450 system, including rifamycins, phenobarbital, phenytoin, and carbamazepine, increase the metabolism of delavirdine and should not be used concurrently (1998g). Conversely, delavirdine, via its inhibition of the P450 system, inhibits the metabolism of many drugs. Consequently, concurrent use of drugs, including cisapride, astemizole, ergot derivatives, terfenadine, simvastatin, lovastatin, and benzodiazepines, with delavirdine can result in serious side effects and should be avoided. Delavirdine may increase the levels of warfarin, dapsone, and quinidine.

J. Nevirapine

1. Pharmacokinetics

Nevirapine is well absorbed orally (>90%) (Lamson, 1993). Food does not alter its absorption. Nevirapine has a wide volume of distribution, can be found in breast milk, and crosses the placenta (1998l). Nevirapine is also metabolized by the P450 system and is also an inducer of the P450 system. Consequently, nevirapine therapy increases its own metabolism. After the first 14 days of therapy, the dose must be increased from 200 mg once per day to 200 mg (one tablet) twice per day. There are no data available on dosing nevirapine in renal or liver failure. The terminal half-life of nevirapine is 25–30 h (Lamson, 1993).

2. Adverse Effects

Rash is again the most common side effect and is seen in up to 35% of patients (D'Aquila *et al.*, 1996; Montaner *et al.*, 1998a; Carr *et al.*, 1996; 1998l). Like delavirdine, nevaripine's rash usually occurs in the first 6 weeks of therapy, is maculopapular, and is distributed over the torso and upper extremities. In clinical trials, 7% of patients stopped taking nevirapine because of the rash. Stevens-Johnson occurs in only 0.3%. Hepatitis occurs in 1%.

3. Drug Interactions

Although nevirapine induces the P450 system, there are few drugs that are absolutely contraindicated (1998l). Rifamycins will also induce the P450 system and can reduce the plasma concentration if used concurrently with nevirapine. When used together, dosage adjustment is necessary. Other inhibitors of the P450 system, such as ketoconazole, cimetidine, and macrolides, are not contraindicated. No dosage adjustment of nevirapine is necessary when these drugs are given together. When used with anticonvulsants, levels of the anticonvulsants should be monitored carefully. Methadone levels are decreased substantially by nevirapine and methadone dose should be titrated to the appropriate effect.

K. Efavirenz

1. Pharmacokinetics

Efavirenz is very well absorbed orally (1999h). Efavirenz has a long half-life and steady state is not reached until day 7 of therapy. Efavirenz is metabolized by the P450 isoenzymes Cyp3A4 and Cyp2B6. The terminal half-life of efavirenz is 40–55 h. The usual adult dose of efavirenz is 600 mg (three 200-mg capsules) orally once per day. A concurrent high-fat meal increases absorption of efavirenz and should be avoided. There are no data on dose adjustment in the setting of renal and hepatic insufficiency.

2. Adverse Effects

Again, skin rash is a frequent side effect of efavirenz and is seen in 27% of patients (1999h; 1998i). Typically, the rash occurs early in therapy, does not require cessation of therapy, and resolves within 1 month. More problematic are the central nervous system (CNS) side effects, which include dizziness, headache, insomnia, nightmares, amnesia, confusion, hallucinations, and euphoria. Though seen in 52% of patients, CNS side effects led to the discontinuation in only 2.6% because the CNS side effects resolve after a few days to weeks of therapy. Efavirenz can also cause mild gastrointestinal symptoms, such as nausea and diarrhea.

3. Drug Interactions

Efavirenz can induce the production of Cyp3A4, which increases the metabolism of many drugs. Conversely, rifampin induces the metabolism of efavirenz. The concurrent rifabutin dose should be increased to 450 mg per day. Certain drugs, including astemizole, terfenadine, cisapride, midazolam, triazolam, and ergot derivatives, should not be used with efavirenz. The prothrombin time should be monitored carefully in patients on warfarin and efavirenz. Efavirenz and saquinavir reduce each other's plasma levels and should not be used concurrently.

L. Protease Inhibitors (PIs)

HIV makes many of its structural and enzymatic proteins as part of a large polyprotein (Watkins *et al.*, 1995). The viral protease cleaves the polyprotein into the active smaller units. Proteolytic cleavage of the polyproteins is essential to virus infectivity. The HIV-1 protease activity occurs after virus budding. In the absence of proteolytic activity, the HIV-1 virus fails to mature and is not infectious. HIV-1 protease inhibitors act by blocking the viral protease. Hence, virions that are released in the presence of a protease inhibitor remain immature and cannot infect other cells. In this way the virus life cycle is terminated.

In the recent past, most antiviral drugs were discovered through empirical testing of many compounds [such as ribavirin (Potter *et al.*, 1976)] or through testing of analogs of necessary substrates (such as ZDV). However, each of the HIV-1 protease inhibitors was developed through a systematic process, which occurred simultaneously at several pharmaceutical companies (Wlodawer *et al.*, 1998). The protease enzyme was first crystallized and the three-dimensional structure of the protein was determined. Protease's structure was analyzed by computer. Inhibitors were designed through this computer analysis. After the chemicals were synthesized, they were tested *in vitro*. Finally, the candidate drugs were tested in humans. In short, the success of this process demonstrates the power of molecular biology in designing novel drugs to combat infectious diseases.

I. Mechanism of Activity

This class of anti-HIV-1 compounds includes ritonavir, saquinavir, indinavir, nelfinavir, and amprenavir (Table III). Each of these has a similar mechanism of action. As above, the protease inhibitors bind to the enzymatic site in the HIV-1 protease and reversibly inhibit the enzymatic activity of the HIV-1 protease (Kaul *et al.*, 1999). Viruses, which are produced in the presence of a protease inhibitor, are defective and cannot infect new cells.

2. Adverse Effects

Each of the protease inhibitors that have been studied to date has a low incidence of side effects that appear to be class related (Kaul *et al.*, 1999). These rare adverse reactions include hyperlipidemia, lipodystrophy, hemolytic anemia, hyperglycemia, and spontaneous bleeding in hemophiliacs. Each reaction has been associated with each of the protease inhibitors. The PI-associated hyperlipidemia is usually an elevation of the triglycerides. In PI-associated lipodystrophy, patients develop aberrant fat deposits around their waists and on the back of their necks ("buffalo humps"). This problem is more cosmetic than medical and may resolve with discontinuation of the PI. PI-associated hyperglycemia can be mild or can present as ketoacidosis with new onset diabetes mellitus.

3. Drug Interactions

Protease inhibitors interact and inhibit the P450 system to varying degrees. Certain drugs should not be coadministered or coadministered with caution in patients taking PIs. Table VI summarizes these contraindicated drugs.

M. Saquinavir

I. Pharmacokinetics

Saquinavir powdered capsules (Invirase) have a low oral bioavailability (4%) (1998e). A new formulation of soft gelatin capsules (Fortovase) has

TABLE VI   Drugs Contraindicated with HIV-1 Protease Inhibitors

Drug Class	Drugs not to be coadministered with protease inhibitors[a]
Antihistamines	Terfenadine, Astemizole
GI motility agents	Cisapride
Antimigraine	Egrot derivatives
Benzodiazepine	Midazolam, Triazolam
Antimycobacterial	Rifampin
Antiarrythmic agents[b]	Amiodarone, Quinidine
Lipid-lowering agents	Simvastatin, Lovastatin
Cardiovascular[b]	Bepridil

[a] List is not necessarily complete; for a complete listing see the product information sheets in the *Physicians Desk Reference*.
[b] Not absolutely contraindicated with each protease inhibitor.

an increased bioavailability of 12% (1998c). Saquinavir has no appreciable penetration of the blood–brain barrier and is 97% protein bound. Saquinavir is metabolized via the P450 system and excreted through the biliary system. The drug's plasma half-life is 1–2 h. The usual dose of Invirase is 400 mg BID and should only be used in combination with ritonavir. The dose of Fortovase is 1200 mg TID (six 200-mg capsules). Fortovase should be taken with food, which increases serum levels.

2. Adverse Effects

Saquinavir has a high frequency of adverse effects (1998c; Perry and Noble, 1998). It commonly causes gastrointestinal symptoms (10–20%), including diarrhea, nausea, and abdominal pain. Rarely, saquinavir usage has been associated with hemolytic anemia, thrombocytopenia, confusion, seizures, renal failure, and pancreatitis.

3. Drug Interactions

Inhibitors of CyP3A4, which metabolizes saquinavir, can increase the saquinavir plasma level. Rifampin and rifabutin, which induce CyP3A4, decrease saquinavir concentration by 80 and 40% respectively. Saquinavir itself inhibits CyP3A4 and therefore should not be administered with cisapride, astemizole, triazolam, and other drugs metabolized by this pathway (1998c).

N. Ritonavir

1. Pharmacology

Ritonavir is well absorbed orally (1998f). Administration with food has only a slight effect on absorption rates. Ritonavir is metabolized by the CyP3A4 enzyme and is also a strong inhibitor of this enzyme. The plasma half-life is 3–5 h. Most (86%) of the drug is excreted in the feces and 11% is excreted in the urine. The usual adult dose is 600 mg BID. However, the dose of ritonavir should be escalated gradually from 300 mg BID to 600 BID over 14 days. Ritonavir is available in an oral solution (80 mg/ml or 600 mg/7.5ml) and a new soft gel capsule (100 mg). Ritonavir should be taken with food, which may decrease ritonavir's tolerability. Ritonavir and ddI doses should be separated by 2 h.

2. Adverse Effects

The major side effects are gastrointestinal, including nausea, vomiting, diarrhea, and abdominal cramping (1998f). Gastrointestinal symptoms develop early in therapy and subside over time. Hepatitis with elevation of the transaminases is seen in 5–6% and is more common in patients with preexisting liver disease. Patients on ritonavir may also develop circumoral paresthesias. Hypertriglyceridemia occurs in 2–8% and hypercholesterolemia occurs in <2%.

3. Drug Interactions

Ritonavir is a potent inhibitor of the P450 isoenzymes (1998f). Coadministration of a drug metabolized by this system results in increased levels of that drug. Additionally, ritonavir may increase the activity of glucuronyl transferase. Consequently, ritonavir may decrease the level of drugs metabolized by this enzyme. Many drugs are contraindicated for patients on ritonavir (Table VI) (1998f).

O. Indinavir

1. Pharmacokinetics

Indinavir is well absorbed in the fasting state (1998a). Administration with food reduces absorption by approximately 60–70% (Yeh *et al.*, 1998). Indinavir is metabolized mainly through the liver and excreted in the feces, but 11% is excreted unchanged in the urine (Balani *et al.*, 1996). The primary pathway responsible for indinavir metabolism is the CyP3A4 enzyme system. The plasma half-life is 1.8–1.9 h. The usual dose of indinavir is 800 mg TID (two 400-mg capsules). In patients with mild to moderate hepatic insufficiency, the recommend dose is 600 mg TID (Table IV). Indinavir doses should be separated from ddI doses by 1 h.

2. Adverse Effects

The most problematic side effect of indinavir is nephrolithiasis, which occurs in 9% (1998a). The stones are made of the drug itself and can be prevented by hydration. Renal function is usually not affected by the nephrolithiasis, but hematuria and flank pain are common. To prevent stone formation, patients are instructed to consume 1.5 liters of liquids per day. Hyperbilirubinemia (indirect bilirubin >2.5 mg/dl) occurs in 10%, is usually asymptomatic, and often resolves spontaneously. Indinavir also causes gastrointestinal symptoms, which are tolerated by most patients.

3. Drug Interactions

Indinavir is metabolized by CyP3A4 enzyme and also inhibits CyP3A4. Consequently, drugs such as cisapride, terfenadine, astemizole, and others metabolized by CyP3A4, should not be administered with indinavir (1998a). Rifampin lowers indinavir levels by 90% and should not be used with indinavir. Coadministration of rifabutin with indinavir results in increased levels of rifabutin and decreased levels of indinavir. If given together, the rifabutin dose should be decreased to 150 mg per day and the indinavir dose should be increased to 1000 mg every 8 h. Efavirenz increases the metabolism of indinavir as well. The dose of indinavir should be increased to 1000 mg every 8 h when coadministered with efavirenz (Ruiz, 1997; Riddler *et al.*, 1998).

P. Nelfinavir

1. Pharmacokinetics

Nelfinavir is well absorbed orally with a bioavailability of 78% (Quart *et al.*, 1995; 1998k). Food enhances nelfinavir absorption. The plasma half-life of nelfinavir is 3.5–5 h. Nelfinavir is metabolized by the CyP3A4 enzyme. Its metabolites and nelfinavir itself (22%) are excreted in the feces. Little of the drug (1–2%) is excreted in the urine. The usual adult dose is 750 mg BID (three 250-mg tablets). Although not approved by the FDA, a regimen of 1250 mg BID has been effective in clinical trials (Johnson *et al.*, 1998). Nelfinavir should be taken with a meal or light snack, which improves absorption two- to threefold.

2. Adverse Effects

The most common side effect of nelfinavir is diarrhea, which is seen in 21% of patients (Powderly *et al.*, 1997; 1998k). The diarrhea is usually worse in the first few weeks of therapy, but may continue. Other side effects, including those discussed above, are infrequent.

3. Drug Interactions

Nelfinavir is metabolized by and also inhibits CyP3A4 enzymes. Drugs, which are metabolized by this system, are contraindicated in a patient taking nelfinavir (1998k). As with indinavir, rifabutin (50%) causes less of a reduction in nelfinavir levels than does rifampin (82%). Other inducers of the CyP3A4 should not be given with nelfinavir.

Q. Amprenavir

1. Pharmacokinetics

Amprenavir has an oral bioavailability of 25–40% and can be taken with or without food (1999a). Amprenavir is metabolized by the P450 system and has many metabolites, which are excreted in urine and feces. The plasma half-life of amprenavir is 7–9.5 h. The usual dose is 1200 mg BID (eight 150-mg capsules).

2. Adverse Effects

In clinical trials to date, the most common side effects are headache, nausea, diarrhea, and rash (1999a). Rash occurs in 18% of patients and develops in the second week of therapy. The rash is usually mild and resolves spontaneously.

3. Drug Interactions

Amprenavir is also an inhibitor of the P-450 CyP3A4 enzyme system. Therefore, patients on amprenavir should not take drugs metabolized by this pathway (1999a). Rifampin coadministration should be avoided as well.

II. Clinical Studies _____

This part of the Chapter summarizes the major clinical studies in chronological order. From the late 1980s to the late 1990s, monotherapy NRTI
therapy was replaced by dual NRTI therapy, which was replaced by triple
therapy. In addition to the development of new drugs, another finding
aided the study of new therapies. In 1989 investigators established that HIV
infection is a chronic, active infection (Coombs *et al.*, 1989; Ho *et al.*, 1989).
However, the concentration HIV in plasma was not well understood. In the
mid-1990s, a technique, branched-DNA or bDNA, was developed at Chiron
(Todd *et al.*, 1995). Branched DNA quantified the number of HIV RNA
genomes in the plasma. For the first time it was shown that HIV virion were
abundant in the plasma. In fact, billions of virions are produced each day
in an infected individual (Perelson *et al.*, 1996). Furthermore, Mellors *et al.*
established that the level of HIV in the plasma was strongly associated with
disease progression (Mellors *et al.*, 1996). Quite simply, the more virus
present in the plasma, the more rapidly an infected individual became immunodeficient. Other investigators and companies developed different techniques, such as quantitative reverse transcription-polymerase chain reaction
(RT-PCR; Roche) and nucleic acid sequence-based amplification (NASBA;
Organon-Teknika), to quantify virus level or load. Studies with these viral
load assays confirmed the relationship between the level of HIV in the
plasma to disease progression rate (Katzenstein *et al.*, 1996b).

While viral load studies established an important prognostic indicator,
these techniques also allowed investigators a way to follow the efficacy of
new therapies. HIV infection is a chronic, insidious disease. Without a good
surrogate marker, trials, especially early in the disease course, can take
years to reach statistical significance (Lange, 1995; DeGruttola *et al.*, 1998).
Plasma viral load measurements allow researchers to study the impact of
new therapies on viral production. In turn (as discussed below), it has been
shown that therapies, which dramatically lower virus levels, effect disease
progression (Katzenstein *et al.*, 1996a; Hammer *et al.*, 1997; Marschner *et
al.*, 1998). Therefore, trials are now designed to evaluate the effect of a new
therapy on the plasma viral load (Tavel *et al.*, 1999). End points, such as
time to an opportunistic infection or acquired immunodeficiency syndrome
(AIDS), are no longer used or are secondary.

Without therapy, most HIV-infected individuals have plasma virus levels
in the order 10^3–10^6 copies/ml (Mellors *et al.*, 1996). Effective antiretroviral
combination therapies can lower this level by several magnitudes (reviewed
in (1999d)). The viral load of many patients can become undetectable on
these therapies. This level of efficacy has been sustained in most of these
patients who continue to take their medications correctly. Current limits of
detection for HIV-1 plasma viral load are in the range of 20–50 copies/ml.
Patients with low or undetectable viral loads have stable CD4 lymphocyte

counts over time. It is felt that if a person's plasma viral load can be suppressed indefinitely, then that individual will not develop immunodeficiency. Consequently, HIV plasma viral load measurements have become the single marker for determining the efficacy of a given drug combination. As discussed below, patients' viral loads are assessed every 3 to 4 months. In those whose viral loads begin to rise, changes in the drug therapy may be made.

A. Zidovudine (ZDV) Monotherapy

In 1983, HIV-1 (then called human T-cell leukemia virus type III) was discovered in patients with AIDS (Gallo *et al.*, 1988). HIV-1 was eventually corroborated as the etiologic agents of AIDS. In 1985, *in vitro* studies demonstrated that ZDV inhibited HIV (Mitsuya *et al.*, 1985). This finding led to the clinical trial BW02, conducted by Fischl and colleagues (Fischl *et al.*, 1987). In this multicenter double-blind, placebo-controlled trial, ZDV was given to 250 mg every 4 h to patients with AIDS or AIDS-related complex (ARC). Patients also were required to have a CD4 lymphocyte count <200 cells/mm^3. The study showed that ZDV improved survival and reduced the incidence of opportunistic infections. ZDV therapy was associated with an increase in CD4 cells and a survival benefit of 21 months. This trial led to FDA approval in 1987.

ACTG 016 was the next trial and included patients with CD4 counts greater than 200 and less than 800 cells/mm^3 (Fischl *et al.*, 1990b). Again, the study was placebo controlled and double blind. The subgroup with CD4 counts less than 500 cells/mm^3 developed AIDS more slowly. However, no benefit was demonstrated to those with CD4 counts greater than 500 cells/mm^3. This finding led to ACTG 019, which directly examined asymptomatic patients with CD4 counts <500 cells/mm^3 (Volberding *et al.*, 1990). ACTG 019 also studied two different dosing regiments of ZDV, 100 mg five times per day or 250 mg every 4 h. Both ZDV regimens were associated with a decrease in the development of AIDS and ARC and an increased CD4 count. The lower dose ZDV group had fewer side effects. These studies led to an NIH-sponsored state-of-the-art conference, which recommended ZDV therapy at 500 mg per day to HIV-infected patients with a CD4 count less than 500 cells/mm^3 (1990).

The VA Study examined the issue delayed versus early therapy for those with a CD4 count between 200 and 500 cells/mm^3 (Hamilton *et al.*, 1992). Patients were initially randomized to the early group (ZDV 250 mg every 4 h) or the late group (placebo). Each patient in the late group was placed on ZDV (same dose) when his CD4 count fell below 200 cells/mm^3 or he had an AIDS-defining event. The study found no difference in the mortality of both groups, but did show a decrease in the development of AIDS in the early group.

The Concorde Study was a major European study (1994). In this study asymptomatic HIV-infected individuals were randomized to receive ZDV immediately or until they developed symptomatic disease. The regimen of ZDV was 250 mg 4 times per day and the trial was placebo controlled. Despite enrolling large numbers of patients, this trial found no statistically significant difference in outcome between the two groups. This trial cast significant doubt on the benefit of treating asymptomatic HIV infection with ZDV monotherapy.

Overall ZDV monotherapy offered a survival advantage to those with advanced disease. However, no consistent or large benefit was seen in patients with early disease. Clearly, there was an enormous need for improved therapies at this time.

B. Didanosine (ddI) Monotherapy

After ZDV, ddI was the second drug approved by the FDA for the treatment of HIV-1 infection. This approval was based primarily on the following three studies. ACTG 116A was a randomized controlled, double blind study which compared ddI to ZDV monotherapy (Dolin *et al.*, 1995). ddI was studied at two doses. Patients were required to have a CD4 lymphocyte count below 200 cells/mm^3 and be asymptomatic, or they were symptomatic and a CD4 count between 200 and 300 cells/mm^3. Patients were allowed to have been on ZDV for up to 16 weeks. The study concluded that ZDV naïve patients did better on ZDV. However, patients who had been on ZDV for longer than 8 weeks did better on ddI. The higher dose (750 mg/day) of ddI was associated with more side effects and no better outcome than the lower dose (500 mg/day).

ACTG 116B/117 was another randomized controlled, double blind study comparing ZDV to ddI monotherapy (Kahn *et al.*, 1992). This study simply addressed the same question as ACTG 116A, except all patients had been on ZDV for over 16 weeks. In this setting of prior ZDV experience, the patients did better on ddI.

BMS 010 again studied ddI versus ZDV monotherapy (Spruance *et al.*, 1994). All patients were on ZDV for greater than 6 months at the time of enrollment and also were clinically deteriorating. The patients were randomized to receive ddI or continue on ZDV. This study confirmed ACTG 116's findings that switching to ddI is superior to continuation of ZDV in patients who are worsening clinically.

C. Zalcitabine (ddC) Monotherapy

ddC was also studied for monotherapy of HIV-1 infection. In ACTG 114, patients who had taken ZDV for less than 3 months or had never taken ZDV were randomized to ZDV or ddC (1998d). All patients had advanced disease with CD4 counts <200 cells/mm^3. Both groups, those with and without a history of ZDV, did better on ZDV therapy. In ACTG 119, ddC monotherapy was again compared to ZDV monotherapy (Fischl *et al.*, 1993). All patients had advanced HIV disease and had been on ZDV for up to 48 weeks. No difference was found in disease progression between the two groups.

In CPCRA 002, ddC was compared to ddI monotherapy in patients with advanced HIV disease (Abrams *et al.*, 1994). Additionally, all had evidence of disease progression despite ZDV therapy. The median CD4 lymphocyte count was 37 cells/mm^3. This study found ddC to be at least as good as ddI and perhaps slightly better in delaying death.

D. Stavudine (d4T) Monotherapy

Only one trial studied d4T monotherapy. BMS 019 compared d4T to ZDV in patients who had been on ZDV treatment for greater than 6 months (Spruance *et al.*, 1997). The study concluded that the patients in the d4T arm progressed more slowly to AIDS and had higher CD4 counts. However, no differences were seen in mortality.

E. NRTI Combination Therapy

Through the clinical trials and other observations it was obvious that patients on monotherapy progressed. While ZDV, ddI, and ddC showed some survival advantage over no therapy, the gains were not great. Consequently, in the early 1990s, investigators began studying combination therapies with two NRTIs.

ACTG 175 had four treatment arms: ZDV alone, ddI alone, ZDV/ddC, and ZDV/ddI (Hammer *et al.*, 1996). Both naïve and drug-experienced patients were included. Patients CD4 counts ranged from 200 to 500 cells/mm^3. The primary endpoint was >50% decline in CD4+ T-cell count. Thirty-two percent of patients on ZDV alone reached the primary endpoint, compared to 18% on ZDV/ddI, 20% on ZDV/ddC, and 22% on ddI alone. However, ZDV/ddC patients only fared better if they were treatment naïve. Overall, ZDV/ddI, ddI alone, and ZDV/ddC were superior to ZDV alone.

ACTG 155 compared ZDV or ddC or both (Fischl *et al.*, 1995). To be eligible, each patient had a CD4 count less than 300 cells/μl and had been treated with ZDV for more than 6 months. The primary end point was time to disease progression or death. In patients with CD4 counts greater than 150 cells/mm^3, ZDV/ddC therapy reduced the disease progression by 50% compared to ZDV monotherapy. No difference was seen in patients with CD4 counts less than 150 cells/mm^3. This study supported the early usage of combination therapy. However, the difference between treatment groups was not large and only found in subgroup analysis.

NUCA 3001 compared ZDV alone, 3TC alone, and ZDV/3TC combination therapy (Eron *et al.*, 1995). Patients had CD4 counts between 200 and 500 cells/mm^3 and had been on ZDV less than 4 weeks. This study examined the viral load and CD4+ cell responses to therapy over 24 weeks. Patients on combination therapy had greater increases CD4 counts and lower plasma viral RNA than did patients on either of the single agent treatment arms. Clinical data were not assessed.

In NUCA 3002, the safety and activity of ZDV/3TC therapy was compared to that of ZDV/ddC therapy (Bartlett *et al.*, 1996). Each patient had been on ZDV for at least 6 months and had a CD4 count of 100–300 cells/mm^3. The study concluded that the 3TC arm had better CD4 response. However, suppression viral RNA was similar in both groups. This study also established the dose of 3TC as 150 mg twice per day.

In the CAESAR trial, the clinical benefit of 3TC combination therapy was assessed (1997). All patients had advanced disease with CD4+ T-cell counts between 25 and 250 cells/mm^3. All patients were on ZDV alone, ZDV/ddC, or ZDV/ddI at the time of enrollment. Patients were then randomized to receive 3TC, placebo, or 3TC and loviride (an investigational NNRTI). The primary end point was progression to an AIDS-defining event or death. The study was prematurely terminated when an interim analysis revealed a highly significant reduction in events in the 3TC treatment arms. Progression to AIDS was lowered from 20 to 9% (>50% reduction) in both 3TC arms. Survival was also improved with 3TC combination therapy. No difference was seen with the addition of loviride. This study demonstrated that the addition of 3TC to a ZDV-containing regimen improved survival and slowed the progression of AIDS. However, the authors noted that these combination therapies were unlikely to produce long-term viral suppression and that many patients still progressed despite the addition of 3TC.

Several other trials involving NRTIs confirmed that combination therapy was superior to monotherapy. By the mid 1990s combination therapy with two NRTIs had become the standard of care for HIV-infected patients (reviewed in Nadler, 1996 and Carpenter *et al.*, 1996). However, even these therapies were not potent or long lasting.

F. Two-Class Combination Therapy

Soon after clinical trials established combination NRTI (single class) therapy as the standard of care for HIV-1 infection, two new classes of HIV-1 therapeutics emerged. Clinical trials with the nonnucleoside reverse transcriptase inhibitors and the protease inhibitors began in 1995–1996. Over the next few years, multiple clinical studies proved that the addition of an NNRTI or PI to a regimen of two NRTIs was more effective than two NRTIs alone. As of early 1999, over 50 trials have been performed with different drug combinations, which employ two or more drug classes in at least one arm. For a summary of these studies the reader is referred to a compilation by Tavel *et al.* (1999). In the following paragraphs we illustrate some of the key trials.

G. Protease Inhibitor Combination Therapy

In ACTG 229, saquinavir was tested with ZDV and ddC (Collier *et al.*, 1996). Patients had a median CD4 count of 156 cells/mm^3 and a median viral load of 4.5 log$_{10}$. Treatment regimens consisted of (1) saquinavir plus ZDV plus ddC, (2) saquinavir plus ZDV, or (3) ZDV plus ddC. The three-drug regimen resulted in a greater CD4 cell count increase and a lower viral load than either two-drug regimen. The study supported the use protease inhibitors in combination with two NRTIs.

The SV 14604 trial compared saquinavir/ZDV/ddC to ZDV/ddC in 3485 patients (Clumeck, 1997). The average HIV-1 RNA concentration (viral load) was approximately 5 log$_{10}$. The median CD4 count was 200 cells/mm^3 (range = 50–350). The study found a 50% reduction in the risk of progression to the first AIDS defining even or death in the three-drug arm compared to the ZDV/ddC arm. This result established the clinical benefit of combination therapy, which includes a protease inhibitor.

In 1997 and 1998, several small studies compared two NRTI regimens with and without ritonavir. These studies showed that ritonavir-containing regimens provided improved suppression of plasma HIV RNA. In 1998, a study by Cameron *et al.* demonstrated that ritonavir added to a NRTI regimen conferred a survival advantage (Cameron *et al.*, 1998). This was the first study to show an improvement in survival in a protease inhibitor-containing regimen.

The third protease inhibitor to be studied, indinavir, was next shown to be equally efficacious to ritonavir and better tolerated (Clumeck *et al.*, 1998). In several studies, indinavir with ZDV and 3TC was shown to be a potent combination (Gulick *et al.*, 1997; Hecht *et al.*, 1998; Perrin *et al.*, 1997). For instance, in AVANTI 2, 75% of patients on indinavir/ZDV/3TC had a plasma viral load below the limit of detection at week 52 of therapy (Gerstoft and Group, 1997). However, only 23% of patients on ZDV/3TC achieved this level of reduction. In naïve or NRTI-experienced patients, this three-drug regimen strongly suppressed viral replication. In most studies the majority of patients achieved viral load reductions to below the limit of detection. This success was seen if the drugs were started together (Gulick *et al.*, 1998). If drugs were added sequentially, they were much less effective. Additionally, in ACTG 343, indinavir was studied with ZDV and 3TC to determine if maintenance therapy had a role in HIV-1 treatment (Havlir *et al.*, 1998). In this study, after an induction period with all three drugs, patients were randomized to take (1) indinavir alone; (2) ZDV and 3TC; or (3) indinavir, ZDV, and 3TC. Patients on all three drugs did better than those on either maintenance regimen. This trial and another with indinavir (Pialoux *et al.*, 1998) have shown that less aggressive maintenance therapies are not as effective as three-drug regimens.

Nelfinavir has also performed well in combination therapy. In AVANTI 3, nelfinavir/ZDV/3TC therapy was compared to ZDV/3TC therapy (Clumeck *et al.*, 1998). At 28 weeks, 83% of the nelfinavir arm had viral loads less than 500 copies/ml, while the ZDV/3TC arm had only 18% below this level. Similarly, in Protocol 511, 79% of patients on nelfinavir (750 mg TID) plus ZDV/3TC had viral loads less than 500 copies at 12 months (Clendeninn *et al.*, 1998).

The most recently approved PI, amprenavir, has worked well in combination therapy. A four-drug regimen, consisting of amprenavir/ABC/ZDV/3TC, produced good viral suppression in both chronically and acutely infected patients (Kost *et al.*, 1998). An interesting combination of only two drugs, amprenavir and ABC, has also demonstrated good short-term antiviral activity (Murphy *et al.*, 1998). This two-drug regimen has a practical advantage in that both drugs are administered twice daily. Long-term data are still needed for amprenavir trials.

Overall, each of the approved protease inhibitors has shown potent anti-HIV-1 activity when used in combination with other drugs. No agent has been clearly shown to be superior to the other PI drugs. However, some appear to be less well tolerated. For instance in one study, 50% of patients on ritonavir stopped taking the drug, secondary to gastrointestinal symptoms or elevation of hepatic transaminases (Mathez *et al.*, 1997). In another study, the rate of drug discontinuation secondary to side effects was 38% in the ritonavir arm compared to 5.4% in the indinavir arm (Clumeck *et al.*, 1998).

H. NNRTI Combination Therapy

Nevirapine was shown to be effective in combination with ZDV and ddI (D'Aquila *et al.*, 1996; Montaner *et al.*, 1998b). This three-drug regimen resulted in HIV plasma RNA below 20 copies per ml in 52% of patients, compared to 12% on ZDV/ddI (Montaner *et al.*, 1998b). Accordingly, the rate of disease in the three-drug arm was 12% compared to 25% in patients taking ZDV/ddI (Montaner *et al.*, 1998b). Similar virologic benefit has been seen in patients on nevirapine/d4T and 3TC (Kaspar *et al.*, 1998). These responses with nevirapine were observed in naïve patients.

Delavirdine has received less attention. No significant benefit when delavirdine was added to ZDV and ddI therapy (Davey *et al.*, 1996). However, a recent trial has reported potent inhibition of HIV-1 with delavirdine/indinavir and ZDV (Daly *et al.*, 1998).

Efivarenz is the newest NNRTI. This compound has been shown to be very effective in combination with ZDV and 3TC in DMP 266–005 (Hicks *et al.*, 1998). In this study of naïve patients, 100% of those receiving the highest dose of efavirenz achieved suppression of HIV RNA levels to below 400 copies/ml. In DMP 266-006, efavirenz/ZDV/3TC therapy achieved similar viral load reductions as indinavir/ZDV/3TC therapy did (Staszewski *et al.*, 1998). Finally, a regimen without a NRTI, efavirenz plus indinavir, has shown good HIV-1 viral load suppression (Riddler *et al.*, 1998). However, because this trial is recent, the data are not long term.

III. Current Recommendations for Treating HIV-1 Infection —

With the advent of potent antiretroviral therapies, the standard of care for treating HIV-1 has changed. In May 1999, the Centers for Disease Control and Prevention published a updated version of the *Guidelines for the Use of Antiretroviral Agents in HIV-Infected Adults and Adolescents* (1999d). This publication is a useful guide for helping a clinician decide which HIV-infected adults and adolescents should be treated and which drugs should be used. Below is a summary of the *Guidelines* with regards to certain questions facing clinicians and patients. The CDC has also published guidelines for treatment of children infected with HIV (1999e). Both *Guidelines* and other valuable information can be found at the HIV/AIDS Treatment Information Service website (**http://www.hivatis.org**).

A. When to Treat?

This is still a controversial area. Some experts feel therapy should be offered to all infected patients with detectable viral loads. However, others feel that therapy should only be offered to patients with viral loads greater than 10,000 copies/ml or a CD4+ T-cell count of less than 500 cells/mm^3. The combination drug therapies, although potent, have significant side effects, which may affect the patient's quality of life. The drug regimens are inconvenient and can be difficult to adhere to. Finally, it is unknown whether HIV-infected individuals, with CD4+ counts greater than 500 cells/mm^3, would clinically benefit. However, theoretically these patients would. In the end, the decision must be made on an individual basis.

B. What to Take?

The *Guidelines* recommend that initial therapy in a patient, who has not been on antiretrovirals previously, consist of two nucleoside reverse transcriptase inhibitors (NRTIs) and a protease inhibitor or efavirenz, which has been recently shown to work well in combination therapy (Staszewski *et al.*, 1998). In naïve patients, these regimens have consistently performed the best. However, potent suppression of viral replication has been observed in other combinations. One such regimen is nevirapine (Montaner *et al.*, 1998a) or delavirdine with two NRTIs. Another is abacavir with ZDV and 3TC. This latter regimen consists of three NRTIs and only short-term data are available. Additionally, success has been achieved with efavirenz and a PI (indinavir) (Staszewski *et al.*, 1998). Other two-drug regimens, such as two protease inhibitors alone, may prove effective in the future. All monotherapies and two NRTIs without a third drug are not recommended.

C. How to Monitor Therapy?

In naïve patients who begin antiretroviral therapy their viral loads should fall to below the level of detection (current limits 20–50 copies/ml). This virologic response should be seen by 12–16 weeks. A detectable viral load past this time point is reason to consider a change in therapy. However, in some patients, the viral load may stabilize at this low, yet detectable level. It is not known whether these patients should stay on their current therapies or should be switched to new therapies. For patients on a stable regimen, viral loads should be assayed every 3–4 months to evaluate the effectiveness of their therapies. A reproducible rise in viral load greater than threefold from the baseline is a strong reason to change therapy. Of note, therapeutic failures have occurred in every drug combination so far tested in large numbers of patients (Tavel *et al.*, 1999). Additionally, it is recommended that a patient's CD4+ T-cell count be monitored every 3–6 months. With strong viral load suppression, a patient's CD4+ T-cell count usually increases and then reaches a plateau level (Powderly *et al.*, 1998). However, in patients whose therapy is failing, the CD4+ T-cell count may fall and a consistent decrease of >30% should be considered as evidence of therapeutic failure.

D. What Drugs to Change to?

This question is perhaps the most difficult. Only a few small studies have addressed the question of "salvage regimens." Although not evaluated, one recommendation is to change to at least two new drugs simultaneously (Saag, 1999). Another question involved in "salvage regimens" is the utility of resistance testing. Drug resistance is a complex issue and is beyond the scope of this article. However, a few simple principles hold. As alluded to above, HIV-1 has become resistant to each drug tested *in vivo* (Hirsch *et al.*, 1998). Resistance to each drug is associated with certain mutations in the viral genes for reverse transcriptase or protease. However, these associations are not universal. Once resistance develops, it is long lasting. Cross-resistance within a drug class is high. For instance, HIV-1 isolates that are resistant to delavirdine are usually resistant to nevirapine (Miller *et al.*, 1998). Currently, the resistance of HIV-1 to drugs can be assessed by genotypic (sequence based) or phenotypic (viral culture based) testing. The utility of resistance testing is unclear. However, many physicians, in making therapeutic changes, are using these tests to guide their choices. A NIH sponsored panel recently produced guidelines for the use of resistance testing (Hirsch *et al.*, 1998). However, the sensitivity and specificity of these assays has not yet been determined. Future studies will determine the utility of these assays in guiding therapeutic changes.

E. Special Situations

Treatment of HIV infection during pregnancy and in acute HIV infection involves unique issues (reviewed in (1999d). We review some of these topics below.

After 48 weeks, monotherapy of primary or acute HIV infection with ZDV resulted in higher CD4+ T-cell counts, but had no impact of HIV viral load (Niu *et al.*, 1998). However, with new potent antiretroviral combination therapy, treatment of primary or acute HIV infection has received much attention. In primary HIV infection virus replicates extensively and immune system abnormalities develop (Schacker *et al.*, 1998; Pantaleo *et al.*, 1998). Acutely infected patients have dramatic decreases in their viral loads on combination therapy (reviewed in Daar, 1998). Several investigators have also noted favorable changes in such patients' immune systems (Carr *et al.*, 1998; McElrath *et al.*, 1998; Sekaly *et al.*, 1998; Sachsenberg *et al.*, 1998; Zaunders *et al.*, 1998). Interestingly, the specific anti-HIV immune response appears attenuated in patients treated aggressively in the acute period (Daar *et al.*, 1998; Lafeuillade *et al.*, 1997). While no clinical benefit has been demonstrated, many experts feel that initiation of therapy during acute infection is appropriate (1999d). Theoretical benefits include the lowering of the viral load set point, which could favorably influence the disease course, and reduction of the overall viral burden, which may reduce the possibility of resistance development. However, once a patient is started on therapy during the acute infection, then that patient must presumably stay on therapy indefinitely. Consequently, a patient and physician must weigh the options before deciding on treatment in the acute setting. Treatment options are the same as those for chronic or established HIV infection.

Vertical transmission is an efficient means of HIV transmission. Without maternal treatment, the rate of HIV vertical transmission is approximately 25% (Fang *et al.*, 1995; Shaffer *et al.*, 1999). Maternal viral load is significantly correlated with transmission; the higher the maternal viral load, the more likely transmission occurs (Fang *et al.*, 1995; Sperling *et al.*, 1996; Katzenstein *et al.*, 1999). In ACTG 076, mothers were randomized to receive either ZDV monotherapy or placebo, starting at 14 weeks. ZDV therapy was associated with a 67.5% reduction in vertical transmission. However, the reduction of the vertical transmission rate was not explained solely by

ZDV's effect on maternal viral load. Recently, a European trial found that elective caesarean section reduced the vertical transmission rate from 10.5% to 1.8% (1999b). However, the mothers in this trial were not treated with potent combination therapy. A consistent, but less striking affect of caesarean section on HIV vertical transmission was found in a meta-analysis (1999g). Again, mothers were not necessarily on potent combination therapy at the time of delivery. The *Guidelines* suggest that pregnant women should be treated in the same manner as nonpregnant women. The teratogenic effect of anti-HIV compounds is not well understood. Consequently, if a woman was not on therapy at the time she became pregnant, a physician should consider holding therapy until after the first trimester. Many anti-HIV compounds have been found to be teratogenic in animal studies (Table VII). Fortunately, major teratogenic effects secondary to anti-HIV drugs have not been seen yet in humans (Culnane *et al.*, 1999). Of note, gross malformations have been observed in cynomologous macaques following intrauterine exposure to efavirenz (1999d). However, these data are limited.

TABLE VII   Anti-HIV Therapeutics in Pregnancy

Drug	FDA pregnancy category[a]
ZDV	C
ddI	B
ddC	C
d4T	C
3TC	C
ABC	C
Nevirapine	C
Delavirdine	C
Efavirenz	C
Saquinavir	B
Ritonavir	B
Indinavir	C
Nelfinavir	B
Amprenavir	C

[a] FDA-defined pregnancy categories: A, human studies showing no risk; B, animal studies showing no risk, but human studies are inadequate or animal risk, but human studies showing no risk; C, animal studies demonstrating toxicity and human studies are inadequate; D, human studies demonstrating some risk, but benefits may outweigh risks; X, fetal toxicity established and risk not greater than benefit.

F. Future Therapeutics

As we have seen since 1993, several new drugs for treating HIV infection will be available in the near future.

Adefovir, which has a phosphate ester, is a nucleotide analog reverse transcriptase inhibitor (De Clercq *et al.*, 1987). Adefovir's mechanism of action is similar to NRTIs. This drug will probably be approved soon by the FDA for the treatment of HIV infection. Adefovir has been studied in small clinical trials and is active against HIV in monotherapy in short-term trials (Deeks *et al.*, 1997; Barditch-Crovo *et al.*, 1997). Adefovir can cause renal dysfunction and can decrease the levels of L-carnitine (1999c).

Hydroyurea is an old drug which has been found to have anti-HIV activity (Lori *et al.*, 1994). Hydroxyurea appears to inhibit the cellular enzyme, ribonucleotide reductase inhibitor. Inhibition of this enzyme reduces the intracellular pool of dATP and thus reduces the efficiency of HIV reverse transcription (Johns and Gao, 1998). Clinical trials have shown that hydroxyurea augments the antiviral effect of ddI, but has little effect alone (Rutschmann *et al.*, 1998; Federici *et al.*, 1998). Hydroxyurea, however, does appear to blunt the CD4+ T-cell increase seen with other potent antiviral therapies. Hydroxyurea is thought to be a teratogen, but its effect on the human fetus is controversial (Diav-Citrin *et al.*, 1999).

As mentioned above, to date all drugs employed against HIV target either reverse transcription or viral protease cleavage. However, in the past few years much has been discovered about viral entry. Drugs which inhibit HIV entry have been developed. One such drug, T20, is a synthetic peptide (Chen *et al.*, 1995). By interacting with a portion of the viral envelope (GP 41), T20 appears to inhibit fusion of the viral membrane with the cellular membrane (Rimsky *et al.*, 1998). This compound has proven to be effective *in vivo* (Kilby *et al.*, 1998) and is currently in phase 2/3 clinical trials. It is expected that other compounds which inhibit HIV entry will enter clinical trials in the near future (Saag, 1999).

Ritonavir, as above, is a potent inhibitor of the P450 enzyme system and all PIs are metabolized by the P450 system. When given with either saquinavir or indinavir, ritonavir increases the half-life of the other PI. Ritonavir with saquinavir or indinavir therapy has already been shown to be effective clinically (Gisolf *et al.*, 1998; Cameron *et al.*, 1999; Workman *et al.*, 1998). In future and ongoing trials, ritonavir at a lower dose (1–200 mg BID) will be used only as a P450 inhibitor to increase the half-life of another PI. In this way the other PI can be given less frequently and at a lower dose.

References ━━━━━━━━━━━━━━━━━━━━━━━━━━━━━━━━━━━━━━

(1990). State-of-the-art conference on azidothymidine therapy for early HIV infection. *Am. J. Med.* **89**, 335–344.

(1994). Concorde: MRC/ANRS randomised double-blind controlled trial of immediate and deferred zidovudine in symptom-free HIV infection. Concorde Coordinating Committee. *Lancet* **343**, 871–881.

(1997). Randomised trial of addition of lamivudine or lamivudine plus loviride to zidovudine-containing regimens for patients with HIV-1 infection: The CAESAR trial. *Lancet* **349**, 1413–1421.

(1998a). Crixivan product information sheet. *In* "HIV Prescribing Guide (PDR)" (R. Arky, Ed.), 1st ed., pp. 51–57. Medical Economics Company, Montvale, NJ.

(1998b). Epivir product information sheet. *In* "HIV Prescribing Guide (PDR)" (R. Arky, Ed.), 1st ed., pp. 7–12. Medical Economics Company, Montvale, NJ.

(1998c). Fortovase product information sheet. *In* "HIV Prescribing Guide (PDR)" (R. Arky, ed.), 1st ed., pp. 58–65. Medical Economics Company, Montvale, NJ.

(1998d). HIVID product information sheet. *In* "HIV Prescribing Guide (PDR)" (R. Arky, Ed.), 1st ed., pp. 13–19. Medical Economics Company, Montvale, NJ.

(1998e). Invirase product information sheet. *In* "HIV Prescribing Guide (PDR)" (R. Arky, Ed.), 1st ed., pp. 66–71. Medical Economics Company, Montvale, NJ.

(1998f). Norvir product information sheet. *In* "HIV Prescribing Guide (PDR)" (R. Arky, Ed.), 1st ed., pp. 72–80. Medical Economics Company, Montvale, NJ.

(1998g). Rescriptor product information sheet. *In* "HIV Prescribing Guide (PDR)" (R. Arky, Ed.), 1st ed., pp. 91–97. Medical Economics Company, Montvale, NJ.

(1998h). Retrovir product information sheet. *In* "HIV Prescribing Guide (PDR)" (R. Arky, Ed.), 1st ed., pp. 20–27. Medical Economics Company, Montvale, NJ.

(1998i). Three new drugs for HIV infection. *Med. Lett. Drugs Ther.* **40**, 114–116.

(1998j). Videx product information sheet. *In* "HIV Prescribing Guide (PDR)" (R. Arky, Ed.), 1st ed., pp. 37–43. Medical Economics Company, Montvale, NJ.

(1998k). Viracept product information sheet. *In* "HIV Prescribing Guide (PDR)" (R. Arky, Ed.), 1st ed., pp. 81–90. Medical Economics Company, Montvale, NJ.

(1998l). Viramune product information sheet. *In* "HIV Prescribing Guide (PDR)" (R. Arky, Ed.), 1st ed., pp. 98–106. Medical Economics Company, Montvale, NJ.

(1998m). Zerit product information sheet. *In* "HIV Prescribing Guide (PDR)" (R. Arky, Ed.), 1st ed., pp. 44–50. Medical Economics Company, Montvale, NJ.

(1999a). "Agenerase Package Insert." Glaxo–Wellcome, Research Triangle Park, NC.

(1999b). Elective caesarean-section versus vaginal delivery in prevention of vertical HIV-1 transmission: a randomised clinical trial: The European Mode of Delivery Collaboration. *Lancet* **353**, 1035–1039.

(1999c). "Expanded Access Program for People with HIV Infection." Gilead Sciences, Foster City, CA.

(1999d). "Guidelines for the Use of Antiretroviral Agents in HIV-Infected Adults and Adolescents." Department of Health and Human Services and the Henry J. Kaiser Foundation, Washington, DC.

(1999e). "Guidelines for the Use of Antiretroviral Agents in Pediatric HIV Infection." Department of Health and Human Services and the Henry J. Kaiser Foundation, Washington, DC.

(1999f). "Letter to Clincians: Ziagen Hypersensitivity Syndrome." Glaxo–Wellcome, Research Triangle Park, NC.

(1999g). The mode of delivery and the risk of vertical transmission of human immunodeficiency virus type 1—A meta-analysis of 15 prospective cohort studies: The International Perinatal HIV Group. *N. Engl. J. Med.* **340**, 977–987.

(1999h). "Sustiva Package Insert." Dupont–Merck, Wilmington, DE.

(1999i). "Ziagen Package Insert." Glaxo–Wellcome, Research Triangle Park, NC.

Abrams, D. I., Goldman, A. I., Launer, C., Korvick, J. A., Neaton, J. D., Crane, L. R., Grodesky, M., Wakefield, S., Muth, K., Kornegay, S., *et al.* (1994). A comparative trial of didanosine or zalcitabine after treatment with zidovudine in patients with human immunodeficiency virus infection: The Terry Beirn Community Programs for Clinical Research on AIDS. *N. Engl. J. Med.* **330,** 657–662.

Balani, S. K., Woolf, E. J., Hoagland, V. L., Sturgill, M. G., Deutsch, P. J., Yeh, K. C., and Lin, J. H. (1996). Disposition of indinavir, a potent HIV-1 protease inhibitor, after an oral dose in humans. *Drug Metab. Dispos.* **24,** 1389–1394.

Barditch-Crovo, P., Toole, J., Hendrix, C. W., Cundy, K. C., Ebeling, D., Jaffe, H. S., and Lietman, P. S. (1997). Anti-human immunodeficiency virus (HIV) activity, Safety, and pharmacokinetics of adefovir dipivoxil (9-[2-(bis-pivaloyloxymethyl)-phosphonylmethox-yethyl]adenine) in HIV-infected patients. *J. Infect. Dis.* **176,** 406-13.

Bareggi, S. R., Cinque, P., Mazzei, M., D'Arminio, A., Ruggieri, A., Pirola, R., Nicolin, A., and Lazzarin, A. (1994). Pharmacokinetics of zidovudine in HIV-positive patients with liver disease. *J. Clin. Pharmacol.* **34,** 782–786.

Batlett, J. A., Benoit, S. L., Johnson, V. A., Quinn, J. B., Sepulveda, G. E., Ehmann, W. C., Tsoukas, C., Fallon, M. A., Self, P. L., and Rubin, M. (1996). Lamivudine plus zidovudine compared with zalcitabine plus zidovudine in patients with HIV infection: A randomized, double-blind, placebo-controlled trial. North American HIV Working Party. *Ann. Intern. Med.* **125,** 161–172.

Buss, N. (1998). Saquinavir soft gel capsule (Fortovase): Pharmacokinetics and drug interactions. *5th Conf. Retrovir. Oppor. Infect.* Abstract 354.

Cameron, D. W., Heath-Chiozzi, M., Danner, S., Cohen, C., Kravcik, S., Maurath, C., Sun, E., Henry, D., Rode, R., Potthoff, A., and Leonard, J. (1998). Randomised placebo-controlled trial of ritonavir in advanced HIV-1 disease: The Advanced HIV Disease Ritonavir Study Group. *Lancet* **351,** 543–549.

Cameron, D. W., Japour, A. J., Xu, Y., Hsu, A., Mellors, J., Farthing, C., Cohen, C., Poretz, D., Markowitz, M., Fallansbee, S., Angel, J. B., McMahon, D., Ho, D., Devanareyan, V., Rode, R., Salgo, M., Kempf, D. J., Granneman, R., Leonard, J. M., and Sun, E. (1999). Ritonavir and saquinavir combination therapy for the treatment of HIV infection. *AIDS* **13,** 213–224.

Carpenter, C. C., *et al.* (1996). Antiretroviral therapy for HIV infection in 1996. Recommendations of an international panel: International AIDS Society—USA. *J. Am. Med. Assoc.* **276,** 146–154.

Carr, A., Vella, S., de Jong, M. D., Sorice, F., Imrie, A., Boucher, C. A., and Cooper, D. A. (1996). A controlled trial of nevirapine plus zidovudine versus zidovudine alone in p24 antigenaemic HIV-infected patients: The Dutch–Italian–Australian Nevirapine Study Group. *AIDS* **10,** 635–641.

Carr, A., Zaunders, J., Cunningham, P., Kaufmann, G., Kelleher, A., and Cooper, D. A. (1998). Immunological and virological effects of combination antiretroviral therapy: Primary HIV infection (PHI) vs asymptomatic, established infection (AEI). *Int. Conf. AIDS* **12,** 50.

Cato, A., III, Qian, J., Hsu, A., Levy, B., Leonard, J., and Granneman, R. (1998). Multidose pharmacokinetics of ritonavir and zidovudine in human immunodeficiency virus-infected patients. *Antimicrob. Agents Chemother.* **42,** 1788–1793.

Chen, C. H., Matthews, T. J., McDanal, C. B., Bolognesi, D. P., and Greenberg, M. L. (1995). A molecular clasp in the human immunodeficiency virus (HIV) type 1 TM protein determines the anti-HIV activity of gp41 derivatives: Implication for viral fusion. *J. Virol.* **69,** 3771–3777.

Cheng, C. L., Smith, D. E., Carver, P. L., Cox, S. R., Watkins, P. B., Blake, D. S., Kauffman, C. A., Meyer, K. M., Amidon, G. L., and Stetson, P. L. (1997). Steady-state pharmacokinetics of delavirdine in HIV-positive patients: effect on erythromycin breath test. *Clin. Pharmacol. Ther.* **61,** 531–543.

Clendeninn, N., Quart, B., Anderson, R., Knowles, M., and Chang, Y. (1998). Analysis of long-term virologic data from the VIRACEPT (nelfinavir, NFV) 511 protocol using 3 HIV-RNA assays. *5th Conf. Retrovir. Oppor. Infect.* Abstract 372.

Clumeck, N. (1997). *In* "Interscience Conference on Antimicrobial Agents and Chemotherapy." American Society of Microbiology, Toronto.

Clumeck, N., Colebunders, B., Vandercam, B., Kabeya, K., Cassano, P., Sommereijns, B., and De Wit, A. (1998). Randomized comparative outcome trial of indinavir (I) and ritonavir (R) in protease inhibitors (PI) naive HIV patients (p) with CD4 below 100 cells/microliter. *5th Conf. Retrovir. Oppor. Infect.* Abstract 386.

Collier, A. C., Coombs, R. W., Schoenfeld, D. A., Bassett, R. L., Timpone, J., Baruch, A., Jones, M., Facey, K., Whitacre, C., McAuliffe, V. J., Friedman, H. M., Merigan, T. C., Reichman, R. C., Hooper, C., and Corey, L. (1996). Treatment of human immunodeficiency virus infection with saquinavir, zidovudine, and zalcitabine. AIDS Clinical Trials Group. *N. Engl. J. Med.* **334,** 1011–1017.

Coombs, R. W., Collier, A. C., Allain, J. P., Nikora, B., Leuther, M., Gjerset, G. F., and Corey, L. (1989). Plasma viremia in human immunodeficiency virus infection. *N. Engl. J. Med.* **321,** 1626–1631.

Cox, S. R., Ferry, J. J., Batts, D. H., Carlson, G. F., Schneck, D. W., Herman, B. D., Della-Coletta, A. A., Chambers, J. H., Carel, B. J., Stewart, F., Buss, N., and Brown, A. (1997). Delavirdine (D) and marketed protease inhibitors (PIs): pharmacokinetic (PK) interaction studies in healthy volunteers. *4th Conf. Retrovir. Oppor. Infect.* Abstract 372.

Cox, S. R., Schneck, D. W., Herman, B. D., Carel, B. J., Gullotti, B. R., Kerr, B. M., and Freimuth, W. W. (1998). Delavirdine (DLV) and nelfinavir (NFV): A pharmacokinetic (PK) drug–drug interaction study in healthy adult volunteers. *5th Conf. Retrovir. Oppor. Infect.* **144,**

Culnane, M., *et al.* (1999). Lack of long-term effects of *in utero* exposure to zidovudine among uninfected children born to HIV-infected women: Pediatric AIDS Clinical Trials Group Protocol 219/076 Teams. *J. Am. Med. Assoc.* **281,** 151–157.

Daar, E. S. (1998). Virology and immunology of acute HIV type 1 infection. *AIDS Res. Hum. Retroviruses* **14**(Suppl. 3), S229–S234.

Daar, E. S., Bai, J., Hausner, M. A., Majchrowicz, M., and Giorgi, J. V. (1998). Lack of CD8(+) cell activation and cytotoxic T-lymphocyte (CTL) activity until after discontinuation of highly active antiretroviral therapy (HAART) for primary HIV infection. *5th Conf. Retrovir. Oppor. Infect.* Abstract 588.

Daly, P., Green, S. L., Freimuth, W. W., Conklin, M. A., Huang, D. C., and Wathen, L. K. (1998). An open-label randomized study of Rescriptor (DLV, delavirdine mesylate) in triple and quadruple combinations with zidovudine (ZDV), indinavir (IDV) and lamivudine (3TC) in HIV-1 infected individuals. *Int. Conf. AIDS* **12,** 84.

D'Aquila, R. T., Hughes, M. D., Johnson, V. A., Fischl, M. A., Sommadossi, J. P., Liou, S. H., Timpone, J., Myers, M., Basgoz, N., Niu, M., and Hirsch, M. S. (1996). Nevirapine, zidovudine, and didanosine compared with zidovudine and didanosine in patients with HIV-1 infection: A randomized, double-blind, placebo-controlled trial: National Institute of Allergy and Infectious Diseases AIDS Clinical Trials Group Protocol 241 Investigators. *Ann. Intern. Med.* **124,** 1019–1030.

Davey, R. T., Jr., *et al.* (1996). Randomized, controlled phase I/II, trial of combination therapy with delavirdine (U-90152S) and conventional nucleosides in human immunodeficiency virus type 1-infected patients. *Antimicrob. Agents Chemother.* **40,** 1657–1664.

De Clercq, E. (1999). Perspectives of non-nucleoside reverse transcriptase inhibitors (NNRTIs) in the therapy of HIV-1 infection. *Farmaco* **54,** 26–45.

De Clercq, E., Sakuma, T., Baba, M., Pauwels, R., Balzarini, J., Rosenberg, I., and Holy, A. (1987). Antiviral activity of phosphonylmethoxyalkyl derivatives of purine and pyrimidines. *Antiviral Res.* **8,** 261–272.

Deeks, S. G., Collier, A., Lalezari, J., Pavia, A., Rodrigue, D., Drew, W. L., Toole, J., Jaffe, H. S., Mulato, A. S., Lamy, P. D., Li, W., Cherrington, J. M., Hellmann, N., and Kahn, J. (1997). The safety and efficacy of adefovir dipivoxil, a novel antihuman immunodeficiency virus (HIV) therapy, in HIV-infected adults: A randomized, double-blind, placebo-controlled trial. *J. Infect. Dis.* **176**, 1517–1523.

DeGruttola, V., *et al.* (1998). Trial design in the era of highly effective antiviral drug combinations for HIV infection. *AIDS* **12**(Suppl. A), S149–S156.

Diav-Citrin, O., *et al.* (1999). Hydroxyurea use during pregnancy: A case report in sickle cell disease and review of the literature. *Am. J. Hematol.* **60**, 148–150.

Dolin, R., Amato, D. A., Fischl, M. A., Pettinelli, C., Beltangady, M., Liou, S. H., Brown, M. J., Cross, A. P., Hirsch, M. S., Hardy, W. D., *et al.* (1995). Zidovudine compared with didanosine in patients with advanced HIV type 1 infection and little or no previous experience with zidovudine: AIDS Clinical Trials Group. *Arch. Intern. Med.* **155**, 961–974.

Dudley, M. N. (1995). Clinical pharmacokinetics of nucleoside antiretroviral agents. *J. Infect. Dis.* **171**(Suppl. 2), S99–S112.

Eron, J. J., Benoit, S. L., Jemsek, J., MacArthur, R. D., Santana, J., Quinn, J. B., Kuritzkes, D. R., Fallon, M. A., and Rubin, M. (1995). Treatment with lamivudine, zidovudine, or both in HIV-positive patients with 200 to 500 CD4+ cells per cubic millimeter: North American HIV Working Party. *N. Engl. J. Med.* **333**, 1662–1669.

Fang, G., Burger, H., Grimson, R., Tropper, P., Nachman, S., Mayers, D., Weislow, O., Moore, R., Reyelt, C., Hutcheon, N., *et al.* (1995). Maternal plasma human immunodeficiency virus type 1 RNA level: A determinant and projected threshold for mother-to-child transmission. *Proc. Natl. Acad. Sci. USA* **92**, 12100–12104.

Federici, M. E., Lupo, S., Cahn, P., Cassetti, I., Pedro, R., Ruiz-Palacios, G., Montaner, J. S., and Kelleher, T. (1998). Hydroxyurea in combination regimens for the treatment of antiretroviral naive, HIV-infected adults. *Int. Conf. AIDS* **12**, 58–59.

Finzi, D., Blankson, J., Siliciano, J. D., Margolick, J. B., Chadwick, K., Pierson, T., Smith, K., Lisziewicz, J., Lori, F., Flexner, C., Quinn, T. C., Chaisson, R. E., Rosenberg, E., Walker, B., Gange, S., Gallant, J., and Siliciano, R. F. (1999). Latent infection of CD4+ T cells provides a mechanism for lifelong persistence of HIV-1, even in patients on effective combination therapy. *Nat. Med.* **5**, 512–517.

Fischl, M. A., Olson, R. M., Follansbee, S. E., Lalezari, J. P., Henry, D. H., Frame, P. T., Remick, S. C., Salgo, M. P., Lin, A. H., Nauss-Karol, C., *et al.* (1993). Zalcitabine compared with zidovudine in patients with advanced HIV-1 infection who received previous zidovudine therapy. *Ann. Intern. Med.* **118**, 762–769.

Fischl, M. A., Parker, C. B., Pettinelli, C., Wulfsohn, M., Hirsch, M. S., collier, A. C., Antoniskis, D., Ho, M., Richman, D. D., Fuchs, E., *et al.* (1990a). A randomized controlled trial of a reduced daily dose of zidovudine in patients with the acquired immunodeficiency syndrome: The AIDS Clinical Trials Group. *N. Engl. J. Med.* **323**, 1009–1014.

Fischl, M. A., Richman, D. D., Grieco, M. H., Gottlieb, M. S., Volberding, P. A., Laskin, O. L., Leedom, J. M., Groopman, J. E., Mildvan, D., Schooley, R. T., *et al.* (1987). The efficacy of azidothymidine (AZT) in the treatment of patients with AIDS and AIDS-related complex. A double-blind, placebo-controlled trial. *N. Engl. J. Med.* **317**, 185–191.

Fischl, M. A., Richman, D. D., Hansen, N., Collier, A. C., Carey, J. T., Para, M. F., Hardy, W. D., Dolin, R., Powderly, W. G., Allan, J. D., *et al.* (1990b). The safety and efficacy of zidovudine (AZT) in the treatment of subjects with mildly symptomatic human immunodeficiency virus type 1 (HIV) infection: A double-blind, placebo-controlled trial: The AIDS Clinical Trials Group. *Ann. Intern. Med.* **112**, 727–737.

Fischl, M. A., Stanley, K., Collier, A. C., Arduino, J. M., Stein, D. S., Feinberg, J. E., Allan, J. D., Goldsmith, J. C., and Powderly, W. G. (1995). Combination and monotherapy with zidovudine and zalcitabine in patients with advanced HIV disease: The NIAID AIDS Clinical Trials Group. *Ann. Intern. Med.* **122**, 24–32.

Fiske, W., Benedek, I. H., Joseph, J. L., Dennis, S., O'Dea, R., Hsu, A., and Kornhauser, D. M. (1998a). Pharmacokinetics of efavirenz (EFV) and ritonavir (RIT) after multiple oral doses in healthy volunteers. *Int. Conf. AIDS* **12**, 827.

Fiske, W. D., Benedek, I. H., White, S. J., Pepperess, K. A., Joseph, J. L., and Kornhauser, D. M. (1998b). Pharmacokinetic interaction between efavirenz (EFV) and nelfinavir mesylate (NFV) in healthy volunteers. *5th Conf. Retrovir. Oppor. Infect.* Abstract 349.

Fletcher, C. V., Rhame, F. S., Beatty, C. C., Simpson, M., and Balfour, H. H., Jr. (1992). Comparative pharmacokinetics of zidovudine in healthy volunteers and in patients with AIDS with and without hepatic disease. *Pharmacotherapy* **12**, 429–434.

Furman, P. A., Fyfe, J. A., St. Clair, M. H., Weinhold, K., Rideout, J. L., Freeman, G. A., Lehrman, S. N., Bolognesi, D. P., Broder, S., Mitsuya, H., *et al.* (1986). Phosphorylation of 3'-azido-3'-deoxythymidine and selective interaction of the 5'-triphosphate with human immunodeficiency virus reverse transcriptase. *Proc. Natl. Acad. Sci. USA* **83**, 8333–8337.

Gallicano, K. D., Tobe, S., Sahai, J., McGilveray, I. J., Cameron, D. W., Kriger, F., and Garber, G. (1992). Pharmacokinetics of single and chronic dose zidovudine in two HIV positive patients undergoing continuous ambulatory peritoneal dialysis (CAPD). *J. Acquir. Immune Defic. Syndr.* **5**, 242–250.

Gallo, R. C., *et al.* (1988). AIDS in 1988. *Sci. Am.* **259**, 41–48.

Gerstoft, J., and Group, T. A. S. (1997). *In* "Interscience Conference on Antimicrobial Agents and Chemotherapy." American Society of Microbiology, Toronto.

Gisolf, E., Colebunders, R., Van Wanzeele, F., Van Der Ende, M., Koopmans, P., Portegies, P., Hoetelmans, R., Japour, A., Ward, P., De Wolf, F., and Danner, S. (1998). Treatment with ritonavir/saquinavir versus ritonavir/saquinavir/stavudine. *5th Conf. Retrovir. Oppor. Infect.* Abstract 389.

Gulick, R. M., Mellors, J. W., Havlir, D., Eron, J. J., Gonzalez, C., McMahon, D., Jonas, L., Meibohm, A., Holder, D., Schleif, W. A., Condra, J. H., Emini, E. A., Isaacs, R., Chodakewitz, J. A., and Richman, D. D. (1998). Simultaneous vs sequential initiation of therapy with indinavir, zidovudine, and lamivudine for HIV-1 infection: 100-week follow-up. *J. Am. Med. Assoc.* **280**, 35–41.

Gulick, R. M., Mellors, J. W., Havlir, D., Eron, J. J., Gonzalez, C., McMahon, D., Richman, D. D., Valentine, F. T., Jonas, L., Meibohm, A., Emini, E. A., and Chodakewitz, J. A. (1997). Treatment with indinavir, zidovudine, and lamivudine in adults with human immunodeficiency virus infection and prior antiretroviral therapy. *N. Engl. J. Med.* **337**, 734–739.

Hamilton, J. D., Hartigan, P. M., Simberkoff, M. S., Day, P. L., Diamond, G. R., Dickinson, G. M., Drusano, G. L., Egorin, M. J., George, W. L., Gordin, F. M., *et al.* (1992). A controlled trial of early versus late treatment with zidovudine in symptomatic human immunodeficiency virus infection: Results of the Veterans Affairs Cooperative Study. *N. Engl. J. Med.* **326**, 437–443.

Hammer, S. M., Katzenstein, D. A., Hughes, M. D., Gundacker, H., Schooley, R. T., Haubrich, R. H., Henry, W. K., Lederman, M. M., Phair, J. P., Niu, M., Hirsch, M. S., and Merigan, T. C. (1996). A trial comparing nucleoside monotherapy with combination therapy in HIV-infected adults with CD4 cell counts from 200 to 500 per cubic millimeter: AIDS Clinical Trials Group Study 175 Study Team. *N. Engl. J. Med.* **335**, 1081–1090.

Hammer, S. M., Squires, K. E., Hughes, M. D., Grimes, J. M., Demeter, L. M., Currier, J. S., Eron, J. J., Jr., Feinberg, J. E., Balfour, H. H., Jr., Deyton, L. R., Chodakewitz, J. A., and Fischl, M. A. (1997). A controlled trial of two nucleoside analogues plus indinavir in persons with human immunodeficiency virus infection and CD4 cell counts of 200 per cubic millimeter or less: AIDS Clinical Trials Group 320 Study Team. *N. Engl. J. Med.* **337**, 725–733.

Havlir, D., Cheeseman, S. H., McLaughlin, M., Murphy, R., Erice, A., Spector, S. A., Greenough, T. C., Sullivan, J. L., Hall, D., Myers, M., *et al.* (1995). High-dose nevirapine: Safety, pharmacokinetics, and antiviral effect in patients with human immunodeficiency virus infection. *J. Infect. Dis.* **171**, 537–545.

Havlir, D. V., Marschner, I. C., Hirsch, M. S., Collier, A. C., Tebas, P., Bassett, R. L., Ioannidis, J. P., Holohan, M. K., Leavitt, R., Boone, G., and Richman, D. D. (1998). Maintenance antiretroviral therapies in HIV infected patients with undetectable plasma HIV RNA after triple-drug therapy: AIDS Clinical Trials Group Study 343 Team. *N. Engl. J. Med.* **339,** 1261–1268.

Hecht, F. M., Chesney, M. A., Busch, M. P., Rawal, B. D., Staprans, S. I. and Kahn, J. O. (1998). Treatment of primary HIV with AZT, 3TC, and indinavir. *5th Conf. Retrovir. Oppor. Infect.* **189.**

Helbert, M., Fletcher, T., Peddle, B., Harris, J. R., and Pinching, A. J. (1988). Zidovudine-associated myopathy. *Lancet* **2,** 689–690.

Hicks, C., Hass, D., Seekins, D., Cooper, R., Gallant, J., Carpenter, C., Ruiz, N. M., Manion, D. J., Ploughman, L. M., and Labriola, D. F. (1998). A phase II, double-blind, placebo-controlled, dose ranging study to assess the antiretroviral activity and safety of DMP 266 (efavirenz, SUSTIVA) in combination with open-label zidovudine (ZDV) with lamivudine (3TC) [DMP 266-005]. *5th Conf. Retrovir. Oppor. Infect.* Abstract 698.

Hilts, A. E., and Fish, D. N. (1998). Dosage adjustment of antiretroviral agents in patients with organ dysfunction. *Am. J. Health Syst. Pharm.* **55,** 2528–2533.

Hirsch, M. S., Conway, B., D'Aquila, R. T., Johnson, V. A., Brun-Vezinet, F., Clotet, B., Demeter, L. M., Hammer, S. M., Jacobsen, D. M., Kuritzkes, D. R., Loveday, C., Mellors, J. W., Vella, S., and Richman, D. D. (1998). Antiretroviral drug resistance testing in adults with HIV infection: Implications for clinical management: International AIDS Society—USA Panel. *J. Am. Med. Assoc.* **279,** 1984–1991.

Ho, D. D., Moudgil, T., and Alam, M. (1989). Quantitation of human immunodeficiency virus type 1 in the blood of infected persons. *N. Engl. J. Med.* **321,** 1621–1625.

Hoetelmans, R. M., van Heeswijk, R. P., Profijt, M., Mulder, J. W., Meenhorst, P. L., Lange, J. M., Reiss, P., and Beijnen, J. H. (1998). Comparison of the plasma pharmacokinetics and renal clearance of didanosine during once and twice daily dosing in HIV-1 infected individuals. *AIDS* **12,** F211–F216.

Horwitz, J., Chua, J., and Noel, M. (1964). The monomesylates of 1-(2'-deoxy-beta-D-lyxofuranosyl) thymine. *J. Org. Chem.* **29,** 2076–2078.

Hsu, A., *et al.* (1998). Pharmacokinetic interaction between ritonavir and indinavir in healthy volunteers. *Antimicrob. Agents Chemother.* **42,** 2784–2791.

Johns, D. G., and Gao, W. Y. (1998). Selective depletion of DNA precursors: An evolving strategy for potentiation of dideoxynucleoside activity against human immunodeficiency virus. *Biochem. Pharmacol.* **55,** 1551–1556.

Johnson, M., Petersen, A., Winslade, J., and Clendeninn, N. (1998). Comparison of BID and TID dosing of Viracept (nelfinavir, NFV) in combination with stavudine (d4T) and lamivudine (3TC). *5th Conf. Retrovir. Oppor. Infect.* Abstract 373.

Kahn, J. O., Lagakos, S. W., Richman, D. D., Cross, A., Pettinelli, C., Liou, S. H., Brown, M., Volberding, P. A., Crumpacker, C. S., Beall, G., et al. (1992). A controlled trial comparing continued zidovudine with didanosine in human immunodeficiency virus infection: The NIAID AIDS Clinical Trials Group. *N. Engl. J. Med.* **327,** 581–587.

Kaspar, J., Werntz, G., and DuBois, D. B. (1998). Early results of combined stavudine, lamivudine, and nevirapine: a twice daily, well-tolerated, protease inhibitor-sparing regimen for the treatment to HIV-1 infection. *5th Conf. Retrovir. Oppor. Infect.* Abstract 696.

Katzenstein, D. A., Hammer, S. M., Hughes, M. D., Gundacker, H., Jackson, J. B., Fiscus, S., Rasheed, S., Elbeik, T., Reichman, R., Japour, A., Merigan, T. C., and Hirsch, M. S. (1996a). The relation of virologic and immunologic markers to clinical outcomes after nucleoside therapy in HIV-infected adults with 200 to 500 CD4 cells per cubic millimeter: AIDS Clinical Trials Group Study 175 Virology Study Team. *N. Engl. J. Med.* **335,** 1091–1098.

Katzenstein, D. A., Mbizvo, M., Zijenah, L., Gittens, T., Munjoma, M., Hill, D., Madzime, S., and Maldonado, Y. (1999). Serum level of maternal human immunodeficiency virus (HIV) RNA, infant mortality, and vertical transmission of HIV in Zimbabwe. *J. Infect. Dis.* **179,** 1382–1387.

Katzenstein, T. L., Pedersen, C., Nielsen, C., Lundgren, J. D., Jakobsen, P. H., and Gerstoft, J. (1996b). Longitudinal serum HIV RNA quantification: Correlation to viral phenotype at seroconversion and clinical outcome. *AIDS* **10**, 167–173.

Kaul, D. R., *et al.* (1999). HIV protease inhibitors: Advances in therapy and adverse reactions, including metabolic complications. *Pharmacotherapy* **19**, 281–298.

Kilby, J. M., Hopkins, S., Venetta, T. M., DiMassimo, B., Cloud, G. A., Lee, J. Y., Alldredge, L., Hunter, E., Lambert, D., Bolognesi, D., Matthews, T., Johnson, M. R., Nowak, M. A., Shaw, G. M., and Saag, M. S. (1998). Potent suppression of HIV-1 replication in humans by T-20, a peptide inhibitor of gp41-mediated virus entry. *Nat. Med.* **4**, 1302–1307.

Kost, R., Cao, Y., Vesanen, M., Talal, A., Hurley, A., Schluger, R., Monard, S., Rogers, M., Johnson, J., Smiley, L., Ho, D., and Markowitz, M. (1998). Combination therapy with abacavir (1592), 141W94, and AZT/3TC in subjects acutely and chronically infected with HIV. *5th Conf. Retrovir. Oppor. Infect.* Abstract 363.

Kremer, D., Munar, M. Y., Kohlhepp, S. J., Swan, S. K., Stinnett, E. A., Gilbert, D. N., Young, E. W., and Bennett, W. M. (1992). Zidovudine pharmacokinetics in five HIV seronegative patients undergoing continuous ambulatory peritoneal dialysis. *Pharmacotherapy* **12**, 56–60.

Lafeuillade, A., *et al.* (1997). Effects of a combination of zidovudine, didanosine, and lamivudine on primary human immunodeficiency virus type 1 infection. *J. Infect. Dis.* **175**, 1051–1055.

Lamson, M. (1993). Pharmacokinetics of Nevirapine following single and multiple doses of 12.5, 50, 200 and 400 mg/d to HIV-infected patients. *Natl. Conf. Hum. Retroviruses Relat. Infect.* Abstract 160.

Lange, J. M. (1995). Current HIV clinical trial design issues. *J. Acquir. Immune Defic. Syndr. Hum. Retrovirol.* **10**(Suppl. 1), S47–S51.

Lea, A. P., and Faulds, D. (1996). Stavudine: A review of its pharmacodynamic and pharmacokinetic properties and clinical potential in HIV infection. *Drugs* **51**, 846–864.

Lin, J. H., Chiba, M., Balani, S. K., Chen, I. W., Kwei, G. Y., Vastag, K. J., and Nishime, J. A. (1996). Species differences in the pharmacokinetics and metabolism of indinavir, a potent human immunodeficiency virus protease inhibitor. *Drug Metab. Dispos.* **24**, 1111–1120.

Lori, F., Malykh, A., Cara, A., Sun, D., Weinstein, J. N., Lisziewicz, J., and Gallo, R. C. (1994). Hydroxyurea as an inhibitor of human immunodeficiency virus-type 1 replication. *Science* **266**, 801–805.

Marschner, I. C., Collier, A. C., Coombs, R. W., D'Aquila, R. T., DeGruttola, V., Fischl, M. A., Hammer, S. M., Hughes, M. D., Johnson, V. A., Katzenstein, D. A., Richman, D. D., Smeaton, L. M., Spector, S. A., and Saag, M. S. (1998). Use of changes in plasma levels of human immunodeficiency virus type 1 RNA to assess the clinical benefit of antiretroviral therapy. *J. Infect. Dis* **177**, 40–47.

Mathez, D., Bagnarelli, P., Gorin, I., *et al.* (1997). Reductions in viral load and increases in T-lymphocyte numbers in treatment-naive patients with advanced HIV-1 infection treated with ritonavir, zidovudine and zalcitabine triple therapy. *Antivir. Ther.* **2**, 175–183.

McElrath, J., Malhotra, U., Musey, L., Berry, M., Huang, Y., and Corey, L. (1998). Improved cellular immunity in acute HIV-1 infection following antiretroviral therapy. *Int. Conf. AIDS* **12**, 530.

McLeod, G. X., and Hammer, S. M. (1992). Zidovudine: Five years later. *Ann. Intern. Med.* **117**, 487–501.

Mellors, J. W., Rinaldo, C. R., Jr., Gupta, P., White, R. M., Todd, J. A., and Kingsley, L. A. (1996). Prognosis in HIV-1 infection predicted by the quantity of virus in plasma. *Science* **272**, 1167–1170.

Merry, C., Barry, M. G., Mulcahy, F., Halifax, K. L., and Back, D. J. (1997). Saquinavir pharmacokinetics alone and in combination with nelfinavir HIV-infected patients. *AIDS* **11**, F117–F120.

Miller, V., de Bethune, M. P., Kober, A., Sturmer, M., Hertogs, K., Pauwels, R., Stoffels, P., and Staszewski, S. (1998). Patterns of resistance and cross-resistance to human immunodeficiency virus type 1 reverse transcriptase inhibitors in patients treated with the nonnucleoside reverse transcriptase inhibitor loviride. *Antimicrob. Agents Chemother.* **42,** 3123–3129.

Mitsuya, H., Weinhold, K. J., Furman, P. A., St. Clair, M. H., Lehrman, S. N., Gallo, R. C., Bolognesi, D., Barry, D. W., and Broder, S. (1985). 3'-Azido-3'-deoxythymidine (BW A509U): An antiviral agent that inhibits the infectivity and cytopathic effect of human T-lymphotropic virus type III/lymphadenopathy-associated virus *in vitro. Proc. Natl. Acad. Sci. USA* **82,** 7096–7100.

Montaner, J., Reiss, P., Cooper, D., Vella, S., Dohnanyi, C., Harris, M., Conway, B., Wainberg, M., Smith, D., Robinson, P., Myers, M., Hall, D., and Lange, J. M. (1998a). Long-term follow-up of patients treated with nevirapine (NVP) based combination therapy within the INCAS trial. *5th Conf. Retrovir. Oppor. Infect.* Abstract 695.

Montaner, J. S., Reiss, P., Cooper, D., Vella, S., Harris, M., Conway, B., Wainberg, M. A., Smith, D., Robinson, P., Hall, D., Myers, M., and Lange, J. M. (1998b). A randomized, double-blind trial comparing combinations of nevirapine, didanosine, and zidovudine for HIV-infected patients: The INCAS Trial: Italy, The Netherlands, Canada and Australia Study. *J. Am. Med. Assoc.* **279,** 930–937.

Moore, K. H., Yuen, G. J., Raasch, R. H., Eron, J. J., Martin, D., Mydlow, P. K., and Hussey, E. K. (1996). Pharmacokinetics of lamivudine administered alone and with trimethoprim-sulfamethoxazole. *Clin. Pharmacol. Ther.* **59,** 550–558.

Morse, G. D., Shelton, M. J., Hewitt, R. G., Adams, J. M., Baldwin, J. R., Della-Coletta, A. A., Freimuth, W. W., and Cox, S. R. (1998). Ritonavir (RIT) pharmacokinetics (PK) during combination therapy with delavirdine (DLV). *5th Conf. Retrovir. Oppor. Infect.* Abstract 343.

Murphy, R., Degruttola, V., Gulick, R., D'Aquila, R., Eron, J., Sommadossi, J. P., Smeaton, L., Currier, J., Tung, R., and Kuritzkes, D. (1998). 141W94 with or without zidovudine/3TC in patients with no prior protease inhibitor or 3TC therapy-ACTG 347. *5th Conf. Retrovir. Oppor. Infect.* Abstract 512.

Murphy, R., Gagnier, P., Lamson, M., Dusek, A., Ju, W., and Hsu, A. (1997). Effect of nevirapine (NVP) on pharmacokinetics (PK) of indinavir (IDV) and ritonavir (RTV) in HIV-1 patients. *4th Conf. Retro. Oppor. Infect.* Abstract 374.

Murphy, R. L., Sommadossi, J. P., Lamson, M., Hall, D. B., Myers, M., and Dusek, A. (1999). Antiviral effect and pharmacokinetic interaction between nevirapine and indinavir in persons infected with human immunodeficiency virus type 1. *J. Infect. Dis.* **179,** 1116–1123.

Nadler, J. P. (1996). Early initiation of antiretroviral therapy for infection with human immunodeficiency virus: Considerations in 1996. *Clin. Infect. Dis.* **23,** 227–230. Review.

Niu, M. T., Bethel, J., Holodniy, M., Standiford, H. C., and Schnittman, S. M. (1998). Zidovudine treatment in patients with primary (acute) human immunodeficiency virus type 1 infection: A randomized, double-blind, placebo-controlled trial: DATRI 002 Study Group: Division of AIDS Treatment Research Initiative. *J. Infect. Dis.* **178,** 80–91.

Ostertag, W., Roesler, G., Krieg, C. J., Kind, J., Cole, T., Crozier, T., Gaedicke, G., Steinheider, G., Kluge, N., and Dube, S. (1974). Induction of endogenous virus and of thymidine kinase by bromodeoxyuridine in cell cultures transformed by Friend virus. *Proc. Natl. Acad. Sci. USA* **71,** 4980–4985.

Pantaleo, G., Cohen, O. J., Schacker, T., Vaccarezza, M., Graziosi, C., Rizzardi, G. P., Kahn, J., Fox, C. H., Schnittman, S. M., Schwartz, D. H., Corey, L., and Fauci, A. S. (1998). Evolutionary pattern of human immunodeficiency virus (HIV) replication and distribution in lymph nodes following primary infection: Implications for antiviral therapy. *Nat. Med.* **4,** 341–345.

Para, M. F., Fischl, M., Meehan, P., Morse, G., Wood, K., Shafer, R., Freimuth, W., Demeter, L., Holden-Wiltse, J., and Nevin, T. (1996). ACTG 260: Randomized phase I/II concentration-controlled trial of the anti-HIV activity of delavirdine. *3rd Conf. Retro. Opport. Infect.* Abstract 163.

Parker, W. B., Shaddix, S. C., Bowdon, B. J., Rose, L. M., Vince, R., Shannon, W. M., and Bennett, L. L., Jr. (1993). Metabolism of carbovir, a potent inhibitor of human immunodeficiency virus type 1, and its effects on cellular metabolism. *Antimicrob. Agents Chemother.* **37**, 1004–1009.

Perelson, A. S., Essunger, P., Cao, Y., Vesanen, M., Hurley, A., Saksela, K., Markowitz, M., and Ho, D. D. (1997). Decay characteristics of HIV-1-infected compartments during combination therapy. *Nature* **387**, 188–191.

Perelson, A. S., Neumann, A. U., Markowitz, M., Leonard, J. M., and Ho, D. D. (1996). HIV-1 dynamics *in vivo*: Virion clearance rate, infected cell life-span, and viral generation time. *Science* **271**, 1582–1586.

Perrin, L., Markowitz, M., Calandra, G., and Chung, M. (1997). An open treatment study of acute HIV infection with zidovudine, lamivudine and indinavir sulfate. *4th Conf. Retro. Opport. Infect.* Abstract 238.

Perry, C. M., and Noble, S. (1998). Saquiavir soft-gel capsule formulation: A review of its use in patients with HIV infection. *Drugs* **55**, 461–486.

Pialoux, G., Raffi, F., Brun-Vezinet, F., Meiffredy, V., Flandre, P., Gastaut, J. A., Dellamonica, P., Yeni, P., Delfraissy, J. F., and Aboulker, J. P. (1998). A randomized trial of three maintenance regimens given after three months of induction therapy with zidovudine, lamivudine, and indinavir in previously untreated HIV-1-infected patients: Trilege (Agence Nationale de Recherches sur le SIDA 072) Study Team. *N. Engl. J. Med.* **339**, 1269–1276.

Piscitelli, S., Vogel, S., Sadler, B., Fiske, W., Metcalf, J., Masur, H., and Falloon, J. (1998). Effect of efavirenz (DMP 266) on the pharmacokinetics of 141 W94 in HIV-infected patients. *5th Conf. Retrovir. Oppor. Infect.* Abstract 346.

Potter, C. W., Phair, J. P., Vodinelich, L., Fenton, R., and Jennings, R. (1976). Antiviral, immunosuppressive and antitumour effects of ribavirin. *Nature* **259**, 496–497.

Powderly, W., Sension, M., Conant, M., Stein, A., and Clendeninn, N. (1997). The efficacy of Viracept (nelfinavir mesylate, NFV) in pivotal phase II/III double-blind randomized controlled trials as monotherapy and in combination with d4T or AZT/3TC. *4th Conf. Retro. Oppor Infect.* Abstract 370.

Powderly, W. G., *et al.* (1998). Recovery of the immune system with antiretroviral therapy: The end of opportunism? *J. Am. Med. Assoc.* **280**, 72–77.

Quart, B. D., Chapman, S. K., Peterkin, J., Webber, S., and Oliver, S. (1995). Phase I safety, tolerance, pharmacokinetics and food effect studies of AG1343—A novel protease inhibitor. *Natl. Conf. Hum. Retroviruses Relat. Infect.*, **167.**

Riddler, S., Kahn, J., Hicks, C., Havlir, D., Stein, D., Horton, J., and Ruiz, N. (1998). Durable clinical anti-HIV-1 activity (72 weeks) and tolerability for efavirenz (DMP 266) in combination with indinavir (IDV) [DMP 266-003, Cohort IV]. *Int. Conf. AIDS* **12**, 85.

Riddler, S., Mayers, D., Wagner, K., Bach, M., Havlir, D., Stein, D., and Kahn, J. (1997). A double-blind pilot study to evaluate the antiviral activity, tolerability of DMP 266 alone and in combination with indinavir. *4th Conf. Retro. Oppor. Infect.*

Rimsky, L. T., Shugars, D. C., and Matthews, T. J. (1998). Determinants of human immunodeficiency virus type 1 resistance to gp41-derived inhibitory peptides. *J. Virol.* **72**, 986–993.

Ruiz, N. (1997). A double-blind pilot study to evaluate the antiretroviral activity, tolerability of DMP 266 in combination with indinavir (cohort III). *4th Conf. Retro. Oppor. Infect.* Abstract L132.

Rutschmann, O. T., Opravil, M., Iten, A., Malinverni, R., Vernazza, P. L., Bucher, H. C., Bernasconi, E., Sudre, P., Leduc, D., Yerly, S., Perrin, L. H., and Hirschel, B. (1998). A placebo-controlled trial of didanosine plus stavudine, with and without hydroxyurea, for HIV infection: The Swiss HIV Cohort Study. *AIDS* **12**, F71–F77.

Saag, M. S. (1999). *In* "HIV Information Network," pp. 1–5.

Sachsenberg, N., Schockmel, G. A., and Perrin, L. (1998). T lymphocyte dynamics in primary HIV infected patients treated with HAART. *Int. Conf. AIDS* **12**, 50.

Sadler, B., Gillotin, C., Chittick, G. E., and Symonds, W. T. (1998). Pharmacokinetic drug interactions with amprenavir. *Int. Conf. AIDS* **12**, 91.

Sahai, J., Cameron, W., Salgo, M., Stewart, F., Myers, M., Lamson, M., and Gagnier, P. (1997). Drug interaction study between saquinavir (SQV) and nevirapine (NVP). *4th Conf. Retro. Oppor. Infect.* Abstract 614.

Schacker, T. W., Hughes, J. P., Shea, T., Coombs, R. W., and Corey, L. (1998). Biological and virologic characteristics of primary HIV infection. *Ann. Intern. Med.* **128**, 613–620.

Sekaly, R. P., Ringuette, N., Lacaille, J., Routy, J. P., Ho, D. D., Markowitz, M., and Conway, B. (1998). Rapid reconstitution of the T-cell receptor (TCR) repertoire following antiretroviral treatment during primary HIV infection. *Int. Conf. AIDS* **12**, 48.

Shaffer, N., Roongpisuthipong, A., Siriwasin, W., Chotpitayasunondh, T., Chearskul, S., Young, N. L., Parekh, B., Mock, P. A., Bhadrakom, C., Chinayon, P., Kalish, M. L., Phillips, S. K., Granade, T. C., Subbarao, S., Weniger, B. G., and Mastro, T. D. (1999). Maternal virus load and perinatal human immunodeficiency virus type 1 subtype E transmission, Thailand: Bangkok Collaborative Perinatal HIV Transmission Study Group. *J. Infect. Dis.* **179**, 590–599.

Singlas, E., Taburet, A. M., Borsa Lebas, F., Parent de Curzon, O., Sobel, A., Chauveau, P., Viron, B., al Khayat, R., Poignet, J. L., Mignon, F., *et al.* (1992). Didanosine pharmacokinetics in patients with normal and impaired renal function: Influence of hemodialysis. *Antimicrob. Agents Chemother.* **36**, 1519–1524.

Skowron, G. (1995). Biologic effects and safety of stavudine: overview of phase I and II clinical trials. *J. Infect. Dis.* **171**(Suppl. 2), S113–S117.

Sperling, R. S., Shapiro, D. E., Coombs, R. W., Todd, J. A., Herman, S. A., McSherry, G. D., O'Sullivan, M. J., Van Dyke, R. B., Jimenez, E., Rouzioux, C., Flynn, P. M., and Sullivan, J. L. (1996). Maternal viral load, zidovudine treatment, and the risk of transmission of human immunodeficiency virus type 1 from mother to infant: Pediatric AIDS Clinical Trials Group Protocol 076 Study Group. *N. Engl. J. Med.* **335**, 1621–1629.

Spruance, S. L., Pavia, A. T., Mellors, J. W., Murphy, R., Gathe, J., Jr., Stool, E., Jemsek, J. G., Dellamonica, P., Cross, A., and Dunkle, L. (1997). Clinical efficacy of monotherapy with stavudine compared with zidovudine in HIV-infected, zidovudine-experienced patients: A randomized, double-blind, controlled trial: Bristol-Myers Squibb Stavudine/019 Study Group. *Ann. Intern. Med.* **126**, 355–363.

Spruance, S. L., Pavia, A. T., Peterson, D., Berry, A., Pollaxd, R., Patterson, T. F., Frank, I., Remick, S. C., Thompson, M., MacArthur, R. D., *et al.* (1994). Didanosine compared with continuation of zidovudine in HIV-infected patients with sings of clinical deterioration while receiving zidovudine: A randomized, double-blind clinical trial: The Bristol-Myers Squibb AI454–010 Study Group. *Ann. Intern. Med.* **120**, 360–368.

Stanley, S. K., and Kaplan, J. E. (1998). Guidelines for the use of antiretroviral agents in HIV-infected adults and adolescents. *Morbid. Mortal. Weekly Rep.* **47** (RR-5), 42–82.

Staszewski, S., Morales-Ramirez, J., Flanigan, T., Hardy, D., Johnson, P., Nelson, M., and Ruiz, N. (1998). A phase II, multicenter, randomized, open-label study to compare the antiretroviral activity and tolerability of efavirenz (EFV) + indinavir (IDV), versus EFV + zidovudine (ZDV) + lamivudine (3TC), versus IDV + ZDV + 3TC at 24 weeks [DMP 266-006]. *Int. Conf. AIDS* **12**, 330.

Taburet, A. M., Naveau, S., Zorza, G., Colin, J. N., Delfraissy, J. F., Chaput, J. C., and Singlas, E. (1990). Pharmacokinetics of zidovudine in patients with liver cirrhosis. *Clin. Pharmacol. Ther.* **47**, 731–739.

Tavel, J. A., Miller, K. D., and Masur, H. (1999). Guide to major clinical trials of antiretroviral therapy in human immunodeficiency virus-infected patients: Protease inhibitors, nonnucleoside reverse transcriptase inhibitors, and nucleotide reverse transcriptase inhibitors. *Clin. Infect. Dis.* **28**, 643–676.

Till, M., and MacDonell, K. B. (1990). Myopathy with human immunodeficiency virus type 1 (HIV-1) infection: HIV-1 or zidovudine? *Ann. Intern. Med.* **113**, 492–494.

Todd, J., Pachl, C., White, R., Yeghiazarian, T., Johnson, P., Taylor, B., Holodniy, M., Kern, D., Hamren, S., Chernoff, D., *et al.* (1995). Performance characteristics for the quantitation of plasma HIV-1 RNA using branched DNA signal amplification technology. *J. Acquir. Immune Defic. Syndr. Hum. Retrovirol.* **10**, S35–S44.

Volberding, P. A., Lagakos, S. W., Koch, M. A., Pettinelli, C., Myers, M. W., Booth. D. K., Balfour, H. H., Jr., Reichman, R. C., Bartlett, J. A., Hirsch, M. S., *et al.* (1990). Zidovudine in asymptomatic human immunodeficiency virus infection. A controlled trial in persons with fewer than 500 CD4-positive cells per cubic millimeter: The AIDS Clinical Trials Group of the National Institute of Allergy and Infectious Diseases. *N. Engl. J. Med.* **322**, 941–949.

Watkins, B. A., Klotman, M. E., and Gallo, R. C. (1995). Human immunodeficiency viruses. *In* "Principles and Practices of Infectious Diseases" (G. L. Mandell, J. E. Bennet, and R. Dolin, Eds.), 4th ed., pp. 1590–1606. Churchill–Livingstone, New York.

Wlodawer, A., *et al.* (1998). Inhibitors of HIV-1 protease: A major success of structure-assisted drug design. *Annu. Rev. Biophys. Biomol. Struct.* **27**, 249–284.

Workman, C., Musson, R., Dyer, W., and Sullivan, J. (1998). Novel double protease combinations-combining indinavir (IDV) with ritonavir (RTV): Results from first study. *Int. Conf. AIDS* **12**, 338.

Yeh, K. C., Deutsch, P. J., Haddix, H., Hesney, M., Hoagland, V., Ju, W. D., Justice, S. J., Osborne, B., Sterrett, A. T., Stone, J. A., Woolf, E., and Waldman, S. (1998). Single-dose pharmacokinetics of indinavir and the effect of food. *Antimicrob. Agents Chemother.* **42**, 332–338.

Zaunders, J., Kelleher, A., Cunningham, P., Smith, D., Carr, A., and Cooper, D. A. (1998). Effect of combination therapy on CD8 T cell activation markers in primary HIV infection. *5th Conf. Retrovir. Oppor. Infect.* Abstract 592.

François Clavel
Esther Race
Fabrizio Mammano

Laboratoire de Recherche Antivirale
IMEA/INSERM
Hôpital Bichat-Claude Bernard
75018 Paris, France

HIV Drug Resistance and Viral Fitness

I. Introduction

Drug resistance is one of the main challenges faced today in the therapy of HIV infection. Soon after the introduction of the first antiretroviral agents active against HIV, it became clear that therapeutic regimens using single or dual combinations of antiviral agents were usually unable to yield stable and satisfactory virological and clinical responses. In all of these cases, a relapse in virus replication was observed, accompanied by immunological and clinical deterioration, and paralleled by emergence of viruses with decreased susceptibility to the treatment drugs. The first documented cases of HIV drug resistance concerned AZT (zidovudine), a nucleoside analog that inhibits HIV replication by virtue of its activity as a DNA chain terminator. Several mutations in the viral reverse transcriptase (RT), the enzyme that transcribes virion RNA into DNA in the newly infected cell, were identified as able to promote significant resistance to AZT (Larder and Kemp, 1989).

Advances in Pharmacology, Volume 49

Thereafter, the introduction of new drugs active against HIV in combined therapeutic regimens went along with the sobering finding that more mutations were readily selected during therapy, leading to HIV resistance to multiple nucleoside analogs (Eron *et al.*, 1993). Subsequently, the use of antiretroviral agents belonging to two new classes, protease inhibitors (PI) and nonnucleosidic RT inhibitors (NNRTI), was also paralleled by the discovery of mutations in the viral protease (PR) or in RT that could confer high-level HIV resistance to these new drugs (Schinazi *et al.*, 1997). Nonetheless, these two new classes of antiretroviral agents appeared to display a stronger overall antiviral activity than the drugs in clinical use before them, and their introduction, along with the availability of precise tools measuring plasma virus load, aided in key discoveries on the dynamics of HIV infection *in vivo*. Analysis of the kinetics of the decrease of viral load in peripheral blood aided in the discovery that most HIV particles detected in peripheral blood samples will have been produced a few hours earlier by cells that had become infected only 1 or 2 days earlier (Coffin, 1995, Perelson, 1996; Ho *et al.*, 1995; Wei *et al.*, 1995). The strong and rapid antiviral response seen with PIs and NNRTIs was attributable to this high virus and cell turnover (Nowak *et al.*, 1997), now a recognized hallmark of HIV infection *in vivo*, which makes this virus potentially remarkably sensitive to any agent that could efficiently inhibit any step of its life cycle. On the other hand, because continuously ongoing cycles of HIV replication are accompanied by repeated rounds of viral DNA synthesis by RT, a notoriously error-prone enzyme (Battula and Loeb, 1976; Bebenek *et al.*, 1989; Preston *et al.*, 1988; Roberts *et al.*, 1988), the introduction of random mutations in the viral genome occurs at a high frequency. Therefore, the probability that virus quasispecies carrying at least one mutation conferring decreased susceptibility to a single antiretroviral agent are present before the introduction of this agent is generally high (Coffin, 1995; Lech *et al.*, 1996; Ribeiro *et al.*, 1998). However, the impact of such single mutations on the level of resistance can be very different from one drug to another. Indeed, while single mutations can confer high-level resistance to a nucleoside analog such as 3TC (lamivudine) or to some of the NNRTIs, it takes the stepwise accumulation of several mutations to confer full resistance to PIs or even to AZT (Boucher *et al.*, 1993; Larder and Kemp, 1989; Richman *et al.*, 1994). In this context, because the accumulation of mutations is only possible following prolonged viral replication under drug selection, it is now recognized that emergence of HIV drug resistance can only be prevented by the use of maximally active drug combinations that completely suppress virus replication (Gulick *et al.*, 1997; Hammer *et al.*, 1997).

The other parameter that is key to the selection of resistance mutations is their effect on the function of the enzyme, PR or RT. It was assumed, for example, that because PR is a relatively small molecule, there might be less leeway for selection of resistance mutations than for RT, a much larger molecule (Debouck, 1992). But even in RT, published reports suggested at one point that HIV may completely lose its replicative capacity following the emergence of combinations of mutations conferring resistance to several nucleoside analogs and to NNRTIs, raising hope that because of evolutionary limitations, HIV resistance to complex drug combinations may be impossible (Chow *et al.*, 1993). In fact, such findings were not confirmed (Larder *et al.*, 1993), and there have been no further observations of viruses in which combinations of mutations conferring resistance to agents used today in HIV therapy could lead to a complete replicative incapacitation of the virus. However, there is increasing evidence that under some circumstances, resistance mutations in HIV RT or PR can indeed reduce the replicative and maybe the pathogenic potential of this virus. Most HIV drug resistance mutations affect amino acid residues that are conserved throughout the HIV-1 lineage, suggesting that they are optimal for the full replicative potential of the virus in the absence of drug (Barrie *et al.*, 1996; Korber *et al.*, 1997; Winslow *et al.*, 1995). Although they all involve structural changes in PR or RT (Erickson, 1995; Huang *et al.*, 1998), the mechanisms of induction of resistance by these mutations appears very different from one drug to another. Resistance to AZT, for example, seems to be the consequence of an increased capacity by RT to excise newly incorporated AZT molecules from the nascent chain of viral DNA (Arion *et al.*, 1998). Resistance to 3TC, which is most often associated with an amino acid substitution at position 184 (methionine into valine: M184V), located next to the catalytic triad of aspartic acids constituting the core of the RT active site (Boucher *et al.*, 1993), involves reduced incorporation of 3TCTP into DNA through its misalignment with the 3' end of nascent DNA (Huang *et al.*, 1998). Resistance to NNRTIs is consecutive to changes in the structural or physical properties of a particular hydrophobic pocket in RT, toward which these molecules display a high affinity (Kohlstaedt *et al.*, 1992; Smerdon *et al.*, 1994). Similarly, reduced susceptibility to PIs is the consequence of mutations that generally reduce the affinity of the inhibitor for the enzyme (Erickson, 1995; Ridky, 1998). Several (but not all) of these changes are located within structures that are essential for a normal function of HIV RT or PR. Therefore, the selection of resistance mutations is always a compromise between the selective advantage conferred by the mutations in the presence of drug and any possible selective disadvantage in the absence of drug.

Here, we must introduce with care the word "fitness" in the context of HIV drug resistance, since several authors may consider the use of this typically Darwinian term inappropriate in some respects. Strictly speaking, the word fitness describes the selective advantage of an organism relative to a competitor in a particular environment. In this case, any experiment on HIV drug-resistant variants that would evaluate their relative competitive advantage or disadvantage relative to wild-type or relative to other mutants, in the presence or in the absence of drug, will indeed describe fitness. On the other hand, several authors have carried out comparative (but not competitive) studies with such mutants, using simplified tissue culture systems that measure, as quantitatively and as reproducibly as possible, the replicative efficiency of the virus. The question of whether viruses with reduced replicative efficiency or "fitness" may also behave as less pathogenic or less virulent viruses further confuses the situation. Indeed, some viruses could be less pathogenic while retaining apparently normal replication parameters in tissue culture systems or the other way around. An additional level of complexity arises with the consideration of whether different tissue culture conditions can affect these measurements. Therefore, for the sake of clarity and simplicity, even if the word may not be always strictly acceptable, we will use here the term "fitness" to describe the replicative capacity of HIV, even when not used within its strictly Darwinian, ecological meaning. As mentioned earlier, the development of HIV drug resistance, as for the selection of any genetic trait in a living organism, is a compromise between resistance in the presence of high concentrations of drug and replication in the absence of drug. Mutant viruses characterized as "resistant" are in fact optimally advantaged relative to their parental strains for limited ranges of drug concentration, which will be different from one mutant to another and from one drug to another. Apart from a few mutants that can generate maximal levels of resistance at a relatively low cost for virus fitness, most drug-resistant HIV variants are still sensitive to high concentrations of drugs. Given that drug concentrations *in vivo* can vary considerably as a function of time and from one tissue compartment to another (Kepler and Perelson, 1998; Wong *et al.*, 1997), the conditions of selection are unstable and uneven in a treated HIV patient. It is therefore easy to imagine how complex the conditions of mutant selection may be in most instances, especially when several drugs with different pharmacokinetic parameters are associated.

The selection of less fit resistant variants may not always be related to the direct impact of resistance mutations on virus infectivity. Other mechanisms, involving the stochastic selection of viruses having survived therapeutic intervention, can lead to a reduction in their replicative capacity. It is well recognized that repeated rounds of bottleneck selection exerted on a heterogeneous population of virus quasispecies can lead to the emergence of viruses with reduced fitness (Clarke *et al.*, 1993; Domingo and Holland, 1997; Novella *et al.*, 1999). This phenomenon, often referred to as "Muller's ratchet," has been thoroughly described following serial low-multiplicity passages of RNA viruses, including retroviruses and HIV (Yuste *et al.*, 1999). The Muller's ratchet phenomenon is related to the marked heterogeneity of RNA virus populations that can be harvested in culture or *in vivo*, where the majority of viral genomes bear random mutations that can be lethal or deleterious to their replicative fitness. *In vivo*, where viral turnover is high with massive production of viral particles, these mutations are efficiently competed out when a sufficient proportion of optimally fit quasispecies is present. Similarly, following virus passage in culture at a high multiplicity of infection, the chance that an optimally fit virus is present in a large inoculum is high. By contrast, when passage is reduced to a very small number of infectious units, that chance is reduced, leading to passive selection of viruses with reduced replicative capacity. In theory, the Muller's ratchet phenomenon could indeed be responsible for transient losses of virus fitness following selection of a resistant mutant *in vivo*. It is likely that a large proportion of viruses with resistance mutations also harbor random deleterious mutations in different viral genes (Meyerhans *et al.*, 1989), leading to a reduction in replicative potential. Marked bottlenecks have indeed been described in populations of HIV quasispecies (Nijhuis *et al.*, 1998) subjected to antiretroviral treatment. However, the size of potential founder virus populations carrying one or more resistance mutations and available for the further selection of a fully resistant virus has not been evaluated with precision. It is likely that the size of these populations is large enough to allow selection of optimally fit viruses in the presence of drug. Therefore, the role of selection bottlenecks in resistance-associated loss of HIV fitness is unclear. Until now, for most of the drug-resistant viruses in which a clear loss of fitness has been described, it has been established that the main factor driving that fitness reduction is related to the resistance mutations themselves.

II. Methods of Assessment of Viral Fitness

A. Mixed Viral Infections

According to the Darwinian definition of fitness, which applies to a selective advantage in conditions of reproductive competition between several variants, one would assume that all fitness assays should be competitive. Indeed, many authors have used culture systems in which mixtures of two or more viral variants are used to infect a population of target cells. In these systems, HIV infection leads to massive production of viral particles in the culture supernatant, reflecting exponential propagation of the infection in the culture following several cycles of virus replication. Virus from this initial peak production is used to inoculate a new culture, and after several such passages, the balance of virus populations is assessed by sequencing cell-associated viral DNA or supernatant viral RNA. Virus fitness is then calculated based on the rate of replacement of one variant by another along these several passages (Harrigan *et al.*, 1998; Martinez-Picado *et al.*, 1999; Tachedjian *et al.*, 1998). The balance between the different variants can be modified by changing their ratios in the initial inoculum. The cultures are usually conducted in the absence of drugs, but some authors have conducted mixed infections in the presence of different concentrations of drug in order to mimick the conditions of virus competition in treated patients (Harrigan *et al.*, 1998).

Clearly, the strength of these competition assays is their high sensitivity. Depending on the size of the inoculum, on the ratio between competing variants, and on the type of cells used in the culture, minute selective advantages can often translate into visible and reproducible shifts in virus populations. However, these assays require several weeks of continuous culture and cannot be easily applied to the study of large series of viruses. In addition, since they require cumulative rounds of virus replication and error-prone reverse transcription, one should not neglect the risk of emergence and further selection of spontaneous changes in the viral sequence in these assays. Finally, because the read-out of the assay is based on sequencing mixtures of viral genomes, quantitative assessment of the balance between variants may not be always very accurate, especially when it is performed by bulk sequencing of PCR products, which is prone to PCR biases and to inaccurate quantitation of sequence minorities.

B. Replication Kinetics Assays

These relatively simple assays are based on infection of separate cell cultures by a normalized inoculum of individual variants (Back *et al.*, 1996; Borman *et al.*, 1996; Zhang *et al.*, 1997). The efficiency of HIV replication is reflected by the time required to achieve mass particle production in the supernatant of parallel cultures. Although simpler than competition assays, they are notoriously insensitive, yielding detectable differences between mutants only for relatively important replicative defects.

C. Single-Cycle Replication Assays

Assuming that the rate of infection of HIV target cells by a virus variant can be reflected by the number of cells infected after a single round of replication, single-cycle assay systems have been developed for the study of HIV fitness. In these systems, virus is produced following transfection of a molecular clone and used to infect target cells in which an indicator gene is turned on by the infection. In some assays, the indicator gene is carried by the viral genome itself. In other assays, the indicator gene is present but silent in target cells, and its expression can be turned on by production of the HIV transactivator Tat following successful infection of the cell (Carron de la Carriere *et al.*, 1999; Mammano *et al.*, 1998; Zennou *et al.*, 1998). The main advantage of such assays is their simplicity and their speed: the rate of infection can be usually measured in less than 2 days. Their other advantage is related to the fact that since there is only one cycle of replication, there is no risk of genetic drift of the virus in the course of the assay. However, their sensitivity is limited, restricting their use for the most prominent fitness changes, such as those observed in viral variants with resistance to protease inhibitors (Fig. 1).

D. Enzymatic Assays

Since all drugs used today in HIV therapy target viral enzymes, several authors have evaluated the effect of resistance mutation on the *in vitro* function of these enzymes. Regarding RT, most assays have focused on the processivity of the enzyme, a parameter that is reflected by the length of individual segments of DNA synthesized by the enzyme in a single round of polymerization (Arion *et al.*, 1996; Back *et al.*, 1996; Miller *et al.*, 1998). In these assays, the template for DNA synthesis by RT can be either the natural template of HIV RT or synthetic DNA or RNA templates. Processivity assays usually require purified enzyme expressed in bacteria, but have also been successfully conducted with RT from concentrated viral particles. Assays of HIV protease activity are always conducted with purified protease expressed in prokaryotic cells (Gulnick *et al.*, 1995; Lin *et al.*, 1995; Sardana *et al.*, 1994). Purified protease is most often used in cleavage assays using synthetic peptides that contain HIV-protease-specific cleavage sites. Although *in vitro* protease assays can be useful for detailed biochemical studies, their results may depend on assay conditions and on the nature of the synthetic peptide substrate used for cleavage and may not be fully reflective of the actual conditions of protease function in the context of the virus (Klabe *et al.*, 1998).

III. Resistance to Nucleoside Analogs

Nucleoside analogs are an essential component of HIV therapy. The antiviral effect of these agents is the consequence of their incorporation into the nascent viral DNA chain, where they act as chain terminators, due to the absence of a 3′ hydroxyl group. Because wild-type RTs lack efficient 3′ exonuclease activity, they cannot efficiently remove incorporated nucleoside analogs, explaining in part the exquisite sensitivity of retroviruses to such compounds. The mechanisms of HIV resistance to nucleoside analogs, although they are not fully understood, can involve two basic processes: excision of the newly incorporated chain terminators (the main mechanism for HIV resistance to AZT and probably to d4T) and reduced incorporation of the nucleoside analog triphosphate (a mechanism best documented for resistance to 3TC). Whatever the basic mechanisms involved in resistance, they imply modifications in key enzyme structures, therefore some loss of drug-free viral fitness should be expected for any resistant virus. However, because these mechanisms are often very different from one analog to another, implying very different sets of mutations in RT, the impact of resis-

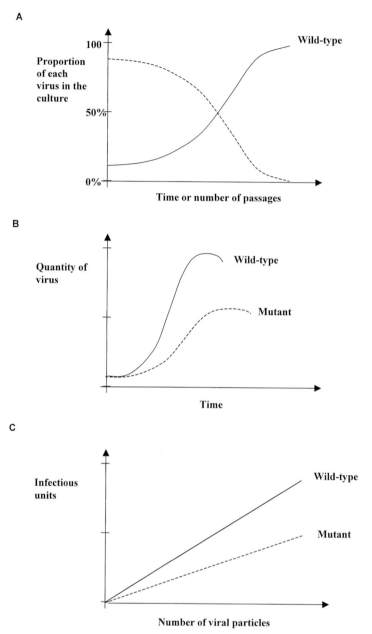

FIGURE I (A) Mixed viral infections. Culture of mixtures of viruses during multiple cycles of infection in the presence or absence of drug. The proportion of each virus in the initial inoculum can be varied. The proportion of each virus in the culture is determined at various time points by sequencing cell-associated viral DNA or viral RNA. (B) Replication kinetics. The quantity of virus produced in cultures is measured at various time points using a suitable assay such as a p24 ELISA. (C) Single-cycle replication assays. The number of target cells infected in a single cycle after inoculation with a known quantity of virus can be measured using various indicator gene systems. Thus a ratio of infectious units to viral particles can be calculated.

tance on virus fitness clearly differs according to the drug used (Schinazi *et al.*, 1997).

A. Resistance to AZT

Resistance to AZT is commonly encountered in treated HIV-infected patients, mainly because AZT was the first available antiviral with some activity against HIV. Resistance to AZT is easy to identify from the RT genotype, which usually bears characteristic amino acid substitutions: M41L, D67N, K70R, L210W, T215Y, and K219Q (Larder and Kemp, 1989). The first HIV-1 variants clearly identified as AZT-resistant carried mutations M41L and T215Y. In spite of a strong resistance phenotype, these viruses appeared to replicate normally in culture in the absence of drug. *In vivo*, reports of transmission of AZT-resistant viruses to treatment-naïve patients in which resistance mutations were maintained for long time periods in the absence of AZT treatment also indicated that HIV resistance to AZT was achieved at the expense of minimal, if any, loss of viral fitness (Yerly *et al.*, 1998). Some authors even proposed that following selection for high-level resistance to AZT, HIV may become more fit than wild-type in the absence of drug, due to an increase in the processivity of RT (Caliendo *et al.*, 1996). In fact, it is clear now that resistance to AZT does not markedly affect HIV replicative capacity in the absence of drug. This was first indicated by calculating fitness based on the rate of reversion of AZT-resistance mutations *in vivo* in the absence of drug. Goudsmit *et al.*, analyzing the stability of mutation T215Y in a treatment-naïve subject newly infected by an AZT-resistant HIV variant, found that the change of tyrosine 215 into a serine, which restored sensitivity to AZT, had a 0.4 to 2.5% selective advantage in the absence of drug compared to its parental AZT-resistant variant (Goudsmit *et al.*, 1996). In further analyses, the same authors observed that, in fact, the range of fitness values could be surprisingly wide among populations of AZT-resistant variants, with differences in fitness between some of the mutants amounting to 25% (Goudsmit *et al.*, 1997).

In other studies, fitness was examined in tissue culture using clonal populations of viruses carrying various combinations of mutations known to confer resistance to AZT. In these experiments, fitness was calculated from the rate of replacement of one mutant by another in mixed viral cultures. They showed that mutation K70R, a mutation often found at early stages of the selection for AZT resistance *in vivo*, is better fit in the absence of drug than mutation T215Y, which itself is better fit than the combination of mutations T215Y and M41L (Harrigan *et al.*, 1998) In this case, the more resistant virus clearly displays the lowest drug-free fitness. However, a mutant with substitution M41L alone, which does not mediate high resistance by itself, is clearly less fit than the T215Y+M41L mutation. These findings are in full agreement with the sequence of the selection of these mutations *in vivo*, where K70R is often selected first, but is usually replaced by combinations of mutations, including T215Y, within a few months of viral escape to AZT (Boucher *et al.*, 1992).

B. Resistance to Other Nucleoside Analogs

Apart from AZT, the impact of resistance on HIV fitness has been examined mainly for two other nucleoside analogs, didanosine (ddl) and lamivudine (3TC). These studies mainly focused on two mutations that appeared to be selectively associated with resistance to either of these two nucleoside analogs, L74V (conferring resistance to ddl) and M184V (conferring resistance to 3TC). Mutation L74V, once considered a signature of resistance to ddl, was in fact mostly observed when ddl was used in monotherapy. This mutation confers only low-level resistance to ddl (usually less than a 10-fold increase in IC50 compared with a reference virus strain) and appears to counteract resistance to AZT of mutants carrying typical AZT-resistance genotypes. Therefore, it is now seldom observed in patients receiving combined antiretroviral therapy. Nonetheless, using mixed culture experiments, this mutation has been clearly shown to reduce the replicative potential of HIV (Sharma and Crumpacker, 1997).

The phenomenon has been more thoroughly documented with mutation M184V, which confers high-level resistance to 3TC. This mutation affects a methionine located next to two of the three key catalytic aspartic acid residues located at positions 185 and 186 of HIV-1 RT, creating the signature motif YMDD (Tisdale *et al.*, 1993). Interestingly, a mutation of the corresponding methionine residue confers 3TC resistance in HBV (Tipples *et al.*, 1996). Although the mechanism of 3TC resistance induced by M184V is not fully delineated, it appears to involve impaired incorporation of 3TCTP, due to misalignment of the nucleotide with the 3′ terminal hydroxyle group of nascent DNA (Huang *et al.*, 1998). Unlike that observed with resistance to other nucleoside analogs, the level of resistance to 3TC conferred by M184V alone is extremely high (Boucher *et al.*, 1993). In fact, in most phenotypic resistance assays, it is close if not equal to the maximal measurable resistance level. Therefore, when an antiretroviral treatment that includes 3TC fails to totally abolish viral replication, M184V mutants have a very strong selective advantage, and the emergence of the mutation is usually rapid. In line with its dramatic effect on resistance, and probably because it affects a highly conserved residue so close to the catalytic core of the enzyme, it is not surprising that mutation M184V can also significantly reduce HIV replication in the absence of drug. *In vitro* experiments conducted on reverse transcriptase from purified viral particles have shown that the mutated enzyme displays a lower processivity than wild type, as measured by the length of DNA segments synthesized during a single round of polymerization (Back *et al.*, 1996; Boyer and Hughes, 1995). In spite of this clear polymerization defect by RT, the replication defect of the M184V mutant is only detectable when the virus is grown in primary human peripheral blood mononuclear cells (PBMC) (Back *et al.*, 1996). The reduced rate of mutant replication in these cells has been attributed to the lower pool of deoxynucleoside triphosphates (dNTPs) present in primary cells compared to cells from established lymphocytic cell lines, where mutant replication is comparable to wild type. In the presence of low concentrations of dNTPs, the processivity defect of mutant enzyme is enhanced (Back and Berkhout, 1997). Interestingly, the enzymatic and replicative defect observed in these studies was found to be even more pronounced for an M184I mutant, often transiently observed in 3TC-treated patients shortly before emergence of the M184V mutant. Because of the mutagenic bias of RT, M184I is often selected before M184V. However, the M184V mutant, although clearly less fit than wild type, is more fit than its parental M184I counterpart, which is therefore rapidly outgrown *in vivo* by M184V (Keulen *et al.*, 1997). One interesting property of the M184V mutation is that it can counteract resistance to AZT induced by classic AZT-resistance mutations such as M41L and T215Y (Larder *et al.*, 1995). This effect is only transient in treated patients (Nijhuis *et al.*, 1997), but may account for the sustained effect of AZT+3TC dual-combination treatments in spite of the presence of resistance mutations to both drugs. The mechanism of this well-documented "resensitization," which has also been observed for other mutations such as L74V (Lacey and Larder, 1994), is not understood. Whether it is related to the individual effect of these mutations on RT processivity or on viral fitness remains to be evaluated.

Patients who fail antiretroviral therapy and in which HIV develops resistance to nucleoside analogs have often received antiretrovirals sequentially before being prescribed combined therapy. Because early antiretrovirals were not used in fully suppressive drug combinations, the emergence of resistance was inevitable and was reinforced along with repeated treatment switches. In such patients, the resistant virus has usually accumulated multiple mutations conferring resistance to a wide range of nucleosides. Some of these viruses display combinations of mutations combining substitutions A62V, V75T, F77L, or F116Y in most cases associated with mutation Q151M, often referred to as the "151 complex" or the MDR (multiple-drug-resistance) mutations (Iversen *et al.*, 1996). Other viruses with multiple nucleoside analog resistance display an insertion of two or more amino acids near serine 69 (Larder *et al.*, 1999; Winters *et al.*, 1998). Although some reports have suggested that MDR viruses may exhibit higher fitness than wild type (Kosalaraksa *et al.*, 1999; Maeda *et al.*, 1998), the full impact of the accumulation of mutations in RT during prolonged treatment failure, leading to resistance to multiple nucleoside analogs, has not yet been carefully examined.

IV. Resistance to Other Reverse Transcriptase Inhibitors ___

A. Nonnucleoside RT Inhibitors (NNRTI)

Nonnucleoside RT inhibitors are noncompetitive, allosteric inhibitors of HIV-1 RT. These molecules display a high affinity for a virtual hydrophobic pocket situated close to the active site of the enzyme (Kohlstaedt *et al.*, 1992). After occupation of the pocket, the mobility of some of the domains of HIV RT is impeded, thereby blocking DNA polymerization. Mutations that confer resistance to these compounds result in amino acid substitutions that change the physical properties of the hydrophobic pocket. Although extensive cross-resistance to most NNRTIs have been described in viruses selected for resistance to any one of these compounds, the number and the positions of the mutations required for resistance often differ from one compound to another: resistance to nevirapine can be achieved with a single Y181C substitution (Richman *et al.*, 1994), resistance to efavirenz and delavirdine is most often characterized by combinations of mutations that include K103N. Recently, it has been shown that mutation P236L, which confers resistance to delavirdine *in vitro* but which is seldom observed in treated patients, reduces viral fitness through a decrease in the RNAse H activity that is associated with HIV RT (Gerondelis *et al.*, 1999). However, there is no clear evidence that resistance to the NNRTIs that are used today in antiviral therapy is accompanied by significant loss of viral fitness. It is suggested that resistance to high concentrations of NNRTI or to novel NNRTI molecules requires the accumulation of large numbers of mutations, resulting in reduction of viral fitness. In nevirapine-treated patients, more mutations seem to be required to achieve high-level resistance and viral escape when patients are prescribed higher drug doses. Although the NNRTI-binding pocket of RT is not directly involved in any particular function of the enzyme, as reflected by the natural NNRTI resistance of HIV-2 and of HIV-1 group 0, one can see the number of mutations that is sufficient for optimal virus replication in the presence of a given concentration of drug as the result of a balance between the effect of the mutations on resistance and the possible deleterious effect of their accumulation on virus fitness in the absence of drug or in the presence of low drug concentrations. Therefore, as for nucleoside analogs, it is possible that viruses exposed to high concentrations of a NNRTI, or to new NNRTI molecules displaying high antiviral activity, need to develop combinations of multiple mutations or novel mutations which may confer higher levels of resistance at the expense of significant loss of viral fitness.

B. Pyrophosphate Analogs

Although pyrophosphate analogs are not used in current antiretroviral therapy regimens, HIV resistance to this third category of RT inhibitors has been documented. The principal molecule that is representative of this category, foscarnet, has not proven active *in vivo* and displays high toxicity. Nevertheless, foscarnet is active against HIV in culture and RT mutations that confer HIV resistance to foscarnet, affecting residues 88 and 89 of RT (Tachedjian *et al.*, 1995), have been shown to significantly affect virus fitness in the absence of drug (Tachedjian *et al.*, 1998).

V. Resistance to Protease Inhibitors

The protease encoded by retroviral genomes is required for proteolytic processing of the large polyproteins that are the precursors of retroviral structural and enzymatic proteins, leading to a second stage of particle assembly that is indispensable for virus infectivity. The HIV-1 protease is an aspartic protease, active as a symmetrically assembled homodimer, with each of the two subunits equally delimiting the active site of the enzyme at the centre of the dimer. Inhibitors of HIV-1 protease have been derived from peptides representing the sites of proteolytic cleavage in the natural substrates of the protease. These molecules bind to the active site of the enzyme with high affinity, but since they cannot be cleaved, they exert a potent and nearly irreversible inhibition of protease activity. Consequently, they behave as extremely potent inhibitors of HIV-1 replication in culture and *in vivo*. Protease inhibitors have proven essential components of the highly active antiretroviral combinations used today in HIV therapy, leading in most cases to complete suppression of detectable virus replication (Gulick *et al.*, 1997; Hammer *et al.*, 1997). In spite of this high activity, early studies conducted on *in vitro* passaged viruses or on treated patients showed that accumulation of specific mutations in the protease can confer high-level resistance to protease inhibitors (Borman *et al.*, 1996; Condra *et al.*, 1995, 1996; Craig *et al.*, 1998; Eastman *et al.*, 1998; Ho *et al.*, 1994; Kaplan *et al.*, 1994; Markowitz *et al.*, 1995; Molla *et al.*, 1996; Otto *et al.*, 1993; Patick *et al.*, 1995, 1998; Roberts, 1995).

In almost all cases, high-level resistance can only be achieved following accumulation of several mutations in the protease (Boden and Markowitz, 1998; Condra *et al.*, 1996; Erickson, 1995; Molla *et al.*, 1996). This accumulation is a stepwise process following which resistance increases gradually: single mutations usually yield low levels of resistance, which becomes significant only after the accumulation of two or more mutations. For some inhibitors, like ritonavir or nelfinavir, the process of accumulation of mutations appears to follow a particular order: resistance to ritonavir almost always starts with selection of mutation V82A and resistance to nelfinavir with mutation D30N (Molla *et al.*, 1996; Patick *et al.*, 1996). For other inhibitors, such as indinavir, selection of authentically resistant viruses appears to be slower and more disordered. The locations of the most common resistance mutations within the protease are shown in Fig. 2. Some mutations, such as V82A, are located directly within the substrate-binding site of the enzyme, while many other mutations are clearly outside of the active site, affecting amino acids that do not contact the inhibitor directly. These mutations, usually termed secondary mutations because they often emerge later in the selection process, are believed to reduce inhibitor efficiency through indirect changes in the tertiary structure of the enzyme, affecting the shape of the substrate-binding site and the corresponding affinity of the inhibitors for the portease (Erickson, 1995).

Early reports on viruses selected for resistance to protease inhibitors in tissue culture have emphasized the effect of resistance on the drug-free fitness of HIV. Some of these mutations, such as R8Q or V32I, which dramatically reduce HIV fitness, are not frequently observed in treated patients. However, many viruses with single or combined resistance mutations in the protease are clearly less fit than wild type, as shown in a variety of assay systems (Borman *et al.*, 1996; Croteau *et al.*, 1997; el-Farrash *et al.*, 1994; Ho *et al.*, 1994; Kaplan *et al.*, 1994; Kuroda *et al.*, 1995; Martinez-Picado *et al.*, 1999; Zennou *et al.*, 1998). Reduced performance is most prominent in enzymatic assays, where loss of enzymatic activity amounting to more than 80% has been described for mutations commonly observed *in vivo* (Gulnick *et al.*, 1995; Ridky *et al.*, 1998; Sardana *et al.*, 1994). In culture assays,

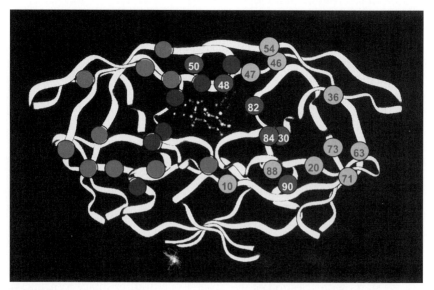

FIGURE 2   Structure of the HIV-1 protease complexed with an inhibitor. Amino acid residues involved in resistance are indicated by a numbered circle, each number indicating the position of the amino acid in the protease sequence. Dark circles indicate primary resistance mutations. Light circles indicate secondary resistance mutations.

including single-cycle systems, the loss of fitness of PI-resistant viruses is generally markedly more prominent than what can be observed with viruses carrying resistance mutations in RT. Some primary protease mutations, which are also those to which the highest level of resistance can be attributed, appear to have a more radical effect on virus fitness. This is the case, in particular, for mutation G48V, which induces high-level resistance to saquinavir at the expense of significant loss of fitness, and for mutation 184V, which confers wide cross-resistance to most known protease inhibitors. In addition, there is now clear evidence that loss of viral fitness can be observed for some particular combinations of mutations: mutations V82A and L90M appear as almost mutually exclusive in the course of selection for resistance to saquinavir (Eastman *et al.*, 1997); similarly, the combination of V82A and I54V is associated with increased resistance to indivavir and ritonavir at the expense of a significant loss of fitness (Nijhuis *et al.*, 1997).

In many cases, however, loss of viral fitness that results from emergence of resistance mutations can be gradually improved by the accumulation of secondary mutations in the protease (Borman *et al.*, 1996; Kaplan *et al.*, 1994; Nijhuis *et al.*, 1997; Rose *et al.*, 1996). Following the introduction of V82A associated with I54V, the emergence of secondary mutations such as A71V and M46I appears to restore virus infectivity to levels that have been described in one report as higher than parental wild-type virus (Nijhuis *et al.*, 1997). In other reports, loss of fitness induced by the V32I or by the V82A mutations has also been found to be partially corrected by the emergence of secondary mutations (Borman *et al.*, 1996; Kaplan *et al.*, 1994). Mutation L10I, often found in viruses with resistance to several protease inhibitors, improves the replicative potential of viruses with various combinations of resistance mutations (Rose *et al.*, 1996). The role of secondary protease mutations in resistance and the mechanism through which they improve fitness is still unclear. These mutations do not seem to be able to induce resistance by themselves, but they are present in most genotypes found in viruses escaping in wide range of protease inhibitors, where the level of resistance clearly increases wtih the number of mutations (Boden and Markowitz, 1998; Erickson, 1995). In parallel, the accumulation of secondary mutations in the protease dramatically widens viral cross-resistance to distinct protease inhibitors, which is not surprising since the same secondary mutations are usually observed in viruses selected for resistance to different molecules. Whether accumulation of secondary changes in the protease can improve viral fitness to the extent that it can recover wild-type levels is still a matter of debate. *In vivo*, the number of mutations required to achieve viral escape at the expense of minimal loss of fitness will depend on the nature of the inhibitor and on the pharmacokinetic properties of the inhibitor. Resistance should be envisioned as a dynamic process within a complex, changing, and uneven environment: in compartments where drug concentrations are high, it is likely that viruses with higher resistance will be favored at the expense of some loss of fitness, while in compartments where drug concentrations are lower, less resistant viruses with optimal fitness will be found. In treated patients, loss of viral fitness can vary dramatically from one virus to another. Although secondary mutations in the protease can in some cases improve fitness, there is now some evidence that viruses with high resistance and numerous primary and secondary mutations often display markedly reduced fitness (Faye *et al.*, 1999). Again, even if viruses with moderate resistance to protease inhibitors can retain wild-type fitness, any increase in drug pressure will create an imbalance that is likely to result in decreased drug-free fitness. Availability of protease inhibitors with improved tolerance, pharmacokinetics, and antiviral activity should lower the risk of emergence of resistant viruses but may also result in the selection of resistant viruses with heavily mutated protease sequences and reduced fitness.

The loss of viral fitness that is consecutive to resistance to protease inhibitors in the direct consequence of abnormal cleavage of some of the structural Gag proteins. Cleavage of Gag and Gag-Pol polyprotein precursors by the protease is mapped on Fig. 3. Accumulation of partially cleaved capsid (CA) and nucleocapsid (NC) proteins in viral particles has been clearly documented in resistant viruses (Rose *et al.*, 1996; Zennou *et al.*, 1998). Since the efficiency of Gag cleavage by the protease is a function of the affinity of the substrate for the enzyme, and since in this system the enzyme and the substrate are both encoded by the virus, it was probably not surprising to observe that changes in the Gag cleavage sites could accompany changes in the protease. This phenomenon was first described for viruses selected for high-level resistance to prototypic protease inhibitors in tissue culture (Doyon *et al.*, 1996). Subsequently, mutations in cleavage sites were observed in indinavir, ritonavir, and saquinavir-treated patients (Mammano

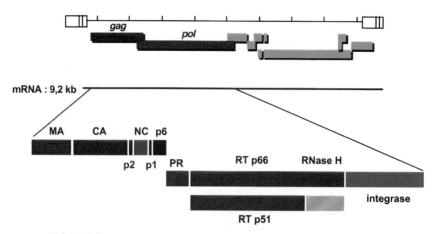

FIGURE 3   Organization and processing of HIV-1 Gag and Pol proteins.

et al., 1998; Zhang *et al.*, 1997). In all of these reports, only two cleavage sites appeared to be involved; one at the C-terminus of the NC protein (NC/p1 site) and another at the N-terminus of the p6 protein (p1/p6 site). The sites of cleavage of HIV proteins by the protease are defined by a sequence of seven amino acids. Although cleavage often occurs at the junction between a phenylalanine and a proline, not all cleavage sites in HIV Gag and Gag-Pol are identical. In fact, most of the cleavage sites in Gag are different and it is believed that their differences determine the order in cleavage events followed in Gag maturation, which appears essential for the formation of infectious viral particles. The mutations in HIV-1 Gag cleavage sites that emerge in parallel to protease resistance mutations are single amino acid substitutions; A432V, located at position P2′ (two amino acids upstream of cleavage) in the NC/p1 cleavage suquence, and F449L at position P1′ (one amino acid downstream of cleavage) in the p1/p6 site. No other Gag or Pol cleavage sequence mutations have been found to be significantly associated with protease resistance. Interestingly, mutation A432V is almost exclusively associated with mutation V82A and appears to be more prominent in indinavir-treated patients. In tissue culture, the introduction of cleavage site mutations significantly corrects loss of viral fitness that results from protease resistance mutations (Mammano *et al.*, 1998). However, that correction is only partial: indeed, some altered cleavage events, in particular those affecting the capsid (CA) protein, are not corrected. It is presumed that this absence of correction is related to structural and evolutionary constraints affecting the CA/p2 cleavage sequence, located at the C-terminus of CA.

The HIV enzymes, including protease itself, RT, and integrase, are also subject to polyprotein cleavage by the protease. Therefore, the effect of protease mutations on virus fitness is not limited to Gag cleavage. Viruses with unfit proteases resulting in abnormal cleavage of RT have been described (Carron de la Carriere *et al.*, 1999). These viruses have reduced particle-associated RT activity and appear to display alterations in their capacity for resistance to AZT. Interestingly, one report reveals that resistance to AZT can partially restore loss of fitness related to abnormal RT cleavage through an improvement in RT processivity (Carron de la Carriere *et al.*, 1999).

VI. *In Vivo* Consequences of Resistance-Associated Loss of HIV Fitness

A. Reversion of Resistant Genotypes after Interruption of Treatment

As long as an antiviral treatment is maintained, resistance mutations confer a selective advantage to the virus. Similarly, compensatory mutations that can correct viral fitness are also subject to positive selection. However, little is known of the behavior of resistant viruses after interruption of the antiretroviral treatment. Preliminary reports have recently shown that in several instances, treatment discontinuation in patients with advanced resistance to multiple antiretrovirals was accompanied by a rapid replacement of the resistant virus with wild-type, drug-sensitive virus. In these cases, most of the resistance mutations seem to disappear simultaneously from the bulk of the viral genomes found in peripheral blood, suggesting that rather than selective reversion of particular mutations, one was witnessing replacement of resistant virus by a competing, more fit, wild-type virus. One explanation of this phenomenon may be that in a virus with multiple resistance and compensatory mutations, which has gone through a series of fitness "troughs" in the course of its evolution up to a satisfactory fitness and resistance balance, no single reversion of resistance mutations is likely to confer a sufficient fitness advantage. Similar to what has been described for some antibiotic-resistant bacteria, resistance mutations and fitness-correcting compensatory mutations are too interdependent to be able to revert one by one. The other explanation is related to the fact that even in patients with evolved HIV drug resistance, it is likely that a large reservoir of wild-type, pretherapy virus is still present but unable to propagate or in a latent form. It has been shown that in patients with complete suppression of active viral replication, a strong reservoir of viral genomes is usually still present in lymphoid cells even after several years of treatment (Finzi *et al.*, 1997; Furtado *et al.*, 1999; Wong *et al.*, 1997; Zhang *et al.*, 1999). The existence of this reservoir, which is immune to the action of antiviral drugs, renders unlikely the possibility of short-term eradication of HIV infection even by the most active antiviral drug combinations. When a resistant virus emerges, it can be assumed that the wild-type virus reservoir is not affected by the colonization of lymphoid organs by resistant genomes. Therefore, depending on the size of this reservoir, and on its difference in fitness with the resistant virus, wild-type virus is able to rapidly replace its resistant counterpart in a matter of a few weeks. In such cases, one could be tempted to reinitiate an antiretroviral treatment based on the new genotype or phenotype, but, again, it is likely that a similar reservoir of latent resistant virus is ready to remerge following treatment reinitiation.

B. Effects of Resistance-Associated Loss of Viral Fitness on Virus Pathogenicity and on Disease Profile

Since fitness is defined in terms of replicative potential, one could predict that loss of viral fitness would result in a reduction in the rate of infection of CD4+ cells, leading to lower amounts of circulating virus and to a reduction in the rate of CD4+ cell depletion, whatever its mechanism. However, even in patients in which viruses with significant loss of viral fitness was documented, the amounts of circulating virus were high, reflecting their strong level of resistance. In parallel, mostly in patients escaping combinations of drugs including protease inhibitors, the number of peripheral CD4+ cells appeared to be "discordant" with the high viral load. Indeed, a significant proportion of patients escaping therapy with at least one protease inhibitor displayed a continued improvement in the number of their circulating CD4+ cells, paralleled by a continued clinical improvement, in spite of high amounts of plasma virus (Deeks *et al.*, 1999; Kaufmann *et al.*, 1998; Piketty *et al.*, 1998). In these situations, however, the discordance between the viral load evolution and the CD4 response is not complete: the best CD4 responses are often found associated with moderate viral rebounds, while rebounds to high viral loads, above baseline, are most often associated with relatively low CD4 responses. Furthermore, preliminary results suggest that the discordance between CD4 and evolution of viremia after the emergence of resistance may be limited in time, with a gradual fall in CD4 numbers after 2–3 years of virological failure of the treatment (Faye *et al.*, 1999). How can this partial disconnection between CD4 and viremia be explained? One possibility is that in these patients, the effect of viral rebound on the number of CD4 cells is delayed. However, such discordance is usually not observed in patients in which resistance to nucleoside analogs or NNRTIs develops. A second possibility is that since viruses considered as "resistant" to protease inhibitors are usually sensitive to high concentrations of the drug, there may be some tissue compartments where treatment activity is maintained, leading to persistence of high CD4 counts. A third possibility relates to an ecological model termed "predator–prey" balance, whereby reduced viral fitness would first reduce viremia and increase the number of HIV-susceptible CD4+ cells, which in turn would lead to a higher production of virus. However, the number of infected cells does not appear to be a limiting factor in HIV disease: even in patients with high viremia and low CD4 counts, the proportion of infected cells is remarkably low. Therefore, even a substantial increase in the number of potential HIV target cells is unlikely to affect viremia. A fourth possibility is based on the assumption that HIV is able to infect a wide range of cell types in which the consequences of a reduction in viral fitness may be very different. In particular, it is possible that reduced fitness could affect selectively HIV infection of a type of cells that could mediate CD4 depletion, while leaving unaffected infection of other cell types. If these latter cell types constituted the majority of HIV-infected cells, their infection would result in high viral load. In this context, it is unclear whether one can envisage the use of resistance-associated loss of viral fitness as an authentic therapeutic objective. Most available data on HIV resistance and fitness, especially in the treated patient, are too preliminary. In particular, the long-term evolution of HIV resistance and fitness under drug pressure or in the absence of drug pressure is not known. A large number of patients harbor viruses with increasing and widening resistance to antiretroviral drugs. Until new molecules are available, either trageting novel viral structures or with improved pharmacokinetics and antiviral properties, the management of HIV drug resistance in these patients will be complex, with many virological parameters that will need to be evaluated in parallel. As much as the resistance genotype, the resistance phenotype and the pharmacokinetics of drugs, the monitoring of viral fitness could prove one of such key parameters.

References

Arion, D., Borkow, G., Gu, Z., Wainberg, M. A., and Parniak, M. A. (1996). The K65R mutation confers increased DNA polymerase processivity to HIV-1 reverse transcriptase. *J. Biol. Chem.* **271,** 19860–19866.

Arion, D., Kaushik, N., McCormick, S., Borkow, G., and Parniak, M. A. (1998). Phenotypic mechanism of HIV-1 resistance to 3'-azido-3'-deoxythymidine (AZT): Increased polymerization processivity and enhanced sensitivity to pyrophosphate of the mutant viral reverse transcriptase. *Biochemistry* **37,** 15908–15917.

Back, N. K., and Berkhout, B. (1997). Limiting deoxynucleoside triphosphate concentrations emphasize the processivity defect of lamivudine-resistant variants of human immunodeficiency virus type 1 reverse transcriptase. *Antimicrob. Agents Chemother.* **41,** 2484–2491.

Back, N. K., Nijhuis, M., Keulen, W., Boucher, C. A., Oude Essink, B. O., van Kuilenburg, A. B., van Gennip, A. H., and Berkhout, B. (1996). Reduced replication of 3TC-resistant HIV-1 variants in primary cells due to a processivity defect of the reverse transcriptase enzyme. *EMBO J.* **15,** 4040–4049.

Barrie, K. A., Perez, E. E., Lamers, S. L., Farmerie, W. G., Dunn, B. M., Sleasman, J. W., and Goodenow, M. M. (1996). Natural variation in HIV-1 protease, Gag p7 and p6, and protease cleavage sites within Gag/Pol polyproteins: Amino acid substitutions in the absence of protease inhibitors in mothers and children infected by Human Immunodeficiency Virus type 1. *Virology* **219,** 407–416.

Battula, N., and Loeb, L. A. (1976). On the fidelity of DNA replication: Lack of exodeoxyribonuclease activity and error-correcting function in avian myeloblastosis virus DNA polymerase. *J. Biol. Chem.* **251,** 982–986.

Bebenek, K., Abbotts, J., Roberts, J. D., Wilson, S. H., and Kunkel, T. A. (1989). Specificity and mechanism of error-prone replication by human immunodeficiency virus-1 reverse transcriptase. *J. Biol. Chem.* **264,** 16948–16956.

Boden, D., and Markowitz, M. (1998). Resistance to human immunodeficiency virus type 1 protease inhibitors. *Antimicrob. Agents Chemother.* **42,** 2775–2783.

Borman, A. M., Paulous, S., and Clavel, F. (1996). Resistance of human immunodeficiency virus type 1 to protease inhibitors: Selection of resistance mutations in the presence and absence of the drug. *J. Gen. Virol.* **77,** 419–426.

Boucher, C. A., Cammack, N., Schipper, P., Schuurman, R., Rouse, P., Wainberg, M. A., and Cameron, J. M. (1993). High-level resistance to (-) enantiomeric 2'-deoxy-3'-thiacytidine in vitro is due to one amino acid substitution in the catalytic site of human immunodeficiency virus type 1 reverse transcriptase. *Antimicrob. Agents Chemother.* **37,** 2231–2234.

Boucher, C. A., O'Sullivan, E., Mulder, J. W., Ramautarsing, C., Kellam, P., Darby, G., Lange, J. M., Goudsmit, J., and Larder, B. A. (1992). Ordered appearance of zidovudine resistance mutations during treatment of 18 human immunodeficiency virus-positive subjects. *J. Infect. Dis.* **165,** 105–110.

Boyer, P. L., and Hughes, S. H. (1995). Analysis of mutations at position 184 in reverse transcriptase of human immunodeficiency virus type 1. *Antimicrob. Agents Chemother.* **39**, 1624–1628.

Caliendo, A. M., Savara, A., An, D., DeVore, K., Kaplan, J. C., and D'Aquila, R. T. (1996). Effects of zidovudine-selected human immunodeficiency virus type 1 reverse transcriptase amino acid substitutions on processive DNA synthesis and viral replication. *J. Virol.* **70**, 2146–2153.

Carron de la Carriere, L., Paulous, S., Clavel, F., and Mammano, F. (1999). effects of human immunodeficiency virus type 1 resistance to protease inhibitors on reverse transcriptase processing, activity, and drug sensitivity. *J. Virol.* **73**, 3455–3459.

Chow, Y. K., Hirsch, M. S., Merrill, D. P., Bechtel, L. J., Eron, J. J., Kaplan, J. C., and D'Aquila, R. T. (1993). Use of evolutionary limitations of HIV-1 multidrug resistance to optimize therapy. *Nature* **361**, 650–654.

Clarke, D. K., Duarte, E. A., Moya, A., Elena, S. F., Domingo, E., and Holland, J. (1993). Genetic bottlenecks and population passages cause profound fitness differences in RNA viruses. *J. Virol,* **67**, 222–228.

Coffin, J. M. (1995). HIV population dynamics *in vivo*: Implications for genetic variation, pathogenesis, and therapy, *Science* **267**, 483–489.

Condra, J. H., Holder, D. J., Schleif, W. A., Blahy, O. M., Danovich, R. M., Gabryelski, L. J., Graham, D. J., Laird, D., Quintero, J. C., Rhodes, A., Robbins, H. L., Roth, E., Shivaprakash, M., Yang, T., Chodakewitz, J. A., Deutsch, P. J., Leavitt, R. Y., Massari, F. E., Mellors, J. W., Squires, K. E., Steigbigel, R. T., Teppler, H., and Emini, E. A. (1996). Genetic correlates of *in vivo* viral resistance to indinavir, a Human Immunodeficiency Virus type 1 protease inhibitor. *J. Virol.* **70**, 8270–8276.

Condra, J. H., Schleif, W. A., Blahy, O. M., Gabryelski, L. J., Graham, D. J., Quintero, J. C., Rhodes, A., Hobbins, H. L., Roth, E., Shivaprakash, M., Titus, D. L., Yang, T., Teppler, H., Squires, K. E., Deutsch, P. J., and Emini, E. A. (1995). *In vivo* emergence of HIV-1 variants resistant to multiple protease inhibitors. *Nature* **374**, 569–571.

Craig, C., Race, E., Sheldon, J., Whittaker, L., Gilbert, S., Moffatt, A., Rose, J., Dissanayeke, S., Chirn, G. W., Duncan, I. B., and Cammack, N. (1998). HIV protease genotype and viral sensitivity to HIV protease inhibitors following saquinavir therapy. *AIDS* **12**, 1611–1618.

Croteau, G., Doyon, L., Thibeault, D., McKercher, G., Pilote, L., and Lamarre, D. (1997). Impaired fitness of Human Immunodeficiency Virus type 1 variants with high-level resistance to protease inhibitors. *J. Virol.* **71**, 1089–1096.

Debouck, C. (1992). The HIV-1 protease as a therapeutic target for AIDS. *AIDS Res. Hum. Retroviruses* **8**, 153–164.

Deeks, S. G., Hecht, F. M., Swanson, M., Elbeik, T., Loftus, R., Cohen, P. T., and Grant, R. M. (1999). HIV RNA and CD4 cell count response to protease inhibitor therapy in an urban AIDS clinic: Response to both initial and salvage therapy. *AIDS* **13**, F35–F43.

Domingo, E., and Holland, J. J. (1997). RNA virus mutations and fitness for survival. *Annu. Rev. Microbiol.* **51**, 151–178.

Doyon, L., Poulin, F., Pilote, L., Clouette, C., Thibeault, D., Croteau, G., and Lamarre, D. (1996). Second locus involved in Human Immunodeficiency Virus type 1 resistance to protease inhibitors. *J. Virol.* **70**, 3763–3769.

Eastman, P. S., Mittler, J., Kelso, R., Gee, C., Boyer, E., Kolberg, J., Urdea, M., Leonard, J. M., Norbeck, D. W., Mo, H., and Markowitz, M. (1998). Genotypic changes in human immunodeficiency virus type 1 associated with loss of suppression of plasma viral RNA levels in subjects treated with ritonavir (Norvir) monotherapy. *J. Virol.* **72**, 5154–5164.

Eastman, S. P., Duncan, I. B., Gee, C., and Race, E. (1997). "Acquisition of Genotypic Mutations Associated with Reduced Susceptibility to Protease Inhibitors during Saquinavir Monotherapy." International Workshop on HIV Drug Resistance, June, 1997, St. Petersburg, FL.

el-Farrash, M. A., Kuroda, M. J., Kitazaki, T., Masuda, T., Kato, K., Hatanaka, M., and Harada, S. (1994). Generation and characterization of a human immunodeficiency virus type 1 (HIV-1) mutant resistant to an HIV-1 protease inhibitor. *J. Virol.* **68**, 233–239.

Erickson, J. W. (1995). The not-so-great escape. *Nat. Struct. Biol.* **2**, 523–529.

Eron, J. J., Chow, Y. K., Caliendo, A. M., Videler, J., Devore, K. M., Cooley, T. P., Liebman, H. A., Kaplan, J. C., Hirsch, M. S., and D'Aquila, R. T. (1993). pol mutations conferring zidovudine and didanosine resistance with different effects in vitro yield multiply resistant human immunodeficiency virus type 1 isolates *in vivo. Antimicrob. Agents Chemother.* **37**, 1480–1487.

Faye, A., Race, E., Paulous, S., Dam, E., Obry, V., Matheron, S., Prevot, M. H., Joly V., and Clavel, F. (1999). "Can Resistance-Associated Loss of Viral Fitness Explain Discordant CD4 and Plasma HIV RNA Evolution Following Protease Inhibitor Failure?" Third International Workshop on HIV Drug Resistance, June, 1999, San Diego.

Finzi, D., Hermankova, M., Pierson, T., Carruth, L. M., Buck, C., Chaisson, R. E., Quinn, T. C., Chadwick, K., Margolick, J., Brookmeyer, R., Gallant, J., Markowitz, M., Ho, D. D., Richman, D. D., and Siliciano, R. F. (1997). Identification of a reservoir for HIV-1 in patients on highly active antiretroviral therapy. *Science* **278**, 1295–1300.

Furtado, M. R., Callaway, D. S., Phair, J. P., Kunstman, K. J., Stanton, J. L., Macken, C. A., Perelson, A. S., and Wolinsky, S. M. (1999). Persistence of HIV-1 transcription in peripheral-blood mononuclear cells in patients receiving potent antiretroviral therapy. *N. Engl. J. Med.* **340**, 1614–1622.

Gerondelis, P., Archer R. H., Palaniappan, C., Reichman, R. C., Fay, P. J., Bambara R. A., and Demeter, L. M. (1999). The P236L delavirdine-resistant human immunodeficiency virus type 1 mutant is replication defective and demonstrates alterations in both RNA 5′-end- and DNA 3′-end-directed RNase H activities. *J. Virol.* **73**, 5803–5813.

Goudsmit, J., de Ronde, A., de Rooij, E., and de Boer, R. (1997). Broad spectrum of *in vivo* fitness of human immunodeficiency virus type 1 subpopulations differing at reverse transcriptase codons 41 and 215. *J. Virol.* **71**, 4479–4484.

Goudsmit, J., De Ronde, A., Ho, D. D., and Perelson, A. S. (1996). Human immunodeficiency virus fitness *in vivo*: Calculations based on a single zidovudine resistance mutation at codon 215 of reverse transcriptase. *J. Virol.* **70**, 5662–5664.

Gulick, R. M., Mellors, J. W., Havlir, D., Eron, J. J., Gonzalez, C., McMahon, D., Richman, D. D., Valentine, F. T., Jonas, L., Meibohm, A., Emini, E. A., and Chodakewitz, J. A. (1997). Treatment with indinavir, zidovudine, and lamivudine in adults with human immunodeficiency virus infection and prior antiretroviral therapy. *N. Engl. J. Med.* **337**, 734–739.

Gulnick, S. V., Suvorov, L. I., Liu, B., Yu, B., Anderson, B., Mitsuya, H., and Erickson, J. W. (1995). Kinetic characterization and cross-resistance patterns of HIV-1 protease mutants selected under drug pressure. *Biochemistry* **34**, 9282–9287.

Hammer, S. M., Squires, K. E., Hughes, M. D., Grimes, J. M., Demeter, L. M., Currier, J. S., Eron, J. J., Jr., Feinberg, J. E., Balfour, H. H., Jr., Deyton, L. R., Chodakewitz, J. A., and Fischl, M. A. (1997). A controlled trial of two nucleoside analogues plus indinavir in persons with human immunodeficiency virus infection and CD4 cell counts of 200 per cubic millimeter or less. AIDS Clinical Trials Group 320 Study Team. *N. Engl. J. Med.* **337**, 725–733.

Harrigan, P. R., Bloor, S., and Larder, B. A. (1998). Relative replicative fitness of zidovudine-resistant human immunodeficiency virus type 1 isolates *in vitro. J. Virol.* **72**, 3773–3778.

Ho, D., Toyoshima, T., Mo, H., Kempf, D., Norbeck, D., Chen, C., Wideburg, N., Burt, S., Erickson, J., and Singh, M. (1994). Characterization of human immunodeficiency virus type 1 variants with increased resistance to a C2-symmetric protease inhibitor. *J. Virol.* **68**, 2016–2020.

Ho, D. D., Neumann, A. U., Perelson, A. S., Chen, W., Leonard, J. M., and Markowitz, M. (1995). Rapid turnover of plasma virions and CD4 lymphocytes in HIV-1 infection. *Nature* **373**, 123–126.

Huang, H., Chopra, R., Verdine, G. L., and Harrison, S. C. (1998). Structure of a covalently trapped catalytic complex of HIV-1 reverse transcriptase: Implications for drug resistance. *Science* **282**, 1669–1675.

Iversen, A. K., Shafer, R. W., Wehrly, K., Winters, M. A., Mullins, J. I., Chesebro, B., and Merigan, T. C. (1996). Multidrug-resistant human immunodeficiency virus type 1 strains resulting from combination antiretroviral therapy. *J. Virol.* 70, 1086–1090.

Kaplan, A. H., Michael, S. F., Wehbie, R. S., Knigge, M. F., Paul, D. A., Everitt, L., Kempf, D. J., Norbeck, D. W., Erickson, J. W., and Swanstrom, R. (1994). Selection of multiple human immunodeficiency virus type 1 variants that encode viral proteases with decreased sensitivity to an inhibitor of the viral protease. *Proc. Natl. Acad. Sci. USA* 91, 5597–5601.

Kaufmann, D., Pantaleo, G., Sudre, P., and Telenti, A. (1998). CD4-cell count in HIV-1-infected individuals remaining viraemic with highly active antiretroviral therapy (HAART): Swiss HIV Cohort Study. *Lancet* 351, 723–724.

Kepler, T. B., and Perelson, A. S. (1998). Drug concentration heterogeneity facilitates the evolution of drug resistance. *Proc. Natl. Acad. Sci. USA* 95, 11514–11519.

Keulen, W., Back, N. K., van Wijk, A., Boucher, C. A., and Berkhout, B. (1997). Initial appearance of the 184Ile variant in lamivudine-treated patients is caused by the mutational bias of human immunodeficiency virus type 1 reverse transcriptase. *J. Virol.* 71, 3346–3350.

Klabe, R. M., Bacheler, L. T., Ala, P. J., Erickson-Viitanen, S., and Meek, J. L. (1998). Resistance to HIV protease inhibitors: A comparison of enzyme inhibition and antiviral potency. *Biochemistry* 37, 8735–8742.

Kohlstaedt, L. A., Wang, J., Friedman, J. M., Rice, P. A., and Steitz, T. A. (1992). Crystal structure at 3.5 A resolution of HIV-1 reverse transcriptase complexed with an inhibitor. *Science* 256, 1783–1790.

Korber, B., Foley, B., Leitner, T., McCutchan, F., Hahn, B., Mellors, J. W., Myers, G., and Kuiken, C. (1997). HIV-1 pol amino-acid alignments. Human Retroviruses and AIDS, a compilation and analysis of nucleic acid and amino acid sequences. NIAID.

Kosalaraksa, P., Kavlick, M. F., Maroun, V., Le, R., and Mitsuya, H. (1999). Comparative fitness of multi-dideoxynucleoside-resistant human immunodeficiency virus type 1 (HIV-1) in an *in vitro* competitive HIV-1 replication assay. *J. Virol.* 73, 5356–5363.

Kuroda, M. J., el-Farrash, M. A., Choudhury, S., and Harada, S. (1995). Impaired infectivity of HIV-1 after a single point mutation in the POL gene to escape the effect of a protease inhibitor *in vitro*. *Virology* 210, 212–216.

Lacey, S. F., and Larder, B. A. (1994). Mutagenic study of codons 74 and 215 of the human immunodeficiency virus type 1 reverse transcriptase, which are significant in nucleoside analog resistance. *J. Virol.* 68, 3421–3424.

Larder, B. A., and Kemp, S. D. (1989). Multiple mutations in HIV-1 reverse transcriptase confer high-level resistance to zidovudine (AZT). *Science* 246, 1155–1158.

Larder, B. A., Bloor, S., Kemp, S. D., Hertogs, K., Desmet, R. L., Miller, V., Sturmer, M., Staszewski, S., Ren, J., Stammers, D. K., Stuart, D. I., and Pauwels, R. (1999). A family of insertion mutations between codons 67 and 70 of human immunodeficiency virus type 1 reverse transcriptase confer multinucleoside analog resistance. *Antimicrob. Agents Chemother.* 43, 1961–1967.

Larder, B. A., Kellam, P., and Kemp, S. D. (1993). Convergent combination therapy can select viable multidrug-resistant HIV-1 *in vitro*. *Nature* 365, 451–453.

Larder, B. A., Kemp, S. D., and Harrigan, P. R. (1995). Potential mechanism for sustained antiretroviral efficacy of AZT-3TC combination therapy. *Science* 269, 696–699.

Lech, W. J., Wang, G., Yang, Y. L., Chee, Y., Dorman, K., McCrae, D., Lazzeroni, L. C., Erickson, J. W., Sinsheimer, J. S., and Kaplan, A. H. (1996). *In vivo* sequence diversity of the protease of human immunodeficiency virus type 1: Presence of protease inhibitor-resistant variants in untreated subject. *J. Virol.* 70, 2038–2043.

Lin, Y., Lin, X., Hong, L., Foundling, S., Heinrickson, R. L., Thaisrivongs, S., Leelamanit, W., Raterman, D., Shah, M., Dunn, B. M., and Tang, J. (1995). Effect of point mutations on the kinetics and the inhibition of Human Immunodeficiency Virus type 1 protease: Relationship to drug resistance. *Biochemistry* 34, 1143–1152.

Maeda, Y., Venzon, D. J., and Mitsuya, H. (1998). Altered drug sensitivity, fitness, and evolution of human immunodeficiency virus type 1 with pol gene mutations conferring multi-dideoxynucleoside resistance. *J. Infect. Dis.* **177**, 1207–1213.

Mammano, F., Petit, C., and Clavel, F. (1998). Resistance-associated loss of viral fitness in human immunodeficiency virus type 1: Phenotypic analysis of protease and gag coevolution in protease inhibitor-treated patients. *J. Virol.* **72**, 7632–7637.

Markowitz, M., Mo, H., Kempf, D., Norbeck D., Bhat, T., Erickson, J., and Ho, D. (1995). Selection and analysis of human immunodeficiency virus type 1 variants with increased resistance to ABT-538, a novel protease inhibitor. *J. Virol.* **69**, 701–706.

Martinez-Picado, J., Savara, A. V., Sutton, L., and D'Aquila, R. T. (1999). Replicative fitness of protease inhibitor-resistant mutants of human immunodeficiency virus type 1. *J. Virol.* **73**, 3744–3752.

Meyerhans, A., Cheynier, R., Albert, J., Seth, M., Kwok, S., Sninsky, J., Morfeldt-Manson, L., Asjo, B., and Wain-Hobson, S. (1989). Temporal fluctuations in HIV quasispecies *in vivo* are not reflected by sequential HIV isolations. *Cell* **58**, 901–910.

Miller, M. D., Lamy, P. D., Fuller, M. D., Mulato, A. S., Margot, N. A., Cihlar, T., and Cherrington, J. M. (1998). Human immunodeficiency virus type 1 reverse transcriptase expressing the K70E mutation exhibits a decrease in specific activity and processivity. *Mol. Pharmacol.* **54**, 291–297.

Molla, A., Kempf, D., Korneyeva, M., Gao, Q., Shipper, P., Mo, H., Markowitz, M., Vasava-nonda, S., Chernyavskyi, T., Niu, P., Lyons, N., Hsu, A., Granneman, G., Ho, D., Boucher, C., Leonard, J., and Norbeck, D. (1996). Ordered accumulation of mutations in HIV protease confers resistance to Ritonavir. *Nat. Med.* **2**, 760–766.

Nijhuis, M., Boucher, C. A., Schipper, P., Leitner, T., Schuurman, R., and Albert, J. (1998). Stochastic processes strongly influence HIV-1 evolution during suboptimal protease-inhibitor therapy. *Proc. Natl. Acad. Sci. USA* **95**, 14441–144416.

Nijhuis, M., Schuurman, R., de Jong, D., van Leeuwen, R., Lange, J., Danner, S., Keulen, W., de Groot, T., and Boucher, C. A. (1997). Lamivudine-resistant human immunodeficiency virus type 1 variants (184V) require multiple amino acid changes to become co-resistant to zidovudine *in vivo*. *J. Infect Dis.* **176**, 398–405.

Nijhuis, M., Schuurman, R., Schipper, P., de Jong, D., van Bommel, T., de Groot, T., Molla, A., Borleffs, J., Danner, S., and Boucher, C. (1997). Reduced replication potential of HIV-1 variants initially selected under Ritonavir therapy is restored upon selection of additional substitutions. *In* "Abstracts of the International Discussion Meeting of HIV Population Dynamics, Variation and Drug Resistance." Edinburgh, UK.

Novella, I. S., Quer, J., Domingo, E., and Holland, J. J. (1999). Exponential fitness gains of RNA virus populations are limited by bottleneck effects. *J. Virol.* **73**, 1668–1671.

Nowak, M. A., Bonhoeffer, S., Shaw, G. M., and May, R. M. (1997). Anti-viral drug treatment: dynamics of resistance in free virus and infected cell populations. *J. Theor. Biol.* **184**, 203–217.

Otto, M., Garber, S., Winslow, D., Reid, C., Aldrich, P., Jadhav, P., Patterson, C., Hodge, C., and Cheng, Y. (1993). *In vitro* isolation and identification of human immunodeficiency virus (HIV) variants with reduced sensitivity to C-2 symmetrical inhibitors of HIV type 1 protease. *Proc. Natl. Acad. Sci. USA* **90**, 7543–7547.

Patick A., Mo, H., Markowitz, M., Appelt, K., Wu, B., Musick, L., Kalish, V., Kaldor, S., Reich, S., Ho, D., and Webber, S. (1996). Antiviral and resistance studies of AG1343, an orally bioavailable inhibitor of human immunodeficiency virus protease. *Antimicrob. Agents Chemother.* **40**, 292–297.

Patick, A., Rose, R., Greytock, J., Bechtold, C., Hermsmeier, M., Chen, P., Barrish, J., Zahler, R., Colonno, R., and Lin, P. (1995). Characterization of a Human Immunodeficiency Virus type 1 variant with reduced sensitivity to an aminodiol protease inhibitor. *J. Virol.* **69**, 2148–2152.

Patick, A. K., Duran, M., Cao, Y., Shugarts, D., Keller, M. R., Mazabel, E., Knowles, M., Chapman, S., Kuritzkes, D. R., and Markowitz, M. (1998). Genotypic and phenotypic characterization of human immunodeficiency virus type 1 variants isolated from patients treated with the protease inhibitor nelfinavir. *Antimicrob. Agents Chemother.* **42**, 2637–2644.

Perelson, A. S., Neumann, A. U., Markowitz, M., Leonard, J. M., and Ho, D. D. (1996). HIV-1 dynamics in vivo: virion clearance rate, infected cell life-span, and viral generation time. *Science* **271**, 1582–1586.

Piketty, C., Castiel, P., Belec, L., Batisse, D., Si Mohamed, A., Gilquin, J., Gonzalez-Canali, G., Jayle, D., Karmochkine, M., Weiss, L., Aboulker, J. P., and Kazatchkine, M. D. (1998). Discrepant responses to triple combination antiretroviral therapy in advanced HIV disease. *AIDS* **12**, 745–750.

Preston, B. D., Poiesz, B. J., and Loeb, L. A. (1988). Fidelity of HIV-1 reverse transcriptase. *Science* **242**, 1168–1171.

Ribeiro, R. M., Bonhoeffer, S., and Nowak, M. A. (1998). The frequency of resistant mutant virus before antiviral therapy. *AIDS* **12**, 461–465.

Richman, D. D., Havlir, D., Corbeil, J., Looney, D., Ignacio, C., Spector, S. A., Sullivan, J., Cheeseman, S., Barringer, K., Pauletti, D., *et al.* (1994). Nevirapine resistance mutations of human immunodeficiency virus type 1 selected during therapy. *J. Virol.* **68**, 1660–1666.

Ridky, T. W., Kikonyogo, A., Leis, J., Gulnik, S., Copeland, T., Erickson, J., Wlodawer, A., Kurinov, I., Harrison, R. W., and Weber, I. T. (1998). Drug-resistant HIV-1 proteases identify enzyme residues important for substrate selection and catalytic rate. *Biochemistry* **37**, 13835–13845.

Roberts, J. D., Bebenek, K., and Kunkel, T. A. (1988). The accuracy of reverse transcriptase from HIV-1. *Science* **242**, 1171–1173.

Roberts, N. A. (1995). Drug-resistance patterns of saquinavir and other HIV proteinase inhibitors. *AIDS* **9**, S27–S32.

Rose, R., Gong, Y., Greytok, J., Bechtold, C., Terry, B., Robinson, B., Alam, M., Colonno, R., and Lin, P. (1996). Human immunodeficiency virus type 1 viral background plays a major role indevelopment of resistance to protease inhibitors. *Proc. Natl. Acad. Sci. USA* **93**, 1648–1653.

Sardana, V. V., Schlabach, A. J., Graham, P., Bush, B. L., Condra, J. H., Culberson, J. C., Gotlib, L., Graham, D. J., Kohl, N. E., LaFemina, R. L., Schneider, C. L., Wolanski, B. S., Wolfgang, J. A., and Emini, E. A. (1994). Human Immunodeficiency Virus type 1 protease inhibitors: Evaluation of resistance engendered by amino-acid substitutions in the enzyme substrate binding site. *Biochemistry* **33**, 2004–2010.

Schinazi, R. F., Larder, B. A., and Mellors, J. W. (1997). Mutations in retroviral genes associated with drug resistance. *Int. Antiviral News* **5**, 129–137.

Sharma, P. L., and Crumpacker, C. S. (1997). Attenuated replication of human immunodeficiency virus type 1 with a didanosine-selected reverse transcriptase mutation. *J. Virol.* **71**, 8846–8851.

Smerdon, S. J., Jager, J., Wang, J., Kohlstaedt, L. A., Chirino, A. J., Friedman, J. M., Rice, P. A., and Steitz, T. A. (1994). Structure of the binding site for nonnucleoside inhibitors of the reverse transcriptase of human immunodeficiency virus type 1. *Proc. Natl. Acad. Sci. USA* **91**, 3911–3915.

Tachedjian, G., Hooker, D. J., Gurusinghe, A. D., Bazmi, H., Deacon, N. J., Mellors, J., Birch, C., and Mills, J. (1995). Characterisation of foscarnet-resistant strains of human immunodeficiency virus type 1. *Virology* **212**, 58–68.

Tachedjian, G., Mellors, J. W., Bazmi, H., and Mills, J. (1998). Impaired fitness of foscarnet-resistant strains of human immunodeficiency virus type 1. *AIDS Res. Hum. Retroviruses* **14**, 1059–1064.

Tipples, G. A., Ma, M. M., Fischer, K. P., Bain, V. G., Kneteman, N. M., and Tyrrell, D. L. (1996). Mutation in HBV RNA-dependent DNA polymerase confers resistance to lamivudine *in vivo*. *Hepatology* **24**, 714–717.

Tisdale, M., Kemp, S. D., Parry, N. R., and Larder, B. A. (1993). Rapid *in vitro* selection of human immunodeficiency virus type 1 resistant to 3′-thiacytidine inhibitors due to a mutation in the YMDD region or reverse transcriptase, *Proc. Natl. Acad. Sci. USA* **90**, 5653–5656.

Wei, X., Ghosh, S. K., Taylor, M. E., Johnson, V. A., Emini, E. A., Deutsch, P., Lifson, J. D., Bonhoeffer, S., Nowak, M. A., Hahn, B. H., *et al.* (1995). Viral dynamics in human immunodeficiency virus type 1 infection. *Nature* **373**, 117–122.

Winslow, D. L., Stack, S., King, R., Scarnati, H., Bincsik, A., and Otto, M. J. (1995). Limited sequence diversity in the HIV type 1 protease gene from clinical isolates and *in vivo* susceptibility to HIV protease inhibitors. *AIDS Res. Hum. Retroviruses* **11**, 107–113.

Winters, M. A., Colley, K. L., Girard, Y. A., Levee, D. J., Hamdan, H., Shafer, R. W., Katzenstein, D. A., and Merigan, T. C. (1998). A 6-basepair insert in the reverse transcriptase gene of human immunodeficiency virus type 1 confers resistance to multiple nucleoside inhibitors. *J. Clin. Invest.* **102**, 1769–1775.

Wong, J. K., Hezareh, M., Gunthard, H. F., Havlir, D. V., Ignacio, C. C., Spina, C. A., and Richman, D. D. (1997). Recovery of replication-competent HIV despite prolonged suppression of plasma viremia. *Science* **278**, 1291–1295.

Wong, J. K., Ignacio, C. C., Torriani, F., Havlir, D., Fitch, N. J., and Richman, D. D. (1997). *In vivo* compartmentalization of human immunodeficiency virus: Evidence from the examination of pol sequences from autopsy tissues. *J. Virol.* **71**, 2059–2071.

Yerly, S., Rakik, A., De Loes, S. K., Hirschel, B., Descamps, D., Brun-Vezinet, F., and Perrin, L. (1998). Switch to unusual amino acids at codon 215 of the human immunodeficiency virus type 1 reverse transcriptase gene in seroconvertors infected with zidovudine-resistant variants. *J. Virol.* **72**, 3520–3523.

Yuste, E., Sanchez-Palomino, S., Casado, C., Domingo, E., and Lopez-Galindez, C. (1999). Drastic fitness loss in human immunodeficiency virus type 1 upon serial bottleneck events. *J. Virol.* **73**, 2745–2751.

Zennou, V., Mammano, F., Paulous, S., Mathez, D., and Clavel, F. (1998). Loss of viral fitness associated with multiple Gag and Gag-Pol processing defects in human immunodeficiency virus type 1 variants selected for resistance to protease inhibitors *in vivo. J. Virol.* **72**, 3300–3306.

Zhang, L., Ramratnam, B., Tenner-Racz, K., He, Y., Vesanen, M., Lewin, S., Talal, A., Racz, P., Perelson, A. S., Korber, B. T., Markowitz, M., and Ho, D. D. (1999). Quantifying residual HIV-1 replication in patients receiving combination antiretroviral therapy. *N. Engl. J. Med.* **340**, 1605–1613.

Zhang, Y., Imamichi, H., Imamichi, T., Lane, H., Falloon, J., Vasudevachari, M., and Salzman, N. (1997). Drug resistance during indinavir therapy is caused by mutations in the protease gene and in its Gag substrate cleavage sites. *J. Virol.* **71**, 6662–6670.

Michael A. Parniak
Nicolas Sluis-Cremer

McGill University AIDS Centre
Lady Davis Institute for Medical Research
Montreal, Quebec H3T 1E2, Canada

Inhibitors of HIV-1 Reverse Transcriptase

I. Introduction

Retroviruses such as HIV carry their genomic information in the form of (+)RNA, but are distinguished from other RNA viruses by the fact that they replicate through a double-stranded DNA (proviral DNA) that is integrated into the infected host cell's genomic DNA (Coffin, 1996). Once integrated, proviral DNA is therapeutically indistinguishable from cellular DNA and results in persistent HIV infection and the establishment of chronically infected cells, which are major contributors to the continued spread of HIV infection in an infected individual. Thus, intervention at stages of viral replication prior to the integration of retroviral DNA, such as the obligate conversion of HIV genomic RNA into double-stranded viral DNA, is a logical therapeutic strategy. While this conversion is a complex process, all chemical steps are catalyzed by the viral enzyme, reverse transcriptase (RT). Since there is no cellular homolog of retroviral RT, the enzyme is an

Advances in Pharmacology, Volume 49

attractive target for the development of HIV therapeutics. Due to intense screening efforts over the past 15 years, thousands of compounds have been identified as inhibitors of HIV RT *in vitro*. However, only a small percentage of these possess the appropriate pharmacological properties, such as a deliverable formulation, bioavailability and attainment of effective plasma concentrations, and low toxicity, that render them useful as anti-HIV therapeutic agents. This chapter discusses the current clinically used RT inhibitors and some promising new inhibitors still in preclinical development, emphasizing the mechanisms of action of RT inhibitors and the mechanisms of viral resistance that develop upon continued exposure of the virus to these compounds.

Inhibitors of HIV-1RT may be classified into three groups:

1. Nucleoside HIV-1 RT inhibitors (NRTIs) such as zidovudine (3'-azido-3'-deoxythymidine; AZT), lamivudine (2',3'-dideoxy-3'-thiacytidine; 3TC), stavudine (2',3'-didehydro-2',3'-dideoxythymidine; d4T), and abacavir ((1S, 4R)-4-[2-amino-6(cyclopropylamino)-9H-purin-9y1]-2-cyclopentene-1-methanol 1592U89; ABC) as well as nucleotide prodrugs such as adefovir dipivoxil.

2. Nonnucleoside HIV-1 RT inhibitors (NNRTIs) such as nevirapine, delavirdine, and efavirenz, which are in current clinical use, and NNRTI in preclinical development such as the quinoxalines and the thiocarboxanilides.

3. Inhibitors with novel mechanisms of action, including *tert*-butyl-dimethylsilyl-*spiro*-aminooxathiole dioxide (TSAOs) derivatives and the N-acylhydrazones.

II. The Target

The conversion of retroviral genomic RNA into double-stranded DNA is a highly complex process. Nonetheless, all chemical steps in this conversion are catalyzed by the retroviral reverse transcriptase. Accordingly, RT must be a multifunctional enzyme. Retroviral RT has two types of DNA polymerase activity, RNA-dependent DNA polymerase (RDDP) activity to prepare a (−)strand DNA copy of the viral genomic RNA template and DNA-dependent DNA polymerase (DDDP) activity to prepare a (+)DNA strand complementary to the newly synthesized (−)DNA. The enzyme also has ribonuclease H (RNaseH) activity in order to degrade the RNA component of the RNA–(−)DNA duplex formed by RT RDDP activity. The released (−)DNA then serves as template for RT DDDP activity to allow completion of retroviral double-stranded DNA that can then be integrated into the host cell genome. Each of the RT activities, DNA polymerase (RDDP, DDDP) and RNaseH, are essential for reverse transcription and thus are appropriate targets for antiviral intervention. However, nearly all RT inhibitors developed to date, and certainly all of the inhibitors in current clinical use, work at the level of RT DNA polymerase; few RNaseH inhibitors have been identified.

HIV-1 RT DNA polymerase activity follows an ordered sequential mechanism that involves three RT mechanistic forms prior to catalysis (Fig. 1). Template/primer(T/P) binds to free RT to form the RT–T/P binary complex. The latter interacts with dNTP to form the RT–T/P–dNTP ternary complex. While free RT can interact with dNTP, this RT–dNTP complex is nonproductive and does not figure in the RT DNA polymerase mechanism (Majumdar *et al.*, 1988; Kati *et al.*, 1992; Reardon, 1992, 1993). Upon binding of the complementary dNTP to the RT–T/P binary complex to form the ternary complex, RT undergoes a biochemically measurable conformational change (Wu *et al.*, 1993). This positions the substrates to enable nucleophilic attack of the 3'-hydroxyl of the primer terminal nucleotide on the α-phosphate of the bound dNTP (Majumdar *et al.*, 1988; Kati *et al.*, 1992; Reardon, 1992, 1993). As is discussed later, the multiple RT mechanistic forms impact significantly on inhibition of RT, especially by NNRTI.

Crystal structures of each of the HIV-1 RT mechanistic forms prior to catalysis have been solved. These include substrate-free (unliganded) RT (Rodgers *et al.*, 1995; Hsiou *et al.*, 1996), the RT–DNA T/P binary complex (Jacobo-Molina *et al.*, 1993; Ding *et al.*, 1998), and most recently an elegant solution of the RT–DNA T/P–dNTP ternary complex (Huang *et al.*, 1998). These data have provided considerable insight into HIV-1 RT molecular structure, the conformational changes associated with DNA polymerization, and the relative position of selected residues during the various stages of catalysis. In addition, numerous structures of RT complexed with a variety of NNRTI are available (Kohlstaedt *et al.*, 1992; Smerdon *et al.*, 1994; Ren *et al.*, 1995a,b, 1998; Ding *et al.*, 1995a,b; Das *et al.*, 1996; Hopkins *et al.*, 1996; Esnouf *et al.*, 1995, 1997; Hsiou *et al.*, 1998). All of these structures are complexes between inhibitor and substrate-free RT. It is interesting to note that to date there has not been a single published example of *de novo* rational drug design based on any of these structures.

The gene for HIV-1 RT encodes a 66-kDa protein; however, the presumed biologically relevant form of HIV-1 RT is a heterodimer consisting

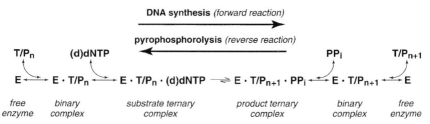

FIGURE I   Schematic of the ordered kinetic mechanism of reverse transcriptase-catalyzed DNA synthesis. The figure also illustrates the relationship between DNA synthesis and pyrophosphorolysis.

of two subunits of 66 and 51 kDa. The latter subunit is derived from the former by proteolytic removal of the 15-kDa C-terminal RNaseH domain during HIV-1 assembly and maturation (reviewed by Le Grice, 1993), a process catalyzed by the HIV-1 protease.

The overall structure of the p66 subunit may be likened to a "right hand" (Kohlstaedt *et al.*, 1992), with subdomains termed "fingers" (residues 1-85 and 118-155), "palm" (86–117 and 156–237), and "thumb" (238–318) that define the polymerase domain. Additional domains include the connection (319–426), involved in T/P interaction and RT subunit interactions, and finally the C-terminal RNaseH domain (427–565). These subdomains have different relative orientations in the p66 and p51 subunits of the HIV-1 RT. The p66 subunit adopts an "open," catalytically competent conformation that can accomodate a nucleic acid template strand, whereas the p51 subunit is in a "closed" conformation and is considered to play a largely structural role. The p51 subunit has also been suggested to play an important role in interaction with the $tRNA^{Lys3}$ primer used for the initiation of reverse transcription of the HIV-1 genomic (+)RNA (Kohlstaedt *et al.*, 1992).

Although the p66/p51 heterodimer is likely to be the species of RT in HIV-1 virions and thus the relevant enzyme to catalyze reverse transcription, this has not been proven unequivocally. Isolated recombinant p51 HIV-1 RT has demonstrable DDDP, activity (Thimmig and McHenry, 1993; Fletcher *et al.*, 1996). In addition, the partially processed p66/p66 HIV-1 RT homodimer has RDDP, DDDP, and RNaseH activities comparable to those of the p66/p51 heterodimer (Fletcher *et al.*, 1996). Since each of these three RT species shows different levels of inhibition by certain RT inhibitors (Arion *et al.*, 1996), additional studies to unequivocally establish the biologically relevant RT quaternary species could be of interest.

III. Nucleoside Reverse Transcriptase Inhibitors (NRTI) _____

Nucleoside reverse transcriptase inhibitors are analogs of naturally occurring deoxyribonucleosides, but lack the 3′-hydroxyl group (Fig. 2). NRTI have been used clinically in the treatment of HIV-1 infection since 1986, and several NRTI are in current clinical use (Table I). NRTI are ineffective as administered and must be metabolically converted by host-cell enzymes to their corresponding 5′-triphosphates for antiviral activity (Mitsuya *et al.*, 1985; Furman *et al.*, 1986). The extent of phosphorylation can vary in different retroviral infectable cell types, depending on the NRTI in question (Dahlberg *et al.*, 1987).

Since NRTIs are analogs of natural deoxynucleosides, there is the potential that they will be used by normal cellular DNA polymerases, with accompanying toxicity. Thus, the selectivity of NRTI use by HIV-1 RT, compared to that by cellular DNA polymerases, is an important issue in the development of NRTI as therapeutic agents.

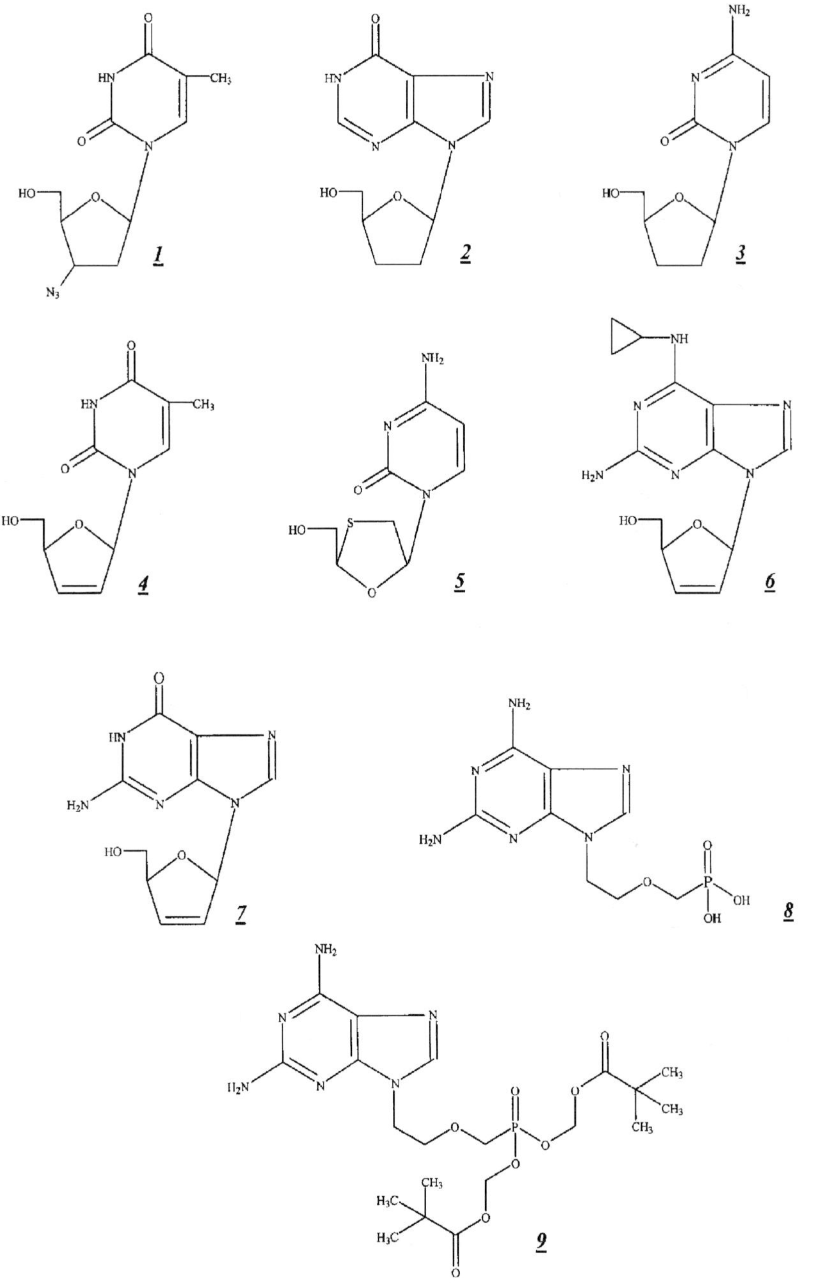

FIGURE 2 Structures of some nucleoside reverse transcriptase inhibitors (NRTIs). Illustrated are AZT (*1*), ddI (*2*), ddC (*3*), d4T (*4*), 3TC (*5*), abacavir (*6*), carbovir (*7*), adefovir (PMEA, (*8*), and adefovir dipivoxil (*9*).

TABLE I RT Inhibitors in Current Clinical Use

Inhibitor	Usual dose	Side effects	Resistance mutations in RT	Cross-resistance
NRTIs				
Zidovudine (AZT)	200 mg tid, or 300 mg bid[a]	Bone marrow suppression: anemia, neutropenia	Generally two or more of M41L, D67N, K70R, L210W, T215Y, K219Q	
Didanosine (ddI)	200 mg bid (>60 kg) 125 mg bid (<60 kg)	Peripheral neuropathy, pancreatitis	K65R, L74V	ddC
Zalcitabine (ddC)	0.75 mg tid	Peripheral neuropathy	K65R, T69D	ddI, 3TC, adefovir
Stavudine (d4T)	40 mg bid (>60 kg)	Peripheral neuropathy	(V75T)[b]	ddI, ddC
Lamivudine (3TC)	150 mg bid[a]	Minimal toxicity	M184V	ddI, ddC
Abacavir (ABC)	300 mg bid	Hypersensitivity (requires discontinuation)	Two or more of K65R, L74V, M184V, Y115F	ddI, ddC, 3TC
NNRTIs				
Nevirapine	200 mg qd (2 weeks) then 200 mg bid	Rash	One of Y181C, L100I, K103N, V106A, V108I, G190A	Many NNRTI
Delavirdine	400 mg tid	Rash	One of K103N, V108I, Y181C, P236L	Many NNRTI
Efavirenz	600 mg qHS	CNS symptoms	One or more of K103N, L100I, V106A, V108I, Y181C, G190A	Most NNRTI[c]

[a] AZT and 3TC exist in a combined formulation, Combivir, taken bid.
[b] Resistance to d4T is generally noted as part of HIV multidrug resistance (two amino acid insertion mutations).
[c] Two or more mutations in the nonnucleoside binding pocket results in cross-resistance to virtually all NNRTIs.

A. NRTIs in Clinical Use

I. Zidovudine (3'-Azido-3'-deoxythymidine; AZT)

AZT (Fig. 2) was first synthesized in 1964 as a potential anticancer drug, but was not further developed for human use because of concerns about toxicity. Twenty years later, the compound was reassessed and found to have potent anti-HIV activity (Mitsuya et al., 1985) and, in 1987, was the first anti-HIV agent approved for clinical use.

AZT is a thymidine analog in which the 3'-OH has been replaced with an azido ($-N_3$) group. Thymidine kinase readily phosphorylates AZT to AZT-MP. However, the rate-limiting step in the metabolic activation of AZT is the conversion of AZT-MP to AZT-DP (Furman et al., 1985; Lavie et al., 1997), thus AZT-MP accumulates in cells, with only small amounts of AZT-TP being formed. However, the potency of AZT-TP against HIV-1 RT is such that these low levels are sufficient to efficiently inhibit HIV replication.

AZT has good oral bioavailability (about 60%) and shows good penetration into the central nervous system (CNS). The toxicity of AZT remains a concern, with bone marrow suppression leading to anemia and/or neutropenia being the most significant dose-limiting effects of AZT therapy.

2. Didanosine (2',3'-Dideoxyinosine; ddI) and Zalcitabine (2',3'-Dideoxycytidine; ddC)

The 2',3'-dideoxynucleosides ddI and ddC (Fig. 2) (Mitsuya and Broder, 1986) were approved for clinical use in 1991 and 1992, respectively. While neither drug has an antiviral potency equivalent to that of AZT, they are active against AZT-resistant strains of HIV and provide an alternative therapy for HIV-infected patients unable to tolerate or failing on AZT therapy.

The active intracellular antiviral metabolite of ddI is ddATP. Didanosine (ddI) is very labile in gastric acid and must be administered in a buffered formulation, without food. Oral bioavailability of ddI is rather less than that of AZT. In contrast, ddC has good oral bioavailability, but absorption is significantly diminished when administered with food. The active metabolite of ddC is ddC triphosphate.

Neither ddI nor ddC penetrates the CNS as effectively as AZT. The most severe side effect associated with ddI or ddC therapy is peripheral neuropathy; this neuropathy is more common with ddC therapy (Simpson and Tagliati, 1995).

3. Stavudine (2',3'-Didehydro-2',3'-dideoxythymidine; d4T)

Stavudine (d4T) (Lin et al., 1987), in 1994, was the fourth NRTI approved for clinical use. Like AZT, d4T (Fig. 2) is a thymidine analog and undergoes metabolic activation by the sequential action of thymidine kinase and thymidylate kinase, as does AZT. However, d4T-MP is a better substrate for the latter enzyme than AZT-MP. In addition, d4T seems to be better tolerated by patients. CNS penetration is less than AZT, but similar to that for ddI/ddC. The major side effect seems to be peripheral neuropathy.

4. Lamivudine ((−)-β-L-2′,3′-Dideoxy-3′-thiacytidine; 3TC)

3TC is a 2′,3′-dideoxynucleoside cytosine analog, with an oxathiolane ring (Fig. 2). 3TC is the β-anomer of the racemate BCH189 and is about 10-fold less potent against HIV than AZT (Soudeyns *et al.*, 1991). Interestingly, both the α- and the β-enantiomers of BCH189 have similar potency against HIV-1 *in vitro;* however, the α-enantiomer is significantly more cytotoxic (Coates *et al.*, 1992).

3TC was approved for clinical use in 1995 and seems to be quite well tolerated, with minimal side effects at the doses used. HIV-1 develops resistance to 3TC very rapidly, but the compound is widely used in coadministration with AZT.

5. Abacavir ((1S, 4R)-4-[2-Amino-6(cyclopropylamino)-9H-purin-9yl]-2-cyclopentene-1-methanol 1592U89; ABC)

Abacavir (Fig. 2) is the most recent NRTI to be approved for clinical use. The *in vitro* antiviral potency of ABC is similar to AZT, and the drug has good bioavailability (Daluge *et al.*, 1997). The active antiviral form of ABC is in fact the triphosphate derivative of carbovir (Fig. 2); however, carbovir itself is not a useful therapeutic because of poor oral bioavailability (Daluge *et al.*, 1998). Metabolic activation of abacavir is unique among the current clinical NRTIs (Faletto *et al.*, 1997). ABC is not a substrate for any of the cellular nucleoside kinases. Instead, the enzyme adenosine phosphotransferase catalyzes the formation of ABC monophosphate (ABC-MP). ABC-MP undergoes elimination of the 6-aminocyclopropyl group, by a specific cytosolic enzyme, to form carbovir monophosphate (CBV-MP). Interestingly, ABC itself is not a substrate for this enzyme. The active antiviral, CBV-TP, is formed via CBV-DP by the action of normal cellular nucleoside kinases.

Side effects associated with abacavir therapy include GI disturbances, rash, and fatigue. The most serious side effect is a hypersensitivity that develops in up to 5% of the treated patients, characterized by fever, nausea/vomiting, and general malaise. These symptoms disappear shortly after discontinuation of ABC. However, hypersensitized patients may no longer receive ABC therapy, since subsequent administration leads to serious, even fatal, reactions.

6. Adefovir Dipivoxil

9-[2-(Phosphonomethoxy)ethyl]adenine (PMEA; Fig. 2) is an acyclic nucleoside phosphonate. These are interesting NRTIs in which the sugar moiety is replaced by an acyclic alkyl side chain linked to a phosphonate group. In nucleoside phosphonates, the labile 5'-phosphodiester bond of normal nucleotides is replaced with a stable carbon–phosphorous linkage. Nucleoside phosphonates are thus analogs of nucleotide monophosphates. PMEA has potent *in vitro* activity against HIV-1 RT (De Clercq *et al.*, 1987; Balzarini *et al.*, 1989) and inhibits reverse transcription by acting as a DNA chain terminator after intracellular conversion to the phosphonate diphosphate form. Charged molecules such as nucleoside phosphonates have poor oral uptake. The diester prodrug adefovir dipivoxil [9-(2-(bispivaloyloxymethyl) phosphonomethoxyethyl)adenine] (Fig. 2) is in late-stage phase III clinical assessment for the treatment of HIV-1 infection. Renal toxicity appears to be the major side effect associated with adefovir dipivoxil treatment.

B. Mechanisms of NRTI Inhibition

NRTIs inhibit HIV-1 RT in two ways. First, after metabolic phosphorylation, the NRTI triphosphate competes with the corresponding dNTP substrate for binding to RT. Although RT tends to have a higher affinity for NRTI-TP than the natural dNTP substrate, competitive inhibition appears to play only a very minor role in NRTI action (Reardon, 1992; Gu *et al.*, 1994). The major mechanism of NRTI antiviral activity is viral DNA chain termination. Since NRTI lack a 3'-hydroxyl group, once incorporated into the nascent viral DNA chain, that chain can no longer be extended.

The base moiety of the nucleoside is the major determinant for selection by RT for incorporation into the growing viral DNA chain. Since HIV-1 genomic RNA is about 10,000 bases long, there are about 5,000 chances for any given NRTI to be incorporated before the completion of HIV-1 double-stranded DNA synthesis by reverse transcription. Each HIV-1 virion carries two copies of genomic RNA (Coffin, 1996); thus in theory only two chain-termination events should suffice to eliminate completion of viral DNA synthesis. In practice, conditions are not nearly this favorable for NRTI incorporation, for a variety of reasons, some of which are discussed later in this chapter. In addition, since NRTI are mimics of the natural nucleosides, there is significant potential for toxicity due to interference with normal cellular nucleoside metabolism; current clinically used NRTIs show a range of toxic side effects.

Clinically, NRTIs are only transiently useful as monotherapy for HIV infection. Drug failure is the result of the appearance of HIV drug resistance. NRTI resistance correlates with specific mutations in HIV-1 RT. It is believed that HIV-1 undergoes facile mutation due to the low transcription fidelity of HIV-1 RT. Because of this, multiple HIV-1 variants exist *in vivo* and drug-resistant mutants may be present at low levels before initiation of therapy (Domingo *et al.*, 1997). Under the selective pressure of anti-HIV therapy, these drug-resistant mutants gain a competitive advantage and eventually become the dominant quasispecies. The time required for this selection to occur is dependent on factors such as mutant frequency at the time of treatment initiation, the replication fitness of the mutant, and the magnitude of the selective pressure (potency of the drug).

Therefore, although NRTI monotherapy is initially effective in reducing the viral load in treatment-naive patients, the rapid emergence of drug-resistant strains of HIV-1 is inevitable under these conditions. Accordingly, it is now considered that monotherapy is inappropriate in the treatment of HIV-infected individuals and that drug combinations must be employed. The most widely used combination therapies involve two NRTIs plus a protease inhibitor, although two NRTIs plus an NNRTI as well as double- and triple-NRTI combinations are also used.

C. Mechanisms of HIV-1 Resistance to NRTIs

The major mechanism for NRTI inhibition of HIV replication is termination of nascent viral DNA elongation due to the lack of a 3'-hydroxyl group on the inhibitor. Competition of the NRTI-TP with the natural dNTP substrate for binding to RT is presumed to be only a minor contributor to the net antiviral effect, even though RT tends to have higher affinity for NRTI-TP than for the corresponding dNTP substrate. Nonetheless, the resistance phenotype for any individual NRTI (with the exception of AZT) seems to involve a decreased binding of the drug to the drug-resistant mutant RT. The recent solution of the RT–T/P–dNTP ternary complex (Huang *et al.*, 1998) supports this probability. The options available for decreased binding are, however, somewhat limited. Mutations that minimize RT interactions with the base component of the NRTI would also decrease interaction with the analogous dNTP, which would be of obvious detriment to HIV replication.

Numerous mutations in RT have been identified in HIV-1 resistance to NRTI (Tantillo *et al.*, 1994). HIV-1 resistance to all individual NRTI except AZT and ABC generally correlates with single point mutations in RT. Mutations that arise in resistance to individual NRTI are listed in Table I. These mutations occur in two distinct subdomains of the p66 catalytic subunit of HIV-1 RT. While the same residues are mutated in the p51 subunit of the RT heterodimer, the significance of these is still unclear. In the p66 catalytic subunit, M41L, I50T, A62V, K65R, D67N, T69D, K70R/E, L74V, V75I, and F77L are situated in the "fingers" subdomain, while Y115F, F116Y, Q151M, M184V, L210W, T215Y/F, and K219Q are in the "palm" subdomain.

Although wild-type HIV-1 RT tends to have a higher affinity for binding NRTI-TP compared to the natural dNTP substrate, the enzyme is considerably less efficient at catalyzing the incorporation of bound NRTI-TP into the nascent DNA compared to the corresponding dNTP substrate (Reardon,

1992; Krebs *et al.*, 1997). Therefore, in many cases, an increase in binding discrimination between the normal dNTP substrate and the corresponding NRTI-TP provided by single point mutations in HIV-1 RT may account for the observed phenotypic resistance. As mentioned above, this altered binding discrimination cannot be completely due to changes in the interaction of RT with the base of the NRTI, since this would lead to significant attenuation in the rate of reverse transcription. Any alterations interactions with the natural dNTP substrate would require some compensatory phenotype to maintain HIV-1 replication efficiency. A good example of this is the K65R mutation, which confers cross-resistance to ddI, ddC, and 3TC. K65R RT shows an eightfold increase in the IC_{50} for ddCTP and ddATP *in vitro*, but only a twofold loss in affinity for the natural substrates dCTP and dATP (Gu *et al.*, 1994). This modest loss of affinity for dNTP substrate is compensated by an increase in the DNA synthesis processivity of the mutant enzyme (Arion *et al.*, 1996a).

For the most part, specific mutations in RT seem to confer specificity in resistance, i.e., the K65R mutation while giving cross-resistance to ddI and ddC, does not confer resistance to ddG nor ddT (Gu *et al.*, 1994). How is this selectivity in resistance conferred? In the HIV-1 RT ternary complex, K65 contacts both the β- and γ-phosphates of the bound dNTP (Huang *et al.*, 1998), positioning the substrate for base pairing with the complementary template residue and for the catalysis of phosphodiester bond formation with the 3'-OH of the primer terminal nucleotide. Crystal structures of *Thermus aquaticus* DNA polymerase I complexed with different dNTPs (Li *et al.*, 1998) show that dNTP binding is also primarily due to interactions of the enzyme with the phosphate component of the nucleotide. However, despite the similarity in binding interactions, the orientations of the nucleoside (base and sugar) of each of the four bound natural dNTP molecules are subtly different, suggesting that mutation of one of the phosphate-contacting residues could lead to a considerable change in specific dNTP positioning for catalysis. If the 3'-OH of the natural dNTP is also crucial for correct positioning of the substrate for catalysis, then loss of this contact (the 3'-OH is missing in all NRTI antivirals) could lead to even more deleterious effects on interaction and thereby to high-level resistance. Recent studies in our laboratory show that when HIV-1 RT K65 is replaced with amino acids other than arginine, the enzyme not only loses catalytic efficiency, but the ability to discriminate between dNTP substrate and the corresponding NRTI-TP inhibitor is lost (Sluis-Cremer *et al.*, 2000).

The recent crystal structure of the RT–T/P–dNTP ternary complex shows that drug-resistance mutations in the "palm" subdomain also cluster around the dNTP binding site (Huang *et al.*, 1998). Interestingly, Q151 makes significant contacts with the nucleotide deoxyribose sugar; mutation of this residue would also influence the binding orientation of dideoxynucleotide triphosphate molecules.

I. HIV-I Resistance to Specific NRTIs

a. Resistance to 3TC   HIV resistance to 3TC develops very rapidly both *in vitro* and *in vivo* (Schinazi *et al.*, 1993; Tisdale *et al.*, 1993; Schuurman *et al.*, 1995). This resistance is due to a single mutation in RT, M184V, which leads to up to 1000-fold resistance to 3TC as well as cross-resistance to ddI and ddC (about 8-fold resistance) (Schinazi *et al.*, 1993; Tisdale *et al.*, 1993; Gao *et al.*, 1993). Importantly, the M184V mutation restores AZT sensitivity to AZT-resistant HIV-1, despite the continued presence of AZT-resistance mutations. This restoration of AZT sensitivity may be one of the reasons for the success of AZT+3TC combination chemotherapy.

In wild-type RT, M184 makes contact with both the sugar and the base of the 3′-terminal nucleotide of the primer (Tantillo *et al.*, 1994; Huang *et al.*, 1998). In the 3TC-resistant M184V mutant, the β-branched side chain of valine can also contact the sugar moiety of the bound dNTP substrate (Huang *et al.*, 1998). Molecular modeling shows that the configuration of the oxathiolane ring of 3TC is such that the valine 184 residue in the mutant sterically hinders 3TC-TP binding and prevents correct positioning of the bound 3TC-TP for catalysis.

b. Resistance to ddI and to ddC   The L74V mutation provides a 6- to 26-fold reduction in sensitivity to ddI (St. Clair *et al.*, 1991). This mutation also leads to cross-resistance to ddC, but partially restores sensitivity to AZT. The reductions in sensitivity conferred by the K65R mutation associated with HIV-1 cross-resistance to ddI, ddC, and 3TC (Gu *et al.*, 1994; Zhang *et al.*, 1994) are also relatively modest (approximately 5-fold for ddI and ddC and 20-fold for 3TC). Also noted is T69D, which provides a 5-fold reduction in sensitivity to ddC but does not impart cross-resistance to other NRTI (Fitzgibbon *et al.*, 1992). *In vitro* selection of HIV-1 resistant to d4T has identified the V75T mutation which confers a 7-fold increase in IC_{50} as well as reduced susceptibility to both ddI and ddC (Lacey and Larder, 1994).

c. Resistance to d4T   Patients previously treated with AZT show a virologic loss of response to d4T monotherapy with increasing length of d4T treatment (Spruance *et al.*, 1997). Although mutations in RT that confer multidrug resistance (see below) also confers resistance to d4T, these mutations are not selected for by d4T therapy alone. Thus, the molecular basis for d4T resistance remains unclear.

d. Resistance to ABC   Abacavir is the most recently approved NRTI and the first guanosine analog NRTI used clinically (Daluge *et al.*, 1997). Resistance to ABC does not develop readily *in vitro* and requires at least two to three mutations in RT to show a significant decrease (about 10-fold) in ABC inhibitory potency (Tisdale *et al.*, 1997). The decrease in antiviral potency correlates with a decreased interaction of inhibitor with RT. The mutations that arise in ABC resistance include K65R, L74V, M184V, and Y115F. ABC-resistant HIV-1 is cross-resistant to ddI, ddC, and 3TC, but not to AZT or d4T (Daluge *et al.*, 1997).

e. *Resistance to Adefovir* *In vitro* experiments show that resistance to adefovir (PMEA) is associated with either of the K70E or the K65R mutations that confer 9- and 16-fold resistance, respectively (Foli *et al.*, 1996; Mulato *et al.*, 1998). These mutants show cross-resistance to ddC, ddI, and 3TC, but not to AZT (Mulato and Cherrington, 1997).

2. The Novel Mechanism of HIV-I Resistance to AZT

HIV-1 resistance to AZT seems to involve a mechanism other than simple alteration of NRTI binding discrimination. HIV-1 mutants with decreased susceptibility to AZT were first reported in 1989 (Larder and Kemp, 1989). Wild-type HIV-1 is very sensitive to inhibition by AZT (50% inhibitory concentrations ranging from 1 to 40 nM), whereas AZT-resistant virus shows up to 200-fold decreased sensitivity to the drug. This resistance correlates with mutations in the viral enzyme reverse transcriptase (RT), namely D67N, K70R, T215F/Y, and K219Q (Larder and Kemp, 1989); an additional mutation (M41L) was subsequently detected (Kellam *et al.*, 1992). In general, high-level resistance requires the simultaneous presence of several of these mutations in RT (Larder *et al.*, 1991; Kellam *et al.*, 1992). Despite the well-validated genotype for AZT resistance and knowledge of the mechanism of AZT inhibition, the phenotype for AZT resistance remained obscure for more than a decade. Unlike the case in resistance to other NRTIs, RT possessing AZT resistance mutations is equally as sensitive to inhibition by AZTTP *in vitro* as wild-type RT (Wainberg *et al.*, 1992; Lacey *et al.*, 1992). However, in the past year we and others have proposed phenotypic mechanisms to account for AZT resistance (Arion *et al.*, 1998; Meyer *et al.*, 1998; Arion and Parniak, 1999). These new insights have interesting implications for future combination chemotherapy, as is discussed later in this chapter.

High-level AZT resistance requires multiple mutations in RT, and these mutations localize to two different subdomains of HIV-1 RT. The D67N/K70R mutations occur in the flexible loop between the $\beta3$-$\beta4$ structures of the "fingers" subdomain, whereas the T215F/Y and K219Q mutations are found in the "palm" subdomain (Tantillo *et al.*, 1994).

Arion *et al.* (1998) have shown that the D67N/K70R and T215F/K219Q mutations impart distinct phenotypic alterations in RT activity and have proposed a model for AZT resistance in which an increased rate of RT-catalyzed pyrophosphorolysis results in removal of chain-terminating AZT from the 3' end of the primer. Enzyme reactions are intrinsically reversible, and pyrophosphorolysis is the reverse reaction of DNA synthesis (Fig. 1). The substrate for pyrophosphorolysis is pyrophosphate, which is the normal product of DNA synthesis and is also present at an intracellular concentration of about 150 μM (Barshop *et al.*, 1991). In this reaction (Fig. 1), the RT–T/P complex binds pyrophosphate (PP$_i$) which carries out nucleophilic attack on the phosphodiester bond between the last two nucleotides of the primer. This results in removal of the 3′-terminal nucleotide from the primer. If the 3′-terminal nucleotide is a chain-terminator such as AZT, pyrophosphorolysis will result in removal of the chain-terminator with the concomitant appearance of a free 3′-hydroxyl, contributed by the nucleotide that was initially penultimate to the 3′-chain-terminating nucleotide.

The D67N/K70R mutations increase the rate of RT-catalyzed pyrophosphorolysis at normal intracellular concentrations of pyrophosphate (Arion *et al.*, 1998). In addition, these mutations increase RT sensitivity to pyrophosphate, leading to decreased binding of AZTTP. This pyrophosphate-mediated attenuation of AZTTP binding is specific for AZTTP, since the binding of the natural substrate TTP and the dideoxynucleotides ddCTP and ddGTP is unaffected by pyrophosphate.

The T215F/K219Q mutations increase the DNA synthesis processivity of RT; this is due to a decreased rate of T/P dissociation from the mutant enzyme (Arion *et al.*, 1998; Canard *et al.*, 1998). The increased mutant RT residence time of the T/P would also provide an increased opportunity for the pyrophosphorolytic removal of chain-terminating AZT. In the presence of dNTPs, pyrophosphorolytic removal of the 3′-chain-terminating AZT is accompanied by renewal of primer extension (viral DNA synthesis). Since the rate of DNA synthesis (forward reaction) is substantially greater than the rate of pyrophosphorolysis (reverse reaction) (Arion *et al.*, 1998), the result is net DNA synthesis. The T215F/K219Q mutation phenotype may provide a compensatory mechanism for the increased rate of pyrophosphorolysis due to the D67N/K70R mutations. Therefore, high-level viral resistance to AZT requires multiple mutations in different subdomains of the viral RT.

The second model for the AZT resistance phenotype is based on a newly discovered reaction carried out by wild-type HIV-1 RT, a ribonucleotide-dependent phosphorolysis reaction that can remove the chain-terminating nucleotides from the 3′ end of the primer, yielding an extendible primer terminus and a dinucleoside tri- or tetraphosphate product (Meyer *et al.*, 1998). The reaction is inhibited by dNTPs that are complementary to the next position on the template, suggesting competition between dinucleoside polyphosphate synthesis and DNA polymerization. Since this reaction occurs *in vitro* at presumed physiological concentrations of ribonucleotide triphosphates, the authors speculate that this reaction may determine the *in vivo* activity of AZT and other NRTIs although this interesting mechanism has not yet been shown to be directly involved in the AZT resistance phenotype at the time of this writing. However, the chemistry involved in this mechanism, nucleophilic attack of a polyphosphate oxygen on the phosphodiester bond between the last two nucleotides of the primer, resulting in removal of the 3′-terminal nucleotide from the primer, is identical to that involved in pyrophosphorolysis. Thus it is quite possible that AZT resistance mutations that enhance the rate of pyrophosphorolysis may also lead to an enhancement of the RT-catalyzed dinucleoside polyphosphate synthesis reaction.[1]

D. NRTI Combination Chemotherapy

Since NRTI monotherapy culminates in the emergence of drug-resistant virus, combinations comprising two or more NRTIs are preferred (Collier *et al.*, 1996; Johnson, 1996; Perrin and Hirschel, 1996). The use of combination therapies has resulted in dramatic improvements in stemming the course of HIV infection in infected patients. It is now recommended that asymptomatic treatment-naïve patients be started with triple drug combinations when plasma HIV RNA is above 5000 copies/ml and/or when CD4 levels drop below 500. Most current combination therapies involve two NRTIs plus a protease inhibitor, although two NRTIs plus an NNRTI as well as double- and triple-NRTI combinations are also used.

AZT and d4T combinations are not used clinically because of adverse effects on CD4+ cell counts. However, this combination is also not advantageous from theoretical biological considerations. Both drugs are thymidine analogs; thus, they would compete with each other and with the natural nucleotide TTP for interaction with HIV-1 RT during reverse transcription and thus would likely act as antagonists to each other. Similarly, although ddC+3TC combinations have been used, they also might be antagonists, since both are cytidine analogs.

The most widely used NRTI combination is AZT+3TC. This combination is now available as a single formulation, Combivir, which reduces the dosage to one pill taken twice daily, a significant improvement over the multiple dosing regimen required for administration of the drugs in individual formulations. AZT+3TC combinations have certain therapeutic advantages. The dual-drug regimen significantly delays the development of AZT resistance compared to that noted with AZT alone (Larder *et al.*, 1995). Perhaps more importantly, the M184V associated with 3TC resistance restores AZT sensitivity to AZT-resistant HIV-1, despite the continued presence of AZT resistance mutations (Larder *et al.*, 1995). Recent studies indicate that the M184V mutation eliminates the increased rate of pyrophosphorolysis associated with the D67N/K70R/T215F/K219Q AZT-resistant RT, thereby restoring chain-terminating activity to AZT (Gotte *et al.*, 2000). This restoration of AZT sensitivity may be one of the reasons for the success of AZT+3TC combination chemotherapy. The M184V mutation also in-

[1] Note added in proof: The correlation between increased ribonucleotide-dependent phosphorolysis and AZT resistance mutations has been demonstrated (Meyer *et al.*, 1999; Arion *et al.*, 2000).

creases the fidelity of HIV-1 RT-catalyzed DNA synthesis relative to the wild-type enzyme (Wainberg *et al.*, 1996). It has been suggested that this increased fidelity may therefore delay resistance development against other antiretroviral agents, such as protease inhibitors (Wainberg and Parniak, 1997).

The newest NRTI to be approved for therapy is abacavir (ABC), which is administered twice daily. The multiple-NRTI regimen, composed of ABC+AZT+3TC, requires two pills twice daily, a significant improvement on other combination regimens and one which might greatly facilitate patient compliance. ABC+AZT+3TC combinations may demonstrate some degree of synergy in inhibiting both wild-type and AZT-resistant HIV replication in peripheral blood mononuclear cells *in vitro* (Tremblay *et al.*, 1998). Ongoing clinical trials suggest that triple NRTI therapy is more effective than two NRTIs alone (Fischl *et al.*, 1999) and that ABC+AZT+3TC triple therapy is as effective as AZT+3TC plus a protease inhibitor (Staszewski *et al.*, 1999). These trials are not complete, and additional studies must be carried out.

Is the use of multiple-NRTI combinations justified from a theoretical biological viewpoint? NRTIs are ineffective as administered and must be phosphorylated for antiviral activity. The extent of phosphorylation varies in different HIV-infectable cell types and differs for each NRTI. Thus, use of multiple NRTIs may maximize the possibility of having a certain amount of at least one chain-terminating NRTI in most, if not all, HIV-infectable cell types. NRTIs are analogs of the natural substrates for DNA polymerases such as HIV RT. The base moiety of the nucleotide (complementarity to the template base) is the major determinant for selection by RT for incorporation into the growing viral DNA chain. Use of multiple NRTIs with different bases would increase the probability of chain termination of viral DNA and may in some cases provide for synergy in inhibition of HIV-1 replication (Villahermosa *et al.*, 1997). However, the use of multiple NRTIs with different bases also increases the potential for incorporation into cellular DNA (in proliferating cells and in mitochondria), which may increase the risk of toxicity. In addition, as discussed below, the emergence of multidrug-resistant HIV-1 strains resulting from the use of multiple NRTIs is an increasing problem.

The extraordinary success of the NRTI+protease inhibitor combination therapies is due in large part to the fact that these combinations are directed at two different viral targets, reverse transcriptase and protease. Multiple-NRTI therapies are directed at a single viral target, the active site of HIV reverse transcriptase. While multiple-NRTI regimens such as ABC+AZT+3TC increase the probability of chain termination of viral DNA, only one NRTI can bind to RT at any given time; thus synergistic inhibition of HIV replication may be less likely. Furthermore, RT is able to overcome the theoretically irreversible blockage of continued viral DNA synthesis once a chain-terminating NRTI has been incorporated by means of pyrophosphorolysis (as discussed above for AZT resistance). This is an intrinsic activity of RT from NRTI-resistant HIV as well as wild-type drug-sensitive virus. The intrinsic pyrophosphorolytic activity of HIV reverse transcriptase may prevent NRTI from being 100% effective in eliminating viral reverse transcription.

I. NRTI Combination Therapy: The Emerging Problem of Multidrug Resistance

Since NRTI monotherapy culminates in the emergence of drug-resistant virus, combinations comprising two or more NRTIs are preferred (Collier *et al.*, 1996; Johnson, 1996; Perrin and Hirschel, 1996). Indeed, triple-NRTI therapy consisting of AZT+3TC+ABC is being considered as an alternative to NRTI+protease inhibitor combinations in certain cases (Fischl *et al.*, 1999; Staszewski *et al.*, 1999). In general, the time to development of HIV-1 resistance to multiple-NRTI treatment *in vitro* is prolonged relative to the development of resistance to any individual drug; this generally translates into a prolonged and sustained antiviral effect *in vivo*. Nonetheless, HIV-1 drug-resistance development remains a significant problem in NRTI combination chemotherapy, all the more so since it is becoming increasingly obvious that HIV-1 strains that replicate under combination-NRTI pressure exhibit multidrug resistance (Shafer *et al.*, 1994; Shirasaka *et al.*, 1995; Iversen *et al.*, 1996).

A chronological study of the HIV-1 variants that developed over length of exposure (up to 48 months) to combinations of AZT with either ddC or ddI revealed the appearance of a series of five mutations; similar results were noted in either simultaneous or sequential therapeutic regimens (Shirasaka *et al.*, 1995). The Q151M mutation appears first and by itself is sufficient to confer multidrug resistance to HIV-1, providing 10-fold resistance to AZT, 20-fold resistance to ddC, and 5-fold resistance to ddI in cell-culture virus-replication assays (Shirasaka *et al.*, 1995). Subsequently, four additional mutations, A62V, V75I, F77L, and F116Y, further increase the level of drug resistance. HIV-1 containing all five mutations shows over 300-fold resistance to AZT and about 40-fold resistance to ddC and to ddI *in vitro* (Shirasaka *et al.*, 1995). Recombinant RT possessing the Q151M mutation exhibited high-level resistance to each of ddATP, ddCTP, ddGTP, ddTTP, and AZTTP (Ueno *et al.*, 1995, 1997).

A disturbing report of the development of an HIV-1 strain resistant to all NRTI and the majority of NNRTI in a patient treated sequentially with AZT, ddI, ddC, 3TC, and d4T, as well as a nonnucleoside inhibitor, has been presented (Schmit *et al.*, 1996). The RT mutations that arose in this multidrug-resistant HIV-1 strain were identical to those described above, but also included K103N, which confers HIV-1 resistance to many NNRTIs.

Multidrug therapy also leads to the generation of insertion mutations in HIV-1 RT, with two serine residues inserted between RT residues 68 and

70 (De Antoni *et al.*, 1997; Tamalet *et al.*, 1998; Winters *et al.*, 1998). These insertion mutants were initially considered to be due to d4T resistance; however, it is now clear that these insertions are not selected by d4T in the absence of prior NRTI therapy, particularly extensive treatment with AZT. HIV-1 with insertion mutations are resistant to most clinically used NRTIs (including AZT, 3TC, d4T, ddC, and ddI), and the multidrug resistance phenotype is generally evident only when the insertion mutations are accompanied by the AZT resistance mutations L210W and/or T215Y.

Thus, resistance development remains a problem even with multiple-NRTI treatment regimens. Use of multiple NRTIs appears to promote the development of multidrug-resistant HIV-1 strains which seriously compromise continued treatment of patients infected with these strains. Current data concerning the benefit of multiple-NRTI treatment, without concomitant administration of either protease inhibitors or NNRTI, are incomplete and some caution should be exercised in promotion of these treatment regimens.

IV. Nonnucleoside HIV-1 RT Inhibitors (NNRTI)

NNRTIs are chemically distinct from nucleosides and, unlike the NRTIs, do not require intracellular metabolism for activity—they work "right out of the box," so to speak. Literally thousands of NNRTIs have been described, including derivatives of 1-(2-hydroxyethoxymethyl)-6-(phenylthio)thymine (HEPT) (Miyasaka *et al.*, 1989), tetrahydroimidazo[4,5,1-*jk*] [1,4]benzodiazepin-2(1*H*)-one and -thione (TIBO) (Pauwels *et al.*, 1990), nevirapine (Merluzzi *et al.*, 1990), pyridinones (Goldman *et al.*, 1991), bis(heteroaryl)piperazine (BHAP) (Romero *et al.*, 1991), α-anilinophenylacetamide (α-APA) (Pauwels *et al.*, 1993), carboxanilides (Bader *et al.*, 1991), and many other subclasses of compounds.

As can be seen for the representative structures illustrated in Fig. 3, NNRTIs are a chemically heterogeneous group and thus may have very different pharmacokinetic and metabolism profiles. Nonetheless, they share a number of common features, including significant hydrophobicity, aromatic structure, and small size (<600 Da). NNRTIs are almost exclusively specific for HIV-1 RT, with little or no activity against even closely related enzymes such as HIV-2 RT, and therefore NNRTIs generally do not exhibit substantial toxicity. Even so, NNRTIs are xenobiotics and as such have the potential to induce unexpected toxicity; this is especially obvious with efavirenz (see below). Despite the apparently improved toxicity properties of NNRTIs compared to NRTIs, the former have not been widely used in treatment of HIV-1 infection, primarily because HIV-1 rapidly develops resistance to early generations of NNRTI, sometimes within 2–3 weeks of therapy.

NNRTIs interact with HIV-1 RT by binding to a single site on the p66 subunit of the HIV-1 RT p66/p51 heterodimer, termed the nonnucleoside

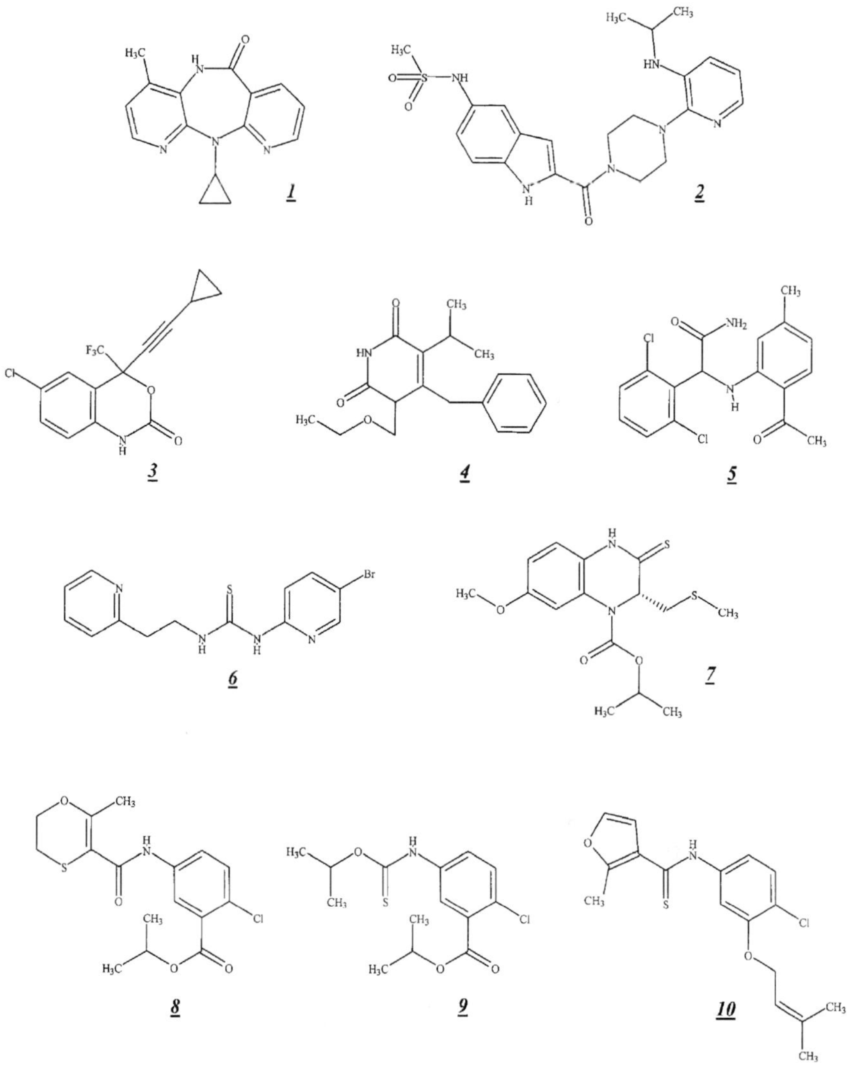

FIGURE 3 Structures of some nonnucleoside reverse transcriptase inhibitors (NNRTIs). Illustrated are nevirapine (*1*); delavirdine (*2*); efavirenz (*3*); the HEPT derivative MKC-442 (*4*); loviride (*5*); trovirdine (*6*), HBY 097 (*7*); and the carboxanilides UC84 (*8*), UC38 (*9*), and UC781 (*10*).

inhibitor binding pocket (NNIBP). This site is in the "palm" subdomain of p66, close to but distinct from the RT polymerase catalytic site. Accordingly, NNRTIs inhibit RT by noncompetitive or uncompetitive mechanisms. The NNIBP has been well-defined by cystallographic characterization of numerous RT–NNRTI complexes (Kohlstaedt *et al.*, 1992; Smerdon *et al.*, 1994;

Ren *et al.*, 1995a,b, 1998; Ding *et al.*, 1995a,b; Das *et al.*, 1996; Hopkins *et al.*, 1996; Esnouf *et al.*, 1995, 1997; Hsiou *et al.*, 1998). The NNRTI binding site is a highly hydrophobic pocket composed primarily of amino acids from secondary-structure elements of the p66 palm subdomain. A probable solvent accessible entrance to the NNIBP is located at the p66/p51 heterodimer interface, ringed by residues L100, K101, K103, V179, and Y181 of the p66 subunit and E138 of the p51 subunit. Most of the amino acid residues that form the NNIBP are hydrophobic and five of them have aromatic side chains. The only hydrophilic residues around the pocket (K101 and K103 of p66 and E138 of p51) are located at the entrance.

Although NNRTIs were discovered nearly a decade ago and have better toxicologic properties than NRTIs, clinical use of these compounds has been minimal due to rapid resistance development, which convinced many pharmaceutical companies to abandon continued development of NNRTIs as anti-HIV therapeutics. However, the increasing failure of HIV-1-infected patients on NRTI+protease inhibitor combination therapies has stimulated renewed interest in NNRTIs, especially in NRTI+NNRTI combination therapy, possibly as a first-line treatment to preserve the more potent protease inhibitors for second-line therapy against HIV-1 infection.

A. NNRTIs in Current Clinical Use

As of mid-1999, only three NNRTIs have been approved for the treatment of HIV-1 infection, namely nevirapine, delavirdine, and efavirenz (Table I). Several other NNRTI trials have been abandoned. Because of the significant delay and expense between discovery of *in vitro* antiviral activity of any given compound and the proof of its pharmacologic (*in vivo*) efficacy, it is difficult to predict which, if any, NNRTIs in preclinical development may be used in future treatment of HIV-infected patients. However, it is certain that future NNRTIs must demonstrate high-potency antiviral activity against wild-type HIV-1 and a broad spectrum of NRTI and NNRTI drug-resistant viral isolates, delayed development of resistance, as well as satisfactory pharmacological characteristics such as oral bioavailability, serum half-life, and minimal side effects.

I. Nevirapine (BI-RG-587; Viramune)

The dipyridodiazepinone nevirapine (Fig. 3) was the first NNRTI approved for clinical use. The compound has a moderate anti-HIV-1 activity *in vitro* ($EC_{50} \approx 100$ nM). Nevirapine has excellent oral bioavailability (>90%) and a significantly prolonged serum half-life compared to NRTI. The major side effect seems to be rash, which can be severe enough to warrant discontinuation of therapy in about 5% of patients.

Nevirapine induces the CYP3A family of cytochrome P450 enzymes. The NNRTI is therefore a potent inducer of its own metabolism, leading to a twofold increase in clearance rate after about 2 weeks of administration.

This induction also creates the potential for clinically significant drug interactions. Nevirapine decreases the plasma levels of rifampin and rifabutin and certain HIV-1 protease inhibitors including saqinavir and indinavir. Thus, nevirapine+protease inhibitor combination therapy is unlikely to be of use.

Resistance to nevirapine develops rapidly both *in vitro* and *in vivo*, and resistant HIV-1 strains are detected within a month after the start of nevirapine monotherapy (Richman *et al.*, 1994). Numerous mutations in the NNIBP have been associated with nevirapine resistance, including L100I, K103N, V106A, Y188C, and Y181C; the latter appears to be the most common. Any one of these single point mutations in the NNIBP confer high-level HIV-1 resistance to nevirapine.

2. Delavirdine (U-90152; Rescriptor)

Delavirdine (Fig. 3) is a bis(heteroaryl)-piperazine (BHAP) derivative, with an antiviral potency similar to nevirapine (Dueweke *et al.*, 1993a). Delavirdine has excellent oral bioavailability (>85%), but a relatively low serum half-life, which requires more frequent dosing than nevirapine.

Delavirdine, like nevirapine, undergoes hepatic metabolism by the CYP3A4 cytochrome P450 system. However, in contrast to nevirapine, delavirdine reduces P450 activity, thereby inhibiting its own metabolism. As well, delavirdine apparently increases the steady-state concentration of the protease inhibitors saquinavir and indinavir and may therefore be useful in combination with protease inhibitors. As for nevirapine, resistance to delavirdine develops rapidly, due to point mutations in the NNIBP such as K103N, V108I, Y181C, and P236L. The latter mutation is not noted in resistance to other NNRTIs and does not confer NNRTI cross-resistance but may actually increase sensitivity to other NNRTI (Dueweke *et al.*, 1993b).

3. Efavirenz (DMP-266; Sustiva)

Efavirenz (Fig. 3) is a benzoxazinone with exceptionally potent antiviral activity against HIV-1 (EC_{50} <1 nM). Efavirenz has a serum half-life similar to nevirapine and crosses the blood–brain barrier, but is less orally available ($\approx$ 40%). The drug has a high degree of protein binding (>99%). Efavirenz also seems to produce more pronounced side effects than either nevirapine or delavirdine. The most serious are a range of CNS symptoms including insomnia, confusion, and abnormal dreams and hallucinations, which may lead to patient-initiated discontinuation of therapy. Efavirenz may induce the CYP3A family of cytochrome P450 enzymes, thereby affecting its own metabolism and that of other drugs.

The value of efavirenz in HIV therapy is its exceptional antiviral potency and activity against many single mutation NNRTI-resistant HIV-1 isolates, particularly those resistant to nevirapine and delavirdine (Young *et al.*, 1995). However, certain mutations such as K101E, K103N, and Y188L confer moderate to high-level resistance to efavirenz.

B. NNRTIs under Clinical Trial Investigation

A number of other NNRTIs are under clinical trial investigation, including the HEPT derivative 6-benzyl-1-ethoxymethyl-5-isopropyluracil (MKC-442; Fig. 3), the α-anilinophenylacetamide Loviride (Fig. 3), and Trovirdine [a phenylethylthiazolyl thiourea (PETT) analog; Fig. 3]. All of these show good to excellent inhibition of HIV-1 replication *in vitro*, with EC_{50} values ranging between 1 to 14 nM.

One of the most promising new NNRTIs is HBY 097 ([S]-4-isopropoxy-carbonyl-6-methoxy-3-methylthiomethyl-3,4-dihydroquinoxaline-2[1H]-thione; Fig. 3), an NNRTI of the quinoxaline class that is a potent inhibitor of HIV-1 replication (EC_{50} <1 nM) (Kleim *et al.*, 1995). Resistance to HBY 097 selects for a G190E/Q mutation, which reduces the catalytic efficiency of RT (Kleim *et al.*, 1994); resistant HIV-1 strains might then demonstrate attenuated replication efficacy, which could assist in maintaining reduced viral loads. Phase 1 clinical trials indicate that HBY 097 is well tolerated and effective at suppressing viral load (Rubsamen-Waigmann *et al.*, 1997).

C. NNRTIs in Preclinical Development:
The Carboxanilides and Thiocarboxanilides

There are numerous NNRTIs that have been identified as having potent inhibitory activity against HIV-1 *in vitro*. However, detailed mechanistic analyses have been carried out primarily with the carboxanilide/thiocarboxanilide class of NNRTI; these studies have provided useful biochemical and virological information, and some surprises, that may provide a foundation for the development of future NNRTIs. Information derived from studies of this class of NNRTIs is provided throughout this and subsequent sections of this chapter.

Carboxanilide NNRTIs are derivatives of oxathiin carboxanilide (2-chloro-5-[[5,6-dihydro-2-methyl-1,4-oxathiin-3-yl)carbonyl]amino]benzoic acid, 1-methylethyl ester; UC84; Fig. 3), which was identified as an antiviral agent in the NIH high-throughput screening program (Bader *et al.*, 1991). Continued design and synthesis of new carboxanilide NNRTIs by chemists at Uniroyal Chemical Company Research Laboratories have provided a series of increasingly potent NNRTIs which have been exceptionally useful in elucidating the mechanism of NNRTI action.

The anti-HIV-1 activity of the first-generation UC84 ($EC_{50} \approx 400$ nM) is about 10-fold less than that of the second-generation compound UC38 (Fletcher *et al.*, 1995b) (Fig. 3); inhibition by the latter is also somewhat less affected by single point mutations in the RT NNIBP (Balzarini *et al.*, 1995). Continued development of UC analogs has resulted in UC781 (N-[4-chloro-3-3-methyl-2-butenyloxy)phenyl]-2-methyl-3-furanocarbothiamide; Fig. 3), one of the most potent NNRTIs yet described (Balzarini *et al.*,

1996b; Barnard *et al.*, 1997). UC781 is active against almost all HIV-1 strains with single point mutations in the NNIBP (Balzarini *et al.*, 1996b) and retains significant antiviral activity against certain double mutants (Motakis *et al.*, 1999). Preliminary studies show that UC781 has good oral bioavailability in a number of animal species (Buckheit *et al.*, 1997).

UC781 was the first NNRTI identified as being a tight-binding inhibitor of HIV-1 RT (Barnard *et al.*, 1997). UC781 binds rapidly to RT, but once bound dissociates very slowly (on the order of minutes to hours). RT is thereby inactivated for extended periods of time following a single "pulse" exposure to UC781. UC781 binds to each of the three RT mechanistic forms, with the strongest interaction to the RT–T/P–dNTP ternary complex (Barnard *et al.*, 1997).

Variable-temperature NMR studies show that UC781 is more flexible than other NNRTIs such as nevirapine and can exist in both cis and trans configurations (specified by the position of the hydrogen on the thioamide nitrogen relative to the sulfur of the thioamide). Both of these configurations can be readily modeled into the NNIBP (Motakis *et al.*, 1999) and likely contribute to the overall tight-binding nature of the inhibition and to the ability to bind to all three RT mechanistic forms. This flexibility of UC781 may also be a factor in the ability of the compound to inhibit many NNRTI-resistant HIV-1 strains (Balzarini *et al.*, 1996b). Importantly, UC781 is a "dead-end" inhibitor of wt RT, a characteristic not described for other NNRTIs. The nevirapine–RT complex catalyzes a low but definite rate of DNA polymerization even when RT is fully saturated with inhibitor. In contrast, The UC781–RT complex has no detectable enzymatic activity. Thus, the high potency of UC781 is due to multiple parameters including interaction with multiple RT mechanistic forms, tight-binding, and the "dead-end" nature of the inhibition (Motakis *et al.*, 1999).

C. Mechanism of NNRTI Inhibition

NNRTIs bind to the same region of RT, thus the basic mechanism of inhibition might be expected to be similar for each compound. The molecular events involved in NNRTI inhibition are, however, still not completely understood and may depend on the mechanistic form(s) of RT to which the inhibitors can bind.

Analysis of the structures of potent NNRTIs indicates a number of features of the inhibitor pharmacophore important for interaction with the NNIBP (Tucker *et al.*, 1996). These include an aromatic ring capable of π-stacking interactions, an amide N-H able to participate in hydrogen bonding, and one or more hydrocarbon-rich regions to participate in hydrophobic contacts. The NNIBP does not exist as a specific binding pocket in RT. Rather, the side chains of the RT residues comprising the NNIBP adapt to each bound NNRTI in a specific manner, closing down around the surface of the drug to make tight van der Waals contacts (Kroeger-Smith *et al.*, 1995).

Several possible mechanisms for the inhibition of HIV-1 RT by NNRTI have been suggested:

1. Conformational changes in the NNIBP induced by NNRTI binding may reposition certain structural elements of RT, thereby distorting the precise geometry of the nearby polymerase catalytic site, especially the highly conserved YMDD motif. There is compelling crystallographic evidence for this possibility (Esnouf *et al.*, 1995).

2. Certain NNIBP residues participate in structural elements that comprise the "primer grip," involved in precise positioning of the primer DNA strand in the polymerase active site (Jacobo-Molina *et al.*, 1993). NNRTI binding deforms these structural elements, thereby inhibiting reaction chemistry by preventing establishment of a catalytically competent ternary complex.

3. The NNIBP may normally function as a hinge between the palm and thumb subdomains. Mobility of the thumb may be important to facilitate template/primer translocation during DNA polymerization (Kohlstaedt *et al.*, 1992; Tantillo *et al.*, 1994). Binding of NNRTIs may restrict the mobility of the thumb subdomain, slowing or preventing T/P translocation and thereby inhibiting facile elongation of nascent viral DNA.

In the absence of inhibitors, the rate-limiting step in RT-catalyzed DNA synthesis is a conformational change in the RT–T/P–dNTP ternary complex that enables a fast rate of nucleotide incorporation (Majumdar *et al.*, 1988; Kati *et al.*, 1992; Reardon, 1992, 1993). NNRTIs appear to change the rate-limiting step so that the chemical step for dNTP incorporation now becomes limiting (Spence *et al.*, 1995), suggesting that NNRTIs may promote nonproductive binding of dNTP substrates.

The various mechanisms suggested are not mutually exclusive and NNRTIs may function by exerting multiple inhibitory effects on RT-catalyzed DNA synthesis.

I. The Role of Multiple RT Mechanistic Forms in NNRTI Inhibition

A number of crystal structures of wild-type and drug-resistant RTs complexed with a variety of NNRTIs have been available for some time (Kohlstaedt *et al.*, 1992; Smerdon *et al.*, 1994; Ren *et al.*, 1995a,b, 1998; Ding *et al.*, 1995a,b; Das *et al.*, 1996; Hopkins *et al.*, 1996; Esnouf *et al.*, 1995, 1997; Hsiou *et al.*, 1998). These structures should be extraordinarily useful in the *de novo* rational design of new NNRTIs. However, to date there has not been a single published example of an RT inhibitor developed in this manner. Why? We suggest that a major reason lies in the fact that all of these crystal structures are of complexes of NNRTIs with substrate-free RT, which is probably not the best RT target for antiviral intervention.

There are three different RT mechanistic forms prior to the catalytic step: substrate-free RT, the RT–T/P binary complex, and the RT–T/P–dNTP ternary complex. These different mechanistic forms have different conformations, as determined biochemically (Wu *et al.*, 1993) and by crystallography (Jacobo-Molina *et al.*, 1993; Rodgers *et al.*, 1994; Hsiou *et al.*, 1996; Huang *et al.*, 1998). While NRTI will bind productively only to the binary complex, NNRTIs have the possibility to bind to any or all of the different RT mechanistic forms.

It has been suggested that certain pyridinones and TIBO derivatives bind more readily to the RT–T/P binary complex than to substrate-free RT (Debyser *et al.*, 1991; Goldman *et al.*, 1991). We showed that UC38 and UC84, which are structurally similar NNRTIs, inhibit via binding to different mechanistic forms of RT (Fletcher *et al.*, 1995a), with UC84 binding to free RT and the RT–T/P binary complex and the more potent UC38 inhibiting only the RT–T/P–dNTP ternary complex. We extended these observations to show that UC38+UC84 combinations of NNRTIs bind to exhibit synergy in inhibiting HIV-1 replication in several cell types (Fletcher *et al.*, 1995b) (Fig. 4). More recent studies in our laboratory show that UC84 and UC38 induce different resistance-related mutations in RT and that RT with high-level resistance to one of these NNRTIs does not necessarily show high-level resistance to that which binds to a different RT mechanistic form. In addition, the development of resistance to synergistic combinations of UC38+UC84 is delayed compared to that for either NNRTI alone, even though they bind to RT in a mutually exclusive manner (unpublished data).

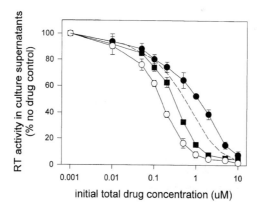

FIGURE 4 Synergistic inhibition of HIV-1 replication in cord blood mononuclear cells by combinations of the structurally similar NNRTIs UC84 and UC38. The data show the concentration dependence for inhibition of wild-type HIV-1 replication by UC84 alone (●), UC38 alone (■), and a 1:1 molar combination of UC84+UC38 (○). The dashed line is the calculated inhibition profile for *additive* inhibition by a 1:1 molar combination of UC84+UC38 using the experimentally determined inhibition constants for each drug. Reproduced with permission from Fletcher *et al.* (1995b, Fig. 5, p. 10110).

These observations provide strong evidence that multiple RT mechanistic forms exist within the HIV-1-infected cell and that NNRTI targeting of different RT mechanistic forms may be an important new strategy for antiviral intervention.

Our recent studies with tight-binding inhibitors of HIV-1 RT suggest that the most potent NNRTIs are those which bind with the highest affinity to the RT–T/P–dNTP ternary complex (Fletcher *et al.*, 1995a; Barnard *et al.*, 1997; Motakis *et al.*, 1999). These biochemical findings are consistent with the differences in antiviral potencies of UC84 (which binds to free RT and RT–T/P binary complex) compared to UC38 (which interacts primarily with the ternary complex).

A recent crystallographic study found few differences in the interaction of different carboxanilide NNRTIs with free RT (Ren *et al.*, 1998). However, we have concerns about the relevance and use of these crystallographic data for future rational drug design, as they are not consistent with biochemical and virological data. For example, the data show that UC84 makes significantly more contacts with NNIBP residues than does UC781, despite the fact that UC781 is more than 100-fold more potent against both RT DNA polymerase activity and HIV-1 replication *in vitro* (Balzarini *et al.*, 1996b; Barnard *et al.*, 1997). Harrison and colleagues have developed an elegant method for trapping the ternary complex for crystallization (Huang *et al.*, 1998). We suggest that crystallographers should now concentrate on solving structures of potent NNRTIs bound to the RT–T/P–dNTP ternary complex in order to provide the basis for future *de novo* rational drug design.

D. Mechanisms of HIV-1 Resistance to NNRTIs

HIV-1 resistance to NNRTIs correlates directly and unequivocally with mutations of one or more RT residues in the NNIBP. Mutations associated with resistance to one or more classes of NNRTIs include L100I, K101E/I/Q, K103N/Q/R/T, V106A/I, V179D/E, Y181C/I, Y188C/L/H, V189I, G190A/E/Q/T, and P236L, all of which are in the p66 RT subunit, and E138K in the p51 RT subunit. As discussed above, the NNIBP residue side chains may adapt to each NNRTI in a specific manner, as confirmed by crystallography and molecular modeling (Kohlstaedt *et al.*, 1992; Smerdon *et al.*, 1994; Ren *et al.*, 1995a,b, 1998; Ding *et al.*, 1995a,b; Kroeger-Smith *et al.*, 1995; Das *et al.*, 1996; Hopkins *et al.*, 1996; Esnouf *et al.*, 1995, 1997; Hsiou *et al.*, 1998). Due to the specificity in these contacts, mutations of some NNIBP residue side chains may confer resistance to one class of NNRTI, while not affecting potency of another (Balzarini *et al.*, 1996, 1997). Nevertheless, mutations of certain residues such as K103 tend to confer broad-spectrum NNRTI resistance. These residues likely form critical contacts with the NNRTI pharmacophore.

Mutations of specific NNIBP residues may lead to NNRTI resistance in a number of ways:

1. Removal of one or more favorable interactions between the inhibitor and NNIBP. For example, the Y181C/I and Y188C/L mutations might eliminate π-stacking interactions between the NNIBP and the aromatic ring of the NNRTI pharmacophore.
2. Introduction of steric barriers to NNRTI binding. For example, the G190A/E/Q/T mutations all introduce bulky side chains, which may prevent binding of NNRTI by sterically interfering with NNRTI functional groups such as the cyclopropyl ring of nevirapine.
3. Introduction or elimination of interresidue contacts in the NNIBP. Such changes could interfere with the ability of other NNIBP side chains to fold down over the NNRTI.

Any or all of these possibilities might diminish the overall affinity of the NNIBP for NNRTI. It has also been suggested that mutations in the NNIBP may not affect the rate of binding of NNRTI such as nevirapine, but rather may lead to increases in the rate of inhibitor dissociation (Spence *et al.*, 1996).

While high-level HIV-1 resistance to many NNRTIs such as nevirapine requires only a single mutation in the NNIBP, significant resistance to more potent NNRTIs such as the tight-binding inhibitor UC781 requires multiple mutations in the NNIBP (Buckheit *et al.*, 1997; Balzarini *et al.*, 1998a). NNRTI conformational "flexibility" may be important in this regard, with more flexible molecules able to bind more readily to mutant RT. The high potency of UC781 is due to multiple parameters including binding to multiple RT mechanistic forms, especially the RT–T/P–dNTP ternary complex, the tight-binding nature of the UC781-RT interaction, and the "dead-end" nature of the inhibition. We have recently determined that multiple mutations in the NNIBP are necessary in order to sequentially eliminate the multiple parameters of UC781 inhibition (Motakis *et al.*, 1999). RT possessing single mutations such as K103T or V106A shows virtually no resistance to UC781, and the NNRTI retains its tight-binding nature against these mutant enzymes. However, these single mutations abolish the ability of UC781 to act as a "dead-end" inhibitor, with the RT–UC781 complex showing a low but definite rate of polymerization activity. The presence of both K103T and V106A abolishes the tight-binding nature of UC781 inhibition and results in greater than 100-fold resistance to UC781. These mutations decrease UC781 binding to each of the RT mechanistic forms to a similar extent. In addition, the DNA polymerase activity of the K103T+V106A RT–UC781 complex is significantly increased compared to that of the single-mutation RT–UC781 complexes.

E. Combinations of NRTIs and NNRTIs as HIV Therapy

NNRTIs and NRTIs do not show cross-resistance, thus NRTI+NNRTI combinations should be well suited for combination chemotherapy. Even though NNRTIs have been known for nearly a decade, NRTI+NNRTI combinations are only now beginning to be used, due to the fact that many pharmaceutical concerns abandoned NNRTI development after the initial clinical observations of rapid resistance development. Thus, there is at present very little clinical information on the prolonged utility of the use of NNRTIs in combination therapy. One of the considerations is that NNRTIs may be used with NRTIs as a first-line treatment of HIV-1-infected patients in order to save the more potent protease inhibitors for use in case of treatment failure.

NRTI+NNRTI combinations are promising from a theoretical biological viewpoint. NRTIs and NNRTIs bind to two different sites on RT, thus binding is not mutually exclusive and has the possibility of synergy. Additive to synergistic antiviral activity *in vitro* was noted with nevirapine in combination with one or more NRTIs (Koup *et al.*, 1993; Merrill *et al.*, 1996). Not surprisingly, significantly greater viral load reductions were seen in patients treated with AZT+ddI+nevirapine compared to patients treated with AZT+ddI only (D'Aquila *et al.*, 1996). Similar results were noted in patients treated with AZT+3TC+delavirdine compared to those treated with AZT+3TC only. However, the AZT+delavirdine combination was much less effective than the AZT+3TC combination (Green *et al.*, 1998).

Clinical trials have shown that the AZT+3TC+efavirenz combination may provide a better virologic response (reduced viral load, improved CD4 levels) than AZT+3TC+indinavir (a protease inhibitor) (Morales-Ramirez, *et al.*, 1998).

The prolonged use of AZT monotherapy has resulted in the generation of AZT-resistant HIV-1 strains, resulting in many newly infected treatment-naïve patients being infected with these resistant strains. Recent studies have shown that certain NNRTIs may be useful in treating such patients when used with current NRTIs including AZT. UC781 not only inhibits RT-catalyzed DNA synthesis, but is also a potent inhibitor of the pyrophosphorolytic cleavage of chain-terminating nucleotides from the 3′ end of the DNA polymerization primer, a process that is important in the AZT resistance phenotype (Arion *et al.*, 1998). Combinations of UC781 and AZT (1:1 molar ratio) show high-level synergy in inhibiting the replication of AZT-resistant virus (Fig. 5), implying that UC781 can restore antiviral activity to AZT-resistant HIV-1 (Borkow *et al.*, 1999). This restoration effect is not noted with other NNRTIs such as nevirapine when combined with AZT. The development of HIV-1 resistance to UC781+AZT combinations is significantly delayed compared to development of resistance to either drug alone. Resistant HIV-1 strains that develop *in vitro* under UC781+AZT drug pressure show multiple mutations, including V118I, which is not seen in resistance to either drug alone (Borkow *et al.*, 1999).

Combinations of NNRTIs might in some cases also be useful in therapy, in a manner somewhat analogous to the use of multiple-NRTI regimens.

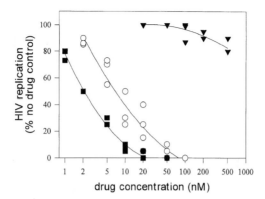

FIGURE 5 The NNRTI UC781 restores antiviral activity to AZT against AZT-resistant HIV-1. The data show the concentration dependence for inhibition of replication of AZT-resistant HIV-1 isolate 691A by AZT alone (▼), UC781 alone (○), and a 1:1 molar combination of AZT+UC781 (■). Reproduced with permission from Borkow *et al.* (1999, Fig. 1, p. 261).

As discussed above, combinations of certain NNRTIs can show synergy inhibiting HIV-1 replication (Fletcher *et al.*, 1995b). Even though the NNRTIs bind to the same site on RT, this synergy is possible because the NNRTIs interact with this site on different RT mechanistic forms. Thus, it may be appropriate to use multiple NNRTIs in a therapeutic regimen, depending on the NNRTI in question.

F. Use of NNRTI as Microbicides

Heterosexual contact is the primary mode of transmission of HIV infection worldwide. Vaginal microbicides that inhibit HIV transmission during sexual activities would help to minimize the spread of HIV from infected to noninfected individuals. Accordingly, the World Health Organization has established a research priority for the development of effective anti-HIV vaginal microbicides. Microbicidal agents should act directly on the virus without the need for prior metabolic activation and should act at replication steps prior to proviral DNA integration. In addition, the ideal retrovirucidal agent should be absorbable by uninfected cells in order to provide a barrier to infection by residual active virus and HIV-infected cells in sexual fluids, and it should be effective at nontoxic concentrations readily attainable by topical application *in vivo*.

Since NNRTIs fulfill many of these parameters, they may be potential microbicides. Biochemical and virological studies (Barnard *et al.*, 1997; Borkow *et al.*, 1997a) combined with preliminary preclinical evaluation (Balzarini *et al.*, 1998b) suggest that UC781 may be very useful in this respect. HIV-1 infectivity is eliminated in a concentration-dependent manner following transient exposure of isolated virions to low concentrations of UC781 ($<$50 nM); this may be due to inhibition of endogenous reverse transcription in these virions, a process which may be important in sexual transmission of HIV-1. In addition, when chronically infected H9 cells or peripheral blood lymphocytes isolated from HIV-1-infected individuals are transiently exposed to UC781, the virus subsequently produced in the absence of the drug are noninfectious. Importantly, transient exposure of uninfected cells to UC781 renders these cells refractory to subsequent HIV-1 infection; this "chemical barrier" to HIV-1 infection persists for extended periods of time following a single low-dose UC781 treatment. The establishment of this barrier suggests that while UC781 is taken up rapidly by cells, it enters some cellular compartment, accessible to HIV-1, in which it may remain for extended periods to exert a continuing protective effect.

Not all NNRTI are useful as microbicides. Transient exposure of HIV-1 virions to nevirapine did not affect viral infectivity. Similarly, no "chemical barrier" to HIV infection was established by transient nevirapine treatment of uninfected cells; these cells were as readily infected by HIV-1 as untreated cells (Borkow *et al.*, 1997a). The tight-binding nature of UC781 inhibition likely figures in the microbicidal activity of the drug.

G. Other Inhibitors of HIV Reverse Transcriptase

I. Inhibitors of RNaseH

While many inhibitors of HIV-1 RT DNA polymerase activity have been developed, very few inhibitors of HIV-1 RT ribonuclease H have been described. These include the natural product illimaquinone (Loya *et al.*, 1990), the metal chelator *o*-phenanthroline (Hizi *et al.*, 1991), and AZT-MP (Tan *et al.*, 1991). However, none of these inhibitors have good potency against HIV-1 RT RNaseH (IC$_{50}$ values $\geqslant$50 μM).

We recently discovered that certain *N*-acyl hydrazones are reasonably potent inhibitors of HIV-1 RT RNaseH. One of these, *N*-(4-*tert*-butylbenzoyl)-2-hydroxy-1-naphthaldehyde hydrazone (BBNH; Fig. 6), inhibits HIV-1 replication *in vitro* with good potency (EC$_{50}$ $\approx$ 1.5 μM) (Borkow *et al.*, 1997b). BBNH is actually a multitarget inhibitor of HIV-1 RT, inhibiting both the RNA-dependent DNA polymerase and the RNaseH activities of the enzyme. A series of biochemical studies established unequivocally that inhibition of RT RDDP activity is due to BBNH binding to the NNIBP, whereas inhibition of RNaseH activity is due to binding of the inhibitor to the RT RNaseH domain (Borkow *et al.*, 1997b).

One of the reasons for the paucity of small molecule inhibitors of HIV-1 RT RNaseH is that the region around the RNaseH active site residues is rather open and unable to provide sufficient contacts with small molecules for resonable binding affinity. X-ray diffraction studies show that BBNH is

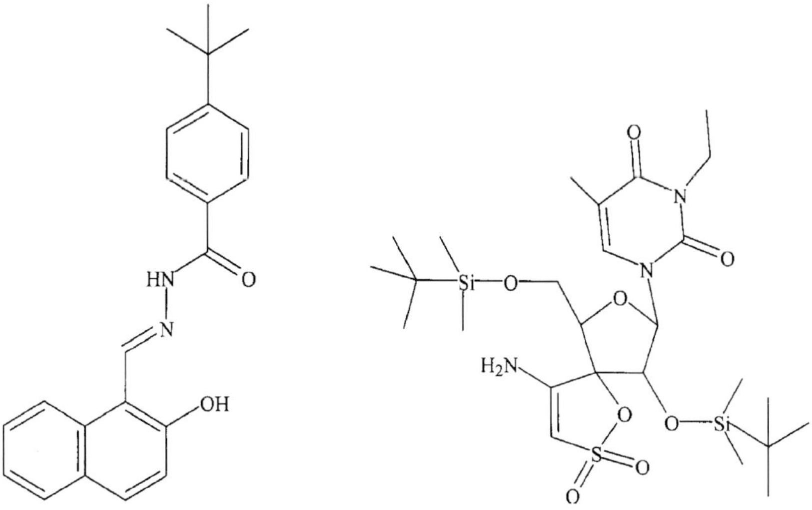

1   *2*

FIGURE 6 Structures of HIV-1 RT inhibitors with novel mechanisms of action. Illustrated are BBNH (*1*) and TSAOe³T (*2*).

a flat molecule (Lanthier *et al.*, 1997), and molecular modeling studies indicate that this planar structure, along with metal interaction, may be important in binding to the RNaseH domain (Parniak *et al.*, 1998a). We have used this information to synthesize a number of BBNH analogs, some of which show significantly improved potency against HIV-1 RT RNaseH and against HIV-1 replication; characterization of these compounds is in progress.

2. Inhibitors of HIV-1 RT Dimer Stability and Assembly

The gene for HIV-1 RT encodes a 66-kDa polypeptide; however, the presumed biologically relevant from of HIV-1 RT is a heterodimer consisting of two subunits of 66 and 51 kDa. The latter subunit is derived from the former by proteolytic processing during HIV-1 assembly and maturation. While the precise kinetics of HIV-1 RT processing and dimerization are still not well characterized, it is clear that the expression of RT RDDP activity requires a dimeric enzyme. Inhibition of RT dimerization during HIV-1 assembly and maturation, or disruption of RT dimer stability, might be novel targets for new RT inhibitor development. Several peptides of 15 to 19 amino acids in length, derived from sequences from the RT connection subdomain, inhibited the reassociation of organic-solvent-denatured RT subunits (Divita *et al.*, 1994). There are, however, problems with the use of peptides as therapeutics, including limited bioavailability, limited cell uptake, and the possibility of immune recognition, all of which would limit accessibility of the peptide to its target.

We have recently identified two small molecules, with very different structures, that disrupt HIV-1 RT p66/p51 heterodimer stability. These "molecular crowbars"[1] may be useful leads in the development of RT inhibitors that function at multiple stages of HIV-1 replication. The first molecule is 1-[2′,5′-bis-O-(t-butyldimethylsilyl)-β-D-ribofuranosyl]-3′-spiro-5″-(4″-amino-1″,2″-oxathiole-2″,2″-dioxide)-3-ethylthymine (TSAOe^{3}T; Fig. 6). The TSAO series of RT inhibitors has been known for some time (Balzarini et al., 1992a, b). Although TSAO derivatives are highly modified nucleosides, they are generally considered to be NNRTIs because they do not require metabolic activation and they inhibit RT DNA polymerase activity in a noncompetitive manner (Balzarini et al., 1992b). However, we have previously shown that the inhibition of RT-catalyzed DNA polymerization by TSAO derivatives differs significantly from that of NNRTIs such as nevirapine analogs and have suggested that TSAO may interact differently with RT than other NNRTIs (Arion et al., 1996b). We have now found that interaction of TSAOe^{3}T with RT results in a 4 kcal mol^{-1} decrease in the strength of the p66-p51 subunit association, a significant destabilization. NNRTIs such as nevirapine and UC781 do not alter the subunit association strength (Sluis-Cremer et al., 2000).

The N-acyl hydrazone BBNH (Fig. 6) is a multitarget RT inhibitor, as described above. N-Acyl hydrazones from very stable complexes with Fe(III). We have found that the Fe(III)BBNH complex retains inhibitory activity against RT DNA polymerase and RNaseH activities and also destabilizes HIV-1 RT p66-p51 subunit interactions, similar to TSAO (Parniak et al., 1998b). Appropriately formulated Fe(III)BBNH inhibits acute HIV-1 infection, as expected. In addition, transient treatment of HIV-1 chronically infected H9 cells with Fe(III)BBNH results in the subsequent production of HIV virions with attenuated infectivity. Fe(III)BBNH thus inhibits HIV replication at both preintegrational and postintegrational stages. We suggest that the postintegrational antiviral activity of Fe(III)BBNH may be due to interference with RT dimerization during HIV assembly and maturation.

V. Conclusions and Future Directions

Reverse transcriptase has been and will continue to be a major target for anti-HIV drug development. NRTIs will continue to be crucial components of anti-HIV combination chemotherapy. Continued development of NRTI prodrugs may circumvent the variability of cellular phosphorylation

---

[1] We thank Dr. E. Arnold, Rutgers University, for suggesting the term "molecular crowbar."

to the active metabolite, providing more consistent antiviral activity in multiple cell types and tissue compartments.

Current NNRTIs in preclinical evaluation are exceptionally potent. Significant *in vitro* inhibition is seen with 1:1 molar ratios of NNRTIs to wild-type RT. It is therefore difficult to imagine future substantive improvements of NNRTI potency against wild-type enzyme. Drug resistance will continue to plague HIV therapy. Unfortunately, this problem is likely to limit current NNRTI long-term efficacy, since just a few mutations in the NNIBP of RT are sufficient to result in high-level cross-resistance to almost all NNRTIs extant. However, appropriate combinations of NRTI+NNRTI may act in synergy to delay the development of resistance, thereby prolonging the clinical utility of NNRTIs. Careful monitoring of patients on such therapies over the next couple of years will allow assessment of this possibility. Rigorous investigations of the mechanisms of NNRTI inhibition and resistance will assist in defining possible new therapeutic combinations. In addition, the recent solution of the RT–T/P–dNTP ternary complex may finally allow *de novo* rational design of new NNRTIs with efficacy against drug-resistant mutants and which are also refractory to resistance development. This design should ensure that the binding of new NNRTIs involves significant contacts with immutable residues of RT in or near the NNIBP, such as W229 and Y318.

HIV-1 RT RNaseH is an underexplored target for antiviral development. The identification of *N*-acyl hydrazones as a class of RNaseH inhibitors with reasonable potency should provide the basis for the development of more potent analogs. Similarly, the concept of multitarget inhibitors of RT, i.e., compounds that inhibit both DNA polymerase and RNaseH activities (such as BBNH) due to multisite binding to RT, needs further exploration. Resistance to such multitarget inhibitors might be difficult to develop, due to the need to introduce mutations at multiple sites in RT.

The development of new RT inhibitors for effective anti-HIV-1 therapy will remain a considerable challenge. RT has proven to be an amazingly plastic enzyme, able to accommodate numerous changes in primary structure while retaining catalytic activity. However, the numerous advances that are being made in our understanding of RT structure and function will certainly figure prominently in future RT inhibitor design and may finally yield a drug with long-term benefit in infected patients.

Acknowledgments

Research in the Parniak laboratory has been supported by grants from the Medical Research Council of Canada, the Natural Sciences and Engineering Research Council of Canada, and the International Research Scholars Program of the Howard Hughes Medical Institute. M.A.P. is an MRC Senior Scientist and HHMI International Research Scholar. N.S.-C. is recipient of an MRC Postdoctoral Fellowship.

References

Arion, D., and Parniak, M. A., (1999). HIV resistance to zidovudine: The role of pyrophosphorolysis. *Drug Res. Updates* **2**, 91–95.

Arion, D., Sluis-Cremer, N., and Parniak, M. A. (2000). Mechanism by which phosphonoformic acid resistance mutations restore 3′-azido-3′-deoxythymidine (AZT) sensitivity to AZT-resistant HIV-1 reverse transcriptase. *J. Biol. Chem.* **275**, 9251–9255.

Arion, D., Borkow, G., Kaushik, N., McCormick, S., and Parniak, M. A. (1998). Phenotypic mechanism of HIV-1 resistance to 3′-azido,3′-deoxythymidine (AZT): Increased DNA polymerization processivity and enhanced sensitivity to pyrophosphate of the mutant viral reverse transcriptase. *Biochemistry* **37**, 15908–15917.

Arion, D., Borkow, G., Gu, Z., Wainberg, M. A., and Parniak, M. A. (1996a). The K65R mutation confers increased DNA polymerase processivity to HIV-1 reverse transcriptase. *J. Biol. Chem.* **271**, 19860–19864.

Arion, D., Fletcher, R. S., Borkow, G., Camarasa, M.-J., Balzarini, J., Dmitrienko, G. I., and Parniak, M. A. (1996b). Difference in the inhibition of HIV-1 reverse transcriptase DNA polymerase activity by analogs of nevirapine and [2′,5′-bis-O-(t-butyldimethylsilyl)-3′-spiro-5″-(4″-amino-1″,2″-oxathiole-2″,2″-dioxide]-thymine (TSAO). *Mol. Pharmacol.* **50**, 1057–1064.

Bader, J. P., McMahon, J. B., Schultz, R. J., Narayanan, V. L., Pierce, J. B., Weislow, O. S., Midelfort, C. F., Stinson, S. F. and Boyd, M. R. (1991). Oxathiin carboxanilide, a potent inhibitor of human immunodeficiency virus reproduction. *Proc. Natl. Acad. Sci. USA* **88**, 6740–6744.

Balzarini, J., Brouwer, W. G., Felauer, E. E., De Clercq, E., and Karlsson, A. (1995). Activity of various thiocarboxanilide derivatives against wild-type and several mutant human immunodeficiency virus type 1 strains. *Antivir. Res.* **27**, 219–236.

Balzarini, J., Naesens, L., Herdewijn, P., Rosenberg, I., Holy, A., Pauwels, R., Baba, M., Johns, D. G., and De Clercq, E. (1989). Marked *in vivo* antiretrovirus activity of 9-(2-phosphonylmethoxyethyl)adenine, a selective anti-human immunodeficiency virus agent. *Proc. Natl. Acad. Sci. USA* **86**, 332–336.

Balzarini, J., Naesens, L., Verbeken, E., Laga, M., Van Damme, L., Parniak, M. A., Van Melaert, L., Anne, J., and De Clercq, E. (1998b). Preclinical studies on thiocarboxanilide UC781 as a virucidal agents. *AIDS* **12**, 1129–1138.

Balzarini, J., Pelemans, H., Aquaro, S., Perno, C.-F., Witvrouw, M., Schols, D., De Clercq, E., and Karlsson, A. (1996b). Highly favorable antiviral activity and resistance profile of the novel thiocarboxanilide pentenyloxy ether derivatives UC781 and UC82 as inhibitors of human immunodeficiency virus type 1 replication. *Mol. Pharmacol.* **50**, 394–401.

Balzarini, J., Pelemans, H., Esnouf, R., and De Clercq, E. (1998a). A novel mutation (F227L) arises in the reverse transcriptase (RT) of human immunodeficiency virus type 1 (HIV-1) upon dose-escalating treatment of HIV-1 infected cell cultures with the non-nucleoside RT inhibitor thiocarboxanilide UC781. *AIDS Res. Hum. Retroviruses.* **14**, 255–260.

Balzarini, J., Perez-Perez, M.-J., San-Felix, A., Schols, D., Perno, C.-F., Vandamme, A.-M., Camarasa, M.-J., and De Clercq, E. (1992a). 2′, 5′-bis-O-(t-butyldimethylsilyl)-3′-spiro-5″ 2″-oxathiole-2″, 2″-dioxide)-pyrimidine (TSAO) nucleoside analogs: Highly selective inhibitors of human immunodeficiency virus type 1 that are targeted at the viral reverse transcriptase. *Proc. Natl. Acad. Sci. USA* **89**, 4392–4396.

Balzarini, J., Perez-Perez, M.-J., San-Felix, A., Camarasa, M.-J., Bathurst, I. C., Barr, P. J., and De Clercq, E. (1992b). Kinetics of inhibition of human immunodeficiency virus type 1 reverse transcriptase by the novel HIV-1 specific nucleoside analogue [2′, 5′-bis-O-(t-butyldimethylsilyl)-3′-spiro-5″ 2″-oxathiole-2″, 2″-dioxide)-thymine (TSAO-T). *J. Biol. Chem.* **267**, 11831–11838.

Barnard, J., Borkow, G., and Parniak, M. A. (1997). The thiocarboxanilide UC781 is a tight-binding nonnucleoside inhibitor of HIV-1 reverse transcriptase. *Biochemistry* **36**, 7786–7792.

Barshop, B. A., Adamson, D. T., Vellom, D. C., Rosen, F., Epstein, B. L., and Seegmiller, J. E. (1991). Luminescent immobolized enzyme test systems for inorganic pyrophosphate: Assays using firefly luciferase and nicotinamide-mononucleotide adenylyl transferase or adenosine-5'-triphosphate sulfurylase. *Anal. Biochem.* **197**, 266–272.

Borkow, G., Arion, D., Wainberg, M. A., and Parniak, M. A. (1999). The thiocarboxanilide nonnucleoside inhibitor UC781 restores antiviral activity of 3'-azido-3'-deoxythymidine (AZT) against AZT-resistant human immunodeficiency virus type 1. *Antimicrob. Agents Chemother.* **43**, 259–263.

Borkow, G., Barnard, J., Nguyen, T. M., Belmonte, A., Wainberg, M. A., and Parniak, M. A. (1997a). Chemical barriers to HIV-1 infection: Retrovirucidal activity of UC781, a thiocarboxanilide nonnucleoside inhibitor of HIV-1 reverse transcriptase. *J. Virol.* **71**, 3023–3030.

Borkow, G., Fletcher, R. S., Barnard, J., Arion, D., Motakis, D., Dmitrienko, G. I., and Parniak, M. A. (1997b). Inhibition of the ribonuclease H and DNA polymerase activities of HIV-1 reverse transcriptase by N-(4-t-butylbenzoyl)-2-hydroxy-1-naphthaldehyde hydrazone. *Biochemistry* **36**, 3179–3185.

Buckheit, Jr., R. W., Hollingshead, M., Stinson, S., Fliakas-Boltz, V., Pallansch, L. A., Roberson, J., Decker, W., Elder, C., Borgel, S., Bonomi, C., Shores, R., Siford, T., Malspeis, L., and Bader, J. P. (1997a). Efficacy, pharmocokinetics, and *in vivo* antiviral activity of UC781, a highly potent, orally bioavailable nonnucleoside reverse transcriptase inhibitor of HIV type 1. *AIDS Res. Hum. Retroviruses.* **13**, 789–796.

Buckheit, R. W., Jr., Snow, M. J., Fliakas-Boltz, V., Kinjerski, T. L., Russell, J. D., Pallansch, L. A., Brouwer, W. G., and Yang, S. S. (1997b). Highly potent oxathiin carboxanilide derivatives with efficacy against nonnucleoside reverse transcriptase inhibitor-resistant human immunodeficiency virus isolates. *Antimicrob. Agents Chemother.* **41**, 831–837.

Canard, B., Sarfati, S. R., and Richardson, C. C. (1998). Enhanced binding of azidothymidine-resistant human immunodeficiency virus 1 reverse transcriptase to the 3'-azido-3'-deoxythymidine 5'-monophosphate-terminated primer. *J. Biol. Chem.* **273**, 14596–14604.

Coates, J. A. V., Cammack, N., Jenkinson, H. J., Mutton, I. M., Pearson, B. A., Storer, R., Cameron, J. M., and Penn, C. R. (1992). The separated enantiomers of 2'-deoxy-3'-thiacytidine (BCH189) both inhibit human immunodeficiency virus replication *in vitro*. *Antimicrob. Agents Chemother.* **36**, 202–205.

Coffin, J. M. (1996). Retroviridae and their replication. In "Virology" (B. N. Fields *et al.* eds.), pp. 1767–1848. Raven, New York.

Collier, A. C., Coombs, R. W., Schoenfeld, D. A., Bassett, R., Baruch, A., and Corey, L. (1996). Combination therapy with zidovudine, didanosine and saquinavir. *Antiviral Res.* **29**, 99.

Dahlberg, J. E., Mitsuya, H., Blam, S. B., Broder, S., and Aaronson, S. A. (1987). Broad spectrum antiretroviral activity of 2',3'-dideoxynucleosides. *Proc. Natl. Acad. Sci. USA* **84**, 2469–2473.

Daluge, S. M., Good, S. S., Faletto, M. B., Miller, W. H., St. Clair, M. H., Boone, L. R., Tisdale, M., Parry, N. R., Reardon, J. E., Dornsife, R. E., Averett, D. R., and Krenitsky, T. A. (1997). 1592U89, a novel carbocyclic nucleoside analog with potent, selective anti-human immunodeficiency virus activity. *Antimicrob. Agents Chemother.* **41**, 1082–1093.

Daluge, S. M., Good, S. S., and Miller, W. H. (1998). Abacavir (1592), a second-generation nucleoside HIV reverse transcriptase inhibitor. *Int. Antiviral News* **6**, 122–124.

D'Aquila, R. T., Hughes, M. D., Johnson, V. A., Fischl, M. A., Sommadossi, J. P., Liou, S. H., Timpone, J., Myers, M., Basgoz, N., Niu, M., and Hirsch, M. S. (1996). Nevirapine, zidovudine, and didanosine compared with zidovudine and didanosine in patients with HIV-1 infection: A randomized, double-blind placebo-controlled trial. *Ann. Intern. Med.* **124**, 1019–1030.

Das, K., Ding, J., Hsiou, Y., Clark, A. D., Jr., Moereels, H., Koymans, L., Andries, K., Pauwels, R., Janssen, P. A. J., Boyer, P. L., Clark, P., Smith, R. H., Jr., Kroeger-Smith, M. B., Michejda, C. J., Hughes, S. H., and Arnold, E. (1996). Crystal structures of 8-Cl and 9-Cl TIBO complexed with wild-type RT and 8-Cl TIBO complexed with the Tyr181 Cys HIV-1 RT drug-resistant mutant. *J. Mol. Biol.* **264**, 1085–1100.

De Antoni, A., Foli, A., Lisziewicz, J., and Lori, F. (1997). Mutations in the pol gene of human immunodeficiency virus type 1 in infected patients receiving didanosine and hydroxyurea combination therapy. *J. Infect. Dis.* **176**, 899–903.

Debyser, Z., Pauwels, R., Andries, K., Desmyter, J., Kukla, M., Janssen, P. A. J., and De Clercq, E. (1991). An antiviral target on reverse transcriptase of human immunodeficiency virus type 1 revealed by tetrahydroimidazo-[4,5,1-*j,k*][1,4]benzodiazepin-2(1*H*)-one and -thione derivatives. *Proc. Natl. Acad. Sci. USA* **88**, 1451–1455.

De Clercq, E., Sakuma, T., Baba, M., Pauwels, R., Balzarini, J., Rosenberg, I., and Holy, A. (1987). Antiviral activity of phosphonylmethoxyalkyl derivatives of purine and pyrimidines. *Antiviral Res.* **8**, 261–272.

Ding, J., Das, K., Hsiou, Y., Sarafianos, S. G., Clark, A. D., Jr., Jacobo-Molina, A., Tantillo, C., Hughes, S. H., and Arnold, E. (1998). Structure and functional implications of the polymerase active site region in a complex of HIV-1 RT with a double-stranded DNA template-primer and an antibody Fab fragment at 2.8 Å resolution. *J. Mol. Biol.* **284**, 1095–1111.

Ding, J., Das, K., Tantillo, C., Zhang, W., Clark, A. D., Jr., Jessen, S., Lu, X., Hsiou, Y., Jacobo-Molina, A., Andries, K., Pauwels, R., Moereels, H., Koymans, L., Janssen, P. A. J., Smith, R. H., Jr., Koepke-Kroeger, M., Michejda, C. J., Hughes, S. H., and Arnold, E. (1995a). Structure of the HIV-1 reverse transcriptase in a complex with the non-nucleoside inhibitor α-APA R95845 at 2.8 Å resolution. *Structure* **3**, 365–379.

Ding, J., Das, K., Moereels, H., Koymans, L., Andries, K., Janssen, P. A. J., Hughes, S. H., and Arnold, E. (1995b). Structure of HIV-1 RT/TIBO R86183 complex reveals similarity in the binding of diverse nonnucleoside inhibitors. *Nature Struct. Biol.* **2**, 407–415.

Divita, G., Restle, T., Goody, R. S., Chermann, J.-C., and Baillon, J. G. (1994). Inhibition of human immunodeficiency virus type 1 reverse transcriptase dimerization using synthetic peptides derived from the connection domain. *J. Biol. Chem.* **269**, 13080–13083.

Domingo, E., Menendez-Arias, L., Quinones-Mateu, M. E., Holguin, A., Gutierrez-Rivas, M., Martinez, M. A., Quer, J., Novella, I. S., and Holland, J. J. (1997). Viral quasispecies and the problem of vaccine-escape and drug-resistant mutants. *Prog. Drug Res.* **48**, 99–128.

Dueweke, T. J., Poppe, S. M., Romero, D. L., Swaney, S. M., So, A. G., Downey, K. M., Althaus, I. W., Reusser, F., Busso, M., Resnick, L., Mayers, D. L., Lane, J., Aristoff, P. A., Thomas, R. C., and Tarpley, W. G. (1993a). U-90152, a potent inhibitor of human immunodeficiency virus type 1 replication. *Antimicrob. Agents Chemother.* **37**, 1127–1131.

Dueweke, T. J., Pushkarskaya, T., Poppe, S. M., Swaney, S. M., Zhao, J. Q., Chen, I. S. Y., Stevenson, M., and Tarpley, W. G. (1993). A mutation in reverse transcriptase of bis(heteroaryl) piperazine-resistant human immunodeficiency virus type 1 that confers increased sensitivity to other nonnucleoside inhibitors. *Proc. Natl. Acad. Sci. USA* **90**, 4713–4717.

Esnouf, R. M., Ren, J., Hopkins, A. L., Ross, C. K., Jones, E. Y., Stammers, D. K., and Stuart, D. I. (1997). Unique features in the structure of the complex between HIV-1 reverse transcriptase and the bis(heteroaryl)piperazine (BHAP) U-90152 explain resistance mutations for this nonnucleoside inhibitor. *Proc. Natl. Acad. Sci. USA* **94**, 3984–3989.

Esnouf, R., Ren, J., Ross, C., Jones, Y., Stammers, D., and Stuart, D. (1995). Mechanism of inhibition of HIV-1 reverse transcriptase by non-nucleoside inhibitors. *Nature Struct. Biol.* **2**, 303–308.

Faletto, M. B., Miller, W. H., Garvey, E. P., St. Clair, M. H., Daluge, S. M., and Good, S. S. (1997). Unique intracellular activation of the potent anti-human immunodeficiency virus agent 1592U89. *Antimicrob. Agents Chemother.* **41**, 1099–1107.

Fischl, M., Greenberg, S., Clumeck, N., Peters, B., Rubio, R., Gould, J., Boone, G., West, M., Spreen, B., and Lafon, S. (1999). "Ziagen (Abacavir, ABC, 1592) Combined with 3TC & ZDV Is Highly Effective and Durable through 48 Weeks in HIV-1 Infected Antiretroviral-Therapy-Naïve Subjects (CNAA3003)," p. 70. 6th Conference on Retroviruses and Opportunistic Infections. Chicago, IL [Abstract 19]

Fitzgibbon, J. E., Howell, R. M., Haberzettl, C. A., Sperber, S. J., Gocke, D. J., and Dubin, D. T. (1992). Human immunodeficiency virus type 1 pol gene mutations which cause decreased susceptibility to 2',3'-dideoxycytidine. *Antimicrob. Agents Chemother.* **36**, 153–157.

Fletcher, R. S., Holleschak, G., Nagy, E., Arion, D., Borkow, G., Gu, Z., Wainberg, M. A., and Parniak, M. A. (1996). Single step purification of HIV-1 recombinant wild type and mutant reverse transcriptase. *Prot. Expression Purif.* **7**, 27–32.

Fletcher, R. S., Syed, K., Mithani, S., Dmitrienko, G. I., and Parniak, M. A. (1995a). Carboxanilide derivative non-nucleoside inhibitors of HIV-1 reverse transcriptase interact with different mechanistic forms of the enzyme. *Biochemistry* **34**, 4036–4042.

Fletcher, R. S., Arion, D., Borkow, G., Wainberg, M. A., Dmitrienko, G. I., and Parniak, M. A. (1995b). Synergistic inhibition of HIV-1 reverse transcriptase DNA polymerase activity and virus replication *in vitro* by combinations of carboxanilide nonnucleoside compounds. *Biochemistry* **34**, 10106–10112.

Foli, A., Sogocio, K. M., Anderson, B., Kavlick, M., Saville, M. W., Wainberg, M. A., Gu, Z., Cherrington, J. M., Mitsuya, H., and Yarchoan, R. (1996). *In vitro* selection and molecular characterization of human immunodeficiency virus type 1 with reduced sensitivity to 9-[2-(phosphonomethoxy)ethyl]adenine (PMEA). *Antiviral Res.* **32**, 91–98.

Furman, P. A., Fyfe, J. A., St. Clair, M. H., Weinhold, K., Rideout, J. L., Freeman, G. A., Nusinoff-Lehrman, S., Bolognesi, D. P., Broder, S., Mitsuya, H., and Barry, D. W. (1986). Phosphorylation of 3'-azido-3'-deoxythymidine and selective interaction of the 5'-triphosphate with human immunodeficiency virus reverse transcriptase. *Proc. Natl. Acad. Sci. USA* **83**, 8333–8337.

Gao, Q., Gu, Z., Parniak, M. A., Cameron, J., Cammack, N., Boucher, C., and Wainberg, M. A. (1993). The same mutation that encodes low-level human immunodeficiency virus type 1 resistance to 2',3'-dideoxyinosine and 2',3'-dideoxycytidine confers high-level resistance to the (−)enantiomer of 2',3'-dideoxy-3'-thiacytidine. *Antimicrob. Agents Chemother.* **37**, 1390–1392.

Goldman, M. E., Nunberg, J. H., O'Brien, J. A., Quintero, J. C., Schlief, W. A., Freund, K. F., Gaul, S. L., Saari, W. S., Wai, J. S., Hoffman, J. M., Anderson, P. S., Hupe, D. J., Emini, E. A., and Stern, A. M. (1991). Pyridinone derivatives: specific human immunodeficiency virus type 1 reverse transcriptase inhibitors with antiviral activity. *Proc. Natl. Acad. Sci. USA* **88**, 6863–6867.

Götte, M., Arion, D., Parniak, M. A., and Wainberg, M. A. (2000). The M184V mutation in the reverse transcriptase of human immunodeficiency virus type 1 impairs rescue of chain-terminated DNA synthesis. *J. Virol.* **74**, 3579–3585.

Green, S., Para, M. F., and Daly P. W. (1998). "Interim Analysis of Plasma Viral Burden Reductions and CD4 Increases in HIV-1 Infected Patients with Rescriptor (DLV) + Retrovir (ZDV) + Epivir (3TC)." 12th World AIDS Conference, Geneva. [Abstract 12219]

Gu, Z., Fletcher, R. S., Wainberg, M. A., and Parniak, M. A. (1994). The K65R mutant reverse transcriptase associated with HIV-1 resistance to 2',3'-dideoxycytidine, 2',3'-dideoxy-3'-thiacytidine and 2',3'-dideoxyinosine shows reduced sensitivity to specific dideoxynucleoside triphosphate inhibitors *in vitro*. *J. Biol. Chem.* **269**, 28118–28122.

Hizi, A., Shaharabany, M., and Loya, S. (1991). Catalytic properties of the reverse transcriptases of human immunodeficiency viruses types 1 and 2. *J. Biol. Chem.* **266**, 6230–6239.

Hopkins, A. L., Ren, J., Esnouf, R. M., Willcox, B. E., Jones, E. Y., Ross, C., Miyasaka, T., Walker, R. T., Tanaka, H., Stammers, D. K., and Stuart, D. I. (1996). Complexes of

HIV-1 reverse transcriptase with inhibitors of the HEPT series reveal conformational changes relevant to the design of potent non-nucleoside inhibitors. *J. Med. Chem.* **39**, 1589–1600.

Hsiou, Y., Das, K., Ding, J., Clark, A. D., Jr., Kleim, J. P., Rosner, M., Winkler, I., Riess, G., Hughes, S. H., and Arnold, E. (1998). Structures of Tyr188Leu mutant and wild-type HIV-1 reverse transcriptase complexed with the non-nucleoside inhibitor HBY 097: Inhibitor flexibility is a useful design feature for reducing drug resistance. *J. Mol. Biol.* **284**, 313–323.

Hsiou, Y., Ding, J., Das, K., Clark, A. D., Jr., Hughes, S. H., and Arnold, E. (1996). Structure of unliganded HIV-1 reverse transcriptase at 2.7 A resolution: Implications of conformational changes for polymerization and inhibition mechanisms. *Structure* **4**, 853–860.

Huang, H., Chopra, R., Verdine, G. L., and Harrison, S. C. (1998). Structure of a covalently trapped catalytic complex of HIV-1 reverse transcriptase: Implications for drug resistance. *Science* **282**, 1669–1675.

Iversen, A. K., Shafer, R. W., Wehrly, K., Winters, M. A., Mullins, J. I., Chesebro, B., and Merigan, T. C. (1996). Multidrug-resistant human immunodeficiency virus type 1 strains resulting from combination antiretroviral therapy. *J. Virol.* **70**, 1086–1090.

Jacobo-Molina, A., Ding, J., Nanni, R. G., Clark, A. D., Jr., Lu, X., Tantillo, C., Williams, R. L., Kamer, G., Ferris, A. L., Clark, P., Hizi, A., Hughes, S. H., and Arnold, E. (1993). Crystal structure of human immunodeficiency virus type 1 reverse transcriptase complexed with double-stranded DNA at 3.0 Å resolution shows bent DNA. *Proc. Natl. Acad. Sci. USA* **90**, 6320–6324.

Johnson, V. A. (1996). Combination therapy for HIV-1 infection-overview: Preclinical and clinical analysis of antiretroviral combinations. *Antiviral Res.* **29**, 35–39.

Kati, W. M., Johnson, K. A., Jerva, L. F., and Anderson, K. S. (1992). Mechanism and fidelity of HIV reverse transcriptase. *J. Biol. Chem.* **267**, 25988–25997.

Kellam, P., Boucher, C. A., and Larder, B. A. (1992). Fifth mutation in human immunodeficiency virus type 1 reverse transcriptase contributes to the development of high-level resistance to zidovudine. *Proc. Natl. Acad. Sci. USA* **89**, 1934–1938.

Kleim, J.-P., Bender, R., Kirsch, R., Meichsner, C., Paessens, A., Rosner, M., and Riess, G. (1994). Mutational analysis of residue 190 of human immunodeficiency virus type 1 reverse transcriptase. *Virology* **200**, 696–701.

Kleim, J.-P., Bender, R., Kirsch, R., Meichsner, C., Paessens, A., Rosner, M., Rubsamen-Waigmann, H., Kaiser, R., Wichers, M., Schneweis, K. E., Winkler, I., and Riess, G. (1995). Preclinical evaluation of HBY 097, a new nonnucleoside reverse transcriptase inhibitor of human immunodeficiency virus type 1 replication. *Antimicrob. Agents Chemother.* **39**, 2253–2257.

Kohlstaedt, L. A., Wang, J., Friedman, J. M., Rice, P. A., and Steitz, T. A. (1992). Crystal structure at 3.5 Å resolution of HIV-1 reverse transcriptase complexed with an inhibitor. *Science* **256**, 1783–1790.

Koup, R. A., Brewster, F., Grob, P., and Sullivan, J. L. (1993). Nevirapine synergistically inhibits HIV-1 replication in combination with zidovudine, interferon or CD4 immunoadhesin. *AIDS* **7**, 1181–1184.

Krebs, R., Immendorfer, U., Thrall, S. H., Wohrl, B. M., and Goody, R. S. (1997). Single-step kinetics of HIV-1 reverse transcriptase mutants responsible for virus resistance to nucleoside inhibitors zidovudine and 3-TC. *Biochemistry* **36**, 10292–10300.

Kroeger-Smith, M. B., Rouzer, C. A., Taneyhill, L. A., Smith, N. A., Hughes, S. H., Boyer, P. L., Janssen, P. A., Moereels, H., Koymans, L., and Arnold, E. (1995). Molecular modeling studies of HIV-1 reverse transcriptase nonnucleoside inhibitors: Total energy of complexation as a predictor of drug placement and activity. *Prot. Sci.* **4**, 2203–2222.

Lacey, S. F., and Larder, B. A. (1994). Novel mutation (V75T) in human immunodeficiency virus type 1 reverse transcriptase confers resistance to 2′,3′-didehydro-2′,3′-dideoxythymidine in cell culture. *Antimicrob. Agents Chemother.* **38**, 1428–1432.

Lacey, S. F., Reardon, J. E., Furfine, E. S., Kunkel, T. A., Bebenek, K., Eckert, K. A., Kemp, S. D., and Larder, B. A. (1992). Biochemical studies on the reverse transcriptase and RNase H activities from human immunodeficiency virus strains resistant to 3′-azido-3′-deoxythymidine. *J. Biol. Chem.* **267**, 15789–15794.

Lanthier, C. M., Parniak, M. A., and Dmitrienko, G. I. (1997). Inhibition of carboxypeptidase A by *N*-(4-*t*-butylbenzoyl)-2-hydroxy-1-naphthaldehyde hydrazone. *Bioorg. Med. Chem. Lett.* **7**, 1557–1562.

Larder, B. A., and Kemp, S. D. (1989). Multiple mutations in HIV-1 reverse transcriptase confer high-level resistance to Zidovudine (AZT). *Science* **246**, 1155–1158.

Larder, B. A., Kellam, P., and Kemp, S. D. (1991). Zidovudine resistance predicted by direct detection of mutations in DNA from HIV-1 infected lymphocytes. *AIDS* **5**, 137–144.

Larder, B. A., Kemp, S. D., and Harrigan, P. R. (1995). Potential mechanism for sustained antiretroviral efficacy of AZT-3TC combination therapy. *Science* **269**, 696–699.

Lavie, A., Vetter, I. R., Konrad, M., Goody, R. S., Reinstein, J., and Schlichting, I. (1997). Structure of thymidylate kinase reveals the cause behind the limiting step in AZT activation. *Nature Struct. Biol.* **4**, 601–604.

Le Grice, S. F. J. (1993). Human immunodeficiency virus reverse transcriptase. In "Reverse Transcriptase" (A. M. Skalka, and S. P. Goff, eds.), pp. 163–191. Cold Spring Harbor Laboratory Press, Cold Spring Harbor, NY.

Li, Y., Kong, Y., Korolev, S., and Waksman, G. (1998). Crystal structures of the Klenow fragment of *Thermus aquaticus* DNA polymerase I complexed with deoxyribonucleoside triphosphates. *Prot. Sci.* **7**, 1116–1123.

Lin, T. S., Schinazi, R. F., and Prusoff, W. H. (1987). Potent and selective *in vitro* activity of 3′-deoxythymidin-2′-ene (3′-deoxy-2′, 3′-didehydrothymidine) against human immunodeficiency virus. *Biochem. Pharmaocol.* **36**, 2713–2718.

Loya, S., Tal, R., Kashman, Y., and Hizi, A. (1990). Illimaquinone, a selective inhibitor of the RNase H activity of human immunodeficiency virus type 1 reverse transcriptase. *Antimicrob. Agents Chemother.* **34**, 2009–2012.

Majumdar, C., Abbotts, J., Broder, S., and Wilson, S. H. (1988). Studies on the mechanism of human immunodeficiency virus reverse transcriptase: Steady-state kinetics, processivity, and polynucleotide inhibition. *J. Biol. Chem.* **263**, 15657–15665.

Merluzzi, V. J., Hargrave, K. D., Labadia, M., Grozinger, K., Skoog, M. T., Wu, J. C., Shih, C.-K., Eckner, K., Hattox, S., Adams, J., Rosenthal, A. S., Faanes, R., Eckner, R. J., Koup, R. A., and Sullivan, J. L. (1990). Inhibition of HIV-1 replication by a nonnucleoside reverse transcriptase inhibitor. *Science* **250**, 1411–1413.

Merrill, D. P., Moonis, M., Chou, T. C., and Hirsch, M. S. (1996). Lamivudine or stavudine in two- and three-drug combination against human immunodeficiency virus type 1 replication *in vitro*. *J. Infect. Dis.* **173**, 355–364.

Meyer, P. R., Matsuura, S. E., So, A. G., and Scott, W. A. (1998). Unblocking of chain-terminated primer by HIV-1 reverse transcriptase through a nucleotide-dependent mechanism. *Proc. Natl. Acad. Sci. USA* **95**, 13471–13476.

Meyer, P. R., Matsuura, S. E., Mian, A. M., So, A. G., and Scott, W. A. (1999). A mechanism of AZT resistance: an increase in nucleotide-dependent primer unblocking by mutant HIV-1 reverse transcriptase. *Mol. Cell* **4**, 35–43.

Mitsuya, H., and Broder, S. (1986). Inhibition of the *in vitro* infectivity and cytopathic effect of human T-lymphotrophic virus type III/lymphadenopathy-associated virus (HTLV-III/LAV) by 2′,3′-dideoxynucleosides. *Proc. Natl. Acad. Sci. USA* **83**, 1911–1915.

Mitsuya, H., Weinhold, K. J., Furman, P. A., St. Clair, M. H., Nusinoff-Lehrman, S., Gallo, R. C., Bolognesi, D., Barry, D. W., and Broder, S. (1985). 3′-azido-3′-deoxythymidine (BW A509U): An antiviral agent that inhibits the infectivity and cytopathic effect of human T-lymphotropic virus type III/lymphadenopathy-associated virus *in vitro*. *Proc. Natl. Acad. Sci. USA* **82**, 7096–7100.

Miyasaka, T., Tanaka, H., Baba, M., Hayakawa, H., Walker, R. T., Balzarini, J., and De Clercq, E. (1989). A new lead for specific anti-HIV-1 agents: 1-[(2-hydroxyethoxy)methyl]-6-(phenylthio)thymine. *J. Med. Chem.* **32**, 2507–2509.

Morales-Ramirez, J., Tashima, K., Hardy, D. *et al.* (1998). "A Phase II, Multi-Center Randomized, Open Label Study to Compare the Antiretroviral Activity and Tolerability of Efavirenz (EFV) + Indinavir (IDV), versus EFV + Zidovudine (ZDV) + Lamivudine (3TC), versus IDV + ZDV + 3TC at >36 Weeks." 38th Interscience Conference on Antimicrobial Agents and Chemotherapy, San Diego. [Abstract I-103]

Motakis, D., Borkow, G., and Parniak, M. A. (1999). Mechanism of resistance of HIV-1 to the tight-binding nonnucleoside inhibitor UC781. Retroviruses, p. 101. Cold Spring Harbor Laboratory, Cold Spring Harbor, NY. [Abstract]

Mulato, A. S., and Cherrington, J. M. (1997). Anti-HIV activity of adefovir (PMEA) and PMPA in combination with antiretroviral compounds: *In vitro* analyses. *Antivir. Res.* **36**, 91–97.

Mulato, A. S., Lamy, P. D., Miller, M. D., Li, W. X., Anton, K. E., Hellmann, N. S., and Cherrington, J. M. (1998). Genotypic and phenotypic characterization of human immunodeficiency virus type 1 variants isolated from AIDS patients after prolonged adefovir dipivoxil therapy. *Antimicrob. Agents Chemother.* **42**, 1620–1628.

Parniak, M. A., Dmitrienko, G. I., Barnard, J., Brown, M. D. R., Borkow, G., and Fletcher, R. S. (1998a). "Small Molecule Inhibitors of HIV Reverse Transcriptase-Associated Ribonuclease H," p. 15. INSERM/NIH Workshop on Ribonucleases H: Tools and Therapeutic Targets. [Abstract]

Parniak, M. A., Borkow, G., Sluis-Cremer, N., Klinsky, E., and Alakhov, V. (1998b). Inhibition of multiple stages of HIV-1 replication *in vitro* by SP1000-Fe(III)BBNH formulations. *Antiviral Ther.* 3(Suppl. 5), 8. [Abstract 5]

Pauwels, R., Andries, K., Desmyter, J., Schols, D., Kukla, M. J., Breslin, H. J., Raeymaeckers, A., Van Gelder, J., Woestenborghs, R., Heykants, J., Schellekens, K., Janssen, M. A., De Clercq. E., and Janssen, P. A. J. (1990). Potent and selective inhibition of HIV-1 replication *in vitro* by a novel series of TIBO derivatives. *Nature* **343**, 470–474.

Pauwels, R., Andries, K., Debyser, Z., Van Daele, P., Schols, D., Stoffels, P., De Vreese, K., Woestenborghs, R., Vandamme, A.-M., Janssen, C. G. M., Anne, J., Cauwenbergh, G., Desmyter, J., Heykants, J., Janssen, M. A. C., De Clercq, E., and Janssen, P. A. J. (1993). Potent and highly selective human immunodeficiency virus type 1 (HIV-1) inhibition by a series of α-anilinophenylacetamide derivatives targeted at HIV-1 reverse transcriptase. *Proc. Natl. Acad. Sci. USA* **90**, 1711–1715.

Perrin, L., and Hirschel, B. (1996). Combination therapy in primary HIV infection. *Antiviral Res.* **29**, 87–89.

Reardon, J. E. (1992). Human immunodeficiency virus reverse transcriptase: Steady-state and presteady-state kinetics of nucleotide incorporation. *Biochemistry* **31**, 4473–4479.

Reardon, J. E. (1993). Human immunodeficiency virus reverse transcriptase: A kinetic analysis of RNA-dependent and DNA-dependent DNA polymerization. *J. Biol. Chem.* **268**, 8743–8751.

Ren, J., Esnouf, R., Garman, E., Somers, D., Ross, C., Kirby, I., Keeling, J., Darby, G., Jones, Y., Stuart, D., and Stammers, D. (1995a). High-resolution structures of HIV-1 RT from four RT–inhibitor complexes. *Nature Struct. Biol.* **2**, 293–302.

Ren, J., Esnouf, R., Hopkins, A., Ross, C., Jones, Y., Stammers, D., and Stuart, D. (1995b). The structure of HIV-1 reverse transcriptase complexed with 9-chloro-TIBO: Lessons for inhibitor design. *Structure* **3**, 915–926.

Ren, J., Esnouf, R. M., Hopkins, A. L., Warren, J., Balzarini, J., Stuart, D. I., and Stammers, D. K. (1998). Crystal structures of HIV-1 reverse transcriptase in complex with carboxanilide derivatives. *Biochemistry* **37**, 14394–14403.

Richman, D. D., Havlir, D., Corbeil, J., Looney, D., Ignacio, C., Spector, S. A., Sullivan, J., Cheeseman, S., Barringer, K., and Pauletti, D. (1994). Nevirapine resistance mutations of human immunodeficiency virus type 1 selected during therapy. *J. Virol.* **68**, 1660–1666.

Rodgers, D. W., Gamblin, S. J., Harris, B. A., Ray, S., Culp, J. S., Hellmig, B., Woolf, D. J., DeBouck, C., and Harrison, S. C. (1995). The structure of unliganded reverse transcriptase from the human immunodeficiency virus type 1. *Proc. Natl. Acad. Sci. USA* **92**, 1222–1226.

Romero, D. L., Busso, M., Tan, C.-K., Reusser, F., Palmer, J. R., Poppe, S. M., Aristoff, P. A., Downey, K. M., So, A. G., Resnick, L., and Tarpley, W. G. (1991). Nonnucleoside inhibitors that potently and specifically block human immunodeficiency virus type 1 replication. *Proc. Natl. Acad. Sci. USA* **88**, 8806–8810.

Rubsamen-Waigmann, R., Huguenel, E., Paessens, A., Kleim, J.-L., Wainberg, M. A., and Shah, A. (1997). Second generation non-nucleosidic reverse transcriptase inhibitor HBY097 and HIV-1 viral load. *Lancet* **349**, 1517.

St. Clair, M. H., Martin, J. L., Tudor-Williams, G., Bach, M. C., Vavro, C. L., King, D. M., Kellam, P, Kemp, S. D., and Larder, B. A. (1991). Resistance to ddI and sensitivity to AZT induced by a mutation in HIV-1 reverse transcriptase. *Science* **253**, 1557–1559.

Schinazi, R. F., Lloyd, R. M., Nguyen, M. H., Cannon, D. L., McMillan, A., Ilksoy, N., Chu, C. K., Liotta, D. C., Bazmi, H. Z., and Mellors, J. W. (1993). Characterization of human immunodeficiency viruses resistant to oxathiolane-cytosine nucleosides. *Antimicrob. Agents Chemother.* **37**, 875–881.

Schmit, J. C., Cogniaux, J., Hermans, P., Van Vaeck, C., Sprecher, S., Van Remoortel, B., Witvrouw, M., Balzarini, J., Desmyter, J., De Clercq, E., and Vandamme, A.-M. (1996). Multiple drug resistance to nucleoside analogues and nonnucleoside reverse transcriptase inhibitors in an efficiently replicating human immunodeficiency virus type 1 patient strain. *J. Infect. Dis.* **174**, 962–968.

Schuurman, R., Nijhuis, M., van Leeuwen, R., Schipper, P., de Jong, D., Collis, P., Danner, S. A., Mulder, J., Loveday, C., and Christopherson, C. (1995). Rapid changes in human immunodeficiency virus type 1 RNA load and appearance of drug-resistant virus populations in persons treated with lamivudine (3TC). *J. Infect. Dis.* **171**, 1411–1419.

Shafer, R. W., Kozal, M. J., Winters, M. A., Iversen, A. K., Katzenstein, D. A., Ragni, M. V., Meyer, W. A., Gupta, P., Rasheed, S., Coombs, R., and Merigan, T. (1994). Combination therapy with zidovudine and didanosine selects for drug-resistant human immunodeficiency virus type 1 strains with unique patterns of pol gene mutations. *J. Infect. Dis.* **169**, 722–729.

Shirasaka, T., Kavlick, M. F., Ueno, T., Gao, W. Y., Kojima, E., Alcaide, M. L., Chokekijchai, S., Roy, B. M., Arnold, E., and Yarchoan, R. (1995). Emergence of human immunodeficiency virus type 1 variants with resistance to ultiple dideoxynucleosides in patients receiving therapy with dideoxynucleosides. *Proc. Natl. Acad. Sci. USA* **92**, 2398–2402.

Simpson, D., and Tagliati, M. (1995). Nucleoside analogue-associated peripheral neuropathy in human immunodeficiency virus infection. *J. Acquir. Immune Defic. Syndr. Hum. Retrovirol.* **9**, 153–161.

Sluis-Cremer, N., Dmitrienko, G. I., Balzarini, J., Camarasa, M.-J., and Parniak, M. A. (2000). Human immunodeficiency virus type 1 reverse transcriptase dimer destabilization by 1-{spiro[4″-amino-2″,2″-dioxo-1″,2″-oxathiole-5″,3′-[2′,5′-bis-o-(tert-butyldimethylsilyl)-β-D-ribofuranosyl]]}-3-ethylthymine. *Biochemistry* **39**, 1427–1433.

Smerdon, S. J., Jager, J., Wang, J., Kohlstaedt, L. A., Chirino, A. J., Friedman, J. M., Rice, P. A., and Steitz, T. A. (1994). Structure of the binding site for nonnucleoside inhibitors of the reverse transcriptase of human immunodeficiency virus type 1. *Proc. Natl. Acad. Sci. USA* **91**, 3911–3915.

Soudeyns, H., Yao, X.-J., Gao, Q., Belleau, B., Kraus, J. L., Nguyen-Bu, N., Spira, B., and Wainberg, M. A. (1991). Anti-humanimmunodeficiency virus type 1 activity and *in vitro* toxicity of 2′-deoxy-3′-thiacytidine (BCH-189), novel heterocyclic nucleoside analog. *Antimicrob. Agents Chemother.* **35**, 1386–1390.

Spence, R. A., Anderson, K. S., and Johnson, K. A. (1996). HIV-1 reverse transcriptase resistance to nonnucleoside inhibitors. *Biochemistry* **35**, 1054–1063.

Spence, R. A., Kati, W. M., Anderson, K. S., and Johnson, K. A. (1995). Mechanism of inhibition of HIV-1 reverse transcriptase by nonnucleoside inhibitors. *Science* **267**, 988–993.

Spruance, S. L., Pavia, A. T., Mellors, J. W., Murphy, R., Gathe, J., Stool, E., Jemsek, J. G., Dellamonica, P., Cross, A., and Dunkle, L. (1997). Clinical efficacy of monotherapy with stavudine compared with zidovudine in HIV-infected, zidovudine-experienced patients: A randomized, double-blind, controlled trial. *Ann. Intern. Med.* **126**, 355–363.

Staszewski, S., Keiser, P., Gathe, J., Haas, D., Montaner, J., Hammer, S., Delfraissy, J. L., Cutrell, A., Lafon, S., Thorborn, D., Pearce, G., Spreen, W., and Tortell, S. (1999). "Ziagen/Combivir Is Equivalent to Indinavir/Combivir in Antiretroviral Therapy (ART) Naïve Adults at 24 Weeks (CNA3005)." 6th Conference on Retroviruses and Opportunistic Infections. [Abstract 20]

Tamalet, C., Izopet, J., Koch, N., Fantini, J., and Yahi, N. (1998). Stable rearrangements of the beta3-beta4 hairpin loop of HIV-1 reverse transcriptase in plasma viruses from patients receiving combination therapy. *AIDS* **12**, F161–F166.

Tan, C.-K., Civil, R., Mian, A. M., So, A. G., and Downey, K. M. (1991). Inhibition of the RNase H activity of HIV reverse transcriptase by azidothymidylate. *Biochemistry* **30**, 4831–4835.

Tantillo, C., Ding, J., Jacobo-Molina, A., Nanni, R. G., Boyer, P. L., Hughes, S. H., Pauwels, R., Andries, K., Janssen, P. A. J., and Arnold, E. (1994). Locations of anti-AIDS drug binding sites and resistance mutations in the three-dimensional structure of HIV-1 reverse transcriptase. *J. Mol. Biol.* **243**, 369–387.

Thimmig, R. L., and McHenry, C. S. (1993). Human immunodeficiency virus reverse transcriptase: Expression in *Escherichia coli*, purification, and characterization of a functionally and structurally asymmetric dimeric polymerase. *J. Biol. Chem.* **268**, 16528–16536.

Tisdale, M., Kemp, S. D., Parry, N. R., and Larder, B. A. (1993). Rapid *in vitro* selection of human immunodeficiency virus type 1 resistant to 3′-thiacytidine inhibitors due to a mutation in the YMDD region of reverse transcriptase. *Proc. Natl. Acad. Sci. USA* **90**, 5653–5656.

Tremblay, C., Merrill, D. P., Chou, T. C., and Hirsch, M. S. (1998). "1529U89 as a Component of 2- and 3-Drug Regimens against Zidovudine-Sensitive and Zidovudine-Resistant HIV Isolates *in Vitro*," p. 198. 5th Conference on Retroviruses and Opportunistic Infections. [Abstract 632]

Tucker, T. J., Lumma, W. C., and Culberson, J. C. (1996). Development of nonnucleoside HIV reverse transcriptase inhibitors. *Methods Enzymol.* **275**, 440–472.

Ueno, T., and Mitsuya, H. (1997). Comparative enzymatic study of HIV-1 reverse transcriptase resistant to 2′,3′-dideoxynucleotide analogs using the single-nucleotide incorporation assay. *Biochemistry* **36**, 1092–1099.

Ueno, T., Shirasaka, T., and Mitsuya, H. (1995). Enzymatic characterization of human immunodeficiency virus type 1 reverse transcriptase resistant to multiple 2′,3′-dideoxynucleoside 5′-triphosphates. *J. Biol. Chem.* **270**, 23605–23611.

Villahermosa, M. L., Martinez-Irujo, J. J., Cabodevilla, F., and Santiago, E. (1997). Synergistic inhibition of HIV-1 reverse transcriptase by combinations of chain-terminating nucleotides. *Biochemistry* **36**, 13223–13231.

Wainberg, M. A., and Parniak, M. A. (1997). Is HIV-1 resistance to 3TC of clinical benefit? *Int. Antiviral News* **5**, 3–5.

Wainberg, M. A., Drosopolous, W. C., Salomon, H., Hsu, M., Borkow, G., Parniak, M. A., Gu, Z., Song, Q., Manne, J., Islam, S., Castriota, G., and Prasad, V. R. (1996). Enhanced fidelity of 3TC-selected mutant HIV-1 reverse transcriptase. *Science* **271**, 1282–1285.

Wainberg, M. A., Tremblay, M., Rooke, R., Blain, N., Soudeyns, H., Parniak, M. A., Yao, X.-J., Li, X., Fanning, M., Montaner, J. S. G., O'Shaughnessy, M., Tsoukas, C., Falutz, J., Dionne, G., Belleau, B., and Ruedy, J. (1991). Characterization of reverse transcriptase activity and susceptibility to other nucleosides of AZT-resistant variants of HIV-1: Results from the Canadian AZT Multicentre Study. *Ann. N. Y. Acad. Sci.* **616,** 346–355.

Winters, M. A., Coolley, K. L., Girard, Y. A., Levee, D. J., Hamdan, H., Shafer, R. W., Katzenstein, D. A., and Merigan, T. C. (1998). A 6-basepair insert in the reverse transcriptase gene of human immunodeficiency virus type 1 confers resistance to multiple nucleoside inhibitors. *J. Clin. Invest.* **102,** 1769–1775.

Wu, J., Amandoron, E., Li, X., Wainberg, M. A., and Parniak, M. A. (1993). Monoclonal antibody-mediated inhibition of HIV-1 reverse transcriptase polymerase activity: Interaction with a possible nucleoside triphosphate binding domain. *J. Biol. Chem.* **268,** 9980–9985.

Young, S. D., Britcher, S. F., Tran, L. O., Payne, L. S., Lumma, W. C., Lyle, T. A., Huff, J. R., Anderson, P. S., Olsen, D. B., and Carroll, S. S. (1995). L-743, 726 (DMP-266): A novel, highly potent nonnucleoside inhibitor of the human immunodeficiency virus type 1 reverse transcriptase. *Antimicrob. Agents Chemother.* **39,** 2602–2605.

Zhang, D., Caliendo, A. M., Eron, J. J., DeVore, K. M., Kaplan, J. C., Hirsch, M. S., and D'Aquila, R. T. (1994). Resistance to 2′, 3′-dideoxycytidine conferred by a mutation in codon 65 of the human immunodeficiency virus type 1 reverse transcriptase. *Antimicrob. Agents Chemother.* **38,** 282–287.

John M. Louis*
Irene T. Weber†
József Tözsér‡
G. Marius Clore*
Angela M. Gronenborn*

*Laboratory of Chemical Physics
National Institute of Diabetes
Digestive and Kidney Diseases
National Institutes of Health
Bethesda, Maryland 20892-0580
†Department of Microbiology and Immunology
Kimmel Cancer Center
Thomas Jefferson University
Philadelphia, Pennsylvania 19107
‡Department of Biochemistry and Molecular Biology
University Medical School of Debrecen
H-4012, Debrecen, Hungary

HIV-1 Protease: Maturation, Enzyme Specificity, and Drug Resistance

I. Introduction

Human immunodeficiency virus type 1 protease (HIV-1 PR) plays an indispensable role in the viral replication cycle. The PR catalyzes the hydrolysis of specific peptide bonds within the HIV-1 Gag and Gag-Pol polyproteins for its own maturation and to produce the other mature structural and functional proteins (Darke *et al.*, 1988; Oroszlan and Luftig, 1990). The active form of the 99-amino-acid-long mature PR is a homodimer. Optimal catalytic activity of the mature PR and ordered processing of the polyproteins are critical for the liberation of infective progeny virus (Oroszlan and Luftig, 1990; DeBouck, 1992; Kaplan *et al.*, 1993; Rose *et al.*, 1995). The initial critical step in the maturation reaction is the formation of an active PR dimer, formed from two Gag-Pol precursors, which is necessary for the release of the mature PR from the precursor. Furthermore, the cleavage at the N terminus of the PR is essential for the liberation of maturelike catalytic

Advances in Pharmacology, Volume 49

activity (Louis *et al.*, 1994) and optimal ordered processing of Gag and viral infectivity (Tessmer and Krausslich, 1998). The level of expression of the Gag-Pol precursor relative to the Gag and the spatial arrangement of the PR domain within the Gag-Pol are also important (Krausslich, 1991). Over-expression of the Gag-Pol precursor results in intracellular activation of the PR and inhibition of virus assembly, suggesting that concentration-dependent dimer formation plays a key role in regulating the autocatalytic maturation of the PR in the viral life cycle (Karacostas *et al.*, 1993).

The mature PR has proven to be a most effective target for antiviral therapy of AIDS. However, the long-term potency of current PR inhibitors as therapeutic agents is limited by the rapid development of drug-resistant variants of the PR. Therefore it is critical to understand the molecular mechanisms of the proteolytic processes of the wild-type and drug-resistant mutants of PR in order to aid the development of new inhibitors and thera-peutic strategies. Viral replication is limited by the relative activity of the PR for the sequential processing of the polyproteins. Up to 400-fold differ-ences in the rates of hydrolysis were shown for different cleavage sites in Gag (Pettit *et al.*, 1994). Thus, it is critical to examine the drug-resistant mutant forms of the precursor and the mature enzyme for changes in the structural properties, kinetics, and specificity for natural substrates as com-pared to the wild-type enzyme.

In this article, we focus on three critical features of the HIV-1 PR, namely autocatalytic maturation from the precursor, substrate specificity, and emergence of drug resistance, utilizing protein and peptide design, en-zyme kinetics, X-ray crystallography, and NMR spectroscopy.

II. Organization of the Gag-Pol Precursor and Molecular Structure of the Mature PR

The structural and functional proteins of retroviruses are produced through translation of polycistronic messenger RNAs into precursor proteins (Oroszlan and Luftig, 1990). In HIV-1, the Gag polyprotein consists of the structural proteins in the arrangement MA-CA-p2-NC-p1-p6 (Fig. 1 [Genbank: HIVHXB2CG]; Henderson *et al.*, 1992; Wondrak *et al.*, 1993). The Gag-Pol polyprotein is translated *via* a mechanism in which a -1 frame-shift of an adenosine residue changes the open reading frame from Gag to Pol at a frequency of 5 to 10% (Jacks *et al.*, 1988; Hatfield *et al.*, 1992). This ribosomal frameshift site corresponds to the second codon within p1, leading to the synthesis of the transframe region (TFR or p6*), which links the structural Gag domain to the functional Pol domain (Fig. 1). Thus, the structure of Gag-Pol is MA-CA-p2-NC-TFR-PR-RT-IN (Fig. 1; Gorelick and Henderson 1994). A protease cleavage site separates each of the sub-domains of the precursors.

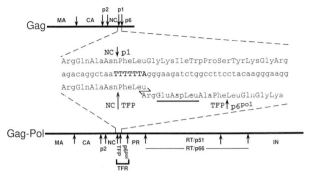

FIGURE I Structural organization of the Gag and Gag-Pol polyproteins of HIV-1 (bold lines). The nucleotide and protein sequence surrounding the translation frameshift site are shown. Arrows indicate the specific sites of cleavage by the viral PR. The DNA and encoded amino acids spanning the NC↓p1 junction in the Gag polyprotein and NC↓TFP and TFP↓p6^pol junctions in the Gag-Pol polyprotein are aligned. Bold letters in the DNA sequence denote the conserved signal sequence required for ribosomal frameshifting. TFP denotes the eight conserved amino acids flanking NC. The remainder of the transframe region (TFR) constitutes p6^pol. The N-terminal dipeptide sequence of TFP is common with Gag p1. The second amino acid of TFP is either Leu or Phe at a ratio of ~7:3, which is dependent upon variation in the frameshifting mechanism and does not alter the remaining sequence of TFP. The underlined amino acid sequence represents the tripeptide core of TFP that inhibits the PR. The coding sequence for the tripeptide Glu-Asp-Leu corresponds to a region in the viral RNA between conserved regions of the signal sequence and the RNA secondary structure. Nomenclature of viral proteins is that of Leis *et al.* (1988).

Retroviral proteases including HIV-1 PR are 99–125 residues in size and contain the characteristic active site triad, Asp-Thr/Ser-Gly (amino acids 25–27 in HIV-1). They are enzymatically active as homodimers in acidic conditions and are structurally similar to pepsinlike proteases which function as monomers. In addition to the catalytic triad, retroviral proteases contain a second highly conserved region (Gly86-Arg87-Asn88 in HIV-1) that is not present in the cellular aspartic proteases (Pearl and Taylor, 1987; Copeland and Oroszlan, 1988). Even a conservative mutation of Arg87→Lys in this region renders the enzyme inactive (Louis *et al.*, 1989). Conserved regions of the PR are important for both catalysis and dimer formation (Weber, 1990; Gustchina and Weber, 1991). Crystal structures have been determined of HIV PR in the presence and absence of inhibitors (reviewed in Wlodawer and Erickson, 1993). The two chemically identical subunits of the PR dimer are in a nearly symmetric arrangement. Each subunit folds into a compact structure of β-strands with a short α-helix near the C terminus (Fig. 2). Residues 44–57 of each subunit form a pair of antiparallel β-strands called the flap. The β-turn at the tip of the flap contains conserved glycines. The flaps are flexible and are thought to fold down over the substrate or inhibitor and act during catalysis both to bind substrate and exclude water from the active site. Mutation of flap residues results in dramatic

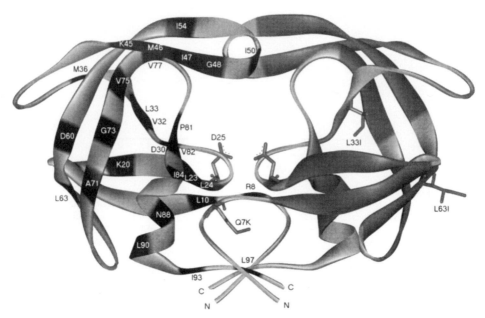

FIGURE 2   Crystal structure of HIV-1 PR. Shaded regions in one monomer indicate the secondary structure of the dimeric enzyme. The three mutations Q7K, L33I, and L63I in the PR to prevent its intrinsic autoproteolysis (degradation) and the active site aspartic acid residues of each monomer are shown in a stick representation. Positions of drug-resistant mutations are indicated in black.

reductions in enzyme activity (Tözsér *et al.*, 1997a). Intersubunit interactions are found between the two catalytic triads (Asp 25-Gly 27); Ile 50 and Gly 51 at the tip of the flaps; the complex salt bridge between Asp 29, Arg 87, and Arg 8′ from the other subunit; and the β-sheet formed by the four termini in the dimer (residues 1–4 and 96–99). These terminal residues form one half of the substrate sequence for cleavage at the p6^{pol} ↓ PR and PR ↓ RT junctions in the Gag-Pol precursor. The substrate binding site is formed by residues 8, 23–30, 32, 45–50, 53, 56, 76, 80–82, and 84. In addition, a number of other PR residues contribute to both the substrate binding and the interface contacts. These important regions are highly conserved in sequence among different isolates of HIV-1 (Gustchina and Weber, 1991) and retroviral proteases share similar structures in these regions (Weber, 1990; Wlodawer and Erickson, 1993).

III. Mechanism and Regulation of PR Maturation

PR-mediated processing of Gag and Gag-Pol and particle maturation are complex events (Vogt, 1996). The TFR, which flanks the N terminus of PR in Gag-Pol (see Fig. 1), is thought to have a role similar to that observed at the N termini of zymogen forms of cellular aspartic proteases. In accordance, early studies have demonstrated that a deletion of the TFR domain leads to a significantly higher rate of processing activity of the Gag-ΔPol precursor, suggesting that TFR negatively regulates PR function (Partin *et al.*, 1991). TFR consists of two domains, the N-terminal transframe octapeptide (TFP), which is conserved in all variants of HIV-1, and a 48- to 60-amino-acid variable region p6^{pol} (Gorelick and Henderson, 1994; Candotti *et al.*, 1994; Louis *et al.*, 1998). TFP and p6^{pol} are separated by a PR cleavage site (Phylip *et al.*, 1992; Louis *et al.*, 1999a). An isolated 68-amino-acid TFR has no overall stable secondary or tertiary structure, although a small potential for helix formation at its N terminus was shown by NMR (Beissinger *et al.*, 1996).

A. Kinetics and Mechanism of Autoprocessing of a Model Precursor

Since kinetic studies to understand the mechanism of the autocatalytic maturation reaction of the full-length Gag-Pol polyprotein are difficult to perform due to the presence of at least nine PR cleavage sites (Fig. 1), a model PR precursor that contains only the two cleavage sites flanking the PR domain was developed. This model polyprotein (MBP-Δp6^{pol}-PR-ΔRT; Fig. 3) was constructed by adding 19 residues of the native reverse transcript-

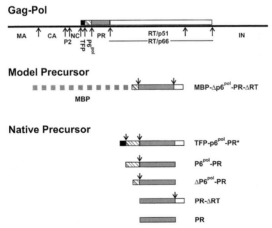

FIGURE 3   Model and native precursor constructs of the PR. The Gag-Pol polyprotein of HIV-1 is shown on top (see legend to Fig. 1). The domains TFP, p6^{pol}, and RT flanking the PR (gray) are shown as closed, hatched, and open boxes, respectively. The dashed gray line denotes the 38-kDa maltose-binding protein (MBP). The catalytic activities of MBP-Δp6^{pol}-PR-ΔRT (referred to as MBP-ΔTF-PR-ΔPol in Louis *et al.*, 1994), TFP-p6^{pol}-PR*, p6^{pol}-PR, and Δp6^{pol}-PR are significantly lower relative to that of the PR-ΔRT and mature PR.

ase sequence (ΔRT) to the C terminus of the PR domain and 12 residues of the native p6^{pol} region (Δp6^{pol}) to its N terminus, which was additionally fused to the maltose binding protein (MBP) of *Escherichia coli* (Louis *et al.*, 1991a). Aggregation of the precursor upon its expression in *E. coli* allowed its isolation in an intact form. Upon refolding the protein from 8 M urea, the model precursor undergoes time-dependent autoprocessing to release the mature PR in two consecutive steps. Initial mutational studies of the cleavage sites in the model polyprotein showed that the N-terminal cleavage is more sensitive to mutations and that it precedes the cleavage at the C terminus of the PR (Louis *et al.*, 1991a,b). Investigation of the time-dependent autocatalytic maturation of MBP-Δp6^{pol}-PR-ΔRT by kinetics led to the mechanism summarized in Fig. 4 (Louis *et al.*, 1994).

I. Processing of the N-Terminal Strands Is Intramolecular

In the mechanism illustrated in Fig. 4, the full-length precursor dimerizes to form tetrapod 1. The dimeric protein binds inhibitors and substrates in a similar manner to the mature enzyme, although its enzymatic activity is significantly lower than that of the mature PR. For autoprocessing to occur, tetrapod 1, which is not an obligatory intermediate, undergoes a conformational change to tetrapod 2, in which one of the two N-terminal strands occupies its active site. Cleavage of the scissile peptide bond at the N terminus is the rate-determining step for the appearance of enzymatic activity. The bipod (PR-ΔRT, see Figs. 3 and 4), which is converted relatively slowly to the mature PR, has catalytic activity comparable to that of the mature enzyme. Restricting the cleavage at the p6^{pol} ↓ PR site in the model precursor leads to reduced cleavage at the C terminus of PR (Louis *et al.*, 1991b). *In*

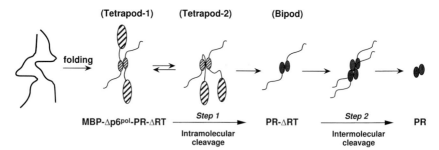

FIGURE 4 Proposed mechanism for the autocatalytic maturation of the HIV-1 PR from the model polyprotein MBP-Δp6^{pol}-PR-ΔRT. The MBP and PR are denoted as large hatched ovals and small hatched ovals, respectively. Lines represent Δp6^{pol} and ΔRT sequences that flank the PR domain. The PR catalyzes the hydrolysis of its N terminus from a transiently dimeric MBP-Δp6^{pol}-PR-ΔRT *via* an intramolecular mechanism to release the intermediate PR-ΔRT (Step 1). Subsequent conversion of the PR-ΔRT (bipod) to the mature PR occurs *via* an intermolecular mechanism (Step 2). The two conformational isomers of the dimeric MBP-Δp6^{pol}-PR-ΔRT are referred to as Tetrapods 1 and 2. Hatched ovals of the PR domain denote the low catalytic activity of the PR precursor as compared to the bipod and mature PR.

vivo, a blocking mutation at the p6^{pol} $\downarrow$ PR site in the Gag-Pol precursor allows maturation of PR to occur at the native TFP $\downarrow$ p6^{pol} and PR $\downarrow$ RT sites to release a p6^{pol}–PR intermediate, which, however, is defective in Gag processing and viral infectivity (Tessmer and Krausslich, 1998).

A similar mechanism for the processing of the PR at its N terminus has been proposed in studies using a mini-precursor in which a mutated PR (Ala28→Ser) was fused to 25 amino acids of the native p6^{pol} sequence (Δp6^{pol}-PR^{Ala28}→^{Ser}; Co *et al.*, 1994). The mutation Ala28→Ser in the mature PR lowers the k_{cat}/K_m by about 250-fold relative to the wild-type mature PR and thus permitted the isolation of the precursor upon its expression in *E. coli* (Ido *et al.*, 1991). This mini-precursor reportedly undergoes time-dependent maturation but with no significant change in the catalytic activity, contrary to our results using the model construct. The proposed mechanism for the maturation of the HIV-1 PR at its N terminus is strikingly similar to that observed for the conversion of pepsinogen to pepsin, a monomeric mammalian aspartic acid protease (Louis *et al.*, 1994).

2. Processing of the C-Terminal Strands Is Intermolecular

Unlike the C terminus of pepsinogen, which is not modified during the activation process, the C terminus of the HIV-1 PR precursor is processed to generate the mature PR. The kinetic order of the reaction involving the bipod monomer (PR-ΔRT, see Fig. 3; Wondrak *et al.*, 1996) was determined by following the initial rates of the reaction. A linear relationship between the rate of conversion (disappearance of the bipod and the appearance of the mature PR) and the square of the protein concentration (varied over sevenfold) indicates that the reaction is bimolecular (intermolecular). Unlike the MBP-Δp6^{pol}-PR-ΔRT, which has very low catalytic activity, both the kinetic parameters for the hydrolysis of the peptide substrate catalyzed by the PR–ΔRT and the inhibition constant determined with a pseudopeptide inhibitor are indistinguishable from those determined for the mature enzyme. In addition, the pH profile for k_{cat}/K_m is similar to that of the mature PR. Activity of the bipod or change in the intrinsic protein fluorescence, both monitored as a function of enzyme dilution, indicated that the majority of the bipod is in the dimeric form under the protein concentrations employed in the kinetic studies.

Our results show that the model precursor exhibits very low catalytic activity prior to the cleavage at the p6^{pol} $\downarrow$ PR junction, the N terminus of PR. In contrast, several mutated PR fusion proteins that either contain a mutation at the p6^{pol} $\downarrow$ PR junction or contain short native or nonnative sequences flanking the N terminus of PR were shown to exhibit catalytic activities comparable to that of the wild-type mature PR (Vogt, 1996; Wondrak and Louis, 1996). This raised the question whether the MBP in our model construct impeded PR folding/dimerization, catalytic activity, and/or the observed kinetics of the maturation reaction. Therefore the autoprocessing reaction was also analyzed using a PR precursor linked to the native transframe region containing the native cleavage sites as described in the next section.

B. Kinetics of Maturation of Native Precursors

I. Maturation of p6^{Pol}-PR Precursor

A native p6^{pol}-PR precursor (see Figs. 1 and 3) containing a native cleavage site at the p6^{pol} ↓ PR junction was expressed and purified. To circumvent aggregation of the precursor associated with Cys thiol oxidation that may lead to anomalous kinetic measurements of the autoprocessing reaction, Cys residues 67 and 95 in the PR were replaced with Ala. These mutations do not alter the kinetic parameters or the structural stability of the mature mutant enzyme (Louis *et al.*, 1999a). The renatured p6^{pol}-PR undergoes a time-dependent maturation reaction concomitant with a large increase in enzymatic activity similar to that of the model precursor (Fig. 5). At ≥pH 5.0, the reaction proceeds in a single step to produce the mature enzyme, whereas at pH <5.0 it is characterized by the appearance and disappearance of a single protein intermediate that migrates between p6^{pol}-PR and PR bands (indicated by arrow in Figs. 5 and 6; see Louis *et al.*, 1999a). This intermediate (termed Δp6^{pol}-PR) is generated via cleavage at the Leu24 ↓ - Gln25 site within p6^{pol} (Fig. 3; Zybarth *et al.*, 1994; Louis *et al.*, 1999a).

The formation of these intermediate precursors under suboptimal conditions for the autoprocessing reaction is similar to that observed for the conversion of the zymogen form of the gastric protease pepsin, which, unlike retroviral proteases, is a monomeric enzyme (Khan and James, 1998). The zymogen pepsinogen differs from mature pepsin by a 44-amino-acid-long positively charged N-terminal proregion. Below pH 2.0, pepsinogen is converted in a single step through an intramolecular maturation process to pepsin with a homogenous N terminus, whereas at pH 4.0 the activation product is heterogeneous with multiple N-terminal products (al-Janabi *et al.*, 1972).

The maturation reaction of p6^{pol}-PR displays good first-order kinetics similar to those of the model precursor between pH 4.0 and pH 6.5, as indicated by a linear relation between the rate of increase in maturelike catalytic activity and protein concentration (inset Fig. 5). Plots of the measured densities corresponding to the starting material and the product (mature PR) and the rate of increase in maturelike catalytic activity versus time are shown superimposed on each other (Fig. 5). The first-order rate constants for the maturation reaction of p6^{pol}-PR and MBP-Δp6^{pol}-PR-ΔPR are similar (Louis *et al.*, 1999a). The first-order rate constant displays a bell-shaped dependency on pH with two ionizable groups having pK_as of 4.9 and 5.1 (Louis *et al.*, 1999a). This pH profile is quite similar to that obtained for the maturation of the mini-precursor Δp6^{pol}-PR$^{Ala28 \rightarrow Ser}$ (Co *et al.*, 1994).

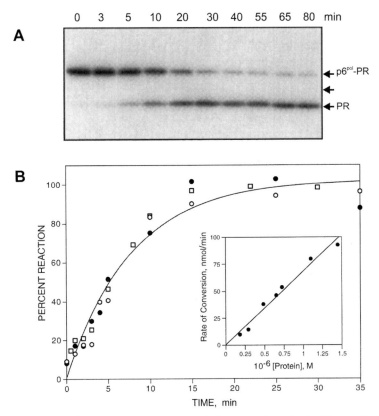

FIGURE 5 (A) Time course of the autocatalytic maturation of p6^{pol}-PR at pH 5.8 in 50 mM sodium acetate at 25°C. Aliquots of the reaction mixture were drawn at the desired time and subjected to 10–20% SDS–PAGE and immunoblotting. The relative mobilities of the full-length precursor and mature PR are indicated by arrows. Band intensities were quantified by densitometry. (B) Time course of the reaction measured by following the increase in enzymatic activity (□), appearance of PR (●), and disappearance of p6^{pol}-PR (○) at pH 4.8. Inset in B is a plot of the dependence of the initial rate of the reaction measured by following the increase in enzymatic activity upon varying the protein concentration. The first-order rate constant measured by following the appearance of enzymatic activity for the maturation of p6^{pol}-PR under the same conditions as described for the model precursor are similar (Louis *et al.*, 1994, 1999a).

The higher pK_a of 5.1 observed for the maturation reaction is in agreement with the pK_a values of 4.8 and 5.2 obtained for the mature PR (Polgar *et al.*, 1994) and PR-ΔPol (Wondrak *et al.*, 1996), respectively, and for those reported using other fusion proteins of the PR (Wondrak and Louis, 1996). The lower pK_a of 4.9 observed for the maturation reaction is about 1.8 pH units higher than that observed for the wild-type mature PR (Polgar *et al.*, 1994).

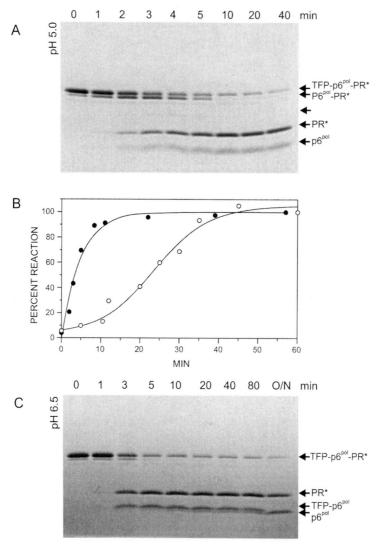

FIGURE 6   Time course of the autocatalytic maturation of TFP-p6^{pol}-PR* at pH 5.0 in 50 mM sodium formate, 0.5 M urea (A), and at pH 6.5 in 25 mM sodium phosphate (C) buffers. Aliquots of the reaction mixture were drawn at the desired time and subjected to 10–20% SDS–PAGE in Tris–Tricine buffer. Protein bands were visualized by Coommasie brilliant G250 staining. The relative mobilities of the full-length precursor and product peaks are indicated by arrows. (B) The time course of the autoprocessing of TFP-p6^{pol}-PR* monitored by following the appearance in enzymatic activity at pH 4.0 (●) and pH 6.5 (○). The lag in the reaction time course coincides to the accumulation of the transient intermediate p6^{pol}-PR* (see A) that is subsequently converted to the mature PR*.

2. Maturation of TFP-p6^{pol}-PR* Precursor

To conduct structural studies of the PR precursor fused to the intact transframe region (TFP-p6^{pol}), a construct which spans the Gag-Pol sequence starting from the NC ↓ TFP junction and ending with the C-terminal amino acid of the PR was expressed (see Figs. 1 and 3). The PR domain in this TFP-p6^{pol}-PR* construct bears the following mutations, Gln7→Lys, Leu33→Ile, Leu63→Ile, Cys67→Ala, and Cys95→A, designed to limit autoprotolysis (degradation of PR) and to prevent Cys thiol oxidation. Importantly, these five mutations do not alter the kinetics or the structural stability of the mature PR* as compared to that of the wild-type mature PR (Rose *et al.*, 1993; Mildner *et al.*, 1994; Mahalingam *et al.*, 1999; Louis *et al.*, 1999a,b). However, the accumulation of a small fraction of the full-length precursor (<5% of the total expressed protein) permitted its purification in just sufficient amounts for kinetic analyses (Louis *et al.*, 1999b).

The full-length precursor TFP-p6^{pol}-PR* has two known native PR cleavage sites, TFP ↓ p6^{pol} and p6^{pol} ↓ PR (Figs. 1 and 3). Since PR-catalyzed hydrolysis of peptides with amino acid sequences corresponding to the two cleavage sites, TFP ↓ p6^{pol} and p6^{pol} ↓ PR, have comparable kinetic parameters (Phylip *et al.*, 1992; Louis *et al.*, 1998), we anticipated complex kinetics and multiple products for the maturation of the PR from TFP-p6^{pol}-PR*. In contrast, TFP-p6^{pol}-PR* undergoes an ordered two-step maturation reaction (Fig. 6A). At pH 5, the first step involves the cleavage of the peptide bond at the TFP/p6^{pol} site to produce the transient intermediate p6^{pol}-PR, which is subsequently converted to the mature PR*. A plot of the rate of appearance of enzymatic activity *versus* time is characterized by a lag period followed by a first-order process indicating that (1) TFP-p6^{pol}-PR* has the same low catalytic activity as that of the intermediate precursor p6^{pol}-PR* and (2) cleavage at the N terminus of the PR (p6^{pol} ↓ PR site) is concomitant with the appearance of maturelike catalytic activity (Fig. 6B). Following the reaction at pH 6.5 no lag period is observed due to preferential cleavage at the p6^{pol} ↓ PR site which precedes that of the TFP ↓ p6^{pol} site releasing the mature PR* (Figs. 6B and 6C).

C. TFP, a Hydrophlic Competitive Inhibitor of the Mature PR

TFP, the N terminus of TFR, may have a regulatory role for the autoprocessing of the PR from the Gag-Pol precursor *in vivo*. TFP and its analogs are competitive inhibitors of the mature PR (Fig. 7A; Louis *et al.*, 1998). The smallest and most potent of the analogs are tripeptides Glu-Asp-Leu and Glu-Asp-Phe with K_is of ca. 50 and 20 μM, respectively. Other substitutions or deletions in the tripeptide Glu-Asp-Leu lead to higher K_is. Substitution of the acidic amino acids in the TFP by neutral amino acids and D- or retro-D configuration of Glu-Asp-Leu results in a >40-fold increase in K_i.

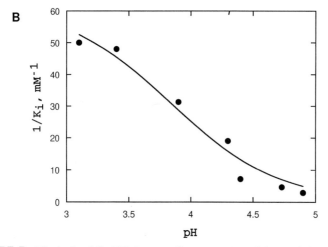

A

PEPTIDE	K_i, μM
Phe-Leu-Arg-Glu-Asp-Leu-Ala-Phe	98 ± 10
Phe-Leu-Arg-Gln-Asn-Leu-Ala-Phe	>2000
Glu-Asp-Leu-Ala	160 ± 20
Glu-Asp-Leu	50 ± 9
Arg-Glu-Asp	3360 ± 400
Glu-Asn-Leu	3070 ± 210
Glu-Asp-Phe	25 ± 3

B

FIGURE 7   The hydrophilic TFP is a specific competitive inhibitor of the mature HIV-1 PR. (A) Inhibition of the action of HIV-1 PR by TFP Phe-Leu-Arg-Glu-Asp-Leu-Ala-Phe and its analogs. The K_i was obtained from a plot of the apparent K_m vs [I] for peptides or from a plot of $1/V$ *vs* [I] at saturating concentration of the substrate. Assays were performed in 50 mM sodium formate at pH 4.25 and 2.5 mM DTT. The final enzyme and substrate concentrations were 150 nM and 390 μM, respectively. All amino acids are of the L-configuration. (B) A plot of inhibition (K_i) by Glu-Asp-Leu for PR-catalyzed hydrolysis of substrate Lys-Ala-Arg-Val-Nle-Phe(NO_2)-Glu-Ala-Nle-NH_2 vs pH. The K_is were obtained from plots of $1/V$ versus [I] at a saturating concentration of the substrate (390 μM). Reactions were carried out in 50 mM sodium formate or 100 mM sodium acetate buffers containing 2.5 mM DTT and 150 nM of PR at 25°C. The line is a calculated curve with a pK_a of 3.8 and a K_i of 20 μM.

Unlike other known inhibitors of the HIV-1 PR which are highly hydrophobic, Glu-Asp-Leu is extremely soluble in water and its binding affinity to the active site of the PR is not enhanced by increasing NaCl concentration. Inhibition of the PR by Glu-Asp-Leu is dependent on the protonated form of a group with a pK_a of 3.8 (Fig. 7B). This result complements the observation of the stepwise maturation of TFP-p6^{pol}-PR* at different pH values. Decreasing pH leads to a higher affinity of TFP to the active site of the PR, promoting the processing at the TFP↓p6^{pol} site prior to the p6^{pol}↓PR site. It was not feasible to investigate the interaction of intact TFP with the PR under conditions of crystal growth due to its hydrolysis, giving rise to two products, Phe-Leu-Arg-Glu-Asp and Leu-Ala-Phe. But the interaction of the tripeptide core of TFP, Glu-Asp-Leu, as studied by X-ray crystallography is similar to those of other product–enzyme complexes (Louis *et al.*, 1998; Rose *et al.*, 1996). Similarly, peptides derived from the proregion of pepsinogen and prorenin have been shown to act as competitive inhibitors of mature pepsin and renin, respectively (Dunn *et al.*, 1978; Richards *et al.*, 1992).

D. Influence of Flanking Sequences on the Structural Stability of Protease Precursors

Loss in enzymatic activity of the mature PR correlates to loss of stable tertiary structure with increasing urea concentration (Wondrak *et al.*, 1996; Wondrak and Louis, 1996). The dimeric form of p6^{pol}-PR, which is mandatory for the formation of an active site capable of supporting a hydrolytic reaction, is highly sensitive to urea denaturation compared to the mature enzyme, which suggests that p6^{pol}-PR is structurally less stable than the mature PR (Louis *et al.*, 1999a). This result is in agreement with earlier studies showing that the PR domain, when fused to the 19 amino acids of the flanking C-terminal RT sequence (PR-ΔRT) or the short native or nonnative sequences at its N terminus, is also less stable toward urea denaturation (Wondrak *et al.*, 1996; Wondrak and Louis, 1996). The above results are also consistent with results showing that the mature PR is largely dimeric above 10 nM (Wondrak *et al.*, 1996), whereas inactive N-terminally extended forms of the PR linked to the TFR fail to dimerize in a qualitative assay (Zybarth and Carter, 1995).

Dissociation of the dimeric form of the mature PR* (autoproteolysis resistant) was also measured by following enzymatic activity as a function of enzyme dilution. Mature PR* exhibits a K_d <5 nM at pH 5.0, similar to that observed for the mature wild-type PR under identical conditions (Louis *et al.*, 1999b). Since TFP-p6^{pol}-PR* precursor undergoes cleavage at the p6^{pol} ↓ PR site as the first step at pH 6.5 (Figs. 6B and 6C), unlike at more acidic pH values, the simultaneous appearance of the maturelike enzymatic activity and the mature PR* protein product was used to evaluate the ability of TFP-p6^{pol}-PR* to undergo dimerization and autocatalytic maturation as a function of precursor concentration. A plot of specific enzymatic activity versus precursor concentration shows that the apparent K_d for TFP-p6^{pol}-PR* is ~680 nM (Louis *et al.*, 1999b). The decreased stability to urea and >130-fold increase in the apparent K_d of TFP-p6^{pol}-PR* as compared to the mature PR* indicates that the structural stability of the PR domain is significantly lower in the precursor.

E. ^{1}H-^{15}N Correlation Spectra of Uniformly Labeled Precursor and Mature Proteases

The fact that the maturation reaction of the TFP-p6^{pol}-PR* and p6^{pol}-PR precursors can be monitored by following the increase in enzymatic activity is clear evidence that there is a large difference between the catalytic activity of the mature PR and that of PR precursor. The low catalytic activity seems to be intrinsic to the PR when linked to the native TFR with the native cleavage sites. This low catalytic activity could be either due to a conformational difference of the dimeric precursor that does not support efficient catalysis or could be an apparent effect of the equilibrium that largely favors the unfolded or partially folded form of the protein relative to the folded enzymatically active dimer. To analyze the PR precursor by NMR requires attaining sufficient quantity of the protein.

In other studies, inhibitor-resistant mutants of the PR fused to the intact TFR were expressed with the aim of understanding the relationship between PR maturation and drug resistance. One among several of these mutants that were analyzed for expression shows elevated level of precursor accumulation as compared to the construct TFP-p6^{pol}-PR*. We chose the R8Q mutant precursor (termed TFP-p6^{pol}-PR*^{Q}) as a good source for preparing sufficient amounts of precursor. Although the mature PR*^{Q} is more sensitive toward urea denaturation than PR*, the dissociation constant (K_d) and kinetic parameters for mature PR*^{Q}-catalyzed hydrolysis of the peptide substrate and the inhibition constant for the hydrolytic reaction with inhibitor are comparable to that of the wild-type PR and PR* (Mahalingham *et al.*, 1999; Louis *et al.*, 1999b). TFP-p6^{pol}-PR*^{Q} undergoes maturation to release the mature PR*^{Q}, similar in kinetics to that of the TFP-p6^{pol}-PR* precursor. We therefore employed uniformly ^{15}N-labeled TFP-p6^{pol}-PR*^{Q} precursor in our subsequent studies of HIV-1 PR maturation.

It appears that for HIV-1 PR, and possibly other viral aspartic proteases, activation is tightly coupled to folding. This is in contrast to most zymogens and their corresponding mature enzymes in which the catalytic machinery is stably preformed and activation is achieved by a conformational change after peptide bond cleavage which involves removing parts of the polypeptide chain protruding into or obstructing access to the active site. The TFP-p6^{pol}-PR precursor protein largely possesses all the hallmarks of an unfolded polypeptide chain and the ^{1}H-^{15}N correlation spectrum of uniformly labeled TFP-p6^{pol}-PR*^{Q} precursor protein at pH 5.0 presented in Fig. 8A exhibits the typical narrow shift dispersion observed for random-coil peptides or proteins (Wishart *et al.*, 1995). This is true for precursor protein in the absence or presence of any tight binding inhibitors. Cleavage at the p6^{pol}/

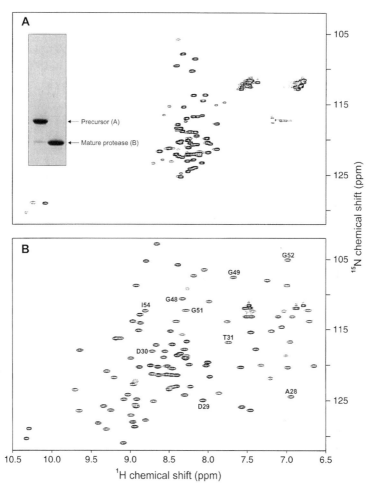

FIGURE 8 ^{1}H-^{15}N correlation spectra of the precursor TFP-p6^{pol}-PR^{*Q} and mature PR^{*Q} proteases in complex with 10-fold excess of DMP323 in 50 mM sodium acetate buffer, pH 5.2, at 25°C. (Inset) An aliquot of the sample was subjected to 20% SDS–PAGE and stained with Coomassie brilliant blue G250. There were no degradation products observed in both preparations but for a minor mature PR product released from TFP-p6^{pol}-PR^{*Q} that occurs at the protein-folding step even in the presence of excess inhibitor DMP 323.

PR site is concomitant with the appearance of stable tertiary structure and enzymatic activity. Figure 8B shows the corresponding ^{1}H-^{15}N correlation spectrum of the mature PR^{*Q} PR complexed with the symmetric tight binding inhibitor DMP323 (Lam *et al.*, 1994), demonstrating a stable three-dimensional structure. A comparison of chemical shifts observed for this mutant complex with previously reported data (Yamazaki *et al.*, 1994) reveals that the structure of the PR^{*Q} is extremely similar to that determined for the wild-type mature enzyme. Viewed in the greater context of zymogen activation, the HIV-1 PR may represent the most extreme case of activation by conformational rearrangement, namely the transition from an unstructured, inactive precursor protein to a stably folded, active mature enzyme.

F. Plausible Mechanism of Regulation of the Protease in the Viral Replication Cycle

The results of studies using different model and native precursor proteins of the HIV-1 PR suggest that in HIV-1 and related retroviruses with similar organization of the Gag-Pol precursor, the transframe region flanking the N terminus of the PR may function as a negative regulator for protein folding and dimerization. The low dimer stability of the PR precursor relative to that of the mature enzyme is an ideal way of preventing enzymatic activity from emerging until the assembly of the viral particle is complete. Depending on the pH of the environment in which Gag-Pol maturation takes place, removal of the transframe region can either occur in two sequential steps or in a single step. Intramolecular cleavage at the p6^{pol}/PR site to release a free N terminus of PR is critical for the formation of a stable tertiary structure of the PR and enzymatic activity (Louis et al., 1999b). Subsequent processing of the other Gag-Pol cleavage sites will occur rapidly via intermolecular processes. Contrary to the TFR, the RT domain flanking the C terminus of PR does not seem to influence the catalytic activity of the PR (Wondrak et al., 1996; Cherry et al., 1998). The Leu5/Trp6 cleavage within PR could be viewed as a final step in the PR-associated cascade of events, resulting in destabilization of the tertiary structure and promoting dissociation of the dimer, thereby down-regulating the catalytic activity of the PR in the viral life cycle (Rose et al., 1993; Louis et al., 1999b).

IV. Substrate/Inhibitor Interactions

Important conserved features of PR–inhibitor interactions have been identified by analysis of crystal structures (Gustchina et al., 1994; Gustchina and Weber, 1991). This information is critical for the design of high-affinity PR inhibitors. Most inhibitors resemble peptides with the scissile peptide bond replaced by a nonhydrolyzable bond, such as the transition-state analogs containing a hydroxyl group instead of the peptide carbonyl oxygen. The PR interactions with peptidelike inhibitors are expected to resemble the interactions with the natural substrates (Fig. 9). Residues 8, 23–32, 45–56, 76, and 80–84 from both subunits form the binding site. About seven residues (P4 to P3′) of peptidelike inhibitors are bound between the catalytic aspartic acids and the two flexible flaps (Gustchina and Weber, 1991). The inhibitor is bound in an extended conformation and forms two short β-sheets with residues 25–29 and flap residues 48–50 from both subunits.

A

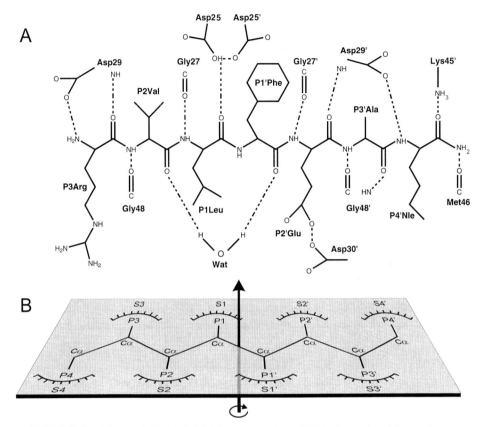

B

FIGURE 9 Scheme of HIV PR–inhibitor interactions. (A) Hydrogen bond interactions between PR residues and the CA↓p2 substrate analog as deduced from the crystal structures (Weber *et al.*, 1997; Wu *et al.*, 1998). Only the PR residues that form hydrogen bond interactions with the peptide are shown. The substrate residues P3–P4′ are shown. Dashed lines indicate hydrogen bonds. Residues in the second subunit of the PR dimer are indicated with a prime. Asp 25 and Asp 25′ are the catalytic aspartates; Asp 25 is protonated in this scheme. The PR hydrogen bond interactions with the substrate main-chain amide and carbonyl oxygen atoms are conserved in different PR–inhibitor crystal structures (Gustchina *et al.*, 1994). The water molecule shown interacting with the carbonyl oxygens of P2 and P1′ is conserved and also interacts with the amides of Ile 50 and Ile 50′. The PR interactions with substrate side chains will depend on the particular substrate sequence. The CA↓p2 peptide shows an interaction between the carboxylates of P2′ Glu and Asp 30′ of the PR. This interaction is expected to involve a proton on either of the acidic groups. (B) Substrate side chains P4–P4′ shown in PR subsites S4–S4′ are indicated by semicircles. This view is approximately perpendicular to the plane of the hydrogen bonds shown in A. The approximate twofold symmetry axis relating the two subunits of the PR dimer is shown.

Analysis of the interactions between HIV PR and peptidic inhibitors showed two major components: (1) Conserved hydrogen bond interactions between the peptidelike NH and C=O groups of the inhibitor and PR residues that are largely independent of the sequence (Fig. 9). The amide and carboxylate oxygen of Asp29, the carbonyl oxygen of Gly27, and the amide and carbonyl of Gly48 form conserved hydrogen bond interactions with the amides and carbonyl oxygens of the substrate or inhibitor in both subunits. Interactions between the Ile50 amide in both flaps and the carbonyl oxygen of P2 and P1' are mediated by a conserved water molecule. The main-chain atoms of the peptidic inhibitors were predicted to make a larger contribution than the side-chain groups to the total binding energy (Gustchina *et al.*, 1994). (2) Each amino acid side chain of the inhibitor (P4 to P3' form a minimum recognition site) fits into successive subsites (S4 to S3') formed by PR residues (Fig. 9). This interaction depends on the nature of the side-chain group at each position of the peptidic inhibitor and is largely hydrophobic. However, polar groups may form specific hydrogen bond or ionic interactions, as observed for P2' Glu in the crystal structure of HIV PR with the CA-p2 analog inhibitor (Weber *et al.*, 1997). The major determinants of PR specificity are located within the subsites, as deduced by mutational and kinetic analysis of HIV-1 and Rous Sarcoma virus proteases (Cameron *et al.*, 1993; Grinde *et al.*, 1992; Cameron *et al.*, 1994). Residues Gly48 and Val82 are critical for substrate selection in the S1 and S1' subsites, while Asp30 and Val32 are important residues in the S2 and S2' subsites.

V. Enzyme Specificity

A. Catalytic Mechanism

The catalytic mechanism of HIV-1 PR (Hyland *et al.*, 1991) is similar to the "push–pull" mechanism of cellular aspartic proteases proposed by Polgár (1987). The two catalytic Asp 25 residues are structurally adjacent in the dimer with 2.5–3.0 Å separating the closest carboxylate oxygen atoms of each residue, suggesting that one Asp 25 is protonated. The two catalytic aspartates mediate a pH-dependent proton transfer from the attacking water molecule to the leaving nitrogen atom of the substrate. The reaction proceeds through a tetrahedral intermediate and involves a physical step (likely the closing down of the flaps on the substrate molecules) as well as chemical steps (the proton transfer). Interestingly, the rate-limiting step is dependent on the amino acid sequence of the substrate (Polgár *et al.*, 1994). Unlike the serine and cysteine proteinases, which form a covalent acyl-enzyme intermediate with the substrate, aspartyl proteinases do not form a covalent intermediate, and therefore they must rely on multiple anchoring of the substrate to the enzyme on both sides of the scissile bond as shown in Fig. 9. Besides the need for correct catalysis, the extended binding site of retroviral proteases may have an evolutionary advantage. These RNA viruses mutate frequently due to the lack of an editing feature of the RT. Since the enzyme recognizes about seven residues, changes in one or in a few binding subsites

Protease **129**

of the enzyme due to mutations may alter, but do not necessarily abolish the enzyme activity or consequently the viral replication.

B. Viral Substrates

The major functional role of the HIV-1 PR is the processing of the viral Gag and Gag-Pol polyproteins in the late phase of replication by catalyzing the hydrolysis of specific peptide bonds in the cleavage sites indicated in Fig. 1. The analysis of naturally occurring cleavage sites, which can be considered as evolutionarily optimized sequences, suggests preferences for certain amino acids at the site of cleavage and in its vicinity (Table I). At P1 and P1′ positions on either side of the cleaved bond hydrophobic amino acids predominate, but Val and Ile are not observed. The occurrence of amino acids at the P1′ position is similar to that found in P1, with the exception that Pro frequently appears in the P1′ position. The P1′ Pro is unique for retroviral PR substrates. No other endopeptidase, except pepsin, is known to hydrolyze cleavage at the imino side of proline. The P2 and P2′ positions are occupied either by hydrophobic or small polar residues, while various types of residues are found at the outer positions. Two major types of cleavage sites were proposed for HIV and related proteases, type 1 having-Tyr(Phe)↓Pro- at P1-P1′ and type 2 having hydrophobic residues (excluding Pro) at P1 and P1′ (Table I; Pettit *et al.*, 1991; Tözsér *et al.*,

TABLE I Protease Cleavage Sites in HIV-1 Gag and Gag-Pol Polyproteins

Location of the cleavage site[a]	P4	P3	P2	P1	P1′	P2′	P3′	Type[b]
In Gag								
MA↓CA	Ser	Gln	Asn	Tyr	Pro	Ile	Val	1
CA↓p2	Ala	Arg	Val	Leu	Ala	Glu	Ala	2
p2↓NC	Ala	Thr	Ile	Met	Met	Gln	Arg	2
NC↓p1	Arg	Gln	Ala	Asn	Phe	Leu	Gly	(2)
p1↓p6	Pro	Gln	Asn	Phe	Leu	Gln	Ser	(2)
in p6	Lys	Glu	Leu	Tyr	Pro	Leu	Thr	1
In Pol								
TFP↓p6^pol	Asp	Leu	Ala	Phe	Leu	Gln	Gly	(2)
p6^pol↓PR	Ser	Phe	Asn	Phe	Pro	Gln	Ile	1
PR↓RT	Thr	Leu	Asn	Phe	Pro	Ile	Ser	1
p66↓p51	Ala	Glu	Thr	Phe	Tyr	Val	Asp	(2)
RT↓IN	Arg	Lys	Ile	Leu	Phe	Leu	Asp	(2)

[a] Notations are according to Schecter and Berger (1967).
[b] Classification of cleavage sites was originally done by Henderson *et al.* (1998) and later modified (Pettit *et al.*, 1991; Tözsér *et al.*, 1992; Griffiths *et al.*, 1992). Oligopeptides corresponding to these cleavage sites were correctly hydrolyzed by the mature PR (Darke *et al.*, 1988; Tözsér *et al.*, 1991; our unpublished results).

1992, Griffiths *et al.*, 1992). These two types of cleavage sites were proposed to have different preferences for the P2 and P2' positions, and these preferences were later confirmed by enzyme kinetics using oligopeptide substrates.

Other viral proteins were later found to be substrates of the HIV-1 PR. Besides its role in the late phase, it was suggested by experiments with equine infectious anemia virus that the viral PR was required in the early phase of viral replication for cleaving the nucleocapsid protein (Roberts *et al.*, 1991). Subsequently, the nucleocapsid of HIV-1 was confirmed to be a substrate of HIV-1 PR (Wondrak *et al.*, 1994). The accessory protein Nef was also found to be a substrate (Freund *et al.*, 1994). However, the significance of any role of the PR in the early stage is not fully understood.

C. Subsite Preference/Substrate Specificity

The extended substrate binding site of HIV-1 consists of six or seven subsites, based on analysis of the crystal structures of enzyme inhibitor complexes (Fig. 9). The extended binding site was also revealed by studies with oligopeptide substrates where a minimal substrate size of six to seven residues was required for optimal catalysis (Darke *et al.*, 1988; Moore *et al.*, 1989; Billich *et al.*, 1988; Tözsér *et al.*, 1991b). The substrate specificity of the HIV-1 PR has been characterized by using either polyproteins or oligopeptides as substrates (for reviews see Dunn *et al.*, 1994; Tomasselli and Heinrickson, 1994). The results are summerized here. The PR cannot hydrolyze oligopeptides with β-branched amino acids substituted in the P1 position (Phylip *et al.*, 1990). Furthermore, introduction of Pro, Ser, and Gly into P1 also prevented hydrolysis (Tözsér *et al.*, 1992; Cameron *et al.*, 1993). Positively charged (Arg) and negatively charged (Glu) amino acids at P1 or P1' positions are also not preferred in substrates (Konvalinka *et al.*, 1990; Cameron *et al.*, 1993). Studies with type 1 oligopeptide substrates indicated a preference at P2 for small residues like Cys or Asn (which is also one of the most frequent amino acid at this position in the polyprotein) and preference for β-branched Val or Ile at P2' position for HIV-1 PR (Margolin *et al.*, 1990, Tözsér *et al.*, 1992). On the other hand, studies of type 2 substrates showed that β-branched residues, especially Val, were favorable at P2, while Glu was preferred at P2' (Phylip *et al.*, 1990; Griffiths *et al.*, 1992). Interestingly, Glu was found to be preferred at P2' in a peptide series in which the P2-P1' sequence of a type 1 cleavage site (-Asn-Tyr $\downarrow$ Pro-) was introduced into a type 2 substrate. Therefore, a general preference for P2' Glu by the HIV-1 PR was suggested (Griffiths *et al.*, 1992). However, P2' Glu was found to be unfavorable in another type 2 series based on a palindromic sequence having two tyrosines at the site of cleavage (Tözsér *et al.*, 1997b). In good agreement with the naturally occurring cleavage site sequences, various residues were acceptable by the S3 and S4' subsites of HIV-1 (Konvalinka *et al.*, 1990, Tözsér *et al.*, 1992; Cameron *et al.*, 1993), while smaller residues were preferred at P4 (Tözsér *et al.*, 1991b). In summary, HIV PR shows a preference for large hydrophobic residues at P1 and P1′, smaller hydrophobic residues at P2, and accommodates a variety of residues at P3 and P3′, while the P2′ preference depends on the peptide sequence.

The different preferences for P2 and P2′ residues as a function of the residues present at P1 and P1′ suggested that the preference for the amino acid at certain positions might strongly depend on the sequence context and conformation of the peptide substrate. The sequence context dependence of the HIV-1 PR was studied for the whole substrate by using doubly and multiply substituted peptides (Ridky *et al.*, 1996; Tözsér *et al.*, 1997b). This structural dependence of the substrate specificity arises from the extended conformation of the bound substrate, as deduced from the crystal structures of PR with peptidelike inhibitors (Fig. 9). The specificity depends on the type of amino acid present in neighboring positions in the sequence (i.e., P2 and P1 or P1 and P1′) as well as amino acid side chains that are adjacent in the extended peptide structure (e.g., the side chain of P2 is next to the side chain of P1′, and P1 is next to P2′). The context dependence may substantially contribute to the high specificity of the retroviral proteases, although this is not apparent from the amino acid sequences of the naturally occurring cleavage sites (Table I). Understanding the strong sequence context dependence of the HIV PR substrates has important implications for both the development of drug resistance and the design of new drugs. Mutation of a PR residue in one subsite will directly influence inhibitor and substrate binding at that subsite and can also indirectly influence the specificity of the other subsites. Conversely, changing substrate residues (or inhibitors) at positions other than those which directly interact with the mutated PR subsite can complement the initial change and restore the efficient substrate processing (or the high potency of the inhibitor).

D. Verification of Knowledge of Specificity

Our understanding of the PR specificity has been verified by engineering Rous sarcoma virus (RSV) PR to recognize the substrates and inhibitors of HIV-1 PR. The wild-type RSV PR and the almost-identical avian myeloblastosis virus (AMV) PR do not hydrolyze most peptides representing the HIV-1 cleavage sites and exhibit low affinity for inhibitors of HIV-1 PR (Ridky *et al.*, 1996; Wu *et al.*, 1998). The crystal structures of RSV and HIV-1 proteases show differences in the substrate binding residues and in the length of the flaps. Differences in substrate selection have been correlated with the differences in 9–10 substrate-binding residues of the two proteases (Cameron *et al.*, 1993; Tözsér *et al.*, 1996). The structurally equivalent residues of HIV-1 PR were substituted into RSV PR. The individual mutations increase the catalytic rate (Thr38→Ser and Ser107→Asn) or alter the substrate specificity (Ile42→Asp, Ile44→Val, Arg105→Pro, Gly106→Val) relative to the wild-type RSV PR (Cameron *et al.*, 1994). The RSV S9 PR designed with nine substitutions of HIV-1 PR residues, Thr38→Ser, Ile42→Asp, Ile44→Val, Met73→Val, Ala100→Leu, Val104→Thr, Arg105→Pro, Gly106→Val, and Ser107→Asn, was shown to hydrolyze all tested substrates of HIV-1 PR and has high affinity for HIV PR inhibitors (Ridky *et al.*, 1996; Wu *et al.*, 1998). The inhibition constants for the CA-p2 analog inhibitor are very similar for RSV S9 and HIV-1 proteases at 20 and 14 n*M*, respectively.

Crystal structures have been determined for both the RSV S9 and HIV-1 proteases with the HIV-1 CA-p2 analog inhibitor (Wu *et al.*, 1998). The RSV S9 interactions with the inhibitor are very similar to those of HIV-1 PR, with the exception of interactions of the flap residues Asn 61, Gln63, and His65 with the distal P3 and P4' positions of the inhibitor. These interactions of the RSV PR flap residues partially substitute for those of Lys45 and Met46 in the flaps of HIV PR (Fig. 9). This comparison suggests that the interactions of P2-P3' of the inhibitor with the PR are most important for affinity since the inhibition constants are very similar for RSV S9 and HIV-1 proteases despite the differences in the flap residues. The engineered RSV S9 PR verifies our knowledge of the critical residues for PR recognition of substrates and inhibitors. The key residues Asp30, Val32, Pro81, and Val82 are mutated in drug-resistant HIV-1 PR, as predicted (Cameron *et al.*, 1993).

E. pH-Dependent Processing of the Gag CA-p2 Cleavage Site

The CA↓p2 cleavage site in the HIV-1 Gag precursor is unique in having a conserved Glu at the P2' position (Table I; Barrie *et al.*, 1996). Cleavage of the CA↓p2 site was shown to be an important regulatory step for the sequential processing of the Gag precursor (Pettit *et al.*, 1994). The CA↓p2 cleavage is also negatively influenced by the p2 domain and is accelerated at low pH. In order to understand the molecular basis for this critical processing step, the crystal structure of HIV-1 PR with a reduced peptide analog of the CA↓p2 site was determined (Weber *et al.*, 1997; Wu *et al.*, 1998). This structure showed a novel proton-mediated interaction between the carboxylates of P2' Glu and Asp30 of the PR (Fig. 9). The proton-mediated interaction was also observed in the crystal structure of a mutant RSV PR with specificity engineered to bind to the same substrate analog (Wu *et al.*, 1998) and for HIV-2 PR crystallized with a different inhibitor containing P2' Glu (Tong *et al.*, 1993). Similar interactions between P2' Glu and Asp30 also occur in the crystal structures of the drug-resistant PR mutants R8Q, K45I, and L90M (see below; Mahalingam *et al.*, 1999); the complex of TFP; the N terminus of the transframe region of Gag-Pol with the HIV-1 PR (Louis *et al.*, 1998); and the product complex with the SIV

PR (Rose *et al.*, 1996). The conservation of the P2′ Glu at the CA ↓ p2 site and its conserved interaction with Asp30 suggest that pH-dependent processing through protonation of P2′ Glu or Asp30 may be a critical regulatory mechanism for PR-mediated protein processing and particle maturation. Consistent with the above observations, the HIV-1 Vpr tethered to the CA-p2 peptide gave nearly complete inhibition of viral replication, unlike the chimera with other peptide substrates (Serio *et al.*, 1997). Also substituting P2′ Glu with Gln of a related chromogenic substrate reduced the catalytic efficiency by ~40-fold at pH 4.0 (Polgar *et al.*, 1994). In accordance with this result, the D30N mutation substantially decreased the replicative capacity of the virus *in vivo* relative to the wild type (Martinez-Picado *et al.*, 1999).

F. Nonviral Substrates

A number of nonviral proteins have been shown to be substrates of HIV-1 PR (for review see Tomasselli and Heinrickson, 1994). One of the first reported nonviral protein substrates of HIV-1 PR was LysPE40, a recombinant derivative of the *Pseudomonas* exotoxin (Tomasselli *et al.*, 1990). However, this protein was hydrolyzed at two unexpected sites and not at the predicted Tyr↓Pro site of the flexible arm of the two domains (Tomasselli *et al.*, 1990). Later, several other nonviral protein substrates of HIV-1 PR were identified, including calmodulin, G-actin, troponin C, prointerleukin 1beta, and lactate dehydrogenase (Tomasselli and Heinrickson, 1994). The intermediate filament proteins vimentin, desmosin, and glial fibrillary acidic protein were found to be cleaved *in vitro* by HIV-1 PR, and microinjection of the enzyme into the cells resulted in the collapse of the vimentin intermediate filament network (Shoeman *et al.*, 1990). The microtubule-associated proteins were also found to be substrates of HIV-1 PR (Wallin *et al.*, 1990). These results indicate that active HIV-1 PR cleavage of cellular proteins may contribute to the pathogenesis associated with the retroviral infection (Kaplan and Swanstrom, 1991). Studies of the nonviral protein substrates provided new cleavage site sequences for databases of HIV PR substrates (Chou *et al.*, 1996). However, many of these sites were not confirmed *in vitro* using peptide substrates. In contrast to the naturally occurring sequences, it is difficult to accommodate the variety of nonviral substrates within any sequence classification and many of these cleavage sites contain charged amino acids, even at P1 or P1′ (Tomasselli and Heinrickson, 1994). A majority of the nonviral substrates have Glu at P2′ (Chou *et al.*, 1996), although only 1 of the 11 HIV-1 polyprotein cleavage sites has P2′ Glu (Table I). Therefore, the HIV-1 PR specificity for nonviral protein substrates appears to differ from its specificity for the viral substrates.

VI. Drug Resistance

Drug resistance is a serious problem for treatment of AIDS infection. Inhibitors of HIV PR are very potent antiviral agents, and several have been approved for treatment of AIDS. The development of antiviral PR inhibitors is a major success of structure-based drug design (Wlodawer and Vondrasek, 1998). The combination therapy using PR inhibitors as well as inhibitors of the reverse transcriptase has had a great impact in extending the lifetime of AIDS patients (Palella *et al.*, 1998). However, the rapid emergence of drug-resistant HIV poses a severe problem for continuous use of PR inhibitors (Korant and Rizzo, 1997). Moreover, multidrug resistance has been observed recently in patients on a combination therapy of RT and PR inhibitors (Shafer *et al.*, 1998). Drug-resistant viruses are selected rapidly due to the high degree of genetic heterogeneity in HIV. The sequence diversity arises due to the error-prone HIV RT (Ji and Loeb, 1992) as well as the high replicative capacity of the virus (Ho *et al.*, 1995; Wei *et al.*, 1995). The RT lacks a $3'$-$5'$-exonuclease proofreading function and has an error rate of ~1 in 10^4. In a fully infected patient as many as 10^9 virus particles are produced daily. Therefore, HIV mutates rapidly.

A. Location of Drug-Resistance Mutations

Resistance to PR inhibitors arises from the selection of mutations in the PR gene (Schinazi *et al.*, 1997). More than 40 mutations have been found in the PR gene of drug-resistant HIV that alter 28 different PR residues (Fig. 2, Table II). Multiple mutations in the PR accumulate over time in response to inhibitor therapy (Molla *et al.*, 1996). In addition, compensating mutations can occur in the PR cleavage sites (Doyon *et al.*, 1996; Zhang *et al.*, 1997). Moreover, the PR gene has extensive natural polymorphisms even in the absence of inhibitors, and these polymorphisms include many amino acid substitutions that contribute to inhibitor resistance (Kozal *et al.*, 1996). Therefore, it is important to understand the molecular basis for drug resistance and to continue the development of new PR inhibitors to overcome the problem of drug resistance.

Mutations are less likely to occur in residues that are essential for the catalytic mechanism, dimer formation, and binding of protein substrates (Table II). About 70% of drug-resistant mutations (Schinazi *et al.*, 1997) and the majority of natural polymorphisms (Kozal *et al.*, 1996) occur in less conserved regions of the PR. The sites of these resistant mutations mapped onto the PR structure are shown in Fig. 2. Many mutations alter residues in the inhibitor binding site and probably act by directly altering the PR affinity for the inhibitor. These include the mutations of V82 arising from exposure to ritonavir or indinavir, Gly48→Val from saquinavir, and Asp30→Asn from nelfinavir. Drug-resistant mutations of these residues were originally predicted on the basis of substrate-specificity studies (Cameron *et al.*, 1993). Other mutations alter residues at the dimer interface, such as Arg8, Ile50, and Leu97; these mutations may exert an effect on dimer formation. Unexpectedly, the majority of resistant mutations alter residues

TABLE II Conserved Regions of HIV-1 Protease and Sites of Drug Resistant Mutations[a]

Note: The table is laid out as three horizontal bands. In each band the "Wild type" sequence is shown with residue numbers, the "Mutants" observed are shown above the relevant residues, and residues involved in substrate binding are marked with an asterisk (*). Region labels (Interface, Active site/interface, Flap/interface, Helix/interface) are indicated below the corresponding residues.

Band 1 (residues 1–33) — regions: Interface; Active site/interface

Pos	Wild type	Mutant(s)	*
1	P		
2	Q		
3	I		
4	T		
5	L		
6	W		
7	Q		
8	R	K, Q	*
9	P		
10	L	I, R, V / F	*
11	V		
12	T		
13	I	V	
14	R		
15	I		
16	G		
17	G		
18	Q		
19	L		
20	K	R, M	*
21	E		*
22	A		
23	L		*
24	L		
25	D		*
26	T		*
27	G		
28	A		*
29	D		*
30	D	N	*
31	T		
32	V	I	*
33	L	F	*

Band 2 (residues 34–66) — regions: Interface; Flap/interface

Pos	Wild type	Mutant(s)	*
34	E		
35	E		
36	M	I	
37	N		
38	L		
39	P		
40	G		
41	K		
42	W		
43	K		
44	P		
45	K		
46	M	F, L, V	*
47	I		*
48	G	V	*
49	G		*
50	I	I, V, V	*
51	G		
52	G		
53	F	M, V	*
54	I	I, V, V	*
55	K		
56	V		
57	R		
58	Q	E	
59	Y		
60	D	E	*
61	Q		
62	I		
63	L	P	
64	V		
65	E		
66	I		

Band 3 (residues 67–99) — regions: Helix/interface; Interface

Pos	Wild type	Mutant(s)	*
67	C		
68	G		
69	H		
70	K		
71	A	T, V	
72	I		
73	G	S	
74	T		
75	V	S, I	*
76	L	I	
77	V		
78	G		
79	P		
80	T		
81	P		
82	V	A, T, F, I, S	*
83	N	D	
84	I	V	*
85	I		
86	G		
87	R		
88	N	S, D	
89	L		
90	L	M, L, V	
91	T		
92	Q		
93	I		
94	G		
95	C		
96	T		
97	L		
98	N		
99	F		

[a] The conserved (−) active site, flap, helix, interface regions of the dimer and residues that are involved in substrate binding (*) are indicated. Protease mutants that are selected upon drug treatment are shown (Schinazi *et al.*, 1997).

that are not part of the inhibitor binding site or the dimer interface. This category includes Leu90→Met and Asn88→Asp, which commonly arise from exposure to saquinavir or nelfinavir, respectively. The molecular basis for the resistance of these distally located mutations is not fully understood.

B. Mutants Exhibit Altered Kinetic Parameters

Resistant PR mutants must act by lowering the PR affinity for the drug while maintaining sufficient catalytic activity for optimal processing of the Gag and Gag-Pol polyproteins leading to the production of infective virions. In order to understand the molecular mechanisms for development of drug resistance, several groups have studied different inhibitor resistant mutants of PR. Most studies have used mutants that were selected by the particular drug of interest and assayed for inhibition and PR activity on a single substrate (Gulnik *et al.*, 1995; Tisdale *et al.*, 1995; Molla *et al.*, 1996). One study has probed the substrate selectivity of mutants Arg8→Lys, Val32→Ile, Val82→Thr, Ile84→Val, Gly48→Val/Leu90→Met, and Val82→Thr/Ile84→Val using a set of peptides based on the HIV-1 CA-p2 cleavage site (Ridky *et al.*, 1998). Recently, a broader range of mutants, Arg8→Gln, Asp30→Asn, Lys45→Ile, Met46→Leu, Gly48→Val, Val82→Ser, Asn88→Asp, and Leu90→Met, were investigated for the hydrolysis of three critical cleavage site peptides and their structural stability (Mahalingam *et al.*, 1999). These studies demonstrate that drug-resistant mutants show alterations in several molecular and enzymatic properties that include lower affinity for inhibitor, altered catalytic rate, substrate specificity, and structural stability.

I. Mutants Exhibit Lower Affinity for Inhibitor

Several studies have shown that when a mutant is selected against a particular inhibitor, the affinity for that inhibitor is reduced (for example, Gulnik *et al.*, 1995; Tisdale *et al.*, 1995; Molla *et al.*, 1996). Cross-resistance has also been observed among different inhibitors. The pattern of mutations and the emergence of cross-resistance are complex and unpredictable (Boden and Markowitz, 1998). A different pattern of mutations is selected by different inhibitors. Exposure to indinavir or ritonavir selects for mutations of Val82 initially, followed by mutations of a number of other residues. In contrast, the single resistant mutations Gly48→Val or Leu90→Met arise from exposure to saquinavir, while Asp30→Asn or Asn88→Asp are commonly selected by nelfinavir (Schinazi *et al.*, 1997).

Many of the individual drug-resistant mutations in the inhibitor binding site have been shown to be critical for substrate specificity. Gly48 and Val82 are important for recognition of the P1 and P1' amino acid side chains and Asp30 and Val32 are critical for recognition of the amino acids at P2 and P2' (Cameron *et al.*, 1994; Lin *et al.*, 1995). These mutations are expected to directly alter the PR affinity for substrates and inhibitors. However, two other common resistant mutations, Leu90→Met and Asp88→Asp, are not part of the substrate or inhibitor binding site, and they must exert an indirect effect on the inhibitor. Crystal structures of PR–inhibitor complexes show few differences from the wild type for the mutants that alter inhibitor binding residues. Direct interactions, mostly changes in van der Waals interactions between the PR and the inhibitor, appear to be responsible for the lowered affinity for the inhibitor (Pazhanisamy et al., 1996; Ala et al., 1997). This analysis again raises the question of how the mutations in regions distant from the binding site cause resistance.

2. Mutants Exhibit Altered Catalytic Activity and Substrate Specificity

Resistant mutants show defects in polyprotein processing and decreased replicative capacity (Zennou et al., 1998; Martinez-Picado et al., 1999). The defects in maturation arise from altered catalytic efficiency for substrate hydrolysis. Many PR mutants show reduced catalytic activity on tested substrates (Gulnik et al., 1995). However, increased catalytic activity on viral substrates has been observed for some mutants (Ridky et al., 1998; Mahalingam et al., 1999). The catalytic activity of the PR mutants Arg8→ Gln, Asp30→Asn, Lys45→Ile, Met46→Leu, Gly48→Val, Val82→Ser, Asn88→Asp, and Leu90→Met was investigated using oligopeptides representing the cleavage sites CA-p2, p6^{pol}-PR, and PR-RT, which are critical for viral maturation (Fig. 10; Mahalingam et al., 1999). These PR mutants include the mutations commonly arising from four inhibitors in clinical use, saquinavir (Gly48→Val and Leu90→Met), indinavir (Val82→Ser and Met46→Leu), ritonavir (Val82→Ser), and nelfinavir (Asp30→Asn and Asn88→Asp) (Boden and Markowitz, 1998). Mutant Arg8→Gln was one of the first inhibitor resistant mutants to be reported (Ho et al., 1994). Lys45→Ile is found in combination with Leu10→Phe and Ile84→Val on exposure to XM323 (Schinazi et al., 1997). The mutants Val82→Ser, Gly48→Val, Asn88→Asp, and Leu90→Met showed reduced catalytic activity compared to the wild-type PR (Fig. 10). Mutant Val82→Ser was the least active, with 2–20% of wild-type PR activity. PR mutants Asn88→Asp, Arg8→Gln, and Leu90→Met exhibited activities ranging from 20 to 40% and Gly48→Val from 50 to 80% of the wild-type activity. In contrast, the D30N mutant showed variable activity on different substrates ranging from 10 to 110% of wild-type activity. Mutants Lys45→Ile and Met46→Leu, usually selected in combination with other mutations, showed activities that are similar to (60–110%) or greater than (110–530%) wild type, respectively.

The substrate preference of PR mutants Arg8→Lys, Val32→Ile, Val82→Thr, Ile84→Val, Gly48→Val/Leu90→Met, and Val82→Thr/ Ile84→Val was studied using peptides with single amino acid substitutions in the CA-p2 cleavage site (Ridky et al., 1998). These inhibitor-resistant

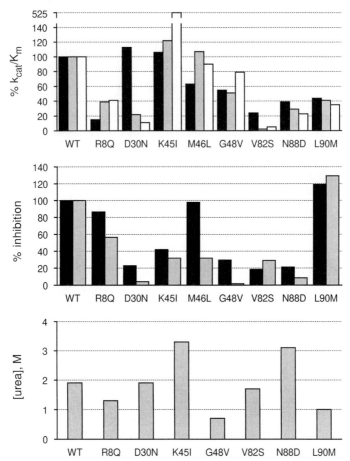

FIGURE 10   Relative inhibition, catalytic efficiency, and structural stability of drug-resistant PR mutants. The PR* and its mutants R8Q, D30N, K45I, G48V, V82S, N88D, and L90M were studied (Mahalingam *et al.*, 1999). The top plot shows the relative k_{cat}/K_m values for hydrolysis of the peptide substrates representing the CA-p2, p6^{pol}-PR, and PR-RT cleavage sites. The middle plot shows the relative inhibition for the CA-p2 and p2-NC reduced substrate analog inhibitors. The bottom plot shows the urea concentration at half maximal activity as a measure of structural stability of the mutants relative to the wild type.

mutations were selected because they alter residues in specific subsites of the PR dimer. Residues Val82 and G48 were predicted to be important for the substrate residue binding in S1 and S1′, and Val32 was predicted to be critical for binding in S2 and S2′ (Cameron *et al.*, 1993). Residue 8 contributes to subsites S1, S1′ S3, and S3′, and I84 forms part of S1 and S1′. Surprisingly, these mutants were similar to the wild-type HIV-1 PR in their substrate specificity. Only the Arg8→Lys and Val32→Ile mutants had significant differences from the wild-type PR. The V32I mutant had significantly enhanced activity on peptides with large hydrophobic residues at P1', while a smaller enhancement was observed for the Arg8→Lys mutant. This increased preference for large hydrophobic P1' residues may explain the observed drug-resistant mutation of P1' Leu to Phe in the Gag p1-p6 cleavage site (Doyon *et al.*, 1996; Zhang *et al.*, 1997). These cleavage site mutations showed improved Gag processing and viral replication in the presence of the drug. Therefore, the altered catalytic rate and altered substrate specificity of the mutant proteases can contribute to drug resistance by enhancing viral replication in the presence of drugs.

C. Mutants Exhibit Altered Structural Stability

Since proper folding and dimer formation are essential for catalytic activity, selected mutants were assayed for their activity as a function of increasing urea concentration. The resistant mutants vary in their structural stability as compared to that of the wild-type PR (Fig. 10; Mahalingam *et al.*, 1999). The mutants Asp30→Asn and Val82→Ser were similar to wild-type PR in their stability toward urea denaturation, while Arg8→Gln, Gly48→Val, and Leu90→Met showed 1.5- to 2.7-fold decreased stability, and Asn88→Asp and Lys45→Ile showed 1.6- to 1.7-fold increased stability. Analysis of the crystal structures of Arg8→Gln, Lys45→Ile, and Leu90→Met mutants complexed with a CA-p2 analog inhibitor showed that the numbers of intersubunit hydrophobic contacts were in good agreement with the relative structural stability of the mutant proteases. The crystal structure of the Arg8→Gln mutant showed changes in the intersubunit interactions as compared to the wild-type PR (Fig. 11). In the wild-type PR, the positively charged side chain of Arg8 forms a strong ionic interaction with the negatively charged side chain of Asp29' from the other subunit in

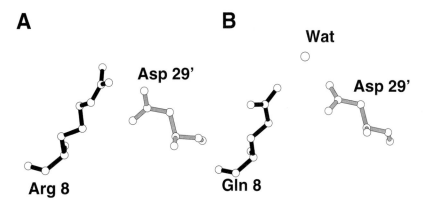

A   **B**

Wat

Asp 29'   Asp 29'

Arg 8   Gln 8

FIGURE 11   The intersubunit interactions of residue 8 in wild-type PR and R8Q mutant. (A) The wild-type ionic interaction of Arg 8 with Asp 29 from the other subunit. (B) The water-mediated hydrogen bond interaction of Gln 8 and Asp 29' in the R8Q mutant.

the dimer. The Arg8→Gln mutant formed a weaker water-mediated hydrogen bond interaction between the uncharged side chain of Gln8 and Asp29', which explained its decreased stability toward urea. Therefore, altered stability of the mutant PR dimers can contribute to drug resistance, since dimer formation is critical for catalytic activity. Both increases and decreases in PR stability were observed relative to that of the wild-type PR. Mutants with decreased stability will show more rapid dissociation of the inhibitor as shown for Gly48→Val and Leu90→Met (Maschera *et al.*, 1996), while mutants with increased stability are likely to show greater catalytic activity even in the presence of inhibitor.

D. Drug Resistance Arise by Multiple Mechanisms

Resistance to PR inhibitors can arise by more than one mechanism. Since no direct relationship was observed between relative catalytic activity, inhibition, and structural stability of the different PR mutants (Fig. 10; Mahalingam *et al.*, 1999), drug resistance can arise from independent changes in any one of these parameters. Prolonged exposure to the drug can result in compensating mutations that act in combination to permit optimal polyprotein processing and replicative capacity in the presence of the PR inhibitor. Selected mutations occurring in PR cleavage sites were shown to also compensate for the reduced activity of the initial mutation in the PR domain, thus giving a growth advantage over the primary PR mutation (Doyon *et al.*, 1996; Zhang *et al.*, 1997). An initial mutation that lowers the affinity for the inhibitor and also lowers the catalytic activity or dimer stability can be combined with additional PR mutations that increase the catalytic activity or dimer stability to confer improved viral replication in the presence of inhibitor. Alternatively, the initial PR mutation can be combined with a mutation in the cleavage sites that restores sufficient catalytic activity for optimal virus maturation.

Acknowledgments

This research was sponsored in part by the Intramural AIDS Targeted Antiviral Program of the Office of the Director of the National Institutes of Health, by the United States Public Health Service Grant AI41380, by the Hungarian Science and Research Fund (OTKA T30092, OTKA T22140), by the Hungarian Ministry of Culture and Education (FKFP 1318/97), and by the Fogarty International Research Collaboration Award (TW01001).

References

al-Janabi, J., Hartsuck, J. A., and Tang, J. (1972). Kinetics and mechanism of pepsinogen activation. *J. Biol. Chem.* **247**, 4628–4632.

Ala, P. J., Huston, E. E., Klabe, R. M., McCabe, D. D., Duke, J. L., Rizzo, C. J., Korant, B. D., DeLoskey, R. J., Lam, P. Y., Hodge, C. N., and Chang, C. H. (1997). Molecular basis of HIV-1 protease drug resistance: Structural analysis of mutant proteases complexed with cyclic urea inhibitors. *Biochemistry* **36**, 1573–1580.

Barrie, K. A., Perez, E. E., Lamers, S. L. *et al.* (1996). Natural variation in HIV-1 protease, Gag p7 and p6, and protease cleavage sites within gag/pol polyproteins: Amino acid substitutions in the absence of protease inhibitors in mothers and children infected by human immunodeficiency virus type 1. *Virology* **219**, 407–416.

Beissinger, M., Paulus, C., Bayer, P., Wolf, H., Rosch, P., and Wagner, R. (1996). Sequence-specific resonance assignments of the 1H-NMR spectra and structural characterization in solution of the HIV-1 transframe protein p6. *Eur. J. Biochem.* **237**, 383–392.

Billich, S., Knoop, M.-T., Hansen, J., Strop, P., Sedlacek, J., Mertz, R., and Moelling, K. (1988). Synthetic peptides as substrates and inhibitors of human immunodeficiency virus-1 protease. *J. Biol. Chem.* **263**, 17905–17908.

Boden, D., and Markowitz, M. (1998). Resistance to human immunodeficiency virus type 1 protease inhibitors. *Antimicrob. Agents Chemother.* **42**, 2775–2783.

Cameron, C. E., Grinde, B., Jacques, P. *et al.* (1993). Comparison of the substrate-binding pockets of the Rous sarcoma virus and human immunodeficiency virus type 1 proteases. *J. Biol. Chem.* **268**, 11711–11720.

Cameron, C. E., Ridky, T. W., Shulenin, S. *et al.* (1994). Mutational analysis of the substrate binding pockets of the Rous sarcoma virus and human immunodeficiency virus-1 proteases. *J. Biol. Chem.* **269**, 11170–11177.

Candotti, D., Chappey, C., Rosenheim, M., M'Pele, P., Huraux, J. M., and Agut, H. (1994). High variability of the gag/pol transframe region among HIV-1 isolates. *C. R. Acad. Sci. III* **317**, 183–189.

Co, E., Koelsch, G., Lin, Y., Ido, E., Hartsuck, J. A., and Tang, J. (1994). Proteolytic processing mechanisms of a miniprecursor of the aspartic protease of human immunodeficiency virus type 1. *Biochemistry* **33**, 1248–1254.

Cherry, E., Liang, C., Rong, L. *et al.* (1998). Characterization of human immunodeficiency virus type-1 (HIV-1) particles that express protease-reverse transcriptase fusion proteins. *J. Mol. Biol.* **284**, 43–56.

Chou, K-C., Tomaselli, A. G., Reardon, I. M., and Heinrikson, R. L. (1996). Predicting human immunodeficiency virus protease cleavage sites in proteins by a discriminant function method. *Proteins* **24**, 51–72.

Copeland, T. D., and Oroszlan, S. (1988). Genetic locus, primary structure, and chemical synthesis of human immunodeficiency virus protease. *Gene Anal. Tech.* **5**, 109–115.

Darke, P. L., Nutt, R. F., Brady, S. F., Garsky, V. M., Ciccarone, T. M., Leu, C.-T., Lumma, P. K., Freidinger, R. M., Veber, D. F., and Sigal, I. S. (1988). HIV-1 protease specificity of peptide cleavage is sufficient for processing of Gag and Pol polyproteins. *Biochem. Biophys. Res. Commun.* **156**, 297–303.

Debouck, C. (1992). The HIV-1 protease as a therapeutic target for AIDS. *AIDS Res. Hum. Retroviruses* **8**, 153–164.

Doyon, L., Croteau, G., Thibeault, D., Poulin, F., Pilote, L., and Lamarre, D. (1996). Second locus involved in human immunodeficiency virus type 1 resistance to protease inhibitors. *J. Virol.* **70**, 3763–3769.

Dunn, B. M., Deyrup, C., Moesching, W. G., Gilbert, W. A., Nolan, R. J., and Trach, M. L. (1978). Inhibition of pepsin by zymogen activation fragments: Spectrum of peptides released from pepsinogen NH2 terminus and solid phase synthesis of two inhibitory peptide sequences. *J. Biol. Chem.* **253**, 7269–7275.

Dunn, B. M., Gustchina, A., Wlodawer, A., and Kay, J. (1994). Subsite preferences of retroviral proteinases. *Methods Enzymol.* **241**, 254–278.

Freund, J., Kellner, R., Konvalinka, J., Wolber, V., Krausslich, H.-G., and Kalbitzer, H. R. (1994). A possible regulation of negative factor (Nef) activity of human immunodeficiency virus type 1 by the viral protease. *Eur. J. Biochem.* **223**, 589–593.

Gorelick, R. J., and Henderson, L. E. (1994). "Human Retroviruses and AIDS: Analyses" (G. Myers, B. Korber, S. Wain-Hobson, K.-T. Jeang, L. Henderson, and G. Pavlakis, Eds.), Part III, pp. 2–5. The Los Alamos National Laboratory, Los Alamos, NM. (http://hiv-web.lanl.gov)

Griffiths, J. T., Phylip, L. H., Konvalinka, J., Strop, P., Gustchina, A., Wlodawer, A., Davenport, R. J., Briggs, R., Dunn, B. M., and Kay, J. (1992). Different requirements for productive interaction between the active site of HIV-1 proteinase and substrates containing hydrophobic*hydrophobic- or- aromatic*Pro-cleavage sites. *Biochemistry* 31, 5193–5200.

Grinde, B., Cameron, C. E., Leis, J., Weber, I. T., Wlodawer, A., Burstein, H., and Skalka, A. M. (1992). Analysis of substrate interactions of the Rous sarcoma virus wild type and mutant proteases and human immunodeficiency virus-1 protease using a set of systematically altered peptide substrates. *J. Biol. Chem.* 267, 9491–9498.

Gulnik, S. V., Suvorov, L. I., Liu, B., Yu, B., Anderson, B., Mitsuya, H., and Erickson, J. W. (1995). Kinetic characterization and cross-resistance patterns of HIV-1 protease mutants selected under drug pressure. *Biochemistry* 34, 9282–9287.

Gustchina, A., and Weber, I. T. (1991). Comparative analysis of the sequences and structures of HIV-1 and HIV-2 proteases. *Proteins* 10, 325–339.

Gustchina, A., Sansom, C., Prevost, M. *et al.* (1994). Energy calculations and analysis of HIV-1 protease-inhibitor crystal structures. *Protein Eng.* 7, 309–317.

Hatfield, D. L., Levin, J. G., Rein, A., and Oroszlan, S. (1992). Translational suppression in retroviral gene expression. *Adv. Virus Res.* 41, 193–239.

Henderson, L. E., Benveniste, R. E., Sowder, R., Copeland, T. D., Schultz, A. M., and Oroszlan, S. (1988). Molecular characterization of *gag* proteins from simian immunodeficiency virus (SIV$_{Mne}$). *J. Virol.* 62, 2587–2595.

Henderson, L. E., Bowers, M. A., Sower, R. C., Serabyn, S. A., Johnson, D. G., Bess, J. W., Jr., Arthur, L. O., Bryant, D. K., and Fenselau, C. (1992). Gag proteins of the highly replicative MN strain of human immunodeficiency virus type 1: Posttranslational modifications, proteolytic processings, and complete amino acid sequences. *J. Virol.* 66, 1856–1865.

Ho, D. D., Neumann, A. U., Perelson, A. S., Chen, W., Leonard, J. M., and Markowitz, M. (1995). Rapid turnover of plasma virions and CD4 lymphocytes in HIV-1 infections. *Nature* 373, 123–126.

Ho, D. D., Toyoshima, T., Mo, H., Kempf, D. J., Norbeck, D., Chen, C. M., Wideburg, N. E., Burt, S. K., Erickson, J. W., and Singh, M. K. (1994). Characterization of human immunodeficiency virus type 1 variants with increased resistance to a C2-symmetric protease inhibitor. *J. Virol.* 68, 2016–2020.

Hyland, L. J., Tomaszek, T. A., and Meek, T. D. (1991). Human immunodeficiency virus-1 protease. 2. Use of pH rate studies and solvent isotope effects to elucidate details of chemical mechanism. *Biochemistry* 30, 8454–8463.

Ido, E., Han, H. P., Kezdy, F. J., and Tang, J. (1991). Kinetic studies of human immunodeficiency virus type 1 protease and its active-site hydrogen bond mutant A28S. *J. Biol. Chem.* 266, 24359–24366.

Jacks, T., Power, M. D., Masiarz, F. R., and Varmus, H. E. (1988). Characterization of ribosomal frameshifting in HIV-1 *gag-pol* expression. *Nature* 331, 280–283.

Ji, J. P., and Loeb, L. A. (1992). Fidelity of HIV-1 reverse transcriptase copying RNA *in vitro*. *Biochem.* 31, 954–958.

Kaplan, A. H., and Swanstrom, R. (1991). Human immunodeficiency virus type 1 Gag proteins are processed in two cellular compartments. *Proc. Natl. Acad. Sci. USA* 88, 4528–4532.

Kaplan, A. H., Zack, J. A., Knigge, M. *et al.* (1993). Partial inhibition of the human immunodeficiency virus type 1 protease results in aberrant virus assembly and the formation of noninfectious particles. *J. Virol.* 67, 4050–4055.

Karacostas, V., Wolffe, E. J., Nagashima, K., Gonda, M. A., and Moss, B. (1993). Overexpression of the HIV-1 gag-pol polyprotein results in intracellular activation of HIV-1 protease and inhibition of assembly and budding of virus-like particles. *Virology* 193, 661–671.

Khan, A. R., and James, M. N. (1998). Molecular mechanisms for the conversion of zymogens to active proteolytic enzymes. *Prot. Sci.* 7, 815–836.

Konvalinka, J., Strop, P., Velek, J., Cerna, V., Kostka, V., Phylip, L. H., Richards, A. D., Dunn, B. M., and Kay, J. (1990). Sub-site preferences of the aspartic proteinase from the human immunodeficiency virus, HIV-1. *FEBS Lett.* 268, 35–38.

Korant, B. D., and Rizzo, C. J. (1997). The HIV protease and therapies for AIDS. *Adv. Exp. Med. Biol.* 421, 279–284.

Kozal, M. J., Shah, N., Shen, N., Yang, R., Fucini, R., Merigan, T. C., Richman, D. D., Morris, D., Hubbell, E., Chee, M., and Gingeras, T. R. (1996). Extensive polymorphisms observed in HIV-1 clade B protease gene using high-density oligonucleotide arrays. *Nature Med.* 2, 753–759.

Krausslich, H. G. (1991). Human immunodeficiency virus proteinase dimer as component of the viral polyprotein prevents particle assembly and viral infectivity. *Proc. Natl. Acad. Sci. USA* 88, 3213–3217.

Lam, P. Y., Jadhav, P. K., Eyermann, C. J. *et al.* (1994). Rational design of potent, bioavailable, nonpeptide cyclic ureas as HIV protease inhibitors. *Science* 263, 380–384.

Leis, J., Baltimore, D., Bishop, J. M. *et al.* (1988). Standardized and simplified nomenclature for proteins common to all retroviruses. *J. Virol.* 62, 1808–1809.

Lin, Y., Lin, X., Hong, L., Foundling, S., Heinrikson, R. L., Thaisrivongs, S., Leelamanit, W., Raterman, D., Shah, M, Dunn, B. M., and Tang, J. (1995). Effect of point mutations on the kinetics and the inhibition of human immunodeficiency virus type 1 protease: Relationship to drug resistance. *Biochemistry* 34, 1143–1152.

Louis, J. M., Dyda, F., Nashed, N. T., Kimmel, A. R., and Davies, D. R. (1998). Hydrophilic peptides derived from the transframe region of Gag-Pol inhibit the HIV-1 protease. *Biochemistry* 37, 2105–2110.

Louis, J. M., McDonald, R. A., Nashed, N. T., *et al.* (1991a). Autoprocessing of the HIV-1 protease using purified wild-type and mutated fusion proteins expressed at high levels in *Escherichia coli. Eur. J. Biochem.* 199, 361–369.

Louis, J. M., Oroszlan, S., and Mora, P. T. (1991b). Studies of the autoprocessing of the HIV-1 protease using cleavage site mutants. *Adv. Exp. Med. Biol.* 306, 499–502.

Louis, J. M., Smith, C. A., Wondrak, E. M., Mora, P. T., and Oroszlan, S. (1989). Substitution mutations of the highly conserved arginine 87 of HIV-1 protease result in loss of proteolytic activity. *Biochem. Biophys. Res. Commun.* 164, 30–38.

Louis, J. M., Nashed, N. T., Parris, K. D., Kimmel, A. R., and Jerina, D. M. (1994). Kinetics and mechanism of autoprocessing of human immunodeficiency virus type 1 protease from an analog of the Gag-Pol polyprotein. *Proc. Natl. Acad. Sci. USA* 91, 7970–7974.

Louis, J. M., Wondrak, E. M., Kimmel, A. R., Wingfield, P. W., and Nashed, N. T. (1999a). Proteolytic processing of the HIV-1 protease precursor: Kinetics and mechanism. *J. Biol. Chem.* 274, 23437–23442.

Louis, J. M., Clore, G. M., and Gronenborn. (1999b). HIV-1 protease regulation: Autoprocessing is tightly coupled to protein folding. *Nat. Struct. Biol.* 6, 868–875.

Mahalingam, B., Louis, J. M., Reed, C. C., Adomat, J. M., Krouse, J., Wang, Y-F., Harrison, R. W., and Weber, I. T. (1999). Structural and kinetic analysis of drug resistant variants of HIV-1 protease. *Eur. J. Biochem.* 263, 238–245.

Maschera, B., Darby, G., Palu, G., Wright, L. L., Tisdale, M., Myers, R., Blair, E. D., and Furfune, E. S. (1996). Human immunodeficiency virus: Mutations in the viral protease that confer resistance to saquinavir increase the dissociation rate constant of the protease–saquinavir complex. *J. Biol. Chem.* 271, 33231–33235.

Margolin, N., Heath, W., Osborne, E., Lai, M., and Vlahos, C. (1990). Substitutions at the P_2' site of *gag* P17-P24 affect cleavage efficiency by HIV-1 protease. *Biochem Biophys Res Commun.* 167, 554–560.

Martinez-Picado, J., Savara, A. V., Sutton, L., and D'Aquila, R. T. (1999). Replicative fitness of protease inhibitor-resistant mutants of human immunodeficiency virus type 1. *J. Virol.* 73, 3744–3752.

Mildner, A. M., Rothrock, D. J., Leone, J. W. *et al.* (1994). The HIV-1 protease as enzyme and substrate: Mutagenesis of autolysis sites and generation of a stable mutant with retained kinetic properties. *Biochemistry* 33, 9405–9413.

Molla, A., Korneyeva, M., Gao, Q., Vasavanonda, S., Schipper, P. J., Mo, H.-M., Markowitz, M., Chernyavskiy, T., Niu, P., Lyons, N., Hsu, A., Granneman, R., Ho, D. D., Boucherm C. A. B., Leonard, J. M., Norbeck, D. W., and Kempf, D. J. (1996). Ordered accumulation of mutations in HIV protease confers resistance to ritonavir. *Nature Med.* 2, 760–766.

Moore, M. L., Bryan, W. M., Fakhoury, S. A., Magaard, V. W., Huffman, W. F., Dayton, B. D., Meek, T. D., Hyland, L., Dreyer, G. B., Metcalf, B. W., Strickler, J. E., Gorniak, J. G., and Debouck, C. (1989). Peptide substrates and inhibitors of the HIV-1 protease. *Biochem. Biophys. Res. Commun.* 159, 420–425.

Oroszlan, S., and Luftig, R. B. (1990). Retroviral proteases. *Curr. Topics Microbiol. Immunol.* 157, 153–185.

Palellar, F. J., Delaney, K. M., Moorman, A. C., Loveless, M. O., Fuhrer, J., Satten, G. A., Aschman, D. J., and Holmberg, S. D. (1998). Declining morbidity and mortality among patients with advanced human immunodeficiency virus infection. *N. Engl. J. Med.* 338, 853–860.

Partin, K., Zybarth, G., Ehrlich, L., DeCrombrugghe, M., Wimmer, E., and Carter, C. (1991). Deletion of sequences upstream of the proteinase improves the proteolytic processing of human immunodeficiency virus type 1. *Proc. Natl. Acad. Sci. USA* 88, 4776–4780.

Pazhanisamy, S., Stuver, C. M., Cullinan, A. B., Margolin, N., Rao, B. G., and Livingston, D. J. (1996). Kinetic characterization of human immunodeficiency virus type-1 protease-resistant variants. *J. Biol. Chem.* 271, 17979–17985.

Pearl, L. H., and Taylor, W. R. (1987). A structural model for the retroviral proteases. *Nature* 329, 351–34.

Pettit, S. C., Moody, M. D., Wehbie, R. S. *et al.* (1994). The p2 domain of human immunodeficiency virus type 1 Gag regulates sequential proteolytic processing and is required to produce fully infectious virions. *J. Virol.* 68, 8017–8027.

Pettit, S. C., Simsic, J., Loeb, D. D., Everitt, L., Hutchinson, C. A., III, and Swanstrom, R. (1991). Analysis of retroviral protease cleavage sites reveals two types of cleavage sites and the structural requirements of the P1 amino acid. *J. Biol. Chem* 266, 14539–14547.

Phylip, L. H., Richards, A. D., Kay, J., Konvalinka, J., Strop, P., Blaha, I., Velek, J., Kostka, V., Ritchie, A. J., Broadhurst, A. V., Farmerie, W. G., Scarborough, P. E., and Dunn, B. M. (1990). Hydrolysis of synthetic chromogenic substrates by HIV-1 and HIV-2 proteinases. *Biochem. Biophys. Res. Commun.* 171, 439–444.

Phylip, L. H., Mills, J. S., Parten, B. F., Dunn, B. M., and Kay, J. (1992). Intrinsic activity of precursor forms of HIV-1 proteinase. *FEBS Lett.* 314, 449–454.

Polgár, L. (1987). The mechanism of action of aspartic proteinases involves "push–pull" catalysis. *FEBS Lett.* 219, 1–4.

Polgár, L., Szeltner, Z., and Boros, I. (1994). Substrate-dependent mechanisms of the catalysis of human immunodeficiency virus protease. *Biochemistry* 33, 9351–9357.

Richards, A. D., Kay, J., Dunn, B. M., Bessant, C. M., and Charlton, P. A. (1992). Inhibition of aspartic proteinases by synthetic peptides derived from the propart region of human prorenin. *Int. J. Biochem.* 24, 297–301.

Ridky, T. W., Cameron, C. E., Cameron, J., Leis, J., Copeland, T. D., Wlodawer, A., Weber, I. T., and Harrison, R. W. (1996). Human immunodeficiency virus, type 1 protease substrate specificity is limited by interactions between substrate amino acids bound in adjacent enzyme subsites. *J. Biol. Chem.* 271, 4709–4717.

Ridky, T. W., Kikonyogo, A., Leis, J. *et al.* (1998). Drug-resistant HIV-1 proteases identify enzyme residues important for substrate selection and catalytic rate. *Biochemistry* 37, 13835–13845.

Roberts, M. M., Copeland, T. D., and Oroszlan, S. (1991). *In situ* processing of a retroviral nucleocapsid protein by the viral proteinase. *Prot. Eng.* 4, 695–700.

Rose, J. R., Babe, L. M., and Craik, C. S. (1995). Defining the level of human immunodeficiency virus type 1 (HIV-1) protease activity required for HIV-1 particle maturation and infectivity. *J. Virol.* **69**, 2751–2758.

Rose, R. B., Craik, C. S., Douglas, N. L., and Stroud, R. M. (1996). Three-dimensional structures of HIV-1 and SIV protease product complexes. *Biochemistry* **35**, 12933–12944.

Rose, J. R., Salto, R., and Craik, C. S. (1993). Regulation of autoproteolysis of the HIV-1 and HIV-2 proteases with engineered amino acid substitutions. *J. Biol. Chem.* **268**, 11939–11945.

Schechter, I., and Berger, A. (1967). On the size of the active site in proteases. I. Papain. *Biochem. Biophys. Res. Commun.* **27**, 157–172.

Schinazi, R. F., Larder, B. A., and Mellors, J. W. (1997). Mutations in retroviral genes associated with drug resistance. *Int. Antivir. News* **5**, 129–142.

Serio, D., Rizvi, T. A., Cartas, M. *et al.* (1997). Development of a novel anti-HIV-1 agent from within: Effect of chimeric Vpr-containing protease cleavage site residues on virus replication. *Proc. Natl. Acad. Sci. USA* **94**, 3346–3351.

Shafer, R. W., Winters, M. A., Palmer, S., and Merigan, T. C. (1998). Multiple concurrent reverse transcriptase and protease mutations and multidrug resistance of HIV-1 isolates from heavily treated patients. *Ann. Intern. Med.* **128**, 906–911.

Shoeman, R. L., Höner, B., Stoller, T. J., Kesselmeier, C., Miedel, M. C., Traub, P., and Graves, M. C. (1990). Human immunodeficiency virus type 1 protease cleaves the intermediate filament proteins vimentin, desmin, and glial fibrillary acidic protein. *Proc. Natl. Acad. Sci. USA* **87**, 6336–6340.

Tessmer, U., and Krausslich, H. G. (1998). Cleavage of human immunodeficiency virus type 1 proteinase from the N- terminally adjacent p6* protein is essential for efficient Gag polyprotein processing and viral infectivity. *J. Virol.* **72**, 3459–3463.

Tisdale, M., Myers, R. E., Maschera, B., Parry, N. R., Oliver, N. M., and Blair, E. D. (1995). Cross-resistance analysis of human immunodeficiency virus type 1 variants individually selected for resistance to five different protease inhibitors. *Antimicrob. Agents Chemother.* **39**, 1704–1710.

Tomasselli, A. G., and Heinrickson, R. L. (1994). Specificity of retroviral proteases: An analysis of viral and nonviral protein substrates. *Methods Enzymol.* **241**, 279–301.

Tomasselli, A. G., Hui, J. O., Sawyer, T. K., Staples, D. J., FitzGerald, D. J., Chaudhary, V. K., Pastan, I., and Heinrikson, R. L. (1990). Interdomain hydrolysis of a truncated *Pseudomonas* exotoxin by the human immunodeficiency virus-1 protease. *J. Biol. Chem.* **265**, 408–413.

Tong, L., Pav, S., Pargellis, C., Do, F., Lamarre, D., and Anderson, P. C. (1993). Crystal structure of human immunodeficiency virus (HIV) type 2 protease in complex with a reduced amide inhibitor and comparison with HIV-1 protease structures. *Proc. Natl. Acad. Sci. USA* **90**, 8387–8391.

Tözsér, J., Bagossi, P., Weber, I. T., Copeland, T. D., and Oroszlan, S. (1996). Comparative studies on the substrate specificity of avian myeloblastosis virus proteinase and lentiviral proteinases. *J. Biol. Chem.* **271**, 6781–6788.

Tözsér, J., Bláha, I., Copeland, T. D., Wondrak, E. M., and Oroszlan, S. (1991a). Comparison of the HIV-1 and HIV-2 proteinases using oligopeptide substrates representing cleavage sites in Gag and Gag-Pol polyproteins. *FEBS Lett.* **281**, 77–80.

Tözsér, J., Gustchina, A., Weber, I. T., Bláha, I., Wondrak, E. M., and Oroszlan, S. (1991b). Studies on the role of the S_4 substrate binding site of HIV proteinases. *FEBS Lett.* **279**, 356–360.

Tözsér, J., Weber, I. T., Gustchina, A., Bláha, I., Copeland, T. D., Louis, J. M., and Oroszlan, S. (1992). Kinetic and modeling studies of S_3-S_3' subsites of HIV proteinases. *Biochemistry* **31**, 4739–4800.

Tözsér, J., Yin, F. H., Cheng, Y. S. *et al.* (1997a). Activity of tethered human immunodeficiency virus 1 protease containing mutations in the flap region of one subunit. *Eur. J. Biochem.* **244**, 235–241.

Tözsér, J., Bagossi, P. Weber, I. T., Louis, J. M., Copeland, T. D., and Oroszlan, S. (1997b). Studies on the symmetry and sequence context dependence of the HIV-1 proteinase specificity. *J. Biol. Chem.* **272**, 16807–16814.

Vogt, V. M. (1996). Proteolytic processing and particle maturation. *Curr. Top. Microbiol. Immunol.* **214**, 95–131.

Wallin, M., Deinum, J., Goobar, L., and Danielson, U. H. (1990). Proteolytic cleavage of microtubule-associated proteins by retroviral proteinases. *J. Gen. Virol.* **71**, 1985–1991.

Weber, I. T. (1990). Comparison of the crystal structures and intersubunit interactions of human immunodeficiency and Rous sarcoma virus proteases. *J. Biol. Chem.* **265**, 10492–10496.

Weber, I. T., Wu, J., Adomat, J. *et al.* (1997). Crystallographic analysis of human immunodeficiency virus 1 protease with an analog of the conserved CA-p2 substrate—Interactions with frequently occurring glutamic acid residue at P2' position of substrates. *Eur. J. Biochem.* **249**, 523–300.

Wei, X., Goshm S. K., Taylor, M. E., Johnson, V. A., Emini, E. A., Deutsch, P., Lifson, J. D., Bonhoeffer, S., Nowak, M. A., Hahn, B. H. *et al.* (1995). Viral dynamics in human immunodeficiency virus type 1 infection. *Nature* **373**, 117–122.

Wishart, D. S., Bigam, C. G., Holm, A., Hodges, R. S., and Sykes, B. D. (1995). ^{1}H, ^{13}C and ^{15}N random coil NMR chemical shifts of the common amino acids. I. Investigations of nearest-neighbor effects. *J. Biomol. NMR* **5**, 67–81.

Wlodawer, A., and Erickson, J. W. (1993). Structure-based inhibitors of HIV-1 protease. *Ann. Rev. Biochem.* **62**, 543–585.

Wlodawer, A., and Vondrasek, J. (1998). Inhibitors of HIV-1 protease: A major succes of structure-assisted drug design. *Annu. Rev. Biophys. Biomol. Struct.* **27**, 249–284.

Wondrak, E. M., and Louis, J. M. (1996). Influence of flanking sequences on the dimer stability of human immunodeficiency virus type 1 protease. *Biochemistry* **35**, 12957–12962.

Wondrak, E. M., Louis, J. M., de Rocquigny, H., Chermann, J. C., and Roques, B. P. (1993). The gag precursor contains a specific HIV-1 protease cleavage site between the NC (P7) and P1 proteins. *FEBS Lett.* **333**, 21–24.

Wondrak, E. M., Nashed, N. T., Haber, M. T., Jerina, D. M., and Louis, J. M. (1996). A transient precursor of the HIV-1 protease: Isolation, characterization, and kinetics of maturation. *J. Biol. Chem.* **271**, 4477–4481.

Wondrak. E. M., Sakaguchi, K., Rice, W. G., Kun, E., Kimmel, A. R., and Louis, J. M. (1994). Removal of zinc is required for processing of the mature nucleocapsid protein of human immunodeficiency virus, type 1, by the viral protease. *J. Biol. Chem.* **269**, 21948–21950.

Wu, J., Adomat, J. M., Ridky, T. W. *et al.* (1998). Structural basis for specificity of retroviral proteases. *Biochemistry* **37**, 4518–4526.

Yamazaki, T., Nicholson, L. K., Torchia, D. A. *et al.* (1994). Secondary structure and signal assignments of human- immunodeficiency-virus-1 protease complexed to a novel, structure-based inhibitor. *Eur. J. Biochem.* **219**, 707–712.

Zennou, V., Mammano, F., Paulous, S., Mathez, D., and Claval, F. (1998). Loss of viral fitness associated with multiple Gag and Gag-Pol processing defects in human immunodeficiency virus type 1 variants selected for resistance to protease inhibitors *in vivo. J. Virol.* **72**, 3300–3306.

Zhang, Y. M., Imamichi, H., Imamichi, T., Lane, H. C., Fallon, J., Vasudevachari, M. B., and Salzman, N. P. (1997). Drug resistance during indinavir therapy is caused by mutations in the protease gene and in its Gag substrate cleavage sites. *J. Virol.* **71**, 6662–6670.

Zybarth, G., and Carter, C. (1995). Domains upstream of the protease (PR) in human immunodeficiency virus type 1 Gag-Pol influence PR autoprocessing. *J. Virol.* **69**, 3878–3884.

Zybarth, G., Krausslich, H. G., Partin, K., and Carter, C. (1994). Proteolytic activity of novel human immunodeficiency virus type 1 proteinase proteins from a precursor with a blocking mutation at the N terminus of the PR domain. *J. Virol.* **68**, 240–250.

Nouri Neamati
Christophe Marchand
Yves Pommier

Laboratory of Molecular Pharmacology
Division of Basic Sciences
National Cancer Institute
Bethesda, Maryland 20892

HIV-1 Integrase Inhibitors: Past, Present, and Future

I. Introduction

The recent success of the highly active antiretroviral therapy (HAART) using combinations of inhibitors of protease and reverse transcriptase have altered the natural course of AIDS by efficiently suppressing viral load for long periods of time. However, issues of patient compliance, drug toxicity, the emergence of multidrug-resistant phenotypes, and the presence of persistent reservoirs of virus replication have highlighted the need to develop alternative therapeutic approaches utilizing other targets in the viral replication cycle. One such replication target is viral integration. Integration of retroviral DNA into host chromosomes is essential for viral replication. For HIV-1, this process is mediated by integrase (IN), a 32-kDa virally encoded protein, and conserved sequences in the HIV long terminal repeats (LTR) (Brown, 1998; Katz and Skalka, 1994; Rice et al., 1996).

Advances in Pharmacology, Volume 49
147

HIV-1 IN contains 288 amino acids and belongs to a superfamily of polynucleotide transferases that include Mu transposase, RNase H, and RuvC (Rice et al., 1996). The functional domains of IN are depicted in Fig. 1. The N terminus contains the highly conserved HHCC motif that binds to a molecule of Zn^{2+} and it is involved in dimerization or multimerization. The catalytic core contains the characteristic three acidic residues known as the DDE motif that is involved in all the catalytic reactions carried out by IN. These residues bind divalent metals and are highly conserved in the polynucleotidyl transferases superfamily. The C terminus is highly charged and is less conserved. It contains nine lysines and seven arginines that are involved in nonspecific DNA binding.

Following reverse-transcription of viral RNA in the cytoplasm of infected cells, IN cleaves two nucleotides from each of the viral DNA ends immediately 3′ to the highly conserved CA dinucleotides. After subsequent migration to the nucleus as a part of a large nucleoprotein complex, IN catalyzes the insertion of the resulting recessed 3′ termini (CA-OH-3′) into a host chromosome by a direct transesterification reaction. These two events are termed 3′ processing and 3′ end-joining (also referred to as integration or strand transfer), respectively (Fig. 2). Structural and biochemical understanding of the enzyme has led to the development of in vitro assays aimed at identifying IN inhibitors. Such assays utilize synthetic oligonucleotides corresponding to the U5 end of the viral LTR, recombinant IN, and divalent metal cofactor (Mn^{2+} or Mg^{2+}). Various assays describing IN activity and DNA binding were reported in detail in two recent reviews (Chow 1997; Mazumder et al., 1999) and are not discussed further in this chapter.

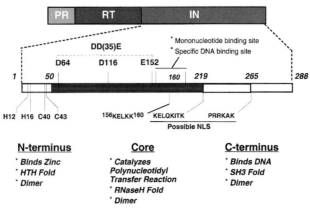

FIGURE I   Functional domains of HIV-1 integrase.

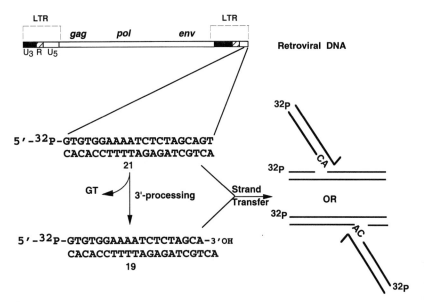

FIGURE 2   Catalytic activities of HIV-integrase. Integrase contains one active site; however, it carries out two different reactions, as explained in the text.

II. Chemistry of Retroviral Integration

The chemistry of viral integration compared to the chemistry of recombination is, in fact, a variation of the same theme. For example, homologus recombinations by bacterial RecA, RecB, RecC, and Ruv (Kogoma, 1997) or VDJ joining by the RAG1 and RAG2 family of proteins in T-cell receptors and immunoglobulins (Gellert, 1997; Lewis and Wu, 1997) or transposition reactions by transposase and resolvases (Craig, 1997; Hallet and Sherratt, 1997) are all carried by very similar types of nucleophilic reactions. In brief, the DNA strand breakage and joining occur by transesterification reactions in which the phosphate of the scissile phosphodiester bond undergoes nucleophilic attack by a hydroxyl group. For 3′ processing (DNA breakage), the hydroxyl group of a water molecule is used as a nucleophile and for DNA strand transfer the 3′-hydroxyl that is generated by 3′ processing becomes the nucleophile (Fig. 3). Interestingly, for the 3′ processing reactions, IN can also use a variety of alcohols and glycerol as nucleophiles (Vink *et al.*, 1991; van Gant *et al.*, 1993; Katzman and Sudol, 1996). This mechanism is unique because one active site is involved in two different reactions (Fig. 3). Consistent with this view, all reported IN mutants have parallel effects on the efficiency of 3′ processing and strand transfer (see Table I) (Cannon *et al.*, 1994; Drelich *et al.*, 1992; Engelman *et al.*, 1993, 1995, 1997; Engel-

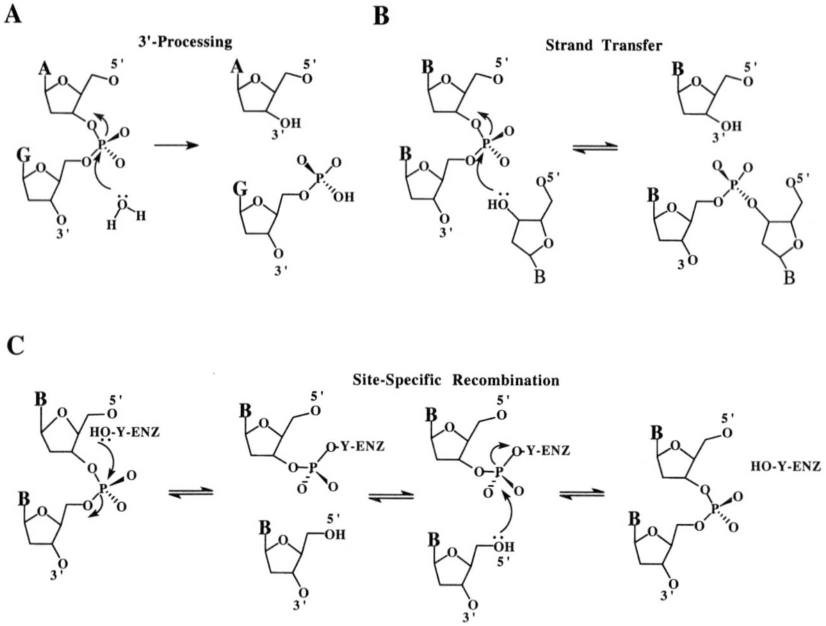

FIGURE 3   Chemistry of retroviral integration and site-specific recombination reactions. (A) The 3′ processing occurs by transesterification reactions in which the phosphate of the scissile phosphodiester bond (CA-p-GT) undergoes nucleophilic attack by an OH group (water) (arrows). (B) The strand-transfer reaction joins the 3′-processed ends to the target DNA. Here the nucleophile is the 3′-OH group from the viral DNA. (C) The strand exchange catalyzed by site-specific recombinases occurs by two-step transesterification involving a covalent protein–DNA intermediate.

man and Craigie, 1992; Kulkosky *et al.*, 1992; LaFemina *et al.*, 1992; Leavitt *et al.*, 1993; Shin *et al.*, 1994; Taddeo *et al.*, 1996; Wiskerchen and Muesing, 1995; Gerton *et al.*, 1998; Lutzke and Plasterk, 1998; Jenkins *et al.*, 1997; Oh *et al.*, 1997).

The dichotomy of the integration reaction in general depends on the active site residues that are intimately involved in interaction with the divalent metal and the substrate DNA. There are two main differences between viral integration and transposition as compared with those from site-specific recombination. First, the IN's active site comprises an invariant triad of acidic residues, termed the DDE motif, while the proteins involved in site-specific recombination contain an invariant RHRY tetrad in which the tyrosine (Y) acts as a nucleophile. Second, integrases do not form covalent intermediates with their DNA substrates, whereas the site-specific recombinases do. In general, activation of the nucleophile requires a divalent metal (Mn^{2+} or Mg^{2+}) as a cofactor. In addition, neither reaction requires ATP as a source of energy. The energy that is conserved in breaking a phosphodiester bond (3' processing) is available later for the formation of a new phospho-diester bond (strand transfer). In accord with classic SN_2-type reactions, both 3' processing and strand transfer reactions proceed by inversion of the configuration, by a single-step substitution, not involving a covalent IN–DNA intermediate (Engelman *et al.*, 1991). On the other hand, reactions involving a covalent intermediate as described above proceed with two inversions to result in an overall retention of configuration. The stereochemical course of the reaction and enzymatic phosphoryl transferase reaction by IN is reminiscent to the mechanism of Mu transposition (Mizuuchi and Adzuma, 1991) and RAG1- and RAG2-mediated VDJ recombination (Van Gent *et al.*, 1996; for a recent report on determining the stereochemistry of DNA cleavage see Mizuuchi *et al.*, 1999).

Throughout evolution, the chemistry of DNA cleavage and recombination (cut and paste) has been well conserved. In many respects integration chemistry is similar to the chemistry of restriction endonucleases. First, in the absence of divalent metal ions, restriction endonucleases, just like the INs, cannot cleave DNA but they can bind to it. Second, in most cases Mn^{2+} but not Ca^{2+} can substitute for Mg^{2+}. Third, mutants in which the acidic residues are replaced by alanine are inactive. Fourth, these enzymes bind comparably to specific and nonspecific DNA sequences, but they cleave the specific sequences preferentially. Fifth, the acidic active site residues form carboxylate-chelates. Sixth, the cleavage of phosphodiester bonds occurs with inversion of configuration at the phosphorus atom as explained above. Seventh, these enzymes recognize a specific site on DNA.

III. Role of Divalent Metals: Stoicheometry and Catalysis ____

It is well established that retroviral integrases and bacterial transposases require divalent metal (Mg^{2+} or Mn^{2+}) for catalysis. The role of divalent metal is not limited to catalysis; they are also important for specific binding of IN to DNA, conformational change, multimerization, and assembly. Therefore, the metal ion binding sites are highly conserved and are essential for enzymatic activity as well as for efficient viral replication. Divalent metal ions orientate and polarize the nonbridging P-O bond, increase the electrophilicity of the phosphorus atom, and subsequently help to stabilize the pentavalent transition state. Under physiological conditions the ratio of the cellular concentrations of Mg^{2+} to Mn^{2+} exceeds several logs; therefore, it is generally believed that Mg^{2+} is the biologically significant metal ion in the integration reaction, although IN functions more efficiently *in vitro* with Mn^{2+} as a cofactor.

There has been a great effort and controversy in establishing the stoicheometry of divalent metal ion in these enzymes. Is there only one metal ion required for catalysis? A literature survey favors a one-metal mecha-

TABLE I Catalytic and Replication Activities of HIV-1 Integrase Mutants

Mutant	Clv.[a]	ST	Dis.	Repl (infec)	References[b]	Mutant	Clv.	ST	Dis	Repl (infec)	References
H12A;N	+[c]	+	+++	−	11,4,5	Q148L	+	++	+		13
H12N/H16N	−	−	+	(+)	4	V151A	−	−	−	(+)	1
H16A					11	V151D/E152Q	−	−	−	−	7
H16C;V	+++	+++	+++	−	8	E152A;C;D;G;H;P;Q	−				1,2,4,6,8,9,11,13,16
C40;S	−	−	+	(−)	1,4	E152A/K156A					11
C40S/C43S	−	−	++		4	S153A;R	+	+	++		4
D41A/K42A	+		+	+	11	S153A	+++		+++	+++	9,16
C43A;S	+	+		(−)	1,7	N155E;K	−	−	−		13
M50A				+	11	N155L	+	+	−	+	13
H51A				+	11	K156A					11
H51A/D55V					11	K156E	−	−	+/−		13,15
D55A				−	11	K156E/K159E	−	−	−	−	15
W61A				(−)	1	K156A/E157A					11
Q62A	+	+	+++	(+/−)	12	E157A/K159A					11
Q62A	−	−	−		13	K159E	−	−	+++		15
Q62E	++	++	+++		12	K159N;S	+	+	+		13
Q62N	+		−		13	K159Q	+++		+++		16
D64C;R					13	K159A/K160A				+	11
D64A;E;N;V	−	−	−	−	1,2,4,8,11	K159A;P;Q:	+++	+++	+++	(+)	1,6,9
D64A/D116A			−	−	11	K160E	+++	+++	+++	−	15
D64R/D116R			−		13	R166A				−	11
D64A/E152A				−	11	R166A/D167A				−	11
D64A/D116A/E152A				−	11	D167A				+	11
T66A	+	+	+++	(+)	1,4	E170A/H171A				+	11
T66A	+	++	++		13	E170A/K173A				+	13
H67S	+++	+++	++		13	H171A/K173A				+	11
E69A/K71A					11	A179P				(−)	1
V75				(−)	1	F185A;K;L;H	+++	+++		(−)	12

Mutation	Clv	ST	Dis	Repl	infec	Ref
S81R;A	+/−	+/−	+/−	(−)		1,8
E92A;Q	+++	+++	+++	+++		12
E92A;N	+	+	++	+++		13
E92K	+	+/−	+++	+++		12
P109A;S	+/−	+/−	−	+		2,10,11
P109S/T125A	+++	+++	+++	(+/−)		10
T112A	+++	+++				2
T115A;S	+++	+++	+++	+++		1,2,4,6,9,11
D116A;E;I;N	−	−	−	−		1,2,4,5,6,8,9,13,16
D116C			+			13
D116A/E152A				−		11
N117K;Q;R	+/++	++	++/+++	Delayed		4,9,13,16
N117S	+	+	+			13
G118A	+++	+++	+++	+++	+/−	9,16
N120Q;S	+++	+++	+++	+++	−	13
F121A	+/−	−	+/−			6
S123A	++	++	++	(+)		1,4
T125A	+++	+++	+++			10
K127A			+	+		11
I135P				(−)		1
K136A	+/−	+/−	+/−	−		11,12
K136E;R	+++	+++	+++	+		12
K136A/E138A	+++	+++	+++			12
E138A			+	+		11
Y143F			Delayed	Delayed		9,11
Y143N	++	+++	+++	Delayed		9,16
S147I	−		−			9,16
G189A	+++	+++				2
E198A/R199A	+++	+++	++		−	11
R199A/D202A	+++	+++	++		−	11
R199T/D202A	+++	+++	++		−	11
R199A;C	+++	+++	+++	+++	+	8,11
K211A/E212A	+++	+++	+++	++	+	11
N222A	++	++	++	++		14
F223A	++	++	++	++		14
R224A	++	++	++	++		14
Y227A	++	+	++	+		14
R231A	++	+	+	+		14
P233A	+++	+++	+++	++	+/−	14
L234A	++	++	++	++		14
W235A;E;F	+++	+++	+++	+++	+/−	1,5,8
K236A/K240A					−	11
L241A	−	−	−	−		14
L242A	−	−	+	+		14
W243A	++	++	++	++		14
K244A/E246A			+		−	11
E246A	++	++	++	++		14
D253A/D256A	++	++	+	++	+	11
N254A	++	++	++	++		14
V260E	−	−	−	−		14
R262G	++	++	++	++		14
R263L	++	++	++	++		14
R269A/D270A					+	11

[a] Abbreviations: Clv, cleavage or 3′ processing; ST, strand transfer; Dis, disintegration; Repl, replication capability; infec, infectivity.

[b] References: (1) Cannon et al. (1994); (2) Drelich et al. (1992); (3) Engelman et al. (1992); (4) Engelman et al. (1993); (5) Engelman et al. (1995); (6) Kulkosky et al. (1992); (7) LaFemina et al. (1992); (8) Leavitt et al. (1993); (9) Shin et al. (1994); (10) Taddeo et al. (1996); (11) Wiskerchen et al. (1995); (12) Engelman et al. (1997); (13) Gerton et al. (1998); (14) Lutzke et al. (1998); (15) Jenkins et al. (1997); (16) Oh et al. (1997).

[c] Notations: −, 0–10%; +, 10–40%; ++, 40–80%; +++, 80–100%.

nism (see, for example, Cowan, 1997). However, a complete structure of a protein–metal–DNA complex is not yet available. Therefore, the two-metal hypothesis is not ruled out for IN.

IV. Inhibitor Development

Extensive compilations of IN inhibitors have been reported previously and recent reviews have dealt at length with the development and the recent progress in design and identification of IN inhibitors (Farnet and Bushman, 1996; Thomas and Brady, 1997; Asante-Appiah and Skalka, 1997; Pommier et al., 1997; Neamati et al., 1997; Robinson, 1998; Hansen et al., 1998; Pommier and Neamati, 1999). Therefore, no further attempt is made here to list all the inhibitors. Herein, we review different classes of inhibitors for their potential as drug leads and give a brief historic perspective for their identification. A series of representative structures is shown in Fig. 5.

A. Early Discoveries

After the initial discovery of IN in 1978 (Grandgenett et al., 1978; Schiff and Grandgenett, 1978) and establishing it as a requirement for viral replication (Hippenmeyer and Grandgenett, 1984), several major discoveries have paved the way for the development of IN inhibitors. These include the in vitro assays for integration (Katzman et al., 1989; Fitzgerald et al., 1991; Sherman and Fyfe, 1990; Craigie et al., 1990, 1991; Bushman and Craigie, 1991), identification of IN domains and highly conserved residues (Engelman and Craigie, 1992, Engelman et al., 1993, van Gent et al., 1993; Vink and Plasterk, 1993; Vink et al., 1993), determination of X-ray structure of the core domain (Dyda et al., 1994; Bujacz et al., 1995, 1996), and the solution structures of the N (Cai et al., 1997) and C termini (Lodi et al., 1995; Eijkelenboom et al., 1995).

The active form of IN that carries out enzymatic activities is multimeric. However, the number of monomers in the active multimer and the stoicheometry of DNA and metal is unknown at present. Both the NMR solution structure of the N and C termini and the X-ray structure of the core region are dimers (Fig. 4). There are extensive dimer–dimer interactions among homologus dimers. It remains to be seen how each domain interacts with the other and with DNA.

Apparent similarities with other DNA binding proteins were the major impetus for testing inhibitors of enzymes such as topoisomerases against IN in vitro. It is reasonable to target the DNA substrates required for IN binding. However, one limitation to such an approach is the difficulty of obtaining highly selective LTR binders. Several classes of DNA binders and well-known inhibitors of other DNA binding proteins such as topoisomerases

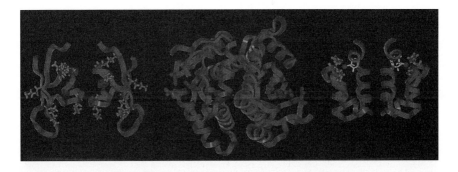

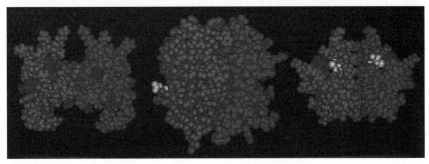

FIGURE 4   Structure of HIV-1 integrase. (Left) The solution structure of the N-terminal zinc binding domain (residues 1–55) of HIV-1 integrase. IN1-55 is dimeric, and each monomer comprises four helices with the zinc tetrahedrally coordinated to His 12, His 16, Cys 40, and Cys 43 (Cai *et al.*, 1997). (Center) The crystal structure of the catalytically active core domain (residues 50 to 212) of HIV-1 integrase. The structure is composed of a five-stranded β-sheet flanked by helical regions. The active site region is identified by the position of two of the conserved D64 and D116 residues and E152 essential for catalysis (Goldgur *et al.*, 1998). (Right) The solution structure of the DNA binding domain of HIV-1 integrase (residues 220–270). The protein is a dimmer in solution, and each subunit is composed of a five-stranded β-barrel with a topology very similar to that of an SH3 domain (Lodi *et al.*, 1995). There are nine lysines and seven arginines in this region. (See also color insert).

have been evaluated against purified IN (Fesen *et al.*, 1993; Billich *et al.*, 1992; Carteau *et al.*, 1993). The topoisomerase II inhibitors, doxorubicin and mitoxantrone, inhibit IN catalytic activity at low micromolar concentrations and in general, the topoisomerase II inhibitors that are not effective DNA binders are not effective inhibitors of IN. In addition, the topoisomerase I poison, camptothecin, was inactive against purified IN. Thus, DNA binding did not correlate closely with IN inhibition (Fesen *et al.*, 1993). A non-DNA binder, dihydroxynaphtaquinone, a moiety present in both doxorubicin and mitoxantrone, inhibited IN at low micromolar concentrations (Fesen *et al.*, 1993). This finding led to the exploration of hydroxylated aromatics (see below) including flavones and caffeic acid phenethyl ester (CAPE). The weak DNA binders, primaquine and chloroquine, were also active against purified IN (Fesen *et al.*, 1993). Moreover, the phenanthroline–cuprous complexes known to bind to the minor groove of DNA only at concentrations of 50 μM or above were also effective inhibitors of IN at the concentration range of 1–25 μM (Mazumder *et al.*, 1995). By contrast, the physiological DNA groove-binders spermine and spermidine had no effect against IN (Fesen *et al.*, 1993; Neamati *et al.*, 1998).

B. LTR-Targeted Inhibitors of IN

Integrase binds to specific sequences located on both extremities of the DNA on the HIV LTR. These sites are highly conserved in all HIV genomes (for alignment of HIV-1 LTRs see Pommier *et al.*, 1997; Neamati *et al.*, 1998) and could provide potential targets for the selective inhibition of integration. The retroviral LTR that recognizes a DNA binding site for IN contains an AT-rich sequence. Presence of this sequence has been exploited as a possible target for the DNA minor groove binder netropsin (Carteau *et al.*, 1994) and triple-helix-forming oligonucleotides (Mouscadet *et al.*, 1994). In an effort to further elucidate the role of AT binding agents in inhibition of IN function, a series of lexitropsins were synthesized and found to inhibit IN catalytic activity at low-nanomolar concentrations (Neamati *et al.*, 1998). Interestingly, many of these lexitropsins are effective inhibitors of HIV-1 replication in CEM cells. However, the monomeric minor groove binders such as distamycin, Hoechst 33258, DAPI, pentamidine, and berenil exhibited poor activity against purified IN (Neamati *et al.*, 1998). Thus, the antiviral activity of some of the IN inhibitors targeting the HIV LTR's may provide a rationale for their further development as anti-HIV drugs (Fig. 5).

C. Nucleotide-Based Inhibitors

HIV-1 IN inhibitors in this class comprise mono-, di-, tri-, and tetranucleotides; single-stranded DNA; double-stranded DNA; triple helices; guanosine quartets; and RNA. A common theme among all these inhibitors is the negative-charge-containing phosphate backbone. Since IN binds to viral DNA and possess extensive DNA binding sites, it is not surprising that a variety of phosphate-containing molecules inhibit IN *in vitro*.

The rational for this kind of inhibitor as drug lead was first illustrated when the guanosine quartet-type structures were found to exhibit antiviral activity in tissue culture. Guanosine quartets are oligonucleotides composed

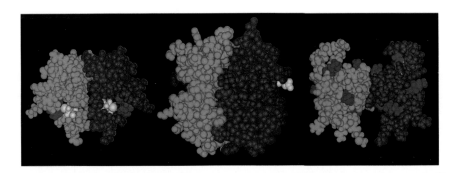

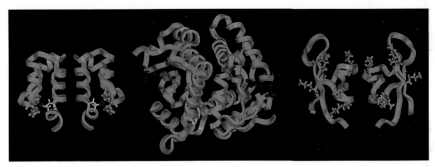

CHAPTER 5, FIGURE 4   Structure of HIV-1 integrase. (Top) The solution structure of the N-terminal zinc binding domain (residues 1–55) of HIV-1 integrase. IN1-55 is dimeric, and each monomer comprises four helices with the zinc (red) tetrahedrally coordinated to His 12, His 16 (blue), Cys 40, and Cys 43 (yellow) (Cai *et al.*, 1997). (Center) The crystal structure of the catalytically active core domain (residues 50 to 212) of HIV-1 integrase. The structure is composed of a five-stranded β-sheet flanked by helical regions. The active site region is identified by the position of two of the conserved D64 and D116 residues (red) and E152 (pink) essential for catalysis (Goldgur *et al.*, 1998). (Bottom) The solution structure of the DNA binding domain of HIV-1 integrase (residues 220–270). The protein is a dimmer in solution, and each subunit is composed of a five-stranded β-barrel with a topology very similar to that of an SH3 domain (Lodi *et al.*, 1995). There are nine lysines (blue) and seven arginines in this region.

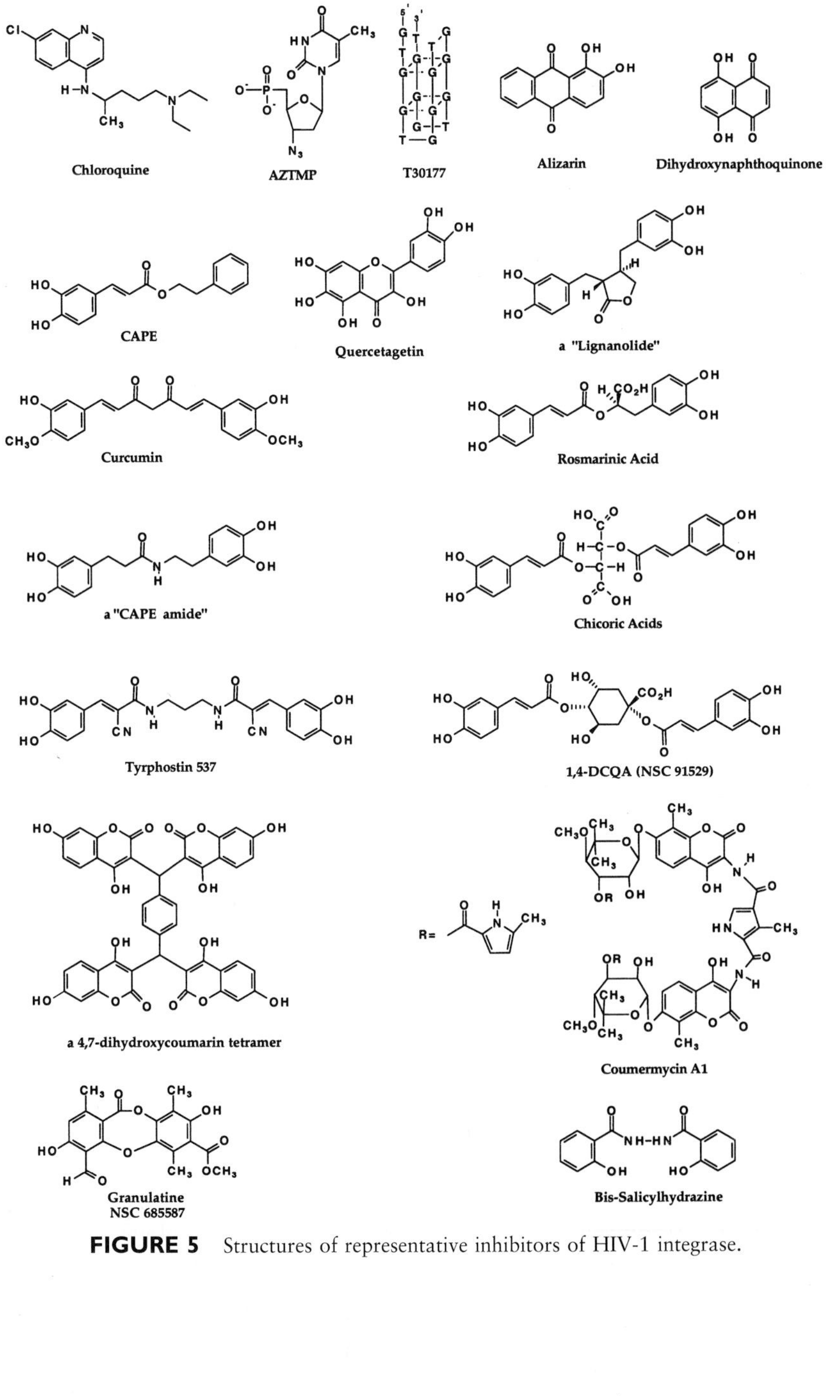

FIGURE 5   Structures of representative inhibitors of HIV-1 integrase.

entirely of deoxyguanosine and thymidine. In addition to guanosine quartets (Mazumder et al., 1996), other nucleotides-based inhibitors of IN have been subsequently reported. For example, phosphorothioate (Tramontano et al., 1998) oligonucleotides and single-stranded DNA (Pemberton et al., 1998) inhibit IN functions at nanomolar concentration. The phosphorothioate also inhibited purified HIV RT (Stein et al., 1991). Although some of these oligonucleotides exhibit antiviral activity both in vitro (Chou et al., 1991; Hotoda et al., 1998) and in vivo (Stoddart et al., 1998; Wallace et al., 1997), their exact mechanism of action remains controversial. For example, it was recently shown that poly (1-methy1-6-thioinosinic acid), or PMTI, which is a single-stranded polyribonucleotide, blocks HIV replication in human cells at its earliest stages by multiple mechanisms including inhibition of virus entry and inhibition of RT (Buckheit et al., 1999). Because these compounds are highly charged, it is expected that they might also inhibit gp-120–CD4 interaction. This mechanism was originally demonstrated for the phosphorothioate (Stein et al., 1991) and more recently for the guanosine quartet structures (Cherepanov et al., 1997; Este et al., 1998). The most potent guanosine quartet inhibitor of IN, now called zintevir, is in phase I clinical trial (for recent reviews see Silvestre et al., 1998; Rando, 1998).

It is generally believed that IN remains tightly bound to viral LTR DNA both in vivo and throughout the course of the integration reaction in vitro. These complexes were shown to be resistant to challenge by poly (Asp50 and Glu50) (Ellison and Brown, 1995), competitor DNA (Ellison and Brown, 1995; Vink et al., 1994), heparin (Ellison et al., 1995), and an exonuclease (Pemberton et al., 1995). In fact, repulsion of charges is an energetically costly process. Therefore, highly charged molecules in general can inhibit a variety of enzymes in vitro. For example, nonspecific inhibition of IN has been reported by suramine, dextran sulfates, and the oligonucleotides (Kvasyuk et al., 1999). However, antiviral activity of such agents is more likely to be due to the inhibition of viral attachment.

D. Peptide-Based Inhibitors

An elegant study using a combinatorial peptide library identified a hexa-peptide (HCKFWW) as a lead inhibitor for IN (Puras Lutzke et al., 1995). This short peptide inhibited HIV-1, HIV-2, FIV, and MoNLV IN, but it was inactive against restriction endonucleases BamHI and EcoRI and DNase I. Extensive modification of this peptide did not lead to a remarkably more potent inhibitor and none of these peptides showed antiviral activity (Nea-mati, Roller, and Pommier, unpublished). However, bicyclo hexapeptides composed entirely of aromatic amino acids such as complestatin and chloro-peptin that were reported to inhibit HIV-1 gp120–CD4 interaction also inhibits HIV-1 IN in vitro (Singh et al., 1998). Selective peptide-based inhibi-tors of IN possessing antiviral activity have not yet been identified. However, there are examples of longer peptides, protein, or antibodies that are briefly discussed.

Monoclonal antibodies (Mabs) against HIV-1 IN were produced in several laboratories (Bizub-Bender et al., 1994; Barsov et al., 1996; Nilsen et al., 1996; Levy-Mintz, et al., 1996; Okui et al., 1998). In general, epitope mapping demonstrated that Mabs raised against the conserved HHCC motif in the N terminus of IN, against the core or the C-terminus clearly inhibited 3′ processing and strand transfer (Bizub-Bender et al., 1994; Nilsen et al., 1996). Inhibition of HIV-1 replication at both early and late stages of the viral life cycle by single-chain antibody against viral IN has also been reported (Kitamura et al., 1999). In addition, peptide antisera against selected regions in HIV-1 and HIV-2 RT and IN were shown to inhibit IN-mediated cleavage of an HIV-1 DNA oligonucleotide substrate in a 3′ processing assay, while anti-RT or normal sera had no effect (Klutch et al., 1998). None of the RT sera inhibited RT activity. These types of studies may be useful for generating HIV peptide vaccines with dual specificity against HIV-1 and HIV-2. Interestingly, monospecific antibodies raised against a synthetic peptide K159 (SQGVVESMNKELKKIIGQVRDQAEHLKTA) reproducing the segment 147–175 of IN exhibited inhibitory activities against IN (Maroun et al., 1999). In contrast, neither P159, a Pro-containing analog of K159 that presents a kink around proline but with intact epitope conformation, nor the truncated analogs encompassing the epitope were inhibitors of IN (Maroun et al., 1999).

E. Natural Products

Natural products have served as a great source for identification of leads. A majority of reported IN inhibitors are from natural products (for a review on natural products as inhibitors of IN see (Eich, 1998). Examples include caffeic acid phenethyl ester (CAPE) (Fesen et al., 1993); anthracyclines (Fesen et al., 1993; Billich et al., 1992); curcumins (Mazumder et al., 1995, 1997); flavones and flavonoids (Fesen et al., 1994); lignans and lignaloids (Eich et al., 1996); depsides and depsidones (Neamati et al., 1997a); conidendrones (LaFemina et al., 1995); caffeoylquinic acids (Neamati et al., 1997a; Robinson et al., 1996a,b); rosmarinic acid (Mazumder et al., 1997) chicoric acids (Neamati et al., 1997a,b; Robinson et al., 1996a,b; Lin et al., 1999; King et al., 1999); coumarins (Mazumder et al., 1996; Zhao et al., 1997); ellagic acid, purpurogallin, and hypericin (Farnet et al., 1998); conformationally restricted cinnamoyl compounds (Artico et al., 1998); and gallic acid flavon-3-yl esters (Desideri et al., 1998). Of particular interest in this class of compounds is that several derivatives inhibit HIV-1 replication in cell-based assays. These include flavones, depsides, caffeoylquinic acids, chicoric acids, coumarins, and hypericin. However, many of these compounds are also known to inhibit other viral targets, such as RT, protease, and gp120 (Pommier et al., 1997; Pommier and Neamati, 1999).

A common structural feature shared by a large number of natural product inhibitors is that they contain one or more catechol moieties (Fig. 1). The most potent inhibitors contain two catechols separated by a linker and this linker can vary in length and geometry. Examples of linkers include short-chain aliphatics, acidic residues, saturated cyclic structures, and aromatics.

The catechol-containing compounds are reported to chelate metals (Fesen et al., 1994; Neamati et al., 1998), cross-link to host proteins (Stanwell et al., 1996), exert remarkable cytotoxicity (Mazumder et al., 1995), and to be relatively nonselective (Neamati et al., 1997; Jagoe et al., 1997; Natarajan et al., 1996). Despite the failure of a large number of catechol-containing inhibitors, the chicoric and caffeoylquinic acids have shown to possess antiviral activity (Lin et al., 1999, King et al., 1999). Structure activity relationships among these compounds have stressed the requirement of the catechol moiety for activity (Lin et al., 1999; King et al., 1999). It was shown that the inhibition of IN by dicaffeoylquinic acid was irreversible, did not require the presence of a divalent cation, and was unaffected by preassembling IN onto viral DNA (Zhu et al., 1999). Moreover, a recent communication has indicated the occurrence of a single glycine-to-serine mutation at position 140 in an HIV-1 strain selected for resistance to L-chicoric acid (King and Robinson, 1998).

The failure of large numbers of natural-product inhibitors of IN in cell-based assays is perhaps due to several factors: poor cellular uptake, formation of inactive metabolites, instability and reactivity in serum-containing medium, and lack of target selectivity. It is important to mention that such inhibitors are great tools to aid in the identification of structural requirements for activity.

V. Perspectives

Like other retroviral targets, identification, optimization, and development of a lead inhibitor against IN requires a multistep approach. A recent publication reviews the strategies for antiviral drug discovery (Jones, 1998).

The combination therapies, or HAART, is now the rule rather than the exception for effective treatment against AIDS (for a recent review see Vandamme et al., 1998). At present, such combinations include nucleoside reverse transcriptase (Mitsuya, 1997), nonnucleoside reverse transcriptase (De Clercq, 1998), and a protease inhibitor. The Food and Drug Administration has approved 16 drugs for the treatment of AIDS and at present there are more than 100 clinical trials ongoing. Many other drugs are in preclinical and phase I studies aimed at keeping viral replication very low. As yet there are no IN-specific drugs under clinical or preclinical investigations. However, it is a matter of time before such an agent will be available.

The major goal of therapy against AIDS is the efficient suppression of viral load for as long as possible. This is the requirement for the prevention of emergence of resistant viruses and for a successful therapy. Therefore, to achieve such a goal, it will probably be necessary to apply combination chemotherapy targeting many viral sites.

With the advent of relatively selective and efficacious inhibitors of HIV replication it is hoped that the prospects for a cure could be improved. In fact, the remarkable clinical efficacy observed with HAART attest to the optimism in the field. Virus-specific events (for a recent review see De Clercq, 1998) are always the attractive targets for chemotherapy due to a possible design for selective inhibitors. In designing inhibitors one can think of an HIV-1 as 15 proteins and an RNA (Frankel and Young, 1998). Therefore, a tremendous amount of effort has been invested in delivering maximum pressure on HIV in hopes of viral eradication. It is generally believed that in due course, the virus is likely to evade all kinds of drug pressure, even combinations of drugs. However, it is hoped that the resistance would emerge inefficiently with combination chemotherapy.

References

1. Artico, M., Di Santo, R., Costi, R., Novellino, E., Greco, G., Massa, S., Tramontano, E., Marongiu, M. E., De Montis, A., and La Colla, P. (1998). *J. Med. Chem.* **41**(21), 3948–3960.
2. Asante-Appiah, E., and Skalka, A. M. (1997). *Antiviral Res.* **36**(3), 139–156.
3. Barsov, E. V., Huber, W. E., Marcotrigiano, J., Clark, P. K., Clark, A. D., Arnold, E., and Hughes, S. H. (1996). *J. Virol.* **70**(7), 4484–4494.
4. Billich, A., Schauer, M., Frank, S., Rosenwirth, B., and Billich, S. (1992). *Antiviral Chem. Chemother.* **3**, 113–119.
5. Bizub-Bender, D., Kulkosky, J., and Skalka, A. M. (1994). *AIDS Res. Hum. Retroviruses* **10**(9), 1105–1115.
6. Brown, P. O. (1998). Integration. In "Retroviruses" (J. M. Coffin, S. H. Hughes, and H. E. Varmus, Eds.). Cold Spring Harbor Press, Cold Spring Harbor, NY.
7. Buckheit, R. W., Lackman-Smith, C., Snow, M. J., Halliday, S. M., White, E. L., Ross, L. J., Agrawal, V. K., and Broom, A. D. (1999). *Antivir. Chem. Chemother.* **10**(1), 23–32.
8. Bujacz, G., Alexandratos, J., Qing, Z. L., Clement-Mella, C., and Wlodawer, A. (1996). *FEBS Lett.* **398**(2–3), 175–178.
9. Bujacz, G., Jaskolski, M., Alexandratos, J., Wlodawer, A., Merkel, G., Katz, R. A., and Skalka, A. M. (1995). *J. Mol. Biol.* **253**(2), 333–346.
10. Bushman, F. D., and Craigie, R. (1991). *Proc. Natl. Acad. Sci. USA* **88**(4), 1339–1343.
11. Cai, M., Zheng, R., Caffrey, M., Craigie, R., Clore, G. M., and Gronenborn, A. M. (1997). *Nat. Struct. Biol.* **4**(7), 567–577.
12. Cannon, P. M., Wilson, W., Byles, E., Kingsman, S. M., and Kingsman, A. J. (1994). *J. Virol.* **68**(8), 4768–4775.
13. Carteau, S., Mouscadet, J. F., Goulaouic, H., Subra, F., and Auclair, C. (1993). *Biochem. Biophys. Res. Commun.* **192**(3), 1409–1414.

14. Carteau, S., Mouscadet, J. F., Goulaouic, H., Subra, F., and Auclair, C. (1994). *Biochem. Pharmacol.* **47**(10), 1821–1826.
15. Cherepanov, P., Este, J. A., Rando, R. F., Ojwang, J. O., Reekmans, G., Steinfeld, R., David, G., De Clercq, E., and Debyser, Z. (1997). *Mol. Pharmacol.* **52**(5), 771–780.
16. Chou, T. C., Zhu, Q. Y., and Stein, C. A. (1991). *AIDS Res. Hum. Retroviruses* **7**(11), 943–951.
17. Chow, S. A. (1997). *Methods* **12**(4), 306–317.
18. Cowan, J. A. (1997). *J. Biol. Inorg. Chem.* **2**(2), 168–176.
19. Craig, N. L. (1997). *Annu. Rev. Biochem.* **66**, 437–474.
20. Craigie, R., Fujiwara, T., and Bushman, F. (1990). *Cell* **62**(4), 829–837.
21. Craigie, R., Mizuuchi, K., Bushman, F. D., and Engelman, A. (1991). *Nucleic Acids Res.* **19**(10), 2729–2734.
22. De Clercq, E. (1998a). *Antivir. Res.* **38**(3), 153–179.
23. De Clercq, E. (1998b). *Pure Appl. Chem.* **70**(3), 567–577.
24. Desideri, N., Sestili, I., Stein, M. L., Tramontano, E., Corrias, S., and La Colla, P. (1998). *Antivir. Chem. Chemother.* **9**(6), 497–509.
25. Drelich, M., Wilhelm, R., and Mous, J. (1992). *Virology* **188**(2), 459–468.
26. Dyda, F., Hickman, A. B., Jenkins, T. M., Engelman, A., Craigie, R., and Davies, D. R. (1994). *Science* **266**(5193), 1981–1986.
27. Eich, E. (1998). *In* "*ACS Symp. Ser.*" Vol. 691, pp. 83–96.
28. Eich, E., Pertz, H., Kaloga, M., Schulz, J., Fesen, M. R., Mazumder, A., and Pommier, Y. (1996). *J. Med. Chem.* **39**(1), 86–95.
29. Eijkelenboom, A. P., Lutzke, R. A., Boelens, R., Plasterk, R. H., Kaptein, R., and Hard, K. (1995). *Nat. Struct. Biol.* **2**(9), 807–810.
30. Ellison, V., and Brown, P. O. (1994). *Proc. Natl. Acad. Sci. USA* **91**(15), 7316–7320.
31. Ellison, V., Gerton, J., Vincent, K. A., and Brown, P. O. (1995). *J. Biol. Chem.* **270**(7), 3320–3326.
32. Engelman, A., and Craigie, R. (1992). *J. Virol.* **66**(11), 6361–6369.
33. Engelman, A., Bushman, F. D., and Craigie, R. (1993). *EMBO J.* **12**(8), 3269–3275.
34. Engelman, A., Englund, G., Orenstein, J. M., Martin, M. A., and Craigie, R. (1995). *J. Virol.* **69**(5), 2729–2736.
35. Engelman, A., Liu, Y., Chen, H., Farzan, M., and Dyda, F. (1997). *J. Virol.* **71**(5), 3507–3514.
36. Engelman, A., Mizuuchi, K., and Craigie, R. (1991). *Cell* **67**(6), 1211–1221.
37. Este, J. A., Cabrera, C., Schols, D., Cherepanov, P., Gutierrez, A., Witvrouw, M., Pannecouque, C., Debyser, Z., Rando, R. F., Clotet, B., Desmyter, J., and De Clercq, E. (1998). *Mol. Pharmacol.* **53**(2), 340–345.
38. Farnet, C. M., and Bushman, F. D. (1996). *AIDS* **10**(Suppl. A), S3–S11.
39. Farnet, C. M., Wang, B., Hansen, M., Lipford, J. R., Zalkow, L., Robinson, W. E., Jr., Siegel, J., and Bushman, F. (1998). *Antimicrob. Agents Chemother.* **42**(9), 2245–2453.
40. Fesen, M. R., Kohn, K. W., Leteurtre, F., and Pommier, Y. (1993). *Proc. Natl. Acad. Sci. USA* **90**(6), 2399–2403.
41. Fesen, M. R., Pommier, Y., Leteurtre, F., Hiroguchi, S., Yung, J., and Kohn, K. W. (1994). *Biochem. Pharmacol.* **48**(3), 595–608.
42. Fitzgerald, M. L., Vora, A. C., and Grandgenett, D. P. (1991). *Anal. Biochem.* **196**(1), 19–23.
43. Frankel, A. D., and Young, J. A. (1998). *Annu. Rev. Biochem.* **67**, 1–25.
44. Gellert, M. (1997). *Adv. Immunol.* **64**, 39–64.
45. Gerton, J. L., Ohgi, S., Olsen, M., DeRisi, J., and Brown, P. O. (1998). *J. Virol.* **72**(6), 5046–5055.
46. Goldgur, Y., Dyda, F., Hickman, A. B., Jenkins, T. M., Craigie, R., and Davies, D. R. (1998). *Proc. Natl. Acad. Sci. USA* **95**(16), 9150–9154.

47. Grandgenett, D. P., Vora, A. C., and Schiff, R. D. (1978). *Virology* **89**(1), 119–132.
48. Hallet, B., and Sherratt, D. J. (1997). *FEMS Microbiol. Rev.* **21**(2), 157–178.
49. Hansen, M. S., Carteau, S., Hoffmann, C., Li, L., and Bushman, F. (1998). *Genet. Eng.* **20**, 41–61.
50. Hippenmeyer, P. J., and Grandgenett, D. P. (1984). *Virology* **137**(2), 358–370.
51. Hotoda, H., Koizumi, M., Koga, R., Kaneko, M., Momota, K., Ohmine, T., Furukawa, H., Agatsuma, T., Nishigaki, T., Sone, J., Tsutsumi, S., Kosaka, T., Abe, K., Kimura, S., and Shimada, K. (1998). *J. Med. Chem.* **41**(19), 3655–3663.
52. Jagoe, C. T., Kreifels, S. E., and Li, J. (1997). *Bioorg. Med. Chem. Lett.* **7**(2), 113–116.
53. Jenkins, T. M., Esposito, D., Engelman, A., and Craigie, R. (1997). *EMBO J.* **16**(22), 6849–6859.
54. Jones, P. S. (1998). *Antiviral. Chem. Chemother.* **9**, 283–202.
55. Katz, R. A., and Skalka, A. M. (1994). *Annu. Rev. Biochem.* **63**, 133–173.
56. Katzman, M., and Sudol, M. (1996). *J. Virol.* **70**(4), 2598–2604.
57. Katzman, M., Katz, R. A., Skalka, A. M., and Leis, J. (1989). *J. Virol.* **63**(12), 5319–5327.
58. Klutch, M., Woerner, A. M., Marcus-Sekura, C. J., and Levin, J. G. (1998). *J. Biomed. Sci.* **5**(3), 192–202.
59. King, P. J., and Robinson, W. E., Jr. (1998). *J. Virol.* **72**(10), 8420–8424.
60. King. P. J., Ma, G. X., Miao, W. F., Jia, Q., McDougall, B. R., Reinecke, M. G., Cornell, C., Kuan, J., Kim, T. R., and Robinson, W. E. (1999). *J. Med. Chem.* **42**(3), 497–509.
61. Kitamura, Y., Ishikawa, T., Okui, N., Kobayashi, N., Kanda, T., Shimada, T., Miyake, K., and Yoshiike, K. (1999). *J. Acquir. Immune Defic. Syndr. Hum. Retrovirol.* **20**(2), 105–114.
62. Kogoma, T. (1997). *Microbiol. Mol. Biol. Rev.* **61**(2), 212–238.
63. Kulkosky, J., Jones, K. S., Katz, R. A., Mack, J. P., and Skalka, A. M. (1992). *Mol. Cell Biol.* **12**(5), 2331–2338.
64. Kvasyuk, E. I., Mikhailopulo, I. A., Homan, J. W., Iacono, K. T., Muto, N. F., Suhadolnik, R. J., and Pfleiderer, W. (1999). *Helvetica Chim. Acta* **82**(1), 19–29.
65. LaFemina, R. L., Graham, P. L., LeGrow, K., Hastings, J. C., Wolfe, A., Young, S. D., Emini, E. A., and Hazuda, D. J. (1995). *Antimicrob. Agents Chemother.* **39**(2), 320–324.
66. LaFemina, R. L., Schneider, C. L., Robbins, H. L., Callahan, P. L., LeGrow, K., Roth, E., Schleif, W. A., and Emini, E. A. (1992). *J. Virol.* **66**(12), 7414–7419.
67. Leavitt, A. D., Shiue, L., and Varmus, H. E. (1993). *J. Biol. Chem.* **268**(3), 2113–2119.
68. Levy-Mintz, P., Duan, L., Zhang, H., Hu, B., Dornadula, G., Zhu, M., Kulkosky, J., Bizub-Bender, D., Skalka, A. M., and Pomerantz, R. J. (1996). *J. Virol.* **70**(12), 8821–8832.
69. Lewis, S. M., and Wu, G. E. (1997). *Cell* **88**(2), 159–162.
70. Lin, Z., Neamati, N., Zhao, H., Kiryu, Y., Turpin, J. A., Aberham, C., Strebel, K., Kohn, K., Witvrouw, M., Pannecouque, C., Debyser, Z., De Clercq, E., Rice, W. G., Pommier, Y., and Burke, T. R., Jr. (1999). *J. Med. Chem.* **42**(8), 1401–1414.
71. Lodi, P. J., Ernst, J. A., Kuszewski, J., Hickman, A. B., Engelman, A., Craigie, R., Clore, G. M., and Gronenborn, A. M. (1995). *Biochemistry* **34**(31), 9826–9833.
72. Maroun, R. G., Krebs, D., Roshani, M., Porumb, H., Auclair, C., Troalen, F., and Fermandjian, S. (1999). *Eur. J. Biochem.* **260**(1), 145–155.
72. Lutzke, R. A., and Plasterk, R. H. (1998). *J. Virol.* **72**(6), 4841–4848.
73. Mazumder, A., Gazit, A., Levitzki, A., Nicklaus, M., Yung, J., Kohlhagen, G., and Pommier, Y. (1995). *Biochemistry* **34**(46), 15111–15122.
74. Mazumder, A., Gupta, M., Perrin, D. M., Sigman, D. S., Rabinovitz, M., and Pommier, Y. (1995). *AIDS Res. Hum. Retroviruses* **11**(1), 115–125.
75. Mazumder, A., Neamati, N., Ojwang, J. O., Sunder, S., Rando, R. F., and Pommier, Y. (1996). *Biochemistry* **35**(43), 13762–13771.
76. Mazumder, A., Neamati, N., Sunder, S., Owen, J., and Pommier, Y. (1999). *In "Methods in Molecular Medicine"* (D. Kinchington and R. Schinazi, Eds.). The Humana Press, Totowa, NJ. pp. 327–338.

77. Mazumder, A., Neamati, N., Sunder, S., Schulz, J., Pertz, H., Eich, E., and Pommier, Y. (1997). *J. Med. Chem.* **40**(19), 3057–3063.
78. Mazumder, A., Raghavan, K., Weinstein, J., Kohn, K. W., and Pommier, Y. (1995). *Biochem. Pharmacol.* **49**(8), 1165–1170.
79. Mazumder, A., Wang, S., Neamati, N., Nicklaus, M., Sunder, S., Chen, J., Milne, G. W., Rice, W. G., Burke, T. R., Jr., and Pommier, Y. (1996). *J. Med. Chem.* **39**(13), 2472–2481.
80. Mitsuya, H. (1997). "Anti-HIV Nucleosides: Past, Present and Future." R. G. Landes, Austin, TX.
81. Mizuuchi, K., and Adzuma, K. (1991). *Cell* **66**(1), 129–140.
82. Mizuuchi, K., and Craigie, R. (1986). *Annu. Rev. Genet.* **20**, 385–429.
83. Mizuuchi, K., Nobbs, T. J., Halford, S. E., Adzuma, K., and Qin, J. (1999). *Biochemistry* **38**(14), 4640–4648.
84. Mouscadet, J. F., Carteau, S., Goulaouic, H., Subra, F., and Auclair, C. (1994). *J. Biol. Chem.* **269**(34), 21635–21638.
85. Natarajan, K., Singh, S., Burke, T. R., Grunberger, D., and Aggarwal, B. B. (1996). *Proc. Natl. Acad. Sci. USA* **93**, 9090–9095.
86. Neamati, N., Hong, H., Mazumder, A., Wang, S., Sunder, S., Nicklaus, M. C., Milne, G. W., Proksa, B., and Pommier, Y. (1997a). *J. Med. Chem.* **40**(6), 942–951.
87. Neamati, N., Hong, H., Owen, J. M., Sunder, S., Winslow, H. E., Christensen, J. L., Zhao, H., Burke, J. T. R., Milne, G. W. A., and Pommier, Y. (1998). *J. Med. Chem.* **41**, 3202–3209.
88. Neamati, N., Hong, H., Sunder, S., Milne, G. W., and Pommier, Y. (1997b). *Mol. Pharmacol.* **52**(6), 1041–1055.
89. Neamati, N., Mazumder, A., Sunder, S., Owen, J. M., Tandon, M., Lown, W., and Pommier, Y. (1998). *Mol. Pharmacol.* **54**, 280–290.
90. Neamati, N., Sunder, S., and Pommier, Y. (1997). *Drug Discov. Today* **2**, 487–498.
92. Nilsen, B. M., Haugan, I. R., Berg, K., Olsen, L., Brown, P. O., and Helland, D. E. (1996). *J. Virol.* **70**(3), 1580–1587.
93. Oh, J. W., Oh, Y. T., Kim, D. J., and Shin, C. G. (1997). *Mol. Cells* **7**(5), 688–693.
94. Okui, N., Kobayashi, N., and Kitamura, Y. (1998). *J. Virol.* **72**(8), 6960–6964.
95. Pemberton, I. K., Buckle, M., and Buc, H. (1996). *J. Biol. Chem.* **271**(3), 1498–1506.
96. Pemberton, I. K., Buc, H., and Buckle, M. (1998). *Biochemistry* **37**(8), 2682–2690.
97. Pommier, Y., and Neamati, N. (1999). *Adv. Virus. Res.* **52**, 427–458.
98. Pommier, Y., Pilon, A., Bajaj, K., Mazumder, A., and Neamati, N. (1997). *Antiviral. Chem. Chemother.* **8**, 483–503.
99. Puras Lutzke, R. A., Eppens, N. A., Weber, P. A., Houghten, R. A., and Plasterk, R. H. (1995). *Proc. Natl. Acad. Sci. USA* **92**(25), 11456–11460.
100. Rando, R. F. (1998). *Curr. Res. Mol. Ther.* **1**(2), 67–75.
101. Rice, P., Craigie, R., and Davies, D. R. (1996). *Curr. Opin. Struct. Biol.* **6**(1), 76–83.
102. Robinson, J. W. E. (1998). *Infect. Med.* **15**, 129–137.
103. Robinson, W. E., Jr., Cordeiro, M., Abdel-Malek, S., Jia, Q., Chow, S. A., Reinecke, M. G., and Mitchell, W. M. (1996a). *Mol. Pharmacol.* **50**(4), 846–855.
104. Robinson, W. E., Jr., Reinecke, M. G., Abdel-Malek, S., Jia, Q., and Chow, S. A. (1996b). *Proc. Natl. Acad. Sci. USA* **93**(13), 6326–6331.
105. Schiff, R. D., and Grandgenett, D. P. (1978). *J. Virol.* **28**(1), 279–291.
106. Sherman, P. A., and Fyfe, J. A. (1990). *Proc. Natl. Acad. Sci. USA* **87**(13), 5119–5123.
107. Shin, C. G., Taddeo, B., Haseltine, W. A., and Farnet, C. M. (1994). *J. Virol.* **68**(3), 1633–1642.
108. Silvestre, J., Graul, A., and Castaner, J. (1998). *Drugs Future* **23**(3), 274–277.
109. Singh, S. B., Jayasuriya, H., Hazuda, D. L., Felock, P., Homnick, C. F., Sardana, M., and Patane, M. A. (1998). *Tetrahedr. Lett.* **39**(48), 8769–8770.
110. Stanwell, C., Ye, B., Yuspa, S. H., and Burke, T. R., Jr. (1996). *Biochem. Pharmacol.* **52**(3), 475–480.

111. Stein, C. A., Neckers, L. M., Nair, B. C., Mumbauer, S., Hoke, G., and Pal, R. (1991). *J. Acquir. Immune Defic. Syndr.* 4(7), 686–693.
112. Stoddart, C. A., Rabin, L., Hincenbergs, M., Moreno, M., Linquist-Stepps, V., Leeds, J. M., Truong, L. A., Wyatt, J. R., Ecker, D. J., and McCune, J. M. (1998). *Antimicrob. Agents Chemother.* 42(8), 2113–2115.
113. Taddeo, B., Carlini, F., Verani, P., and Engelman, A. (1996). *J. Virol.* 70(12), 8277–8284.
114. Thomas, M., and Brady, L. (1997). *Trends Biotechnol.* 15(5), 167–172.
115. Tramontano, E., Colla, P. L., and Cheng, Y. C. (1998). *Biochemistry* 37(20), 7237–7243.
116. Vandamme, A.-M., Van Vaerenbergh, K., and De Clercq, E. (1998). *Antivir. Chem. Chemother.* 9(3), 187–203.
117. van Gent, D. C., Mizuuchi, K., and Gellert, M. (1996). *Science* 271(5255), 1592–1594.
118. van Gent, D. C., Oude Groeneger, A. A., and Plasterk, R. H. (1993). *Nucleic Acids Res.* 21(15), 3373–3377.
119. van Gent, D. C., Vink, C., Groeneger, A. A., and Plasterk, R. H. (1993). *EMBO J.* 12(8), 3261–3267.
120. Vink, C., and Plasterk, R. H. (1993). *Trends Genet.* 9(12), 433–438.
121. Vink, C., Lutzke, R. A., and Plasterk, R. H. (1994). *Nucleic Acids Res.* 22(20), 4103–4110.
122. Vink, C., Oude Groeneger, A. M., and Plasterk, R. H. (1993). *Nucleic Acids Res.* 21(6), 1419–1425.
123. Vink, C., Yeheskiely, E., van der Marel, G. A., van Boom, J. H., and Plasterk, R. H. (1991). *Nucleic Acids Res.* 19(24), 6691–6698.
124. Wallace, T., Kahn, J., Kennedy, B., Bazemore, S., and Cossum, P. (1997). "37th Interscience Conference for Antimicrobiol Agents and Chemotherapy," **Abstract I-70**. September 28–October 1, Toronto.
125. Wiskerchen, M., and Muesing, M. A. (1995). *J. Virol.* 69(1), 597–601.
126. Zhao, H., Neamati, N., Hong, H., Mazumder, A., Wang, S., Sunder, S., Milne, G. W., Pommier, Y., and Burke, T. R., Jr. (1997). *J. Med. Chem.* 40(2), 242–249.
127. Zhu, K., Cordeiro, M. L., Atienza, J., Robinson, W. E., and Chow, S. A. (1999). *J. Virol.* 73(4), 3309–3316.

Seongwoo Hwang
Natarajan Tamilarasu
Tariq M. Rana
Department of Pharmacology
Robert Wood Johnson Medical School
Piscataway, New Jersey 08854

Selection of HIV Replication Inhibitors: Chemistry and Biology

I. Introduction

Triple-drug therapies for HIV have produced spectacular results in reducing the number of virus particles and have facilitated remarkable recoveries in AIDS patients. However, current AIDS therapies face three major problems: (1) first-line drugs are not effective in some patients, (2) newer medications have major side effects, and (3) new drug-resistant strains of HIV are emerging. Therefore, there is a great need to find new drugs and treatment strategies. Available HIV drugs inhibit two key enzymes of the virus, reverse transcriptase and protease. Given the pathogenesis of HIV mutants capable of resisting triple-drug therapies, the identification of drugs that target HIV proteins other than reverse transcriptase and protease is a high priority for the development of new drugs.

HIV-1 is a complex retrovirus that encodes six regulatory proteins, including Tat and Rev, essential for viral replication. Inhibition of Tat and Rev function provides attractive targets for new antiviral therapies.

Advances in Pharmacology, Volume 49

A. Tat Protein

The Tat protein is a potent transcriptional activator of the HIV-1 long terminal repeat promoter element. A regulatory element between +1 and +60 in the HIV-1 long terminal repeat, which is capable of forming a stable stem-loop structure designated TAR, is critical for Tat function. In the absence of Tat, RNA polymerase II (pol II) terminates transcription prematurely. Tat–TAR interactions convert pol II into its processive form and lead to the efficient production of full-length viral transcripts (Cullen, 1998).

Tat proteins are small arginine-rich RNA-binding proteins. HIV-1 Tat is encoded by two exons containing 86 to 101 amino acids in different HIV-1 isolates. Amino acids encoded by the first exon are both necessary and functional for TAR RNA binding and transactivation *in vivo*. Tat protein is composed of several functional regions (Fig. 1). A cysteine-rich region (amino acids 22–37) contains seven cysteine residues; a "core" sequence (amino acids 37–48) contains hydrophobic amino acids; a basic RNA-binding region (amino acids 48–59) contains six arginines and two lysines and is a characteristic of a family of sequence-specific RNA-binding proteins; and a glutamine-rich region at the carboxyl terminus of the first exon contains several regularly spaced glutamines. In lentiviral proteins, only the basic and core regions are conserved. Although the integrity of the Cys-rich region is essential for transactivation, this region does not appear to be directly involved in TAR RNA recognition. Based on mutational analysis, Tat can be divided into two functional domains. The first domain is the activation domain (amino acids 1 to 47), or cofactor binding domain, which is functionally autonomous and is active when recruited to the HIV-1 long terminal repeat (LTR) via a heterologous RNA-binding protein (Selby and Peterlin, 1990). The second functional domain contains the basic region and is required for both RNA-binding and nuclear localization activities of Tat (Dingwall *et al.*, 1990).

HIV-1 Tat protein acts by binding to the TAR (*trans*-activation responsive) RNA element, a 59-base stem-loop structure located at the 5′ ends of all nascent HIV-1 transcripts (Berkhout *et al.*, 1989). Upon binding to the TAR RNA sequence, Tat causes a substantial increase in transcript levels (Jones and Peterlin, 1994). TAR RNA was originally localized to nucleotides +1 to +80 within the viral long terminal repeat (Rosen *et al.*, 1985). Subsequent deletion studies have established that the region from +19 to +42 incorporates the minimal domain that is both necessary and sufficient for Tat responsiveness *in vivo* (Fig. 1) (Jakobovits *et al.*, 1988). TAR RNA contains a six-nucleotide loop and a three-nucleotide pyrimidine bulge which separates two helical stem regions (Berkhout and Jeang, 1989).

Tat protein recognizes the trinucleotide bulge in TAR RNA. Key elements required for TAR recognition by Tat have been defined by extensive mutagenesis, chemical probing, and peptide-binding studies (Berkhout and

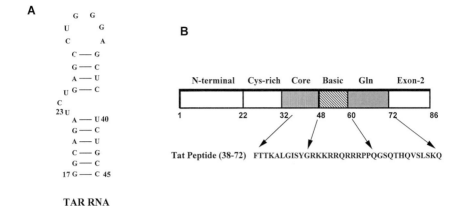

TAR RNA

FIGURE I (A) Sequence and secondary structure of TAR RNA used in structural studies. TAR RNA spans the minimal sequences that are required for Tat responsiveness *in vivo* (Jakobovits *et al.*, 1988) and for *in vitro* binding of Tat-derived peptides (Cordingley *et al.*, 1990). Wild-type TAR contains two non-wild-type base pairs to increase transcription by T7 RNA polymerase (Wang *et al.*, 1996). (B) Regions of the HIV-1 Tat protein and sequence of the Tat (37–72) peptide that recognizes Tat with high affinity and specificity.

Jeang, 1991; Calnan *et al.*, 1991; Churcher *et al.*, 1993; Cordingley *et al.*, 1990; Weeks and Crothers, 1991). Tat interacts with U23 and two other bulge residues, C24 and U25, which act as spacers because they can be replaced by other nucleotides or linkers (Churcher *et al.*, 1993; Sumner-Smith *et al.*, 1991). In addition to the trinucleotide bulge region, two base pairs above and below the bulge also contribute significantly to Tat binding (Churcher *et al.*, 1993; Weeks and Crothers, 1991). Phosphate contacts below the bulge at positions 22, 23, and 40 are critical for Tat interactions (Calnan *et al.*, 1991; Churcher *et al.*, 1993; Hamy *et al.*, 1993). Chemical crosslinking studies showed that a TAR duplex containing a trisubstituted pyrophosphate replacing the phosphate at 38–39 reacted specifically with Lys51 in the basic region of Tat(37–72) peptide (Naryshkin *et al.*, 1997). Site-specific photocrosslinking experiments on a Tat–TAR complex using 4-thioUracil as a photoactive nucleoside showed that Tat interacts with U23, U38, and U40 in the major groove of TAR RNA (Wang and Rana, 1996). In a recent study, 6-thio-G was incorporated at specific sites in the TAR RNA sequence and Tat–TAR photocrosslinking experiments were performed (Wang and Rana, 1998). Results of these experiments provide direct evidence that during RNA–protein recognition Tat is in close proximity to O^6 of G21 and G26 in the major groove of TAR RNA. Taken together, these studies establish that Tat binds TAR RNA at the trinucleotide bulge region and interacts with two base pairs above and below the bulge in the major groove of RNA.

B. Rev Protein

The Tat protein functions as a transcriptional activator, whereas Rev acts as a sequence-specific nuclear RNA export factor. Rev is involved in efficient nuclear export and hence expression of the various incompletely spliced viral transcripts (Cullen, 1998). The target RNA sequence required for Rev function is called the Rev response element (RRE) and it is located within the *env* reading frame (Cochrane *et al.*, 1990; Daly *et al.*, 1989; Dayton *et al.*, 1989; Felber *et al.*, 1989; Malim *et al.*, 1989; Rosen *et al.*, 1988; Zapp and Green, 1989). It has been suggested previously that binding of Rev to RRE RNA is important in regulation of HIV mRNA splicing (Chang and Sharp, 1989; Kjems *et al.*, 1991; Lu *et al.*, 1990), in facilitating the nuclear export of the incompletely spliced mRNA (Emerman *et al.*, 1989; Felber *et al.*, 1989; Malim *et al.*, 1989), and in increasing the translational efficiency of the structural proteins (D'Agostino *et al.*, 1992; Lawrence *et al.*, 1991). More recent studies show that Rev induces the sequence-specific nuclear export of pre-mRNAs and mRNAs containing RRE sequence (Bogerd *et al.*, 1995; Fischer *et al.*, 1995; Fischer *et al.*, 1994; Stutz *et al.*, 1995; Wen *et al.*, 1995).

Rev protein contains $\sim$116 amino acids and an arginine-rich sequence located near the amino terminus that serves as both an RNA-binding motif and as a nuclear localization signal. Initially Rev binds to an RNA bulge sequence as a monomer and serves as a nucleation site to recruit additional Rev monomers to the RRE in a multimerization process that requires both RNA–protein and protein–protein interactions. The second functional domain of Rev is located between residues 75 and 84 and contains an $\sim$10-amino-acid leucine-rich sequence. This leucine-rich domain functions as a nuclear export signal both in Rev and when attached to other substrate proteins (Fischer *et al.*, 1995). Since Rev contains both a nuclear localization signal and a nuclear export sequence, it rapidly shuttles back and forth between the nucleus and the cytoplasm of expressing cells.

RRE has been mapped to 234-residue-long RNA and it has an elaborate secondary structure (Felber *et al.*, 1989; Holland *et al.*, 1992; Malim *et al.*, 1989; Rosen *et al.*, 1988). Subsequent studies show that a complete biologically active RRE RNA is 351 nucleotides (nt) in length (Mann *et al.*, 1994). This extended RRE contains an extra 58 nucleotides on the 5′ end and 59 nucleotides on the 3′ end beyond the sites included in the original models for the RRE secondary structure (Mann *et al.*, 1994). Purified Rev protein binds to RRE RNA with a dissociation constant of $\approx$1 nM (Cochrane *et al.*, 1990; Daly *et al.*, 1989; Heaphy *et al.*, 1990; Malim *et al.*, 1989; Zapp and Green, 1989). Mutagenesis and RNase protection experiments determined that a 66-nucleotide-long RNA fragment, domain II of the RRE RNA, is sufficient to form high-affinity Rev–RRE complexes *in vitro* (Heaphy *et al.*, 1990; Holland *et al.*, 1990; Malim *et al.*, 1990). Rev binding to RRE is a complex reaction and involves an initial interaction of Rev with a high-affinity binding site in the RNA that is followed by the addition of Rev molecules to the lower affinity binding sites on the flanking RNA sequences (Heaphy *et al.*, 1990, 1991; Malim and Cullen, 1991). The high-affinity binding site of Rev on RRE has been localized to a relatively small stem-loop structure containing nucleotides 45–75 of the RRE RNA by *in vitro* selection (Bartel *et al.*, 1991), mutational analysis (Heaphy *et al.*, 1991), chemical protection and modification (Kjems *et al.*, 1992; Tiley *et al.*, 1992), and nucleotide analog studies (Iwai *et al.*, 1992). Figure 2 shows the RRE structure containing the high-affinity binding site of Rev.

II. RNA–Protein Interactions as a Target for Therapeutic Interventions

Protein–nucleic acid interactions are involved in many cellular functions including transcription, RNA splicing, and translation. Readily accessible synthetic molecules that can bind with high affinity to specific sequences of single- or double-stranded nucleic acids have the potential to interfere with these interactions in a controllable way, making them attractive tools for molecular biology and medicine. Successful approaches used thus far include duplex-forming (antisense) (Miller, 1996) and triplex-forming (antigene) oligonucleotides (Beal and Dervan, 1991; Helene *et al.*, 1992; Maher *et al.*, 1991), peptide nucleic acids (PNA) (Nielsen, 1999), and pyrrole–imidazole polyamide oligomers (Gottesfeld *et al.*, 1997; White *et al.*, 1998). Each

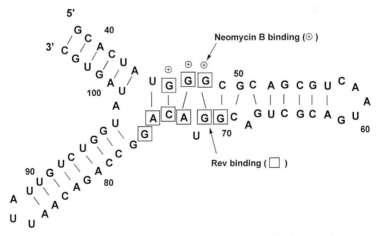

FIGURE 2   Secondary structure of RRE RNA which contains binding site for *in vitro* Rev binding. Numbering of nucleotides in the RNA is derived from Malim *et al.* (1989). Nucleotides involved in Rev and neomycin binding are highlighted.

class of compounds employs a readout system based on simple rules for recognizing the primary or secondary structure of a linear nucleic acid sequence. Another approach employs carbohydrate-based ligands, caliche-amicin oligosaccharides, which interfere with the sequence-specific binding of transcription factors to DNA and inhibit transcription *in vivo* (Ho *et al.*, 1994; Liu *et al.*, 1996). While antisense oligonucleotides and PNA employ the familiar Watson–Crick base pairing rules, two others, the triplex-forming oligonucleotides and the pyrrole–imidazole polyamides, take advantage of straightforward rules to read the major and minor grooves, respectively, of the double helix itself.

In addition to its primary structure, RNA has the ability to fold into complex tertiary structures consisting of such local motifs as loops, bulges, pseudoknots, and turns (Chastain and Tinoco, 1991; Chow and Bogdan, 1997). It is not surprising that when they occur in RNAs that interact with proteins these local structures are found to play important roles in protein–RNA interactions (Weeks and Crothers, 1993). This diversity of local and tertiary structure, however, makes it impossible to design synthetic agents with general, simple-to-use recognition rules analogous to those for the formation of double- and triple-helical nucleic acids. Since RNA–RNA and protein–RNA interactions can be important in viral and microbial disease progression, it would be advantageous to have a general method for rapidly identifying synthetic compounds for targeting specific RNA structures. A particular protein-binding RNA structure can be considered as a molecular receptor not only for the protein with which it interacts but also for synthetic compounds, which may prove to be antagonistic to the protein–RNA interaction.

Two examples of such interactions are recognition of Tat–TAR and Rev–RRE, which are essential elements in the mechanism of HIV-1 gene expression. The following sections describe recent developments in identification of ligands for inhibition of Tat–TAR and Rev–RRE interactions.

III. TAR RNA Ligands as Tat Antagonists

A. TAR RNA Bulge Binders

As discussed above, Tat protein binds a trinucleotide bulge sequence in TAR RNA. Therefore, it is obvious that a Tat antagonist should be able to recognize the bulge structure and, ideally, should be able to bind TAR bulge with affinities higher than that of Tat protein. Hamy and co-workers identified a peptidic compound, CGP64222, that was able to bind TAR RNA with high affinities (Hamy *et al.*, 1997). NMR studies suggested that CGP64222 binds the bulge region of TAR and induces conformational change in TAR, resulting in a structure very similar to that of a Tat-peptide-bound TAR RNA (Aboul-ela *et al.*, 1995; Hamy *et al.*, 1997). This nine-residue oligomer, a hybrid peptide/peptoid, was screened and identified by a deconvolution combinatorial library method. The Tat activity in a cellular Tat-dependent transactivation assay was inhibited with 10–30 μM CGP64222. The structure of CGP64222 is shown in Fig. 3.

In another study, Hamy *et al.* (1998) reported the identification of low-molecular-weight Tat antagonists and their affinities for TAR RNA and biological activities. A series of compounds on the basis of published structural data of the molecular interactions between TAR and Tat-derived peptide was synthesized. This new class of Tat antagonists contains two different functional motifs, a polyaromatic motif for stacking interactions with TAR RNA and a polycationic anchor for contacts with the phosphate backbone of RNA. A varying linker to connect the stacking motif with the RNA binding motif was used. The most active compound competed with Tat–TAR complex formation with a competition dose (CD_{50}) of 22 nM *in vitro* and blocked Tat activity in a cellular system with an IC_{50} of 1.2 μM. Figure 3 shows the structure of the active compound. From structure–activity relationship studies, two new features of Tat–TAR inhibitors became clear: (1) Modification of the linker length has a mild effect on activity and the structure of polyamine moiety is critical for Tat–TAR inhibition and (2) the position of polyaromatic ring for a substitution of the linker is important and the type of chemical bond between the linker and the polyaromatic motif is also crucial for activity.

B. TAR RNA Targeting at Multiple Sites

To identify small organic molecules that inhibit HIV-1 replication by blocking Tat–TAR interactions, Mei *et al.* (1998) screened their research

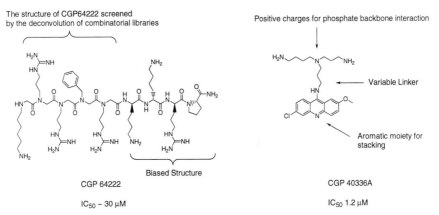

FIGURE 3   Structures of TAR RNA bulge-binding molecules that inhibit Tat–TAR interactions (Hamy *et al.*, 1997, 1998).

compound libraries and reported three inhibitors of Tat–TAR interactions that target TAR RNA and not the protein. These three Tat–TAR inhibitors include neomycin, quinoxaline, and aminoquinozaline. Chemical structures of these compounds, IC_{50}, and their binding sites on TAR RNA are outlined in Fig. 4. Each of these inhibitors recognizes a different structural region in TAR RNA such as the bulge, lower stem, and the loop sequence.

In another attempt to discover Tat–TAR inhibitors, Mei *et al.* (1997) screened their corporate compound library containing ~150,000 compounds. Selective Tat–TAR inhibitors were screened by *in vitro* high-throughput screening assays and inhibitory activities were determined by gel mobility shift assays, scintillation proximity assays, filtration assays, and electrospray ionization mass spectrometry (ESI-MS). After *in vitro* assays, Tat-activated reporter gene analyses were employed to investigate the cellular activities of the primary Tat–TAR inhibitors. Approximately 500 Tat–TAR inhibitors were selected from *in vitro* assays and 50 compounds exhibited dose-dependent cellular activities with IC_{50} values ≤ 50 μM. Among them, approximately 20 compounds were relatively nontoxic (therapeutic index, TC_{50}/IC_{50}, ≥ 5) and considered selective for Tat-dependent transcription.

C. Unnatural Peptides as Ligands for TAR RNA

I. Backbone Modification

We have recently begun to examine TAR RNA recognition by unnatural biopolymers (Tamilarasu *et al.*, 1999; Wang *et al.*, 1997). We synthesized oligocarbamates and oligourea containing the basic arginine-rich region of Tat by solid-phase synthesis methods and tested for TAR RNA binding.

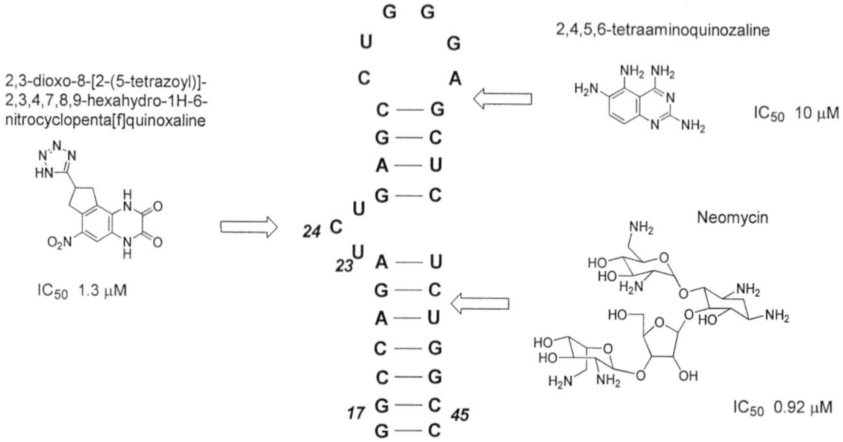

FIGURE 4 Structures and IC_{50} values of three TAR RNA ligands. Putative RNA-binding sites are indicated (Mei *et al.*, 1998).

The oligocarbamate backbone consists of a chiral ethylene backbone linked through relatively rigid carbamate groups (Cho *et al.*, 1993). Oligoureas have backbones with hydrogen bonding groups, chiral centers, and a significant degree of conformational restriction. Introducing additional side chains at the backbone NH sites can further modify biological and physical properties of these oligomers (Fig. 5).

A tat-derived oligourea binds specifically to TAR RNA with affinities significantly higher than the wild-type Tat peptide. To synthesize Tat-derived oligourea on solid support, we used activated *p*-nitrophenyl carbamates and protected amines in the form of azides, which were reduced with SnCl₂-thiophenol-triethylamine on solid support (Kick and Ellman, 1995; Kim *et al.*, 1996). After HPLC purification and characterization by mass spectrometry, the oligourea was tested for TAR RNA binding. The tat-derived oligourea was able to bind TAR RNA and failed to bind a mutant TAR RNA without the bulge residues (Fig. 6).

Equilibrium dissociation constants of the oligourea–TAR RNA complexes were measured using direct and competition electrophoretic mobility assays. Dissociation constants were calculated from multiple sets of experiments which showed that the oligourea binds TAR RNA with a K_d of $0.11 \pm 0.07 \ \mu M$. To compare the RNA-binding affinities of the oligourea to natural peptide, we synthesized a tat-derived peptide (Tyr47 to Arg57)

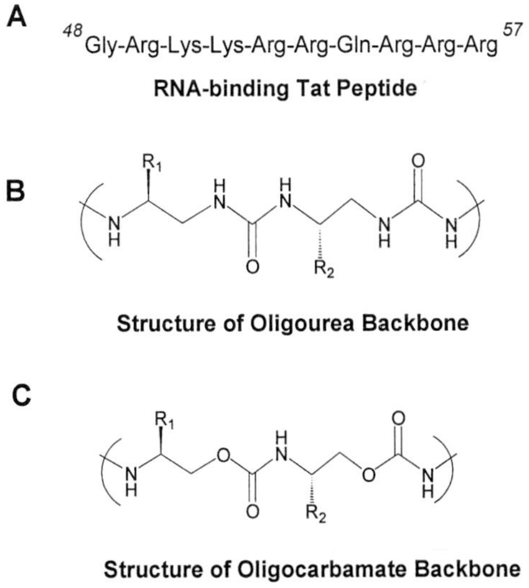

A

^{48}Gly-Arg-Lys-Lys-Arg-Arg-Gln-Arg-Arg-Arg^{57}

RNA-binding Tat Peptide

B

Structure of Oligourea Backbone

C

Structure of Oligocarbamate Backbone

FIGURE 5   (A) The Tat-derived peptide, amino acids 48 to 57, contains the RNA-binding region of Tat protein. Strucutre of the oligourea (B) and oligocarbamate (C) backbone. Sequence of the oligourea and oligocarbamate correspond to the tat peptide shown in A, except the addition of an L-Tyr amino acid at the carboxyl terminus of oligourea (Tamilarasu *et al.*, 1999; Wang *et al.*, 1997).

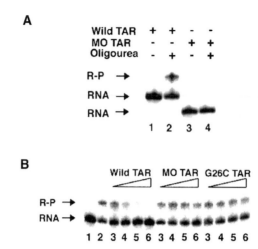

FIGURE 6   (A) Electrophoretic mobility shift analysis for the Tat-derived oligourea binding to wild-type and trinucleotide bulge mutant (M0) TAR RNA. RNA–oligourea complexes are indicated R-P. The ^{32}P-5′-end-labeled TAR RNAs were heated to 85°C for 3 min and then cooled to room temperature in TK buffer (50 mM Tris–HCl (pH 7.4), 20 mM KCl, 0.1% Triton X-100). The oligourea was added to wild-type or mutant TAR and incubated at room temperature for 1 h. After adding 30% glycerol, the oligourea–RNA complexes were resolved on a nondenaturing 12% acrylamide gel and visualized by autoradiography or phosphor-imaging. (B) Specificity of the oligourea–TAR complex formation determined by competition experiments. Oligourea–RNA complexes were formed between ^{32}P-5′-end-labeled TAR RNA (40 nM) and the oligourea (150 nM) in the presence of increasing concentrations of unlabeled wild-type or mutant TAR RNAs. Mutant M0 TAR contained no bulge residue in its sequence. In mutant G26C, a base pair in the upper stem of TAR RNA, G26-C39, was substituted by C26-G39. Concentrations of the competitor RNAs in lanes 3, 4, 5, 6 were 50, 100, 150, and 200 nM, respectively. Lanes 1 and 2 were marker lanes showing RNA and oligourea–RNA complexes. Oligourea–RNA complexes are labeled as R-P. Adapted from Tamilarasu et al. (1999).

containing the RNA-binding domain of Tat protein. Dissociation constants of the Tat peptide–RNA complexes were determined from multiple sets of experiments under the same conditions used for oligourea–TAR RNA complexes. These experiments showed that the Tat peptide (47–57) binds TAR RNA with a K_d of 0.78 ± 0.05 μM. A relative dissociation constant (K_{rel}) can be determined by measuring the ratios of wild-type Tat peptide to the oligourea dissociation constants for TAR RNA. Our results demonstrate that the calculated value for K_{rel} was 7.09, indicating that the urea backbone structure significantly enhanced the TAR binding affinities of the unnatural biopolymer, and this difference in k_d values could be more dramatic because gel mobility shift methods severely underestimate absolute peptide–RNA binding affinities (Long and Crothers, 1995).

Specificity of the oligourea–TAR RNA complex formation was addressed by competition experiments (Fig. 6). Oligourea–RNA complex formation was inhibited by the addition of unlabeled wild-type TAR RNA and not by mutant TAR RNAs. Mutant TAR RNA without a trinucleotide bulge or with one base bulge was not able to compete for oligourea binding to wild-type TAR RNA. Two base pairs immediately above the pyrimidine bulge are critical for Tat recognition (Churcher et al., 1993). To determine whether the oligourea recognizes specific base pairs in the stem region of TAR RNA or only a trinucleotide bulge containing RNA, we synthesized a TAR mutant where the G26-C39 base pair was substituted by a C26-G39 base pair. Competition experiments showed that this mutant TAR (G26C) did not inhibit oligourea binding to TAR RNA. These results indicate that the tat-derived oligourea can specifically recognize TAR RNA.

To probe the oligourea–RNA interactions and determine the proteolysis stability of oligourea, we synthesized TAR RNA containing 4-thioU at position 23 and performed photocrosslinking experiments (Fig. 7) (Wang et al., 1997; Wang and Rana, 1996). Irradiation of the oligourea–RNA complex yields a new band with electrophoretic mobility less than that of the RNA. Both the oligourea and UV (360 nm) irradiation are required for the formation of this crosslinked RNA–oligourea complex (see lanes 3 and 4). Since the crosslinked oligourea–RNA complex is stable to alkaline pH (9.5), high temperature (85°C), and denaturing conditions (8 M urea, 2% SDS), we conclude that a covalent bond is formed between TAR RNA and the oligourea during the crosslinking reaction. To test the protease stability of the oligourea–RNA complexes, we subjected the oligourea–RNA crosslink products to vigorous proteinase K digestion, which showed that the complexes were completely stable and there were no signs of oligourea degradation (lanes 5 and 6). Under similar proteinase K treatment, a degradation of the RNA–protein crosslink products was observed (see lanes 7 and 8).

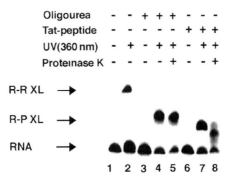

FIGURE 7   Site-specific photocrosslinking reaction of TAR RNA labeled with 4-thioUracil at position 23 with the oligourea (Tamilarasu et al., 1999). For photochemical reactions, RNA duplex was prepared by hybridizing two strands (Wang and Rana, 1996). Strand 1 of the duplex was 5′-end-labeled with ^{32}P. Preformed RNA duplexes (40 nM) in the presence of the oligourea (100 nM) (lanes 4–5) and the Tat peptide (lanes 7–8) were irradiated (360 nm) and analyzed by denaturing gels as described earlier (Wang and Rana, 1996). Proteinase K digestion was performed at 55°C for 15 min after UV irradiation. R-R and R-P XL indicate the RNA–RNA and RNA–oligourea crosslink, respectively. Reprinted from Tamilarasu et al. (1999).

These findings show that a small tat-derived oligourea binds TAR RNA specifically with high affinity and interacts in the major groove of TAR RNA similar to Tat peptides. Due to the difference in backbone structure, oligoureas may differ from peptides in hydrogen-bonding properties, lipophilicity, stability, and conformational flexibility. Moreover, oligoureas are resistant to proteinase K degradation. These characteristics of oligoureas may be useful in improving pharmacokinetic properties relative to peptides. RNA recognition by an oligourea provides a new approach for the design of drugs, which will modulate RNA–protein interactions.

2. D-Peptides

Due to the difference in chirality, D-peptides are resistant to proteolytic degradation and cannot be efficiently processed for major histocompatibility complex class II-restricted presentation to T helper cells (T_H cells). Consequently, D-peptides would not induce a vigorous humoral immune response that impairs the activity of L-peptide drugs (Dintzis *et al.*, 1993). The D-peptide ligands may provide useful starting points for the design or selection of novel drugs.

Can D-peptides recognize naturally occurring nucleic acid structures? To test this hypothesis, we synthesized a D-Tat peptide, Tat(37–72), containing the basic arginine-rich region of Tat by solid-phase peptide synthesis methods. After HPLC purification and characterization by mass spectrometry, the D-Tat peptide was tested for TAR RNA binding (Fig. 8). Similar to L-Tat, the D-Tat peptide was able to bind TAR RNA and failed to bind a mutant TAR RNA without the bulge residues. Equilibrium dissociation constants of the D-Tat–TAR RNA complexes were measured using direct electrophoretic mobility shift assays (Fried and Crothers, 1981; Wang and Rana, 1995). Dissociation constants were calculated from eight sets of experiments which showed that the D-Tat peptide binds TAR RNA with a K_d of 0.22 μM. Under similar experimental conditions, L-Tat(37–72) binds TAR RNA with a K_d of 0.13 μM.

To test the effects of D-Tat peptide on HIV-1 transcription in a cell free system, *in vitro* transcription reactions were performed using HeLa cell nuclear extract and linearized HIV-1 DNA template (Wang and Rana, 1997). Tat produced a large increase in the synthesis of correctly initiated 530-nucleotide runoff RNA transcripts (Huq *et al.*, 1999). Tat stimulated transcription at concentrations ranging from 50 to 300 ng per 10-μL reactions. Quantitation revealed that 100 ng of Tat per reaction produced a 10-12-fold stimulation of HIV-1 transcription. Control experiments showed that Tat did not significantly increase transcription from an HIV-1 promoter with a mutated TAR element either in the stem region (G26 to C26) or in the loop sequence (U31 to G31). Increasing amounts of D-Tat resulted in a significant decrease in Tat-mediated transcriptional activation. In the presence of 1 μg D-Tat ($\approx$ 3$\times$ to the wild-type Tat), more than 80% Tat transactivation was inhibited (Fig. 9A) (Huq *et al.*, 1999). The amount of

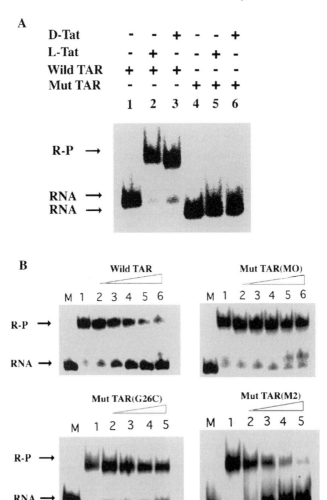

FIGURE 8   (A) Electrophoretic mobility shift analysis for the D-Tat peptide binding to wild-type (wild) and mutant M0 (mut) TAR RNA. RNA–peptide complexes are indicated as R-P. (B) Specificity of the D-Tat–TAR complex formation determined by competition experiments. RNA–peptide complexes were formed between ^{32}P-5'-end-labeled TAR RNA and the D-Tat peptide in the presence of increasing concentrations of unlabeled wild-type or mutant TAR RNAs. Mutant TAR RNAs (M0 and M2) contained different numbers of mismatch bases in the pyrimidine bulge: M0, without bulge bases; M2, two-uridine bulge. In mutant G26C, a base pair in the upper stem of TAR RNA, G26-C39, was substituted by C26-G39. Lane M is a marker lane and RNA–peptide complexes are labeled as R-P. Adapted from Huq *et al.* (1997).

A

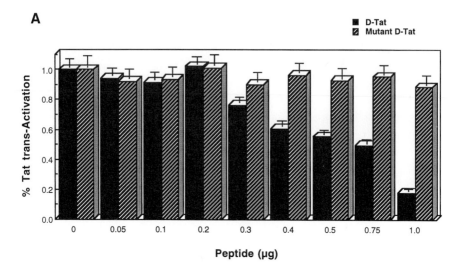

B

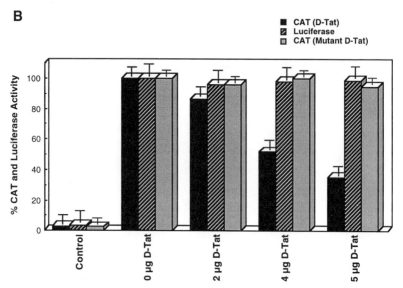

FIGURE 9 (A) Inhibition of Tat transactivation by the D-Tat peptide *in vitro*. Transcriptional activities were measured from three sets of experiments and normalized to Tat activation without the D-Tat peptide inhibitor (100%). To address the specificity of D-Tat inhibitory activities, Tat trans-activation was carried out in the presence of a mutant D-Tat peptide, Gly44-Gln72, where all Arg residues in the RNA-binding region were replaced with Ala. (B) Inhibition of Tat transactivation by the D-Tat peptide *in vivo*. CAT activity expressed from the integrated HIV-1 LTR of HL3T1 cells with increasing amounts of D-Tat or mutant D-Tat peptide is shown. Mutant D-Tat peptide contains Ala in place of Arg at positions 49, 52, 53, 55, 56, and 57 in the RNA-binding region of the peptide. Luciferase activity was a control

recovered transcripts and the efficiency of transcription were normalized by including a labeled RNA not originating from HIV-1 LTR. To determine the specificity of transactivation inhibition by D-Tat, a mutant D-Tat peptide, Gly44-Gln72, where all Arg residues in the RNA-binding region were substituted with Ala, was synthesized. The mutant D-Tat was unable to bind TAR RNA in electrophoretic mobility-shift experiments and did not inhibit Tat *trans*-activation *in vitro* (Fig. 9A). These results indicate that D-Tat is able to specifically inhibit Tat transactivation *in vitro*.

It has been previously established that Tat peptides containing the basic domain are taken up by cells within less than 5 min and accumulate in the cell nucleus (Vives *et al.*, 1997). Since the D-Tat peptide also contains the basic domain of Tat, we reasoned that this peptide would be rapidly taken up by HeLa cells and accumulate in the nucleus. Once D-Tat peptide reaches the nucleus, it would compete with Tat for TAR binding and lead to inhibition of Tat function. To test this hypothesis, we added D-Tat during transfection of pSV2-Tat (Frankel and Pabo, 1988) and pAL (Nordeen, 1988) plasmids into HeLa cells containing an integrated LTR-CAT reporter (Felber and Pavlakis, 1988). Plasmids pSV2Tat and pAL express the first exon of Tat protein and luciferase enzyme, respectively. Transfection of pSV2Tat enhanced transcription, as determined by CAT activity. As shown in Fig. 9B, increasing amounts of the D-Tat resulted in a decrease of CAT activity while luciferase activity was not affected. Tat *trans*-activation was inhibited more than 60% by 5 μg ($\approx$ 0.5 μM) D-Tat peptide. Further addition of D-Tat did not further inhibit Tat transactivation, probably because maximum peptide uptake efficiency is reached at 5 μg of D-Tat. To rule out the possibility that the observed inhibition of transactivation is due to some nonspecific toxicity of the D-peptide or reduction of the pSV2Tat plasmid uptake, transcription of luciferase gene was monitored (Fig. 9B). Transcription of luciferase gene was not affected by D-Tat peptide, as measured by luciferase enzymatic activity assays. Cell viability assays showed that cells were not killed by D-Tat treatment. Specificity of the inhibition was tested by adding a mutant D-Tat peptide, Gly44-Gln72, where all Arg residues in the RNA-binding region were substituted with Ala during transfection of plasmids and the CAT and luciferase activities were analyzed as described above for D-Tat. This mutant D-Tat peptide did not inhibit Tat *trans*-activation. Thus, these results indicate that the D-Tat peptide specifically inhibits *trans*-activation by Tat protein *in vivo*.

---

experiment to monitor the transfection inhibition of pSV2Tat by the addition of D-Tat. Transfection and enzymatic activity (CAT and luciferase) assyas were performed as described previously (Frankel and Pabo, 1988; Nordeen, 1988). CAT and luciferase activities were measured from five experiments and normalized to 100%. Control lane does not contain pSV2Tat or pAL and shows basal level of transcription. (Reprinted from Huq *et al.*, 1999.)

These findings show that a small Tat-derived D-peptide binds TAR RNA and selectively inhibits Tat *trans*-activation. It remains to be determined whether a broad range of RNA–protein interactions can be selectively targeted. These results present an example of the application of D-peptides as artificial regulators of cellular processes involving RNA–protein interactions *in vivo*.

IV. Encoded One-Bead/One-Compound Combinatorial Library Approach in Discovery of HIV Replication Inhibitors

A. Combinatorial Chemistry

The demand for a variety of chemical compounds for identifying and optimizing new drug candidates has increased dramatically. In the past, traditional mass screening of natural products from plants, marine organisms, and synthetic compounds has been successful in identifying a lead chemical structure. In order to support this demand, chemists have developed new methodologies that are accelerating the drug-discovery process. Combinatorial synthesis is one of the most promising approaches to the synthesis of a large collection of diverse molecules because vast libraries of molecules having different chemical identities are synthesized in a short period of time (Armstrong *et al.*, 1996; Brown, 1996; DeWitt and Czarnik, 1996; Ellman, 1996; Still, 1996; Wentworth, Jr. and Janda, 1998). There is considerable evidence to show that combinatorial chemistry plays an important role in the lead-discovery process. For example, a variety of biological targets such as proteases (Kick and Ellman, 1995; Lam *et al.*, 1994; Li *et al.*, 1998), protein kinases (Gray *et al.*, 1998; Lam *et al.*, 1998; Norman *et al.*, 1996), cathepsin D (Kick *et al.*, 1997), and SH3 domain (Feng *et al.*, 1994; Kapoor *et al.*, 1998; Morken *et al.*, 1998) have been screened with this new technology.

Combinatorial synthesis has been primarily facilitated by the application of solid-phase synthesis (Gravert and Janda, 1997; Hermkens *et al.*, 1997). Each substrate is linked to a solid support (a polymer bead), and it is possible to synthesize a variety of products that are spatially separated, and thus reagents and by-products not bound to the beads may be removed simply by filtration. In short, the combinatorial synthesis is faster and thus more efficient and much cheaper than classic organic synthesis and can give up to thousands or even millions of products simultaneously.

The combinatorial drug-discovery process has three major parts: (1) the generation of a large collection of diverse molecules, known as combinatorial libraries, by systematic synthesis of a variety of building blocks; (2) screening of such libraries with biological targets to identify novel lead compounds; and (3) determining the chemical structures of active compounds. Therefore, the combinatorial discovery process requires not only the rational design and synthesis of combinatorial libraries of molecular diversity, but also the development of screening methodologies for library evaluation.

A variety of libraries have been prepared by synthetic and biological methods such as phage display, polysomes, and plasmids (Cortese, 1996; Osborne and Ellington, 1997). Two major synthetic methods are the iterative deconvolution approach (Puras Lutzke *et al.*, 1995; Wilson-Lingardo *et al.*, 1996) and the one-bead/one-compound method (Lam *et al.*, 1991). First, iterative deconvolution is a chemical method in which the chemical structure of the active compound in a combinatorial library is characterized in an iterative manner. It involves screening of combinatorial library pools, identi-fication of the active sublibrary pool, resynthesis of sublibraries, and re-screening of the resynthesized sublibraries. This method has the advantage that it affords fully characterizable, nonmodified structure, solution-phase libraries which afford more realistic interaction results than solid-support bound libraries. However, the time-consuming nature of having to resynthe-size and reassay sublibraries and the potential inconsistencies during resyn-thesis have led to a search for alternative combinatorial methods. Recently, Hamy *et al.* (1997) have used this method to identify inhibitors of HIV-1 replication. The one-bead/one-compound method is based on the fact that the combinatorial bead library contains single beads displaying only one type of compound although there may be up to $\sim$100 pmole ($\sim$10^{13} molecules) of the same molecules on a single 90-μm-diameter polymer bead. This approach has several unique features: (1) a large combinatorial library is synthesized by a split synthesis method; (2) each library member (compound) is spatially separated in the solution and all the library compounds can be screened independently at the same time; and (3) once active beads are screened, the chemical structure of the active beads may be determined directly by using NMR, mass spectrometry, and HPLC (Ni *et al.*, 1996; Youngquist *et al.*, 1995) or by an encoding method (Czarnik, 1997; Ohlmeyer *et al.*, 1993).

B. Split Synthesis

The beauty of split synthesis is that it is a simple and very efficient synthetic method (Fig. 10) (Furka *et al.*, 1991). A sample of support material (bead) is divided into a number of equal portions (*n*) and each reaction vessel is individually reacted with a specific substrate. After completion of the substrate reaction, it is subsequently washed to remove by-products and excess reagent. The individual reaction products are recombined and the whole is thoroughly mixed and divided again into portions. Further reaction with a set of reagents gives complete sets of possible dimer combinatorial libraries, and this whole process may then be repeated as necessary (total *x* times). After this split synthesis, the number of library compounds obtained arises from the exponential increase in molecular diversity, in this case *n* to

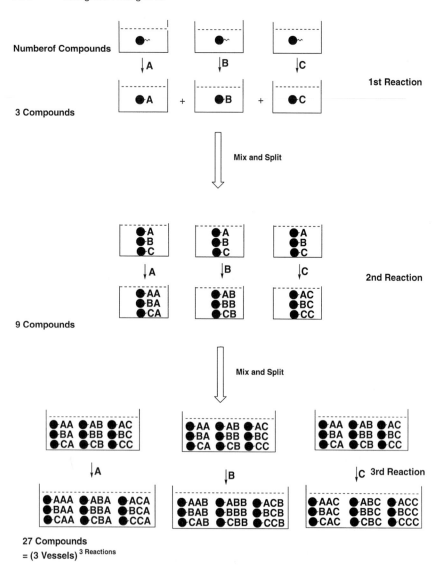

FIGURE 10 Split synthesis using three monomers would produce as many as 27 different timers.

the power of x (n^x). For example, split synthesis using 20 different D and L amino acids at each site of a pentapeptide would produce as many as 20^5 (3,200,000) different compounds.

C. Encoding

The idea of encoding synthetic information with a chemical tag was first proposed by Brenner and Lerner (1992). Encoding involves attaching unique arrays of readily analyzable chemical tags to each bead that designate the particular set of reagents used in the split synthesis of that specific bead (Czarnik, 1997; Lam *et al.*, 1997). Thus analyzing any bead for its tag content yields the history for the synthesis of that specific bead (Fig. 11). Although one could use a different tag for each reagent, it is much simpler to use a mixture of tags because tag mixtures of N different tags can encode 2^N different reactions. For example, only six tags are needed to encode as many as 64 different reactions. While almost any kind of chemical can be used for encoding, there are practical problems because tags need to be chemically inert to library synthesis and reliably analyzed in picogram quantity from a single positive bead. Numerous tagging methods have been developed such as oligonucleotide, peptide, and halophenyl tag approaches (Lam *et al.*, 1997; Maclean *et al.*, 1997; Still, 1996). Among these methods, halophenyl tagging was chosen because it fulfills the following requirements: (1) An assurance of fast structural determination by GC/ECD (gas chromatrography using electron capture detector). In early stages of encoding strategies, biopolymer tags were used (Janda, 1994). The problem with molecular structure determination using biopolymers relates to the stability of the tag to the vigorous organic reaction conditions, and the amount of compound on one bead may not allow for characterization by Edman degradation or PCR. This problem can be solved using photocleavable halophenyl tags that can afford the structural determination in 10 min using femtomole quantities of tags on the bead. (2) Affordability of a variety of tagging compounds: Using different spacers and halophenyl compounds, 30–40 tagging compounds are easily obtained with commercially available starting materials. (3) Simplicity of the tag cleavage reactions: The photocleavable halophenyl tag is cleaved by UV irradiation and it does not require any chemical reactions. (4) Sensitivity of halophenyl derivatives: Halophenyl functionality on the level of only a few femtomoles has been detected with GC/ECD analysis.

D. Identification of Tat–TAR Inhibitors from an Encoded Combinatorial Library

Still's laboratory has recently been involved in the design and combinatorial testing of synthetic receptors for homo- and heterochiral peptides

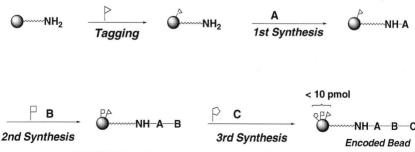

FIGURE 11   Schematic representation of encoding process.

(Still, 1996). This method for finding receptors for specific peptides has led us to approach the problem of RNA recognition, and in particular the recognition of protein-binding RNAs, by treating a particular RNA structure as a receptor for an unknown small-molecule ligand. Peptides are well suited to this task not only because they are made from the same building blocks as the natural protein ligand but also because they can be coupled in high yield on a solid-phase resin, which might then be used directly in screening. In lieu of a reliable set of nucleic acid recognition rules, then, the combinatorial synthesis of many diverse potential antagonists might be used to find new lead compounds for disruption of a particular RNA–protein interaction.

Previous studies using combinatorial chemistry to identify new ligands to block the TAT–TAR interaction have relied on a variety of complex methods that are labor intensive or require expensive robotics equipment (Mei et al., 1997). For the most part, these methods originated in the study of individual protein–nucleic acid interaction experiments. Moreover, in some cases time-consuming deconvolution strategies are also needed to iden- tify the individual compounds responsible for the properties found in a mixture of compounds tested together (Hamy et al., 1997, 1998). We decided to try methods that have previously been used with success on small organic receptors. This entailed covalently attaching the dye Disperse red to the TAR RNA and incubating it in a suspension of library bends made from the split-synthesis method. Diffusion of low-molecular-weight receptors into a bead of Tentagel resin is known to be rapid, whereas one might expect that a macromolecule such as a protein or large nucleic acid might be excluded from the bead interior where the bulk of the peptide is displayed. Nevertheless, we have found that the dye–TAR conjugate was able to enter the beads and bind in a structure-dependent manner. Peptides specific for portions of TAR other than the bulge region were blocked by using a relatively large concentration of an unlabeled TAR analog lacking the natural 3-nt bulge. Using a small amount of detergent and a low RNA concentration (250 nM) also minimizes nonspecific binding. Another advantage to our method is the use of chemically encoded beads (Ohlmeyer et al., 1993). Once a dye-stained bead is selected, the identification of the peptide sequence is rapid and straightforward (Fig. 12). Although many binding experiments are conducted simultaneously, the compounds remain discrete, each in its own assay vessel, the bead produced by the split-synthesis method.

We have made an encoded combinatorial tripeptide library of 24,389 possible members from D- and L-alpha amino acids on Tentagel resin (Ohl- meyer et al., 1993). Using on-bead screening we have identified a small family of mostly heterochiral tripeptides capable of structure-specific binding to the bulge loop of TAR RNA. *In vitro* binding studies reveal stereospecific discrimination when the best tripeptide ligand is compared to diastereomeric

A

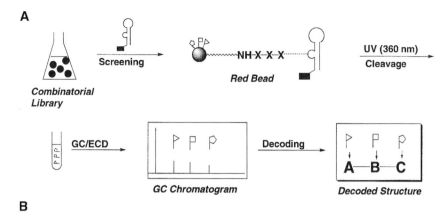

B

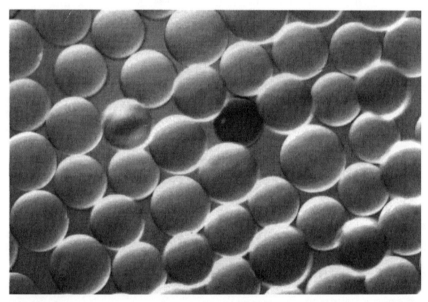

FIGURE 12 (A) Schematic presentation of screening and decoding of the combinatorial library. TAR RNA was labeled with a red dye, Disperse red, at its 5′ end by chemical synthesis and incubated with the trimer library as described in the text. (B) A portion of the beads in the library after incubating with red-dye-labeled TAR RNA. The dark bead in the center was identified as a TAR ligand that specifically binds TAR RNA with high affinties (Hwang *et al.*, 1999).

peptide sequences. In addition, the most strongly binding tripeptide was shown to repress Tat-induced *trans*-activation *in vivo* with an IC_{50} of approximately 50 nM (Hwang *et al.*, 1999).

V. Ligands for RRE RNA as Inhibitors of Rev–RRE Interactions

A. Aminoglycosides

Aminoglycoside antibiotics are antimicrobial agents that block protein synthesis by interacting with ribosomal RNA (Moazed and Noller, 1987). These drugs have also been shown to inhibit *in vitro* self-splicing of group I introns (von Ahsen *et al.*, 1991; von Ahsen and Noller, 1993). Zapp *et al.* (1993) reported that aminoglycoside antibiotics directly interact with RRE RNA and specifically blocked Rev–RRE interactions *in vivo*. A number of aminoglycoside antibiotics such as tobramycin B, kanamycin A, and neomycin B were tested and it was found that neomycin B was the best inhibitor of Rev function with IC_{50} = 0.1 to 1 μM (Zapp *et al.*, 1993).

Due to toxicity of neomycin B, it cannot be used systemically as a successful drug to inhibit viral replication. Neomycin B is also relatively unstable toward acidic conditions and is a good substrate for *in vivo* enzymatic modifications, which leads to drug inactivation and resistance. To overcome these problems associated with the use of neomycin and identify more potent RRE RNA ligands, Wong and coworkers synthesized a library of neomycin B mimetics (Park *et al.*, 1996). Wong's group developed effective and rapid solution-phase synthetic methodologies for the preparation of library compounds containing neamine as a common core structure (Fig. 13). A library was synthesized using a neamine-derived aldehyde and a polyethyleneglycol-linked amine as key substances in the multiple component Ugi condensation. The library of neomycin B mimetics was screened for binding to RRE RNA and several compounds were found to be more active than neamine with IC_{50} in the micromolar range.

B. Diphenylfuran Derivatives

Wilson and colleagues studied the interaction of diphenylfuran derivatives with DNA model systems and showed that these compounds bind

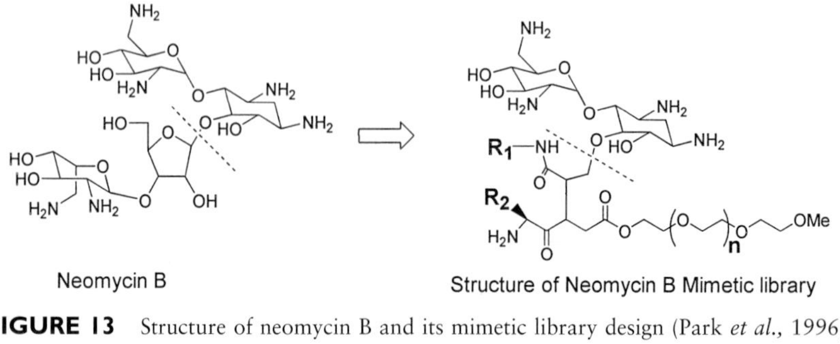

Neomycin B   Structure of Neomycin B Mimetic library

FIGURE 13   Structure of neomycin B and its mimetic library design (Park *et al.*, 1996).

DNA and that the binding mode was dependent on both the sequence and structure of DNA. Subsequent studies showed that diphenylfurans were able to bind duplex RNA with a wide range of binding affinities and modes (Li *et al.*, 1997; Ratmeyer *et al.*, 1996). Based on these observations, a series of synthetic diphenylfuran cations were tested for RRE RNA binding and inhibition of Rev–RRE complex formation. Structure of a diphenylfuran derivative that binds RRE with high affinity is shown in Fig. 14. Fluorescence titrations and gel-shift results showed that diphenylfurans bind to RRE and inhibited Rev–RRE complex formation in a structure-dependent manner (Li *et al.*, 1997; Ratmeyer *et al.*, 1996). The compounds with greatest affinity for RRE inhibited Rev–RRE complex formation at concentrations below 1 μM. The best inhibitors of Rev–RRE interactions contain tetracation structures and appear to interact with RNA through a threading intercalation mode. Spectroscopic experiments suggest that the active diphenylfurans bind at the structured internal loop of RRE and cause a conformational change that is responsible for inhibition of Rev binding.

C. Natural Products

Researchers at Bristol-Myers Squibb Pharmaceutical Research Institute have an extensive natural-product screening program to identify inhibitors of Rev–RRE interactions. During the screening of the natural products, two fungal metabolites, harziphilone and fleephilone, were isolated from the alcohol extract of the fermentation broth of *Trichoderma hazianum* (Qian-Cutrone *et al.*, 1996). Harziphilone and fleephilone inhibited Rev–RRE complex formation with IC$_{50}$ values of 2.0 and 7.6 μM, respectively. However, both compounds did not protect CEM-SS cells from acute HIV-1 infection at concentration levels up to 200 μg/mL. Harziphilone was also cytotoxic at 38 μM against murine tumor cell line M-190.

Additional screening of microbial fermentation extracts was carried out to identify more potent Rev–RRE binding inhibitors. An extract of the fungal solid fermentation culture of *Epicoccum nigrum* WC47880 was found to inhibit the binding of Rev–RRE. After bioassay-guided fractionation, a novel oxopolyene, orevactaene, was isolated as an inhibitor of Rev–RRE complex formation. Orevactaene inhibited Rev–RRE binding with an IC$_{50}$ of 3.6 μM and showed moderate cytotoxicity on murine cell line M-109 (Shu

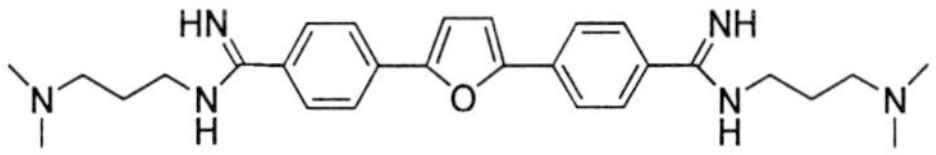

FIGURE 14   Structure of a Rev-RRE inhibitor, a diphenylfuran derivative (Ratmeyer *et al.*, 1996).

et al., 1997). The structures of these natural-product Rev–RRE inhibitors are shown in Fig. 15.

VI. Summary and Perspective

   Tat and Rev proteins and other auxiliary factors intricately control regulation of HIV-1 gene expression. The relevance of RNA structure and RNA–protein interactions to the regulation of gene expression suggests the design of drugs that specifically target regulatory RNA sequences. There has been a considerable effort to discover specific inhibitors of Tat–TAR and Rev–RRE interactions over the past few years and a number of high-affinity RNA ligands have been identified. Combinatorial chemical synthesis is one of the most promising approaches to the synthesis of large collections of diverse molecules. The use of encoded combinatorial chemistry to screen ligands for specific RNA targets would allow the discovery of novel molecules containing diverse structures. To avoid the problem of rapid hydrolysis by host enzymes, peptides and drug molecules containing unnatural linkages can be designed. Structural and activity analysis of the lead molecules will provide new insights into designing drugs with improved properties. The discovery of highly selective and cell-permeable inhibitors of RNA–protein interactions would greatly assist us in understanding the functional significance of RNA–protein interactions *in vivo*.

   An important consideration in identification of inhibitors of RNA–protein interactions, which is often overlooked, is that the inhibitor does not always have to bind RNA at the binding site of the protein. Although

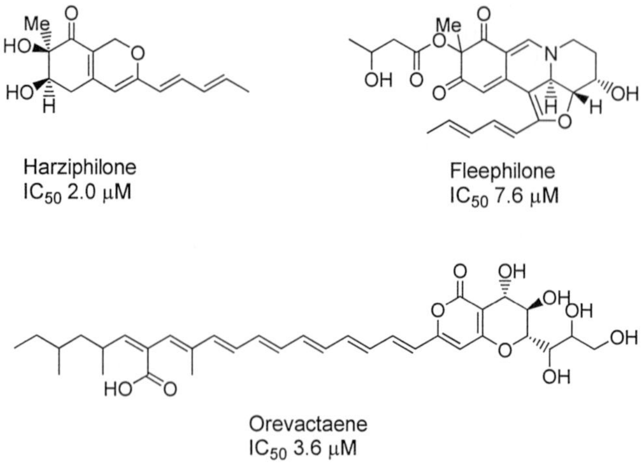

Harziphilone
IC$_{50}$ 2.0 μM

Fleephilone
IC$_{50}$ 7.6 μM

Orevactaene
IC$_{50}$ 3.6 μM

FIGURE 15   Structures and IC$_{50}$ values of Rev-RRE inhibitors screened from natural products (Qian-Cutrone *et al.*, 1996; Shu *et al.*, 1997).

it is true that flexibility of RNA structure makes it difficult to design molecules for specific RNA sequences, however, this flexibility of RNA folding can be exploited to lock an RNA into a nonfunctional structure. For example, a TAR ligand can bind the sequences below or above the bulge or loop region and cause conformational changes in TAR RNA that are not recognized by Tat and cellular proteins involved in transcriptional activation. Similarly, TAR bulge-binding ligands can be identified that interact with the RNA in a manner different from that of Tat. Locking RNA into a nonfunctional structure could also be kinetically more favorable than a ligand–RNA interaction that competes with protein for the same site.

With the increasing wealth of knowledge being generated in the field of RNA–protein recognition and identification of new RNA targets, it would be exciting to see how chemists and biologists use this information to discover drugs that target specific RNA structures and manipulate RNA–protein interactions to control biological processes.

Acknowledgments

This work was supported by grants from the National Institutes of Health (AI 41404 and AI 43198). T.M.R. is a recipient of a Research Career Development Award from NIH.

References

Aboul-ela, F., Karn, J., and Varani, G. (1995). The structure of the human immunodeficiency virus type-1 TAR RNA reveals principles of RNA recognition by Tat protein. *J. Mol. Biol.* **253**, 313–332.

Armstrong, R., Combs, A., Tempest, P., Brwon, S., and Keating, T. (1996). Multiple-component condensation strategies for combinatorial library synthesis. *Acc. Chem. Res.* **29**, 123–131.

Bartel, D. P., Zapp, M. L., Green, M. R., and Szostak, J. W. (1991). HIV-1 Rev regulation involves recognition of non-Watson–Crick base pairs in viral RNA. *Cell* **67**, 529–536.

Beal, P. A., and Dervan, P. B. (1991). Second structural motif for recognition of DNA by oligonucleotide-directed tripl-helix formation. *Science* **251**, 1360–1363.

Berkhout, B., and Jeang, K.-T. (1991). Detailed mutational analysis of TAR RNA: Critical spacing between the bulge and loop recognition domains. *Nucl. Acids Res.* **19**, 6169–6176.

Berkhout, B., and Jeang, K.-T. (1989). Trans-activation of human immunodeficiency virus type 1 is sequence specific for both the single-stranded bulge and loop of the trans-acting-responsive hairpin: A quantitative analysis. *J. Virol.* **63**, 5501–5504.

Berkhout, B., Silverman, R. H., and Jeang, K. T. (1989). Tat trans-activates the human immunodeficiency virus through a nascent RNA target. *Cell* **59**, 273–282.

Bogerd, H. P., Fridell, R. A., Madore, S., and Cullen, B. R. (1995). Identification of a novel cellular cofactor for the Rev/Rex class of retroviral regulatory proteins. *Cell* **82**, 485–494.

Brenner, S., and Lerner, R. A. (1992). Encoded combinatorial chemistry. *Proc. Natl. Acad. Sci. USA* **89**, 5381–5183.

Brown, D. (1996). Future pathway for combinatorial chemistry. *Mol. Div.* **2**, 217–222.

Calnan, B. J., Biancalana, S., Hudson, D., and Frankel, A. D. (1991). Analysis of arginine-rich peptides from the HIV Tat protein reveals unusual features of RNA protein recognition. *Genes Dev.* **5**, 201–210.

Chang, D. A., and Sharp, P. A. (1989). Regulation by HIV rev depends upon recognition of splice site. *Cell* **59**, 789–795.

Chastain, M., and Tinoco, I., Jr. (1991). Structural elements in RNA. *Progr. Nucl. Acid Res. Mol. Biol.* **41**, 131–177.

Cho, C. Y., Moran, E. J., Cherry, S. R., Stephans, J. C., Fodor, S. P., Adams, C. L., sundaram, A., Jacobs, J. W., and Schultz, P. G. (1993). An unnatural biopolymer. *Science* **261**, 1303–1305.

Chow, C. S., and Bogdan, F. M. (1997). A structural basis for RNA–ligand interactions. *Chem. Rev.* **97**, 1489–1514.

Churcher, M. J., Lamont, C., Hamy, F., Dingwall, C., Green, S. M., Lowe, A. D., Butler, P. J. C., Gait, M. J., and Karn, J. (1993). High affinity binding of TAR RNA by the human immunodeficiency virus type-1 *tat* protein requires base-pairs in the RNA stem and amino acid residues flanking the base region. *J. Mol. Biol.* **230**, 90–110.

Cochrane, A. W., Chen, C.-H., and Rosen, C. A. (1990). Specific interaction of the human immunodeficiency virus rev protein with a structured region in the *env* mRNA. *Proc. Natl. Acad. Sci. USA* **87**, 1198–1202.

Cordingley, M. G., La Femina, R. L., Callahan, P. L., Condra, J. H., Sardana, V. V., Graham, D. J., Nguyen, T. M., Le Grow, K., Gotlib, L., Schlabach, A. J., and Colonno, R. J. (1990). Sequence-specific interaction of Tat protein and Tat peptides with the transactivation-responsive sequence element of human immunodeficiency virus type 1 *in vitro. Proc. Natl. Acad. Sci. USA* **87**, 8985–8989.

Cortese, R. (1996). "Combinatorial Libraries: Synthesis, Screening and Application." de Gruyter, New York.

Cullen, B. R. (1998). HIV-1 auxiliary proteins: Making connections in a dying cell. *Cell* **93**, 685–692.

Czarnik, A. (1997). Encoding strategies in combinatorial chemistry. *Proc. Natl. Acad. Sci. USA* **94**, 12738–12739.

D'Agostino, D. M., Felber, B. K., Harrison, J. E., and Pavlakis, G. N. (1992). The *rev* protein of human immunodeficiency virus type 1 promotes polysomal association and translation of *gag/pol* and *vpu/env* mRNA. *Mol. Cell. Biol.* **12**, 1375–1386.

Daly, T. J., Cook, K. S., Gary, G. S., Maione, T. E., and Rusche, J. R. (1989). Specific binding of HIV-1 recombinant *rev* protein to the *rev*-responsive element *in vitro. Nature* **342**, 816–819.

Dayton, E. T., Konings, D. A. M., Powell, D. M., Shapiro, B. A., Butini, L., Maizel, J. V., and Dayton, A. I. (1989). Functional analysis of CAR, the target sequence for the *rev* protein if HIV-1. *Science* **246**, 1625–1629.

DeWitt, S., and Czarnik, A. (1996). Combinatorial organic synthesis using Parke-Davis's DIVERSOMER method. *Acc. Chem. Res.* **29**, 114–122.

Dingwall, C., Ernberg, I., Gait, M. J., Green, S. M., Heaphy, S., Karn, J., Lowe, A. D., Singh, M., and Skinner, M. A. (1990). HIV-1 Tat protein stimulates transcription by binding to the stem of the TAR RNA structure. *EMBO J.* **9**, 4145–4153.

Dintzis, H. M., Symer, D. E., Dintzis, R. Z., Zawadzke, L. E., and Berg, J. M. (1993). A comparison of the immunogenicity of a pair of enantiomeric proteins. *Proteins* **16**, 306–308.

Ellman, J. (1996). Design, synthesis, and evaluation of small-molecule libraries. *Acc. Chem. Res.* **29**, 132–143.

Emerman, M., Vazeaux, R., and Peden, K. (1989). The *rev* gane product of the human immunodeficiency virus affects specific RNA localization. *Cell* **57**, 1155–1165.

Felber, B. K., Hadzopoulou-Cladaras, M., Cladaras, C., Copeland, T., and Pavlakis, G. N. (1989). *Rev* protein of human immunodeficiency virus type 1 affects the stability and transport of the viral mRNA. *Proc. Natl. Acad. Sci. USA* **86**, 1495–1499.

Felber, B. K., and Pavlakis, G. N. (1988). A quantitative bioassay for HIV-1 based *trans*-activation. *Science* **239**, 184–187.

Feng, S., Chen, J., Yu, H., Simon, J., and Schreiber, S. (1994). Two binding orientations for peptides to the Src SH3 domain: Development of a general model for SH3-ligand interactions. *Science* 266, 1241–1247.

Fischer, U., Huber, J., Boelens, W. C., Mattaj, I. W., and Lührmann, R. (1995). The HIV-1 Rev activation domains is a nuclear export signal that accesses an export pathway used by specific cellular RNAs. *Cell* 82, 475–483.

Fischer, U., Meyer, S., Teufel, M., Heckel, C., Lührmann, R., and Rautmann, G. (1994). Evidence that HIV-1 rev directly promotes the nuclear export of unspliced RNA. *EMBO J.* 13, 4105–4112.

Frankel, A. D., and Pabo, C. O. (1988). Cellular uptake of the tat protein from human immunodeficiency virus. *Cell* 55, 1189–1194.

Fried, M., and Crothers, D. M. (1981). Equilibria and kinetics of lac repressor-operator interactions by polyacrylamide gel electrophoresis. *Nucl. Acids Res.* 9, 6505–6525.

Furka, A., Sebestyen, F., Asgedom, M., and Dibo, G. (1991). General method for rapid synthesis of multicomponent peptide mixtures. *Int. J. Pept. Prot. Res.* 37, 487–493.

Gottesfeld, J. M., Neely, L., Trauger, J. W., Baird, E. E., and Dervan, P. B. (1997). Regulation of gene expression by small molecules. *Nature* 387, 202–205.

Gravert, D., and Janda, K. (1997). Organic synthesis on soluble polymer supports: Liquid-phase methodologies. *Chem. Rev.* 97, 489–509.

Gray, N. S., Wodicka, L., Thunnissen, A.-M. W. H., Norman, T. C., Kwon, S., Espinoza, F. H., Morgan, D. O., Barnes, G., LeClerc, S., Meijer, L., Kim, S.-H., Lockhart, D. J., and Schultz, P. G. (1998). Exploiting chemical libraries, structure, and genomics in the search of kinase inhibitors. *Science* 281, 533–538.

Hamy, F., Asseline, U., Grasby, J., Iwai, S., Pritchard, C., Slim, G., Butler, P. J. G., Karn, J., and Gait, M. J. (1993). Hydrogen-bonding contacts in the major groove are required for human immunodeficiency virus type-1 *tat* protein recognition of TAR RNA. *J. Mol. Biol.* 230, 111–123.

Hamy, F., Brondani, V., Florscheimer, A., Stark, W., Blommers, M. J. J., and Klimkait, T. (1998). A new class of HIV-1 Tat antagonist acting through Tat-TAR inhibition. *Biochemistry* 37, 5086–5095.

Hamy, F., Felder, E., Heizmann, G., Lazdins, J., Aboulela, F., Varani, G., Karn, J., and Klimkait, T. (1997). An inhibitor of the TAT/TAR RNA interaction that effectively suppresses HIV-1 replication. *Proc. Natl. Acad. Sci. USA* 94, 3548–3553.

Heaphy, S., Dingwall, C., Ernberg, I., Gait, M. J., Green, S. M., Karn, J., Lowe, A. D., Singh, M., and Skinner, M. A. (1990). HIV-1 regulator of virion expression (Rev) protein binds to an RNA stem-loop structure located within the Rev response element region. *Cell* 60, 685–693.

Heaphy, S., Finch, J. T., Gait, M. J., Karn, J., and Singh, M. (1991). Human immunodeficiency virus type 1 regulator of virion expression, rev, forms nucleoprotein filaments after binding to a purine-rich "bubble" located within the rev-responsive region of viral mRNAs. *Proc. Natl. Acad. Sci. USA* 88, 7366–7370.

Helene, C., Thuong, N. T., and Harel-Bellan, A. (1992). Control of gene expression by triple helix-forming oligonucleotides: The antigene strategy. *Ann. N. Y. Acad. Sci.* 660, 27–36.

Hermkens, P., Ottenheijm, H., and Rees, D. (1997). Solid-phase organic reactions II: A review of the literature Nov 95–Nov 96. *Tetrahedron* 53, 5643–5678.

Ho. S. N., Boyer, S. H., Schreiber, S. L., Danishefsky, S. J., and Crabtree, G. R. (1994). Specific inhibition of formation of transcription complexes by a calicheamicin oligosaccharide: A paradigm for the development of transcriptional antagonists. *Proc. Natl. Acad. Sci. USA* 91, 9203–9207.

Holland, S. M., Ahmad, N., Maitra, R. K., Wingfield, P., and Venkatesan, S. (1990). Human immunodeficiency virus Rev protein recognize a target sequence in rev-responsive element RNA within the context of RNA secondary structure. *J. Virol.* 64, 5966–5975.

Holland, S. M., Chavez, M., Gerstberger, S., and Venkatesan, S. (1992). A specific sequence within a bulged guanosine residues(s) in a stem-bulge-stem structure *rev*-responsive element RNA is required for trans-activation by human immunodeficiency virus type 1 rev. *J. Virol.* **66**, 3699–3706.

Huq, I., Ping, Y.-H., Tamilarasu, N., and Rana, T. M. (1999). Controlling human immunodeficiency virus type 1 gene expression by unnatural peptides. *Biochemistry* **38**, 5172–5177.

Huq, I., Wang, X., and Rana, T. M. (1997). Specific recognition of HIV-1 TAR RNA by a D-Tat peptide. *Nat. Struct. Biol.* **4**, 881–882.

Hwang, S., Tamilarasu, N., Ryan, K., Huq, I., Still, W. C., and Rana, T. M. (1999). Inhibition of gene expression in human cells through small molecule-RNA interactions. *Proc. Natl. Acad. Sci. USA,* **96**, 12997–13002.

Iwai, S., Pritchard, C., Mann, D. A., Karn, J., and Gait, M. J. (1992). Recognition of the high affinity binding site in rev-response element RNA by the human immunodeficiency virus type-1 rev protein. *Nucl. Acids Res.* **20**, 6465–6472.

Jakobovits, A., Smith, D. H., Jakobovits, E. B., and Capon, D. J. (1988). A discrete element 3 of human immunodeficiency virus 1 (HIV-1) and HIV-2 mRNA initiation sites mediates transcriptional activation by an HIV trans activator. *Mol. Cell. Biol.* **8**, 2555–2561.

Janda, K. D. (1994). Tagged versus untagged libraries: Methods for the generation and screening of combinatorial chemical libraries. *Proc. Natl. Acad. Sci. USA* **91**, 10779–10785.

Jones, K. A., and Peterlin, B. M. (1994). Control of RNA initiation and elongation at the HIV-1 promoter. *Annu. Rev. Biochem.* **63**, 717–743.

Kapoor, T., Andreotti, A., and Schreiber, S. (1998). Exploring the specificity pockets of two homologous SH3 domains using structure-based, split-pool synthesis and affinity-based selection. *J. Am. Chem. Soc.* **120**, 23–29.

Kick, E., and Ellman, J. (1995). Expedient method for the solid-phase synthesis of aspartic acid protease inhibitors directed toward the generation of libraries. *J. Med. Chem.* **38**, 1427–1430.

Kick, E., Roe, D., Skillman, A., Liu, G., Ewing, T., Sun, Y., Kuntz, I., and Ellman, J. (1997). Structure-based design and combinatorial chemistry yield low nanomolar inhibitors of cathepsin D. *Chem. Biol.* **4**, 297–307.

Kim, J. M., Bi, Y. Z., Paikoff, S. J., and Schultz, P. G. (1996). The solid phase synthesis of oligoureas. *Tetrahedron. Lett.* **37**, 5305–5308.

Kjems, J., Calnan, B. J., Frankel, A. D., and Sharp, P. A. (1992). Specific binding of a basic peptide from HIV-1 *rev*. *EMBO J.* **11**, 1119–1129.

Kjems, J., Frankel, A. D., and Sharp, P. A. (1991). Specific regulation of mRNA splicing *in vitro* by a peptide from HIV-1 Rev. *Cell* **67**, 169–178.

Lam, K., Lebl, M., and Krchnak, V. (1997). The one-bead-one compound combinatorial library method. *Chem. Rev.* **97**, 411–448.

Lam, K., Salmon, S., Hersh, E., Hruby, V., Kazmierski, W., and Knapp, R. (1991). A new type of synthetic peptide library for identifying ligand-binding activity. *Nature* **354**, 82–84.

Lam, K., Sroka, T., Chen, M., Zhao, Y., Lou, Q., Wu, J., and Zhao, Z. (1998). Application of "one-bead one-compound" combinatorial library methods in signal trasduction research. *Life Sci.* **62**, 1577–1583.

Lam, P., Jadhav, P., Eyermann, C., Hodge, C., Ru, Y., Bacheler, L., Meek, J., Otto, M., Rayner, M., Wong, Y. *et al.* (1994). Rational design of potent, bioavailable, nonpeptide cyclic ureas as HIV protease inhibitors. *Science* **263**, 380–384.

Lawrence, J. B., Cochrane, A. W., Johnson, C. V., Perkin, A., and Rosen, C. A. (1991). The HIV-1 *rev* protein: A model system for the coupled RNA transport and translation. *New Biol.* **3**, 1220–1232.

Li, J., Murray, C., Waszkowycz, B., and Young, S. (1998). Targeted molecular deversity in drug discovery: Integration of structure-based design and combinatorial chemistry. *Drug Discov. Today* **3**, 105–112.

Li, K., Fernandez-Saiz, M., Rigl, C., Kumar, A., Ragunathan, K., McConnaughie, A., Boykin, D., Schneider, H.-J., and Wilson, W. (1997). Design and analysis of molecular motifs for specific recognition of RNA. *Bioorg. Med. Chem.* **5**, 1157–1172.

Liu, C., Smith, B. M., Ajito, K., Komatsu, H., Gomez-Paloma, L., Li, T., Theodorakis, E. A., Nicolaou, K. C., and Vogt, P. K. (1996). Sequence-selective carbohydrate-DNA interaction: Dimeric and monomeric forms of the calicheamicin oligosaccharide interfere with transcription factor function. *Proc. Natl. Acad. Sci. USA* **93**, 940–944.

Long, K. S., and Crothers, D. M. (1995). Interaction of human immunodeficiency virus type 1 Tat-derived peptides with TAR RNA. *Biochemistry* **34**, 8885–8895.

Lu, X., Heimer, J., Rekosh, D., and Hammarskjöld, M.-L. (1990). *Proc. Natl. Acad. Sci. USA* **87**, 7598–7602.

Maclean, D., Schullek, J., Murphy, M., Ni, Z., Gordon, E., and Gallop, M. (1997). Encoded combinatorial chemistry: Synthesis and screening of a library of highly functionalized pyrrolidines. *Proc. Natl. Acad. Sci. USA* **94**, 2805–2810.

Maher, L. J. D., Wold, B., and Dervan, P. B. (1991). Oligonucleotide-directed DNA triple-helix formation: An approach to artificial repressors? *Antisense Res. Dev.* **1**, 277–281.

Malim, M. H., and Cullen, B. R. (1991). HIV-1 structural gene expression requires the binding of multiple Rev monomers to the viral RRE: Implications for HIV-1 latency. *Cell* **65**, 241–248.

Malim, M. H., Hauber, J., Le, S.-Y., Maizel, J. W., and Cullen, B. R. (1989). The HIV-1 *rev* *trans*-activator acts through a structured target sequence to activate nuclear export of unspliced viral mRNA. *Nature* **338**, 254–257.

Malim, M. H., Tiley, L. S., McCarn, D. F., Rusche, J. R., Hauber, J., and Cullen, B. R. (1990). HIV-1 structural gene expression requires binding of the *rev* trans-activator to its RNA target sequence. *Cell* **60**, 675–683.

Mann, D. A., Mikaelian, I., Zemmel, R. W., Green, S. M., Lowe, A. D., Kimura, T., Singh, M., Butler, P. J., Gait, M. J., and Karn, J. (1994). A molecular rheostat: Co-operative rev binding to stem I of the rev-response element modulates human immunodeficiency virus type-1 late gene expression. *J. Mol. Biol.* **241**, 193–207.

Mei, H., Mack, D., Galan, A., Halim, N., Heldsinger, A., Loo, J., Moreland, D., Sannes-Lowery, K., Sharmeen, L., Truong, H., and Czarnik, A. (1997). Discovery of selective, small-molecule inhibitors of RNA complexes. I. The Tat protein/TAR RNA complexes required for HIV-1 transcription. *Bioorg. Med. Chem.* **5**, 1173–1184.

Mei, H.-Y., Cui, M., Heldsinger, A., Lemrow, S., Loo, J., Sannes-Lowery, K., Sharmeen, L., and Czarnik, A. (1998). Inhibitors of protein–RNA complexation that target the RNA: Specific recognition of human immunodeficiency virus type I TAR TNA by small organic molecules. *Biochemistry* **37**, 14204–14212.

Miller, P. S. (1996). Development of antisense and antigene oligonucleotide analogs. *Progr. Nucl. Acid Res. Mol. Biol.* **52**, 261–291.

Moazed, D., and Noller, H. F. (1987). Chloramphenicol, erythromycin, carbomycin, and varnamycin B protect overlapping sites in the peptidyl transferase region of 23S ribosomal RNA. *Biochimie* **69**, 879–884.

Moazed, D., and Noller, H. F. (1987). Interaction of antibiotics with functional sites in 16S ribosomal RNA. *Nature* **327**, 389–394.

Morken, J., Kapoor, T., Feng, S., Shirai, F., and Schreiber, S. (1998). Exploring the leucine-proline binding pocket of the src SH3 domain using structure-based, split-pool synthesis and affinity-based selection. *J. Am. Chem. Soc.* **120**, 30–36.

Naryshkin, N. A., Farrow, M. A., Ivanovskaya, M. G., Oretskaya, T. S., Shabarova, Z. A., and Gait, M. J. (1997). Chemical cross-linking of the human immunodeficiency virus type 1 Tat protein to synthetic models of the RNA recognition sequence TAR containing site-specific trisubstituted pyrophosphate analogues. *Biochemistry* **36**, 3496–3505.

Ni, Z., Maclean, D., Holmes, C., Murphy, M., Ruhland, B., Jacobs, J., Gordon, E., and Gallop, M. (1996). Versatile approach to encoding combinatorial organic syntheses using chemically robust secondary amine tags. *J. Med. Chem.* **1996**, 1601–1608.

Nielsen, P. E., (1999). Applications of peptide nucleic acids. *Curr. Opin. Biotechnol.* **10**, 71–75.

Nordeen, S. K. (1988). Luciferase reporter gene vectors for analysis of promoters and enhancers. *Biotechniques* **6**, 454–457.

Norman, T., Gray, N., Koh, J., and Schultz, P. (1996). A structure-based library approach to kinase inhibitors. *J. Am. Chem, Soc.* **118**, 7430–7431.

O'Donnell, M. J., Zhou, C., and Scott, W. L. (1996). Solid-phase unnatural peptide synthesis (UPS). *J. Am. Chem. Soc.* **118**, 6070–6071.

Ohlmeyer, M. H. J., Swanson, R. N., Dillard, L. W., Reader, J. C., Asouline, G., Kobayashi, R., Wigler, M., and Still, W. C. (1993). Complex synthetic chemical libraries indexed with molecular tags. *Proc. Natl. Acad. Sci. USA* **90**, 10922–10926.

Osborne, S., and Ellington, A. (1997). Nucleic acid selection and the challenge of combinatorial chemistry. *Chem. Rev.* **97**, 349–370.

Park, W., Auer, M., Jaksche, H., and Wong, C. (1996). Rapid combinatorial synthesis of amioglycoside antibiotic mimetics: Use of a polyethylene glycol-linked amine and a neamine-derived aldehyde in multiple component condensation as a strategy for the discovery of new inhibitors of the HIV RNA Rev responsive element. *J. Am. Chem. Soc.* **118**, 10150–10155.

Puras Lutzke, R., Eppens, N., Weber, P., Houghten, R., and Plasterk, R. (1995). Identification of a hexapeptide inhibitor of a human immunodeficiency virus integrase protein by using a combinatorial chemical library. *Proc. Natl. Acad. Sci. USA* **92**, 11456–11460.

Qian-Cutrone, J., Huang, S., Chang, L.-P., Pirnik, D., Klohr, S., Dalterio, R., Hugill, R., Lowe, S., Alam, M., and Kadow, K. (1996). Harziphilone and fleephilone, two new HIV Rev/RRE binding inhibitors produced by *Trichoderma harzianum*. *J. Antibiot.* **49**, 990–997.

Ratmeyer, L., Zapp, M., Green, M., Vinayak, R., Kumar, A., Boykin, D., and Wilson, W. (1996). Inhibition of HIV-1 Rev–RRE interaction by diphenylfuran derivatives. *Biochemistry* **35**, 13689–13696.

Rosen, C. A., Sodroski, J. G., and Haseltine, W. A. (1985). Location of cis-acting regulatory sequences in the human T cell lymphotropic virus type III (HTLV-III/LAV) long terminal repeat. *Cell* **41**, 813–823.

Rosen, C. A., Terwillinger, E., Dayton, A. I., Sodrowski, J. G., and Haseltine, W. A. (1988). Intragenic *cis*-acting *art*-responsive sequences of the human immunodeficiency virus. *Proc. Natl. Acad. Sci. USA* **85**, 2071–2075.

Selby, M. J., and Peterlin, B. M. (1990). Trans-activation by HIV-1 Tat via a heterologous RNA binding protein. *Cell* **62**, 769–776.

Shu, Y.-Z., Ye, Q., Li, H., Kadow, K., Hussain, R., Huang, S., Gustavson, D., Lowe, S., Chang, L.-P., Pirnik, D., and Kodukula, K. (1997). Orevactaene, a novel binding inhibitor of HIV-1 REV protein to REV response element (RRE) from Epicoccum Nigrum WC47880. *Bioorg. Med. Chem. Lett.* **7**, 2295–2298.

Still, W. C. (1996). Discovery of sequence-selective peptide binding by synthetic receptors using encoded combinatorial libraries. *Acc. Chem. Res.* **29**, 155–163.

Stutz, F., Neville, M., and Rosbash, M. (1995). Identification of a novel nuclear pore-associated protein as a functional target of the HIV-1 rev protein in yeast. *Cell* **82**, 495.

Sumner-Smith, M., Roy, S., Barnett, R., Reid, L. S., Kuperman, R., Delling, U., and Sonenberg, N. (19991). Critical chemical features in *trans*-acting-responsive RNA are required for interaction with human immunodeficiency virus tye 1 tat protein. *J. Virol.* **65**, 5196–5202.

Tamilarasu, N., Huq, I., and Rana, T. M. (1999). High affinity and specific binding of HIV-1 TAR RNA by a Tat-derived oligourea. *J. Am. Chem. Soc.* **121**, 1597–1598.

Tiley, L. S., Malim, M. H., Tewary, H. K., Stockley, P. G., and Cullen, B. R. (1992). Identification of a high-affinity RNA-binding site for the human immunodeficiency virus type 1 Rev protein. *Proc. Natl. Acad. Sci. USA* **89**, 758–762.

Vives, E., Broden, P., And Lebleu, B. (1997). A truncated HIV-1 Tat protein basic domain rapidly translocates through the plasma membrane and accumulates in the cell. *J. Biol. Chem.* **272**, 16010–16017.

von Ahsen, U., Davies, J., and Schroeder, R. (1991). Antibiotic inhibition of group I ribozyme function. *Nature* **353**, 368–370.

von Ahsen, U., and Noller, H. F. (1993). Footprinting the sites of interaction of antibiotics with catalytic group I intron RNA. *Science* **260**, 1500–1503.

Wang, X., Huq, I., and Rana, T. M. (1997). HIV-1 TAR RNA recognition by an unnatural biopolymer. *J. Am. Chem. Soc.* **119**, 6444–6445.

Wang, Z., and Rana, T. M. (1995). Chemical conversion of a TAR RNA-binding fragment of HIV-1 Tat protein into a site-specific crosslinking agent. *J. Am. Chem. Soc.* **117**, 5438–5444.

Wang, Z., and Rana, T. M. (1997). DNA damage-dependent transcriptional arrest and termination of RNA polymerase II elongation complexes in DNA template containing HIV-1 promoter. *Proc. Natl. Acad. Sci. USA* **94**, 6688–6693.

Wang, Z., and Rana, T. M. (1996). RNA conformation in the Tat–TAR complex determined by site-specific photo-cross-linking. *Biochemistry* **35**, 6491–6499.

Wang, Z., and Rana, T. M. (1998). RNA-protein interactions in the Tat-*trans*-activation response element complex determined by site-specific photo-cross-linking. *Biochemistry* **37**, 4235–4243.

Wang, Z., Wang, X., and Rana, T. M. (1996). Protein orientation in the Tat-TAR complex determined by psoralen photocross-linking. *J. Biol. Chem.* **271**, 16995–16998.

Weeks, K. M., and Crothers, D. M. (1993). Major groove accessibility of RNA. *Science* **261**, 1574–1577.

Weeks, K. M., and Crothers, D. M. (1991). RNA recognition by Tat-derived peptides: Interaction in the major groove? *Cell* **66**, 577–588.

Wen, W., Meinkoth, J. L., Tsien, R. Y., and Taylor, S. S. (1995). Identification of a signal for rapid export of proteins from the nucleus. *Cell* **82**, 463–473.

Wentworth, P., Jr., and Janda, K. (1998). Generating and analyzing combinatorial chemistry libraries. *Curr. Opin. Biotechnol.* **9**, 109–115.

White, S., Szewczyk, J. W., Turner, J. M., Baird, E. E., and Dervan, P. B. (1998). Recognition of the four Watson–Crick base pairs in the DNA minor groove by synthetic ligands. *Nature* **391**, 468–471.

Wilson-lingardo, L., Davis, P., Ecker, D., Hebert, N., Acevedo, O., Sprankle, K., Brennan, T., Schwarcz, L., Freier, S., and Wyatt, J. (1996). Deconvolution of combinatorial libraries for drug discovery: Experimental comparison of pooling strategies. *J. Med. Chem.* **39**, 2720–2726.

Youngquist, R., Fuentes, G., Lacey, M., and Keough, T. (1995). Generation and screening of combinatorial peptide libraries designed for rapid sequencing by mass spectrometry. *J. Am. Chem. Soc.* **117**, 3900–3906.

Zapp, M. L., and Green, M. R. (1989). Sequence-specific binding by the HIV-1 *rev* protein. *Nature* **342**, 714–716.

Zapp, M. L., Stern, S., and Green, M. R. (1993). Small molecules that selectively block RNA binding of HIV-1 Rev protein inhibit Rev function and viral production. *Cell* **74**, 969–978.

Andrew I. Dayton
Ming Jie Zhang
Laboratory of Molecular Virology
Division of Transfusion Transmitted Disease
Office of Blood Research and Review
Center for Biologics Evaluation and Research
Food and Drug Administration
Rockville, Maryland 20852-1448

Therapies Directed against the Rev Axis of HIV Autoregulation

I. Introduction

The Rev axis of HIV autoregulation is a key axis required for viral replication and is consequently an attractive target for anti-HIV gene therapy strategies. The axis contains two positively acting elements, the Rev protein and its RNA target sequence, the RRE (see Fig. 1). Scattered throughout the HIV genome in the genes coding for virion structural proteins are CRSs (constitutive repressor sequences), which act in cis to constitutively downregulate the expression of the mRNAs which contain them. During early HIV infection, in the absence of Rev, only small, regulatory viral mRNAs which are multiply spliced and lack CRSs are expressed as protein. Later in infection, after levels of Rev have built up, Rev binds to the RRE, which is retained in the mRNAs for genomic and virion structural RNAs, and overcomes the CRS repression to promote the cytoplasmic expression of RRE-containing viral RNAs. Rev may have multiple mechanisms of action, but by far the

Advances in Pharmacology, Volume 49

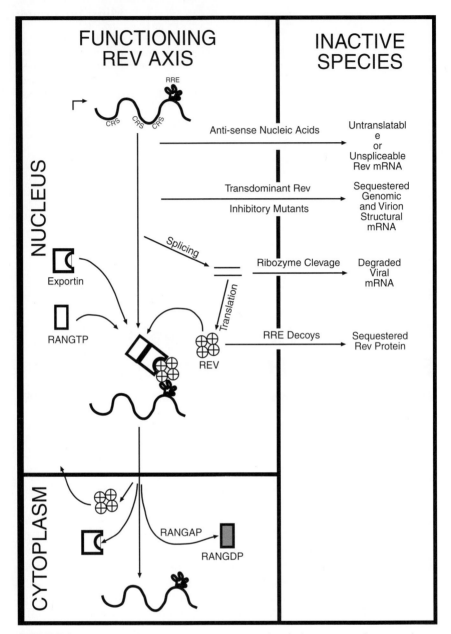

FIGURE I   Summary of therapeutic targets associated with the Rev axis. The Rev pathway is diagrammed on the left. The results of targeting indicated steps and intermediates are summarized on the right.

best established is its role in promoting the transport of RRE-containing RNAs from the nucleus to the cytoplasm. Rev binds to the RRE as a multimer. Evidence indicates that the Rev protein, through its nuclear export signal (NES), binds exportin/CRM1 and Ran GTP cooperatively when it is bound to RRE-containing RNA. The complex undergoes export to the cytoplasm where Ran GAP (Ran GTPase-activating protein) stimulates GTP hydrolysis, inducing the dissociation of the complex. The RNA is freed to the cytoplasm and Rev shuttles back into the nucleus to start the cycle over again. Eukaryotic translation initiation factor eIF5A has also been implicated in contributing to Rev's action, but the mechanism remains unclear. Cytoplasmic effects of Rev have been described, but their mechanisms of action also remain to be determined. (For reviews of Rev's mechanism of action, see Dayton, 1996; Cullen, 1998 a,b; Pollard and Malim, 1998; and Emerman and Malim, 1998.)

Numerous steps in the Rev pathway are appropriate targets for gene therapy. Most efforts have focused on blocking Rev access to the RRE by transdominant Rev inhibitors which bind to the RRE without promoting expression. Major efforts have also focused on sequestering Rev protein with various RRE decoys. Efforts have also been made to direct antisense RNA, antisense oligonucleotides, and ribozymes against Rev mRNA. Many of these approaches have been combined with one another and with other anti-HIV genes (for recent reviews of anti-Rev-based gene therapies, see Nakaya *et al.*, 1997a; Heguy, 1997; Bunnell and Morgan, 1997.)

II. Transdominant Rev Mutants and Modified Rev Proteins ___

The early discovery of several Rev mutations, which not only inactivated the Rev protein but conferred upon it the property of trans-dominant inhibition of wild-type Rev (Malim *et al.*, 1998, 1992; Zapp *et al.*, 1991), paved the way for the development of anti-HIV therapies based on trans-dominant protein inhibitors. The size and nature of such inhibitors implies they require intracellular introduction via gene therapy and to date this remains the exclusive mode of their introduction. The chief advantage of transdominant inhibitors, such as the well-known RevM10, is that they already have the required signals to ensure their transport to the nucleus where Rev's actions occur initially and probably predominantly. Rev-based transdomiant inhibitors (RevTD) may have the potential to act not only by binding to the RRE and blocking access of wild-type Rev, but also by forming inactive mixed multimers with wild-type Rev. In the latter case, one Rev might effectively inactivate a multimer containing a number of wild-type Rev molecules, depending on the size of the functional multimers and on the dependence of their function on the number of wild-type Rev molecules they incorporate. At least in the case of RevTDs, acting by competitively binding to the RRE and blocking the binding of wild-type Rev, there is an additional theoretical advantage. In this situation, viral resistance should require at least two viral mutations rather than one for the evolution of full resistance. Escape mutants should need to have acquired one mutation in the RRE to interfere with RRE binding to the inhibitor and a second mutation, in the viral Rev, to compensate for the altered structure of the RRE. RevTDs are the most prominent of the antivirals directed against the Rev axis if for no other reason than they have been the first to advance to clinical trials. The workhorse of Rev transdominant inhibitors has uniformly been the RevM10 mutant, in which the LERL at the core of the nuclear export signal has been mutated to DLRL (Malim *et al.*, 1988). In general, RevM10 is transduced into cells to form stable, Rev-M10-expressing derivatives which are then tested for activity.

Early results with RevTDs were mixed, though not without promise. Malim *et al.* (1992) used retroviral transduction to stably express RevM10 in CEM cells. Consistent with Rev's acting late in infection, all of the RevM10-expressing clones could be infected by an HIV-1 provirus construct with the CAT (chloramphenicol acetyl transferase) gene substituted for the Nef gene, which is expressed early and is not under Rev control, when infection was monitored by CAT expression. However, the RevM10-expressing clones would not support productive replication when infected by a clone of HIV-1(IIIB). Unfortunately, only two of three RevM10-expressing CEM cell lines were resistant to highly productive infection by a heterogeneous HIV-1(IIIB) pool. On the other hand, these studies demonstrated the basic feasibility of transdominant Rev inhibition and demonstrated that constitutive RevM10 expression did not grossly interfere with cellular physiology, having no effect on the secretion of IL-2 in response to mitogen stimulation of EL-4 and Jurkat cells. In similar studies, Bevec *et al.* (1992) reported that HIV-1(IIIB) infection of CEM cells, as measured by PCR of proviral DNA in high-molecular-weight DNA, was not inhibited by constitutive RevM10 expression. In contrast to infection of CEM cells or CEM cells transduced with wild-type REV, HIV-1 infection of CEM cells transduced with RevM10 failed to produce significant levels of p24 antigen released into the media. Further support for this general approach was offered by a later extension of this early work. Stable expression of RevM10 was shown to inhibit HIV-1 production in chronically infected A3.01 cell lines, suggesting the feasibility of using this type of gene therapy to treat patients already infected with HIV (Escaich *et al.*, 1995).

Subsequent to these early studies, work in the area tended to follow along three different lines, focusing on improving the efficiency of expression of the transdominant inhibitor, on combining its expression with other anti-HIV agents, and on efficiently transducing it into primary cells to lay the groundwork for eventual introduction into HIV patients, using gene therapy techniques.

Much of the effort at boosting the efficiency of RevM10 expression has been directed at choosing appropriate transcriptional promoters. Bahner *et al.* (1993) retrovirally transduced RevM10 into CEM cells under the control of various promoters. Expression of RevM10 from a CMV promoter was the most effective at inhibiting HIV-1 production (p24) subsequent to challenge by HIV-1 infection. Slightly less effective was expression of RevM10 from a MoMLV LTR promoter. Minimal inhibition was seen with expression from an HIV-1 LTR promoter. Plavec *et al.* (1997), on the other hand, using vectors of various structures, reported that steady-state RevM10 levels in transduced CEMSS cells were uniformly higher when expressed from MoMLV, MPSV (myeloproliferative sarcoma virus), or MESV (mouse embryonic stem cell virus) promoters located in the LTR than when expressed from internal promoters, such as CMV or PGK. LTR expression of RevM10 with these retroviral promoters also yielded high steady-state protein levels in activated primary T cells and low levels in resting primary T cells, correlating, respectively, with strong and weak inhibition of HIV replication in these cell types. However, in the experiments presented, it was not demonstrated that these transgenic constructs specifically affected the Rev axis. Liu *et al.* (1994) boosted RevM10-endowed resistance to HIV-1 infection by expressing it from a strong constitutive enhancer and including in its mRNA a TAR (HIV Tat response) element. Ranga *et al.* (1997) boosted the efficiency of RevM10 by placing it under the control of CD4-specific regulatory elements and an upstream CMV(IE) enhancer containing KB and AP1 regulatory elements in addition to TAR. CD4+, peripheral blood mononuclear cells were rendered resistant to productive HIV-1 infection by several such constructs.

Some laboratories have focused on improving transdominant Rev inhibition of HIV replication by searching for improved Rev transdominant inhibitor mutants or by modifying Rev or its derivatives with functional groups from other molecules. Furuta *et al.* (1995) constructed dRev, a mutant of Rev in which the nucleolar targeting signal and RNA binding domain are deleted (Δ38–44). Even though the intracellular distribution of dRev is predominantly cytoplasmic, when transduced virally by either of two viral systems into HeLa CD4 cells to make stable expressors (with transcription either from an HIV-1 LTR or from a CMV promoter), dRev inhibited virus replication, syncytium formation, and cell death caused by HIV-1 infection. Stable expression of dRev also conferred upon CEM cells an HIV-1-resistant phenotype. Inhibition obtained by expressing the dRev from the HIV-1 LTR was generally modest. In CEM cells, however, expressing the dRev from a CMV promoter resulted in approximately 1 log inhibition in virus production. Ragheb *et al.* (1995) tested the relative efficiency of several transdominant inhibitors and found no major, consistent advantages over RevM10 of any of five other inhibitors tested.

Success in fusing RevTDs to functional regions of other molecules has occurred both with fusing RevTDs to other HIV regulatory gene mutants and with fusing RevTDs to non-HIV viral gene fragments. Liem *et al.* (1993) retrovirally transduced either Tat-inducible or Tat/Rev-inducible transdominant mutants of Tat and/or Rev into MT4 cells. As measured by p24 levels, a high level of resistance to infection by HIV-1(NL4-3) was observed in cells expressing a double Tat/Rev mutant in which the transgenes were expressed from overlapping open reading frames (ORFs) in a Tat-inducible manner. Tat+Rev inducibility did not work as efficiently, presumably because the expected late expression of such a construct would be relatively less effective against an early viral gene system such as the Tat/TAR axis. Similarly, Aguilar-Cordova *et al.* (1995) constructed a fusion protein, tRev, containing transdominant mutants of both Tat and Rev (Δ80–82), demonstrating that the tRev protein entered the nucleus and independently inhibited both Tat and Rev functions (as measured by separate, specific indicator assays). Using this construct they achieved approximately sixfold inhibition of the Rev axis (compared to about eightfold inhibition of the Tat axis). Fusion of Tat and Rev synergistically inhibited HIV replication in the Jurkat-derived 1G5 cells. Fusion of Tat and Rev mutants synergeistically inhibited HIV-1 replication when measured 10 days postchallenge. However, when measured at 4 or 7 days postchallenge, the fusion construct was no more inhibitory than Tat or Rev mutants expressed separately. Despite the simplicity of coexpressing Tat and Rev transdominant inhibitors as a fusion construct, fusion is not absolutely necessary. Ulich *et al.* (1996) separately cotransduced RevM10 and a Tat transdominant inhibitor into H9 cells and demonstrated that the two synergized in inhibiting HIV-1(HXB2) replication (as measured by p24 production) despite the relatively moderate inhibitory effect of the Tat mutant expressed alone.

At least two groups have attempted to harness the influenza virus NS1 protein's multiple properties of inhibiting splicing, transport, and 3′-end formation of pre-mRNAs (Qian *et al.*, 1994; Qiu and Krug, 1994; Lu *et al.*, 1994; Nemeroff *et al.*, 1998) by targeting the protein specifically to the HIV RRE via the RNA binding domain of Rev. The Krug laboratory reported that a fusion construct of wild-type Rev and full-length NS1 is equivalent to RevM10 alone in inhibiting the Rev axis in transient transfection assays. A different NS1/Rev fusion bearing full-length NS1 with a mutation in the RNA binding domain also inhibits the Rev axis in HIV infection and transfection experiments, but does not block the nuclear export of free Rev protein. Curiously, in their system, a RevM10/NS1 fusion construct is less inhibitory than RevM10 alone (Qian *et al.*, 1996; Chen *et al.*, 1998)! Our own laboratory (Zhang and Dayton, 1997) fused a peptide containing the NS1 effector domain, but no polyA-RNA binding domain, to RevM10. In transient transfection assays we demonstrated that the effector domain, when fused to RevM10, improves the inhibitory effects of RevM10 in a manner dependent on an RRE in the target RNA. Furthermore, the improved efficiency was most noticeable at lower levels of the fusion construct. These studies demonstrated the validity of targeting general inhibitory molecules specifically to viral RNAs.

Considerable effort has been made to improve the transduction of Rev transdominant inhibitors into various primary cells as a first step toward the clinic. Woffendin *et al.* (1994) used gold microparticle introduction of RevM10 into human PBL to convey HIV resistance *in vitro*. Vandendriessche *et al.* (1995) transduced CD4 T lymphocytes (CD8-depleted PBMC selected with G418 after transduction) from uninfected donors and reported greater inhibition of HIV-1 replication with transdominant Rev (Leu78→Asn78) than with antisense TAR or antisense Rev RNA when challenged with either an ordinary primary HIV-1 isolate or an AZT-resistant, primary isolate. Bauer *et al.* (1997) used neomycin-resistant retroviral vectors to transduce various anti-HIV genes into GCSF-mobilized CD34 cells from patients already infected with HIV-1. After 3 days' transduction on allogenic stroma with IL-6 and IL-3, followed by G418 selection for neomycin resistant cells, cultures were resistant to challenge with HIV-1(JR-FL) and a primary culture of HIV-1. In this system, RevM10 was only marginally less inhibitory than a double hammerhead ribozyme directed against Tat/Rev mRNA and considerably more inhibitory than an RRE decoy when cells were challenged with HIV-1(JR-FL) or a primary HIV-1 isolate. Davis *et al.* (1998) measured the transduction efficiency of transdominant Rev (Leu78→Asn78) into CD34 cord blood cells from uninfected individuals and obtained, without selection, transduction efficiencies of from 7 to 85% (average, 28%) in CFU-GM. Macrophages derived from RevTD-transduced CFU-GM colonies supported infection by HIV-1 Bal in four of eight cultures. Though nominally better than the vector-alone controls, which supported infection in only six of eight controls, these results suggested considerable room for improvement. To improve transduction efficiency by another selection method, Kaneshima and coworkers expressed RevM10 in a MoMLV-based retroviral vector which coexpressed the murine CD8-α chain (Lyt2) to allow for immuno-enrichment and immuno-tagging. They demonstrated that RevM10-expressing, CD34-enriched hematopoetic stem cell progenitors could differentiate normally (after immunoselection for Lyt2) into T cells and CD14(+) cells which showed decreased susceptibility to HIV infection. In myeloid cells, they demonstrated Rev inhibition of JR-FL infection. In T cells they demonstrated inhibition of JR-CSF and a primary HIV-1 isolate (Su *et al.*, 1997; Bonyhadi *et al.*, 1997). Extending earlier work which showed that a fusion construct of transdominant inhibitor mutants of Tat and Rev in Jurkat cells effectively inhibited HIV infection (Aguilar-Cordova *et al.*, 1995 and see above), Chinen *et al.* (1997) transduced their tRev fusion construct into peripheral blood lymphocytes, with an efficiency in the range of 50 to 60% and challenged the cells with two primary HIV isolates and two cloned isolates. In cultures of tRev-CD4+ T cells, tRev inhibited primary HIV strains 4- to 15-fold (with respect to untransduced parental cells) and completely inhibited NL4-3 and SF2.

Although most of the work on Rev transdominant inhibitors has focused on efficacy, safety issues are also of concern. In general there have been few reported toxicities associated with transdominant Rev inhibitors (however, see Nosaka *et al.*, 1993) and they are generally considered inert except for their antiviral properties. For instance, Fox *et al.*, (1995) used gold-particle gene transfer to transduce human lymphocytes *in vitro*. They observed no obvious pathologic defects following adoptive transfer of the genetically modified human T cells into SCID mice. T-cell function was not impaired in PBL of HIV-seropositive donors transduced with RevM10 compared to cells transduced with a ΔRevM10 control mutant. Furthermore, RevM10 had no malignant transformation ability using *in vitro* IL-2 dependence and fibroblast transformation assays. There has been a report that RevM10, as well as wild-type Rev, enhances HTLV-I gene expression in a manner not directed to the RxRE. This enhancement effect seems to be directed at the RU5 region, but, surprisingly, seems not to involve binding of RNA (Kubota *et al.*, 1998).

III. Antisense Oligonucleotides

A. Antisense Oligonucleotides against Rev mRNA

The comparatively simple process of delivering to cells oligonucleotides directed against various parts of the Rev axis prompted early investigation into the effects of oligonucleotides directed against Rev mRNA as well as against the RRE. The overlap of the Rev, Tat, and Env coding sequences and the overlap of the Env and RRE sequences offered the possibility of simultaneously targeting multiple viral elements by each approach.

In early work on targeting Rev mRNA, Matsukura *et al.* (1989) designed 28-base phosphorothioate antisense oligonucleotides complementary to the translation initiation site of Rev. Applied free in solution to H9 cells chronically infected with HIV-1, these oligonucleotides inhibited Rev-dependent protein expression and drastically reduced unspliced viral mRNA, consistent with an impairment of the Rev pathway. Controls which had no effect included the same oligo but with normal phosphodiester linkages; phosphorothioate oligonucleotides containing sense, random, or homopolymeric sequences; and antisense sequences with four thymidine residues substituted by N3-methyl thymidine. In other early work, the Gallo laboratory determined that this same 28-base phosphorothioate oligonucleotide and a second 28-base overlapping phorphorothioate oligonucleotide directed against Rev mRNA blocked HIV-1 replication in MOLT-3 cells in a sequence-specific manner during long culture periods, in contrast to an oligo directed against an HIV splice acceptor for Tat which allowed viral breakthrough by 25 days in culture (Lisziewicz et al., 1992, 1993). Sequence-dependent antiviral activity of the anti-Rev mRNA oligonucleotides was limited to a window of specificity at concentrations between 0.25 and 1.0 μM, a window which overlaps the nonspecific toxicity window. Cell survival was 100% at 0.1 μM, but decreased to 5% at 10 μM (Weichold et al., 1995). Although longer oligonucleotides should have greater specificity, Kim et al. (1995) reported that a 15-mer phosphorothioate against Rev mRNA had higher activity than phosphorothioate oligodeoxynucleotides with sense, random, mismatched or homo-20-mer sequences. Using the same 28-base sequence described by Matsukura (1989; and see above), Lund and Hansen (1998) reported that substitution of the normal phosphodiester linkages with two, four, or six phorphorothioate linkages at each end only resulted in molecules with non-sequence-specific HIV inhibitory properties.

Efforts have been made to improve the effectiveness of antisense oligonucletides by encapsulating them in antibody-targeted, immuno-liposomes and by chemically modifying them. Zelphati et al. (1993) reported that liposome immuno-targeting to HLA class I molecules enhanced the inhibitory effects of an unmodified, anti-Rev phorphorothioate oligonucleotide by 60-fold in acutely infected CEM cells. In further work with immunoliposomes, they showed that β- (normal conformation), but not α- (abnormal conformation), phosphodiester oligonucleotides complementary to the translation initiation site of Rev mRNA interfere with viral mRNA in a sequence-specific manner (Zelphati et al., 1994). Selvam et al. (1996) targeted liposomes with monoclonal antibody to CD4 to deliver a 20-base phorphorothioate oligonucleotide directed against the Rev mRNA initiation site. They achieved 85% inhibition of HIV production in acutely infected H9 cells and in acutely infected peripheral blood lymphocytes from normal donors. Empty immunoliposomes, or immunoliposomes containing phosphorothioate oligonucleotides with scrambled sequences, were inactive. Lavigne and Thierry (1997) complexed phosphorothioate oligonucleotides directed against Rev mRNA in liposomes, using the lipid DLS, without using immuno-targeting. Compared to the results without DLS, using DLS allowed 90–95% inhibition with 100- to 1000-fold lower concentrations of oligonucleotide in cultures of acutely infected MOLT-3 and at 1.6×10^6 lower concentrations in cultures of PBMC from normal donors.

A wide range of chemical modifications of oligonucleotides has been assessed in anti-Rev therapy. Svinarchuk et al. (1993) studied the effects of conjugating lipophilic groups to antisense oligonucleotides directed against the Rev initiation site (as well as against other HIV targets). Conjugation with lipophilic groups stimulated binding of oligonucleotides to cells, protected the oligonucleotides from cellular nucleases, and enhanced their activity with respect to unmodified oligonucleotides. The greatest inhibition of

HIV replication in their hands (90% inhibition in MT4 cells) was obtained with cholesterol-conjugated oligonucleotides. The lipophilic derivatives of oligonucleotides containing an ester bond in the linker structure were cleaved by cellular esterases yielding oligonucleotides protected from 5′ nuclease degradation by the glycine residue. Lund *et al.* (1995) reported that a 15-base C-5-propyne-modified phosphorothioate oligodeoxynucleotide complementary to the Rev initiation site was approximately fivefold more effective in providing viral inhibition when added to a culture of H9 chronically infected with HIV-1 LAI than was a 28-mer unmodified phosphorothioate oligodeoxynucleotide targeted to the same sequence and previously shown to inhibit HIV in a sequence-dependent manner. Kuwasaki *et al.* (1996) achieved moderate improvements in efficacy by introducing 2′ methoxy nucleotides and hairpin structures at the 3′ end of antisense phosphorothioates directed against REV initiation and splice acceptor sites.

B. Antisense Oligonucleotides against the RRE

Less advanced than the efforts to target oligonucleotides to Rev mRNA are efforts to target oligonucleotides to the RRE. In parallel with their studies on oligonucleotides directed against Rev mRNA, Lisziewicz *et al.* (1992) reported that 28-base phosphorothioate oligonucleotides against the RRE inhibited HIV-1 replication in MOLT-3 cells, in a sequence-specific manner, at 1 μM over long times in culture, in contrast to an oligo directed against an HIV splice acceptor, which allowed breakthrough by 25 days of culture. Li *et al.* (1993) demonstrated that antisense phosphorothioate oligonucleotides against the high-affinity site of the RRE inhibit Rev activity and HIV-1 replication better than comparable oligonucleotides against Env mRNA in long-term cultures of HeLa-Tat and MOLT-3 cells after acute infection. Fenster *et al.* (1994) tested *in vitro* oligonucleotides against the RRE and described a series of oligonucleotides complementary to stem/loop V (VI, according to Dayton *et al.*, 1989). Expression of HIV-1 Env in COS-7 cells was blocked by nuclear microinjection of ODNs with C-5-propyne-modified pyrimidines and phosphorothioate linkages. Inhibition was highly dependent upon RNA target position, internucleotide chemistry, oligodeoxynucleotide sequence, and concentration. Unmodified phosphodiester or phosphorothioate oligodeoxynucleotides were inactive. In the future, the field may see the development of TOPs (tethered oligonucleotide probes), which have so far only been tested in *in vitro* chemical binding experiments. Moses *et al.* (1997) reported that TOPs containing two functional regions, one complementary to single-stranded RRE regions and the other containing sequences which form Hoogsteen base pairs with a double-stranded RRE region, the two tethered by a short linker, can effectively inhibit Rev/RRE binding *in vitro*.

IV. Antisense RNA ———————————————————————

A. Anti-Rev Antisense RNA

As has been the case for oligonucleotide therapies, antisense RNA therapies against Rev have been directed both at the Rev mRNA and at the RRE. As is the case for Rev transdominant inhibitors, large antisense RNAs need to be introduced into cells by gene therapy technology. Because of their size, antisense RNAs complementary to Rev mRNA are typically complementary to parts of Env and, in most cases, to parts of Tat (because of the desirability of targeting the initiation site for Rev which overlaps the Tat coding sequence). Thus they usually posses the theoretical advantage of targeting multiple genes. Additionally, it should be more difficult for viruses to evolve to antisense-RNA resistance if the antisense RNA is large and individual subregions are sufficient for inhibition.

Studies reported by Gyotoku *et al.* (1991) demonstrated that a 2.7-kb HIV fragment covering Tat, Rev, and a part of Env could, in the antisense orientation, inhibit HIV infection. When the transgene was stably transduced into CEM cells under the control of an SV40 promoter, the antisense construct inhibited HIV replication by 90–95% with respect to vector-alone controls and by 50–60% with respect to CEM cells. In this system, transduction of cells with the retroviral vector alone enhanced HIV-1 replication considerably, suggesting a possible disadvantage to retroviral transduction. Work by Sczakiel *et al.* (1992) underscored the significance of the m.o.i used in these *in vitro* model systems. When they stably expressed a 562-base antisense RNA against Rev/Tat mRNA in Jurkat cells, the transgene could inhibit HIV-1 infection for up to 2 weeks at high m.o.i. and for up to 5 weeks at low m.o.i.

Numerous attempts have been made to improve the efficiency of expression of antisense RNAs directed against Rev. Cagnon *et al.* (1995) inserted a 28-base sequence complementary to the Rev mRNA into a Pol III transcribed, adenovirus VA1 RNA gene in a projecting loop of the VA1 RNA central domain. Stable expression (postelectroporation and postcloning) of this construct in human CEM cells specifically inhibited HIV-1, but not HIV-2, replication for at least 3 months after challenge by infection. The antisense transgene constituted up to 3% of total cellular RNA! Peng *et al.* (1996) tested three different promoters in the context of double- or single-copy retroviral vectors to express a transcript complementary to full-length Rev mRNA in Jurkat cells. The best long-term protection to infection by pNL4-3, measured after 20 weeks or more, was obtained with antisense RNA expression driven by a tRNA promoter in the context of a double-copy vector. Less effective was expression driven by HIV LTR or MLV LTR promoters. These investigators also reported that expression of the antisense RNA resulted in no alteration in cell proliferation or surface CD4 expression.

Using the promonocytic cell line U937, Liu *et al.* (1997) investigated the effects of coexpressing antisense RNAs against the Rev mRNA initiation site, against TAR, and against the Rev/Tat splice acceptor, each sequence expressed as a part of an SnRNA gene, independently expressed from within the same plasmid. After stable transfection and G418 selection, cultures were challenged twice with HIV-1 Bal and the surviving cells were analyzed. All surviving cells produced normal amounts of CD4 and all produced U1/ HIV antisense RNA by *in situ* hybridization and with no detectable remnant HIV infection as determined by p24 assay and PCR. On rechallenge with HIV-1(III B), the cells were resistant to HIV infection even at high m.o.i. Further encouragement for this approach comes from the work of Donahue *et al.* (1998), who obtained CD34-enriched lymphocytes from Rhesus macaques and retrovirally introduced antisense RNA against Tat/Rev mRNA, using repeated rounds of an optimized gene transfer protocol, which resulted in 30–40% gene transfer efficiencies into primary T cells *in vitro*, before reintroducing the lymphocytes into the host. All three animals so treated were challenged by subsequent infection. Only one of the three anti-HIV gene-transduced animals developed measurable p27 serum levels compared to three of three control animals. The degree of protection was correlated with normal lymph node architecture.

B. Anti-RRE Antisense RNA

Moderately encouraging results have also been reported for antisense RNAs directed against the RRE. Kim *et al.* (1996) transiently contransfected RRE antisense and RRE decoy constructs into COS and SUPT1 cells. Both constructs inhibited Rev-dependent expression from an indicator plasmid, pgTat, in which Tat expression is Rev dependent (Malim *et al.*, 1998). Using various cell lines stably transduced by retroviral vectors, Tagieva and Vaquero (1997) obtained similar results with antisense RRE constructs based on an HIV antisense RNA which naturally occurs during HIV infection in tissue culture and PBL. In all cases they achieved from 3-fold to over 10-fold inhibition of HIV-1(IIIB), HIV-1(NDK), and HIV-1(BRU), but not HIV-2. Cells studied included HeLa-CD4; A301; CEM; Jurkat; the chronically HIV-1-infected A301 derivative ACH2; and transduced PBL, selected in G418.

V. Ribozymes

The excitement generated by the possibility of using ribozymes in gene therapy has gone neither unnoticed nor unaddressed in the field of anti-HIV therapies directed against the Rev axis. As is the case with the other therapies, ribozyme approaches can be divided into two categories, those directed against Rev mRNA and those directed against the RRE. As with antisense RNA and oligonucleotides directed against these same targets, the ribozyme approach theoretically benefits from the advantage of multiple targets being hit simultaneously because of the overlap of many coding and regulatory regions in HIV. As with oligonucleotides, however, ribozymes can be rendered ineffective by single base changes in HIV. But if the ribozymes are directed against critical RRE elements, presumably the virus would also have to mutate the RRE binding domain of Rev to compensate for RRE-coded ribozyme escape mutations.

A. Ribozymes against Rev mRNA

Yamada *et al.* (1994) targeted a hairpin ribozyme to the Rev mRNA of HIV-1(HXB2), which successfully inhibited HIV-1(HXB2) when infection was done at a m.o.i. of <0.1 in retrovirally transduced MOLT-4. This same ribozyme exhibited a marginal inhibition of the SF2 strain, which bears a single base substitution at the ribozyme cleavage site. However, the inhibition of SF2 was equivalent to that achieved by a disabled ribozyme which may have residual antisense effects. The data suggested that ribozymes against Rev mRNA of HIV can function catalytically. Zhou *et al.* (1994) expressed a hammerhead ribozyme targeted to the Tat/Rev common mRNA from the LTR of retroviral vectors coding for neomycin resistance and transduced it into CEM cells. Following selection of pools in G418, transduced cells showed resistance to HIV-1(IIIB) when compared to ribozyme controls with point mutations in the catalytic domain of the ribozyme.

As with the other gene therapies directed against the Rev axis, attempts have been made to boost the efficiencies of anti-Rev ribozymes by boosting their expression and by coexpressing them with other anti-HIV molecules either by fusion or separate expression. Zhou *et al.* (1996) retrovirally transduced CEM cells with tandem ribozymes transcribed from various promoters. One ribozyme was directed against Rev/Tat mRNA and the other against Tat mRNA only. Although the MoMLV LTR transcribed transgene RNA to levels 10- to 30-fold higher than the human CMV promoter or tRNA^{Met} promoter, all constructs inhibited infection with HIV-1(IIIB), as measured by p24 levels, about equally. The authors interpreted the results as indicating that less flanking RNA in the ribozyme constructs and higher levels of transgene RNA both correlated with inhibitory activity. Gervaix *et al.* (1997a; see also Section VI) fused an RRE decoy to a ribozyme directed against U5 and a ribozyme directed against Rev mRNA in double- and triple-copy vectors which they transduced into MOLT-4/8 cells. Their strongest inhibition was obtained with a vector containing two separate cassettes, the first cassette being an anti-U5 ribozyme fused to an RRE stem/loop II RRE decoy and the second cassette being the same RRE decoy fused to a ribozyme directed against Rev/Env mRNA. The first cassette was inserted into the 3' LTR of the vector, with transcription being driven by a tRNA^{Val} promoter. During reverse transcription and insertion, this region is duplicated and the final integrated construct has two such cassettes, one at the 5' end and the other at the 3' end. Expression of the second cassette was driven by an internal tRNA^{Val} promoter. Michenzi et al. (1998) combined a high-expressing system (U1SnRNA) with a targeting system which depended either on U1SnRNA's localization in pre-mRNA-containing spliceosomes in general or on a derivatized SnRNA which paired perfectly with the suboptimal 5' splice site of the Rev pre-mRNA. Using either of these two systems to target ribozymes directed against Tat/Rev mRNA, they studied several clones of Jurkat cells selected after transduction. Constructs bearing the derivatives specifically targeted to the 5' splice site of Rev pre-mRNA showed moderately better inhibition than constructs based on the unmodified SnRNA targeting. However, the improvement was variable. As mentioned above in the section on Rev transdominant inhibitors, Bauer et al. (1997) used neomycin-resistant retroviral vectors to transduce anti-HIV ribozymes into G-CSF-mobilized CD34 cells from patients already infected with HIV-1. After 3 days' transduction on allogenic stroma, with IL-6 and IL-3, followed by G418 selection for neomycin-resistant cells, cultures were resistant to challenge with HIV-1(JR-FL) and a primary culture of HIV-1. RevM10 reduced infection by two to three logs in monocytic cells derived from these cultures. Similar results were obtained with a retroviral construct containing a message for two ribozymes, one directed against Rev mRNA and the other against Tat mRNA.

B. Ribozymes Directed against the RRE

Ribozymes directed against the RRE have been somewhat less actively studied than those against Rev mRNA. Duan et al. (1997) studied several ribozymes directed against the RRE, retrovirally transducing them into SUPT1 cells or PHA/IL2-activated PBMC. Constructs included a ribozyme directed against stem/loop V of the RRE and transcribed from an internal CMV promoter; a construct with tandem ribozymes against stem loops IIB and V, also transcribed from an internal CMV promoter; a ribozyme against stem/loop V transcribed from a tRNA promoter in the 3' LTR of the vector; and a derivative of this latter construct containing an additional protein SFv derived from a monoclonal antibody against Rev and transcribed from an internal CMV promoter. Strongest inhibition of HIV-1(NL4-3) infection was obtained with the tRNA/ribozyme against stem/loop V expressed in conjunction with the D8SFv. Best results were obtained when viral infection was at a m.o.i. in the range of 0.01–0.22. The inhibitory effects were minimized at high m.o.i.

VI. RRE Decoys

RRE decoys rank among the most promising and most heavily studied of the anti-Rev therapies. Although large-molecular-weight RRE decoys have received most of the attention to date and generally require introduction by gene therapy techniques, small-molecular-weight aptamers capable of binding the RRE should be able to be introduced extracellularly, either free in solution or encapsulated in liposomes, and merit close study. RRE decoys also possess the theoretical advantage of requiring at least two viral mutations for full escape: one in the viral Rev so it no longer binds therapeutic RRE and another in the viral RRE to compensate for the altered viral Rev. As with other gene therapy approaches, approaches to RRE decoys have included attempts to achieve improved efficiency by boosting their intracellular levels and attempts to achieve improved efficiency by coexpressing them with other anti-HIV molecules.

Early results were obtained by Lee *et al.* (1992). Using a double-copy murine retroviral vector to transduce a 43-base tRNA–RRE fusion construct into CEM-SS cells, they obtained 90% inhibition of HIV replication postinfection, as determined by *in situ* IF and p24 EIA. However, it is doubtful that the inhibition seen occurs because of interference with the Rev axis. The decoy used involves deletion of a major portion of one side of the high-affinity site and the claimed accumulation of relatively greater amounts of small viral mRNA in cells expressing the decoy seems tenuous at best, considering the extremely faint high-molecular-weight viral bands seen on the published Northern blots.

Much subsequent work in the field addressed the development of high intracellular levels of RRE decoys. Bevec *et al.* (1994) expressed various intracellular levels of an RRE decoy by fusing two, three, or six tandem copies of the RRE to the neomycin phosphotransferase gene of a retroviral vector. After selecting clonal and mass cell populations derived from transduced CEM cells for G418 resistance, they challenged them with HIV-1(IIIB). Forty-three days after challenge with virus, they scored cells for HIV antigens and CD4 expression. Only 10% of the cells in populations from "three-copy" RRE construct transfections and 20% of the cells from "two-copy" RRE construct transfections displayed HIV antigens. The "three-copy" and "two-copy" populations were 47 and 64% positive for CD4. Control cell populations transduced with retroviral vector only or, surprisingly, transduced with the "six-copy" construct were 80% positive for HIV antigens, but displayed no CD4 antigen. Choli *et al.* (1994) used retroviral transduction into MT4 cells to express one or two RRE copies, in the sense orientation, from the TK (thymidine kinase) promoter. They found that two RRE copies, but not one, were sufficient to inhibit HIV-1 replication, but again, only very moderately, though still more than the single RRE decoy.

In contrast, one antisense RRE copy sufficed to inhibit HIV replication. Smith *et al.* (1996) designed a construct in which a 13-base minimal Rev binding domain was embedded in a 23-base fragment and expressed from a tRNA-Pol III promoter. By stably transducing this construct into CEM-SS cells via adeno-associated vectors, they achieved 70 to 99% inhibition of HIV gene expression postinfection, as determined by intracellular and extracellular HIV p24. As mentioned above in section V,B, Kim *et al.* (1996) cloned full-length RRE decoys (and RRE antisense RNA) into HIV-based retroviral vectors. When the RRE decoy constructs were transiently cotransfected into SUPT1 cells together with HIV-1(HXB2), expression of p24 was inhibited only 5% at a 4:1 ratio of inhibitor to provirus and by 91% at a 16:1 ratio. In contrast, antisense RRE constructs only inhibited replication by 21 and 65% at the 4:1 and 16:1 ratios respectively. Good *et al.* (1997) expressed a variety of RNAs directed against the Rev axis in tRNA^{Met} or U6 SnRNA expression cassettes to achieve high intranuclear levels of stable Rev-binding RNA aptamers which had been obtained by *in vitro* evolution of Rev binding RNAs (Tuerk *et al.*, 1993). These aptamer constructs were effective when contransfected with HIV proviral DNA into human 293 cells and infection was followed by reverse transcriptase production. Similarly contransfected hairpin and hammerhead ribozymes directed against the U5 region of RNA or antisense RNA constructs also directed against the U5 region of RNA were ineffective.

As with the transdominant inhibitors, some work has addressed the expression of RRE decoys in primary cell types suitable for eventual gene therapy in humans. Bahner *et al.* (1996) transduced retroviral vectors expressing RRE decoy sequences as part of an LTR-directed transcript expressing the neo gene into CD34 progenitor cells which were allowed to differentiate into mature myelomonocytic cells normally permissive for HIV-1(JR-FL) replication. Compared to vector alone control inhibition of HIV-1(JR-FL), inhibition in unselected, decoy-expressing cultures was 50 to 77%. In G418-selected cultures, the inhibition was 99.4–99.9%. Transduced cells grew normally and were able to differentiate into mature myelomonocytic cells. Similarly, Bauer *et al.* (1997) transduced various anti-HIV genes expressed on retroviral vectors containing neomycin phosphotransferase into G-CSF-mobilized CD34 cells from patients already HIV-1 positive. When cells expressing a 41-base RRE subfragment decoy were challenged with HIV-1(JR-FL) infection, replication was inhibited by three logs. When challenged by a primary HIV-1 isolate, replication was inhibited by almost two logs. When the cells expressed a ribozyme against Rev/Tat mRNA, two logs of inhibition were seen with HIV-1(JR-FL) and greater than three logs were seen with the primary HIV-1 isolate. When the cells expressed the RevM10 gene, greater than three logs inhibition was seen with both HIV-1(JR-FL) and the primary isolate.

In the future, new ways to boost intracellular expression of RRE decoys may include expressing them in a novel fashion. Puttaraju and Been (1995), for instance, modified the Anabaena pre-tRNA group I self-splicing PIE (permuted intron–exon) sequence to generate circular forms of the RRE high-affinity site, which were resistant to nuclease degradation in cellular extracts. These circles specifically bind Rev and are candidates for gene therapy molecules, but have not yet been tested in culture.

As is the case with other anti-Rev therapies, efforts have been made to increase the efficiency of RRE decoys by coexpressing them with other antiviral agents. As early as 1994, Yuyama et al. (1994) described a hybrid RRE decoy/ribozyme construct. In this multifunctional construct, an RRE decoy consisting of the primary Rev binding domain was piggybacked onto flanking, cis-acting ribozymes which trimmed the flanking regions of a centrally located ribozyme, which was in turn targeted against Tat mRNA. A similar piggybacked TAR decoy was located at the 5′ end of the construct in a cis-acting ribozyme which trimmed 5′ flanking sequences from the central trans-acting ribozyme. All of the functional groups were demonstrated to have the desired ribozyme and/or decoy functions in vitro, but the vector remains to be optimized for in vivo function. Caputo et al. (1997) transduced into Jurkat cells an RRE decoy and a Tat transdominant inhibitor (Tat22/37) coexpressed on the same vector either tandemly from the same promoter or from separate promoters. Cell clones expressing both agents were resistant to HIV replication at low m.o.i. but not at high m.o.i. Synergism between the RRE decoy and the Tat trans-dominant inhibitor was obtained only when the RRE was expressed 3′ of and in tandem with the Tat construct. Control cultures expressing either agent alone were ineffective at inhibiting viral replication. Yamada et al. (1996) fused an RRE decoy to either one of two ribozymes, one directed against the U5 region of viral RNA and the other against Rev mRNA, expression being driven from an internal tRNA^{Val} promoter in a retroviral construct. These constructs were transfected into MOLT-4 cells and G418-resistant cells were selected. The stem/loop II decoy linked to the anti-Rev ribozyme gave more inhibition of HIV-1(SF2) than the anti-Rev ribozyme alone. The anti-U5 ribozyme-expressing cells were strongly inhibitory and almost as inhibitory as the stem/loop II decoy fused to the anti-U5 ribozyme. Gervaix et al. (1997a) used various combinations of an RRE stem/loop II decoy fused to a ribozyme targeted to U5 RNA or to a ribozyme targeted against Rev/Env mRNA in single-, double-, and triple-copy vectors transduced into MOLT-4 cells which were then clonally selected in G418 after transduction. Across clades A, B, C, D, and E of HIV-1, the most effective inhibition was obtained with coexpression of both ribozyme fusions. In general, expression of ribozymes in double-copy vectors was more effective than expression in single-copy vectors. In later work, Gervaix et al. (1997b) demonstrated they could introduce a triple-copy retroviral construct with two stem/loop II RRE decoys into CD34(+)-enriched cells from G-CSF treated, HIV-1-infected patients with 69 to 100% efficiency. One stem/loop II RRE decoy in the construct was fused to an anti-U5 ribozyme and the other was fused to an anti-Env/Rev ribozyme. CD34(+)-derived macrophages expressed the transgenes and were resistant to HIV challenge *in vitro*. Inouye *et al.* (1998) expressed a Rev binding domain, optimized by Given *et al.* (1993), using adeno-associated virus vectors. With these constructs, inhibition of clinical and laboratory HIV-1 isolates could be obtained in human alveolar macrophages and in CD4-positive lymphocytes.

RRE decoy effects may come from lentiviral vectors useful for targeting resting cells and cells which are natural targets for HIV. Corbeau and Wong-Staal (1998) reported that minimal, HIV-based vectors containing HIV LTRs and extended packaging signals including the RRE could act as TAR and RRE decoys. These anti-HIV properties may represent an added benefit in anti-HIV therapies using HIV-derived vectors.

The report by Lee *et al.* (1994) that expression of a 23-base minimal Rev binding element (Tiley *et al.*, 1992) containing a 13-nucleotide RRE decoy (in a tRNA cassette) could inhibit HIV replication in CEM cells (after retroviral transduction and clonal selection) opened the way for developing small RRE decoys, which could then be introduced by direct application to cells rather than through gene therapy technology. Nakaya *et al.* (1997a,b) developed 29- and 31-base synthetic RRE decoys that were RNA–DNA chimeras and bound Rev with high affinity. When added to the culture medium, one decoy inhibited 90% of viral production from chronically infected MOLT-8 LAI cells. Inhibition of acute infection of healthy donor-derived, peripheral blood mononuclear cells challenged with HIV-1(NL4-3) or a clinical isolate was also seen. These chimeric RRE decoys only moderately inhibited HIV-1 production after TNFα stimulation of the latently infected ACH2 cell line. Symensma *et al.* (1996) further bolstered the promise of this approach by demonstrating that effective RRE decoys could be selected from random sequence pools. They isolated two such RNA aptamers which competed with the RRE for Rev binding but which had primary sequences and structures very different from the RRE. These aptamers could functionally replace the RRE when substituted for it in a Rev-dependent indicator plasmid, mediating the Rev response as well as the RRE when Rev was saturating and better than the RRE when Rev was limiting. These results suggest that in the future RRE decoys will be improved by combinatorial techniques combined with large-scale screening methodology. Kanopka *et al.* (1998), using Lipofectin, cotransfected into HeLa cells an HIV proviral clone and an anti-HIV Rev binding aptamer to obtain specific inhibition of HIV production by the aptamer. Fusion of an anti-HIV-Env-targeted ribozyme to the aptamer resulted in no additional inhibition.

VII. Intracellular Antibodies against Rev _____

A literal, if not unimaginative, interpretation of "intracellular immuniza-tion" against the Rev axis has been to introduce into cells, mostly by gene therapy, intracellular antibodies targeted against the axis. The Pomerantz group reported the introduction into HeLa T4 cells (by calcium phosphate-mediated transfection) of a genetic fusion of light- and heavy-chain variable regions of an anti-Rev monoclonal antibody, D8, targeted to the nuclear export signal of Rev. Expressed intracellularly in SUPT1, CEM, and human PBMC, this SFv construct potently inhibited HIV replication (Duan *et al.*, 1994, 1995), even though the heavy-chain region of the construct later proved to be aberrant and possessing unknown specificity. The VI region, however, was correctly cloned and presumably contains the correct specific-ity. Despite the questionable origin of the heavy-chain region, this SFv does seem mechanistically to inhibit the Rev axis (Duan *et al.*, 1995) in that it induces formation of relatively large amounts of spliced HIV proviral mes-sages. This same construct moderately inhibited PMA, IL-6, and TNFα induction of HIV in the latently infected U1 cell line and PMA or TNFα induction of HIV in the latently infected ACH2 cell line (Ho *et al*, 1998).

As with other therapies, coexpression of intracellular antibodies with other anti-HIV genes is expected to be of value. Along these lines, Inouye *et al.* (1998) coexpressed an RRE decoy and a single-chain antibody, H10, targeted to a C-terminal region of Rev distinct from the nuclear export signal and having a verified sequence. Expressed in an adeno-associated virus vector, the two transgenes together were more inhibitory than either one alone.

In the future, intracellular antibodies against Rev may be improved by designing more efficient antibodies. Pilkington *et al.* (1996) produced a Fab phage display library from the PBL of an HIV-infected, long-term nonprog-ressor. The library was screened against Rev and several clones were obtained with epitopes and binding constants in a range appropriate for intracellular immunization. Such libraries may be future sources for effective antibodies.

There is one report in the literature of a method of introducing anti-Rev antibodies intracellularly without genetic transduction (Pardridge *et al.*, 1994). Affinity purified, polyclonal antibodies directed against a 16-amino-acid sequence in the Rev protein (35–50, encompassing the RRE binding domain) were "cationized" by covalently linking surface carboxyl groups to polyamines via carbodiimide. This raises the isoelectric point and promotes cellular uptake. Cationized, anti-Rev antibodies applied extracellularly at a concentration which would result in 50% of the antibodies binding to the Rev epitope *in vitro* resulted in a very modest, though specific, 20% inhibi-tion of HIV-1 replication in human PBL. However, no data were presented to demonstrate that these cationized antibodies have a mechanistic effect related to the Rev axis.

VIII. Small Molecules Directed against the Rev Axis ─────────

The report by Zapp *et al.* (1993) that certain aminoglycosides inhibit Rev/RRE binding by binding to the RRE and that neomycin B can antagonize Rev function in cultured cells and inhibit HIV production opened the possibility that nonnucleotide, small-molecular-weight compounds might be developed that could serve as anti-Rev therapeutics. Ratmeyer *et al.* (1996) designed and characterized a set of diphenylfuran cations that selectively inhibit Rev/RRE binding. Their best derivatives bound the RRE with a K_a of $10^7\ M^{-1}$ and inhibited binding at levels as low as 1 μM. The binding involves intercalation, probably in the structured internal loop. Zapp *et al.* (1997) reported that a tetracationic diphenylfuran, AK.A, can competitively bind to the RRE high-affinity site and block binding of Rev to the high-affinity site in the RRE at concentrations as low as 0.1 μM. The molecular basis for the AK.A–RNA interaction as well as the mode of RNA binding differs from the aminoglycoside Rev inhibitors. It remains to be seen how well such molecules, other than the aminoglycosides, can function in living cells.

IX. Therapies against Rev Cofactors ─────────────────────

Although the role of eIF-5A in the Rev axis is not as well established as that of CRM-1, it is of interest that Bevec *et al.* (1996) and Junker *et al.* (1996) have reported inhibition of HIV-1 replication in lymphocytes by mutants of eIF-5A. They expressed, by retroviral transduction into human T cell lines, mutants of eIF-5A which still bind Rev/RRE complexes and successfully inhibited replication of several strains of HIV-1 in several cell types. Furthermore, each of the mutant eIF-5A proteins could block nuclear export of Rev in microinjection studies. Junker *et al.* (1996) also showed that expression of eIF5A mutant M13Δ in PBLs resulted in no difference in proliferation and metabolic activity as determined in a 3-(4,5 dimethylthiazole-2-yl)-2, 5-diphenyl tetrazolium bromide (MTT) assay, suggesting that this mutant causes no cellular toxicity and is an appropriate candidate for gene therapy.

X. Rev-Activated Suicide and Antiviral Therapies ─────────

Among the more creative uses of the Rev axis for antiviral strategies has been the use of Rev regulation of toxic genes so that they are specifically expressed only in HIV-infected cells. Rev regulation, of course, involves two levels of regulation. First, CRSs introduced in cis are expected to lower or eliminate constitutive expression of the toxic transgene. Second, an RRE, also in cis, is expected to render the toxic transgene Rev-responsive so that it will be expressed in response to HIV infection. An additional theoretical advantage of this approach is that given a sufficiently toxic transgene such as diphtheria toxin A (DTA) chain, conferring upon it Rev regulation should induce its expression with the earliest production of virion structural proteins, allowing early interdiction of the process of virion production.

Although this strategy was pioneered in the Haseltine laboratory shortly after the discovery of Rev (E. Terwilliger, personal communication), Harrison et al. (1991, 1992a,b) were the first to demonstrate its implementation. In their first approach, a luciferase reporter construct which constitutively expressed luciferase in a manner independent of Rev and Tat was cotransfected with a DTA construct that was Rev/Tat inducible, with baseline expression of DTA downregulated by HIV CRS elements. Introduction of a source of Tat and Rev reduced the luciferase expression by the reporter gene by almost 75%. Similar results were observed upon stable transduction of the Rev-/Tat-inducible DTA into HeLa cells. In this format, the DTA construct impaired HIV production from transfected proviral DNA by as much as 80–90%. Stable introduction of the Rev-/Tat-inducible DTA construct into H9 cells afforded up to two logs of inhibition of HIV-1(IIIB) replication postinfection. Smythe et al. (1994) pursued a variation of this approach by placing a dominant negative mutant Gag under the regulation of Rev. Although this construct would not be expected to kill cells upon induction, theoretically it should inhibit virus production. In uncloned CEM-SS cells transduced with this construct and selected as G418-resistant pools, HIV-1 replication was inhibited 94% and was associated with decreased CPE. SIV replication was also inhibited.

More recent work on the DTA concept has made the system seem less attractive. Regheb et al. (1999) constructed a retroviral vector in which an HIV promoter drove transcription in a sense opposite to the retroviral LTR (which drives neo). From the HIV promoter is transcribed a cassette containing an RRE and a gene of interest. Thus, genes inserted into the cassette are under the regulation of Rev and Tat (with CRSs coming from the cloned region of Env containing the RRE). CD4-enriched, G418-selected PBLs were challenged after transduction with this vector containing in its expression cassette either IFN-α2 or diphtheria toxin. Whereas the DTA transduction only inhibited HIV replication about threefold, and that only after 2 weeks in culture, transduction of IFN-α2 inhibited HIV replication by about 1.5 logs, even at 3 days postinfection, the earliest time point studied. The inhibition of viral replication compared favorably to inhibition by a retrovirally transduced, constitutively expressed, Rev transdominant mutant when the cells were challenged with HIV-1 RF. Inhibition by DTA transduction, however, was almost 2 logs less efficient than inhibition by the Rev transdominant mutant when the cells were challenged with HIV-1(IIIB).

XI. Miscellaneous Anti-Rev Effects

One of the more curious observations to have been reported concerning the Rev axis is its interaction with I-κBα (Wu *et al.*, 1995, 1997), which inhibited HIV replication, in part, through inhibition of Rev function, possibly acting through a cellular factor involved in Rev transactivation. IKBα inhibited by fivefold or less the Rev stimulation of a Rev-dependent reporter plasmid and had similar effects on Rex. However, no Rev effect was noticeable in Northern blots of HIV-1(BRU)-infected 293HEK-CD4 cells after 70% transduction of experimental and control cells. Possibly a fivefold inhibition is too small to be noticeable at the level of relative RNA species. Interestingly, the N terminus of I-κBα is required for inhibition of Rev function but not for inhibition of NF-κBα. Another NF-κB family member, IKBβ, with a distinct amino-terminal sequence, inhibits NF-κBα, but not Rev. Conversely, the C terminus of I-κBα is required for full NF-κB inhibition, but not for Rev inhibition. Furthermore, the nuclear export signal of I-κBα is not required for inhibition of either Rev or NF-κB. Surprisingly, mutants with defects in the inhibition of either Rev or NF-κB could inhibit HIV replication. As intriguing as these observations are, given the small effects observed, more work has to be done to rule out transcriptional effects which might lower steady-state levels of Rev protein.

XII. Clinical Trials

Despite the many exciting possibilities for Rev-directed gene therapy, few such therapies have advanced to clinical trials. By far the most advanced of these strategies, in terms of progression through clinical trials, is RevM10. First results were reported by Woffendin *et al.* (1996), who placed the RevM10 gene, or an inactive deletion mutant, RevM10Δ, under the control of the RSV promoter and the HIV TAR element. These constructs were introduced separately into the CD4-enriched PBLs of three HIV-infected patients using gold-particle-mediated gene transfer and neomycin-resistant vectors. RevM10 and RevM10Δ were placed under the transcriptional control of the RSV promoter. For *in vitro* experiments, cells containing the two constructs were selected in G418 and challenged by HIV-1(BRU) or HIV-1(IIIB) infection. In each case, RevM10 inhibited p24 production by at least three- to fivefold. For patient treatment, transfected cells were reinfused together, after separate *in vitro* expansion without selection, into the donor patient. The numbers of RevM10 and RevM10Δ cells were monitored by

PCR. Although equivalent numbers of both cell types were present in blood 1 h after infusion, by 1 week levels of circulating RevM10Δ cells were down to less than 5% of the levels of RevM10 cells, suggesting preferential HIV resistance of the RevM10 cells. Predominance of the RevM10 cells was seen as long as 56 days postinfusion though both cell types decreased in number over time, with even the RevM10 cells disappearing after 8 weeks. In more recent work along similar lines, Ranga *et al.* (1998) inserted the RevM10 and control genes into retroviral vectors derived from the pLJ plasmid (Korman *et al.*, 1987). The results were generally similar to the gold-particle-mediated gene-transfer experiments except that the relative selection for RevM10 cells was much slower and the persistence of detectable transgenic cells was much more prolonged, with RevM10 cells lasting for an average of 6 months in the three patients studied. Presumably, determinants expressed by the vector and possibly the transgenes are eliciting an immune reaction to the engineered cells, clearing both transgene-containing and vector-alone retrovirus insertions.

Morgan *et al.* (1996, 1998) are pursuing similar studies using identical twins discordant for HIV-1 infection. In these experiments they retrovirally transduce into CD4-enriched lymphocytes from the HIV-negative twin antisense Tar and/or transdominant Rev transgenes in parallel with vector-alone controls. The genetically engineered T cells are then infused into the HIV-infected twin and monitored. Preliminary results indicate that transgenic vectors can be detected for at least 30 weeks postinfusion and, in six of eight patients, a 2- to 20-fold enrichment in the ratio of transgene to vector-alone can also be detected.

XIII. Conclusion

Despite the promise afforded anti-Rev-directed therapies by encouraging *in vitro* results, considerable hurdles remain before the attainment of effective *in vivo* therapies is likely to be obtained. Even the elegant clinical studies establishing that RevTDs introduced by gene therapy can have a protective effect are far from demonstrating any clinical benefit. As is the case for much of gene therapy, the field of gene therapy directed against the Rev axis is in need of better vectors, particularly ones which avoid the *in vivo* decay of transgenic cells. As can be said for much of gene therapy in general, the approaches to date have probably contributed more to our understanding of basic viral, cellular, and organismal mechanisms than to our quest to cure disease.

References

Aguilar-Cordova, E., Chinen, J., Donehower, L. A., and Harper, J. W. (1995). Inhibition of HIV-1 by a double transdominant fusion gene. *Gene Ther.* 2(3), 181–186.

Bahner, I., Kearns, K., Hao, Q. L., Smogorzewska, E. M., and Kohn, D. B. (1996). Transduction of human CD34(+)hematopoietic progenitor cells by a retroviral vector expressing an RRE decoy inhibits human immunodeficiency virus type 1 replication in myelomonocytic cells produced in long-term culture. *J. Virol.* **70**(7), 4352–4360.

Bahner, I., Zhou, C., Yu, X. J., Hao, Q. L., Guatelli, J. C., and Kohn, D. B. (1993). Comparison of transdominant inhibitory mutant human immunodeficiency virus type 1 genes expressed by retroviral vectors in human T lymphocytes. *J. Virol.* **67**(6), 3199–3207.

Bauer, G., Valdez, P., Kearns, K., Bahner, I, Wen, S. F., and Zaia, J. A. (1997). Inhibition of human immunodeficiency virus-1 (HIV-1) replication after transduction of granulocyte colony-stimulating factor-mobilized CD34+ cells from HIV-1-infected donors using retroviral vectors containing anti-HIV-1 genes. *Blood* **89**(7), 2259–2267.

Bevec, D., Jaksche, H., Oft, M., Wohl, T., Himmelspach, M., Pacher, A., Schebesta, M., Koettnitz, K., Dobrovnik, M., Csonga, R., Lottspeich, F., and Hauber, J. (1996). Inhibition of HIV-1 replication in lymphocytes by mutants of the Rev cofactor eIF-5A. *Science* **271**(5257), 1858–1860.

Bevec, D., Volc-Platzer, B., Zimmermann, K., Dobrovnik, M., Hauber, J., Veres, G., and Bohnlein, E. (1994). Constitutive expression of chimeric neo-Rev response element transcripts suppresses HIV-1 replication in human CD4+ T lymphocytes. *Hum. Gene Ther.* **5**(2), 193–201.

Bonyhadi, M. L., Moss, K., Voytovich, A., Auten, J., Kalfoglou, C., Plavec, I., Forestell, S., Su, L., Bohnlein, E., and Kaneshima, H. (1997). Human hematopoietic stem-progenitor cells inhibit human immunodeficiency virus replication. *J. Virol.* **71**(6), 4707–4716.

Bunnell, B. A., and Morgan, R. (1998). Gene therapy for infectious diseases. *Clin. Microbiol. Rev.* **11**(1), 42–56.

Cagnon, L., Cucchiarini, M., Lefebyre, J. C., and Doglio, A. (1995). Protection of a T-cell line from human immunodeficiency virus replication by the stable expression of a short antisense RNA sequence carried by a shuttle RNA molecule. *J. Acq. Immune Defic. Syndr. Hum. Retrovirol.* **9**(4), 349–358.

Caputo, A., Rossi, C., Balboni, P. G., Bozzini, R., Grossi, M. P., Betti, M., and Barbanti-Brodano, G. (1996). The HIV-1 regulatory genes Tat and Rev as targets for gene therapy. *Antibiot. Chemother.* **48**, 205–216.

Caputo, A., Rossi, C., Bozzini, R., Betti, M., Grossi, M. P., Barbanti-Brodano, G., and Balboni, P. G. (1997). Studies on the effect of the combined expression of anti-Tat and anti-Rev genes on HIV-1 replication. *Gene Ther.* **4**(4), 288–295.

Chan, S. Y., Louie, M. C., Piccotti, J. R., Iyer, G., Ling, X., Yang, Z. Y., Nabel, G. J., and Bishop, D. K. (1998). Genetic vaccination-induced immune responses to the human immunodeficiency virus protein Rev: Emergence of the interleukin 2- producing helper T lymphocyte. *Hum. Gene Ther.* **9**(15), 2187–2196.

Chen, Z., Li, Y., and Krug, R. M. (1998). Chimeras containing influenza NS1 and HIV-1 Rev protein sequences: Mechanism of their inhibition of nuclear export of Rev protein-RNA complexes. *Virology* **241**(2), 234–250.

Chinen, J., Aguilar-Cordova, E., Ng-Tang, D., and Lewis, D. E. (1997). Protection of primary human T cells from HIV infection by TRev: A transdominant fusion gene. *Hum. Gene Ther.* **8**(7), 861–868.

Cohli, H., Fan, B., Joshi, R. L., Ramezani, A., Li, X., Joshi, S. (1994). Inhibition of HIV-1 multiplication in a human CD4+ lymphocytic cell line expressing antisense and sense RNA molecules containing HIV-1 packaging signal and Rev response element(s). *Antisense Res. Dev.* **4**(1), 19–26.

Corbeau, P., and Wong-Staal, F. (1998). Anti-HIV effects of HIV vectors. *Virology* **243**(2), 268–274.

Cullen, B. R. (1998a). HIV-1 auxiliary proteins: Making connections in a dying cell. *Cell* **93**(5), 685–692.

Cullen, B. R. (1998b). Retroviruses as model systems for the study of nuclear RNA export pathways. *Virology* **249**(2), 203–210.

Davis, B. R., Saitta, F. P., Bauer, G., Bunnell, B. A., Morgan, R. A., and Schwartz, D. H. (1998). Targeted transduction of CD34(+) cells by transdominant negative Rev-expressing retrovirus yields partial anti-HIV protection of progeny macrophages. *Hum. Gene Ther.* **9**(8), 1197–1207.

Dayton, A. I. (1996). The Rev axis of HIV-1 and its associated host cofactors: A viral window on the workings of eukaryotic post-transcriptional RNA processing. *J. Biomed. Sci.* **3**, 69–77.

Dayton, E. T., Powell, D. M., and Dayton, A. I. (1989). Functional analysis of CAR, the target sequence for the Rev protein of HIV-1. *Science* **246**, 1625–1629.

Donahue, R. E., Bunnell, B. A., Zink, M. C., Metzger, M. E., Westro, R. P., Kirby, M. R., Unangst, T., Clements, J. E., and Morgan, R. A. (1998). Reduction in SIV replication in rhesus macaques infused with autologous lymphocytes engineered with antiviral genes. *Nat. Med.* **4**(2), 181–186.

Duan, L., Zhu, M., Ozaki, I., Zhang, H., Wei, D. L., and Pomerantz, R. J. (1997). Intracellular inhibition of HIV-1 replication using a dual protein- and RNA-based strategy. *Gene Ther.* **4**(6), 533–543.

Duan, L., Bagasra, O., Laughlin, M. A., Oakes, J. W., and Pomerantz, R. J. (1994). Potent inhibition of human immunodeficiency virus type 1 replication by an intracellular anti-Rev single-chained antibody. *Proc. Natl. Acad. Sci. USA* **91**(11), 5075–5079.

Duan, L., Zhu, M., Bagasra, O., and Pomerantz, R. J. (1995). Intracellular immunization against HIV-1 infection of human T lymphocytes: Utility of anti-Rev single-chain variable fragments. *Hum. Gene Ther.* **6**(12), 1561–1573.

Emerman, M., and Malim, M. (1998). HIV-1 regulatory/accessory genes: Keys to unraveling viral and host cell biology. *Science* **280**(5371), 1880–1884.

Escaich, S., Kalfoglou, C., Plavec, I., Kaushal, S., Mosca, J. D., and Bohnlein, E. (1995). RevM10-mediated inhibition of HIV-1 replication in chronically infected T cells. *Hum. Gene Ther.* **6**(5), 625–634.

Fenster, S. D., Wagner, R. W., Froehler, B. C., and Chin, D. J. (1994). Inhibition of human immunodeficiency virus type-1 Env expression by C-5 propyne oligoncleotides specific for Rev-response element stem-loop V. *Biochemistry* **33**(28), 8391–8398.

Fox, B. A., Woffendin, C., Yang, Z. Y., San, H., Ranga, U., Gordon, D., Osterholzer, J., and Nabel, G. J. (1995). Genetic modification of human peripheral blood lymphocytes with a transdominant negative form of Rev: Safety and toxicity. *Hum. Gene Ther.* **6**(8), 997–1004.

Furuta, R. A., Kubota, S., Maki, M., Miyazaki, Y., Hattori, T., and Hatanaka, M. (1995). Use of a human immunodeficiency virus type 1 Rev mutant without nucleolar dysfunction as a candidate for potential AIDS therapy. *J. Virol.* **69**(3), 1591–1599.

Gervaix, A., Li, X., Kraus, G., and Wong-Staal, F. (1997a). Multigene antiviral vectors inhibit diverse human immunodeficiency virus type 1 clades. *J. Virol.* **71**(4), 3048–3053.

Gervaix, A., Schwarz, L., Law, P., Ho, A. D., Looney, D., Lane, T., and Wong-Staal, F. (1997b). Gene Therapy targeting peripheral blood CD34(+) hematopoietic stem cells of HIV-infected individuals. *Hum. Gene Ther.* **8**(18), 2229–2238.

Giver, L., Bartel, D., Zapp, M., Pawul, A., Green, M., and Ellington, A. D. (1993). Selective optimization of the Rev-binding element of HIV-1. *Nucleic Acids Res.* **21**(23), 5509–5516.

Good, P. D., Krikos, A. J., Li, S. X., Bertrand, E., Lee, N. S., Giver, L., Ellington, A., Zaia, J. A., Rossi, J. J., and Engelke, D. R. (1997). Expression of small, therapeutic RNAs in human cell nuclei. *Gene Ther.* **4**(1), 45–54.

Gyotoki, J., el-Farrash, M. A., Fujimoto, S., Germeraad, W. T., Watanabe, Y., Teshigawara, K., Harada, S., and Katsura, Y. (1991). Inhibition of human immunodeficiency virus replication in a human T cell line by antisense RNA expressed in the cell. *Virus Genes* **5**(3), 189–202.

Harrison, G. S., Long, C. J., Maxwell, F., Glode, L. M., and Maxwell, I. H. (1992). Inhibition of HIV production in cells containing an integrated, HIV-regulated diphtheria toxin A chain gene. *AIDS Res. Hum. Retroviruses* 8(1), 39–45.

Harrison, G. S., Long, C. J., Curiel, T. J., Maxwell, F., and Maxwell, I. H. (1992). Inhibition of human immunodeficiency virus-1 production resulting from transduction with a retrovirus containing an HIV-regulated diphtheria toxin A chain gene. *Hum. Gene Ther.* 3(5), 461–469.

Harrison, G. S., Maxwell, F., Long, C. J., Rosen, C. A., Glode, L. M., and Maxwell, I. H. (1991). Activation of a diphtheria toxin A gene by expression of human immunodeficiency virus-1 Tat and Rev proteins in transfected cells. *Hum. Gene Ther.* 2(1), 53–60.

Heguy, A. (1997). Inhibition of the HIV Rev transactivator: A new target for therapeutic intervention. *Front. Biosci.* 2, 283–289.

Ho, W. Z., Lai, J. P., Bouhamdan, M., Duan, L., Pomerantz, R. J., and Starr, S. E. (1998). Inhibition of HIV type 1 replication in chronically infected monocytes and lymphocytes by retrovirus-mediated gene transfer of anti-Rev single-chain variable fragments. *AIDS Res. Hum. Retroviruses* 14(17), 1573–1580.

Inouye, R. T., Du, B., Boldt-Houle, D., Ferrante, A., Park, I. W., Hammer, S. M., Duan, L., Groopman, J. E., Pomerantz, R. J., and Terwilliger, E. F. (1997). Potent inhibition of human immunodeficiency virus type 1 in primary T cells and alveolar macrophages by a combination anti-Rev strategy delivered in an adeno-associated virus vector. *J. Virol.* 71(5), 4071–4078.

Junker, U., Bevec, D., Barske, C., Kalfoglou, C., Escaich, S., Dobrovnik, M., Hauber, J., and Bohnlein, E. (1996). Intracellular expression of cellular eIF-5A mutants inhibits HIV-1 replication in human T cells: A feasibility study. *Hum. Gene Ther.* 7(15), 1861–1869.

Kim, J. H., McLinden, R. J., Mosca, J. D., Vahey, M. T., Greene, W. C., and Redfield, R. R. (1996). Inhibition of HIV replication by sense and antisense Rev response elements in HIV-based retroviral vectors. *J. Acq. Immune Defic. Syndr. Hum. Retrovirol.* 12(4), 343–351.

Kim, S. G., Hatta, T., Tsukahara, S., Nakashima, H., Yamamoto, N., Shoji, Y., Takai, K., and Takaku, H. (1995). Antiviral effect of phosphorothioate oligodeoxyribonucleotides complementary to human immunodeficiency virus. *Bioorg. Med. Chem.* 3(1), 49–54.

Konopka, K., Duzgunes, N., Rossi, J., and Lee, N. S. (1998). Receptor ligand-facilitated cationic liposome delivery of anti-HIV-1 Rev-binding aptamer and ribozyme DNAs. *J. Drug Target* 5(4), 247–259.

Korman, A. J., Frantz, J. D., Strominger, J. L., and Mulligan, R. C. (1987). Expression of human class II major histocompatibility complex antigens using retrovirus vectors. *Proc. Natl. Acad. Sci. USA* 84, 2150–2154.

Kubota, S., Furuta, R. A., Hatanaka, M., and Pomerantz, R. J. (1998). Modulation of HTLV-1 gene expression by HIV-1 Rev through an alternative RxRE-independent pathway mediated by the RU5 portion of the 5′-LTR. *Biochem. Biophys. Res. Commun.* 243(1), 79–85.

Kuwasaki, T., Hosono, K., Takai, K., Ushijima, K., Nakashima, H., Saito, T., Yamamoto, N., and Takaku, H. (1996). Hairpin antisense oligonucleotides containing 2′-methoxynucleosides with base-pairing in the stem region at the 3′-end: Penetration, localization, and Anti-HIV activity. *Biochem. Biophys. Res. Commun.* 228(2), 623–631.

Lavigne, C., and Thierry, A. R. (1997). Enhanced antisense inhibition of human immunodeficiency virus type 1 in cell cultures by DLS delivery system. *Biochem. Biophys. Res. Commun.* 237(3), 566–571.

Lee, S. W., Gallardo, H. F., Gilboa, E., and Smith, C. (1994). Inhibition of human immunodeficiency virus type 1 in human T cells by a potent Rev response element decoy consisting of the 13-nucleotide minimal Rev-binding domain. *J. Virol.* 68(12), 8254–8264.

Lee, T. C., Sullenger, B. A., Gallardo, H. F., Ungers, G. E., and Gilboa, E. (1992). Overexpression of RRE-derived sequences inhibits HIV-1 replication in CEM cells. *New Biol.* 4(1), 66–74.

Li, G., Lisziewicz, J., Sun, D., Zon, G., Daefler, S., Wong-Staal, F., Gallo, R. C., and Klotman, M. E. (1993). Inhibition of Rev activity and human immunodeficiency virus type 1 replication by antisense oligodeoxynucleotide phosphorothioate analogs directed against the Rev-responsive element. *J. Virol.* **67**(11), 6882–6888.

Liem, S. E., Ramezani, A., Li, X., and Joshi, S. (1993). The development and testing of retroviral vectors expressing trans-dominant mutants of HIV-1 proteins to confer anti-HIV-1 resistance. *Hum. Gene Ther.* **4**(5), 625–634.

Lisziewicz, J., Sun, D., Klotman, M., Agrawal, S., Zamecnik, P., and Gallo, R. (1992). Specific inhibition of immunodeficiency virus type 1 replication by antisense oligonucleotides: An *in vitro* model for treatment. *Proc. Natl. Acad. Sci. USA* **89**(23), 11209–11213.

Lisziewicz, J., Sun, D., Metelev, V., Zamecnik, P., Gallo, R. C., and Agrawal, S. (1993). Long-term treatment of human immunodeficiency virus-infected cells with antisense oligonucleotide phosphorothioates. *Proc. Natl. Acad. Sci. USA* **90**(9), 3860–3864.

Liu, D., Donegan, J., Nuovo, G., Mitra, D., and Laurence, J. (1997). Stable human immunodeficiency virus type 1 (HIV-1) resistance in transformed CD4+ monocytic cells treated with multitargeting HIV-1 antisense sequences incorporated into U1 snRNA. *J. Virol.* **71**(5), 4079–4085.

Liu, J., Woffendin, C., Yang, Z. Y., and Nabel, G. J. (1994). Regulated expression of a dominant negative form of Rev improves resistance to HIV replication in T cells. *Gene Ther.* **1**(1), 32–37.

Lu, Y., Qian, X. Y., and Krug, R. M. (1994). The influenza virus NS1 protein: A novel inhibitor of pre-mRNA splicing. *Genes Dev.* **8**(15), 1817–1828.

Lund, O. S., and Hansen, J. (1998). Inhibition of HIV-1 replication by chimeric phosphorothioate oligodeoxynucleotides applied in free solution. *Intervirology* **41**(2–3), 63–68.

Lund, O. S., Nielsen, J. O., and Hansen, J. E. (1995). Inhibition of HIV-1 *in vitro* by C-5 propyne phosphorothioate antisense to Rev. *Antiviral Res.* **28**(1), 81–91.

Malim, M. H., Freimuth, W. W., Liu, J., Boyle, T. J., Lyerly, H. K., Cullen, B. R., and Nabel, G. J. (1992). Stable expression of transdominant Rev protein in human T cells inhibits human immunodeficiency virus replication. *J. Exp. Med.* **176**(4), 1197–1201.

Malim, M. H., Hauber, J., Fenrick, R., and Cullen, B. R. (1988). Immunodeficiency virus Rev transactivator modulates the expression of the viral regulatory genes. *Nature* **335**, 181–183.

Matsukura, M., Zon, G., Shinozuka, K., Robert-Guroff, M., Shimada, T., Stein, C. A., Mitsuya, H., Wong-Staal, F., Cohen, J. S., and Broder, S. (1989). Regulation of viral expression of human immunodeficiency virus *in vitro* by an antisense phosphorothioate oligodeoxynucleotide against Rev (art/trs) in chronically infected cells. *Proc. Natl. Acad. Sci. USA* **86**(11), 4244–4248.

Michienzi, A., Conti, L., Varano, B., Prislei, S., Gessani, S., and Bozzoni, I. (1998). Inhibition of human immunodeficiency virus type 1 replication by nuclear chimeric anti-HIV ribozymes in a human T lymphoblastoid cell line. *Hum. Gene Ther.* **9**(5), 621–628.

Morgan, R. A. (1997). Gene therapy for HIV infection. *Clin. Exp. Immunol.* **107**(Suppl. 1), 41–44.

Morgan, R. A., and Walker, R. (1996). Gene therapy for AIDS using retroviral mediated gene transfer to deliver HIV-1 antisense TAR and transdominant Rev protein genes to syngeneic lymphocytes in HIV-1 infected identical twins. *Hum. Gene Ther.* **7**(10), 1281–1306.

Morgan, R. A., Walker, R., Carter, C., Bachtel, C., Bunnell, B., Fellowes, V., Muul, L., Leitman, S., Lane, C., and Blaese, M. (1998). Intracellular immunity: Preferential survival of CD4 T-cells engineered with anti-HIV genes in a gene therapy trial in HIV-1 discordant identical twins. In "Abstracts of Papers Presented at the 1998 Meeting on Gene Therapy," p. 145. Cold Spring Harbor Laboratory Press, Cold Spring Harbor, NY.

Moses, A. C., Huang, S. W., and Schepartz, A. (1997). Inhibition of Rev–RRE complexation by triplex tethered oligonucleotide probes. *Bioorg. Med. Chem.* **5**(6), 1123–1129.

Nabel, G. J., Fox, B. A., Post, L., Thompson, C. B., and Woffendin, C. (1994). A molecular genetic intervention for AIDS—Effects of a transdominant negative form of Rev. *Hum. Gene Ther.* **5**(1), 79–92.

Nakaya, T., Iwai, S., Fujinaga, K., Otsuka, E., and Ikuta, K. (1997a). Inhibition of HIV-1 replication by targeting the Rev protein. *Leukemia* **11**(Suppl. 3), 134–137.

Nakaya, T., Iwai, S., Fujinaga, K., Sato, Y., Otsuka, E., and Ikuta, K. (1997b). Decoy approach using RNA–DNA chimera oligonucleotides to inhibit the regulatory function of human immunodeficiency virus type 1 Rev protein. *Antimicrob. Agents Chemother.* **41**(2), 319–325.

Nemeroff, M. E., Barabino, S. M., Keller, W., and Krug, R. M. (1998). Influenza virus NS1 protein interacts with the cellular 30 kDa subunit of CPSF and inhibits 3′-end formation of cellular pre-mRNAs. *Mol. Cell* **1**(7), 991–1000.

Nosaka, Ö., and Hatanaka (1993). Cytotoxic activity of the Rev protein of HIV-1 by nucleolar dysfunction. *Exp. Cell Res.* **209**, 89–102.

Pardridge, W. M., Bickel, U., Buciak, J., Yang, J., and Diagne, A. (1994). Enhanced endocytosis and anti-human immunodeficiency virus type 1 activity of anti-Rev antibodies after cationization. *J. Infect. Dis.* **169**(1), 55–61.

Peng, H., Callison, D., Li, P., and Burrell, C. (1996). Long-term protection against HIV-1 infection conferred by Tat or re antisense RNA was affected by the design of the retroviral vector. *Virology* **220**(2), 377–389.

Pilkington, G. R., Duan, L., Zhu, M., Keil, W., and Pomerantz, R. J. (1996). Recombinant human Fab antibody fragments to HIV-1 Rev and Tat regulatory proteins: Direct selection from a combinatorial phage display library. *Mol. Immunol.* **33**(4–5), 439–450.

Plavec, I., Agarwal, M., Ho, K. E., Pineda, M., Auten, J., Baker, J., Matusuzaki, H., Escaich, S., Bonyhadi, M., and Bohnlein, E. (1997). High transdominant RevM10 protein levels are required to inhibit HIV-1 replication in cell lines and primary T cells: Implication for gene therapy of AIDS. *Gene Ther.* **4**(2), 128–139.

Pollard, V. W., and Malim, M. (1998). The HIV-1 Rev protein. *Annu. Rev. Microbiol.* **52**, 491–532.

Puttaraju, M., and Been, M. D. (1995). Generation of nuclease resistant circular RNA decoys for HIV-Tat and HIV-Rev by autocatalytic splicing. *Nucleic Acids Symp. Ser.* **33**, 49–51.

Qian, X. Y., Alonso-Caplen, F., and Krug, R. M. (1994). Two functional domains of the influenza virus NS1 protein are required for regulation of nuclear export of mRNA. *J. Virol.* **68**(4), 2433–2441.

Qui, Y., and Krug, R. M. (1994). The influenza virus NS1 protein is a poly (A)-binding protein that inhibits nuclear export of mRNAs containing poly (A). *J. Virol.* **68**(4), 2425–2432.

Ragheb, J. A., Bressler, P., Daucher, M., Chiang, L., Chuah, M. K., Vandendriessche, T., and Morgan, R. A. (1995). Analysis of trans-dominant mutants of the HIV type 1 Rev protein for their ability to inhibit Rev function, HIV type 1 replication, and their use as anti-HIV gene therapeutics. *AIDS Res. Hum. Retroviruses* **11**(11), 1343–1353.

Ragheb, J. A., Couture, L., Mullen, C., Ridgway, A., and Morgan, R. A. (1999). Inhibition of human immunodeficiency virus type 1 by Tat/Rev-regulated expression of Cytosine Deaminase, Interferon a2, or diphtheria toxin compared with inhibition by transdominant Rev. *Hum. Gene Ther.* **10**(1), 103–112.

Ranga, U., Woffendin, C., Verma, S., Xu, L., June, C. H., Bishop, D. K., and Nabel, G. J. (1998). Enhanced T cell engraftment after retroviral delivery of an antiviral gene in HIV-infected individuals. *Proc. Natl. Acad. Sci. USA* **95**(3), 1201–1206.

Ranga, U., Woffendin, C., Yang, Z. Y., Xu, L., Verma, S., Littman, D. R., and Nabel, G. J. (1997). Cell and viral regulatory elements enhance the expression and function of a human immunodeficiency virus inhibitory gene. *J. Virol.* **71**(9), 7020–7029.

Ratmeyer, L., Zapp, M. L., Green, M. R., Vinayak, R., Kumar, A., Boykin, D. W., and Wilson, W. D. (1996). Inhibition of HIV-1 Rev–RRE interaction by diphenylfuran derivatives. *Biochemistry* **35**(42), 13689–13696.

Rosenberg, S. A., Blaese, R. M., Brenner, M. K., Deisseroth, A. B., Ledley, F. D., Lotze, M. T., Wilson, J. M., Nabel, G. J., Cornetta, K., Economou, J. S., Freeman, S. M., Riddell, S. R., Oldfield, E., Gansbacher, B., Dunbar, C., Walker, R. E., Schuening, F. G., Roth, J. A., Crystal, R. G., Welsh, M. J., Culver, K., Heslop, H. E., Simons, J., Wilmott, R. W., Boucher, R. C., et al. (1997). Human gene marker/therapy clinical protocols. Hum. Gene Ther. 8(18), 2301–2338.

Rosenzweig, M., Marks, D. E., Hempel, D., Heusch, M., Kraus, G., Wong-Staal, F., and Johnson, R. P. (1997). Intracellular immunization of rhesus CD34+ hematopoietic progenitor cells with a hairpin ribozyme protects T cells and macrophages from simian immunodeficiency virus infection. Blood 90(12), 4822–4831.

Sczakiel, G., Oppenlander, M., Rittner, K., and Pawlita, M. (1992). Tat- and Rev-directed antisense RNA expression inhibits and abolishes replication of human immunodeficiency virus type 1: A temporal analysis. J. Virol. 66(9), 5576–5581.

Selvam, M. P., Buck, S. M., Blay, R. A., Mayner, R. E., Mied, P. A., and Epstein, J. S. (1996). Inhibition of HIV replication by immunoliposomal antisense oligonucleotide. Antiviral Res. 33(1), 11–20.

Smith, C., Lee, S. W., Wong, E., Gallardo, H., Page, K., Gaspar, O., Lebkowski, J., and Gilboa, E. (1996). Transient protection of human T-cells from human immunodeficiency virus type 1 infection by transduction with adeno-associated viral vectors which express RNA decoys. Antiviral Res. 32(2), 99–115.

Smythe, J. A., Sun, D., Thomson, M., Markham, P. D., Reitz, M. S., Jr., Gallo, R. C., and Lisziewicz, J. (1994). A Rev-inducible mutant gag gene stably transferred into T lymphocytes: An approach to gene therapy against human immunodeficiency virus type 1 infection. Proc. Natl. Acad. Sci. USA 91(9), 3657–3661.

Su, L., Lee, R., Bonyhadi, M., Matsuzaki, H., Forestell, S., Escaich, S., Bohnlein, E., and Kaneshima, H. (1997). Hematopoietic stem cell-based gene therapy for acquired immunodeficiency syndrome: Efficient transduction and expression of RevM10 in myeloid cells in vivo and in vitro. Blood 89(7), 2283–2290.

Svinarchuk, F. P., Konevetz, D. A., Pliasunova, O. A., Pokrovsky, A. G., and Vlassov, V. V. (1993). Inhibition of HIV proliferation in MT-4 cells by antisense oligonucleotide conjugated to lipophilic groups. Biochimie 75(1–2), 49–54.

Symensma, T. L., Giver, L., Zapp, M., Takle, G. B., and Ellington, A. D. (1996). RNA aptamers selected to bind human immunodeficiency virus type 1 Rev in vitro are Rev responsive in vivo. J. Virol. 70(1), 179–187.

Tagieva, N. E., and Vaquero, C. (1997). Expression of naturally occurring antisense RNA inhibits human immunodeficiency virus type heterologous strain replication. J. Gen. Virol. 78(10), 2503–2511.

Tiley, L. S., Malim, M. H., Tewary, H. K., Stockley, P. G., and Cullen, B. R. (1992). Identification of a high-affinity RNA-binding site for the human immunodeficiency virus type 1 Rev protein. Proc. Natl. Acad. Sci. USA 89(2), 758–762.

Tuerk, C., and MacDougal-Waugh, S. (1993). In vitro evolution of functional nucleic acids: High affinity RNA ligands of HIV-1 proteins. Gene 137, 33–39.

Ulich, C., Harrich, D., Estes, P., and Gaynor, R. B. (1996). Inhibition of human immunodeficiency virus type 1 replication is enhanced by a combination of transdominant Tat and Rev proteins. J. Virol. 70(7), 4871–4876.

Vandendriessche, T., Chuah, M. K., Chiang, L., Chang, H. K., Ensoli, B., and Morgan, R. A. (1995). Inhibition of clinical human immunodeficiency virus (HIV) type 1 isolates in primary CD4+T lymphocytes by retroviral vectors expressing anti-HIV genes. J. Virol. 69(7), 4045–4052.

Weichold, F. F., Lisziewicz, J., Zeman, R. A., Nerurkar, L. S., Agrawal, S., Reitz, M. S., Jr., and Gallo, R. C. (1995). Antisense phosphorothioate oligodeoxynucleotides alter HIV type 1 replication in cultured human macrophages and peripheral blood mononuclear cells. AIDS Res. Hum. Retroviruses 11(7), 863–867.

Woffendin, C., Ranga, U., Yang, Z., Xu, L., and Nabel, G. J. (1996). Expression of a protective gene prolongs survival of T cells in human immunodeficiency virus-infected patients. *Proc. Natl. Acad. Sci. USA* **93**(7), 2889–2894.

Woffendin, C., Yang, Z. Y., Udaykumar, Xu, L., Yang, N. S., Sheehy, M. J., and Nabel, G. J. (1994). Nonviral and viral delivery of a human immunodeficiency virus protective gene into primary human T cells. *Proc. Natl. Acad. Sci. USA* **91**(24), 11581–11585.

Wong-Staal, F., Poeschla, E. M., and Looney, D. J. (1998). A controlled, Phase 1 clinical trial to evaluate the safety and effects in HIV-1 infected humans of autologous lymphocytes transduced with aribozyme that cleaves HIV-1 RNA. *Hum. Gene Ther.* **9**(16), 2407–2425.

Wu, B., Woffendin, C., Duckett, C. S., Ohno, T., and Nabel, G. J. (1995). Regulation of human retroviral latency by the NF-kappa B family: Inhibition of human immunodeficiency virus replication by I kappa B through a Rev-dependent mechanism. *Proc. Natl. Acad. Sci. USA* **92**(5), 1480–1484.

Wu, B., Woffendin, C., MacLachlan, I., and Nabel, G. J. (1997). Distinct domains of IkappaB-alpha inhibit human immunodeficiency virus type 1 replication through NF-kappaB and Rev. *J. Virol.* **71**(4), 3161–3167.

Yamada, O., Kraus, G., Leavitt, M. C., Yu, M., and Wong-Staal, F. (1994). Activity and cleavage site specificity of an anti-HIV-1 hairpin ribozyme in human T cells. *Virology* **205**(1), 121–126.

Yamada, O., Kraus, G., Luznik, L., Yu, M., and Wong-Staal, F. (1996). A chimeric human immunodeficiency virus type 1 (HIV-1) minimal Rev response element-ribozyme molecule exhibits dual antiviral function and inhibits cell–cell transmission of HIV-1. *J. Virol.* **70**(3), 1596–1601.

Yuyama, N., Ohkawa, J., Koguma, T., Shirai, M., and Taira, K. (1994). A multifunctional expression vector for an anti-HIV-1 ribozyme that produces a 5'- and 3'-trimmed trans-acting ribozyme, targeted against HIV-1 RNA, and cis-acting ribozymes that are designed to bind to and thereby sequester trans-activator proteins such as Tat and Rev. *Nucleic Acids Res.* **22**(23), 5060–5067.

Zapp, M. L., Hope, T. J., Parslow, T. G., and Green, M. R. (1991). Oligomerization and RNA binding domains of the type 1 human immunodeficiency virus Rev protein: A dual function for an arginine-rich binding motif. *Proc. Natl. Acad. Sci. USA* **88**, 7734–7738.

Zapp, M. L., Stern, S., and Green, M. R. (1993). Small molecules that selectively block RNA binding of HIV-1 Rev protein inhibit Rev function and viral production. *Cell* **74**(6), 969–978.

Zapp, M. L., Young, D. W., Kumar, A., Singh, R., Boykin, D. W., Wilson, W. D., and Green, M. R. (1997). Modulation of the Rev–RRE interaction by aromatic heterocyclic compounds. *Bioorg. Med. Chem.* **5**(6), 1149–1155.

Zelphati, O., Imbach, J. L., Signoret, N., Zon, G., Rayner, B., and Leserman, L. (1994). Antisense oligonucleotides in solution or encapsulated in immunoliposomes inhibit replication of HIV-1 by several different mechanisms. *Nucleic Acids Res.* **22**(20), 4307–4314.

Zelphati, O., Zon, G., and Leserman, L. (1993). Inhibition of HIV-1 replication in cultured cells with antisense oligonucleotides encapsulated in immunoliposomes. *Antisense Res. Dev.* **3**(4), 323–338.

Zhang, M. J., Dayton, A. I. (1997). Targeting to the HIV-1 RRE of the influenza virus NS1 protein effector domain as a potent, specific anti-HIV agent. *J. Biomed. Sci.* **4**, 35–38.

Zhou, C., Bahner, I. C., Larson, G. P., Zaia, J. A., Rossi, J. J., and Kohn, E. B. (1994). Inhibition of HIV-1 in human T-lymphocytes by retrovirally transduced anti-Tat and Rev hammerhead ribozymes. *Gene* **149**(1), 33–39.

Zhou, C., Bahner, I., Rossi, J. J., and Kohn, D. B. (1996). Expression of hammerhead ribozymes by retroviral vectors to inhibit HIV-1 replication: Comparison of RNA levels and viral inhibition. *Antisense Nucleic Acid Drug Dev.* **6**(1), 17–24.

Ralph Dornburg
Roger J. Pomerantz

The Dorrance H. Hamilton Laboratories
Center for Human Virology
Division of Infectious Diseases
Jefferson Medical College
Thomas Jefferson University
Philadelphia, Pennsylvania 19107

HIV-1 Gene Therapy: Promise for the Future

I. Introduction

The systematic investigation of steps and enzymes involved in the replication of the human immunodeficiency virus type I (HIV-1) has led to the discovery of several potential targets for anti-HIV-1 drugs. Indeed, since the discovery that the acquired immunodeficiency syndrome (AIDS) is caused by HIV-1, several billion dollars have been invested worldwide to develop and test new pharmaceutical agents to control the spread of this epidemic disease. New conventional drugs were specifically designed to block the action of HIV-1-specific enzymes, such as the reverse transcriptase or the protease.

However, as a result of the high mutation rate of the virus, new virus variants continuously emerge which are resistant to such conventional therapies. Thus, great efforts are currently being made in many laboratories to develop alternative genetic approaches to inhibit the replication of this virus.

Advances in Pharmacology, Volume 49

With growing insight into the mechanism and regulation of HIV-1 replication, in the past decade a large arsenal of genetic antivirals has been developed.

Gene expression vectors have been constructed which express various anti-HIV-1 products (RNAs or proteins) which attack basically every step in the viral life cycle. Tissue-culture cells have been transduced with such genes, and it has been shown that such gene-transduced cells can become rather resistant to HIV-1 infection. However, although such antivirals have been proven to be very effective *in vitro*, their beneficiary effect *in vivo* is very difficult to evaluate and still remains to be shown. In particular, the long latent period from infection to the onset of AIDS (up to 10 years or even longer) makes it very difficult to evaluate the efficacy of a new drug. To make things even more complicated, HIV-1 can infect various nonlymphoid tissues (e.g., cells of the central nervous system) and can become dormant in such cells and thus "invisible" to the immune system. Thus, it is unknown whether such dormant viruses can become active again late in the life of an HIV-1-infected individual and whether cells carrying anti-HIV-1 antivirals would still be available and active to block reemerging viruses and reoccurring infections.

Another problem is still the delivery of genes encoding for such genetic antivirals. Although a series of gene-transfer tools exist which enable efficient transduction of genes in tissue culture, it becomes more and more evident that *ex vivo* transduced cells do not survive long *in vivo*. No efficient gene-delivery tools are available at this point which would enable robust delivery to the actual target cell *in vivo*. This chapter summarizes the experimental genetic approaches to block HIV-1 replication and current gene-delivery techniques to transduce therapeutic genes into the precise target cells.

II. HIV-I Infection and Conventional Pharmaceutical Agents

HIV-1 primarily infects and destroys cells of the human immune system, in particular CD4+ T lymphocytes and macrophages. The destruction of such cells leads to a severe immunodeficiency, e.g., the inability to fight other infectious agents or tumor cells. Thus AIDS patients usually die from secondary infections (e.g., tuberculosis or pneumonia) or cancer (e.g., Kaposi's sarcoma). Enormous efforts have been made to study the life cycle and pathogenesis of HIV-1 in order to find potential targets to block the replication of this virus. For a description of the life cycle of HIV-1, see below and Fig. 1. Some viral proteins, such as the protease and reverse transcriptase, have been crystallized and their three-dimensional structures have been determined. These studies were performed to design specific compounds which would irreversibly bind to the active sites of such enzymes

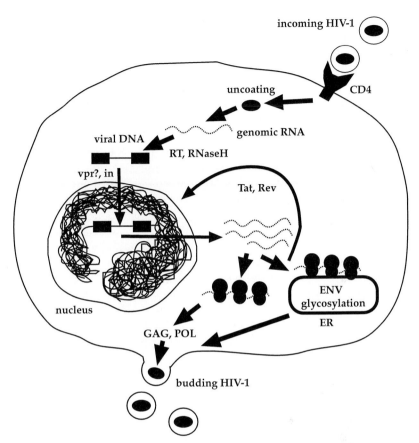

FIGURE 1   Life cycle of HIV-1. The life cycle of the human immunodeficiency virus type I is similar to that of all retroviruses studied. HIV-1 attaches to the target cell mainly by binding to the CD4 molecule. After fusion of the viral and cellular membranes, retroviral core particles are released into the cytoplasm. The RNA genome is converted into a double-stranded DNA by the viral reverse transcriptase (RT) and ribonuclease H (RNaseH) and actively transported into the nucleus, probably aided by the viral protein Vpr. The viral DNA is integrated into the genome of the host cell by the viral integrase (in). The integrated DNA form of the virus is called the provirus. In contrast to other retroviruses, transcription and RNA splicing of the provirus is regulated by viral accessory proteins. For example, the viral protein Tat must to bind to a specific sequence in the HIV genome (termed TAR) to enable highly efficient transcription of the provirus. Rev is required to control RNA splicing and the transport of RNAs into the cytoplasm. Finally, in the cytoplasm, virus core particles are assembled by encapsidating full-length genomic viral RNAs (recognized by specific encapsidation sequences). At the cell membrane virus particle assembly is completed by the interaction of the core with the viral membrane proteins and new particles "bud" (are released) from the infected cell. For more details regarding regulatory proteins, see also the legend to Fig. 5. Abbreviations: Env, envelope; ER, endoplasmatic reticulum.

and therefore inhibit their function. Indeed, specific chemical compounds which efficiently block the activity or function of these viral proteins are now commercially available and in use worldwide (Johnston *et al.*, 1989; Erickson and Burt, 1996; de Clercq, 1996; Sandstrom and Folks, 1996; Fulcher and Clezy, 1996).

However, treatment with just one enzyme inhibitor led to the emergence of drug-resistant viruses in the patient. Thus, a few years ago, therapies were initiated using three antiviral compounds (called combination chemo-therapy or highly active antiretroviral therapy, HAART) simultaneously. Recent studies have demonstrated that the administration of this mixture can lead to significant reduction of viral load *in vivo*. Utilizing two reverse transcriptase and one protease inhibitor in treatment-naive patients, the serum HIV-1 RNA levels could be reduced to an undetectable level in many patients. However, how long this response will last in these patients remains an open issue (Chun *et al.*, 1997; Finzi *et al.*, 1997; Wong *et al.*, 1997). Most recently, it has been shown that very low amounts of the HIV-1 virus can still be detected in lymphocytes isolated from semen samples of patients that had received HAART for several years (Zhang *et al.*, 1998). Even worse, these dormant viruses could be reactivated in *ex vivo* experiments. Thus, even if patients have extremely low or no detectable levels of viral RNA in their blood specimen, they may still be capable of transmitting the virus to uninfected individuals.

Furthermore, reports of patients have emerged describing variant strains of HIV-1 which are resistant to this combination chemotherapy. In particu-lar, patients who have been treated with one antiviral inhibitor alone in the past appear to already carry a virus strain which is resistant to one of these compounds and therefore are partially resistant to combination therapy. Such virus strains have a higher chance of further mutating and of evading the inhibitory effects of the other chemical compounds. In particular, the viral enzyme reverse transcriptase, which converts the viral RNA genome into a double-stranded DNA, does not have proofreading capabilities. On average, it inserts at least one incorrect nucleotide into the viral genome per replication cycle. This error rate is orders of magnitude higher than that of the cellular DNA polymerase I, which is the main enzyme for the replication of the eukaryotic genome. This high mutation rate explains why drug-resistant virus mutants emerge rapidly in HIV-1-infected patients (Richman, 1992). This high mutation rate also explains why new mutant viruses contin-uously arise which are "new" to and therefore not recognized and inactivated by the immune system. Consequently, in the clinically latent stage of HIV-1 infection, high virus loads persist which consistently change their genetic "outfit" to escape drugs that inhibit virus replication and the immune system (Baltimore, 1995; Embretson *et al.*, 1993; Pantaleo *et al.*, 1993).

In summary, the clinical application of all conventional drugs for the treatment of AIDS has not led to a cure of the disease and even the new combination therapy may only halt the development of AIDS in infected people temporarily, as new drug resistant variants of the HIV-1 virus start to emerge. Thus, efforts are underway in many laboratories to develop alternative therapeutics.

III. Genetic Antivirals to Block HIV-I Replication

The primary target cells for HIV-1 are cells of the hematopoietic system, in particular CD4+ T lymphocytes and macrophages. During HIV-1 infection these cells are destroyed by the virus, leading to immunodeficiency of the infected individual. In order to prevent the destruction of the cells of the immune system, a diverse array of efforts is now underway to make such cells resistant to HIV-1 infection. This approach has been termed "intracellular immunization" (Baltimore, 1988). In particular, the development of genetic agents (also termed "genetic antivirals") which attack the virus at several points simultaneously inside the cell and/or which are independent from viral mutations has gained great attention.

HIV-1 replicates via a classic retroviral life cycle. Initially, it binds to the high-affinity receptor, CD4, on CD4+ T lymphocytes and certain monocyte/macrophage populations. Recently, it has been determined that chemokine coreceptors, including CXCR-4 and CCR5, are also critical in viral entry. Thus, efforts are being made in several laboratories to develop agents which make T lymphocytes and/or macrophages resistant against virus entry (for details see below).

After entry and disassembly the viral RNA is reverse transcribed to viral DNA by reverse transcriptase (RT), like other retroviruses. However, in contrast to most other retroviruses, the resulting preintegration complex is then actively transported across the nuclear membrane. Thus, HIV-1 is capable of infecting quiescent cells. Many attempts are now also underway to endow immune cells with genes which prevent reverse transcription and/or integration (see below).

Besides the structural proteins, which form the virus particle, lentiviruses express a number of critical regulatory genes from multiple-spliced mRNAs. Thus, a series of studies are currently underway to test the potential of genetic antivirals directed not only against the structural Core and Envelope proteins, which are found in all retroviruses (e.g., matrix proteins, reverse transcriptase, integrase, and protease), but also against some regulatory proteins, which are specific and essential for the life cycle of lentiviruses. HIV-1 contains six regulatory genes which are involved in the complex pathogenesis (Table I).

First, the Tat gene is the major transcriptional transactivator of HIV-1 and essential for the activity of the long terminal repeat (LTR) promoter. The Tat protein stimulates HIV-1 transcription via an RNA intermediate

TABLE I   HIV Proteins and Their Functions

Protein	Size	Function
Gag	p25 (p24)	Capsid (CA) structural protein
	p17	Matrix (MA) protein–myristoylated
	p9	RNA–binding protein (?)
	p6	RNA–binding protein (?) helps virus budding
Polymerase (Pol)	p66,p51	reverse transcriptase (RT); RNase H-inside core
Protease (PR)	p10	Post-translation processing of viral proteins
Integrase (IN)	p32	Integration of viral DNA
Envelope (Env)	gp160	Envelope precursor protein, proteolytically cleaved
	gp120	Envelope surface protein, virus binding to cell surface
	gp41 (gp36)	Envelope transmembrane protein, membrane fusion
Tat*	p14	Transactivation
Rev*	p19	Regulation of viral RNA expression
Nef*	p27	Pleiotropic, including virus suppression–myristoylated
Vif	p23	Increases virus infectivity and cell-to-cell transmission; helps in proviral DNA synthesis and/or virion assembly
Vpr	p15	Helps in virus replication; nuclear import (?) transactivation (?)
Vpu*	p16	Helps in virus release; disrupts gp160-CD4 complexes

* not associated with virion

called the TAR region, which is found just downstream of the 5′ LTR. The product of the Rev gene rescues the unspliced viral RNA from the nucleus of infected cells by increasing transport through the nuclear pore. These two regulatory proteins are absolutely essential for HIV-1 replication and therefore became major targets for the development of genetic inhibitors of virus replication. These inhibitors attack the virus after integration into the chromosomes of the host and aim to prevent or reduce particle formation and/or release from infected cells.

Other critical accessory proteins include the Vpr gene, which leads to G_2 arrest in the cell cycle of infected cells; Nef, which stimulates viral production and activation of infected cells; Vpu, which stimulates viral release; and Vif, which seems to augment viral production in either early or late steps in the viral life cycle. These regulatory proteins may be less crucial to viral load and replication in comparison to Tat and Rev. Consequently, antiviral agents which attack these proteins are less likely to significantly prevent infection and/or the spread of the virus.

Potential genetic inhibitors of virus replication should have four features to overcome the shortcomings of conventional treatment. First, they should be directed against a highly conserved moiety in HIV-1 which is absolutely essential for virus replication, eliminating the chance that new mutant variants arise which can escape this attack. Second, they have to be highly effective, greatly reducing or, ideally, completely blocking the production of progeny virus. Third, they have to be nontoxic. A fourth criterion, which should not be overlooked, is that the antiviral agent has to be tolerated by the immune system. It would not make much sense to endow immune cells with an antiviral agent which elicits an immune response against itself, leading to the destruction of the HIV-1-resistant cell after a short period of time.

In the past few years many strategies have been developed and proposed for clinical application to block HIV-1 replication inside the cell (see Fig. 2). Such strategies use either antiviral RNAs or proteins. They include antisense

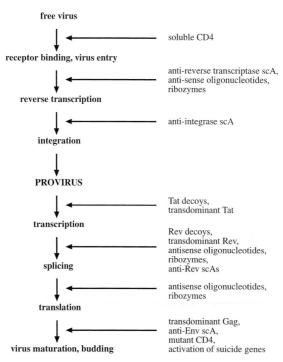

FIGURE 2   Overview of possible genetic targets to block HIV-1 replication. The different approaches are described in detail in the text.

oligonucleotides, ribozymes, RNA decoys, transdominant negative mutant proteins, products of toxic genes, and single-chain antigen binding proteins (for reviews, see Dropubic and Jeang, 1994; Yu *et al.*, 1994a; Kaplan and Hirsch, 1994; Gilboa and Smith, 1994; Anderson, 1994; Sarver and Rossi, 1993; Bahner *et al.*, 1993; Blaese, 1995; Bridges and Sarver, 1995; Bunnell and Morgan, 1996a; Cohen, 1996; Ho *et al.*, 1995; Pomerantz and Trono, 1995). Antiviral strategies that employ RNAs have the advantage that they are less likely to be immunogenic than protein-based antiviral agents. However, protein-based systems have been engineered that use inducible promoters that only become active upon HIV-1 infection.

A. RNA-Based Inhibitors of HIV-I Replication

I. Antisense RNAs and Ribozymes

Prokaryotes and bacteriophages express antisense RNAs, which provide regulatory control over gene expression by hybridizing to specific RNA sequences (Stein and Cheng, 1993). In animal cells, artificial antisense oligonucleotides (RNAs or single-standed DNAs) have been successfully used to selectively prevent expression of various genes (e.g., oncogenes, differentiation genes, and viral genes) (Stein and Cheng, 1993; Agrawal, 1992). Furthermore, the presence of double-stranded RNA inside the cell can induce the production of interferon and/or other cytokines, stimulating an immune response. Indeed, it has also been reported that the expression of RNAs capable of forming a double-stranded RNA molecule with the HIV-1 RNA (antisense RNAs) can significantly reduce the expression of HIV-1 proteins and consequently the efficiency of progeny virus production (Castanotto *et al.*, 1994; Rossi and Sarver, 1992; Rothenberg *et al.*, 1989; Cohen, 1991; Stein and Cheng, 1993; Agrawal, 1992).

Ribozymes are very similar to antisense RNAs, e.g., they bind to specific RNA sequences, but they are also capable of cleaving their target at the binding site catalytically. Thus, they have the advantage that they may not need to be overexpressed in order to fulfill their function (Bahner *et al.*, 1996a; Biasolo *et al.*, 1996; Larsson *et al.*, 1996; Heusch *et al.*, 1996; Poeschla and Wong-Staal, 1995; Chuah *et al.*, 1994; Yu *et al.*, 1995; Leavitt *et al.*, 1994; Poeschla and Wong-Staal, 1994; Sullivan, 1994; Yu *et al.*, 1993; Weerasinghe *et al.*, 1991; Leavitt *et al.*, 1996; Wong-Staal, 1995). Certain ribozymes (e.g., hairpin and hammerhead ribozymes, Fig. 3) require only a GUC sequence (Fig. 2). Thus, many sites in the HIV-I genome can be targeted. However, several questions remain unanswered. For example, it is unclear whether efficient subcellular colocalization can be obtained, in particular *in vivo*. This problem has been addressed recently by inserting anti-HIV-1 (anti-Rev) ribozymes into the U1 small nuclear RNA. Such RNAs were found to be retained in the nucleus and to exhibit anti-HIV-1 activity *in vitro* (Michienzi *et al.*, 1998). Another question is whether the target

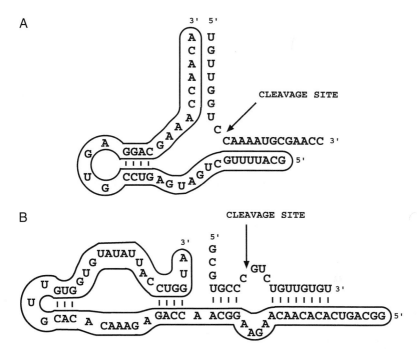

FIGURE 3 Schematic representation of two ribozymes to block HIV-1 replication. The structures shown are paired with actual HIV-1 target sequences. (A) A hammerhead ribozyme pairs specifically with a sequence in the gag region of the HIV-1 genome. (B) A hairpin ribozyme designed to bind to and cleave the 5' end of the viral genome, abolishing the reverse transcription and integration of progeny virus.

RNA will be efficiently recognized due to secondary and tertiary folding of the target RNA or whether RNA binding proteins would prevent efficient binding. Experimentation in several laboratories has addressed these problems (Gervaix *et al.*, 1997b; Earnshaw and Gait, 1997; Rossi *et al.*, 1997; Hampel, 1998; Ramezani *et al.*, 1997; Duarte *et al.*, 1997; Campbell and Sullenger, 1995) and clinical trial have been initiated to test the therapeutic effect of ribozymes in AIDS patients (Li *et al.*, 1998).

2. RNA Decoys

In contrast to ribozymes and antisense RNAs, RNA decoys do not attack the viral RNAs directly. RNA decoys are mutant RNAs that resemble authentic viral RNAs that have crucial functions in the viral life cycle. They mimic such RNA structure and decoy viral and/or cellular factors required for the propagation of the virus (Bahner *et al.*, 1996a; Chang *et al.*, 1994; Liu *et al.*, 1994; Yu *et al.*, 1994b; Lisziewicz *et al.*, 1995; Cesbron *et al.*, 1994; Bahner *et al.*, 1996b; Lee *et al.*, 1994; Smythe *et al.*, 1994; Matsuda *et al.*, 1993; Lisziewicz *et al.*, 1993; Brother *et al.*, 1996; Ilves *et al.*, 1996; Bunnell and Morgan, 1996b; Sullenger *et al.*, 1990). For example, HIV-1 replication largely depends on the two regulatory proteins Tat and Rev. These proteins bind to specific regions in the viral RNA, the transactivation response (TAR) loop and the Rev response element (RRE), respectively. Tat binding to TAR is crucial in the initiation of RNA transcription, and Rev binding to RRE is essential in controlling splicing, RNA stability, and the transport of the viral RNA from the nucleus to the cytoplasm. These two complex secondary RNA structure within the HIV-1 genome appear to be unique for the HIV-1 virus and no cellular homologous structures have been identified. Thus, such structures appear to be valuable targets for attack with genetic antivirals.

The strategy here is to endow HIV-1 target cells with genes that overexpress short RNAs containing TAR or RRE sequences. The rationale for this is to have RNA molecules within the cells in abundance which will capture Tat or Rev proteins, preventing the binding of such proteins to their actual targets. Consequently, HIV-1 replication is markedly impaired. This strategy has the advantage over antisense RNAs and ribozymes in that mutant Tat or Rev, which will not bind to the RNA decoys, will also not bind to their actual targets. Thus, the likelihood that mutant strains would arise which would bypass the RNA decoy trap is low. However, it still remains to be elucidated whether cellular factors also bind to Tat or Rev decoys and whether overexpression of decoy RNAs would lead to the sequestering of the resulting protein–RNA complexes in the cell.

B. Protein-Based Inhibitors of HIV-1 Replication

1. Transdominant Negative Mutant Proteins

Efficient replication of the HIV-1 virus depends on several regulatory proteins which are involved in viral gene expression and gene regulation. Mutant forms of such proteins greatly reduce the efficiency of viral replication. Transdominant (TD) negative mutants are genetically modified viral proteins that still bind to their targets but are unable to perform their actual function. They compete with the corresponding native, wild-type protein inside the cell. The competition of several TD proteins with the wild-type counterpart has been shown to greatly reduce virus replication, especially when such TD mutants are expressed from strong promoters (e.g., the cytomegalovirus immediate early promoter, CMV-IE) (Baltimore, 1988; Trono *et al.*, 1988; Buchschacher *et al.*, 1992; Freed *et al.*, 1992; Wu *et al.*, 1996; Regheb *et al.*, 1995; Liu *et al.*, 1994; Lori *et al.*, 1994; Liem *et al.*, 1993; Bahner *et al.*, 1993; Caputo *et al.*, 1996).

For example, mutations in the Gag proteins, which lead to abnormal virus core assembly, have been shown to inhibit HIV-1 replication (Lori *et al.*, 1994; Cara *et al.*, 1998). Furthermore, transcription from the HIV-1 long terminal repeat promoter is dependent on the Tat protein. Mutant Tat proteins, which still bind to the nascent viral RNA, but which are unable to further trigger RNA elongation of transcription, greatly reduce the production of HIV-I-RNAs and consequently the production of progeny virus (Caputo *et al.*, 1996, 1997; Rossi *et al.*, 1997). In a similar way, mutant Rev proteins interfere with regulated posttranscriptional events and also greatly reduce the efficiency of virus replication in an infected cell. In particular, one Rev mutant, termed RevM10, has been shown to efficiently block HIV-1 replication and has become a standard to measure the inhibitory effect of other anti-HIV-1 genetic antivirals (Liu *et al.*, 1994; Woffendin *et al.*, 1994; Woffendin *et al.*, 1996; Ranga *et al.*, 1998a; Junker *et al.*, 1997; Bonyhadi *et al.*, 1997; Plavec *et al.*, 1997; Caputo *et al.*, 1997). Clinical trials are now underway to test long-term expression and therapeutic effects of RevM10 in AIDS patients (see Section VI).

Although TD mutants have been shown to be effective *in vitro*, it still remains unclear how long cells endowed with such proteins will survive *in vivo*. There is a significant possibility that peptides of such proteins will be displayed via HLA leading to the destruction of the HIV-I-resistant cell by the patient's own immune system. There is, however, the potential to express such proteins from the HIV-1 LTR promoter, which only becomes activated upon HIV-1 infection. However, this would rule out that a TD Tat can be used because TD Tat may also abolish its own expression. Even if other TD proteins are expressed from inducible promoters, it still remains unclear whether such inducible promoters are really silent enough so that no protein is made (and no immune response) as long as there is no viral infection (e.g., a "leakiness problem").

2. Toxic Genes

Another approach to reduce the production of progeny virus is to endow the target cells of HIV-1 with toxic genes which become activated immediately after virus infection. The activation of the toxic gene leads to immediate cell death and therefore no new progeny virus particles can be produced. Theoretically, this would lead to an overall reduction of the virus load in the patient. *In vitro* experiments have shown that the production of HIV-1 virus particles was indeed reduced if target cells had been endowed with genes coding for the herpes simplex virus (HSV) thymidine kinase or a mutant form of the bacterial diphtheria toxin protein. Such genes were inserted downstream of the HIV-1 LTR promoter, which only becomes activated upon HIV-1 infection when the viral Tat protein is expressed (Dinges *et al.*, 1995b; Curiel *et al.*, 1993; Caruso and Klatzmann, 1992; Caruso, 1996; Dinges *et al.*, 1995a; Brady *et al.*, 1994).

Besides the question regarding the "silence" of the HIV-1 LTR promoter without Tat (discussed above), the main problem with this approach is the actual number of cells that carry a toxic gene present in the patient. Since the HIV-1 virus will not only infect cells that carry the toxic gene, but also many other cells of its host, it remains to be shown whether this approach is effective enough to have a therapeutic effect or whether it only slows down virus replication for a short period of time until all cells that carry the toxic genes undergo self-destruction upon infection.

3. CD4 and HIV-I Coreceptors as Decoy

The CD4 molecule is the major receptor for the HIV-1 virus for entry into T lymphocytes. Besides CD4, entry of HIV-1 into T lymphocytes and/or macrophages also depends on the presence of chemokine coreceptors, mainly CXCR-4 and CCR-5. Thus efforts have been made to block HIV-1 replication by modifying the main envelope docking molecules CD4, CXCR-4, or CCR-5 (Vandendriessche *et al.*, 1995; Morgan *et al.*, 1990; 1994; San Jose *et al.*, 1998; Yang *et al.*, 1997).

In a similar way to RNA decoys, the therapeutic effect of mutant CD4, which stays inside the endoplasmatic reticulum (ER), has been evaluated. In such experiments a chimeric CD4 coding gene was constructed which contained an ER retention sequence derived from the TCR CD3-epsilon chain (San Jose *et al.*, 1998). Thus this mutant CD4 molecule was able to prevent HIV-1 envelope maturation by retaining the HIV-1 envelope in the ER and blocking its transport to the cell surface. Consequently, formation of infectious particles could not take place. Indeed, it has been shown that human-tissue-culture T-lymphocyte lines (e.g., Jurkat or MT-2 cells) as well as primary T-cells from asymptomatic HIV-1-infected individuals which express this chimeric CD4 molecule became resistant to HIV-1 infection.

In another approach, soluble CD4 has been used to block the envelope of free extracellular virus particles and to prevent binding to fresh target cells. However, the question remains if soluble and/or mutant CD4 in the blood of the patient will also serve as a trap for natural CD4 ligands, leading to the impairment of important physiological functions (Vandendriessche *et al.*, 1995; Morgan *et al.*, 1990, 1994).

It has been found that people who have a mutant form of CCR-5, a coreceptor used by macrophage-tropic strains of HIV-1, are naturally protected against HIV-1 infection (Lyman *et al.*, 1994). Such mutant CCR-5 appears to be retained in the ER and therefore is not available at the cell surface for HIV-1 entry. Thus, efforts are being made to phenotypically knock out wild-type CCR-5 in HIV-1-infected individuals to make their macrophages resistant to CCR-5-tropic HIV-1 infection (Yang *et al.*, 1997). Macrophages have been transduced with a genetically modified chemokine gene which expresses a modified protein targeted to the ER. In the ER, the modified chemokine binds to CCR-5, preventing its transport to the cell surface.

4. Single-Chain Antibodies

Single-chain antibodies (scA) have originally been developed for *Escherichia coli* expression to bypass the costly production of monoclonal antibodies in tissue culture or mice (Bird *et al.,* 1988; Whitlow and Filpula, 1991). They comprise only the variable domains of both the heavy and the light chain of an antibody. These domains are expressed from a single gene, in which the coding region for these domains are separated by a short spacer sequence coding for a peptide bridge, which connects the two variable domain peptides. The resulting single-chain antibody (scA, also termed single-chain variable fragment, scFv) can bind to its antigen with similar affinity as an Fab fragment of the authentic antibody molecule.

scFvs have been developed by our group and others to combat HIV-1 replication. sFvs, which lack a hydrophobic signal peptide, are expressed intracellularly and are retained in the cytoplasm and are capable of binding to specific domains of HIV-1 proteins. Many proteins of HIV-1 have been targeted to prevent integration of the HIV-1 into the host chromosome (e.g., sFv against the integrase, reverse transcriptase, or matrix protein) or to block virus maturation (e.g., sFvs against Rev, Tat, and Env). Some intracellular sFvs have been found to greatly reduce the ability of HIV-1 to replicate, while other sFvs have produced only moderate success (Duan and Pomerantz, 1996; Morgan *et al.,* 1994; Chen *et al.,* 1994a,b; Buonocore and Rose, 1993; Duan *et al.,* 1994a,b; Pomerantz *et al.,* 1992; Pomerantz and Trono, 1995; Marasco *et al.,* 1998; Rondon and Marasco, 1997).

The ability to inhibit levels of HIV-1 infection, which may occur in the interstices of lymphoid tissue, remains a problem for anti-HIV gene therapy. New approaches which may target anti-HIV SFvs to escaping virion particles themselves have been developed. Of note, a new finding utilizing a specific motif which binds to the intravirion HIV-1 regulatory protein, Vpr, has demonstrated the ability to target anti-HIV-1 SFvs to the virion particle themselves. This motif, entitled WXXF, docks to Vpr and is able, when using chimeric fusion proteins, to target HIV-1 virions with inhibitory SFvs (BouHamdan *et al.,* 1998).

C. Combination Therapies

As described above, many strategies have been developed to block HIV-1 replication with genetic antivirals. As with conventional drugs, the question has to be asked whether mutant forms of the virus would arise which would eventually escape the blocking genetic antiviral. Little information is available regarding this issue because none of the genetic approaches have been tested on a broad clinical basis. Generally, most genetic antivirals attack HIV-1 at points at which mutations are expected to be fatal for the virus. Thus, there is significant hope that resistant virus will not emerge, at least not as rapidly as in the case of conventional drugs. In fact, a recent *in vitro* study showed the absence of emerging mutations within the specific viral ribozyme target sequences (Wang *et al.*, 1998).

Another problem which still remains for gene therapy to combat HIV-1 infection is the fact that increased multiplicities of infection seem to overwhelm most gene-therapeutic approaches to inhibit HIV-1 replication. As such, new approaches to augment the anti-HIV-1 activities of single genetic antivirals are being developed and experiments are underway in many laboratories to test the therapeutic effect of a combination of two or more anti-HIV-1 antivirals (Cara *et al.*, 1998; Bauer *et al.*, 1997; Gervaix *et al.*, 1997a). Furthermore, it has been reported that a combination of conventional drugs with genetic antivirals inhibits HIV-1 replication 10 times more efficiently than either therapy alone (Junker *et al.*, 1997).

For example, combinations of different ribozymes, i.e., a combination of an RRE decoy with RevM10 and a ribozyme, were more efficient blockers of HIV-1 replication than either genetic antiviral alone (Bauer *et al.*, 1997; Gervaix *et al.*, 1997a). In a similar way, expression of the transdominant mutant REV-M10 (see Section III,B,1) together with a gag-antisense RNA has been reported to increase the inhibition of HIV-1 replication (Cara *et al.*, 1998). Our laboratory is now developing combination single-chain variable fragments in a single construct. These use the internal ribosome entry site (IRES) to express proteins from a single promoter (manuscript in preparation).

IV. Genetic "Guns" to Deliver Genetic Antivirals

In all therapeutic approaches listed above, the therapeutical agent cannot be delivered directly to the cell. Instead, the corresponding genes have to be transduced to express the therapeutic agent of interest within the target cell. Genes can be delivered using a large variety of molecular tools. Such tools range from nonviral delivery agents (liposomes or even naked DNA) to viral vectors. Since HIV-1 remains and replicates in the body of an infected person for many years, it will be essential to stably introduce therapeutic genes into the genome of target cells for either continuous expression or for availability upon demand. Thus, gene delivery tools such as naked DNA, liposomes, or adenoviruses (AV), which are highly effective for transient expression of therapeutic genes, may not be useful for gene therapy of HIV-1 infection.

Adeno-associated virus (AAV), a nonpathogenic single-stranded DNA virus of the parvovirus family, has recently gained a great deal of attention as a vector because it is not only capable of inserting its genome specifically at one site at chromosome 19 in human cells, but it is also capable of infecting nondividing cells. However, vectors derived from AAV are much less efficient and loose their ability to target chromosome 19 (Muzyczka,

1992; Hallek and Wendtner, 1996). Another shortcoming of AAV is the need for it to be propagated with replication-competent AV, since AAV alone is replication-defective. It also remains to be shown, how efficiently AAV vectors transduce genes into human hematopoietic stem cells and/or mature T lymphocytes and macrophages. Since gene therapy of HIV-1 infection may also require multiple injections of the vector (*in vivo* gene therapy, see below) or of *ex vivo* manipulated cells, it also remains to be shown whether even small amounts of contaminating AV, which is used as a helper agent to grow AAV, will cause immunity problems.

The most efficient tools for stable gene delivery are retroviral vectors (Miller, 1990; Morgan and Andserson, 1993; Dornburg, 1995; Temin, 1986; Gilboa, 1990; Eglitis and Anderson, 1988; Gunzburg and Salmons, 1996), which stably integrate into the genome of the host cell, as this is a part of the retroviral life cycle (Fig. 1). This is why the first virus-based gene-delivery systems have been derived from this class of viruses. This is also why they are being used in almost all current human gene-therapy trials, including on-going clinical AIDS trials.

Retroviral vectors are basically retroviral particles that contain a genome in which all viral protein coding sequences have been replaced with the gene(s) of interest. As a result, such viruses cannot further replicate after one round of infection. Furthermore, infected cells do not express any retroviral proteins, which makes cells that carry a vector provirus (the integrated DNA form of a retrovirus) invisible to the immune system (Miller, 1990; Morgan and Anderson, 1993; Dornburg, 1995; Temin, 1986; Gilboa, 1990; Eglitis and Anderson, 1988; Gunzburg and Salmons, 1996). A large variety of genes have been inserted in retroviral vectors and highly efficient, long-term gene expression has been reported for many applications.

However, efficient gene expression may not always be achieved and different genes behave in different ways. For example, our laboratory encountered some problems in robustly expressing sFvs from murine leukemia virus (MLV)-derived gene delivery vectors. As such, new approaches utilizing various vectors have been developed to express SFvs. In particular, SV40, which infects a wide range of human and mammalian cells, has been utilized with an anti-integrase SFv (BouHamdan *et al.*, 1999).

A. Retroviral Vectors Derived from C-Type Retroviruses

All current retroviral vectors used in clinical trials have been derived from murine leukemia virus (MLV), a C-type retroviruses with a rather simple genomic organization (Fig. 4). MLV contains only two gene units, which code for the inner core structure proteins and the envelop protein, respectively. It does not contain regulatory genes like HIV-1. Thus, the construction of safe gene-delivery systems is rather simple and straightfor-

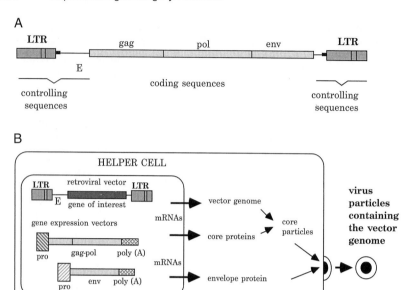

FIGURE 4   Retroviral helper cells derived from C-type retroviruses. (A) A C-type retroviral provirus (the DNA intermediate of a retrovirus is shown on the top). The protein coding genes (*gag-pol* and *env*) are flanked by *cis*-acting or controlling sequences, which play essential roles during replication. (B) In a retroviral helper cell, the retroviral protein coding genes, which code for all virion proteins, are expressed (ideally) from heterologous promoters (pro) and polyadenylated via a heterologous polyadenylation signal sequence (poly A). To minimize reconstitution of a full-length provirus by recombination, the *gag-pol* and *env* genes are split to different gene expression vectors. In the retroviral vector, the viral protein coding sequences are completely replaced by the gene(s) of interest. Since the vector contains specific encapsidation sequences (E), the vector genome is encapsidated into retroviral vector particles, which bud from the helper cell. The virion contains all proteins necessary to reverse transcribe and integrate the vector genome into that of a newly infected target cell. However, since there are no retroviral protein coding sequences in the target cell, vector replication is limited to one round of infection. Abbreviation: LTR, long terminal repeat.

ward. Such delivery systems consist of two components: the retroviral vector, which is a genetically modified viral genome which contains the gene of interest replacing retroviral protein coding sequences, and a helper cell that supplies the retroviral proteins for the encapsidation of the vector genome into retroviral particles (Fig. 4). Modern helper cells contain separate plasmid constructs which express all retroviral proteins necessary for replication. After transfection of the vector genome into such helper cells, the vector genome is encapsidated into virus particles (due to the presence of specific encapsidation sequences). Virus particles are released from the helper cell carrying a genome containing only the gene(s) of interest (Fig. 4). Thus, once established, retrovirus helper cells can produce gene transfer particles for very long time periods (e.g., several years). In the past decade, several retroviral vector systems have also been derived from other C-type chicken retroviruses (Miller, 1990; Dornburg, 1995; Gunzburg and Salmons, 1996).

B. Retroviral Vectors Derived from HIV-I

Retroviral vectors derived from MLV have been shown to be very useful in transfering genes into a large variety of human cells. However, they poorly infect human hematopoietic cells because such cells lack the receptor, which is recognized by the MLV envelope protein (von Laer *et al.*, 1998). Furthermore, retroviral vectors derived from C-type retroviruses are unable to infect quiescent cells: such viruses (and their vectors) can only establish a provirus after one cell division, during which the nuclear membrane is temporarily dissolved (Bukrinsky *et al.*, 1993; Lewis and Emerman, 1994; Schwedler *et al.*, 1999; Miyake *et al.*, 1999; Uchida *et al.*, 1998). Thus, efforts are underway in many laboratories to develop retroviral vectors from lentiviruses, e.g., HIV-1 or the simian immunodeficiency virus, SIV, which are able to establish a provirus in nondividing cells (Naldini *et al.*, 1996; Jiang *et al.*, 1998; Kaul *et al.*, 1998; Mochizuki *et al.*, 1998; Dull *et al.*, 1998; Kim *et al.*, 1998; Warner *et al.*, 1995; Parolin and Sodroski, 1995). However, the fact that lentiviruses contain several regulatory proteins which are essential for virus replication makes the construction of lentiviral packaging cells more complicated. Furthermore, the fact that the lentiviral envelope proteins (e.g., that of HIV-1) can cause syncytia and/or that some viral regulatory proteins are toxic to the cells further hampers the development of stable packaging lines.

The "envelope problem" has been solved by generating packaging cells which express the envelope protein of MLV or the envelope of vesicular stomatitis virus (VSV). Such envelope proteins are efficiently incorporated into lentiviral particles. The second and major problem for generating stable packaging lines is the toxicity of some retroviral regulatory proteins to the cell. Thus, retroviral vectors can only be generated in transient systems: 293T cells (human-tissue-culture cells highly susceptible for transfecting DNAs) are simultaneously transfected with all plasmids constructs to express the particle proteins and the vector genome. Figure 5 shows plasmid constructs used to make HIV-1-derived packaging cells. Vector virus can be harvested from the transfected cells for a limited time period and can be used to infect fresh target cells. Although this gene-transfer system has been shown to be functional, initially it was not highly efficient and efforts are underway to develop better packaging to make it suitable for broad clinical applications (Jiang *et al.*, 1998; Mochizuki *et al.*, 1998; Dull *et al.*, 1998).

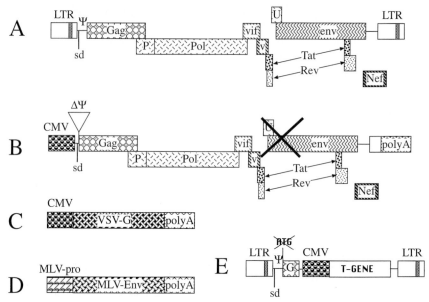

FIGURE 5 Retroviral packaging system derived from HIV-1. (A) A provirus of HIV-1 is shown on the top. (B–D) Plasmid constructs that express pseudotyped HIV-1 retroviral particles; (E) a plasmid construct that encapsidate and transduce genes with a HIV-1 vector (the plasmid sequences that propagate such constructs in bacteria are not shown). Besides the genes encoding for HIV-1 proteins, which form the core of the virus (Gag, structural core proteins; P, protease; Pol, reverse transcriptase and integrase) and the envelope (env, envelope protein), the HIV-1 genome also codes for several regulatory proteins (termed Vif; U = Upu; V = Vpr; Tat; Rev; Nef), which are expressed from spliced mRNAs and which have important functions in the viral life cycle. (B) Plasmid construct that expresses the core and regulatory proteins. To avoid encapsidation and transduction of genes coding for such proteins, the following modifications have been made: the 5′ LTR promoter of the HIV-1 provirus has been replaced with the promoter of cytomegalovirus (CMV) to enable constitutive gene expression; the 3′ LTR has been partially replaced with the polyadenylation signal sequence of simian virus 40 (polyA); the encapsidation signal has been deleted (ΔΨ); and the reading frames for the envelope and vpu genes have been blocked. (C and D) Plasmid constructs used to express the envelope proteins of the vesicular stomatitis virus (VSV-G) or the envelope protein of murine leukemia virus (MLV), respectively. In the absence of HIV-1 envelope proteins, which are rather toxic to the cell, HIV-1 efficiently incorporates the envelope proteins of VSV or MLV into virions. The use of such envelopes also further reduces the risk of the reconstitution of a replication-competent HIV-1 by homologous recombination between the plasmid constructs. (E) A retroviral vector used to package and transduce a gene of interest (T-gene) with HIV-1-derived vectors. Since the encapsidation sequence extends into the Gag region, part of the gag gene (G) has been conserved in the vector. However, the ATG start codon has been mutated. The gene of interest is expressed from an internal promoter, since the HIV-1 LTR promoter is silent without Tat (sd, splice donor site).

In addition, may questions regarding the safety of such vectors still need to be addressed (Putman, 1998).

C. Cell-Type-Specific Retroviral Vectors

All retroviral vectors currently used in human gene therapy trials contain the envelope protein of amphotropic (ampho) MLV or that of VSV. Ampho-MLV as well as VSV have a very broad host range and can infect various tissues of many species including humans. Thus, the use of vectors containing such envelope proteins enables transduction into many different human tissues. However, due to this broad host range, gene transfer has to be performed *ex vivo*. If injected directly into the blood stream, the chances that the vector particles would infect their actual target cells are very low. Furthermore, such vector particles may infect germ line cells (which are continuously dividing). Thus, the target cells have to be isolated and the gene transfer is performed in tissue culture. Gene-transduced cells are then selected and reintroduced into the patient.

However, this protocol has major shortcomings in regard to gene therapy of HIV-1 infection. First, it is very expensive and requires highly trained personnel to administer it. Second, human cells, which are kept in tissue culture, change their physiological behavior and/or take up fetal bovine proteins (a component of the tissue culture medium) and display bovine peptides via HLA on the cell surface. Consequently, such cells become immunogenic and are eliminated by the immune system of the patient. To bypass such *ex vivo* protocols, efforts are now underway in many laboratories to develop cell-type-specific gene-delivery systems which would enable to inject the gene delivery vehicle directly into the patient's bloodstream or tissue of interest. In the past few years, several attempts have been made to develop cell-type-specific gene-delivery tools, again, with retrovirus-derived vectors leading the field.

The cell-type specificity of a virus particle is determined by the nature of the retroviral envelope protein which mediates the binding of the virus to a receptor of the target cell (Hunter and Swanstrom, 1990). Thus, experiments have been initiated in several laboratories to modify the envelope protein of retroviruses in order to alter the host range of the vector. One of the first attempts to specifically deliver genes into distinct target cells has been performed in the laboratory of Dr. H. Varmus. Using retroviral vectors derived from avian leukosis virus (ALV), these investigators incorporated the human CD4 molecule into virions to specifically transduce genes into HIV-1-infected cells (Young *et al.*, 1990). However, such particles were not infectious for unknown reasons. Recent reports indicate that MLV particles that carry CD4 can infect HIV-1-infected cells, although at very low efficiencies (Matano *et al.*, 1995).

In another attempt to target retroviral particles to specific cells, Roux *et al.* have shown that human cells could be infected with eco-MLV if they added two different antibodies to the virus particle solution. The antibodies were connected at their carboxy-termini by streptavidine. One antibody was directed against a cell-surface protein, the other antibody was directed against the retroviral envelope protein (Etienne-Julan *et al.*, 1992; Roux *et al.*, 1989). Although this approach was not practical (infectivity was very inefficient and was performed at 4°C), these experiments showed that cells which do not have an appropriate receptor for a particular virus can be infected with that virus if binding to the cell surface had been facilitated. These data also indicated that antibody-mediated cell targeting with retroviral vectors was possible.

To overcome the technical problems of creating an antibody bridge, it was logical to incorporate the antibody directly into the virus particle. However, complete antibodies are very bulky and are not suitable for this approach. The problem has been solved using single-chain antibody technology (Chu and Dornburg, 1995; Chu *et al.*, 1994; Russell *et al.*, 1993; Chu and Dornburg, 1997). Using hapten model systems, it has been shown that retroviral vectors that contain scAs fused to the envelope are competent for infection (Chu *et al.*, 1994; Russell *et al.*, 1993). Retroviral vector particles derived from spleen necrosis virus (SNV, an avian retrovirus) that display various scAs against human cell-surface proteins are competent for infection on human cells that express the antigen recognized by the antibody (Chu and Dornburg, 1995; Chu *et al.*, 1994; Chu and Dornburg, 1997; Jiang *et al.*, 1998). Recent experiments in our laboratory have shown that a cell-type-specific gene delivery can be obtained *in vivo* as well (Jiang and Dornburg, 1999).

Most recently, it became possible to use scA-displaying SNV to introduce genes into human T lymphocytes with the same high efficiency obtained with vectors containing wild-type envelope. In all such experiments, the wild-type envelope of SNV had to be copresent in the virus particle to enable efficient infection of human cells. However, since SNV vector particles with wild-type SNV envelope do not infect human cells at all, this requirement is not a drawback for using such vector particles for human gene therapy (Dougherty *et al.*, 1989; Purchase and Witter, 1975).

Although successful gene transfer using scA-displaying MLV vector particles has been reported from one laboratory (Somia *et al.*, 1995), further experimentation in the laboratories of several other investigators revealed that MLV-derived vector particles that display various scAs are not competent for infection in human cells (Etienne-Julan *et al.*, 1992; Roux *et al.*, 1989; Marin *et al.*, 1996; Valsesia-Wittmann *et al.*, 1996; Nilson *et al.*, 1996; Schnierle *et al.*, 1996). The difference between MLV and SNV cell-targeting vectors is certainly based on the different features of the wild-type envelope and the mode of virus entry (reviewed in Dornburg, 1997). Moreover, wild-type ampho-MLV infects human hematopoietic cells extremely poorly, most likely due to the absence of an ampho-MLV receptor on such cells (Orlic *et al.*, 1996). Thus, MLV-derived vectors are certainly not the best candidates for human gene therapy of AIDS and alternative vectors need to be developed.

D. Other New Potential Vector Systems

Most recently, other very interesting attempts have been made to combat HIV-1-infected cells. Recombinant vesicular stomatitis virus (VSV) has been engineered, which lacks its own glycoprotein gene. Instead, genes coding for the HIV-1 receptor CD4 and a chemokine coreceptor, CXCR4, have been inserted. The corresponding virus was able to efficiently infect HIV-1-infected cells, which display the HIV-1 glycoprotein on the cell surface. Since VSV is a virus which normally kills infected cells, the engineered virus only infects and kills HIV-1-infected cells. It has been reported that this virus indeed reduced HIV-1 replication in tissue culture cells up to 10,000-fold (Schnell *et al.*, 1997). This novel approach to combat one virus with another will certainly gain a great deal of further attention. However, it remains to be shown how effective this approach will be *in vivo*. Will the "antivirus" succeed in eliminating a large load of HIV-infected cells before it will be cleared by the immune system? On the other hand, since HIV-1 preferentially kills activated immune cells, will it destroy the immune cells which are attempting to clear the body from its own "enemy"? How will the body tolerate a virus that does not look like one because it carries human cell-surface proteins on the viral surface? The answers to these and other questions are eagerly awaited (Nolan, 1997).

E. Other Potential Problems

Even if we find a gene-transfer system that can transduce enough cells within the body to inhibit virus replication significantly, there are many other questions that still remain to be answered. For example, it is not clear whether a cell which has been endowed with a certain HIV-1 resistance gene will be able to fulfill its normal biological function *in vivo*. Will the body be able to eliminate all HIV-1-infected cells or will the infected person become a lifelong carrier of the virus, which is still replicating in his or her body although at levels that cause no clinical symptoms due to the presence of HIV-1-resistance genes? Will the patient be capable of infecting new individuals? Finally, as genetic therapies can all be overcome with *in vitro* challenges of very high m.o.i.s of HIV-1, will there be a difference in antiviral effects in peripheral blood versus lymphoid tissues?

V. Animal Model Systems

One of the major problems with any therapeutic agent against HIV-1 infection is the lack of an appropriate and inexpensive animal model system to test the efficiency of an antiviral agent. Since HIV-1 only causes AIDS in humans, it is very difficult to test and evaluate the therapeutic effect of novel antiviral agents *in vivo*. Furthermore, the evaluation of the efficacy of a new drug is further complicated by the very long clinical latency period of the virus until the onset of AIDS (which can be 10 years or more). Although a virus similar to HIV-1 has been found in monkeys, the simian immunodeficiency virus (SIV) results obtained with this virus do not necessarily reflect the onset of AIDS in humans caused by HIV-1. Furthermore, many antivirals that block HIV-1 are ineffective in blocking SIV. Thus, other animal model systems need to be developed to study the effect of anti-HIV-1 therapies.

In the past decade, many strains of laboratory mice have been bred which lack components of the immune system. Severe combined immunodeficient (SCID) mice are deficient in functional B and T lymphocytes. Thus, they are unable to reject allogeneic organ grafts (Uckun, 1996; Dick, 1991, 1994; Dick *et al.*, 1992; Mueller and Reisfeld, 1991; Shen *et al.*, 1990). SCID mice have been used extensively to study human leukemia and other malignancies and for modeling human retroviral pathogenesis including antiviral gene therapy. Furthermore, in the past few years, much progress has been made to transplant hematopoietic stem cells into SCID mice to mimic and study human hematopoiesis. It has been shown that transplantation of human hematopoietic cells into such mice can lead to the repopulation of the mouse's blood with human CD4 and CD8-positive T lymphocytes. Thus, SCID mice appear also to be very good candidates to develop mouse model systems for HIV-1 infection.

At present, two different SCID mice model systems are used to study the effects of anti-HIV-1 antiviral agents. The two systems are somewhat different because they represent different components of the human immune system. One is called the hu-PBL SCID model; the other is termed SCID-hu mouse (Akkina *et al.*, 1994; Chen and Zack, 1994; Gauduin *et al.*, 1997, 1998; Jamieson *et al.*, 1996).

In the hu-PBL Scid mouse system, human peripheral blood leukocytes are injected into the peritoneum of the animal. Thus, cells residing in such animals are mature CD4+ and CD8+ T lymphocytes. The presence of activated and/or memory T cells (CD45RO+ cells) has also been demonstrated. Such human cells can be recovered from various organs in the mouse, e.g., the spleen or lymph nodes. Since these animals contain human CD4+ T lymphocytes, they can be infected easily with HIV-1. Even more, HIV-1 virus replicates in the animal, leading to the depletion of CD4+ T lymphocytes over a period of several weeks after infection. Thus, this experimental system is very valuable in testing the effect of antiviral agents (Gauduin *et al.*, 1997, 1998; Jamieson *et al.*, 1996). For example, experiments have been performed to test the effect of monoclonal antibodies against the HIV-1 envelope protein (neutralizing the receptor binding V3 domain). It has been shown that this anitbody indeed could block the replication of the viral strain for which the antibody is specific. This model system has also been used to test the efficiency of a vaccinia virus-derived vaccine.

In the SCID-hu mouse model system various human fetal hematopoietic tissues (e.g., liver, lymph nodes, and/or thymus) are transplanted into the mouse. For example, human fetal thymus and liver tissues are engrafted under the murine kidney capsule. It has been shown that normal thymopoiesis takes place for up to 1 year after implantation and human CD4+ and CD8+ T lymphocytes are found in the mouse blood at low levels. In contrast to the hu-PBL-SCID mouse, human cells are also found which express the CD45RA antigen, which is considered a marker for "naive" T lymphocytes. Such mice can also be infected with HIV-1, although the injection of a high virus dose directly into the implant is necessary to establish an infection. However, once an infection has been established, the pathologic effects observed are very similar to those observed in the thymuses of infected human adults, children, and fetuses. Moreover, a depletion of CD4+/CD8− T lymphocytes is observed.

Both SCID mice model systems are very useful to study pathogenesis and the effect of anti-HIV-1 drugs and to test the effect of anti-HIV antiviral genes. However, one has to keep in mind that such model systems only represent a portion of the human immune system and HIV-1 also infects other cells in humans, such as dendritic cells, microvascular endothelial cells, and sometimes neurons in the brain. Thus, the pathogenesis observed in SCID mice certainly does not accurately reflect the pathogenesis in humans. Moreover, it is not clear whether SCID mice transplanted with human immune cells do have a functional immune system.

VI. Clinical Trials

Recently certain initial *in vivo* studies have been conducted for intracellular immunization against primate lentiviruses. A transdominant negative Rev protein (RevM10) has been studied in humans infected with HIV-1 by Dr. G. Nabel's group. In these initial studies, it was demonstrated that cells transduced with RevM10 had a significant longer half-life as compared to control cells when reinfused into patients in different stages of disease (Ranga *et al.*, 1998a). These early initial Phase I trials were performed using murine retroviral vectors (MLV) as well as microparticulate bombardment using a "gene gun" (Ranga *et al.*, 1998b). In another approach, clinical trials have just begun to test the therapeutic effect of anti-HIV-1 ribozymes in AIDS patients (Gervaix *et al.*, 1997a).

In addition, a very exciting study has recently been reported by R. Morgan's group in which an antisense construct to Tat and Rev genes in SIV was used to transduce T lymphocytes from rhesus macaques (Donahue *et al.*, 1998). The monkeys were then challenged with SIV intravenously. Of note, the animals with the transduced cells had significantly lower viral loads and higher CD4 counts compared to control monkeys. Thus this suggests, for the first time, that gene therapy against lentiviruses may have significant efficacy *in vivo*. Clearly, these are both very preliminary studies in humans and in primates which require more detailed evaluation. Other trials using a variety of different approaches are on-going in initial Phase I studies.

In summary, considering all facts and problems of current gene-transfer technologies and considering our lack of knowledge regarding many functions of the immune system, how should we move forward with Phase I gene-therapy trials to combat HIV-1? In spite of the lack of knowledge of many aspects of this disease, we have to remain optimistic and can only hope that one approach or the other will lead to measurable success toward viral eradication or at least will significantly prolong the life expectancy of the infected person. Clearly, in addition to further exploring novel molecular therapeutics *in vivo*, significant attention must be placed toward answering critical basic science questions pertaining to "intracellular immunization."

References

Agrawal, S. (1992). Antisense oligonucleotides as antiviral agents. *Trends Bioctechnol.* **10**, 152–158.

Akkina, R. K., Rosenblatt, J. D., Campbell, A. G., Chen, I. S., and Zack, J. A. (1994). Modeling human lymphoid precursor cell gene therapy in the SCID-hu mouse. *Blood* **84**, 1393–1398.

Anderson, W. F. (1994). Gene therapy for AIDS. *Hum. Gene Ther.* **5**, 149–150.

Bahner, I., Kearns, K., Hao, Q. L., Smogorzewska, E. M., and Kohn, D. B. (1996a). Transduction of human CD34+ hematopoietic progenitor cells by a retroviral vector expressing an RRE decoy inhibits human immunodeficiency virus type 1 replication in myelomonocytic cells produced in long-term culture. *J. Virol.* **70**, 4352–4360.

Bahner, I., Kearns, K., Hao, Q. L., Smogorzewska, E. M., and Kohn, D. B. (1996b). Transduction of human CD34(+) hematopoietic progenitor cells by a retroviral vector expressing an rre decoy inhibits human immunodeficiency virus type 1 replication in myelomonocytic cells produced in long-term culture. *J. Virol.* **67**, 3199–3207.

Bahner, I., Zhou, C., Yu, X. J., Hao, Q. L., Guatelli, J. C., and Kohn, D. B. (1993). Comparison of trans-dominant inhibitory mutant human immunodeficiency virus type 1 genes expressed by retroviral vectors in human T lymphocytes. *J. Virol.* **67**, 3199–3207.

Baltimore, D. (1988). Intracellular immunization. *Nature* **235**, 395–396.

Baltimore, D. (1995). The enigma of HIV infection. *Cell* **82**, 175–176.

Bauer, G., Valdez, P., Kearns, K., Bahner, I., Wen, S. F., Zaia, J. A., and Kohn, D. B. (1997). Inhibition of human immunodeficiency virus-1 (HIV-1) replication after transduction of granulocyte colony-stimulating factor-mobilized CD34+ cells from HIV-1-infected donors using retroviral vectors containing anti-HIV-1 genes. *Blood* **89**, 2259–2267.

Biasolo, M. A., Radaelli, A., Del Pup, L., Franchin, E., De Giuli-Morghen, C., and Palu, G. (1996). A new antisense tRNA construct for the genetic treatment of human immunodeficiency virus type 1 infection. *J. Virol.* **70**, 2154–2161.

Bird, R. E., Hardman, K. D., Jacobson, J. W., Johnson, S., Kaufmann, B. M., Lee, S.-M., Lee, T., Pope, S. H., Riordan, G. S., and Whitlow, M. (1988). Single-chain antigen-binding proteins. *Science* **242**, 423–426.

Blaese, R. M. (1995). Steps toward gene therapy. 2. Cancer and AIDS. *Hosp. Pract.* **30**, 37–45.

Bonyhadi, M. L., Moss, K., Voytovich, A., Auten, J., Kalfoglou, C., Plavec, I., Forestell, S., Su, L., Bohnlein, E., and Kaneshima, H. (1997). Revm10-expressing t cells derived *in vivo* from transduced human hematopoietic stem-progenitor cells inhibit human immunodeficiency virus replication. *J. Virol.* **71**, 4707–4716.

BouHamdan, M., Xue, Y., Baudat, Y., Hu, B., Sire, J., Pomerantz, R. J., and Duan, L. X. (1998). Diversity of HIV-1 vpr interactions involves usage of the wxxf motif of host cell proteins. *J. Biol. Chem.* **273**, 8009–8016.

Brady, H. J., Miles, C. G., Pennington, D. J., and Dzierzak, E. A. (1994). Specific ablation of human immunodeficiency virus Tat-expressing cells by conditionally toxic retroviruses. *Proc. Natl. Acad. Sci. USA* **91**, 365–369.

Bridges, S. H., and Sarver, N. (1995). Gene therapy and immune restoration for HIV disease. *Lancet* **345**, 427–432.

Brother, M. B., Chang, H. K., Lisziewicz, J., Su, D., Murty, L. C., and Ensoli, B. (1996). Block of tat-mediated transactivation of tumor necrosis factor beta gene expression by polymeric-tar decoys. *Virology* **222**, 252–256.

Buchschacher, G. L., Freed, E. O., and Panganiban, A. T. (1992). Cells induced to express a human immunodeficiency virus type 1 envelope mutant inhibit the spread of the wild-type virus. *Hum. Gene Ther.* **3**, 391–397.

Bukrinsky, M. I., Haggerty, S., Dempsey, M., Sharova, N., Adzhubel, A., Spitz, L., Lewis, P., Goldfarb, D., Emerman, M., and Stevenson, M. (1993). A nuclear localization signal within HIV-1 matrix protein that governs infection of non-dividing cells. *Nature* **365**, 666–670.

Bunnell, B. A., and Morgan, R. A. (1996a). Gene therapy for aids. *Mol. Cells* **6**, 1–12.

Bunnell, B. A., and Morgan, R. A. (1996b). Gene therapy for HIV infection. *Drugs Today* **32**, 209–224.

Buonocore, L., and Rose, J. K. (1993). Blockade of human immunodeficiency virus type 1 production in CD4+ T cells by an intracellular CD4 expressed under control of the viral long terminal repeat. *Proc. Natl. Acad. Sci. USA* **90**, 2695–2699.

Campbell, T. B., and Sullenger, B. A. (1995). Alternative approaches for the application of ribozymes as gene therapies for retroviral infections. *Adv. Pharmacol.* **33**, 143–178.

Caputo, A., Grossi, M. P., Bozzini, R., Rossi, C., Betti, M., Marconi, P. C., Barbantibrodano, G., and Balboni, P. G. (1996). Inhibition of HIV-1 replication and reactivation from latency by tat transdominant negative mutants in the cysteine rich region. *Gene Ther.* **3**, 235–245.

Caputo, A., Rossi, C., Bozzini, R., Betti, M., Grossi, M. P., Barbanti-Brodano, G., and Balboni, P. G. (1997). Studies on the effect of the combined expression of anti-tat and anti-rev genes on HIV-1 replication. *Gene Ther.* **4**, 288–295.

Cara, A., Rybak, S. M., Newton, D. L., Crowley, R., Rottschafer, S. E., Reitz, M. S., Jr., and Gusella, G. L. (1998). Inhibition of HIV-1 replication by combined expression of Gag dominant negative mutant and a human ribonuclease in a tightly controlled HIV-1 inducible vector. *Gene Ther.* **5**, 65–75.

Caruso, M., and Klatzmann, D. (1992). Selective killing of CD4+ cells harboring a human immunodeficiency virus-inducible suicide gene prevents viral spread in an infected cell population. *Proc. Natl. Acad. Sci. USA* **89**, 182–186.

Caruso, M. (1996). Gene therapy against cancer and HIV infection using the gene encoding herpes simplex virus thymidine kinase. *Mol. Med. Today* **2**, 212–217.

Castanotto, D., Rossi, J. J., and Sarver, N. (1994). Antisense catalytic RNAs as therapeutic agents. *Adv. Pharmacol.* **25**, 289–317.

Cesbron, J. Y., Agut, H., Gosselin, B., Candotti, D., Raphael, M., Puech, F., Grandadam, M., Debre, P., Capron, A., and Autran, B. (1994). SCID-hu mouse as a model for human lung HIV-1 infection. *C. R. Acad. Sci. III* **317**, 669–674.

Chang, H. K., Gendelman, R., Lisziewicz, J., Gallo, R. C., and Ensoli, B. (1994). Block of HIV-1 infection by a combination of antisense tat RNA and Tar decoys: A strategy for control of HIV-1. *Gene Ther.* **1**, 208–216.

Chen, I. S., and Zack, J. A. (1994). Modeling human retroviral pathogenesis and antiretroviral gene therapy in the scid mouse. *Res. Immunol.* **145**, 385–392.

Chen, S. Y., Bagley, J., and Marasco, W. A. (1994a). Intracellular antibodies as a new class of therapeutic molecules for gene therapy. *Hum. Gene Ther.* **5**, 595–601.

Chen, S. Y., Khouri, Y., Bagley, J., and Marasco, W. A. (1994b). Combined intra- and extracellular immunization against human immunodeficiency virus type 1 infection with a human anti-gp120 antibody. *Proc. Natl. Acad. Sci. USA* **91**, 5932–5936.

Chu, T.-H., and Dornburg, R. (1995). Retroviral vector particles displaying the antigen-binding site of an antibody enable cell-type-specific gene transfer. *J. Virol.* **69**, 2659–2663.

Chu, T.-H., and Dornburg, R. (1997). Towards highly-efficient cell-type-specific gene transfer with retroviral vectors that display a single chain antibody. *J. Virol* **71**, 720–725.

Chu, T.-H., Martinez, I., Sheay, W. C., and Dornburg, R. (1994). Cell targeting with retroviral vector particles containing antibody-envelope fusion proteins. *Gene Ther.* **1**, 292–299.

Chuah, M. K., Vandendriessche, T., Chang, H. K., Ensoli, B., and Morgan, R. A. (1994). Inhibition of human immunodeficiency virus type-1 by retroviral vectors expressing antisense-Tar. *Hum. Gene Ther.* **5**, 1467–1475.

Chun, T. W., Stuyver, L., Mizell, S. B., Ehler, L. A., Mican, J. A., Baseler, M., Lloyd, A. L., Nowak, M. A., and Fauci, A. S. (1997). Presence of an inducible HIV-1 latent reservoir during highly active antiretroviral therapy. *Proc. Natl. Acad. Sci.* **94**, 13193–13197.

Cohen, J. S. (1991). Antisense oilgonucleotides as antiviral agents. *Antiviral Res.* **16**, 121–133.

Cohen, J. (1996). Gene therapy-new role for HIV—A vehicle for moving genes into cells. *Science* **272**, 195.

Curiel, T. J., Cook, D. R., Wang, Y., Hahn, B. H., Ghosh, S. K., and Harrison, G. S. (1993). Long-term inhibition of clinical and laboratory human immunodeficiency virus strains in human T-cell lines containing an HIV-regulated diphtheria toxin a chain gene. *Hum. Gene. Ther.* **4**, 741–747.

de Clercq, E. (1996). Non-nucleoside reverse transcriptase inhibitors (nnrtis) for the treatment of human immunodeficiency virus type 1 (HIV-1) infections: strategies to overcome drug resistance development. *Med Res Rev* **16**, 125–157.

Dick, J. E. (1991). Immune-deficient mice as models of normal and leukemic human hematopoiesis. *Cancer Cells* **3**, 39–48.

Dick, J. E., Sirard, C., Pflumio, F., and Lapidot, T. (1992). Murine models of normal and neoplastic human haematopoiesis. *Cancer Surv.* **15**, 161–181.

Dick, J. E. (1994). Future prospects for animal models created by transplanting human haematopoietic cells into immune-deficient mice. *Res. Immunol.* **145**, 380–384.

Dinges, M. M., Cook, D. R., King, J., Curiel, T. J., Zhang, X. Q., and Harrison, G. S. (1995a). HIV-regulated diphtheria toxin a chain gene confers long-term protection against HIV type 1 infection in the human promonocytic cell line u937. *Hum. Gene Ther.* **6**, 1437–1445.

Dinges, M. M., Cook, D. R., King, J., Curiel, T. J., Zhang, X. Q., and Harrison, G. S. (1995b). HIV-regulated diphtheria toxin a chain gene confers long-term protection against HIV type 1 infection in the human promonocytic cell line U937. *Hum. Gene Ther.* **6**, 1437–1445.

Donahue, R. E., Bunnell, B. A., Zink, M. C., Metzger, M. E., Westro, R. P., Kirby, M. R., Unangst, T., Clements, J. E., and Morgan, R. A. (1998). Reduction in SIV replication in rhesus macaques infused with autologous lymphozytes engineered with antiviral genes. *Nat. Med.* **4**, 181–186.

Dornburg, R. (1995). Reticuloendotheliosis viruses and derived vectors. *Gene Ther.* **2**, 301–310.

Dornburg, R. (1997). From the natural evolution to the genetic manipulation of the host range of retroviruses. *Biol. Chem.* **378**, 457–468.

Dougherty, J. P., Wisniewski, R., Yang, S., Rhode, B. W., and Temin, H. M. (1989). New retrovirus helper cells with almost no nucleotide sequence homology to retrovirus vectors. *J. Virol.* **63**, 3209–3212.

Dropubic, B., and Jeang, K.-T. (1994). Gene therapy for human immunodeficiency virus infection: Genetic antiviral strategies and targets for intervention. *Hum. Gene Ther.* **5**, 927–939.

Duan, L., and Pomerantz, R. J. (1996). Intracellular antibodies for HIV-1 gene therapy. *Sci. Med.* **3**, 24–36.

Duan, L., Bagasra, O., Laughlin, M. A., Oakes, J. W., and Pomerantz, R. J. (1994a). Potent inhibition of human immunodeficiency virus type 1 replication by an intracellular anti-rev single-chain antibody. *Proc. Natl. Acad. Sci. USA* **91**, 5075–5079.

Duan, L., Zhang, H., Oakes, J. W., Bagasra, O., and Pomerantz, R. J. (1994b). Molecular and virological effects of intracellular anti-rev single-chain variable fragments on the expression of various human immunodeficiency virus-1 strains. *Hum. Gene Ther.* **5**, 1315–1324.

Duarte, E. A., Leavitt, M. C., Yamada, O., and Yu, M. (1997). Hairpin ribozyme gene therapy for aids. *Methods Mol. Biol.* **74**, 459–468.

Dull, T., Zufferey, R., Kelly, M., Mandel, R. J., Nguyen, M., Trono, D., and Naldini, L. (1998). A third-generation lentivirus vector with a conditional packaging system. *J. Virol.* **72**, 8463–8471.

Earnshaw, D. J., and Gait, M. J. (1997). Progress toward the structure and therapeutic use of the hairpin ribozyme. *Antisense Nucleic Acid Drug. Dev.* **7**, 403–411.

Eglitis, M. A., and Anderson, W. F. (1988). Retroviral vectors for introduction of genes into mammalian cells. *Biotechniques* **6**, 608–614.

Embretson, J. E., Zupanic, M., Ribas, J. L., Burke, A., Racz, P., Tenner-Racz, K., and Haase, A. T. (1993). Massive covert infection of helper T lymphocytes and macrophages by HIV during incubation period of AIDS. *Nature* **362**, 359–362.

Erickson, J. W., and Burt, S. K. (1996). Structural mechanisms of HIV drug resistance. *Annu. Rev. Pharmacol. Toxicol.* **36**, 545–571.

Etienne-Julan, M., Roux, P., Carillo, S., Jeanteur, P., and Piechaczyk, M. (1992). The efficiency of cell targeting by recombinant retroviruses depends on the nature opf the receptor and the composition of the artificial cell-virus linker. *J. Gen. Virol.* **73**, 3251–3255.

Finzi, D., Hermankova, M., Pierson, T., Carruth, L. M., Buck, C., Chaisson, R. E., Quinn, T. C., Chadwick, K., Margolick, J., Brookmeyer, R., Gallant, J., Markowitz, M., Ho, D. D., Richman, D. D., and Siliciano, R. F. (1997). Identification of a reservoir for HIV-1 in patients on highly active antiretroviral therapy. *Science* **278**, 1295–1300.

Freed, E. O., Delwart, E. L., Buchschacher, G. L., and Panganiban, A. T. (1992). A mutation in the human immunodeficiency virus type 1 transmembrane glycosylation gp41 dominantly interferes with fusion and infectivity. *Proc. Natl. Acad. Sci. USA* **89**, 70–74.

Fulcher, D., and Clezy, K. (1996). Antiretroviral therapy. *Med. J. Aust.* **164**, 607.

Gauduin, M.-C., Allaway, G. P., Olson, W. C., Weir, R., Maddon, P. J., and Koup, R. A. (1998). CD4-immunoglobulin G2 protects hu-PBL-SCID mice against challenge by primary human immunodeficiency virus type 1 isolates. *J. Virol.* **72**, 3475–3478.

Gauduin, M.-C., Parren, P. W. H. I., Weir, R., Barbas, C. F., Burton, D. R., and Koup, R. A. (1997). Passive immunization with a human monoclonal antibody protects hu-PBL-SCID mice against challenge by primary isolates of HIV-1. *Nat. Med.* **3**, 1389–1393.

Gervaix, A., Li, X., Kraus, G., and Wong-Staal, F. (1997a). Multigene antiviral vectors inhibit diverse human immunodeficiency virus type 1 clades. *J. Virol.* **71**, 3048–3053.

Gervaix, A., Schwarz, L., Law, P., Ho, A. D., Looney, D., Lane, T., and Wong-Staal, F. (1997b). Gene therapy targeting peripheral blood CD34+ hematopoietic stem cells of HIV-infected individuals. *Hum. Gene Ther.* **8**, 2229–2238.

Gilboa, E. (1990). Retroviral gene transfer: Applications to human gene therapy. *Prog. Clin. Biol. Res.* **352**, 301–311.

Gilboa, E., and Smith, C., (1994). Gene therapy for infectious diseases: The AIDS model. *Trends Genet.* **10**, 139–144.

Gunzburg, W. H., and Salmons, B. (1996). Development of retroviral vectors as safe, targeted gene delivery systems. *J. Mol. Med.* **74**, 171–182.

Hallek, M., and Wendtner, C. M. (1996). Recombinant adeno-associated virus (raav) vectors for somatic gene therapy—Recent advances and potential clinical applications. *Cytokines Mol. Ther.* **2**, 69–79.

Hampel, A. (1998). The hairpin ribozyme: Discovery, two-dimensional model, and development for gene therapy. *Prog. Nucleic Acid Res. Mol. Biol.* **58**, 1–39.

Heusch, M., Kraus, G., Johnson, P., and Wong-Staal, F., (1996). Intracellular immunization against SIVmac utilizing a hairpin ribozyme. *Virology* **216**, 241–244.

Ho, A. D., Li, X. Q., Lane, T. A., Yu, M., Law, P., and Wongstaal, F. (1995). Stem cells as vehicles for gene therapy—Novel strategy for HIV infection. *Stem Cells* **13**, 100–105.

Hunter, E., and Swanstrom, R. (1990). Retrovirus envelope glycoproteins. *Curr. Top. Microbiol. Immunol.* **157**, 187–253.

Ilves, H., Barske, C., Junker, U., Bohnlein, E., and Veres, G. (1996). Retroviral vectors designed for targeted expression of rna polymerase III-driven transcripts—A comparative study. *Gene* **171**, 203–208.

Jamieson, B. D., Aldrovandi, G. M., and Zack, J. A. (1996). The SCID-hu mouse: An *in-vivo* model for HIV-1 pathogenesis and stem cell gene therapy for AIDS. *Semin. Immunol.* **8**, 215–221.

Jiang, A., Chu, T.-H. T., Nocken, F., Cichutek, K., and Dornburg, R., (1998). Cell-type-specific gene transfer into human cells with retroviral vectors that display single-chain antibodies. *J. Virol.* **72**, 10148–10156.

Jiang, A. and Dornburg, R. (1999). *In vivo* cell-type-specific gene delivery with retroviral vectors that display single chain antibodies. *Gene Ther.* **6**, 1982–1987.

Johnston, M. I., Allaudeen, H. S., and Sarver, N. (1989). HIV proteinase as a target for drug action. *Trends Pharmacol. Sci.* **10**, 305–307.

Junker, U., Baker, J., Kalfoglou, C. S., Veres, G., Kaneshima, H., and Bohnlein, E., (1997). Antiviral potency of drug-gene therapy combinations against human immunodeficiency virus type 1. *AIDS Res. Hum. Retroviruses* **13**, 1395–1402.

Kaplan, J. C., and Hirsch, M. S., (1994). Therapy other than reverse transcriptase inhibitors for HIV infection. *Clin. Lab. Med.* **14**, 367–391.

Kaul, M., Yu, H., Ron, Y., and Dougherty, J. P., (1998). Regulated lentiviral packaging cell line devoid of most viral cis-acting sequences. *Virology* **249**, 167–174.

Kim, V. N., Mitrophanous, K., Kingsman, S. M., and Kingsman, A. J. (1998). Minimal requirement for a lentivirus vector based on human immunodeficiency virus type 1. *J. Virol.* **72**, 811–816.

Larsson, S., Hotchkiss, G., Su, J., Kebede, T., Andang, M., Nyholm, T., Johansson, B., Sonnerborg, A., Vahlne, A., Britton, S., and Ahrlund-Richter, L. (1996). A novel ribozyme target site located in the HIV-1 nef open reading frame. *Virology* **219**, 161–169.

Leavitt, M. C., Yu, M., Yamada, O., Kraus, G., Looney, D., Poeschla, E., and Wong-Staal, F. (1994). Transfer of an anti-HIV-1 ribozyme gene into primary human lymphocytes. *Hum. Gene Ther.* **5**, 1115–1120.

Leavitt, M. C., Yu, M., Wongstaal, F., and Looney, D. J. (1996). *Ex vivo* transduction and expansion of CD4(+) lymphocytes from HIV+ donors—Prelude to a ribozyme gene therapy trial. *Gene Ther.* **3**, 599–606.

Lee, S. W., Gallardo, H. F., Gilboa, E., and Smith, C. (1994). Inhibition of human immunodeficiency virus type 1 in human T cells by a potent Rev response element decoy consisting of the 13-nucleotide minimal Rev-binding domain. *J. Virol.* **68**, 8254–8264.

Lewis, P., and Emerman, M. (1994). Passage through mitosis is required for oncoretroviruses but not for the human immunodeficiency virus. *J. Virol* **68**, 510–516.

Li, X., Gervaix, A., Kang, D., Law, P., Spector, S. A., Ho, A. D., and Wong-Staal, F. (1998). Gene therapy targeting cord blood-derived CD34+ cells from HIV-exposed infants: Preclinical studies. *Gene Ther.* **5**, 233–239.

Liem, S. E., Ramezani, A., Li, X., and Joshi, S. (1993). The development and testing of retroviral vectors expressing trans-dominant mutants of HIV-1 proteins to confer anti-HIV-1 resistance. *Hum. Gene Ther.* **4**, 625–634.

Lisziewicz, J., Sun, D., Lisziewicz, A., and Gallo, R. C. (1995). Antitat gene therapy: A candidate for late-stage AIDS patients. *Gene Ther.* **2**, 218–222.

Lisziewicz, J., Sun, D., Smythe, J., Lusso, P., Lori, F., Louie, A., Markham, P., Rossi, J., Reitz, M., and Gallo, R. C. (1993). Inhibition of human immunodeficiency virus type 1 replication by regulated expression of a polymeric Tat activation response RNA decoy as a strategy for gene therapy in AIDS. *Proc. Natl. Acad. Sci. USA* **90**, 8000–8004.

Liu, J., Woffendin, C., Yang, Z. Y., and Nabel, G. J. (1994). Regulated expression of a dominant negative form of Rev improves resistance to HIV replication in T cells. *Gene Ther.* **1**, 32–37.

Lori, F., Lisziewicz, J., Smythe, J., Cara, A., Bunnag, T. A., Curiel, D., and Gallo, R. C. (1994). Rapid protection against human immunodeficiency virus type 1 (HIV-1) replication mediated by high efficiency non-retroviral delivery of genes interfering with HIV-1 Tat and Gag. *Gene Ther.* **1**, 27–31.

Lyman, S. D., Brasel, K., Rousseau, A. M., and Williams, D. E. (1994). The flt3 ligand: A hematopoietic stem cell factor whose activities are distinct from steel factor. *Stem Cells* **12**(Suppl. 1), 99–107.

Marasco, W. A., Chen, S., Richardson, J. H., Ramstedt, U., and Jones, S. D. (1998). Intracellular antibodies against HIV-1 envelope protein for AIDS gene therapy. *Hum. Gene Ther.* **9**, 1627–1642.

Marin, M., Noel, D., Valsesia-Wittman, S., Brockly, F., Etienne-Julan, M., Russell, S., Cosset, F.-L., and Piechaczyk, M. (1996). Targeted infection of human cells via major histocompatibility complex class I molecules by Moleney leukemia virus-derived viruses displaying single chain antibody fragment-envelope fusion proteins. *J. Virol.* **70**, 2957–2962.

Matano, T., Odawara, T., Iwamoto, A., and Yoshikura, H. (1995). Targeted infection of a retrovirus bearing a CD4-env chimera into human cells expressing human immunodeficiency virus type 1. *J. Gen. Virol.* **76**, 3165–3169.

Matsuda, Z., Yu, X., Yu, Q. C., Lee, T. H., and Essex, M. (1993). A virion-specific inhibitory molecule with therapeutic potential for human immunodeficiency virus type 1. *Proc. Natl. Acad Sci. USA* **90**, 3544–3548.

Michienzi, A., Conti, L., Varano, B., Prislei, S., Gessani, S., and Bozzoni, I. (1998). Inhibition of human immunodeficiency virus type 1 replication by nuclear chimeric anti-HIV ribozymes in a human T lymphoblastoid cell line. *Hum. Gene Ther.* **9**, 621–628.

Miller, A. D. (1990). Retrovirus packaging cells. *Hum. Gene. Ther.* **1**, 5–14.

Miyake, K., Suzuki, N., Matsuoka, H., Tohyama, T., and Shimada, T. (1999). Stable integration of human immunodeficiency virus-based retroviral vectors into the chromosomes of nondividing cells. *Hum. Gene Ther.* **9**, 467–475.

Mochizuki, H., Schwartz, J. P., Tanaka, K., Brady, R. O., and Reiser, J. (1998). High-titer human immunodeficiency virus type 1-based vector systems for gene delivery into nondividing cells. *J. Virol.* **72**, 8873–8883.

Morgan, R. A., and Anderson, W. F. (1993). Human gene therapy. *Annu. Rev. Biochem.* **62**, 191–217.

Morgan, R. A., Baler-Bitterlich, G., Ragheb, J. A., Wong-Staal, F., Gallo, R. C., and Anderson, W. F. (1994). Further evaluation of soluble CD4 as an anti-HIV type 1 gene therapy: Demonstration of protection of primary human peripheral blood lymphocytes from infection by HIV type 1. *AIDS Res. Hum. Retroviruses* **10**, 1507–1515.

Morgan, R. A., Looney, D. J., Muenchau, D. D., Wong-Staal, F., Gallo, R. C., and Anderson, W. F. (1990). Retroviral vectors expressing soluble CD4: A potential gene therapy for AIDS. *AIDS Res. Hum. Retroviruses* **6**, 183–191.

Mueller, B. M., and Reisfeld, R. A. (1991). Potential of the SCID mouse as a host for human tumors. *Cancer Metastasis Rev.* **10**, 193–200.

Muzyczka, N. (1992). Use of adeno-associated virus as a general transduction vector for mammalian cells. *Curr. Top. Microbiol. Immunol.* **158**, 97–129.

Naldini, L., Blomer, U., Gage, F. H., Trono, D., and Verma, I. M. (1996). Efficient transfer, integration, and sustained long-term expression of the transgene in adult rat brains injected with a lentiviral vector. *Proc. Natl. Acad. Sci. USA* **93**, 11382–11388.

Nilson, B. H. K., Morling, F. J., Cosset, F.-L., and Russell, S. J. (1996). Targeting of retroviral vectors through protease–substrate interactions. *Gene Ther.* **3**, 280–286.

Nolan, G. P. (1997). Harnessing viral devices as pharmaceuticals: Fighting HIV-1's fire with fire. *Cell* **90**, 821–824.

Orlic, D., Girard, L. J., Jordan, C. T., Anderson, S. M., Cline, A. P., and Bodine, D. M. (1996). The level of mRNA encoding the amphotropic retrovirus receptor in mouse and human hematopoietic stem cells is low and correlates with the efficiency of retrovirus transduction. *Proc. Natl. Acad. Sci. USA* **93**, 11097–11102.

Pantaleo, G., Graziosi, C., Demarest, J. F., Butini, L., Montroni, M., Fox, C. H., Orenstein, J. M., Kotler, D. P., and Fauci, A. S. (1993). HIV infection is active and progressive in lymphoid tissue during the clinically latent stage of disease. *Nature* **362**, 355–358.

Parolin, C., and Sodroski, J. (1995). A defective HIV-1 vector for gene transfer to human lymphocytes. *J. Mol. Med.* **73**, 279–288.

Plavec, I., Agarwal, M., Ho, K. E., Pineda, M., Auten, J., Baker, J., Matsuzaki, H., Escaich, S., Bonyhadi, M., and Bohnlein, E. (1997). High transdominant revm10 protein levels are required to inhibit HIV-1 replication in cell lines and primary T cells: Implication for gene therapy of aids. *Gene Ther.* **4**, 128–139.

Poeschla, E., and Wong-Staal, F. (1994). Antiviral and anticancer ribozymes. *Curr. Opin. Oncol.* **6**, 601–606.

Poeschla, E. M., and Wong-Staal, F. (1995). Gene therapy and HIV disease. *AIDS Clin. Rev.* 1–45.

Pomerantz, R. J., and Trono, D. (1995). Genetic therapies for HIV infections: Promise for the future. *AIDS* **9**, 985–993.

Pomerantz, R. J., Bagasra, O., and Baltimore, D. (1992). Cellular latency of human immunodeficiency virus type 1. *Curr. Opin. Immunol.* **4**, 475–480.

Purchase, H. G., and Witter, R. L. (1975). The reticuloendotheliosis viruses. *Curr. Top. Microbiol. Immunol.* **71**, 103–124.

Putman, L. (1998). Debate grows on safety of gene-therapy vectors. *Lancet* **351**, 808.

Ragheb, J. A., Bressler, P., Daucher, M., Chiang, L., Chuah, M. K., Vandendriessche, T., and Morgan, R. A. (1995). Analysis of trans-dominant mutants of the HIV type 1 Rev protein for their ability to inhibit Rev function, HIV type 1 replication, and their use as anti-HIV gene therapeutics. *AIDS Res. Hum. Retroviruses* **11**, 1343–1353.

Ramezani, A., Ding, S. F., and Joshi, S. (1997). Inhibition of HIV-1 replication by retroviral vectors expressing monomeric and multimeric hammerhead ribozymes. *Gene Ther.* **4**, 861–867.

Ranga, U., Woffendin, C., Verma, S., Xu, L., June, C. H., Bishop, D. K., and Nabel, G. J. (1998a). Enhanced T cell engraftment after retroviral delivery of an antiviral gene in HIV-infected individuals. *Proc. Natl. Acad. Sci. USA* **95**, 1201–1206.

Ranga, U., Woffendin, C., Verma, S., Xu, L., June, C. H., Bishop, D. K., and Nabel, G. J. (1998b). Enhanced T cell engraftment after retroviral delivery of an antiviral gene in HIV-infected individuals. *Proc. Natl. Acad. Sci. USA* **95**, 1201–1205.

Richman, D. (1992). HIV drug resistance. *AIDS Res. Hum. Retroviruses* **8**, 1065–1071.

Rondon, I. J., and Marasco, W. A. (1997). Intracellular antibodies (intrabodies) for gene therapy of infectious diseases. *Annu. Rev. Microbiol.* 51, 257–283.

Rossi, C., Balboni, P. G., Betti, M., Marconi, P. C., Bozzini, R., Grossi, M. P., Barbanti-Brodano, G., and Caputo, A. (1997). Inhibition of HIV-1 replication by a tat transdominant negative mutant in human peripheral blood lymphocytes from healthy donors and HIV-1-infected patients. *Gene Ther.* 4, 1261–1269.

Rossi, J. J., and Sarver, N. (1992). Catalytic antisense RNA (ribozymes): Their potential and use as anti-HIV-1 therapeutic agents. *Adv. Exp. Med. Biol.* 312, 95–109.

Rothenberg, M., Johnson, G., Laughlin, C., Green, I., Cradock, J., Sarver, N., and Cohen, J. S. (1989). Oligodeoxynucleotides as anti-sense inhibitors of gene expression: Therapeutic implications. *J. Natl. Cancer Inst.* 81, 1539–1544.

Roux, P., Jeanteur, P., and Piechaczyk, M. (1989). A versatile and potentially general approach to the targeting of specific cell types by retroviruses: Application to the infection of human by means of major histocompatibility complex class I and class II antigens by mouse ecotropic murine leukemia virus-derived viruses. *Proc. Natl. Acad. Sci. USA* 86, 9079–9083.

Russell, S. J., Hawkins, R. E., and Winter, G. (1993). Retroviral vectors displaying functional antibody fragments. *Nucleic Acid. Res.* 21, 1081–1085.

San Jose, E., Munoz-Fernandez, M. A., and Alarcon, B. (1998). Retroviral vector-mediated expression in primary human T cells of an endoplasmic reticulum-retained CD4 chimera inhibits human immunodeficiency virus type-1 replication. *Hum. Gene Ther.* 9, 1345–1357.

Sandstrom, P. A., and Folks, T. M. (1996). New strategies for treating aids. *Bioessays* 18, 343–346.

Sarver, N., and Rossi, J. (1993). Gene therapy: A bold direction for HIV-1 treatment. *AIDS Res. Hum. Retroviruses* 9, 483–487.

Schnell, M. J., Johnson, J. E., Buonocore, L., and Rose, J. K. (1997). Construction of a novel virus that targets HIV-1-infected cells and controls HIV-1 infection. *Cell* 90, 849–857.

Schnierle, B. S., Moritz, D., Jeschke, M., and Groner, B. (1996). Expression of chimeric envelope proteins in helper cell lines and integration into Moloney murine leukemia virus particles. *Gene Ther.* 3, 334–342.

Schwedler, U., Kornbluth, R. S., and Trono, D. (1999). The nuclear localization signal of the matrix protein of human immunodeficiency virus type 1 allows the establishment of infection in macrophages and quiescent T lymphocytes. *Proc. Natl. Acad. Sci. USA* 91, 6992–6996.

Shen, R. N., Lu, L., and Broxmeyer, H. E. (1990). New therapeutic strategies in the treatment of murine diseases induced by virus and solid tumors: Biology and implications for the potential treatment of human leukemia, aids, and solid tumors. *Crit. Rev. Oncol. Hematol.* 10, 253–265.

Smythe, J. A., Sun, D., Thomson, M., Markham, P. D., Reitz, M. S., Jr., Gallo, R. C., and Lisziewicz, J. (1994). A Rev-inducible mutant Gag gene stably transferred into T lymphocytes: An approach to gene therapy against human immunodeficiency virus type 1 infection. *Proc. Natl. Acad. Sci. USA* 91, 3657–3661.

Somia, N. V., Zoppe, M., and Verma, I. M. (1995). Generation of targeted retroviral vectors by using single-chain variable fragment: An approach to *in vivo* gene delivery. *Proc. Natl. Acad. Sci. USA* 92, 7570–7574.

Stein, C. A., and Cheng, Y. C. (1993). Antisense oligonucleotides as therapeutic agents: Is the bullet really magical? *Science* 261, 1004–1012.

Sullenger, B. A., Gallardo, H. F., Ungers, G. E., and Gilboa, E. (1990). Overexpression of TAR sequences renders cells resistant to HIV replication. *Cell* 63, 601–608.

Sullivan, S. M. (1994). Development of ribozymes for gene therapy. *J. Invest. Dermatol.* 103, 85S–89S.

Temin H. M. (1986). Retrovirus vectors for gene transfer: Efficient integration into and expression of exogenous DNA in vertebrate cell genomes. *In* "Gene Transfer" (R. Kucherlapati., Ed.), pp. 144–187. Plenum, New York.

Trono, D., Feinberg, M. B., and Baltimore, D. (1988). HIV-1 gag mutants can dominantly interfere with the replication of wild-type virus. *Cell* **59**, 113–120.

Uchida, N., Sutton, R. E., Friera, A. M., He, D., Reitsma, M. J., Chang, W. C., Veres, G., Scollay, R., and Weissman, I. L. (1998). HIV, but not murine leukemia virus, vectors mediate high efficiency gene transfer into freshly isolated g0/g1 human hematopoietic stem cells. *Proc. Natl. Acad. Sci. USA* **95**, 11939–11944.

Uckun, F. M. (1996). Severe combined immunodeficient mouse models of human leukemia. *Blood* **88**, 1135–1146.

Valsesia-Wittmann, S., Morling, F., Nilson, B., Takeuchi, Y., Russell, S., and Cosset, F.-L. (1996). Improvement of retroviral retargeting by using acid spacers between an additional binding domain and the N terminus of Moloney leukemia virus SU. *J. Virol.* **70**, 2059–2064.

Vandendriessche, T., Chuah, M. K., Chiang, L., Chang, H. K., Ensoli, B., and Morgan, R. A. (1995). Inhibition of clinical human immunodeficiency virus (HIV) type 1 isolates in primary CD4+ T lymphocytes by retroviral vectors expressing anti-HIV genes. *J. Virol.* **69**, 4045–4052.

von Laer, D., Thomsen, S., Vogt, B., Donath, M., Kruppa, J., Rein, A., Ostertag, W., and Stocking, C. (1998). Entry of amphotropic and 10a1 pseudotyped murine retroviruses is restricted in hematopoietic stem cell lines. *J. Virol.* **72**, 1424–1430.

Wang, L., Witherington, C., King, A., Gerlach, W. L., Carr, A., Penny, R., Cooper, D., Symonds, G., and Sun, L. Q. (1998). Preclinical characterization of an anti-tat ribozyme for therapeutic application. *Hum. Gene Ther.* **9**, 1283–1291.

Warner, J. F., Jolly, D., Mento, S., Galpin, J., Haubrich, R., and Merritt, J. (1995). Retroviral vectors for HIV immunotherapy. *Ann. N.Y. Acad. Sci.* **772**, 105–116.

Weerasinghe, M., Liem, S. E., Asad, S., Read, S. E., and Joshi, S. (1991). Resistance to human immunodeficiency virus type 1 (HIV-1) infection in human CD4+ lymphocyte-derived cell lines conferred by using retroviral vectors expressing an HIV-1 RNA-specific ribozyme. *J. Virol.* **65**, 5531–5534.

Whitlow, M., and Filpula, D. (1991). Single-chain Fv proteins and their fusion proteins. *Methods Enzymol.* **2**, 1–9.

Woffendin, C., Ranga, U., Yang, Z. Y., Xu, L., and Nabel, G. J. (1996). Expression of a protective gene prolongs survival of T cells in human immunodeficiency virus-infected patients. *Proc. Natl. Acad. Sci. USA* **93**, 2889–2894.

Woffendin, C., Yang, Z. Y., Udaykumar, Xu, L., Yang, N. S., Sheehy, M. J., and Nabel, G. J. (1994). Nonviral and viral delivery of a human immunodeficiency virus protective gene into primary human T cells. *Proc. Natl. Acad. Sci. USA* **91**, 11581–11585.

Wong, J. K., Hezareh, M., Gunthard, H. F., Havlir, D. V., Ignacio, C. C., Spina, C. A., and Richman, D. D. (1997). Recovery of replication-competent HIV despite prolonged suppression of plasma viremia. *Science* **278**, 1291–1295.

Wongstaal, F. (1995). Ribozyme gene therapy for HIV infection—Intracellular immunization of lymphocytes and CD34+ cells with an anti-HIV-1 ribozyme gene. *Adv. Drug Deliv. Rev.* **17**, 363–368.

Wu, X., Liu, H., Xiao, H., Conway, J. A., and Kappes, J. C. (1996). Inhibition of human and simian immunodeficiency virus protease function by targeting vpx-protease-mutant fusion protein into viral particles. *J. Virol.* **70**, 3378–3384.

Yang, A. G., Bai, X., Huang, X. F., Yao, C., and Chen, S. (1997). Phenotypic knockout of HIV type 1 chemokine coreceptor ccr-5 by intrakines as potential therapeutic approach for HIV-1 infection. *Proc. Natl. Acad. Sci. USA* **94**, 11567–11572.

Young, J. A. T., Bates, P., Willert, K., and Varmus, H. E. (1990). Efficient incorporation of human CD4 protein into avian leukosis virus particles. *Science* **250**, 1421–1423.

Yu, M., Ojwang, J., Yamada, O., Hampel, A., Rapapport, J., Looney, D., and Wong-Staal, F. (1993). A hairpin ribozyme inhibits expression of diverse strains of human immunodeficiency virus type 1. *Proc. Natl. Acad. Sci. USA* **90,** 6340–6344.

Yu, M., Poeschla, E., and Wong-Staal, F. (1994a). Progress towards gene therapy for HIV infection. *Gene Ther.* **1,** 13–26.

Yu, M., Poeschla, E., and Wong-Staal, F. (1994b). Progress towards gene therapy for HIV infection. *Gene Ther.* **1,** 13–26.

Yu, M., Poeschla, E., Yamada, O., Degrandis, P., Leavitt, M. C., Heusch, M., Yees, J. K., Wong-Staal, F., and Hampel, A. (1995). *In vitro* and *in vivo* characterization of a second functional hairpin ribozyme against HIV-1. *Virology* **206,** 381–386.

Zhang, H., Dornadula, G., Beumont, M., Livornese, L., Jr., Van Uitert, B., Henning, K., and Pomerantz, R. J. (1998). Human immunodeficiency virus type 1 in the semen of men receiving highly active antiretroviral therapy. *N. Engl. J. Med.* **339,** 1803–1809.

Michael W. Cho

AIDS Vaccine Research and Development Unit
Laboratory of Molecular Microbiology
National Institute of Allergy and Infectious Diseases
National Institutes of Health
Bethesda, Maryland 20892

Assessment of HIV Vaccine Development: Past, Present, and Future

I. Introduction

It's been over 15 years since the discovery of human immunodeficiency virus-1 (HIV-1), the virus that causes acquired immunodeficiency syndrome (AIDS) in humans. Although this epidemic resulted in the deaths of millions worldwide and the spread of the virus is still rampant, especially in Africa and Southeast Asia, there is yet no cure in sight and the efforts in vaccine development has been thus far mostly unsuccessful. It is a widely held belief that a vaccine is the only cost-effective measure that will contain the AIDS epidemic. Currently available multiple-drug therapies are too expensive for most AIDS patients in the world and require regimens that are too difficult to follow. Increasing reports of drug-resistant variants and debilitating side effects of the drugs are creating serious doubts about the long-term success of these treatments.

The slow progress of AIDS vaccine development has been frustrating for government officials, health organizations, social activists, and perhaps

Advances in Pharmacology, Volume 49

above all, the scientists. The lack of major advances in this endeavor is not due to underfunding by the government, lack of leadership, or the absence of capable scientists. The principal difficulty lies from the fact that HIV-1 is drastically different from other human pathogens in many respects. It includes the lack of documented natural immunity (i.e., full recovery subsequent to infection), the ability of the virus to integrate its genetic codes into host's genome, and the ability of the virus to infect and kill immune cells that play a central role in eliciting immune response against pathogens. High antigenic variation of its surface envelope glycoprotein, the lack of a good animal model, and unknown correlates of protection make the vaccine development even more difficult.

HIV-1 probably is the most thoroughly studied virus. The biochemical and structural properties of many viral gene products are well understood and have led to the development of many antiviral agents. There are equally impressive amounts of data in the literature on viral pathogenesis, virus–cell interactions, and immunological responses against the virus and viral proteins both in human and in nonhuman primates. Many aspects of AIDS vaccine research and development have been reviewed elsewhere, including various vaccine strategies (Bagarazzi *et al.*, 1998; Barnett *et al.*, 1998; Excler and Plotkin, 1997; Kim and Weiner, 1997; Robinson, 1997; Ruprecht *et al.*, 1996; Stott and Schild, 1996; Tartaglia *et al.*, 1998; Wagner *et al.*, 1996), nonhuman primate models (Almond and Heeney, 1998; Heeney, 1996; Hulskotte *et al.*, 1998; Lamb-Wharton *et al.*, 1997; Murthy *et al.*, 1998; Schultz, 1998; Schultz and Stott, 1994; Stott *et al.*, 1998), immune response and correlates of protection (Ada and McElrath, 1997; Burton, 1997; Gotch, 1998; Haigwood and Zolla-Pazner, 1998; Haynes *et al.*, 1996; Heeney *et al.*, 1997; Hu and Norrby, 1997; McElrath *et al.*, Weinhold, 1997; Parren *et al.*, 1997; Poignard *et al.*, 1996; Steinman and Germain, 1998), human clinical trials (Dolin, 1995), and general progress and problems (Baltimore and Heilman, 1998; Johnston, 1997; Letvin, 1998; Verani *et al.*, 1997). In this review, our current understanding of biochemical, structural, and immunological properties of the viral envelope glycoprotein that are pertinent to vaccine development is summarized. Subsequently, a summary of various vaccine strategies, assessment of the past and recent progress on HIV-1 vaccine development, and some of the factors that must be considered in designing new vaccine candidates are discussed. A brief overview of interactions between HIV-1 and the host immune system is first presented.

II. Virus–Host Interaction

Understanding the complexity of the interactions between HIV-1 and the host immune system is invaluable to designing an effective vaccine against the virus. A simple overview of the virus replication cycle and the interactions between the virus, an infected cell, and immune cells is shown in Fig. 1. HIV-1 infects CD4$^+$ T lymphocytes, macrophages, dendritic cells, and brain microglial cells through a series of interactions between the viral envelope

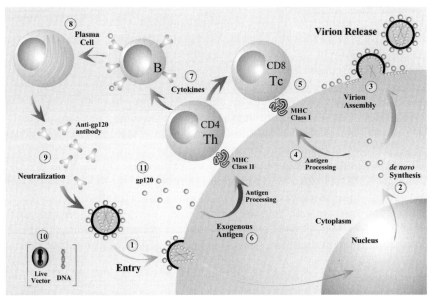

FIGURE I   A simplified overview of interactions between HIV-1 and immune systems. (1) HIV-1 enters cells that express CD4 and one of the coreceptor molecules on the surface. (2) Integrated provirus directs synthesis of viral gene products *de novo*. (3) Structural proteins assemble into virus particles on the plasma membrane, viral genome is encapsidated, and progeny virions are released. (4) Some of the *de novo* synthesized viral proteins are degraded into peptides via the Class I antigen-processing pathway. CTL epitopes, which bind to MHC Class I molecules, are presented on the cell surface. (5) The virus-infected cell is killed by CD8$^+$ CTL (when activated) that recognize the viral epitope-Class I molecule complex. (6) Exogenous viral antigens (e.g., gp120 or its debris) that are endocytosed are processed *via* the Class II antigen-processing pathway. The peptides (Th epitopes) associate with MHC Class II molecules and the complex is recognized by CD4$^+$ Th cells. (7) Stimulated virus-specific Th cells secrete various cytokines to signal virus-specific CTL and B cells to undergo clonal expansion. (8) B cells (either naïve- or memory-) that recognize viral antigens (e.g., gp120) mature into plasma cells, which secrete large amounts of gp120-specific antibodies. (9) Antibodies that can recognize certain epitopes on the native quaternary structure of the envelope glycoprotein bind virus particles and inhibit viral entry, thus neutralizing the virus. (10) Vaccine candidates, such as recombinant live vector or plasmid DNA encoding viral genes, work by expressing viral antigens *de novo* subsequent to its entry into cells, mimicking the natural virus infection. Thus, these vaccine approaches can potentially elicit CTL as well as the humoral immune response. (11) In contrast, vaccines that directly deliver viral antigens (e.g., whole-inactivated or subunit vaccine) are likely to elicit mostly humoral immune response. However, because of large amount of antigen than can be administered, subunit vaccine can greatly enhance immune response elicited by either live vector or DNA vaccine (e.g., prime-boost vaccine approach).

glycoprotein and the cellular receptor CD4 and coreceptors, which include several members of the chemokine receptor family (see below). When the virus enters a susceptible cell, the viral RNA genome is reverse transcribed into double-stranded DNA, which subsequently integrates into the host genome. Viral gene expression is highly regulated through complex interactions between cellular transcriptional and posttranscriptional machinery and viral gene products (for reviews, see Freed and Martin, 1999; Luciw, 1996). Late in the virus replication cycle, viral structural proteins and envelope glycoproteins are made, virions are assembled at the cell surface, and mature progeny viruses are released.

As a part of a natural immune surveillance system, proteins in the body are continuously "sampled" by immune cells. When foreign proteins are synthesized *de novo* (e.g., in virus-infected cells), some are proteolytically degraded into peptides via the Class I antigen-processing pathway. Some of these peptides associate with the major histocompatibility complex (MHC) Class I molecules and are presented on the cell surface for immune surveillance by $CD8^+$ cytotoxic T lymphocytes (Tc or CTL). These peptides are called CTL epitopes and are about 8 to 10 amino acids in lengths. In contrast, exogenous viral antigens that are internalized by endocytosis are degraded via Class II antigen-processing pathway and are presented on the cell surface in association with the MHC Class II molecules. These viral antigens, which are 10 to 12 amino acids in length, are recognized by $CD4^+$ helper T lymphocytes (Th). Therefore, both Th and CTL epitopes are linear and are more sensitive to the primary sequence rather than the conformation. While most of the cells in a body can present foreign antigens to CTL, only a few professional antigen-presenting cells (e.g., macrophage and dendritic cells) can present antigens to Th cells. Th cells play a central role in the body's immune system because they stimulate propagation of antigen-specific CTL and antibody-producing B cells through a complex signal transduction using cytokines. CTL are responsible for killing virus-infected cells, which limit production, and therefore the spreading, of progeny viruses. When B cells are stimulated, they mature into plasma cells, which produce high levels of virus antigen-specific antibodies. When these antibodies bind virus, the antibodies neutralize, or block, the virus from infecting susceptible cells. Although some B cell epitopes are linear, many are nonlinear and are highly conformation-dependent. Th and CTL make up the cellular immune response arm of the immune system while B cells are a part of the humoral immune response.

Based on this simple overview, several problems are already evident in developing a vaccine against AIDS. First, if HIV-1 enters a resting $CD4^+$ T cell, viral gene expression is silenced and immune cells do not recognize the infected cells as being invaded by a foreign entity. Second, HIV-1 integrates its own genetic information into the host's genome. Thus, unless the infected cells are killed, HIV-1 continuously persists in the body. Third, HIV-1 infects and kills CD4$^+$ helper T cells that play a central role in the immune system. Thus, once all of HIV-1-specific Th cells are killed, the body is unable to mount any immune response against the virus.

III. Envelope Glycoprotein

The envelope glycoprotein of HIV-1 has been the focus of the majority of AIDS vaccine development efforts during the past decade. It is the only viral antigen on the surface of a virion that can be recognized by the body's immune system. It is the interaction between this viral protein and cellular receptors that allows HIV-1 to enter cells. Understanding the structural, biochemical, functional, and immunological properties of this protein would undoubtedly be a crucial step in designing an effective vaccine against HIV-1. Because this protein has been a focus of many reviews in the past (Fenouillet *et al.*, 1994; Freed and Martin, 1995; Putney *et al.*, 1990; Wyatt *et al.*, 1998; Wyatt and Sodroski, 1998), only those subjects that are pertinent to vaccine development are discussed.

A. General Features

The precursor form of HIV-1 envelope glycoprotein (gp160) is divided into three segments: the signal peptide (SP), the surface glycoprotein gp120, and the transmembrane glycoprotein gp41 (Fig. 2). The SP is required for directing protein synthesis on rough endoplasmic reticulum (ER), trafficking the protein through the ER-Golgi pathway, and eventual cell surface expression. The SP is cleaved by the signal peptidase in ER and is not present in the mature form of the protein on virion. The junction between gp120 and gp41 is cleaved by furin or furinlike protease in Golgi at the carboxy-terminus of a highly conserved Lys/Arg-X-Lys/Arg-Arg motif (Anderson *et al.*, 1993; Decroly *et al.*, 1994; Hallenberger *et al.*, 1992). Subsequent to cleavage, gp120 and gp41 are associated noncovalently to form a loose heterodimer. The gp120–gp41 complex is anchored to the membrane via the transmembrane domain of gp41, which spans the lipid bilayer. This gp120–gp41 heterodimer is thought to exist as an oligomer on virion surface. Although the results from some of the chemical cross-linking studies suggested that the complex might exist as a dimer/tetramer (Doms *et al.*, 1991; Earl *et al.*, 1990; Pinter *et al.*, 1989; Schawaller *et al.*, 1989), other studies including more recent X-ray crystal structure analyses of the ectodomain of gp41 indicated that the complex is likely to exist as a trimer (Weiss *et al.*, 1990; Chan *et al.*, 1997; Tan *et al.*, 1997; Weissenhorn *et al.*, 1997).

Gp120 is a heavily glycosylated protein with 23 to 24 potential N-linked glycosylation sites (Leonard *et al.*, 1990) as well as some O-linked modifications (Bernstein *et al.*, 1994) throughout the length of the protein.

A

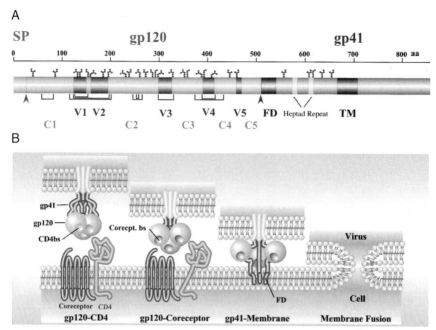

FIGURE 2 A schematic diagram of important features on HIV-1 envelope glycoprotein and proposed mechanism of virus entry into cells. (A) HIV-1 envelope glycoprotein (gp160) is processed into three fragments: signal peptide (SP), surface glycoprotein gp120, and trans-membrane glycoprotein gp41. Nonglycosylated forms of the two subunits are 54 and 39 kDa, respectively. Approximate positions of the cleavage sites, glycosylation sites, the variable (V1-V5) and conserved regions (C1–C5), fusion domain (FD), transmembrane domain (TM), disulfide bonds (brackets), and heptad repeat sequences are indicated. (B) Sequential interactions between gp120 and CD4, gp120 and coreceptor, and gp41 and cellular membrane, ultimately leading to fusion between viral and cellular membranes (see text).

Greater than one-half of the protein's total mass is from carbohydrate moieties, as fully deglycosylated protein has a molecular weight of approximately 54 kDa. Gp41 is also glycosylated, but much less extensively compared to gp120. One of the striking features of envelope glycoprotein is the great genetic diversity of gp120 from isolate-to-isolate. Comparative sequence analyses of the protein have revealed five hypervariable regions (V1–V5) with conserved regions in between. Adjacent to most of these variable regions, there are highly conserved cystein residues. Intramolecular disulfide linkages between these cystein residues form loops for the first four variable regions (Fig. 2; Leonard et al., 1990).

While gp120 is the only known viral protein that directly interacts with the cellular receptors (see below), gp41 also plays an important role in the virus entry process. Two structural features important for the function of gp41 are a long stretch of hydrophobic amino acid residues at the amino-terminus of gp41, called the fusion domain (FD), and two heptad repeat sequences downstream of the FD in the ectodomain of the protein. The FD is speculated to be the determinant that inserts into (or closely associate with) the plasma membrane of a target cell, inducing a fusion between cellular and viral membranes. X-ray crystallographic studies of regions containing the heptad repeats of HIV-1 (Chan *et al.*, 1997; Tan *et al.*, 1997; Weissenhorn *et al.*, 1997) and SIV (Malashkevich *et al.*, 1998) have revealed that they form a six-helix (trimer of an antiparallel heterodimer) coiled-coil structure. This structure, by analogy with other viral envelope glycoproteins, has been proposed to function as a spring that interjects the FD into the plasma membrane of target cells.

B. Genetic Diversity

HIV-1, as with other retroviruses, is subject to a high degree of genetic variation due to the error-prone nature of the viral reverse transcriptase, which lacks an editing function. Based on the phylogenetic analyses of all the viruses isolated worldwide, HIV-1 is divided into two groups, the major (M) and the outlier (O) groups. The viruses within the M group are further divided into nine subtypes (A–H and J), or clades, based on genetic clustering (Leitner *et al.*, 1997). Compared to other regulatory, structural, and enzymatic proteins, the envelope glycoprotein of HIV-1 exhibits higher sequence diversity, particularly in the hypervariable regions. It is speculated that the combination of immune selection pressure and the ability of the regions to tolerate a great deal of amino acid changes, without significantly altering the functional capacity of the protein, are responsible for such a genetic variation. The ability of the virus to utilize multiple coreceptors (see below) is also likely to contribute. This extreme genetic diversity of gp120, which most likely translates into antigenic diversity, is one of the major problems in developing a vaccine for HIV-1.

C. Cellular Tropism and Viral Entry

The cellular tropism of HIV-1 is determined largely at the level of viral entry. HIV-1 enters cells through a series of interactions between its envelope glycoprotein and cellular receptors. In addition to CD4, the primary receptor, HIV-1 requires a coreceptor that includes several members of the G protein-coupled chemokine receptor family (for reviews, see Berger, 1997; D'Souza and Harden, 1996; Doranz *et al.*, 1997; Littman, 1998; Moore *et al.*, 1997; Wells *et al.*, 1996). HIV-1 isolates are divided into three general groups based on their cellular tropism. Macrophage-tropic (M-tropic; R5) strains that are able to infect macrophages but not transformed T-cell lines, utilize the β-chemokine receptor CCR5. T-cell-line-tropic (T-tropic; X4) isolates utilize the α-chemokine receptor CXCR4 and infect T-cell lines, but not macrophages. Dual tropic viruses (X4R5) can utilize both chemokine receptors and are able to infect both cell types. Regardless of the cellular tropism, all isolates can infect primary CD4$^+$ T lymphocytes. Although there are many other chemokine receptors identified that could serve as a coreceptor for HIV-1 (e.g., CCR2b, CCR3, and CCR8), their significance and biological relevance are not yet clear. Existing clinical, epidemiological, and molecular data suggest that CCR5 and CXCR4 are the most physiologically important coreceptors for the pathogenesis of HIV-1.

A working model of HIV-1 entry into a target cell is shown in the lower panel of Fig. 2. First, gp120 is thought to bind to CD4. This interaction between gp120 and CD4 induces a conformational change in gp120 from a pre-CD4 binding, "closed" conformation to a post-CD4-binding, "open" conformation (Fig. 3). The "open" conformation of gp120 allows subsequent interaction between gp120 and coreceptors. The interaction between gp120 and a coreceptor is thought to induce even more drastic changes in gp120 and gp41 conformation such that the fusion domain of gp41 can interact with the plasma membrane, allowing a fusion between viral and

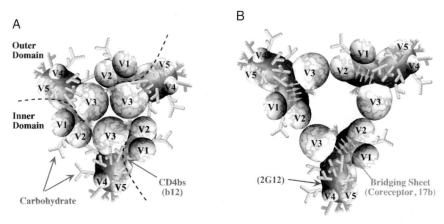

FIGURE 3   A hypothetical model of trimeric arrangement of HIV-1 gp120 in a pre-CD4 binding closed conformation (A) and a post-CD4 binding open conformation (B). The model is from the prospective of a target cell. The five variable domains are indicated as V1–V5. As discussed in the text, the gp120 molecule is divided into two domains (inner and outer), divided by dashed lines. The surface of much of outer domain is covered by carbohydrate structures. The CD4 binding region lies in the border between the two domains, recessed in a pocket that is partially covered by the V1 loop. Subsequent to CD4 binding, gp120 undergoes a conformational shift, exposing the bridging sheet where coreceptor binds. Neutralizing monoclonal antibodies b12 and 17b binds to the CD4 binding site (CD4bs) and coreceptor binding site (CD4-induced), respectively. Monoclonal antibody 2G12, which uniquely targets the glycosylation-rich outer domain, has a broad neutralizing activity against many HIV-1 isolates. Interaction between gp120 and coreceptors is thought to induce even more dramatic conformational change in the envelope glycoprotein that allows the fusion domain of gp41 to interject into the plasma membrane of a target cell.

cellular membranes to occur. The exact mechanism of membrane fusion is still not fully elucidated.

While the determinants of HIV-1 envelope glycoprotein that interact with CD4 are clearly identified from a recent X-ray crystallographic analysis (see below), those that interact with the coreceptors are identified using a multitude of indirect experimental approaches. They include site-directed mutational analyses, functional and infectivity analyses with chimeric envelopes and viruses, and biochemical competition analyses with site-specific antibodies and chemokines. Results from earlier studies using either exclusively T-tropic, tissue-culture laboratory-adapted (TCLA) HIV-1 isolates or M-tropic viruses indicated that the V3 region of gp120 is the principal, if not the only, determinant that confers cellular tropism of the virus (Cann *et al.*, 1992; Chesebro *et al.*, 1991; Choe *et al.*, 1996; Cocchi *et al.*, 1996; De Jong *et al.*, 1992; Hwang *et al.*, 1991; O'Brien *et al.*, 1990; Pleskoff *et al.*, 1997; Shioda *et al.*, 1992; Speck *et al.*, 1997; Trkola *et al.*, 1996a; Westervelt *et al.*, 1991). However, more recent studies with dual tropic virus isolates clearly demonstrated that the V1/V2 region, in addition to the V3 region, interacts with coreceptors and plays an important role in viral tropism (Cho *et al.*, 1998; Lee *et al.*, 1999; Ross and Cullen, 1998; Smyth *et al.*, 1998).

D. Glycosylation

Although many cellular and viral membrane proteins are glycosylated, the extent of glycosylation of HIV-1 gp120 is highly unusual (Myers and Lenroot, 1992). Oligosaccharide composition analyses of HIV-1 gp120 produced in Chinese hamster ovary (CHO) cells show that the protein is modified with high-mannose, complex, and hybrid types of N-linked glycans (Leonard *et al.*, 1990). Potential functions of the carbohydrate structures of HIV-1 gp120, and that of simian immunodeficiency virus (SIV), have been examined using variety of approaches, including the use of glycosylation inhibitors, enzymatic deglycosylation, and site-directed mutagenesis. Carbohydrate moieties seem to have two major roles that are important in the viral life cycle and pathogenesis. First, they are important for proper folding and processing of envelope glycoprotein, as well as its interaction with CD4 and/or coreceptors (Bandres *et al.*, 1998; Fennie and Lasky, 1989; Gruters *et al.*, 1987; Huang *et al.*, 1997; Li *et al.*, 1993; Nakayama *et al.*, 1998; Ogert *et al.*, in preparation; Walker *et al.*, 1987; for reviews, see Feizi and Larkin, 1990; Fenouillet *et al.*, 1994). Not all glycosylation sites, however, contribute equally in the functionality of the protein. Site-directed mutagenesis analyses show that many glycosylation sites of HIV-1 and SIV can be mutated without affecting virus infectivity or ability of the envelope to interact with cellular receptors (Bolmstedt *et al.*, 1991; Gram *et al.*, 1994; Hemming *et al.*, 1994; Lee *et al.*, 1992; Ogert *et al.*, in preparation; Ohgi-

moto *et al.*, 1998; Reitter and Desrosiers, 1998). The relative importance
of an individual glycosylation site likely varies depending on the virus isolate.
A certain density of carbohydrates appears to be necessary at a given region
of gp120 to maintain the functional protein conformation, since the removal
of multiple, but not individual, glycosylation sites is detrimental (Ohgimoto
et al., 1998). The loss of function by the removal of glycosylation sites can
be restored by a second-site mutation in other regions of gp120 without
introducing a new glycosylation site (Ogert *et al.*, in preparation; Reitter
and Desrosiers, 1998; Willey *et al.*, 1988). These observations strongly
indicate that carbohydrate moieties contribute to the overall tertiary or
quaternary structure of envelope glycoprotein.

The second important role of carbohydrates is in evasion of immune
response. They have been shown to reduce immunogenicity by masking
potential antigenic epitope and antigenicity of the protein by hindering
recognition by antibodies (Back *et al.*, 1994; Davis *et al.*, 1990; Huang *et
al.*, 1997; Papandreou and Fenouillet, 1998; Reitter *et al.*, 1998; Schonning
et al., 1996). Fully glycosylated gp120 also have been reported to elicit less
CTL response compared to deglycosylated protein, suggesting negative effect
of glycans on antigen processing and/or presentation (Doe *et al.*, 1994). It
is clear from a number of studies that the carbohydrate composition and
the extent of modification of HIV-1 and HIV-2 gp120 can vary significantly,
depending on the origin of the cell the protein was produced in (Butters *et
al.*, 1998; Liedtke *et al.*, 1994, 1997; Pal *et al.*, 1993 and references therein).
Moreover, the sensitivity of HIV-1 to neutralizing antibodies depends on
the cell type in which the virus was propagated (Sawyer *et al.*, 1994; Willey
et al., 1996).

E. Detailed Structural Analyses

Until very recently, structural analyses of HIV-1 gp120 have been limited
to site-directed mutagenesis and topological antigenic mapping studies with
monoclonal antibodies. The crystal structure of HIV-1 gp120 core has finally
been determined at a high resolution (Kwong *et al.*, 1998, 1999). Attempts
to crystallize gp120 have been unsuccessful for many years due to high
degree of glycosylation, irregular glycosylation pattern, and high flexibility
of some of the regions of the protein (e.g., variable loops). Three approaches
were taken to overcome these problems (Kwong *et al.*, 1999). First, the
variable loops V1/V2 and V3 were removed (in addition to a large fragment
in the N terminus). Second, the protein was produced in the *Drosophila*
Schneider 2 cell line in which the glycosylation is limited to high-mannose,
which were subsequently removed enzymatically. Finally, the crystal lattice
was stabilized by binding the gp120 core with two N-terminal domains of
CD4 and a Fab fragment of a human monoclonal antibody, 17b. Detailed
analyses of the structure have been described elsewhere (Turner and Sum-

mers, 1999; Wyatt *et al.*, 1998; Wyatt and Sodroski, 1998) and only a brief summary is presented.

Although the structural information of such an artificial product should be interpreted with some caution, the fact that this structure binds CD4 and a neutralizing monoclonal antibody likely indicates that it closely resembles the core structure of the native, fully glycosylated, and untruncated gp120. The core structure of gp120 is composed of two domains, the inner and the outer (Kwong *et al.*, 1998). The two domains are bridged by a minidomain called a "bridging sheet," a four-stranded, antiparallel β-sheet. A model of a trimeric form of gp120 is depicted in Fig. 3. The inner domain, which is more conserved than the outer domain, is largely buried internally where intermolecular interactions occur between gp120 and gp41 molecules in the trimer. As the surface of the inner domain is not exposed on the trimeric, quaternary structure, the antibodies directed against this region do not neutralize the virus, hence a "nonneutralizing face." The outer domain, which is more variable than the inner domain, is extensively glycosylated and does not elicit antibody response (an "immunosilent face").

The CD4 binding pocket is formed at a region that borders the two domains and the bridging sheet (Kwong *et al.*, 1998). Although the CD4 binding site (CD4bs) is conserved and devoid of glycosylation, only a limited number of neutralizing antibodies have been generated against CD4bs because it is highly recessed and partially masked by the V1 loop. Coreceptors interact with conserved residues that make up part of the bridging sheet (Rizzuto *et al.*, 1998) as well as the variable regions V3 and/or V1/V2. This coreceptor binding site, which is normally buried under the variable regions V1/V2, becomes exposed only after the conformational shift in gp120 that occurs subsequent to CD4 binding (Wyatt *et al.*, 1995). Thus, the binding of gp120 to CCR5 is highly CD4-dependent (Hill *et al.*, 1997; Trkola *et al.*, 1996a; Wu *et al.*, 1996). The monoclonal antibody 17b, which binds to this rather conserved CD4-induced epitope (CD4i) (Sullivan *et al.*, 1998; Thali *et al.*, 1993) and competes with CCR5 (Rizzuto *et al.*, 1998; Trkola *et al.*, 1996a; Wu *et al.*, 1996), neutralizes a number of divergent HIV-1 isolates (Thali *et al.*, 1993). However, antibodies with such specificity are not common, suggesting that this epitope is either nonimmunogenic, not exposed well, or both. Although the infection by TCLA T-tropic HIV-1 isolates is CD4-dependent, binding of T-tropic gp120 to CXCR4 is less CD4-dependent than binding to CCR5 (Hesselgesser *et al.*, 1997; Bandres *et al.*, 1998). TCLA viruses are, in general, more easily neutralized than primary HIV-1 isolates (Cohen, 1993; Golding *et al.*, 1994; Mascola *et al.*, 1996; Matthews, 1994; Moore and Ho, 1995). The V3 loop has been shown to be the principal neutralization determinant for TCLA viruses (Goudsmit *et al.*, 1988; Javaherian *et al.*, 1989; Palker *et al.*, 1988; Rusche *et al.*, 1988). These observations suggest that the V3 loop of gp120 on these viruses might extrude out and partially expose the bridging sheet, making it a better target for neutralizing antibodies.

IV. Viral Vaccine Strategies

Vaccines can be classified into six general categories as shown in Fig. 4: live attenuated, whole-inactivated, subunit, live vector, DNA, and multi-

Immunogen	Advantages	Disadvantages
Live Attenuated	CTL Native conformation Extended antigen presentation Low cost Long-lasting immunity	Safety Special handling
Whole Inactivated	"Native" conformation	Safety No CTL Cellular protein
Virus-Like Particles (Subunit)	Native conformation Safe	Difficult to make No CTL Cellular protein
Peptide Protein (Subunit)	CTL (peptide) High antigen delivery Safe	Native conformation? Multiple boost No CTL
Live Vector	CTL Native conformation	Immunity against the vector Side effects? Special handling
DNA	CTL Native conformation Ease of preparation Low cost	Low antigen delivery Multiple boost Immune tolerance? Less experience
Multimodal (Prime-Boost)	CTL High antigen delivery Native conformation	Complex vaccine schedule

FIGURE 4   Some of the advantages and disadvantages of various vaccine strategies against HIV-1. Vaccine candidates can be generally divided into six different categories: live attenuated, whole-inactivated, subunit, live vector, DNA, and multimodal (prime-boost). Subunit vaccine candidates can include viruslike particles, recombinant proteins, or chemically synthesized peptides.

modal. Historically, the most successful viral vaccines have been either live attenuated vaccines or whole-inactivated vaccines. These include vaccines against smallpox, poliovirus, measles, and influenza. Live attenuated vaccine is composed of a live, but substantially weakened pathogen that contains a single mutation or multiple mutations in its genome. Examples include Sabin vaccine against poliovirus, chicken pox, and measles. Although an attenuated pathogen can infect cells and complete its life cycle, the replication kinetic of the virus is significantly delayed and/or the virus cannot cause disease. During this limited and non-life-threatening replication, the body is able to mount a strong immune response that is able to eliminate the pathogen from the body. The immune response is long lasting and is able to protect the body from future encounters with the same pathogen. There are several advantages of a live attenuated vaccine compared to others: (1) the vaccine actually mimics the natural course of virus infection (e.g., the same route of entry and target cells); (2) the viral antigens presented to the immune system are in native conformation and therefore have the potential to elicit a good humoral immune response; (3) the viral antigens are synthesized *de novo* and therefore can elicit a CTL response; and (4) the period of antigen presentation from a single vaccine dose is longer and therefore elicits a strong immune response.

Whole-inactivated vaccine has also been used quite successfully against a number of human pathogens (e.g., Salk poliovirus vaccine, influenza). In this case, a pathogen is killed by either heat or chemical treatment so the pathogen is no longer infectious. Although various treatment processes can alter the conformation and immunogenicity of the viral antigens, it is assumed that the process retains the native structure of epitopes critical for eliciting protective immune responses. CTL responses generally are not elicited since there is no *de novo* synthesis of viral antigens subsequent to immunization. Because the viruses used to make the vaccine are fully infectious, it is highly critical to completely kill the pathogen during the inactivation process.

Subunit vaccines are prepared from protein components of a virus with no genetic materials. The vaccine is usually composed of a protein or a mixture of proteins that can elicit humoral and Th immune response. A set of peptides that are known to contain B- and/or T-cell epitopes can also be used. Large amounts of viral protein can be produced in bacteria, yeast, insect cells, or in mammalian cells. Peptides can be chemically synthesized also in large quantities. Because these antigens are not infectious, safety is one of the advantages of this vaccine approach. Except for peptides, subunit vaccines are not efficient in eliciting a CTL response. Furthermore, these antigens do not elicit a strong immune response by themselves, requiring some form of adjuvant. In addition to proteins and peptides, noninfectious viruslike particles (or pseudovirions) can be used, although preparing them is substantially more difficult.

With the advent of molecular biology techniques, more novel approaches of vaccine development are being explored. These include live vector and DNA vaccine strategies. In case of a live vector vaccine, a gene or set of genes encoding proteins of a pathogen are inserted into other nonpathogenic viruses or bacteria. Once these vectors enter a body, they replicate and express the protein(s) of interest and the immune responses are elicited against them. Some of the vectors currently being explored include enteric bacteria (e.g., salmonella), poxviruses (e.g., vaccinia and canarypox), small RNA viruses (e.g., poliovirus and semliki forest virus), and other DNA viruses (e.g., adenovirus or adeno-associated viruses). Although currently there are no licensed vaccines using this approach, they have good potential for future vaccine development. In case of a DNA vaccine, plasmids encoding a pathogen's gene(s) are directly injected into the body where they are taken up by cells and proteins are expressed. Because proteins are expressed *de novo* for both live vector and DNA vaccines, a CTL response can be elicited. One disadvantage of these vaccine strategies is that it is difficult to assess how much of antigen is actually being expressed *in vivo*. With a live vector vaccine, the vector itself elicits an immune response. Thus successive vaccine administration is likely to deliver less antigen and therefore be less effective. As for DNA vaccines, the delivery of DNA into cells using currently available techniques is highly inefficient. Because DNA vaccine strategy is a relatively new concept with little experience, there are many unknowns and some safety concerns. Combining with a subunit vaccine, a multimodal vaccine (or prime-boost) approach has been explored. This vaccine approach includes "priming" with either live vector or DNA vaccine and a subsequent "boost" with a subunit vaccine (e.g., recombinant protein). Interestingly, this prime-boost approach has been shown to be far superior to live vector, DNA, or subunit vaccine alone for HIV and SIV (see below).

V. Progress in AIDS Vaccine Development

All of the strategies described above have been explored as a potential AIDS vaccine candidates in a number of small laboratory animals, nonhuman primates, and even some in humans. This section presents a summary of past efforts and most recent studies with respect to each of these vaccine strategies.

A. Animal Model for HIV-1

Many small laboratory animals (e.g., mice, guinea pigs, or rabbits) have been used to examine the safety and the immunogenicity of various vaccine candidates. They also have been used to evaluate the potency of adjuvants and immune stimulatory molecules (e.g., TNF-α, IL-2, or IL-12). However, they cannot be used for efficacy studies because these animals are not susceptible to infection by HIV-1. One of the major problems in HIV-1 vaccine development has been the lack of a good animal model for HIV-1 pathogenesis. Most nonhuman primates are not suitable because HIV-1 does not infect them or infects only at extremely low levels and does not cause disease (Fig. 5). Chimpanzees (*Pan troglodytes*), from which HIV-1 is thought to have originated (Gao *et al.*, 1999), is the only nonhuman species that can be consistently infected with HIV-1. However, chimpanzees, except for a single reported case (Novembre *et al.*, 1997), do not develop the AIDS-like syndrome and the ability of HIV-1 to replicate in these animals varies significantly depending on the virus isolate, making it very difficult to interpret vaccine efficacy. That is, the protective efficacy often inversely correlates with the replicative capacity of the HIV-1 strain used for the challenge in chimpanzees, which raises questions as to the potential efficacy of the vaccines against primary HIV-1 isolates in humans.

Much of our knowledge in primate lentiviral pathogenesis has been acquired from studying SIV infections in various macaque species. Because of many similarities between HIV-1 and SIV (e.g., overall genome organization, receptor usage and cellular targets, and viral pathogenesis), the SIV/macaque system has been used extensively as a model system for HIV-1 vaccine

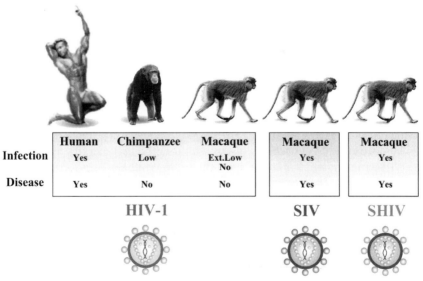

	Human	Chimpanzee	Macaque	Macaque	Macaque
Infection	Yes	Low	Ext.Low No	Yes	Yes
Disease	Yes	No	No	Yes	Yes

HIV-1   SIV   SHIV

FIGURE 5 Host range and primate animal model for HIV-1. HIV-1 does not reliably infect nonhuman primates and it can cause immunodeficiency syndrome only in humans. SIV, a close relative of HIV-1, can infect and cause simian AIDS in macaques. A chimeric virus, SHIV, which encodes *tat, rev, vpr, vpu,* and *env* genes in the background of SIV genome, can infect and cause AIDS-like symptoms in macaques.

studies, in particular, to understand the correlates of protection. However, because SIV proteins are significantly different from those of HIV-1, it is impossible to examine the protective efficacy of any specific HIV-1 vaccine candidate using the SIV/macaque system. More recently, chimeric viruses were generated between SIV and HIV-1 (SHIV) in which the HIV-1 *env* gene (as well as *tat, rev, vpr,* and *vpu*) is inserted into the backbone of an SIV genome. The resulting virus can replicate in macaques to varying degrees. There are currently three major SHIVs being examined, encoding *env* genes of HIV-1 isolates HXBc2 (Joag *et al.,* 1997), 89.6 (Li *et al.,* 1995), and DH12 (Shibata *et al.,* 1997a). Although initial constructs did not cause disease, either serial *in vivo* passages of the viruses in monkeys (Joag *et al.,* 1996; Reimann *et al.,* 1996) or treatment of virus-infected monkeys with anti-CD8 antibodies (Igarashi *et al.,* 1999) resulted in pathogenic variants. These pathogenic SHIVs can cause rapid CD4 depletion, immune dysfunction, and eventual death of the animals. Because SHIVs encode HIV-1 envelope glycoprotein, any vaccine candidate based on envelope glycoprotein can be tested for protective efficacy. Attempts to generate SHIV constructs with more HIV-1 genes have been unsuccessful. The SHIV construct with HIV-1 *nef* gene replicates less well compared to the one with SIV *nef in vivo* (Shibata *et al.,* 1997a). SHIV constructs containing HIV-1 LTR, *gag-pol,* and/or *vif* does not replicate in macaque PBMC, suggesting that they are the determinants of host-range restriction (Shibata *et al.,* 1991). Regardless of the system, results from animal studies should be interpreted with caution, as immune systems of human and nonhumans may respond differently to any antigen.

B. Live Attenuated Vaccine

Live attenuated vaccines for HIV-1 have not been explored due to serious potential risks in humans and the lack of a good animal model. Much of our understanding on attenuated vaccines for primate lentiviruses is based on the SIV/macaque system. Although it is generally agreed that this vaccine strategy can elicit a strong and long-lasting immune response, the correlates of protection are still largely unknown, the degree of protection varies significantly from study to study (and from animal to animal), and, most importantly, the safety issues remain highly controversial.

Various combinations of vaccine/challenge viruses have been examined, including SIV/SIV, HIV-2/SIV, SHIV/SIV, SIV/SHIV, and SHIV/SHIV in different macaque species. Immunization of either rhesus (Daniel *et al.,* 1992) or cynomolgus (Almond *et al.,* 1995) macaques with attenuated SIV were reported to completely protect against pathogenic SIV. Other studies, however, showed only a partial protection with reduced virus load and delay of disease (Lohman *et al.,* 1994; Marthas *et al.,* 1990). The degree of protection had an inverse relationship with the degree of attenuation (i.e., the best protection from the least attenuated virus), suggesting the requirement for viral replication for eliciting a good immune response (Johnson *et al.,* 1999; Lohman *et al.,* 1994). Strong protection was observed in HIV-2-infected cynomolgus macaques challenged with SIV, although sterilizing immunity was not achieved (Putkonen *et al.,* 1990). Macaques infected with attenuated SIV (Bogers *et al.,* 1995; Dunn *et al.,* 1997; Shibata *et al.,* 1997b) or SHIV (Igarashi *et al.,* 1997; Joag *et al.,* 1998) exhibited strong immune response and various levels of protection from subsequent challenge with SHIV. Conversely, cynomolgus macaques immunized with SHIV were also protected from SIV_{sm} challenge (Quesada-Rolander *et al.,* 1996). In contrast, protection was not observed when cynomolgus macaques immunized with the same SHIV were challenged with a different SIV isolate (J5 clone of $SIV_{mac\ 32H}$; Letvin *et al.,* 1995).

Safety of a vaccine is of utmost importance. Because HIV-1 and SIV integrate into the host genome, it is absolutely crucial to make sure that the attenuated vaccine strain does not revert and become pathogenic even long after the vaccination. Recent observations that attenuated SIV with triple deletions (in *nef, vpr,* and negative regulatory element) has potential to become pathogenic in neonate (Baba *et al.,* 1995) as well as in adult macaques during a prolonged observation period (Baba *et al.,* 1999) raise serious safety concerns and doubts in HIV-1 vaccine development using this approach. It has been reported that the incidents of disease occurrence in neonate macaques are low and the pathogenic potential depends on the initial dosage (Wyand *et al.,* 1997). Nevertheless, the potential danger of attenuated vaccine remains real, especially when it was shown that, in some animals immunized with attenuated SIV, the virus reactivated and caused CD4 depletion (Bogers *et al.,* 1995).

It is unknown whether the lessons learned from attenuated SIV vaccine studies can be applied to HIV-1 vaccine development. In this regard, it might be more relevant to analyze human patients who were infected with naturally occurring attenuated HIV-1 isolates with a deletion in the *nef* gene. These individuals are long-term, nonprogressors who remain healthy substantially longer than those patients infected with viruses with an intact *nef* gene (Deacon *et al.,* 1995; Kirchhoff *et al.,* 1995) and have been shown to have high levels of CTL activity against HIV-1 (Dyer *et al.,* 1999). Although they were once considered to indicate safety of an attenuated HIV-1 vaccine, recent reports of slow decline in the number of $CD4^+$ T cells in one such patient demonstrates a potential problem with attenuated HIV-1 vaccine (Greenough *et al.,* 1999). In this regard, the potential health risk of any attenuated HIV-1 vaccine is demonstrated in the recent observation that an attenuated HIV-1 with triple deletions is genetically unstable and can evolve into a fast-replicating variant (Berkhout *et al.,* 1999).

C. Whole-Inactivated Vaccine

Due to the potential danger of incompletely inactivated HIV-1 (real or perceived), most of the vaccine studies using this approach were done in macaques using either SIV or HIV-2. While some of the early studies using whole-inactivated SIV vaccine demonstrated only partial protection (Desrosiers *et al.*, 1989; Sutjipto *et al.*, 1990), others reported complete protection in most of the vaccinated animals (Murphey-Corb *et al.*, 1989). Complete protection was also observed in animals immunized with an inactivated SIV-infected cell vaccine (Stott *et al.*, 1990). However, a protective immune response was also elicited using uninfected cells in a subsequent study (Stott, 1991). Later studies by several other groups showed that the immunized animals were protected if they were challenged with SIV grown in human cells, but not in macaque cells (Heeney *et al.*, 1992; Hirsch *et al.*, 1994; Le Grand *et al.*, 1992; Mills *et al.*, 1992).

In all of these studies, the virus preparations used for the immunization were produced in human cells. High levels of antibodies against human cellular proteins were detected in all of these animals and the degree of protection was shown to correlate with the level of the antibody titer (Langlois *et al.*, 1992; Mills *et al.*, 1992). The presence of cellular proteins has been observed in sucrose gradient-purified HIV-1 preparations (Arthur *et al.*, 1992). The likelihood that the protective immune response was directed against the xenoantigen rather than the viral antigens was later demonstrated when macaques immunized with purified human HLA-DR were protected from SIV grown in human cell, but not from the virus grown in macaque PBMC (Arthur *et al.*, 1995). These results were a great disappointment to efforts in AIDS vaccine development using whole-inactivated virus.

When immunized animals were challenged with virus stocks prepared in macaque PBMC, variable protection was observed. While two cynomolgus monkeys immunized with whole-inactivated HIV-2 in incomplete Freund's adjuvant (IFA) were protected from HIV-2 challenge, two other monkeys immunized in immune-stimulating complexes (iscoms) were not (Putkonen *et al.*, 1991). Whether the difference in protection was due to the adjuvant used could not be determined because different antigen dose and immunization schedule was used. In a similar study using various adjuvants (IFA, iscoms, Alum, Ribi, and muramyl dipeptide), 7 of 21 animals were protected from HIV-2 challenge (Putkonen *et al.*, 1994). The correlates of protection, however, could not be determined and the immune response was not broad enough to protect against SIV. Efficacy of whole-inactivated vaccine approach against HIV-1 was examined in a limited scale using chimeric SHIV (Lu *et al.*, 1996). Rhesus macaques immunized with whole-inactivated HIV-1 and two subsequent boosts with recombinant HIV-1 gp120 were protected from SHIV expressing a highly related HIV-1 envelope. Recently, targeted immunization of the iliac lymph nodes with inactivated SIV has been examined in rhesus macaques in effort to elicit stronger local immunity. Although high titers of anti-SIV antibodies were detected in serum and cervicovaginal secretions, animals were not protected from intravaginal challenge with SIV (Lu *et al.*, 1998).

Although some vaccine studies using whole-inactivated virus may have elicited a protective immune response, the approach, in general, received very little attention due to difficulty and the potential danger with growing large amounts of virus, possible side effects of alloantigen immunization, and statistically questionable levels of protection, which sometimes seem sporadic. Whether the native antigenic conformation is preserved (especially for viral envelope glycoprotein) after conventional virus inactivation processes (formalin, detergent, β-propiolactone, and psoralen cross-linking) and various adjuvant treatments needs to be further evaluated. In this regard, recent reports of the use of aldrithiol-2 to inactivate the essential zinc fingers in the viral nucleocapsid seem promising since the conformational and functional properties of the envelope glycoprotein are preserved (Arthur et al., 1998; Rossio et al., 1998). However, many safety concerns mentioned above still need to be addressed.

D. Subunit Vaccine

Because subunit vaccines are generally perceived to be safe, this approach has been a focus of many vaccine studies from early on, including a number of human clinical trials testing the safety and immunogenicity of several vaccine candidates. Although small laboratory animals have been invaluable and used extensively in AIDS vaccine research, the discussion in this chapter is largely limited to vaccine studies done with humans and nonhuman primates due to the overwhelming amount of data available.

I. Nonhuman Primates

Chimpanzees immunized with p55 *gag* protein generated a high titer of antibodies against the protein. However, these antibodies did not neutralize HIV-1 and the animals were not protected from the virus challenge (Emini et al., 1990b). While some of the rhesus macaques immunized with an envelope-enriched fraction of SIV virion preparation were protected, animals immunized with an envelope-depleted fraction got infected (Murphey-Corb et al., 1991). These results strongly suggested the importance of envelope glycoprotein in eliciting a protective immune response. The degree of protection observed in chimpanzees against HIV-1, following immunization with various envelope glycoprotein-based subunit vaccine candidates, has been highly variable (or inconsistent) to a degree where it is difficult to make any strong conclusions. Chimpanzees immunized with either recombinant gp120 (Berman et al., 1988) or the protein purified from virus-infected cells (Arthur et al., 1989) were not protected from subsequent challenges with HIV-1. Although a strong T-cell proliferative response was demonstrated, only low levels of neutralizing antibodies were detected (Arthur et al., 1989). In contrast, several other studies were able to show good protection against HIV-1 challenges (Berman *et al.*, 1996; Fultz *et al.*, 1992; Girard *et al.*, 1991; Girard *et al.*, 1995).

A number of parameters could potentially modulate the immune response, which consequently affects the protective efficacy of a vaccine. One factor that is becoming increasingly clear to have an influence on the immune response is adjuvants. Alum has been used in many studies, primarily because it is the only adjuvant licensed for use in humans at the present time. Immune responses to gp120 or gp160 subunit vaccines, administered with alum, were generally weak (Arthur *et al.*, 1989; Berman *et al.*, 1988). Higher levels of immune response were elicited with the use of lipid-based adjuvant (Barrett *et al.*, 1991), incomplete Freund's adjuvant (IFA), mixture of IFA/ MTP-PE, or other emulsion-based adjuvants (Haigwood *et al.*, 1992) compared to alum. However, animals immunized with gp120 conjugated to alum have been protected against challenges with HIV-1 with either homologous or heterologous envelope glycoprotein (Berman *et al.*, 1990, 1996). These results suggest that while adjuvants play an important role, it is not the only factor that determines the quality of an immune response.

Another factor that may influence the protective efficacy of an immune response is the nature (or the source) of the antigen used to immunize animals. In one study, chimpanzees immunized with gp120, but not with gp160, were protected (Berman *et al.*, 1990). Both recombinant proteins were produced in CHO cells, suggesting that these two proteins are conformationally different and elicit different types (or qualities) of immune response. In another study, the inverse was true; one of the two animals immunized with gp160 was protected but two animals immunized with gp120 got infected (Bruck *et al.*, 1994). In the latter study, however, gp160 was derived from monkey kidney cells (CV-1), whereas gp120 was prepared from *Drosophila* cells. In both of these studies, protection correlated with the level of neutralizing antibody, but not the level of T-cell proliferative response against the immunogen. It has been reported that "native," fully glycosylated gp120 from CHO cells elicited high levels of antibodies that could neutralize both homologous and heterologous HIV-1 isolates but not nonglycosylated protein from yeast (Haigwood *et al.*, 1992). Similarly, mammalian-derived gp120 has been reported to produced better humoral response than insect- or yeast-derived proteins (Graham, 1994). It remains to be examined whether subtle differences in glycosylation of envelope glycoprotein in different cell types alter the immunogenicity of the protein.

One of the major challenges in developing an effective vaccine against HIV-1 is the enormous genetic diversity of the envelope glycoprotein. Early vaccine studies showed that neutralizing antibodies generated against gp120 from HIV-1$_{IIIB}$ strains were highly isolate-specific (Berman *et al.*, 1988; Weiss *et al.*, 1986). However, chimpanzees immunized with gp120 (Berman *et al.*, 1996) or the combination of the V3 peptide and hybrid gp160 (Girard *et al.*, 1995) derived from HIV-1$_{MN}$ isolate were protected from a challenge with a heterologous isolate (HIV-1$_{SF2}$) of the same subtype (clade B). It was suggested that the gp120 from the MN isolate was better able to elicit cross-neutralizing antibodies because the principal neutralization determinant (PND) of the MN isolate was more representative of the PND of the clade B viruses than the IIIB isolate (LaRosa et al., 1990). These studies appeared promising, since a monomeric gp120 of a single TCLA HIV-1 isolate was able to protect from a challenge with a primary, heterologous isolate. However, the major caveat of these studies is that HIV-1$_{SF2}$ is considered to be relatively weak and, as with other HIV-1 isolates, it does not cause disease in chimpanzees. An important question that was left unanswered is whether these animals could be protected from other clade B viruses that replicate better than the SF2 isolate or have more divergent gp120. Two chimpanzees that were protected from intraclade challenge, however, were not protected from a subsequent challenge with an isolate from clade E (Girard et al., 1996), suggesting that a simple monovalent vaccine candidate would not be successful in a field test.

HIV-1 can be transmitted via a multiple route of entry, through intravenous drug use as well as homo- and heterosexual activities. It is a formidable challenge to develop a vaccine candidate that would elicit both systemic and mucosal immunity against the virus. Targeted immunization of iliac lymph node in rhesus macaques with recombinant *gag* and *env* proteins successfully protected the animals against a rectal challenge with SIV (Lehner et al., 1996). However, the same immunization approach did not work against a vaginal challenge (Lu et al., 1998). To date, there has not been a systematic, comparative analysis of various immunization routes for protection against different routes of virus challenge.

2. Human Clinical Trials

More than 20 vaccine candidates have been tested in over 2000 healthy human volunteers since 1988 through the NIAID-sponsored AIDS Vaccine Evaluation Group. Although most of the subunit vaccine candidates (gp120 or gp160 from HIV-1 isolates IIIB, MN, and SF2), in combination with various adjuvants, have been shown to be safe (Keefer et al., 1997), efficacy results from these studies, by and large, have been disappointing. Low levels of neutralizing antibodies were elicited against the homologous virus strain, with only a weak cross-neutralization activity against closely related isolates and none against primary isolates (Gorse et al., 1996; Keefer et al., 1996; Salmon-Ceron et al., 1995). Immunization with either gp120 or gp160 from IIIB isolates elicited antibodies that preferentially reacted to epitopes present on the denatured form of the protein (VanCott et al., 1995). In contrast, antibodies from virus-infected individuals preferentially bound native protein. There was also a lack of antibodies specific for epitopes that are conserved across divergent HIV-1 strains in vaccinated individuals. As in chimpanzees, human subjects immunized with gp120 from the MN isolate developed better cross-reactive antibodies to heterologous gp120 than IIIB gp120-immunized individuals (Gorse *et al.*, 1996). Interestingly, priming with IIIB gp120 followed by boosting with MN gp120 resulted in increase in cross-reactive antibodies to heterologous clade B HIV-1 isolates. Whether this observation is generally applicable (i.e., the higher cross-reactivity with the increase in multivalency of gp120) is not yet known and requires further evaluation.

The biggest disappointment came when some of the vaccinated individuals became infected during Phase I and II clinical trials (Connor *et al.*, 1998; Graham *et al.*, 1998). Subjects were immunized with gp120/gp160 from either MN or SF2. Most individuals developed antibody response from the vaccine, but only transiently. No neutralizing antibodies were present against the infecting strain prior to infection. All of the subjects were infected with clade B virus, the same subtype as the vaccine strain. There were no significant differences in the frequency of infection between the vaccinated and placebo groups. Vaccine-induced immune response apparently did not exert any selection pressure, since neither genotypic nor phenotypic characteristics of the transmitted virus were affected. In summary, no obvious beneficial or adverse effects were observed from the vaccine. These results led many people to conclude that monomeric gp120 or soluble gp160, at least by themselves or without any modification to improve immunogenicity, would not be sufficient to elicit protective immune response against highly divergent field isolates.

Recently, the first large-scale Phase III human clinical trial was approved by the U.S. FDA for bivalent monomeric gp120 vaccines (AIDS VAX B/B and B/E from Vaxgen). The vaccine is composed of gp120 from either two clade B isolates or one each from clades B and E, formulated in alum (Berman, 1998; Francis *et al.*, 1998). They are currently being tested in the United States and Thailand, respectively. Many critics of the study argue that the vaccine does not address many of the important issues (e.g., eliciting a CTL response or protection against heterologous isolates) and that protection from the vaccine, if any, would be very limited (see Boily *et al.*, 1999 and references therein). In addition, the study is very similar to previously failed vaccine trials and would only use up valuable resources. However, proponents argue that (1) the vaccine candidate has been shown to be safe and immunogenic; (2) even if the candidate is only partially protective, the benefits to some communities can be significant; and (3) a vaccine against HIV-1 would have to be developed empirically and lessons could be learned even if the trial fails (Boily *et al.*, 1999).

E. DNA Vaccine

DNA vaccine is relatively a new concept. Immunization is performed by directly injecting naked plasmid DNA encoding viral gene(s) intramuscularly, intradermally, or by a gene gun that projects DNA-coated gold particles into the epidermis. Various parameters of immunization and the types of immune responses elicited with DNA vaccine have been characterized extensively in mice for HIV-1 as well as other pathogens. The general approach of DNA vaccine has been reviewed elsewhere (Bagarazzi *et al.*, 1998; Kim and Weiner, 1997; Levitsky, 1997; Lu, 1998; Robinson, 1997; Suhrbier, 1997; Wahren and Brytting, 1997) and the discussions here are mostly limited to vaccine studies done in nonhuman primates.

The amount of DNA that is administered with a gene gun is much less (5–10 μg) than by intramuscular injection (0.1–3 mg) because the gene gun allows direct intracellular delivery. Intramuscular immunization elicited a higher lymphoproliferative response than intradermal immunization in mice (Shiver *et al.*, 1997). A better CTL response was observed in rhesus macaques immunized with a combination route (gene gun, intramuscular, and intravenous) than with the gene gun alone (Yasutomi *et al.*, 1996), suggesting a potential benefit of a diversified immunization route. Regardless of the route of administration, both cell-mediated and humoral immune responses are generated in macaques and chimpanzees immunized with plasmids encoding viral gene products. The cytokine profile of helper T cells generated from immunization of HIV-1 envelope gene indicated that a Th1-like immune response was elicited in rhesus macaques (Lekutis *et al.*, 1997). A Th1 response is important for augmenting a CTL response, which may be essential for successful containment of HIV-1 infection. Unfortunately, the humoral immune response elicited from DNA vaccine is relatively weak, generating only low titers of ELISA and neutralizing antibodies even after multiple administrations (Boyer *et al.*, 1997; Fuller *et al.*, 1996, 1997a; Haigwood *et al.*, 1999). However, the strength of the immune response is strongly influenced by the choice of the vector system as well as the vaccine schedule. A higher antibody titer was observed with a reduced number of immunizations and a longer resting period (Fuller *et al.*, 1996, 1997a). Although DNA vaccination itself does not elicit a strong immune response, it appears to prime the immune system quite effectively, since a strong anemnestic response is observed subsequent to boosting with recombinant protein or vaccinia virus (Fuller *et al.*, 1996, 1997a,b; Haigwood *et al.*, 1999; Letvin *et al.*, 1997).

The protective efficacy of DNA vaccines against a virus challenge has been highly variable, similar to subunit vaccines. Rhesus macaques immunized with DNA vaccine encoding SIV *gag* and *env* genes were not protected from SIV_{mac} challenge despite modest levels of CTL response generated in the animals (Yasutomi *et al.*, 1996). Humoral immune response was not measured in the study. Protection was also not observed in rhesus macaques immunized with DNA encoding gp160 and gp120 of SIV_{mac239} from a heterologous challenge with $SIV_{Delta/B670}$ (Fuller *et al.*, 1997b). Similar results were obtained in animals primed with DNA and subsequently boosted with recombinant vaccinia virus encoding SIV_{mac239} gp160. However, vaccinated animals exhibited lower virus load and p27 antigenemia. Recently, macaques (*Macaca fascicularis*) were immunized with a multigenic DNA vaccine encoding *gag, pol,* and *env* genes, with or without subsequent boost with recombinant gp160 and gag-pol viruslike particles (Haigwood *et al.,* 1999). Unfortunately, all of the animals got infected when challenged with SIV_{mne}. Interestingly, a significantly lower viral load was detected only in animals immunized with DNA, although the animals primed with DNA and boosted with protein exhibited higher antibody levels. Lymphoproliferative responses to gp160 were similar in both groups. However, the animals immunized with DNA only and those immunized with DNA and protein elicited Th1- and Th2-like T-cell responses, respectively. The significance of this observation was not determined.

In contrast to the dismal rate of protection seen in macaques immunized with DNA vaccine against SIV challenges, a higher rate of protection was observed in animals challenged with SHIV or HIV-1. One of four cynomolgus macaques immunized with DNA encoding $HIV-1_{Z6}$ gp160 was protected from SHIV that encodes envelope protein of a heterologous $HIV-1_{HIVB2}$ strain (Boyer *et al.,* 1996). Complete protection was reported in two rhesus macaques primed with HIV-1 env DNA and boosted with recombinant gp160 against SHIV-HXBc2 that encodes homologous envelope (Letvin *et al.,* 1997). In contrast, no protection was observed in animals immunized with the recombinant gp160 only. Envelope-specific lymphoproliferative response and CTL activity were detected only in animals immunized with both DNA and recombinant gp160. Chimpanzees immunized with DNA vaccine only ($HIV-1_{MN}$ gp160 and $HIV-1_{IIIB}$ *gag-pol*) were similarly protected from heterologous $HIV-1_{SF2}$ (Boyer *et al.,* 1997). However, because of relatively low levels of antibody titer, neutralizing activity, and CTL response that were elicited from the vaccine, no clear correlates of protection could be determined. Furthermore, the use of a relatively weakly replicating challenging virus, which only resulted in a virus load of 10^4 particles/ml in the control animal, raised some concerns as to the stringency of the efficacy test. Despite apparent lack of protection against SIV and protection only in suboptimal animal model system for HIV-1, the DNA vaccine approach clearly shows significant promise and additional parameters of the vaccine strategy (e.g., the use of immune stimulatory molecules and optimization of gene delivery and expression) must be examined in nonhuman primates. The Phase I clinical trial with HIV-1-infected human patients clearly demonstrated the safety of DNA-based vaccines (MacGregor *et al.,* 1998; Ugen *et al.,* 1998). Although some signs of increased anti-HIV-1 immune response were reported, the long-term benefits of postinfection therapeutic immunization have yet to be clearly demonstrated.

F. Live Vector Vaccine

The overwhelming majority of live vector vaccine or live vector-prime followed by subunit-boost vaccine research during the past decade have made the use of pox viruses (different strains of vaccinia and canarypox virus). As with other vaccine strategies discussed above, the level of the immune response elicited and protective efficacy vary significantly from study to study depending on the immunogen, the animal species, and the challenge virus strain used in the study. Most of the efficacy studies were performed in macaques, but some also in chimpanzees. Quite a few clinical trials have also been conducted in humans, testing to the safety and immunogenicity of various candidates. Although it is difficult to clearly identify correlates of protection due to many inconsistent and sometimes discrepant results, some general conclusions can be made about the vaccine approach.

Studies in humans show that recombinant vaccinia or canarypox viruses encoding HIV-1 envelope or capsid proteins are safe and well tolerated. Stronger immune responses to HIV-1 proteins are elicited in vaccinia-naive subjects than in vaccinia-primed subjects (those who were vaccinated against smallpox) when recombinant vaccinia virus was used (Cooney et al., 1991, 1993; Graham et al., 1992; McElrath et al., 1994). In contrast, the vaccinia immune status does not affect the outcome when using recombinant canarypox (ALVAC) (Clements-Mann et al., 1998). Both humoral and cellular immune responses can be elicited in subjects vaccinated with recombinant pox viruses. In general, relatively low levels of lymphoproliferative response and anti-HIV-1 protein antibody titer are elicited using recombinant virus alone. Boosting with subunit protein has consistently shown to enhance immune response dramatically (Clements-Mann et al., 1998; Cooney et al., 1993; Corey et al., 1998; Graham et al., 1993; Montefiori et al., 1992; Pialoux et al., 1995). This is especially true in eliciting neutralizing antibodies. However, the neutralizing activity is largely limited toward the homologous virus strain (i.e., the same strain the immunogen is derived from; Graham et al., 1993; Montefiori et al., 1992; Pialoux et al., 1995). A low CTL response is elicited in only a fraction of the subjects and is usually transient (Egan et al., 1995; Fleury et al., 1996). However, it has been reported that the response can also be enhanced by a subunit boost (Clements-Mann et al., 1998). A CTL response from subjects immunized against the envelope is more restricted (i.e., cannot kill cells infected with heterologous primary isolates) than those immunized with both envelope and gag proteins, suggesting that gag proteins can present more CTL epitopes, which are HLA haplotype-dependent (Ferrari et al., 1997).

Analyses of the humoral immune response revealed that the level of neutralizing antibody correlated with a longer interval between vaccinia priming and gp160 boosting; lesion formation after the first, but not the second, vaccinia inoculation; and induction of envelope-specific antibodies after the priming (Graham et al., 1994). Antibody epitope analyses showed that the immune response is determined by priming rather than the boost, which supports the view of "original antigenic sin" (Fazekas de St. Groth and Webster, 1966). The observation also indicates that there is a fundamental difference between a live vector-expressed gp160 and a recombinant subunit gp160. It is likely that the presentation of a structurally native immunogen during the priming may be critical for eliciting a functional immune response. Long-term follow-up of vaccinated subjects revealed that although both cellular and antibody response were detected after a year, they were significantly diminished (McElrath et al., 1994). Good anemnestic responses were observed after a boost. However, neutralizing antibody was detected in only a few vaccinees and the level remained extremely low. The CTL activity was not observed. Analyses of a person who became infected while receiving vaccinia prime-gp160 boost vaccine regimen revealed that the patient mounted a strong B- and T-cell responses within 3 weeks of infection (McElrath et al., 1996). Although the neutralizing antibody titer against the homologous vaccine strain (HIV-1$_{LAI}$) increased, none was detected against the virus the patient was infected with. The patient exhibited progressive CD4$^+$ T-cell decline and immune dysfunction within 2 years. These results suggest that although the vaccine was able to elicit an immune response that could be recalled upon infection, the limited breath of the response was not able to prevent infection by a heterologous virus. In this regard, it should be noted that a chimpanzee immunized with ALVAC expressing HIV-1$_{LAI}$ gp160, in addition to gag and protease, was protected from a homologous cell-associated virus challenge (Girard et al., 1997). However, the same chimpanzee and another animal immunized with antigens from HIV-1$_{MN}$ isolate were not protected from a heterologous challenge with a HIV-1$_{DH12}$. Neither animal had any neutralizing antibodies against HIV-1$_{DH12}$. The observation that chimpanzees with no or low levels of neutralizing antibodies are not protected from homologous virus challenges suggests a possible importance of neutralizing antibodies in protection (Girard et al., 1997; Hu et al., 1987).

The inability of the vaccine candidates to elicit a protective immune response against a heterologous virus challenge is more evident in the SIV/macaque system. Animals primed with recombinant vaccinia virus expressing gp160 and boosted recombinant gp160 protein were either completely protected or had only a transient viremia against a cloned homologous virus SIV$_{mneE11S}$ (Hu et al., 1992, 1996; Polacino et al., 1999). A dramatic increase in antibody titer and neutralizing antibodies were observed subsequent to boosting with recombinant subunit protein. However, only limited protection was observed against the parental uncloned SIV$_{mne}$ (Hu et al., 1996; Polacino et al., 1999). Sequence analyses of the uncloned virus stock showed 15% of the viruses had a divergent V1 region. Characterization of the viruses in immunized but persistently infected animals showed three of four animals had the variant virus, suggesting that the protection was directed against homologous virus and that immune response against the V1 region may have played an important role (Polacino *et al.*, 1999).

One of the reasons for the inconsistent results of vaccine studies could be attributed to the relative virulence of the virus isolate and/or the relative susceptibility of macaque species used in the study. In contrast to macaques (*M. fascivularis*) challenged with SIV$_{mneE11S}$ (Hu *et al.*, 1992, 1996; Polacino *et al.*, 1999), most rhesus macaques (*Macaca mulatta*) challenged with SIV$_{mac}$ became infected (Ahmad *et al.*, 1994; Daniel *et al.*, 1994; Giavedoni *et al.*, 1993). Furthermore, the virus load in the immunized animals was significantly lower compared to control when challenged with SIV$_{mac251}$, but not with SIV$_{mac239}$ even when the immunogen was derived from SIV$_{mac239}$ (Ahmad *et al.*, 1994). Additional variables that could also have affected the outcome of the virus challenge include differences in the immunization schedule (e.g., duration of resting period between priming and boosting) and the nature of the immunogen used (e.g., the use of gp130; Giavedoni *et al.*, 1993; inactivated SIV particles or pseudovirus particles; Daniel *et al.*, 1994). All of 12 macaques immunized with recombinant NYVAC expressing SIV *gag*, *pol*, and *env* gene products became infected when the animals were challenged intravenously (Benson *et al.*, 1998). However, only a transient viremia was observed in most animals subsequent to intrarectal challenge, suggesting that the immunization may have elicited a strong mucosal but not systemic immunity. Alternatively, the results could represent inefficient virus entry via intrarectal route. Protection could also depend on the strain of vaccinia virus used to generate the recombinant virus since animals immunized with modified vaccinia virus Ankara-based recombinant exhibited significantly lower levels of viral replication than those immunized with the Wyeth-based recombinant (Hirsch *et al.*, 1996). Although most studies observed enhanced immune response by priming the immune system with recombinant vaccinia virus prior to subunit protein immunization, Isreal *et al.* (1994) did not see any benefit. In some cases, vaccinia-primed animals did much worse than those animals that received subunit only. In this study, however, animals were primed only once, whereas other studies immunized with live vectors two or more times prior to subunit boosting. The importance of multiple priming with live vectors has not been carefully examined.

Recombinant Semliki Forest virus (rSFV) and adenovirus have also been used in HIV-1 vaccine development, although much less extensively than the poxviruses. Pigtail macaques immunized with rSFV expressing SIV$_{smmPBj14}$ gp160 or purified recombinant subunit gp120 developed envelope-specific antibodies (Mossman *et al.*, 1996). However, neither neutralizing antibodies nor a T-cell proliferative response was detected. Although all of the animals got infected subsequent to SIV$_{smmPBj14}$ challenge, symptoms in vaccinated animals were much less severe and exhibited lower plasma viremia compared to the unimmunized animals. Immunization of cynomolgus macaques with rSFV expressing HIV-1$_{IIIB}$ gp160 also did not protect from a challenge with SHIV encoding HIV-1$_{IIIB}$ envelope (Berglund et al., 1997). However, three of four vaccinated animals had no viral antigenemia and lower viral load compared to the control animals. A protective immunity, however, was elicited in chimpanzees primed with a recombinant adenovirus expressing HIV-1$_{MN}$ gp160 and boosted with HIV-1$_{SF2}$ gp120 against a low-dose challenge with HIV-1$_{SF2}$ (Lubeck et al., 1997). Both neutralizing antibodies and CTL responses could be elicited with the vaccine. The results from a rechallenge with a higher dose of the same virus showed that the protection correlated with the presence of neutralizing antibodies (Lubeck et al., 1997; Zolla-Pazner et al., 1998). Whether the same vaccine regimen can protect animals from a more rigorous challenge by other HIV-1 strains or in macaque-SIV or -SHIV systems needs to be further examined.

VI. Factors to Consider in Designing an Envelope-Based Vaccine

One of the difficulties in developing a prophylactic vaccine against AIDS has been the lack of a clear understanding of the immune correlates of protection. That is, it is not known what types of immune responses a vaccine should elicit in order for it to be protective against a virus challenge. No patient has ever been cured of the virus subsequent to an unequivocal infection and the results from the animal vaccine studies have been highly inconsistent, as alluded to above. The typical immune parameters measured are antigen-specific antibody titers (binding or neutralizing) and cell-mediated immunity (lymphoproliferative or cytotoxic response). Antigen-binding antibody titer, which is often determined by ELISA or Western blot, and lymphoproliferative responses do not correlate with either protection in vivo or inhibition of virus replication in vitro. What appears to be important are neutralizing antibody titer and CTL activity. Several lines of observations have demonstrated the importance of CTL in containing the early spread of HIV-1 and in the long-term survival of infected patients (Borrow et al., 1994; Koup et al., 1994; Klein et al., 1995; Pinto et al., 1995; Rinaldo et al., 1995; Rowland-Jones et al., 1998a,b, 1995, 1999). The presence of neutralizing antibody against a challenging virus strain often, although not always, resulted in some levels of protection in animal experiments (Berman et al., 1990; Bruck et al., 1994; Girard et al., 1991; Zolla-Pazner et al., 1998). The protective efficacy of neutralizing antibodies, when present in adequate levels, has been directly demonstrated in several passive immunization studies (Conley et al., 1996; Emini et al., 1990a, 1992; Gauduin et al., 1997; Mascola et al., 1999; Prince et al., 1991; Shibata et al., 1999; also see Moore and Burton, 1999). It is generally viewed that a successful AIDS vaccine should elicit both CTL and neutralizing antibodies.

To date, live attenuated vaccines have elicited the strongest and perhaps the most consistent protection against SIV in macaques. However, additional studies must be performed to thoroughly assess the potential long-term risks and rigorously address all of the safety concerns prior to any human clinical trials. Multimodal vaccine approaches (e.g., priming with either DNA or live vectors followed by subunit recombinant protein boost) have also elicited strong immune responses, albeit less protection compared to live attenuated vaccines. Immunizing with either DNA or live vectors alone can elicit CTL response. However, only a weak humoral immune response is elicited. Although subsequent boosting with recombinant proteins can dramatically enhance antigen-binding antibody titers, the ability of the antibodies to neutralize viruses has been relatively poor, especially against heterologous or clinical isolates. Obtaining sterilizing immunity (i.e., total neutralization of incoming viruses prior to the completion of the first infection cycle) has been the general goal of AIDS vaccine development. The reasoning behind the effort is the prediction that it would be increasingly more difficult (perhaps impossible) to prevent the virus from spreading, evading the immune system, and eventually establishing a persistent infection the longer the virus replication is allowed to occur. Sterilizing immunity can be achieved, at least in theory, if there is a high-enough level of preexisting antibodies (at the mucosal surface or in the blood stream) that could neutralize all of the incoming viruses prior to entering cells in a newly infected person. Alternatively, sterilizing immunity could be achieved if HIV-1-specific CTL can kill all virus-infected cells prior to producing progeny viruses. Since neutralizing antibody is the first line of immunological defense, the ability of a vaccine to elicit a potent neutralizing antibody is crucial.

Unlike T-cell epitopes, which are linear, many B-cell epitopes are highly conformation-specific. Thus, special care should be given in designing, producing, and administering envelope-based recombinant subunit vaccine candidates to elicit neutralizing antibodies. Some of the structural properties of the envelope glycoprotein that must be taken into consideration are shown in Fig. 6. As described above, HIV-1 gp120 is a heavily glycosylated protein and carbohydrate structures affect both immunogenicity and antigenicity of the protein. Macaques infected with SIV with certain glycosylation site mutations in the V1 region of gp120 produced a higher titer of neutralizing antibodies against the mutant, as well as the wild-type strain, compared to animals infected with the wild-type virus (Reitter et al., 1998). One possible interpretation of the result is that the removal of bulky carbohydrate chains revealed a hidden neutralization epitope. Whether this phenomenon is generally applicable to other strains of SIV or HIV-1, and to other glycosylation sites, remains to be examined. This is important since another study showed that immunization of guinea pigs with glycosylation site-mutated HIV-1

Glycosylation.

Hypo or non-glycosylated   **Glycosylated**   **Hyperglycosylated**

Conformation.

Monomeric gp120   **Oligomeric gp140**   **Virus-like particles**

Sequence Diversity.

Single isolate   **Multiple isolates**   **Multiple isolates**
                        Single clade   **Multiple clades**

FIGURE 6   Factors to consider in designing an envelope-based HIV-1 vaccine. Three of the most important factors to consider in designing an envelope-based vaccine are the status of gp120 glycosylation, structural conformation of the protein, and genetic/antigenic diversity of gp120. These factors should be more important for developing a subunit vaccine candidate and for eliciting humoral immune response. As discussed in the text, glycans affect both structural and immunological properties of gp120. Removal of certain carbohydrate moieties could potentially uncover hidden neutralizing epitopes that are conserved between different isolates. The sites to be mutated must be chosen carefully since the removal of glycans can also modify local antigenic structure or even completely misfold the protein. Although monomeric gp120 is easiest to produce, its antigenic properties could be somewhat different from those of native envelope glycoprotein on virion surface. The challenge lies in producing viruslike particles in sufficiently large quantities or oligomeric gp140 molecules (or other oligomers)

gp160 elicited antibodies that preferentially neutralized the virus with the mutations over the wild-type virus and vice versa (Bolmstedt *et al.*, 1996). Such a pattern of preferred homologous neutralization suggested that the elimination of carbohydrate moieties alters local antigenic conformation, creating new epitopes rather than uncovering hidden ones.

The removal of certain carbohydrate moieties can be detrimental to the protein function because of altered structure, defective processing, or a gross misfolding which ultimately leads to degradation. Thus, one has to make sure that elimination of glycosylation sites does not modify the structure drastically from that of the native protein. Eliciting CTL response is considered to be critical for developing a successful AIDS vaccine. There is lack of sufficient information on the effects of carbohydrate moieties on HIV-1 envelope glycoprotein processing and the efficiency of association between viral peptides and MHC molecules. Whether a nonglycosylated protein enhances cellular immune response should be examined further. HIV-1 gp120 on macrophage-derived viruses has been shown to be hyperglycosylated with lactoseaminoglycans (Willey *et al.*, 1996). More importantly, these viruses were 8- to 10-fold more resistant to neutralizing antibodies compared to PBMC-derived viruses. Although immunization with hyperglycosylated envelope protein may sound counterintuitive, it is possible that gp120 with lactoseaminoglycans can elicit higher levels of neutralizing antibody against macrophage-derived viruses compared to gp120 without lactoseaminoglycans. This could be important for containing HIV-1 from spreading during the primary infection stage, since one of the first cell types to be infected by the virus at the mucosal surface is macrophage.

Another important factor is the overall conformation of gp120 (i.e., tertiary and quaternary structure of the protein). HIV-1 envelope glycoprotein is thought to exist as a trimer on the virion surface. Among various subunit envelope vaccine candidates examined thus far, monomeric gp120 has been used most widely. It is relatively easy to produce and it can elicit high levels of antibody response in most circumstances. Unfortunately, only low levels of neutralizing antibodies are elicited, suggesting that its structure may not be entirely identical to that of the trimer on the virion surface. However, monomeric gp120 likely possesses many of the structural and immunological features present on the native trimer since it can bind CD4

---

that have structural and antigenic properties similar to those of the native trimer on virions. If it is not possible to elicit broadly neutralizing antibodies that can recognize conserved epitopes on envelope glycoprotein from a wide variety of isolates, serious consideration must be given to a polyvalent envelope vaccine candidate. A polyvalent envelope vaccine candidate could comprise multiple isolates from a single clade (e.g., for North America) or multiple isolates from multiple clades (e.g., Africa or Southeast Asia). Similar to the flu vaccine, annual inoculation of an AIDS vaccine containing different sets of envelope glycoprotein could be a possibility.

and neutralizing antibodies. More convincingly, the monomeric form of the protein can recall anemnestic response to neutralizing antibodies when subjects have been primed with either DNA or live vector vaccine. It is possible that many of the conformational epitopes on the surface of gp120, which are critical for eliciting neutralizing antibodies, are partially denatured when the protein is formulated with adjuvants. As a result, much of the antibodies elicited are against linear epitopes that are not present on the external surface of the trimer. The effects of adjuvants on the structural features critical for eliciting neutralizing antibodies have not yet been examined rigorously.

The ability of antibodies to bind oligomeric envelope glycoprotein strongly correlated with their ability to neutralize virus (Fouts *et al.*, 1997; Parren *et al.*, 1998). The envelope structures on noninfectious viruslike particles (VLP) probably best resemble the ones on virions. Unfortunately, however, the production of VLP in quantities required for a vaccine use has been very difficult due to problems of inefficient gag protein processing, particle assembly and release, and/or envelope incorporation in recombinant systems. Furthermore, it has been very difficult to purify VLP away from contaminating cellular proteins, which could pose potential problems. As an alternative to producing VLP, a noncleavable gp140 construct, which could form into an oligomeric structure, was expressed (Earl *et al.*, 1994). The antigenic characteristics of the oligomeric gp140 closely resembled those of the envelope protein complex on virions (Richardson *et al.*, 1996). The oligomer was able to elicit antibodies to conserved, highly conformational epitopes in mice (Broder *et al.*, 1994; Earl *et al.*, 1994). Immunization of macaques with oligomeric gp140, however, did not seem to improve the protective efficacy of the vaccine compared to monomeric gp120 (Pat Earl, personal communication). The results raise possibilities that the structure of oligomeric gp140, while better than that of monomeric gp120, may not be identical to the trimeric structure of the envelope protein on virions or that they are also partially denatured upon formulating with an adjuvant.

A recent observation that immunization of mice with fusion intermediates, captured by formaldehyde fixation of cells undergoing a gp120/CD4/coreceptor-dependent fusion reaction, elicited broad neutralizing antibodies against a number of primary HIV-1 isolate is promising (LaCasse *et al.*, 1999). It remains to be seen whether a similar immunogen can protect against a viral challenge *in vivo* in primates. Furthermore, the safety of a whole-cell vaccine must also be carefully evaluated. The results from the study raise a possibility that the "native antigen" may not necessarily be the best immunogen to elicit neutralizing antibodies. Perhaps an antigen with limited modifications (e.g., removal of certain combination of glycosylation sites, deletion of certain regions of the envelope, or partial denaturation of the protein) might elicit more potent neutralizing antibodies with broader reactivity. Answers to such a question could be determined only by an empirical approach.

As alluded to earlier, one of the major problems of developing an effective AIDS vaccine is the extreme genetic diversity of the envelope glycoprotein. Two classes of neutralizing antibodies are elicted against gp120: (1) antibodies directed against the variable regions that are highly strain-specific (i.e., reactive only against the same virus strain the immunogen is derived from; Arthur *et al.*, 1987; Goudsmit, 1988; Nara *et al.*, 1988; Rusche *et al.*, 1988; Weiss *et al.*, 1986); and (2) those that are directed against the conserved regions, which are more broadly cross-reactive (Berkower *et al.*, 1989; Katzenstein *et al.*, 1990; Steimer and Haigwood, 1991; Steimer *et al.*, 1991; Weiss *et al.*, 1986). Most of the neutralizing antibodies are virus isolate-specific. These antibodies are directed primarily against the V3 loop (principal neutralizing determinant; Goudsmit *et al.*, 1988; Javaherian *et al.*, 1989; Palker *et al.*, 1988; Rusche *et al.*, 1988) for TCLA T-tropic envelopes or against the V1/V2 region for the primary isolates (Gorny *et al.*, 1994; Pinter *et al.*, 1998). Neutralizing antibodies with broad specificity are rare, with only two human monoclonal antibodies (2G12 and b12) identified against gp120 worldwide (Mo *et al.*, 1997; Roben *et al.*, 1994; Trkola *et al.*, 1995; Trkola *et al.*, 1996b; Fig. 3). One way to elicit broadly cross-reactive neutralizing antibodies is to identify and target conserved, yet accessible, epitopes (e.g., by using modified gp120 with some of the variable regions and/or glycosylation sites removed). Alternatively, a pool of envelope proteins, which represent the majority of "antigenic subtypes" prevalent in a given geographic area (if one could be established), can be prepared as a polyvalent vaccine. One potential advantage of a polyvalent vaccine is that it might reduce any variations in immune response to a given immunogen by individuals in the population (e.g., due to HLA haplotype). Both of these tasks are equally difficult and would require years of trial-and-error.

VII. Concluding Remarks

Vaccine development is an enormous task. It involves a number of scientific disciplines, multiple stages of development, and countless variables. Completing a single nonhuman primate vaccine study, for example, which includes designing and producing a vaccine candidate, planning, administering, analyzing the immune response, virus challenge, and postchallenge monitoring and analysis, can take over 2 to 3 years. A typical animal study will examine a limited number of vaccine candidates, utilize one adjuvant, follow one vaccine dose and schedule by one immunization route, challenge with one virus strain by a singe route of infection with a given dose, and in one animal species. All of these variables make it extremely difficult to perform every experiment systematically and well controlled with the number of animals that would yield statistically significant results. What is worse is that each and every one of these variables can potentially alter the outcome of the study. To make things even more problematic, different studies often use different methodologies to monitor immune parameters (e.g., virus neutralization assay; see Vujcic and Quinnan, 1995), which makes comparison of vaccine efficacy between different studies even more difficult. Every assay has both advantages and disadvantages, making it difficult for different research groups to agree and standardize assays.

AIDS research during the past decade has made a significant contribution toward our understanding of the basic virology and pathogenesis as well as the biochemical, structural, and immunological properties of HIV-1 and its gene products. Developing an AIDS vaccine has been a significant scientific challenge. Despite the absence of an effective vaccine, what has been learned thus far will be the foundation for the next decade of research in developing a successful vaccine. Continued collaborations between research groups, division of labor, and development and standardization of new, faster, and more quantitative assays would help us achieve our goal.

References

Ada, G. L., and McElrath, M. J. (1997). HIV type 1 vaccine-induced cytotoxic T cell responses: Potential role in vaccine efficacy. *AIDS Res. Hum. Retroviruses* **13**, 205–210.

Ahmad, S., Lohman, B., Marthas, M., Giavedoni, L., el-Amad, Z., Haigwood, N. L., Scandella, C. J., Gardner, M. B., Luciw, P. A., and Yilma, T. (1994). Reduced virus load in rhesus macaques immunized with recombinant gp160 and challenged with simian immunodeficiency virus. *AIDS Res. Hum. Retroviruses* **10**, 195–204.

Almond, N., Kent, K., Cranage, M., Rud, E., Clarke, B., and Stott, E. J. (1995). Protection by attenuated simian immunodeficiency virus in macaques against challenge with virus-infected cells. *Lancet* **345**, 1342–1344.

Almond, N. M., and Heeney, J. L. (1998). AIDS vaccine development in primate models. *AIDS* **12**, S133–S140.

Anderson, E. D., Thomas, L., Hayflick, J. S., and Thomas, G. (1993). Inhibition of HIV-1 gp160-dependent membrane fusion by a furin-directed alpha 1-antitrypsin variant. *J. Biol. Chem.* **268**, 24887–24891.

Arthur, L. O., Bess, J. W., Jr., Chertova, E. N., Rossio, J. L., Esser, M. T., Benveniste, R. E., Henderson, L. E., and Lifson, J. D. (1998). Chemical inactivation of retroviral infectivity by targeting nucleocapsid protein zinc fingers: A candidate SIV vaccine. *AIDS Res. Hum. Retroviruses* **14**(Suppl. 3), S311–S319.

Arthur, L. O., Bess, J. W., Jr., Sowder, R. C. D., Benveniste, R. E., Mann, D. L., Chermann, J. C., and Henderson, L. E. (1992). Cellular proteins bound to immunodeficiency viruses: Implications for pathogenesis and vaccines. *Science* **258**, 1935–1938.

Arthur, L. O., Bess, J. W., Jr., Urban, R. G., Strominger, J. L., Morton, W. R., Mann, D. L., Henderson, L. E., and Benveniste, R. E. (1995). Macaques immunized with HLA-DR are protected from challenge with simian immunodeficiency virus. *J. Virol.* **69**, 3117–3124.

Arthur, L. O., Bess, J. W., Jr., Waters, D. J., Pyle, S. W., Kelliher, J. C., Nara, P. L., Krohn, K., Robey, W. G., Langlois, A. J., Gallo, R. C. *et al.* (1989). Challenge of chimpanzees (*Pan troglodytes*) immunized with human immunodeficiency virus envelope glycoprotein gp120. *J. Virol.* **63**, 5046–5053.

Arthur, L. O., Pyle, S. W., Nara, P. L., Bess, J. W., Jr., Gonda, M. A., Kelliher, J. C., Gilden, R. V., Robey, W. G., Bolognesi, D. P., Gallo, R. C. *et al.* (1987). Serological responses in chimpanzees inoculated with human immunodeficiency virus glycoprotein (gp120) subunit vaccine. *Proc. Natl. Acad. Sci. USA* **84**, 8583–8587.

Baba, T. W., Jeong, Y. S., Pennick, D., Bronson, R., Greene, M. F., and Ruprecht, R. M. (1995). Pathogenicity of live, attenuated SIV after mucosal infection of neonatal macaques. *Science* **267**, 1820–1825.

Baba, T. W., Liska, V., Khimani, A. H., Ray, N. B., Dailey, P. J., Penninck, D., Bronson, R., Greene, M. F., McClure, H. M., Martin, L. N., and Ruprecht, R. M. (1999). Live attenuated, multiply deleted simian immunodeficiency virus causes AIDS in infant and adult macaques. *Nat. Med.* **5**, 194–203.

Back, N. K., Smit, L., De Jong, J. J., Keulen, W., Schutten, M., Goudsmit, J., and Tersmette, M. (1994). An N-glycan within the human immunodeficiency virus type 1 gp120 V3 loop affects virus neutralization. *Virology* **199**, 431–438.

Bagarazzi, M. L., Boyer, J. D., Ayyavoo, V., and Weiner, D. B. (1998). Nucleic acid-based vaccines as an approach to immunization against human immunodeficiency virus type-1. *Curr. Top. Microbiol. Immunol.* **226**, 107–143.

Baltimore, D., and Heilman, C. (1998). HIV vaccines: Prospects and challenges. *Sci. Am.* **279**, 98–103.

Bandres, J. C., Wang, Q. F., O'Leary, J., Baleaux, F., Amara, A., Hoxie, J. A., Zolla-Pazner, S., and Gorny, M. K. (1998). Human Immunodeficiency virus (HIV) envelope binds to CXCR4 independently of CD4, and binding can be enhanced by interaction with soluble CD4 or by HIV envelope deglycosylation. *J. Virol.* **72**, 2500–2504.

Barnett, S. W., Klinger, J. M., Doe, B., Walker, C. M., Hansen, L., Duliege, A. M., and Sinangil, F. M. (1998). Prime-boost immunization strategies against HIV. *AIDS Res. Hum. Retroviruses* **14**(Suppl. 3), S299–S309.

Barrett, N., Eder, G., and Dorner, F. (1991). Characterization of a vaccinia-derived recombinant HIV-1 gp160 candidate vaccine and its immunogenicity in chimpanzees. *Biotechnol. Ther.* **2**, 91–106.

Benson, J., Chougnet, C., Robert-Guroff, M., Montefiori, D., Markham, P., Shearer, G., Gallo, R. C., Cranage, M., Paoletti, E., Limbach, K., Venzon, D., Tartaglia, J., and Franchini, G. (1998). Recombinant vaccine-induced protection against the highly pathogenic simian immunodeficiency virus SIV (mac251): Dependence on route of challenge exposure. *J. Virol.* **72**, 4170–4182.

Berger, E. A. (1997). HIV entry and tropism: The chemokine receptor connection. *AIDS* **11**, S3–S16.

Berglund, P., Quesada-Rolander, M., Putkonen, P., Biberfeld, G., Thorstensson, R., and Liljestrom, P. (1997). Outcome of immunization of cynomolgus monkeys with recombinant Semliki Forest virus encoding human immunodeficiency virus type 1 envelope protein and challenge with a high dose of SHIV-4 virus. *AIDS Res. Hum. Retroviruses* **13**, 1487–1495.

Berkhout, B., Verhoef, K., van Wamel, J. L., and Back, N. K. (1999). Genetic instability of live, attenuated human immunodeficiency virus type 1 vaccine strains. *J. Virol.* **73**, 1138–1145.

Berkower, I., Smith, G. E., Giri, C., and Murphy, D. (1989). Human immunodeficiency virus 1: Predominance of a group-specific neutralizing epitope that persists despite genetic variation. *J. Exp. Med.* **170**, 1681–1695.

Berman, P. W. (1998). Development of bivalent rgp120 vaccines to prevent HIV type 1 infection. *AIDS Res. Hum. Retroviruses* **14**(Suppl. 3), S277–S289.

Berman, P. W., Gregory, T. J., Riddle, L., Nakamura, G. R., Champe, M. A., Porter, J. P., Wurm, F. M., Hershberg, R. D., Cobb, E. K., and Eichberg, J. W. (1990). Protection of chimpanzees from infection by HIV-1 after vaccination with recombinant glycoprotein gp120 but not gp160. *Nature* **345**, 622–625.

Berman, P. W., Groopman, J. E., Gregory, T., Clapham, P. R., Weiss, R. A., Ferriani, R., Riddle, L., Shimasaki, C., Lucas, C., Lasky, L. A. *et al.* (1998). Human immunodeficiency virus type 1 challenge of chimpanzees immunized with recombinant envelope glycoprotein gp120. *Proc. Natl. Acad. Sci. USA* **85**, 5200–5204.

Berman, P. W., Murthy, K. K., Wrin, T., Vennari, J. C., Cobb, E. K., Eastman, D. J., Champe, M., Nakamura, G. R., Davison, D., Powell, M. F., Bussiere, J., Francis, D. P., Matthews, T., Gregory, T. J., and Obijeski, J. F. (1996). Protection of MN-rgp120-immunized chimpanzees from heterologous infection with a primary isolate of human immunodeficiency virus type 1. *J. Infect. Dis.* **173**, 52–59.

Bernstein, H. B., Tucker, S. P., Hunter, E., Schutzbach, J. S., and Compans, R. W. (1994). Human immunodeficiency virus type 1 envelope glycoprotein is modified by O-linked oligosaccharides. *J. Virol.* **68**, 463–468.

Bogers, W. M., Niphuis, H., ten Haaft, P., Laman, J. D., Koornstra, W., and Heeney, J. L. (1995). Protection from HIV-1 envelope-bearing chimeric simian immunodeficiency virus (SHIV) in rhesus macaques infected with attenuated SIV: Consequences of challenge. *AIDS* **9**, F13–F18.

Boily, M. C., Masse, B. R., Desai, K., Alary, M., and Anderson, R. M. (1999). Some important issues in the planning of phase III HIV vaccine efficacy trials. *Vaccine* **17**, 989–1004.

Bolmstedt, A., Hemming, A., Flodby, P., Berntsson, P., Travis, B., Lin, J. P., Ledbetter, J., Tsu, T., Wigzell, H., Hu, S. L. *et al.* (1991). Effects of mutations in glycosylation sites and disulphide bonds on processing, CD4-binding and fusion activity of human immunodeficiency virus envelope glycoproteins. *J. Gen. Virol.* **72**, 1269–1277.

Bolmstedt, A., Sjolander, S., Hansen, J. E., Akerblom, L., Hemming, A., Hu, S. L., Morein, B., and Olofsson, S. (1996). Influence of N-liked glycans in V4-V5 region of human immunodeficiency virus type 1 glycoprotein gp160 on induction of a virus-neutralizing humoral response. *J. Acq. Immune Defic. Syndr. Hum. Retrovirol.* **12**, 213–220.

Borrow, P., Lewicki, H., Hahn, B. H., Shaw, G. M., and Oldstone, M. B. (1994). Virus-specific CD8+ cytotoxic T-lymphocyte activity associated with control of viremia in primary human immunodeficiency virus type 1 infection. *J. Virol.* **68**, 6103–6110.

Boyer, J. D., Ugen, K. E., Wang, B., Agadjanyan, M., Gilbert, L., Bagarazzi, M. L., Chattergoon, M., Frost, P., Javadian, A., Williams, W. V., Refaeli, Y., Ciccarelli, R. B., McCallus, D., Coney, L., and Weiner, D. B. (1997). Protection of chimpanzees from high-dose heterologous HIV-1 challenge by DNA vaccination. *Nat. Med.* **3**, 526–532.

Boyer, J. D., Wang, B., Ugen, K. E., Agadjanyan, M., Javadian, A., Frost, P., Dang, K., Carrano, R. A., Ciccarelli, R., Coney, L., Williams, W. V., and Weiner, D. B. (1996). *In vivo* protective anti-HIV immune responses in non-human primates through DNA immunization. *J. Med. Primatol.* **25**, 242–250.

Broder, C. C., Earl, P. L., Long, D., Abedon, S. T., Moss, B., and Doms, R. W. (1994). Antigenic implications of human immunodeficiency virus type 1 envelope quaternary structure: Oligomer-specific and -sensitive monoclonal antibodies. *Proc. Natl. Acad. Sci. USA* **91**, 11699–11703.

Bruck, C., Thiriart, C., Fabry, L., Francotte, M., Pala., P., Van Opstal, O., Clup, J., Rosenberg, M., De Wilde, M., Heidt, P. *et al.* (1994). HIV-1 envelope-elicited neutralizing antibody titres correlate with virus load in chimpanzees. *Vaccine* **12**, 1141–1148.

Burton, D. R. (1997). A vaccine for HIV type 1: The antibody perspective. *Proc. Natl. Acad. Sci. USA* **94**, 10018–10023.

Butters, T. D., Yudkin, B., Jacob, G. S., and Jones, I. M. (1998). Structural characterization of the N-linked oligosaccharides derived from HIV gp120 expressed in lepidopteran cells. *Glycoconj. J.* **15**, 83–88.

Cann, A. J., Churcher, M. J., Boyd, M., O'Brien, W., Zhao, J. Q., Zack, J., and Chen, I. S. (1992). The region of the envelope gene of human immunodeficiency virus type 1 responsible for determination of cell tropism. *J. Virol.* **66**, 305–309.

Chan, D. C., Fass, D., Berger, J. M., and Kim, P. S. (1997). Core structure of gp41 from the HIV envelope glycoprotein *Cell* **89**, 263–273.

Chesebro, B., Nishio, J., Perryman, S., Cann, A., O'Brien, W., Chen, I. S., and Wehrly K. (1991). Identification of human immunodeficiency virus envelope gene sequences influencing viral entry into CD4-positive HeLa cells, T-leukemia cells, and macrophages. *J. Virol.* **65**, 5782–5789.

Cho, M. W., Lee, M. K., Carney, M. C., Berson, J. F., Doms, R. W., and Martin, M. A. (1998). Identification of determinants on a dualtropic human immunodeficiency virus type 1 envelope glycoprotein that confer usage of CXCR4. *J. Virol.* **72**, 2509–2515.

Choe, H., Farzan, M., Sun, Y., Sullivan, N., Rollins, B., Ponath, P. D., Wu, L., Mackay, C. R., LaRosa, G., Newman, W., Gerard, N., Gerard, C., and Sodroski, J. (1996). The beta-chemokine receptors CCR3 and CCR5 facilitate infection by primary HIV-1 isolates. *Cell* **85**, 1135–1148.

Clements-Mann, M. L., Weinhold, K., Matthews, T. J., Graham, B. S., Gorse, G. J., Keefer, M. C., McElrath, M. J., Hsieh, R. H., Mestecky, J., Zolla-Pazner, S., Mascola, J., Schwartz, D., Siliciano, R., Corey, L., Wright, P. F., Belshe, R., Dolin, R., Jackson, S., Xu, S., Fast, P., Walker, M. C., Stablein, D., Excler, J. L., Tartaglia, J., Paoletti, E. *et al.* (1998). Immune responses to human immunodeficiency virus (HIV) type 1 induced by canarypox expressing HIV-1MN gp120, HIV-1SF2 recombinant gp120, or both vaccines in seronegative adults. NIAID AIDS Vaccine Evaluation Group. *J. Infect. Dis.* **177**, 1230–1246.

Cocchi, F., DeVico, A. L., Garzino-Demo, A., Cara, A., Gallo, R. C., and Lusso, P. (1996). The V3 domain of the HIV-1 gp120 envelope glycoprotein is critical for chemokine-mediated blockade of infection. *Nat. Med.* **2**, 1244–1247.

Cohen, J. (1993). Jitters jeopardize AIDS vaccine trials. *Science* **262**, 980–981.

Conley, A. J., Kessler, J. A., II, Boots, L. J., McKenna, P. M., Schleif, W. A., Emini, E. A., Mark, G. E., III, Katinger, H., Cobb, E. K., Lunceford, S. M., Rouse, S. R., and Murthy, K. K. (1996). The consequence of passive administration of an anti-human immunodeficiency virus type 1 neutralizing monoclonal antibody before challenge of chimpanzees with a primary virus isolate. *J. Virol.* **70**, 6751–6758.

Connor, R. I., Korber, B. T., Graham, B. S., Hahn, B. H., Ho, D. D., Walker, B. D., Neumann, A. U., Vermund, S. H., Mestecky, J., Jackson, S., Fenamore, E., Cao, Y., Gao, F., Kalams, S., Kunstman, K. J., McDonald, D., McWilliams, N., Trkola, A., Moore, J. P., and Wolinsky, S. M. (1998). Immunological and virological analyses of persons infected by human immunodeficiency virus type 1 while participating in trials of recombinant gp120 subunit vaccines. *J. Virol.* **72**, 1552–1576.

Cooney, E. L., Collier, A. C., Greenberg, P. D., Coombs, R. W., Zarling, J., Arditti, D. E., Hoffman, M. C., Hu, S. L., and Corey, L. (1991). Safety of and immunological response to a recombinant vaccinia virus vaccine expressing HIV envelope glycoprotein. *Lancet* **337**, 567–572.

Cooney, E. L., McElrath, M. J., Corey, L., Hu, S. L., Collier, A. C., Arditti, D., Hoffman, M., Coombs, R. W., Smith, G. E., and Greenberg, P. D. (1993). Enhanced immunity to human immunodeficiency virus (HIV) envelope elicited by a combined vaccine regimen consisting of priming with a vaccinia recombinant expressing HIV envelope and boosting with gp160 protein. *Proc. Natl. Acad. Sci. USA* **90**, 1882–1886.

Corey, L., McElrath, M. J., Weinhold, K., Matthews, T., Stablein, D., Graham, B., Keefer, M., Schwartz, D., and Gorse, G. (1998). Cytotoxic T cell and neutralizing antibody responses to human immunodeficiency virus type 1 envelope with a combination vaccine regimen. AIDS Vaccine Evaluation Group. *J. Infect. Dis.* **177**, 301–309.

D'Souza, M. P., and Harden, V. A. (1996). Chemokines and HIV-1 second receptors: Confluence of two fields generates optimism in AIDS research. *Nat. Med.* **2**, 1293–1300.

Daniel, M. D., Kirchhoff, F., Czajak, S. C., Sehgal, P. K., and Desrosiers, R. C. (1992). Protective effects of a live attenuated SIV vaccine with a deletion in the nef gene. *Science* **258**, 1938–1941.

Daniel, M. D., Mazzara, G. P., Simon, M. A., Sehgal, P. K., Kodama, T., Panicali, D. L., and Desrosiers, R. C. (1994). High-titer immune responses elicited by recombinant vaccinia virus priming and particle boosting are ineffective in preventing virulent SIV infection. *AIDS Res. Hum. Retroviruses* 10, 839–851.

Davis, D., Stephens, D. M., Willers, C., and Lachmann, P. J. (1990). Glycosylation governs the binding of antipeptide antibodies to regions of hypervariable amino acid sequence within recombinant gp120 of human immunodeficiency virus type 1. *J. Gen. Virol.* 71, 2889–2898.

De Jong, J. J., Goudsmit, J., Keulen, W., Klaver, B., Krone, W., Tersmette, M., and de Ronde, A. (1992). Human immunodeficiency virus type 1 clones chimeric for the envelope V3 domain differ in syncytium formation and replication capacity. *J. Virol.* 66, 757–765.

Deacon, N. J., Tsykin, A., Solomon, A., Smith, K., Ludford-Menting, M., Hooker, D. J., McPhee, D. A., Greenway, A. L., Ellett, A., Chatfield, C. *et al.* (1995). Genomic structure of an attenuated quasi species of HIV-1 from a blood transfusion donor and recipients. *Science* 270, 988–991.

Decroly, E., Vandenbranden, M., Ruysschaert, J. M., Cogniaux, J., Jacob, G. S., Howard, S. C., Marshall, G., Kompelli, A., Basak, A., Jean, F. *et al.* (1994). The convertases furin and PC1 can both cleave the human immunodeficiency virus (HIV)-1 envelope glycoprotein gp160 into gp120 (HIV-1 SU) and gp41 (HIV-I TM). *J. Biol. Chem.* 269, 12240–12247.

Desrosiers, R. C., Wyand, M. S. Kodama, T., Ringler, D. J., Arthur, L. O., Sehgal, P. K., Letvin, N. L., King, N. W., and Danial, M. D. (1989). Vaccine protection against simian immunodeficiency virus infection. *Proc. Natl. Acad. Sci. USA* 86, 6353–6357.

Doe, B., Steimer, K. S., and Walker, C. M. (1994). Induction of HIV-1 envelope (gp120)-specific cytotoxic T lymphocyte responses in mice by recombinant CHO cell-derived gp120 is enhanced by enzymatic removal of N-linked glycans. *Eur. J. Immunol.* 24, 2369–2376.

Dolin, R. (1995). Human studies in the development of human immunodeficiency virus vaccines. *J. Infect. Dis.* 172, 1175–1183.

Doms, R. W., Earl, P. L., and Moss, B. (1991). The assembly of the HIV-1 env glycoprotein into dimers and tetramers. *Adv. Exp. Med. Biol.* 300, 203–219.

Doranz, B. J., Berson, J. F., Rucker, J., and Doms, R. W. (1997). Chemokine receptors as fusion cofactors for human immunodeficiency virus type 1 (HIV-1). *Immunol. Res.* 16, 15–28.

Dunn, C. S., Hurtrel, B., Beyer, C., Gloeckler, L., Ledger, T. N., Moog, C., Kieny, M. P., Mehtali, M., Schmitt, D., Gut, J. P., Kirn, A., and Aubertin, A. M. (1997). Protection of SIVmac-infected macaque monkeys against superinfection by a simian immunodeficiency virus expressing envelope glycoproteins of HIV type 1. *AIDS Res. Hum. Retroviruses* 13, 913-922.

Dyer, W. B., Ogg, G. S., Demoitie, M. A., Jin, X., Geczy, A. F., Rowland-Jones, S. L., McMichael, A. J., Nixon, D. F., and Sullivan, J. S. (1999). Strong human immunodeficiency virus (HIV)-specific cytotoxic T-lymphocyte activity in Sydney Blood Bank Cohort patients infected with nef-defective HIV type 1. *J. Virol.* 73, 436–443.

Earl, P. L., Broder, C. C., Long, D., Lee, S. A., Peterson, J., Chakrabarti, S., Doms, R. W., and Moss, B. (1994). Native oligomeric human immunodeficiency virus type 1 envelope glycoprotein elicits diverse monoclonal antibody reactivities. *J. Virol.* 68, 3015–3026.

Earl, P. L., Doms, R. W., and Moss, B. (1990). Oligomeric structure of the human immunodeficiency virus type 1 envelope glycoprotein. *Proc. Natl. Acad. Sci. USA* 87, 648–652.

Egan, M. A., Pavlat, W. A., Tartaglia, J., Paoletti, E., Weinhold, K. J., Clements, M. L., and Siliciano, R. F. (1995). Induction of human immunodeficiency virus type 1 (HIV-1)-specific cytolytic T lymphocyte responses in seronegative adults by a nonreplicating, host-range-restricted canarypox vector (ALVAC) carrying the HIV-1MN env gene. *J. Infect. Dis.* 171, 1623–1627.

Emini, E. A., Nara, P. L., Schleif, W. A., Lewis, J. A., Davide, J. P., Lee, D. R., Kessler, J., Conley, S., Matsushita, S., Putney, S. D. *et al.*, (1990a). Antibody-mediated *in vitro* neutralization of human immunodeficiency virus type 1 abolishes infectivity for chimpanzees. *J. Virol.* **64**, 3674–3678.

Emini, E. A., Schleif, W. A., Nunberg, J. H., Conley, A. J., Eda, Y., Tokiyoshi, S., Putney, S. D., Matsushita, S., Cobb, K. E., Jett, C. M. *et al.*, (1992). Prevention of HIV-1 infection in chimpanzees by gp120 V3 domain-specific monoclonal antibody. *Nature* **355**, 728–730.

Emini, E. A., Schleif, W. A., Quintero, J. C., Conard, P. G., Eichberg, J. W., Vlasuk, G. P., Lehman, E. D., Polokoff, M. A., Schaeffer, T. F., Schultz, L. D. *et al.*, (1990b). Yeast-expressed p55 precursor core protein of human immunodeficiency virus type 1 does not elicit protective immunity in chimpanzees. *AIDS Res. Hum. Retroviruses* **6**, 1247–1250.

Excler, J. L., and Plotkin, S. (1997). The prime-boost concept applied to HIV preventive vaccines. *AIDS* **11**, S127–S137.

Fazekas de St Groth, B., and Webster, R. G. (1966). Disquisitions on original antigenic sin. I. Evidence in man. *J. Exp. Med* **124**, 331–345.

Feizi, T., and Larkin, M. (1990). AIDS and glycosylation. *Glycobiology* **1**, 17–23.

Fennie, C., and Lasky, L. A. (1989). Model for intracellular folding of the human immunodeficiency virus type 1 gp120. *J. Virol.* **63**, 639–646.

Fenouillet, E., Gluckman, J. C., and Jones, I. M. (1994). Functions of HIV envelope glycans. *Trends Biochem. Sci.* **19**, 65–70.

Ferrari, G., Humphrey, W., McElrath, M. J., Excler, J. L., Duliege, A. M., Clements, M. L., Corey, L. C., Bolognesi, D. P., and Weinhold, K. J. (1997). Clade B-based HIV-1 vaccines elicit cross-clade cytotoxic T lymphocyte reactivities in uninfected volunteers. *Proc. Natl. Acad. Sci. USA* **94**, 1396–1401.

Fleury, B., Janvier, G., Pialoux, G., Buseyne, F., Robertson, M. N., Tartaglia, J., Paoletti, E., Kieny, M. P., Excler, J. L., and Riviere, Y. (1996). Memory cytotoxic T lymphocyte responses in human immunodeficiency virus type 1 (HIV-1)-negative volunteers immunized with a recombinant canarypox expressing gp160 of HIV-1 and boosted with a recombinant gp160. *J. Infect. Dis.* **174**, 734–738.

Fouts, T. R., Binley, J. M., Trkola, A., Robinson, J. E., and Moore, J. P. (1997). Neutralization of the human immunodeficiency virus type 1 primary isolate JR-FL by human monoclonal antibodies correlates with antibody binding to the oligomeric form of the envelope glycoprotein complex. *J. Virol.* **71**, 2779–2785.

Francis, D. P., Gregory, T., McElrath, M. J., Belshe, R. B., Gorse, G. J., Migasena, S., Kitayaporn, D., Pitisuttitham, P., Matthews, T., Schwartz, D. H., and Berman, P. W. (1998). Advancing AIDSVAX to phase 3. Safety, immunogenicity, and plans for phase 3. *AIDS Res. Hum. Retroviruses* **14**(Suppl. 3), S325–S331.

Freed, E. O., and Martin, M. A. (in press). The role of human immunodeficiency virus type 1 envelope glycoproteins in virus infection. *J. Biol. Chem.* **270**, 23883–23886.

Freed, E. O., and Martin, M. A. (1999). The molecular and biological properties of the human immunodeficiency virus. *In* "The Molecular Basis of Blood Diseases" (G. Stamatoyanno-poulos, A. Neinhuis, P. Majerus, and M. Varmus, Eds.), 3rd ed. W. B. Saunders, New York.

Fuller, D. H., Corb, M. M., Barnett, S., Steimer, K., and Haynes, J. R. (1997a). Enhancement of immunodeficiency virus-specific immune responses in DNA- immunized rhesus macaques. *Vaccine* **15**, 924–926.

Fuller, D. H., Murphey-Corb, M., Clements, J., Barnett, S., and Haynes, J. R. (1996). Induction of immunodeficiency virus-specific immune responses in rhesus monkeys following gene gun-mediated DNA vaccination. *J. Med. Primatol.* **25**, 236–241.

Fuller, D. H., Simpson, L., Cole, K. S., Clements, J. E., Panicali, D. L., Montelaro, R. C., Murphey-Corb, M., and Haynes, J. R. (1997b). Gene gun-based nucleic acid immunization alone or in combination with recombinant vaccinia vectors suppresses virus burden in rhesus macaques challenged with a heterologous SIV. *Immunol. Cell. Biol.* **75**, 389–396.

Fultz, P. N., Nara, P., Barre-Sinoussi, F., Chaput, A., Greenberg, M. L., Muchmore, E., Kieny, M. P., and Girard, M. (1992). Vaccine protection of chimpanzees against challenge with HIV-1-infected peripheral blood mononuclear cells. *Science* **256**, 1687–1690.

Gao, F., Bailes, E., Robertson, D. L., Chen, Y., Rodenburg, C. M., Michael, S. F., Cummins, L. B., Arthur, L. O., Peeters, M., Shaw, G. M., Sharp, P. M., and Hahn, B. H. (1999). Origin of HIV-1 in the chimpanzee *Pan troglodytes troglodytes*. *Nature* **397**, 436–441.

Gauduin, M. C., Parren, P. W., Weir, R., Barbas, C. F., Burton, D. R., and Koup, R. A. (1997). Passive immunization with a human monoclonal antibody protects hu-PBL-SCID mice against challenge by primary isolates of HIV-1. *Nat. Med.* **3**, 1389–1393.

Giavedoni, L. D., Planelles, V., Haigwood, N. L., Ahmad, S., Kluge, J. D., Marthas, M. L., Gardner, M. B., Luciw, P. A., and Yilma, T. D. (1993). Immune response of rhesus macaques to recombinant simian immunodeficiency virus gp130 does not protect from challenge infection. *J. Virol.* **67**, 577–583.

Girard, M., Kieny, M. P., Pinter, A., Barre-Sinoussi, F., Nara, P., Kolbe, H., Kusumi, K., Chaput, A., Reinhart, T., Muchmore, E. *et al.* (1991). Immunization of chimpanzees confers protection against challenge with human immunodeficiency virus. *Proc. Natl. Acad. Sci. USA* **88**, 542–546.

Girard, M., Meignier, B., Barre-Sinoussi, F., Kieny, M. P., Matthews, T., Muchmore, E., Nara, P. L., Wei, Q., Rimsky, L., Weinhold, K. *et al.* (1995). Vaccine-induced protection of chimpanzees against infection by a heterologous human immunodeficiency virus type 1. *J. Virol.* **69**, 6239–6248.

Girard, M., van der Ryst, E., Barre-Sinoussi, F., Nara, P., Tartaglia, J., Paoletti, E., Blondeau, C., Jennings, M., Verrier, F., Meignier, B., and Fultz, P. N. (1997). Challenge of chimpanzees immunized with a recombinant canarypox-HIV-1 virus. *Virology* **232**, 98–104.

Girard, M., Yue, L., Barre-Sinoussi, F., van der Ryst, E., Meignier, B., Muchmore, E., and Fultz, P. N. (1996). Failure of a human immunodeficiency virus type 1 (HIV-1) subtype B-derived vaccine to prevent infection of chimpanzees by an HIV-1 subtype E strain. *J. Virol.* **70**, 8229–8233.

Golding, H., D'Souza, M. P., Bradac, J., Mathieson, B., and Fast, P. (1994). Neutralization of HIV-1. *AIDS Res. Hum. Retroviruses* **10**, 633–643.

Gorny, M. K., Moore, J. P., Conley, A. J., Karwowska, S., Sodroski, J., Williams, C., Burda, S., Boots, L. J., and Zolla-Pazner, S. (1994). Human anti-V2 monoclonal antibody that neutralizes primary but not laboratory isolates of human immunodeficiency virus type 1. *J. Virol.* **68**, 8312–8320.

Gorse, G. J., Patel, G. B., Newman, F. K., Belshe, R. B., Berman, P. W., Gregory, T. J., and Matthews, T. J. (1996). Antibody to native human immunodeficiency virus type 1 envelope glycoproteins induced by IIIB and MN recombinant gp120 vaccines: The NIAID AIDS Vaccine Evaluation Group. *Clin. Diagn. Lab. Immunol.* **3**, 378–386.

Gotch, F. (1998). Cross-clade T cell recognition of HIV.1. *Curr. Opin. Immunol.* **10**, 388–392.

Goudsmit, J. (1988). Immunodominant B-cell epitopes of the HIV-1 envelope recognized by infected and immunized hosts. *AIDS* **2**, S41–S45.

Goudsmit, J., Thiriart, C., Smit, L., Bruck, C., and Gibbs, C. J. (1988). Temporal development of cross-neutralization between HTLV-IIIB and HTLV-III RF in experimentally infected chimpanzees. *Vaccine* **6**, 229–232.

Graham, B. S. (1994). Serological responses to candidate AIDS vaccines. *AIDS Res. Hum. Retroviruses* **10**, S145–S148.

Graham, B. S., Belshe, R. B., Clements, M. L., Dolin, R., Corey, L., Wright, P. F., Gorse, G. J., Midthun, K., Keefer, M. C., Roberts, N. J., Jr. *et al.* (1992). Vaccination of vaccinia-naive adults with human immunodeficiency virus type 1 gp160 recombinant vaccinia virus in a blinded, controlled, randomized clinical trial: The AIDS Vaccine Clinical Trials Network. *J. Infect. Dis.* **166**, 244–252.

Graham, B. S., Gorse, G. J., Schwartz, D. H., Keefer, M. C., McElrath, M. J., Matthews, T. J., Wright, P. F., Belshe, R. B., Clements, M. L., Dolin, R. *et al.* (1994). Determinants of antibody response after recombinant gp160 boosting in vaccinia-naive volunteers primed with gp160-recombinant vaccinia virus: The National Institute of Allergy and Infectious Diseases AIDS Vaccine Clinical Trials Network. *J. Infect. Dis.* **170**, 782–786.

Graham, B. S., Matthews, T. J., Belshe, R. B., Clements, M. L., Dolin, R., Wright, P. F., Gorse, G. J., Schwartz, D. H., Keefer, M. C., Bolognesi, D. P. *et al.* (1993). Augmentation of human immunodeficiency virus type 1 neutralizing antibody by priming with gp160 recombinant vaccinia and boosting with rgp160 in vaccinia-naive adults: The NIAID AIDS Vaccine Clinical Trials Network. *J. Infect. Dis.* **167**, 533–537.

Graham, B. S., McElrath, M. J., Connor, R. I., Schwartz, D. H., Gorse, G. J., Keefer, M. C., Mulligan, M. J., Matthews, T. J., Wolinsky, S. M., Montefiori, D. C., Vermund, S. H., Lambert, J. S., Corey, L., Belshe, R. B., Dolin, R., Wright, P. F., Korber, B. T., Wolff, M. C., and Fast, P. E. (1998). Analysis of intercurrent human immunodeficiency virus type 1 infections in phase I and II trials of candidate AIDS vaccines: AIDS Vaccine Evaluation Group, and the Correlates of HIV Immune Protection Group. *J. Infect. Dis.* **177**, 310–319.

Gram, G. J., Hemming, A., Bolmstedt, A., Jansson, B., Olofsson, S., Akerblom, L., Nielsen, J. O., and Hansen, J. E. (1994). Identification of an N-linked glycan in the V1-loop of HIV-1 gp120 influencing neutralization by anti-V3 antibodies and soluble CD4. *Arch. Virol.* **139**, 253–261.

Greenough, T. C., Sullivan, J. L., and Desrosiers, R. C. (1999). Declining CD4 T-cell counts in a person infected with nef-deleted HIV-1. *N. Engl. J. Med.* **340**, 236–237.

Gruters, R. A., Neefjes, J. J., Tersmette, M., de Goede, R. E., Tulp, A., Huisman, H. G., Miedema, F., and Ploegh, H. L. (1987). Interference with HIV-induced syncytium formation and viral infectivity by inhibitors of trimming glucosidase. *Nature* **330**, 74–77.

Haigwood, N. L., Nara, P. L., Brooks, E., Van Nest, G. A., Ott, G., Higgins, K. W., Dunlop, N., Scandella, C. J., Eichberg, J. W., and Steimer, K. S. (1992). Native but not denatured recombinant human immunodeficiency virus type 1 gp120 generates broad-spectrum neutralizing antibodies in baboons. *J. Virol.* **66**, 172–182.

Haigwood, N. L., Pierce, C. C., Robertson, M. N., Watson, A. J., Montefiori, D. C., Rabin, M., Lynch, J. B., Kuller, L., Thompson, J., Morton, W. R., Benveniste, R. E., Hu, S. L., Greenberg, P., and Mossman, S. P. (1999). Protection from pathogenic SIV challenge using multigenic DNA vaccines. *Immunol. Lett.* **66**, 183–188.

Haigwood, N. L., and Zolla-Pazner, S. (1998). Humoral immunity to HIV, SIV, and SHIV *AIDS* **12**, S121–S132.

Hallenberger, S., Bosch, V., Angliker, H., Shaw, E., Klenk, H. D., and Garten, W. (1992). Inhibition of furin-mediated cleavage activation of HIV-1 glycoprotein gp160. *Nature* **360**, 358–361.

Haynes, B. F., Pantaleo, G., and Fauci, A. S. (1996). Toward an understanding of the correlates of protective immunity of HIV infection. *Science* **271**, 324–328.

Heeney, J. L. (1996). Primate models for AIDS vaccine development. *AIDS* **10**, S115–S122.

Heeney, J. L., Bruck, C., Goudsmit, J., Montagnier, L., Schultz, A., Tyrrell, D., and Zolla-Pazner, S. (1997). Immune correlates of protection from HIV infection and AIDS. *Immunol. Today* **18**, 4–8.

Heeney, J. L., de Vries, P., Dubbes, R., Koornstra, W., Niphuis, H., ten Haaft, P., Boes, J., Dings, M. E., Morein, B., and Osterhaus, A. D. (1992). Comparison of protection from homologous cell-free vs cell-associated SIV challenge afforded by inactivated whole SIV vaccines. *J. Med. Primatol.* **21**, 126–130.

Hemming, A., Bolmstedt, A., Jansson, B., Hansen, J. E., Travis, B., Hu, S. L., and Olofsson, S. (1994). Identification of three N-linked glycans in the V4-V5 region of HIV-1 gp 120, dispensable for CD4-binding and fusion activity of gp 120. *Arch. Virol.* **134**, 335–344.

Hesselgesser, J., Halks-Miller, M., Del Vecchio, V., Peiper, S. C., Hoxie, J., Kolson, D. L., Taub, D., and Horuk, R. (1997). CD4-independent association between HIV-1 gp120 and CXCR4: Functional chemokine receptors are expressed in human neurons. *Curr. Biol.* **7**, 112–121.

Hill, C.M., Deng, H., Unutmaz, D., Kewalramani, V. N., Bastiani, L., Gorny, M. K., Zolla-Pazner, S., and Littman, D. R. (1997). Envelope glycoproteins from human immunodeficiency virus types 1 and 2 and simian immunodeficiency virus can use human CCR5 as a coreceptor for viral entry and make direct CD4-dependent interactions with this chemokine receptor. *J. Virol.* **71**, 6296–6304.

Hirsch, V. M., Fuerst, T. R., Sutter, G., Carroll, M. W., Yang, L. C., Goldstein, S., Piatak, M., Jr., Elkins, W. R., Alvord, W. G., Montefiori, D. C., Moss, B., and Lifson, J. D. (1996). Patterns of viral replication correlate with outcome in simian immunodeficiency virus (SIV)-infected macaques: Effect of prior immunization with a trivalent SIV vaccine in modified vaccinia virus Ankara. *J. Virol.,* **70**, 3741–3752.

Hirsch, V. M., Goldstein, S., Hynes, N. A., Elkins, W. R., London, W. T., Zack, P. M., Montefiori, D., and Johnson, P. R. (1994). Prolonged clinical latency and survival of macaques given a whole inactivated simian immunodeficiency virus vaccine. *J. Infect. Dis.* **170**, 51–59.

Hu, S., and Norrby, E. (1997). Vaccines and immunology: Overview. *AIDS* **11**, S85–S86.

Hu, S. L., Abrams, K., Barber, G. N., Moran, P., Zarling, J. M., Langlois, A. J., Kuller, L., Morton, W. R., and Benveniste, R. E. (1992). Protection of macaques against SIV infection by subunit vaccines of SIV envelope glycoprotein gp160. *Science* **255**, 456–459.

Hu, S. L., Fultz, P. N., McClure, H. M., Eichberg, J. W., Thomas, E. K., Zarling, J., Singhal, M. C., Kosowski, S. G., Swenson, R. B., Anderson, D. C. *et al.* (1987). Effect of immunization with a vaccinia-HIV env recombinant on HIV infection of chimpanzees. *Nature* **328**, 721–723.

Hu, S. L., Polacino, P., Stallard, V., Klaniecki, J., Pennathur, S., Travis, B. M., Misher, L., Kornas, H., Langlois, A. J., Morton, W. R., and Benveniste, R. E.(1996). Recombinant subunit vaccines as an approach to study correlates of protection against primate lentivirus infection. *Immunol. Lett.* **51**, 115–119.

Huang, X., Barchi, J. J., Jr., Lung, F. D., Roller, P. P., Nara, P. L., Muschik, J., and Garrity, R. R. (1997). Glycosylation affects both the three-dimensional structure and antibody binding properties of the HIV-1IIIB GP120 peptide RP135. *Biochemistry* **36**, 10846–10856.

Hulskotte, E. G., Geretti, A. M., and Osterhaus, A. D. (1998). Towards an HIV-1 vaccine: Lessons from studies in macaque models. *Vaccine* **16**, 904–915.

Hwang, S. S., Boyle, T. J., Lyerly, H. K., and Cullen, B. R. (1991). Identification of the envelope V3 loop as the primary determinant of cell tropism in HIV-1. *Science* **253**, 71–74.

Igarashi, T., Ami, Y., Yamamoto, H., Shibata, R., Kuwata, T., Mukai, R., Shinohara, K., Komatsu, T., Adachi, A., and Hayami, M. (1997). Protection of monkeys vaccinated with vpr- and/or nef-defective simian immunodeficiency virus strain mac/human immunodeficiency virus type 1 chimeric viruses: A potential candidate live-attenuated human AIDS vaccine. *J. Gen. Virol.* **78**, 985–989.

Igarashi, T., Endo, Y., Englund, G., Sadjadpour, R., Matano, T., Buckler, C., Buckler-White, A., Plishka, R., Theodore, T., Shibata, R., and Martin, M. (1999). Emergence of a highly pathogenic simian/human immunodeficiency virus in a rhesus macaque treated with anti-CD8 mAb during a primary infection with a nonpathogenic virus. *Proc. Natl. Acad. Sci. USA* **96**, 14049–14054.

Israel, Z. R., Edmonson, P. F., Maul, D. H., O'Neil, S. P., Mossman, S. P., Thiriart, C., Fabry, L., Van Opstal, O., Bruck, C., Bex, F. *et al.* (1994). Incomplete protection, but suppression of virus burden, elicited by subunit simian immunodeficiency virus vaccines. *J. Virol.* **68**, 1843–1853.

Javaherian, K., Langlois, A. J., McDanal, C., Ross, K. L., Eckler, L. I., Jellis, C. L., Profy, A. T., Rusche, J. R., Bolognesi, D. P., Putney, S. D. *et al.* (1989). Principal neutralizing domain of the human immunodeficiency virus type 1 envelope protein. *Proc. Natl. Acad. Sci. USA* **86**, 6768–6772.

Joag, S. V., Li, Z., Foresman, L., Pinson, D. M., Raghavan, R., Zhuge, W., Adany, I., Wang, C., Jia, F., Sheffer, D., Ranchalis, J., Watson, A., and Narayan, O. (1997). Characterization of the pathogenic KU-SHIV model of acquired immunodeficiency syndrome in macaques. *AIDS Res. Hum. Retroviruses* **13**, 635–645.

Joag, S. V., Li, Z., Foresman, L., Stephens, E. B., Zhao, L. J., Adany, I., Pinson, D. M., McClure, H. M., and Narayan, O. (1996). Chimeric simian/human immunodeficiency virus that causes progressive loss of CD4+ T cells and AIDS in pig-tailed macaques. *J. Virol.* **70**, 3189–3197.

Joag, S. V., Liu, Z. Q., Stephens, E. B., Smith, M. S., Kumar, A., Li, Z., Wang, C., Sheffer, D., Jia, F., Foresman, L., Adany, I., Lifson, J., McClure, H. M., and Narayan, O. (1998). Oral immunization of macaques with attenuated vaccine virus induces protection against vaginally transmitted AIDS. *J. Virol.* **72**, 9069–9078.

Johnson, R. P., Lifson, J. D., Czajak, S. C., Cole, K. S., Manson, K. H., Glickman, R., Yang, J., Montefiori, D. C., Montelaro, R., Wyand, M. S., and Desrosiers, R. C. (1999). Highly attenuated vaccine strains of simian immunodeficiency virus protect against vaginal challenge: Inverse relationship of degree of protection with level of attenuation. *J. Virol.* **73**, 4952–4961.

Johnston, M. I. (1997). HIV vaccines: Problems and prospects. *Hosp. Pract.* **32**, 125–128, 131–140.

Katzenstein, D. A., Vujcic, L. K., Latif, A., Boulos, R., Halsey, N. A., Quinn, T. C., Rastogi, S. C., and Quinnan, G. V., Jr. (1990). Human immunodeficiency virus neutralizing antibodies in sera from North Americans and Africans. *J. Acq. Immune Defic. Syndr.* **3**, 810–816.

Keefer, M. C., Graham, B. S., McElrath, M. J., Matthews, T. J., Stablein, D. M., Corey, L., Wright, P. F., Lawrence, D., Fast, P. E., Weinhold, K., Hsieh, R. H., Chernoff, D., Dekker, C., and Dolin, R. (1996). Safety and immunogenicity of Env 2-3, a human immunodeficiency virus type 1 candidate vaccine, in combination with a novel adjuvant, MTP-PE/MF59: NIAID AIDS Vaccine Evaluation Group. *AIDS Res. Hum. Retroviruses* **12**, 683–693.

Keefer, M. C., Wolff, M., Gorse, G. J., Graham, B. S., Corey, L., Clements-Mann, M. L., Verani-Ketter, N., Erb, S., Smith, C. M., Belshe, R. B., Wagner, L. J., McElrath, M. J., Schwartz, D. H., and Fast, P. (1997). Safety profile of phase I and II preventive HIV type 1 envelope vaccination: Experience of the NIAID AIDS Vaccine Evaluation Group. *AIDS Res. Hum. Retroviruses* **13**, 1163–1177.

Kim, J. J., and Weiner, D. B. (1997). DNA gene vaccination for HIV. *Springer Semin. Immunopathol.* **19**, 175–194.

Kirchhoff, F., Greenough, T. C., Brettler, D. B., Sullivan, J. L., and Desrosiers, R. C. (1995). Brief report: Absence of intact nef sequences in a long-term survivor with nonprogressive HIV-1 infection. *N. Engl. J. Med.* **332**, 228–232.

Klein, M. R., van Baalen, C. A., Holwerda, A. M., Kerkhof Garde, S. R., Bende, R. J., Keet, I. P., Eeftinck-Schattenkerk, J. K., Osterhaus, A. D., Schuitemaker, H., and Miedema, F. (1995). Kinetics of Gag-specific cytotoxic T lymphocyte responses during the clinical course of HIV-1 infection: A longitudinal analysis of rapid progressors and long-term asymptomatics. *J. Exp. Med.* **181**, 1365–1372.

Koup, R. A., Safrit, J. T., Cao, Y., Andrews, C. A., McLeod, G., Borkowsky, W., Farthing, C., and Ho, D. D. (1994). Temporal association of cellular immune responses with the initial control of viremia in primary human immunodeficiency virus type 1 syndrome. *J. Virol.* **68**, 4650–4655.

Kwong, P. D., Wyatt, R., Desjardins, E., Robinson, J., Culp, J. S., Hellmig, B. D., Sweet, R. W., Sodroski, J., and Hendrickson, W. A. (1999). Probability analysis of variational crystallization and its application to gp120, the exterior envelope glycoprotein of type 1 human immunodeficiency virus (HIV-1). *J. Biol. Chem.* **274**, 4115–4123.

Kwong, P. D., Wyatt, R., Robinson, J., Sweet, R. W., Sodroski, J., and Hendrickson, W. A. (1998). Structure of an HIV gp120 envelope glycoprotein in complex with the CD4 receptor and a neutralizing human antibody. *Nature* **393**, 648–659.

LaCasse, R. A., Follis, K. E., Trahey, M., Scarborough, J. D., Littman, D. R., and Nunberg, J. H. (1999). Fusion-competent vaccines: Broad neutralization of primary isolates of HIV. *Science* 283, 357–362.

Lamb-Wharton, R. J., Joag, S. V., Stephens, E. B., and Narayan, O. (1997). Primate models of AIDS vaccine development. *AIDS* 11, S121–S126.

Langlois, A. J., Weinhold, K. J., Matthews, T. J., Greenberg, M. L., and Bolognesi, D. P. (1992). Detection of anti-human cell antibodies in sera from macaques immunized with whole inactivated virus. *AIDS Res. Hum. Retroviruses* 8, 1641–1652.

LaRosa, G. J., Davide, J. P., Weinhold, K., Waterbury, J. A., Profy, A. T., Lewis, J. A., Langlois, A. J., Dreesman, G. R., Boswell, R. N., Shadduck, P. *et al.* (1990). Conserved sequence and structural elements in the HIV-1 principal neutralizing determinant. *Science* 249, 932–935.

Le Grand, R., Vogt, G., Vaslin, B., Roques, P., Theodoro, F., Aubertin, A. M., and Dormont, D. (1992). Specific and non-specific immunity and protection of macaques against SIV infection. *Vaccine* 10, 873–879.

Lee, M. K., Heaton, J., and Cho, M. W. (1999). Identification of determinants of interaction between CXCR4 and gp120 of a dual-tropic HIV-1DH12 isolate. *Virology* 257, 290–296.

Lee, W. R., Syu, W. J., Du, B., Matsuda, M., Tan, S., Wolf, A., Essex, M., and Lee, T. H. (1992). Nonrandom distribution of gp120 N-linked glycosylation sites important for infectivity of human immunodeficiency virus type 1. *Proc. Natl. Acad. Sci. USA* 89, 2213–2217.

Lehner, T., Wang, Y., Cranage, M., Bergmeier, L. A., Mitchell, E., Tao, L., Hall, G., Dennis, M., Cook, N., Brookes, R., Klavinskis, L., Jones, I., Doyle, C., and Ward, R. (1996). Protective mucosal immunity elicited by targeted iliac lymph node immunization with a subunit SIV envelope and core vaccine in macaques. *Nat. Med.* 2, 767–775.

Leitner, T., Korber, B., Robertson, D., Gao, F., and Hahn, B. (1997). Updated proposal of reference sequences of HIV-1 genetic subtypes. *In* "Human Retroviruses and AIDS" (B. Kober, B. Hahn, B. Foley, J. W. Mellors, T. Leitner, G. Myers, F. McCutchan, and C. Kuiken, Eds.), pp. III19–III24. Los Alamos National Laboratory, Los Alamos.

Lekutis, C., Shiver, J. W., Liu, M. A., and Letvin, N. L. (1997). HIV-1 env DNA vaccine administered to rhesus monkeys elicits MHC class II-restricted CD4+ T helper cells that secrete IFN-gamma and TNF-alpha. *J. Immunol.* 158, 4471–4477.

Leonard, C. K., Spellman, M. W., Riddle, L., Harris, R. J., Thomas, J. N., and Gregory, T. J. (1990). Assignment of intrachain disulfide bonds and characterization of potential glycosylation sites of the type 1 recombinant human immunodeficiency virus envelope glycoprotein (gp120) expressed in Chinese hamster ovary cells. *J. Biol. Chem.* 265, 10373–10382.

Letvin, N. L. (1998). Progress in the development of an HIV-1 vaccine. *Science* 280, 1875–1880.

Letvin, N. L., Li, J., Halloran, M., Cranage, M. P., Rud, E. W., and Sodroski, J. (1995). Prior infection with a nonpathogenic chimeric simian-human immunodeficiency virus does not efficiently protect macaques against challenge with simian immunodeficiency virus. *J. Virol.* 69, 4569–4571.

Letvin, N. L., Montefiori, D. C., Yasutomi, Y., Perry, H. C., Davies, M. E., Lekutis, C., Alroy, M., Freed, D. C., Lord, C. I., Handt, L. K., Liu, M. A., and Shiver, J. W. (1997). Potent, protective anti-HIV immune responses generated by bimodal HIV envelope DNA plus protein vaccination. *Proc. Natl. Acad. Sci. USA* 94, 9378–9383.

Levitsky, H. I. (1997). Accessories for naked DNA vaccines. *Nat. Biotechnol.* 15, 619–620.

Li, J. T., Halloran, M., Lord, C. I., Watson, A., Ranchalis, J., Fung, M., Letvin, N. L., and Sodroski, J. G. (1995). Persistent infection of macaques with simian–human immunodeficiency viruses. *J. Virol.* 69, 7061–7067.

Li, Y., Luo, L., Rasool, N., and Kang, C. Y. (1993). Glycosylation is necessary for the correct folding of human immunodeficiency virus gp120 in CD4 binding. *J. Virol.* 67, 584–588.

Liedtke, S., Adamski, M., Geyer, R., Pfutzner, A., Rubsamen-Waigmann, H., and Geyer, H. (1994). Oligosaccharide profiles of HIV-2 external envelope glycoprotein: Dependence on host cells and virus isolates. *Glycobiology* 4, 477–484.

Liedtke, S., Geyer, R., and Geyer, H. (1997). Host-cell-specific glycosylation of HIV-2 envelope glycoprotein. *Glycoconj. J.* **14**, 785–793.

Littman, D. R. (1998). Chemokine receptors: Keys to AIDS pathogenesis? *Cell* **93**, 677–680.

Lohman, B. L., McChesney, M. B., Miller, C. J., McGowan, E., Joye, S. M., Van Rompay, K. K., Reay, E., Antipa, L., Pedersen, N. C., and Marthas, M. L. (1994). A partially attenuated simian immunodeficiency virus induces host immunity that correlates with resistance to pathogenic virus challenge. *J. Virol.* **68**, 7021–7029.

Lu, S. (1998). Developing DNA vaccines against immunodeficiency viruses. *Curr. Top. Microbiol. Immunol.* **226**, 161–173.

Lu, X., Kiyono, H., Lu, D., Kawabata, S., Torten, J., Srinivasan, S., Dailey, P. J., McGhee, J. R., Lehner, T., and Miller, C. J. (1998). Targeted lymph-node immunization with whole inactivated simian immunodeficiency virus (SIV) or envelope and core subunit antigen vaccines does not reliably protect rhesus macaques from vaginal challenge with SIV-mac251. *AIDS* **12**, 1–10.

Lu, Y., Salvato, M. S., Pauza, C. D., Li, J., Sodroski, J., Manson, K., Wyand, M., Letvin, N., Jenkins, S., Touzjian, N., Chutkowski, C., Kushner, N., LeFaile, M., Payne, L. G., and Roberts, B. (1996). Utility of SHIV for testing HIV-1 vaccine candidates in macaques. *J. Acq. Immune Defic. Syndr. Hum. Retrovirol.* **12**, 99–106.

Lubeck, M. D., Natuk, R., Myagkikh, M., Kalyan, N., Aldrich, K., Sinangil, F., Alipanah, S., Murthy, S. C., Chanda, P. K., Nigida, S. M., Jr., Markham, P. D., Zolla-Pazner, S., Steimer, K., Wade, M., Reitz, M. S., Jr., Arthur, L. O., Mizutani, S., Davis, A., Hung, P. P., Gallo, R. C., Eichberg, J., and Robert-Guroff, M. (1997). Long-term protection of chimpanzees against high-dose HIV-1 challenge induced by immunization. *Nat. Med.* **3**, 651–658.

Luciw, P. A. (1996). Human immunodeficiency viruses and their replication. *In* "Fields Virology" (B. N. Fields, D. M. Knipe, and P. M. Howley, Eds.), 3rd ed., Vol. 2, pp. 1881–1952. Lippincott–Raven, Philadelphia.

MacGregor, R. R., Boyer, J. D., Ugen, K. E., Lacy, K. E., Gluckman, S. J., Bagarazzi, M. L., Chattergoon, M. A., Baine, Y., Higgins, T. J., Ciccarelli, R. B., Coney, L. R., Ginsberg, R. S., and Weiner, D. B. (1998). First human trial of a DNA-based vaccine for treatment of human immunodeficiency virus type 1 infection: Safety and host response. *J. Infect. Dis.* **178**, 92–100.

Malashkevich, V. N., Chan, D. C., Chutkowski, C. T., and Kim, P. S. (1998). Crystal structure of the simian immunodeficiency virus (SIV) gp41 core: Conserved helical interactions underlie the broad inhibitory activity of gp41 peptides. *Proc. Natl. Acad. Sci. USA* **95**, 9134–9139.

Marthas, M. L., Sutjipto, S., Higgins, J., Lohman, B., Torten, J., Luciw, P. A., Marx, P. A., and Pedersen, N. C. (1990). Immunization with a live, attenuated simian immunodeficiency virus (SIV) prevents early disease but not infection in rhesus macaques challenged with pathogenic SIV. *J. Virol.* **64**, 3694–3700.

Mascola, J. R., Lewis, M. G., Stiegler, G., Harris, D., VanCott, T. C., Hayes, D., Louder, M. K., Brown, C. R., Sapan, C. V., Frankel, S. S., Lu, Y., Robb, M. L., Katinger, H., and Birx, D. L. (1999). Protection of Macaques against pathogenic simian/human immunodeficiency virus 89.6PD by passive transfer of neutralizing antibodies. *J. Virol.* **73**, 4009–4018.

Mascola, J. R., Snyder, S. W., Weislow, O. S., Belay, S. M., Belshe, R. B., Schwartz, D. H., Clements, M. L., Dolin, R., Graham, B. S., Gorse, G. J., Keefer, M. C., McElrath, M. J., Walker, M. C., Wagner, K. F., McNeil, J. G., McCutchan, F. E., and Burke, D. S. (1996). Immunization with envelope subunit vaccine products elicits neutralizing antibodies against laboratory-adapted but not primary isolates of human immunodeficiency virus type 1: The National Institute of Allergy and Infectious Diseases AIDS Vaccine Evaluation Group. *J. Infect. Dis.* **173**, 340–348.

Matthews, T. J. (1994). Dilemma of neutralization resistance of HIV-1 field isolates and vaccine development. *AIDS Res. Hum. Retroviruses* **10**, 631–632.

McElrath, M. J., Corey, L., Berger, D., Hoffman, M. C., Klucking, S., Dragavon, J., Peterson, E., and Greenberg, P. D. (1994). Immune responses elicited by recombinant vaccinia-human immunodeficiency virus (HIV) envelope and HIV envelope protein: Analysis of the durability of responses and effect of repeated boosting. *J. Infect. Dis.* **169**, 41–47.

McElrath, M. J., Corey, L., Greenberg, P. D., Matthews, T. J., Montefiori, D. C., Rowen, L., Hood, L., and Mullins, J. I. (1996). Human immunodeficiency virus type 1 infection despite prior immunization with a recombinant envelope vaccine regimen. *Proc. Natl. Acad. Sci. USA* **93**, 3972–3977.

McElrath, M. J., Siliciano, R. F., and Weinhold, K. J. (1997). HIV type 1 vaccine-induced cytotoxic T cell responses in phase I clinical trials: Detection, characterization, and quantitation. *AIDS Res. Hum. Retroviruses* **13**, 211–216.

Mills, K. H., Page, M., Chan, W. L., Kitchin, P., Stott, E. J., Taffs, F., Jones, W., Rose, J., Ling, C., Silvera, P. *et al.* (1992). Protection against SIV infection in macaques by immunization with inactivated virus from the BK28 molecular clone, but not with BK28-derived recombinant env and gag proteins. *J. Med. Primatol.* **21**, 50–58.

Mo, H., Stamatatos, L., Ip, J. E., Barbas, C. F., Parren, P. W., Burton, D. R., Moore, J. P., and Ho, D. D. (1997). Human immunodeficiency virus type 1 mutants that escape neutralization by human monoclonal antibody IgG1b12. *J. Virol.* **71**, 6869–6874.

Montefiori, D. C., Graham, B. S., Kliks, S., and Wright, P. F. (1992). Serum antibodies to HIV-1 in recombinant vaccinia virus recipients boosted with purified recombinant gp160: NIAID AIDS Vaccine Clinical Trials Network. *J. Clin. Immunol.* **12**, 429–439.

Moore, J. P., and Burton, D. R. (1999). HIV-1 neutralizing antibodies: How full is the bottle? *Nat. Med.* **5**, 142–144.

Moore, J. P., and Ho, D. D. (1995). HIV-1 neutralization: the consequences of viral adaptation to growth on transformed T cells. *AIDS* **9**, S117–S136.

Moore, J. P., Trkola, A., and Dragic, T. (1997). Co-receptors for HIV-1 entry. *Curr. Opin. Immunol.* **9**, 551–562.

Mossman, S. P., Bex, F., Berglund, P., Arthos, J., O'Neil, S. P., Riley, D., Maul, D. H., Bruck, C., Momin, P., Burny, A., Fultz, P. N., Mullins, J. I., Liljestrom, P., and Hoover, E. A. (1996). Protection against lethal simian immunodeficiency virus SIVsmmPBj14 disease by a recombinant Semliki Forest virus gp160 vaccine and by a gp120 subunit vaccine. *J. Virol.* **70**, 1953–1960.

Murphey-Corb, M., Martin, L. N., Davison-Fairburn, B., Montelaro, R. C., Miller, M., West, M., Ohkawa, S., Baskin, G. B., Zhang, J. Y., Putney, S. D. *et al.* (1989). A formalin-inactivated whole SIV vaccine confers protection in macaques. *Science* **246**, 1293–1297.

Murphey-Corb, M., Montelaro, R. C., Miller, M. A., West, M., Martin, L. N., Davison-Fairburn, B., Ohkawa, S., Baskin, G. B., Zhang, J. Y., Miller, G. B. *et al.* (1991). Efficacy of SIV/deltaB670 glycoprotein-enriched and glycoprotein-depleted subunit vaccines in protecting against infection and disease in rhesus monkeys. *AIDS* **5**, 655–662.

Murthy, K. K., Cobb, E. K., Rouse, S. R., McClure, H. M., Payne, J. S., Salas, M. T., and Michalek, G. R. (1998). Active and passive immunization against HIV type 1 infection in chimpanzees. *AIDS Res. Hum. Retroviruses* **14**(Suppl. 3.), S271–S276.

Myers, G., and Lenroot, R. (1992). HIV glycosylation: What does it portend? *AIDS Res. Hum. Retroviruses* **8**, 1459–1460.

Nakayama, E. E., Shioda, T., Tatsumi, M., Xin, X., Yu, D., Ohgimoto, S., Kato, A., Sakai, Y., Ohnishi, Y., and Nagai, Y. (1998). Importance of the N-glycan in the V3 loop of HIV-1 envelope protein for CXCR-4- but not CCR-5-dependent fusion. *FEBS Lett.* **426**, 367–372.

Nara, P. L., Robey, W. G., Pyle, S. W., Hatch, W. C., Dunlop, N. M., Bess, J. W., Jr., Kelliher, J. C., Arthur, L. O., and Fischinger, P. J. (1988). Purified envelope glycoproteins from human immunodeficiency virus type 1 variants induce individual, type-specific neutralizing antibodies. *J. Virol.* **62**, 2622–2628.

Novembre, F. J., Saucier, M., Anderson, D. C., Klumpp, S. A., O'Neil, S. P., Brown, C. R., 2nd, Hart, C. E., Guenthner, P. C., Swenson, R. B., and McClure, H. M. (1997). Development of AIDS in a chimpanzee infected with human immunodeficiency virus type 1. *J. Virol.* **71**, 4086–4091.

O'Brien, W. A., Koyanagi, Y., Namazie, A., Zhao, J. Q., Diagne, A., Idler, K., Zack, J. A., and Chen, I. S. (1990). HIV-1 tropism for mononuclear phagocytes can be determined by regions of gp120 outside the CD4-binding domain. *Nature* **348**, 69–73.

Ogert, R., Lee, M. K., Ross, W., White-Buckler, A., Martin, M. A., and Cho, M. W. (manuscript in preparation).

Ohgimoto, S., Shioda, T., Mori, K., Nakayama, E. E., Hu, H., and Nagai, Y. (1998). Location-specific, unequal contribution of the N glycans in simian immunodeficiency virus gp120 to viral infectivity and removal of multiple glycans without disturbing infectivity. *J. Virol.* **72**, 8365–8370.

Pal, R., di Marzo Veronese, F., Nair, B. C., Rittenhouse, S., Hoke, G., Mumbauer, S., and Sarngadharan, M. G. (1993). Glycoprotein of human immunodeficiency virus type 1 synthesized in chronically infected Molt3 cells acquires heterogeneous oligosaccharide structures. *Biochem. Biophys. Res. Commun.* **196**, 1335–1342.

Palker, T. J., Clark, M. E., Langlois, A. J., Matthews, T. J., Weinhold, K. J., Randall, R. R., Bolognesi, D. P., and Haynes, B. F. (1988). Type-specific neutralization of the human immunodeficiency virus with antibodies to env-encoded synthetic peptides. *Proc. Natl. Acad. Sci. USA* **85**, 1932–1936.

Papandreou, M. J., and Fenouillet, E. (1998). Effect of changes in the glycosylation of the human immunodeficiency virus type 1 envelope on the immunoreactivity and sensitivity to thrombin of its third variable domain. *Virology* **241**, 163–167.

Parren, P. W., Gauduin, M. C., Koup, R. A., Poignard, P., Fisicaro, P., Burton, D. R., and Sattentau, Q. J. (1997). Relevance of the antibody response against human immunodeficiency virus type 1 envelope to vaccine design. *Immunol. Lett.* **57**, 105–112.

Parren, P. W., Mondor, I., Naniche, D., Ditzel, H. J., Klasse, P. J., Burton, D. R., and Sattentau, Q. J. (1998). Neutralization of human immunodeficiency virus type 1 by antibody to gp120 is determined primarily by occupancy of sites on the virion irrespective of epitope specificity. *J. Virol.* **72**, 3512–3519.

Pialoux, G., Excler, J. L., Riviere, Y., Gonzalez-Canali, G., Feuillie, V., Coulaud, P., Gluckman, J. C., Matthews, T. J., Meignier, B., Kieny, M. P. *et al.* (1995). A prime-boost approach to HIV preventive vaccine using a recombinant canarypox virus expressing glycoprotein 160 (MN) followed by a recombinant glycoprotein 160 (MN/LAI). *AIDS Res. Hum. Retroviruses* **11**, 373–381.

Pinter, A., Honnen, W. J., Kayman, S. C., Trochev, O., and Wu, Z. (1998). Potent neutralization of primary HIV-1 isolates by antibodies directed against epitopes present in the V1/V2 domain of HIV-1 gp120. *Vaccine* **16**, 1803–1811.

Printer, A., Honnen, W. J., Tilley, S. A., Bona, C., Zaghouani, H., Gorny, M. K., and Zolla-Pazner, S. (1989). Oligomeric structure of gp41, the transmembrane protein of human immunodeficiency virus type 1. *J. Virol.* **63**, 2674–2679.

Pinto, L. A., Sullivan, J., Berzofsky, J. A., Clerici, M., Kessler, H. A., Landay, A. L., and Shearer, G. M. (1995). ENV-specific cytotoxic T lymphocyte responses in HIV seronegative health care workers occupationally exposed to HIV-contaminated body fluids. *J. Clin. Invest.* **96**, 867–876.

Pleskoff, O., Sol, N., Labrosse, B., and Alizon, M. (1997). Human immunodeficiency virus strains differ in their ability to infect CD4+ cells expressing the rat homolog of CXCR-4 (fusin). *J. Virol.* **71**, 3259–3262.

Poignard, P., Klasse, P. J., and Sattentau, Q. J. (1996). Antibody neutralization of HIV-1. *Immunol. Today* **17**, 239–246.

Polacino, P., Stallard, V., Montefiori, D. C., Brown, C. R., Richardson, B. A., Morton, W. R., Benveniste, R. E., and Hu, S. L. (1999). Protection of macaques against intrarectal infection by a combination immunization regimen with recombinant simian immunodeficiency virus SIVmne gp160 vaccines. *J. Virol.* **73**, 3134–3146.

Prince, A. M., Reesink, H., Pascual, D., Horowitz, B., Hewlett, I., Murthy, K. K., Cobb, K. E., and Eichberg, J. W. (1991). Prevention of HIV infection by passive immunization with HIV immunoglobulin. *AIDS Res. Hum. Retroviruses* **7**, 971–973.

Putkonen, P., Nilsson, C., Walther, L., Ghavamzadeh, L., Hild, K., Broliden, K., Biberfeld, G., and Thorstensson, R. (1994). Efficacy of inactivated whole HIV-2 vaccines with various adjuvants in cynomolgus monkeys. *J. Med. Primatol.* **23**, 89–94.

Putkonen, P., Thorstensson, R., Albert, J., Hild, K., Norrby, E., Biberfeld, P., and Biberfeld, G. (1990). Infection of cynomolgus monkeys with HIV-2 protects against pathogenic consequences of a subsequent simian immunodeficiency virus infection. *AIDS* **4**, 783–789.

Putkonen, P., Thorstensson, R., Walther, L., Albert, J., Akerblom, L., Granquist, O., Wadell, G., Norrby, E., and Biberfeld, G. (1991). Vaccine protection against HIV-2 infection in cynomolgus monkeys. *AIDS Res. Hum. Retroviruses* **7**, 271–277.

Putney, S. D., Rusche, J., Javaherian, K., Matthews, T., and Bolognesi, D. (1990). Structural and functional features of the HIV envelope glycoprotein and considerations for vaccine development. *Biotechnology* **14**, 81–110.

Quesada-Rolander, M., Makitalo, B., Thorstensson, R., Zhang, Y. J., Castanos-Velez, E., Biberfeld, G., and Putkonen, P. (1996). Protection against mucosal SIVsm challenge in macaques infected with a chimeric SIV that expresses HIV type 1 envelope. *AIDS Res. Hum. Retroviruses* **12**, 993–999.

Reimann, K. A., Li, J. T., Veazey, R., Halloran, M., Park, I. W., Karlsson, G. B., Sodroski, J., and Letvin, N. L. (1996). A chimeric simian/human immunodeficiency virus expressing a primary patient human immunodeficiency virus type 1 isolate env causes an AIDS-like disease after *in vivo* passage in rhesus monkeys. *J. Virol.* **70**, 6922–6928.

Reitter, J. N., and Desrosiers, R. C. (1998). Identification of replication-competent strains of simian immunodeficiency virus lacking multiple attachment sites for N-linked carbohydrates in variable regions 1 and 2 of the surface envelope protein. *J. Virol.* **72**, 5399–5407.

Reitter, J. N., Means, R. E., and Desrosiers, R. C. (1998). A role for carbohydrates in immune evasion in AIDS. *Nat. Med.* **4**, 679–684.

Richardson, T., Jr., Stryjewski, B. L., Broder, C. C., Hoxie, J. A., Mascola, J. R., Earl, P. L., and Doms, R. W. (1996). Humoral response to oligomeric human immunodeficiency virus type 1 envelope protein. *J. Virol.* **70**, 753–762.

Rinaldo, C., Huang, X. L., Fan, Z. F., Ding, M., Beltz, L., Logar, A., Panicali, D., Mazzara, G., Liebmann, J., Cottrill, M. *et al.* (1995). High levels of anti-human immunodeficiency virus type 1 (HIV-1) memory cytotoxic T-lymphocyte activity and low viral load are associated with lack of disease in HIV-1-infected long-term nonprogressors. *J. Virol.* **69**, 5838–5842.

Rizzuto, C. D., Wyatt, R., Hernandez-Ramos, N., Sun, Y., Kwong, P. D., Hendrickson, W. A., and Sodroski, J. (1998). A conserved HIV gp120 glycoprotein structure involved in chemokine receptor binding. *Science* **280**, 1949–1953.

Roben, P., Moore, J. P., Thali, M., Sodroski, J., Barbas, C. F., 3rd, and Burton, D. R. (1994). Recognition properties of a panel of human recombinant Fab fragments to the CD4 binding site of gp120 that show differing abilities to neutralize human immunodeficiency virus type 1. *J. Virol.* **68**, 4821–4828.

Robinson, H. L. (1997). DNA vaccines for immunodeficiency viruses. *AIDS* **11**, S109–S119.

Ross, T. M., and Cullen, B. R. (1998). The ability of HIV type 1 to use CCR-3 as a coreceptor is controlled by envelope V1/V2 sequences acting in conjunction with a CCR-5 tropic V3 loop. *Proc. Natl. Acad. Sci. USA* **95**, 7682–7686.

Rossio, J. L., Esser, M. T., Suryanarayana, K., Schneider, D. K., Bess, J. W., Jr., Vasquez, G. M., Wiltrout, T. A., Chertova, E., Grimes, M. K., Sattentau, Q., Arthur, L. O.,

Henderson, L. E., and Lifson, J. D. (1998). Inactivation of human immunodeficiency virus type 1 infectivity with preservation of conformational and functional integrity of virion surface proteins. *J. Virol.* **72,** 7992–8001.

Rowland-Jones, S., Dong, T., Krausa, P., Sutton, J., Newell, H., Ariyoshi, K., Gotch, F., Sabally, S., Corrah, T., Kimani, J., MacDonald, K., Plummer, F., Ndinya-Achola, J., Whittle, H., and McMichael, A. (1998a). The role of cytotoxic T-cells in HIV infection. *Dev. Biol. Stand.* **92,** 209–214.

Rowland-Jones, S., Sutton, J., Ariyoshi, K., Dong, T., Gotch, F., McAdam, S., Whitby, D., Sabally, S., Gallimore, A., Corrah, T. *et al.* (1995). HIV-specific cytotoxic T-cells in HIV-exposed but uninfected Gambian women *Nat. Med.* **1,** 59–64.

Rowland-Jones, S. L., Dong, T., Dorrell, L., Ogg, G., Hansasuta, P., Krausa, P., Kimani, J., Sabally, S., Ariyoshi, K., Oyugi, J., MacDonald, K. S., Bwayo, J., Whittle, H., Plummer, F. A., and McMichael, A. J. (1999). Broadly cross-reactive HIV-specific cytotoxic T-lymphocytes in highly-exposed persistently seronegative donors. *Immunol. Lett.* **66,** 9–14.

Rowland-Jones, S. L., Dong, T., Fowke, K. R., Kimani, J., Krausa, P., Newell, H., Blanchard, T., Ariyoshi, K., Oyugi, J., Ngugi, E., Bwayo, J., MacDonald, K. S., McMichael, A. J., and Plummer, F. A. (1998b). Cytotoxic T cell responses to multiple conserved HIV epitopes in HIV-resistant prostitutes in Nairobi. *J. Clin. Invest.* **102,** 1758-1765.

Ruprecht, R. M., W., B. T., Li, A., Ayehunie, S., Hu, Y., Liska, V., Rasmussen, R., and Sharma, P. L. (1996). Live attenuate HIV as a vaccine for AIDS: Pros and cons. *Semin. Virol.* **7,** 147–155.

Rusche, J. R., Javaherian, K., McDanal, C., Petro, J., Lynn, D. L., Grimaila, R., Langlois, A., Gallo, R. C., Arthur, L. O., Fischinger, P. J. *et al.* (1988). Antibodies that inhibit fusion of human immunodeficiency virus-infected cells bind a 24-amino acid sequence of the viral envelope, gp120. *Proc. Natl. Acad. Sci. USA* **85,** 3198–3202.

Salmon-Ceron, D., Excler, J. L., Sicard, D., Blanche, P., Finkielstzjen, L., Gluckman, J. C., Autran, B., Matthews, T. J., Meignier, B., Kieny, M. P. *et al.* (1995). Safety and immunogenicity of a recombinant HIV type 1 glycoprotein 160 boosted by a V3 synthetic peptide in HIV-negative volunteers. *AIDS Res. Hum. Retroviruses* **11,** 1479–1486.

Sawyer, L. S., Wrin, M. T., Crawford-Miksza, L., Potts, B., Wu, Y., Weber, P. A., Alfonso, R. D., and Hanson, C. V. (1994). Neutralization sensitivity of human immunodeficiency virus type 1 is determined in part by the cell in which the virus is propagated. *J. Virol.* **68,** 1342–1349.

Schawaller, M., Smith, G. E., Skehel, J. J., and Wiley, D. C. (1989). Studies with crosslinking reagents on the oligomeric structure of the env glycoprotein of HIV. *Virology* **172,** 367–369.

Schonning, K., Jansson, B., Olofsson, S., Nielsen, J. O., and Hansen, J. S. (1996). Resistance to V3-directed neutralization caused by an N-linked oligosaccharide depends on the quaternary structure of the HIV-1 envelope oligomer. *Virology* **218,** 134–140.

Schultz, A. (1998). Encouraging vaccine results from primate models of HIV type 1 infection. *AIDS Res. Hum. Retroviruses* **14**(Suppl. 3), S261–S263.

Schultz, A. M., and Stott, E. J. (1994). Primate models for AIDS vaccines. *AIDS* **8,** S203–S212.

Shibata, R., Igarashi, T., Haigwood, N., Buckler-White, A., Ogert, R., Ross, W., Willey, R., Cho, M. W., and Martin, M. A. (1999). Neutralizing antibody directed against the HIV-1 envelope glycoprotein can completely block HIV-1/SIV chimeric virus infections of macaque monkeys. *Nat. Med.* **5,** 204–210.

Shibata, R., Kawamura, M., Sakai, H., Hayami, M., Ishimoto, A., and Adachi, A. (1991). Generation of a chimeric human and simian immunodeficiency virus infectious to monkey peripheral blood mononuclear cells. *J. Virol.* **65,** 3514–3520.

Shibata, R., Maldarelli, F., Siemon, C., Matano, T., Parta, M., Miller, G., Fredrickson, T., and Martin, M. A. (1997a). Infection and pathogenicity of chimeric simian-human immunodeficiency viruses in macaques: Determinants of high virus loads and CD4 cell killing. *J. Infect. Dis.* **176,** 362–373.

Shibata, R., Siemon, C., Czajak, S. C., Desrosiers, R. C., and Martin, M. A. (1997b). Live, attenuated simian immunodeficiency virus vaccines elicit potent resistance against a challenge with a human immunodeficiency virus type 1 chimeric virus. *J. Virol.* **71**, 8141–8148.

Shioda, T., Levy, J. A., and Cheng-Mayer, C. (1992). Small amino acid changes in the V3 hypervariable region of gp120 can affect the T-cell-line and macrophage tropism of human immunodeficiency virus type 1. *Proc. Natl. Acad. Sci. USA* **89**, 9434–9438.

Shiver, J. W., Davies, M. E., Yasutomi, Y., Perry, H. C., Freed, D. C., Letvin, N. L., and Liu, M. A. (1997). Anti-HIV env immunities elicited by nucleic acid vaccines. *Vaccine* **15**, 884–887.

Smyth, R. J., Yi, Y., Singh, A., and Collman, R. G. (1998). Determinants of entry cofactor utilization and tropism in a dualtropic human immunodeficiency virus type 1 primary isolate. *J. Virol.* **72**, 4478–4484.

Speck, R. F., Wehrly, K., Platt, E. J., Atchison, R. E., Charo, I. F., Kabat, D., Chesebro, B., and Goldsmith, M. A. (1997). Selective employment of chemokine receptors as human immunodeficiency virus type 1 coreceptors determined by individual amino acids within the envelope V3 loop. *J. Virol.* **71**, 7136–7139.

Steimer, K. S., and Haigwood, N. L. (1991). Importance of conformation on the neutralizing antibody response to HIV- 1 gp120. *Biotechnol. Ther.* **2**, 63–89.

Steimer, K. S., Scandella, C. J., Skiles, P. V., and Haigwood, N. L. (1991). Neutralization of divergent HIV-1 isolates by conformation-dependent human antibodies to Gp120. *Science* **254**, 105–108.

Steinman, R. M., and Germain, R. N. (1998). Antigen presentation and related immunological aspects of HIV-1 vaccines. *AIDS* **12**, S97–S112.

Stott, E. J. (1991). Anti-cell antibody in macaques. *Nature* **353**, 393.

Stott, E. J., and Schild, G. C. (1996). Strategies for AIDS vaccines. *J. Antimicrob. Chemother.* **37**(Suppl. B), 185–198.

Stott, E. J., Chan, W. L., Mills, K. H., Page, M., Taffs, F., Cranage, M., Greenaway, P., and Kitchin, P. (1990). Preliminary report: Protection of cynomolgus macaques against simian immunodeficiency virus by fixed infected-cell vaccine. *Lancet* **336**, 1538–1541.

Stott, J., Hu, S. L., and Almond, N. (1998). Candidate vaccines protect macaques against primate immunodeficiency viruses. *AIDS Res. Hum. Retroviruses* **14**(Suppl. 3), S265–S270.

Suhrbier, A. (1997). Multi-epitope DNA vaccines. *Immunol. Cell. Biol.* **75**, 402–408.

Sullivan, N., Sun, Y., Sattentau, Q., Thali, M., Wu, D., Denisova, G., Gershoni, J., Robinson, J., Moore, J., and Sodroski, J. (1998). CD4-Induced conformational changes in the human immunodeficiency virus type 1 gp120 glycoprotein: Consequences for virus entry and neutralization. *J. Virol.* **72**, 4694–4703.

Sutjipto, S., Kodama, T., Yee, J., Geetie, A., Jennings, M., Desrosiers, R. C., and Marx, P. A. (1990). Characterization of monoclonal antibodies that distinguish simian immunodeficiency virus isolates from each other and from human immunodeficiency virus types 1 and 2. *J. Gen. Virol.* **71**, 247–249.

Tan, K., Liu, J., Wang, J., Shen, S., and Lu, M. (1997). Atomic structure of a thermostable subdomain of HIV-1 gp41. *Proc. Natl. Acad. Sci. USA* **94**, 12303–12308.

Tartaglia, J., Excler, J. L., El Habib, R., Limbach, K., Meignier, B., Plotkin, S., and Klein, M. (1998). Canarypox virus-based vaccines: Prime-boost strategies to induce cell-mediated and humoral immunity against HIV. *AIDS Res. Hum. Retroviruses* **14**(Suppl. 3), S291–S298.

Thali, M., Moore, J. P., Furman, C., Charles, M., Ho, D. D., Robinson, J., and Sodroski, J. (1993). Characterization of conserved human immunodeficiency virus type 1 gp120 neutralization epitopes exposed upon gp120-CD4 binding. *J. Virol.* **67**, 3978–3988.

Trkola, A., Dragic, T., Arthos, J., Binley, J. M., Olson, W. C., Allaway, G. P., Cheng-Mayer, C., Robinson, J., Maddon, P. J., and Moore, J. P. (1996a). CD4-dependent, antibody-sensitive interactions between HIV- 1 and its co-receptor CCR-5. *Nature* **384**, 184–187.

Trkola, A., Pomales, A. B., Yuan, H., Korber, B., Maddon, P. J., Allaway, G. P., Katinger, H., Barbas, C. F., 3rd, Burton, D. R., Ho, D. D. *et al.* (1995). Cross-clade neutralization of primary isolates of human immunodeficiency virus type 1 by human monoclonal antibodies and tetrameric CD4-IgG. *J. Virol.* **69**, 6609–6617.

Trkola, A., Purtscher, M., Muster, T., Ballaun, C., Buchacher, A., Sullivan, N., Srinivasan, K., Sodroski, J., Moore, J. P., and Katinger, H. (1996b). Human monoclonal antibody 2G12 defines a distinctive neutralization epitope on the gp120 glycoprotein of human immunodeficiency virus type 1. *J. Virol.* **70**, 1100–1108.

Turner, B. G., and Summers, M. F. (1999). Structural biology of HIV. *J. Mol. Biol.* **285**, 1–32.

Ugen, K. E., Nyland, S. B., Boyer, J. D., Vidal, C., Lera, L., Rasheid, S., Chattergoon, M., Bagarazzi, M. L., Ciccarelli, R., Higgins, T., Baine, Y., Ginsberg, R., Macgregor, R. R., and Weiner, D. B. (1998). DNA vaccination with HIV-1 expressing constructs elicits immune responses in humans. *Vaccine* **16**, 1818–1821.

VanCott, T. C., Bethke, F. R., Burke, D. S., Redfield, R. R., and Birx, D. L. (1995). Lack of induction of antibodies specific for conserved, discontinuous epitopes of HIV-1 envelope glycoprotein by candidate AIDS vaccines. *J. Immunol.* **155**, 4100–4110.

Verani, P., Titti, F., Corrias, F., Belli, R., Di Fabio, S., Geraci, A., Konga-Mogtomo, M., Maggiorella, M. T., and Sernicola, L. (1997). Vaccines against HIV. *J. Biol. Regul. Homeost. Agents* **11**, 82–87.

Vujcic. L. K., and Quinnan, G. V., Jr. (1995). Preparation and characterization of human HIV type 1 neutralizing reference sera. *AIDS Res. Hum. Retroviruses* **11**, 783–787.

Wagner. R., Deml, L., Teeuwsen, V., Heeney, J., Yiming, S., and Wolf, H. (1996). A recombinant HIV-1 virus-like particle vaccine: From concepts to a field study. *Antibiot. Chemother.* **48**, 68–83.

Wahren, B., and Brytting, M. (1997). DNA increases the potency of vaccination against infectious diseases. *Curr. Opin. Chem. Biol.* **1**, 183–189.

Walker, B. D., Kowalski, M., Goh, W. C., Kozarsky, K., Krieger, M., Rosen, C., Rohrschneider, L., Haseltine, W. A., and Sodroski, J.(1987). Inhibition of human immunodeficiency virus syncytium formation and virus replication by castanospermine. *Proc. Natl. Acad. Sci. USA* **84**, 8120–8124.

Weiss, C. D., Levy, J. A., and White, J. M. (1990). Oligomeric organization of gp120 on infectious human immunodeficiency virus type 1 paticles. *J. Virol.* **64**, 5674–5677.

Weiss, R. A., Clapham, P. R., Weber, J. N., Dalgleish, A. G., Lasky, L. A., and Berman, P. W. (1986). Variable and conserved neutralization antigens of human immunodeficiency virus. *Nature* **324**, 572–575.

Weissenhorn, W., Dessen, A., Harrison, S. C., Skehel, J. J., and Wiley, D. C. (1997). Atomic structure of the ectodomain from HIV-1 gp41. *Nature* **387**, 426–430.

Wells, T. N., Proudfoot, A. E., Power, C. A., and Marsh, M. (1996). Chemokine receptors—The new frontier for AIDS research. *Chem. Biol.* **3**, 603–609.

Westervelt, P., Gendelman, H. E., and Ratner, L. (1991). Identification of a determinant within the human immunodeficiency virus 1 surface envelope glycoprotein critical for productive infection of primary monocytes. *Proc. Natl. Acad. Sci. USA* **88**, 3097–3101.

Willey, R. L., Shibata, R., Freed, E. O., Cho, M. W., and Martin, M. A. (1996). Differential glycosylation, virion incorporation, and sensitivity to neutralizing antibodies of human immunodeficiency virus type 1 envelope produced from infected primary T-lymphocyte and macrophage cultures. *J. Virol.* **70**, 6431–6436.

Willey, R. L., Smith, D. H., Lasky, L. A., Theodore, T. S., Earl, P. L., Moss, B., Capon, D. J., and Martin, M. A. (1998). *In vitro* mutagenesis identifies a region within the envelope gene of the human immunodeficiency virus that is critical for infectivity. *J. Virol.* **62**, 139–147.

Wu, L., Gerard, N. P., Wyatt, R., Choe, H., Parolin, C., Ruffing, N., Borsetti, A., Cardoso, A. A., Desjardin, E., Newman, W., Gerard, C., and Sodroski, J. (1996). CD4-induced interaction of primary HIV-1 gp120 glycoproteins with the chemokine receptor CCR-5. *Nature* **384**, 179–183.

Wyand, M. S., Manson, K. H., Lackner, A. A., and Desrosiers, R. C. (1997). Resistance of neonatal monkeys to live attenuated vaccine strains of simian immunodeficiency virus. *Nat. Med.* **3**, 32–36.

Wyatt, R., and Sodroski, J. (1998). The HIV-1 envelope glycoproteins: Fusogens, antigens, and immunogens. *Science* **280**, 1884–1888.

Wyatt, R., Kwong, P. D., Desjardins, E., Sweet, R. W., Robinson, J., Hendrickson, W. A., and Sodroski, J. G. (1998). The antigenic structure of the HIV gp120 envelope glycoprotein. *Nature* **393**, 705–711.

Wyatt, R., Moore, J., Accola, M., Desjardin, E., Roginson, J., and Sodroski, J. (1995). Involvement of the V1/V2 variable loop structure in the exposure of human immunodeficiency virus type 1 gp120 epitopes induced by receptor binding. *J. Virol.* **69**, 5723–5733.

Yasutomi, Y., Robinson, H. L., Lu, S., Mustafa, F., Lekutis, C., Arthos, J., Mullins, J. I., Voss, G., Manson, K., Wyand, M., and Letvin, N. L. (1996). Simian immunodeficiency virus-specific cutotoxic T-lymphocyte induction through DNA vaccination of rhesus monkeys. *J. Virol.* **70**, 678–681.

Zolla-Pazner, S., Lubeck, M., Xu, S., Burda, S., Natuk, R. J., Sinangil, F., Steimer, K., Gallo, R. C., Eichberg, J. W., Matthews, T., and Robert-Guroff, M. (1998). Induction of neutralizing antibodies to T-cell line-adapted and primary human immunodeficiency virus type 1 isolates with a prime-boost vaccine regimen in chimpanzees. *J. Virol.* **72**, 1052–1059.

Fred C. Krebs
Heather Ross
John McAllister
Brian Wigdahl
The Pennsylvania State University
College of Medicine
Department of Microbiology and Immunology
Hershey, Pennsylvania 17033

HIV-1-Associated Central Nervous System Dysfunction

I. Introduction

In June of 1981, the Centers for Disease Control reported five cases of *Pneumocystis carinii* pneumonia in five men in the Los Angeles area (Centers for Disease Control, 1981). In less than 2 years, hundreds of cases with similar diagnoses were reported throughout the United States and around the world. Most of the patients afflicted with this new syndrome shared several distinguishing characteristics, including immune system dysfunction and a history of homosexual activity, intravenous drug abuse, or hemophilia. Efforts to identify the infectious agent responsible for this rapidly spreading immunodeficiency syndrome culminated in 1983 and 1984 with the isolation of a human retrovirus which would later be designated the human immunodeficiency virus type 1 (HIV-1). In the years that followed, HIV-1 infection and the acquired immunodeficiency syndrome (AIDS), a consequence of infection, spread globally to millions of individuals.

Advances in Pharmacology, Volume 49

As more HIV-1-infected individuals were identified and treated, clinicians recognized an association between HIV-1 infection and the appearance of central nervous system (CNS) functional abnormalities. Patients infected with HIV-1 presented with numerous cognitive and motor-skill deficiencies concurrent with progressive immune system failure. Postmortem analyses identified specific pathological hallmarks within the brain as a consequence of HIV-1 invasion. As more cases of CNS involvement were identified and documented, the progressive spectrum of neurologic abnormalities linked to HIV-1 infection was collectively referred to as the AIDS dementia complex (ADC) or, more recently, HIV-1 dementia (HIVD).

At our current level of understanding, development of HIVD following HIV-1 infection of the CNS is the result of complex interactions between infected and uninfected cell types residing within the brain. Further comprehension of HIVD will require answers to numerous questions. For example, what are the correlates of HIVD? Can these markers be used to diagnose the onset of HIVD or predict the level of risk for developing HIV-1-related neurological complications? What are the exact interactions between uninfected and infected neuroglial cells that lead to identifiable neuropathogenesis and the appearance of clinical symptoms? Which viral and cellular soluble mediators are most important to the development of HIVD? Can treatments be developed to interrupt the complex cascade of cellular interactions that lead to HIV-1-associated CNS dysfunction? How well do the currently available antiretroviral drugs penetrate the blood–brain barrier and are they effective in the CNS? Can the CNS serve as a hidden HIV-1 reservoir that can reseed the periphery and confound the ability of antiretroviral therapies to lower the viral load below detectable levels? How will the extended life span of patients under treatment affect the incidence or severity of HIVD? These questions are just a sampling of the issues that have yet to be addressed.

This chapter presents a three-part overview of investigations surrounding HIV-1 infection of the CNS and the appearance of HIVD. In the first part, our contemporary understanding of the clinical and pathological consequences of HIV-1 within the CNS are outlined. In the second part, selected cellular aspects of HIV-1 CNS infection and HIVD are examined. In the third and final part, current findings in the chemotherapeutic treatment of HIVD are discussed.

II. HIV-1-Associated CNS Disease

Over the course of HIV-1-associated disease, the presence of the virus within the CNS becomes manifested both functionally and physically. Numerous studies have examined and documented the clinical aspects of HIV-1 CNS infection as well as the underlying pathological consequences of viral replication within the brain and periphery of the nervous system. Although a great deal of information has been accumulated, ongoing clinical and laboratory studies continue to contribute to our incomplete understanding of HIV-1-associated CNS disease.

A. Clinical Manifestations of HIV-1 CNS Infection

As more adult patients presented with clinical symptoms characteristic of what would later be called AIDS, reports began to appear which described neurologic abnormalities in approximately 40% of these patients (Britton and Miller, 1984; Snider *et al.*, 1983). Initially, these abnormalities were thought to be related to the numerous opportunistic infections that arise in the CNS during the progression of AIDS (Britton and Miller, 1984; Navia *et al.*, 1986c; Snider *et al.*, 1983). However, as more patients were examined, it was soon apparent that opportunistic infections could be linked to only a third of those with CNS abnormalities (Navia *et al.*, 1986b). Furthermore, patients that presented with symptoms characteristic of both AIDS and neurologic involvement frequently showed no indication of the presence of opportunistic infectious agents within the CNS. When HIV-1 was identified in 1984 as the putative etiologic agent of AIDS, an association between the presence of HIV-1 in the CNS and the onset of neurologic dysfunction was postulated.

1. Prevalence of HIVD in HIV-1-Infected Individuals

Of those patients infected with HIV-1, approximately 30 to 60% exhibit symptoms indicative of neurologic dysfunction (McArthur *et al.*, 1994). Of those, about 20% present with symptoms characteristic of HIVD (Janssen *et al.*, 1992; McArthur *et al.*, 1994). Postmortem examinations of AIDS-patient CNS tissue revealed that an even greater percentage (approximately 90%) of HIV-1-infected individuals had some form of peripheral or CNS abnormalities (Geleziunas *et al.*, 1992; Ho *et al.*, 1989; Hollander, 1991; Navia *et al.*, 1986a,b). The association between the presence of HIV-1 and neurologic dysfunction was strongly indicated by detection of HIV-1 gene products in the CNS of AIDS patients as well as similarities between the neurologic implications of infection by HIV-1 and infection of sheep by visna virus, the prototypical lentiviral model for retroviral-induced neurodegenerative disorders (Gonzalez-Scarano *et al.*, 1995; Spencer and Price, 1992). The terms ADC or HIVD were coined to collectively describe specific clinical symptoms associated with HIV-1 CNS infection (Spencer and Price, 1992). HIVD may appear as the only indication of HIV-1 infection and may occur in the absence of any immune system degeneration (Navia and Price, 1987).

Unfortunately, the appearance of HIVD is not limited to the adult population. As many as 75% of HIV-1-infected infants and children present with symptoms, including a static or progressive encephalopathy associated with HIV-1 infection of the CNS (Belman *et al.*, 1986; Epstein *et al.*, 1985, 1986; Oleske *et al.*, 1983). Neurologic symptoms of pediatric HIV-1 infection include impaired brain growth, progressive motor function deterioration, and loss of developmental milestones and cognitive abilities (Belman, 1994; Belman *et al.*, 1986; Epstein *et al.*, 1985, 1986, 1988; Navia *et al.*, 1986b; Oleske *et al.*, 1983; Tardieu, 1998). However, the rate and time of disease onset and involvement of the CNS is variable and appears to be related to the progression of immunologic disease. In a longitudinal study of 94 HIV-1-infected children, approximately one-third suffered from severe encephalopathy correlated with the early onset of immunodeficiency. The remainder of the infected children progressed both immunologically and neurologically at a much slower rate, suggesting a relationship between a pediatric patient's immunologic state and the appearance of neurologic symptoms (Blanche *et al.*, 1990). Neurologic disease severity may also be linked to virus load, which, in cases of vertical transmission, is tightly correlated to maternal viral loads at delivery (Tardieu, 1998). Before the advent of combination drug therapies, approximately 20% of infected children suffered from severe immunodeficiency and encephalopathy during the first 3 years of life, while the remainder experienced a much slower progression (Tardieu, 1998). New drug therapies may be effective against the slower progressing pediatric cases, but may be generally ineffective against the rapid onset of neurologic disease (Tardieu, 1998). In a 1997 retrospective study of pediatric AIDS cases over the preceding 7 years, extended pediatric life spans associated with improved medical care and more effective antiretroviral therapies were accompanied by increased involvement of organ systems, including the central nervous system, in the progression of disease (Johann-Liang *et al.*, 1997). These data suggest that continued advancements in the treatment of pediatric HIV-1 and opportunistic infections might lead to increases in the frequency and severity of pediatric CNS disease.

2. Onset of HIVD Relative to HIV-1-Associated Immune Disease

Although HIVD may appear at any time during the progression of HIV-1 infection, it most often appears late in the course of disease concurrent with low CD4-positive cell numbers and high viral titers in the peripheral circulation (McArthur *et al.*, 1993). Although clinical symptoms of neurologic impairment may appear prior to the onset of AIDS-related immunologic abnormalities, it is not a common observation (McArthur and Selnes, 1997). In patients who do not demonstrate early neurologic impairment, progression through the asymptomatic phase of infection is not accompanied by a decline in neurologic abilities (Selnes *et al.*, 1990). The time of onset and rate of HIVD progression, which is variable and may depend on currently undefined host and viral factors (Bouwman *et al.*, 1998), are not necessarily temporally linked to the progression of systemic immunologic dysfunction.

The apparent lag between initial seeding of the cerebrospinal fluid (and, presumably, the brain) and onset of demonstrable neurologic abnormalities may be explained by clearance of the initial infection by the immune system and subsequent reinfection of the CNS as the immune system deteriorates. Evidence of a host immune response to infection includes increased cerebrospinal fluid (CSF) levels of immunoglobulins, proteins, and whole cells as well as anti-HIV-1 cytotoxic T lymphocytes and intrathecally produced anti-HIV-1 antibodies. Alternatively, selective infection of the CNS by less virulent strains may result in a more chronic spread of HIV-1 through the CNS. The general prognosis for patients diagnosed with HIVD is poor (Mayeux *et al.*, 1993), since the mean survival of demented patients (approximately 6 months) is less than half the remaining life span of nondemented HIV-1-infected patients (McArthur, 1987; Portegies, 1994; Rothenberg *et al.*, 1987).

3. Definition and Classification of HIV-1-Associated Dementia

The definition of HIVD encompasses a broad constellation of symptoms characteristic of cognitive, behavioral, and motor abnormalities associated with the presence of HIV-1 in the CNS. The most distinct clinical aspect of HIVD is, as the name implies, dementia with progressive loss of cognitive functions. Specifically, HIVD can be classified as a subcortical dementia, which distinguishes it from dementia associated with Alzheimer's disease (Kolson *et al.*, 1998). Other common clinical symptoms of HIVD include decreased memory, difficulty concentrating, psychomotor retardation, and headache. Less frequent manifestations include motor deficits, seizures, and psychiatric dysfunction. Early symptoms of HIVD include a slowing of mental capacity; impairment of motor control and coordination; and behavioral changes that include apathy, social withdrawal, and alterations in personality. Later in the progression of the disease, nearly all aspects of cognition, motor function, and behavior are affected. Severe cases of HIVD may be characterized by near or absolute mutism, incontinence, and severe dementia (Price *et al.*, 1988). Ultimately, the diagnosis of HIVD is based on the symptoms observed, the time course of the disease, and the application of laboratory diagnostic criteria (Navia *et al.*, 1986b; Price and Sidtis, 1992; Worley and Price, 1992).

To provide a common language for discussing HIVD diagnoses, a numerical staging system has been formulated for defining the severity and progression of HIVD based on an assessment of the patient's cognitive and motor functions. Within the Memorial Sloan-Kettering (MSK) scale of HIV-1-associated dementia (Fig. 1), patients are classified as unaffected (Stage 0); subclinically or equivocally affected (Stage 0.5); or suffering mild (Stage 1), moderate (Stage 2), severe (Stage 3), or end-stage disease (Stage 4) (Price and Brew, 1988). Patients with no neurologic symptoms, usually in the clinical latency period of systemic infection, are generally classified as Stage

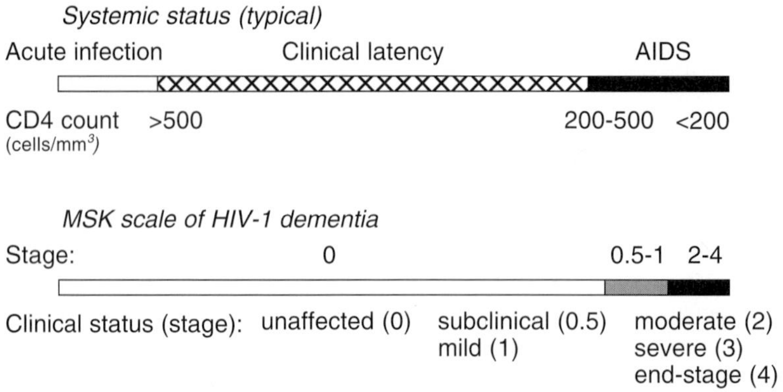

Systemic status (typical)

Acute infection   Clinical latency   AIDS

CD4 count   >500   200-500   <200
(cells/mm³)

MSK scale of HIV-1 dementia

Stage: 0   0.5-1   2-4

Clinical status (stage): unaffected (0)   subclinical (0.5)   moderate (2)
                                            mild (1)   severe (3)
                                                                end-stage (4)

FIGURE I   The Memorial Sloan-Kettering (MSK) scale of dementia severity provides a common reference for assessing the progression of HIVD. Patients diagnosed with HIVD are classified symptomatically as unaffected (Stage 0); subclinically or equivocally affected (Stage 0.5); or suffering mild (Stage 1), moderate (Stage 2), severe (Stage 3), or end-stage disease (Stage 4) (Price and Brew, 1988). The progression from Stage 0 through Stage 4 generally corresponds with the advancement of immune system disease.

0. Patients with HIVD Stages 0.5 and 1 have been generally nonprogressive for several years with CD4-positive cell counts between 200 and 500 cells/mm³. Neurologic deficits may include difficulty with concentration, forgetfulness, the need to keep lists for personal organization, and motor abnormalities characterized by slowing of rapid eye and extremity movements. HIVD Stage 2 (or greater) generally coincides with CD4-positive cell counts below 200 cells/mm³, immunosuppression, increased viral titers, and opportunistic infections of both the immune and central nervous systems. Patients who have progressed to and beyond Stage 2 have generally more severe motor, behavioral, and cognitive deficits, which may include quadraparesis, generalized spasticity, seizures, incontinence, or mutism.

There are, however, patients who cannot be classified according to the guidelines of the MSK system. Some patients exhibit normal CNS function despite the presence of numerous systemic opportunistic infections. Conversely, other HIV-1-infected patients develop neurologic impairment in the absence of any immunologic dysfunction or symptoms related to the systemic presence of HIV-1 (Navia and Price, 1987). Progressive HIVD does not always correlate with increased systemic compromise; patients have been classified as HIVD Stage 2 or greater with CD4-positive counts over 200 cells/mm³ (Price, 1994). Furthermore, the severity of HIVD does not always correlate with the viral burden in the brain, as demonstrated by inconsistent correlations between high levels of viral gene expression in the brain and neurologic dysfunction (Dickson et al., 1991; Kure et al., 1990a; Wiley and Achim, 1994). This phenomenon is particularly evident when considering milder forms of HIVD. Studies by Brew and coworkers have demonstrated that viral load is often less than would be expected given the severity of the clinical abnormalities (Brew *et al.*, 1995b). Similar observations have been reported by Vazeux and coworkers in pediatric patients with HIV-1-associated neurologic dysfunction (Vazeux *et al.*, 1992).

B. Pathology Associated with HIV-1 CNS Infection and HIVD

A complete understanding of HIV-1-associated disease and HIVD must include data from clinical diagnoses of neurologic impairment and the onset of HIVD as well as pathological findings compiled from physical examinations of brains infected with HIV-1. Early investigations were limited primarily by available technologies to postmortem examinations of brain tissues from HIV-1-infected patients, which demonstrated that HIV-1 infection results in specific structural and cellular abnormalities. These studies provided information about terminal CNS neuropathology but shed little light on events that occur early in HIV-1-associated CNS disease. However, recent advances in imaging technology have facilitated investigations at all stages of HIVD. These and other studies continue to contribute toward a greater understanding of HIVD pathology.

1. HIV-1-Associated Neuropathology in Adults

Gross pathological changes are clearly evident in the postmortem brains of many HIV-1-infected patients who have succumbed as a consequence of the infection. Of those who die from AIDS, roughly 20–30% exhibit postmortem CNS disease (Anders *et al.*, 1986). A diagnosis of HIVD does not necessarily guarantee the presence of detectable neuropathology, since up to half of all patients diagnosed with HIVD have no definitive pathologic abnormalities (Budka, 1991; Ho *et al.*, 1989; Sharer, 1992). Among the numerous heterogeneous pathologies associated with HIVD (Rosenblum, 1990), three common pathologies are evident: white matter pallor and gliosis, multinucleated-cell encephalitis, and vacuolar myelopathy (Price, 1994). White matter pallor and gliosis are the most common but least specific pathologies identified in HIVD patients. Multinucleated-cell encephalitis, also observed in HIV-1-infected patients with CNS disease, is characterized by the appearance of multinucleated giant cells (MGC or syncytia). These multinuclear cell complexes, composed of infected macrophages and microglial cells, are sometimes clustered around blood vessels (Price, 1994). However, MGCs are only found in approximately 50% of patients diagnosed with HIVD (Kato *et al.*, 1987; Wiley and Achim, 1994). The presence of MGCs is indicative of active HIV-1 replication, which has been confirmed by the detection of viral particles, virus-specific antigens, and HIV-1-specific nucleic acids within multinuclear cells (Budka, 1990; Budka *et al.*, 1987;

Gabuzda et al., 1986; Koenig et al., 1986; Kure et al., 1990a,b; Michaels et al., 1988; Peudenier et al., 1991; Price et al., 1988; Pumarola-Sune et al., 1987; Shaw et al., 1985; Stoler et al., 1986; Vazeux et al., 1987; Wiley et al., 1986). Multinucleated-cell encephalitis is usually observed against a background of gliosis and white matter pallor and occurs in the subcortical regions of the brain (Chrysikopoulos et al., 1990; Epstein et al., 1984). Vacuolar myelopathy, the third pathologic finding, often affects the cervical and thoracic sections of the spinal cord more severely than other regions (Petito et al., 1985).

Other, more diverse pathologies have also been associated with HIV-1 CNS infection. Wiley and coworkers, as well as other groups, have reported atrophy and neuronal loss within the neocortical regions of the HIV-1-infected brain (Wiley and Achim, 1994). Volume loss in the basal ganglia also correlates with the presence of dementia (Aylward et al., 1995; Dal Pan et al., 1992). Neuronal dropout has been demonstrated in a number of studies (Everall et al., 1994; Janssen et al., 1991; Ketzler et al., 1990; Masliah et al., 1992b) but has not been correlated with severity of disease (Everall et al., 1994). Neuronal apoptosis, demonstrated in both HIV-1-infected adults and children, is associated with microglial activation and axonal damage, but not necessarily with the occurrence of HIVD (Adle-Biassette et al., 1999; Adle-Biassette et al., 1995; Gelbard et al., 1995). However, a correlation between HIV-1-mediated immune system alterations and progressive peripheral and central sensory tract lesions has been observed (Husstedt et al., 1998). Morphological alterations in the neuritic processes of neurons in both the frontal cortex and hippocampus (Masliah et al., 1992b) and damage to dendritic projections in cortical pyramidal neurons have been reported (Everall et al., 1993). Abnormalities in oligodendrocytes have also been observed (Esiri et al., 1991). Other studies have implicated neuronal apoptosis as a factor in both HIVD and encephalopathy found in pediatric cases of HIV-1 CNS infection (Adle-Biassette et al., 1995; Gelbard et al., 1995). Apoptosis associated with HIV-1 CNS infection has also been demonstrated in vascular endothelial cells, suggesting that programmed cell death may play a role in compromising the blood–brain barrier (BBB) (Shi et al., 1996).

2. Correlates of HIVD

Since the recognition of neurologic disease associated with HIV-1 CNS infection, numerous groups have attempted to identify correlates of HIVD development such as viral load in the brain or immune system. To date, these links have not been definitively established. In the brain, detection of proviral DNA has been associated with neurologic deficits as well as HIV-1-associated histopathology (Boni et al., 1993). However, results from a similar study that used quantitative PCR to detect proviral DNA indicated no correlation between CNS viral load and HIVD (Johnson et al., 1996).

HIV-1 p24 expression in the brain was also found to be linked to HIVD severity (Brew et al., 1995b). Studies of brain viral load and HIVD may be complicated by a lag between development of HIV-1 encephalitis within the CNS and manifestation of clinical dementia (Wiley and Achim, 1994). The severity of dementia has also been associated with the extent of CNS monocyte/macrophage activation (Glass et al., 1995) as well as expression of gp41 and inducible nitric oxide synthase (iNOS) (Adamson et al., 1999; Rostasy et al., 1999). iNOS expression is particularly interesting as a marker for HIVD since iNOS synthesizes nitric oxide (NO), a compound implicated as a potential neurotoxin involved in the development of HIV-1-associated disease.

Unintegrated DNA, a subset of the total viral DNA load within the brain, may also be associated with HIVD. Following conversion of the viral RNA genome to double-stranded DNA through the enzymatic activity of reverse transcriptase, HIV-1 DNA is integrated as a linear molecule into the host genome by the HIV-1-encoded integrase. Circular DNA structures containing either one or two LTRs are thought not to serve as preintegration intermediates and are considered a nonproductive offshoot of the replication cycle. Because these circular molecules are not integrated, they accumulate within the infected cell. In vitro, unintegrated HIV-1 DNA accumulates in HIV-1-infected monocyte/macrophage cells to levels lower that those found in productively infected T cells (Sonza et al., 1994). In macrophages and microglial cells, lower levels of unintegrated DNA may be indicative of resistance to superinfection or a more chronic level of replication. One study of HIV-1-infected patients demonstrated that levels of unintegrated linear and circular viral DNA are elevated in postmortem brains of patients with HIV-1 encephalitis (Pang et al., 1990). A similar study found that the presence of unintegrated single LTR proviral structures and increased HIV-1 replication were associated with the appearance of MGCs and dementia (Teo et al., 1997). These results parallel a link between the presence of unintegrated viral DNA in the blood and the severity of immunologic disease (Nicholson et al., 1996). Within infected macrophages and microglia in the brain, unintegrated DNA may serve as transcription templates for production of viral RNA and, subsequently, proteins such as gp120 and Tat. These proteins may function as neurotoxins when released into the CNS (Stevenson et al., 1990). In addition, a low level of viral replication from unintegrated templates in macrophages may contribute to the viral load within the brain, as has been demonstrated in vitro in HeLa cells (Cara et al., 1996).

Other groups have identified associations between HIVD and markers in the CSF or peripheral circulation. Correlations have been demonstrated between HIV-1 nucleic acids or protein products (RNA or p24) in the CSF and the appearance of clinical disease (Brew et al., 1997; McArthur et al., 1997; Royal et al., 1994). In pediatric HIV-1 CNS disease, viral RNA in the CSF correlates with cognitive dysfunction (Sei et al., 1996). In adult

HIV-1-infected patients, levels of HIV-1 RNA in the CSF correlate with viral burden in the brain, indicating that CSF viral RNA may be a surrogate marker for studies of viral load in HIVD (Wiley *et al.*, 1998). However, elevated levels of CSF HIV-1 may also be indicative of other CNS infections (Brew *et al.*, 1997). Peripheral blood p24 levels have also been linked to the appearance of CNS dysfunction (Royal *et al.*, 1994).

3. Neuropathology during Pediatric HIV-1 CNS Infection

HIV-1 infection of the developing pediatric brain results in neuropathology that differs somewhat from that reported in HIV-1-infected adults. HIV-1 infection of the pediatric brain may result in development of either static or progressive encephalopathy (Epstein *et al.*, 1985). In a study of 49 cases of pediatric HIV-1 CNS infection, over half of the patients examined demonstrated some form of clinical neurological dysfunction and over two-thirds presented with progressive encephalopathy (Roy *et al.*, 1992). In contrast to adult CNS infection, pediatric HIV-1 infection of the CNS results in increased frequencies of inflammatory infiltrates and multinucleated giant cells, and the relatively rare appearance of vacuolar myelopathy (Elder and Sever, 1988; Price *et al.*, 1988). Brain weights of HIV-1-infected infants and children are lower than normal (Sharer, 1992) and increases in head circumference are correspondingly reduced (Tardieu, 1998). Volumetric reductions in all cortical regions, subcortical gray matter, and cerebral white matter have also been observed (Kozlowski *et al.*, 1997) along with enlargement of subarachnoidal spaces (Tardieu, 1998). As was noted above, neuronal apoptotic death has been reported as one consequence of pediatric HIV-1 infection (Gelbard *et al.*, 1995). Children infected with HIV-1 acquire fewer opportunistic infections than adults, providing evidence that supports a direct causal relationship between the presence of HIV-1 and onset of neurologic abnormalities in both children and adults.

4. In vivo Imaging of the HIV-1-Infected Brain

Technological advances have facilitated studies of the living brain under normal as well as diseased states. Numerous imaging techniques have been employed in the study of HIV-1 CNS infection and have made valuable contributions to understanding HIV-1 CNS infection and HIVD. These techniques have been used to examine postmortem CNS tissues to corroborate findings documented during more traditional postmortem examinations. More importantly, they have been used to examine living patients to diagnose HIV-1-associated neurologic dysfunction and to rule out other conditions, such as opportunistic infections or brain tumors. Structural neuroimaging techniques used to study HIVD include computed tomography (CT) and magnetic resonance imaging (MRI). Imaging techniques designed to demonstrate patterns of brain perfusion or metabolic function, such as single photon emission computed tomography (SPECT) or positron emission tomography (PET), have also been employed.

The most common CT and MRI finding is cerebral atrophy, indicated by increased ventricular CSF volume (Gelman and Guinto, 1992; Grafe *et al.*, 1990; Jakobsen *et al.*, 1989; Jarvik *et al.*, 1988). Cerebral atrophy is identified at the greatest frequency in patients with advanced immunosuppression or in patients who have been infected longer than 4 years (Bencherif and Rottenberg, 1998). The degree of cerebral atrophy has been correlated with the extent of neurologic impairment (Jakobsen *et al.*, 1989; Levin *et al.*, 1990), implying that regional atrophy of the brain may be reflected in the appearance of specific neurologic symptoms. The next most common finding is the appearance of bilateral white matter lesions, which appear in CT scans as hypodense regions and in MRI scans as hyperintense areas on T2-weighted images (Bencherif and Rottenberg, 1998; Chrysikopoulos *et al.*, 1990; Jarvik *et al.*, 1988). Lesions of this type become more prominent with disease progression and may be transiently reduced (as documented by MRI) by treatment with zidovudine (Tozzi *et al.*, 1993). MRI scans, which are superior to CT scans in spatial resolution and image contrast, have also documented increased deep gray matter cerebral blood volumes, which may signal impending structural changes concomitant with HIVD and CNS HIV-1 infection (Tracey *et al.*, 1998). Decreases in posterior cortical and basal ganglia volumes have also been observed in demented patients (Aylward *et al.*, 1995). These techniques have highlighted a correlation between decreased brain volume, increased CSF volume, and the severity of HIVD (Dal Pan *et al.*, 1992).

PET and SPECT scanning have been employed to document functional irregularities associated with HIV-1 CNS infection (Bencherif and Rottenberg, 1998). Dementia-associated metabolic abnormalities identified by PET include basal ganglia and thalamic hypermetabolism during early HIVD and cortical and subcortical hypometabolism as the severity of HIVD increases (Rottenberg *et al.*, 1987). PET scanning has also been used effectively to monitor brain metabolic responses to antiviral treatments (Bencherif and Rottenberg, 1998). SPECT scanning, which is used to monitor brain perfusion, has also been used to diagnose HIVD-related abnormalities. Numerous SPECT studies have reported perfusion irregularities associated with HIVD and detectable improvements in perfusion following treatment with zidovudine (Bencherif and Rottenberg, 1998).

Various imaging techniques have also been employed to view the HIV-1-infected pediatric brain. CT scans have demonstrated that central and cortical atrophy is also found frequently in HIV-1-infected pediatric brains (Roy *et al.*, 1992) and that administration of zidovudine results in partial reversal of brain atrophy and improved cognitive abilities (DeCarli *et al.*, 1991). White matter lesions are also present, as are basal ganglia calcifications, an observation unique to the pediatric brain (Roy *et al.*, 1992). Another

MRI study of pediatric patients with encephalopathy noted a pattern of central atrophy in the subcortical white matter and the basal ganglia (Scarmato *et al.*, 1996). Advanced imaging techniques may also be used to detect and diagnose abnormal CNS pathologies before the presentation of clinical symptoms. Using magnetic resonance spectroscopy (MRS), Cortey and coworkers demonstrated CNS abnormalities in the brains of newborn children specifically born to HIV-1-infected mothers (Cortey *et al.*, 1994).

III. Cellular Aspects of HIV-1 CNS Infection

The clinical symptoms and neuropathology associated with HIV-1 infection of the CNS have their foundation in HIV-1 replication within susceptible CNS cells and the complex interplay between infected and uninfected cells of the brain. Indeed, the molecular basis of HIV-1-associated CNS impairment is supported by the observation that HIV-1-related neuropathology is undetectable by macroscopic or microscopic analysis in 30–50% of HIVD patients examined (Budka, 1991; Ho *et al.*, 1989; Sharer, 1992). This section highlights relevant studies that have examined cellular and molecular aspects of HIV-1 neuropathogenesis, including HIV-1 entry into the CNS, chemokine receptor and chemokine expression within the CNS, susceptibility of resident CNS cells to HIV-1 infection, transcriptional control in cells of monocyte/macrophage lineage (including brain microglial cells) and the complex interplay between cells during HIV-1 CNS disease and the progression of HIVD.

A. HIV-1 Entry into the CNS

The observation that neurologic dysfunction can be diagnosed in the absence of detectable immunologic compromise in a small number of HIV-1-infected patients suggests that HIV-1 may gain access to the brain relatively soon after initial infection. In support of this hypothesis, a number of studies have shown that the CSF is apparently seeded with virus soon after systemic infection (Davis *et al.*, 1992; Levy *et al.*, 1985; Price and Brew, 1988). Although a definitive correlation between the presence of HIV-1 in the CSF and eventual onset of HIVD has not been established, HIV-1 has been observed in the CSF as early as seroconversion and during the period of clinical latency, as demonstrated by detection of HIV-1-specific antibodies, nucleic acids, and infectious virus (Buffet *et al.*, 1991; Chiodi *et al.*, 1988; Goswami *et al.*, 1991; Ho *et al.*, 1985). Similar observations have been made in rhesus macaques infected with simian immunodeficiency virus (SIV). Following systemic inoculation, SIV was detectable in the brain within 2 weeks (Sharer *et al.*, 1991).

Infiltration of HIV-1 into the brain and initiation of events that lead to CNS dysfunction requires that the virus circumvent the BBB, a selective barrier which separates the CNS from the rest of the body. Several entry mechanisms have been described (Fig. 2). One means by which HIV-1 may enter the CNS is by crossing the BBB as cell-free virus (Fig. 2A). Although the cell-free mode of entry is attractive for its simplicity, the Trojan horse theory, in which HIV-1-infected cells transit the BBB into the CNS, is more generally accepted as the means of HIV-1 infiltration into the CNS (Haase, 1986). The likely vehicle for this method of entry is the infected monocyte, which can carry HIV-1 in its proviral DNA form or as viral particles (Fig. 2B). Systemic infection and immune cell activation may serve to enhance this mechanism (Hickey, 1999). There is similar precedent for immune cell circulation in the CNS in observations made in animal models (Perry and Gordon, 1988; Wekerle *et al.*, 1986).

HIV-1 infection of monocytes and macrophages may serve to facilitate CNS entry of both cell-free and cell-associated virus through the BBB. Macrophage infection by HIV-1 leads to a state in which the cell is considered "primed" (Nottet and Gendelman, 1995). Subsequent activation of the "primed" macrophage results in increased expression of pro-inflammatory molecules such as eicosinoids (Nottet *et al.*, 1995), platelet-activating factor (PAF) (Gelbard *et al.*, 1994), and tumor necrosis factor-α (TNFα) (D'Addario *et al.*, 1990, 1992; Nottet *et al.*, 1995). TNFα, in particular, is upregulated through its promoter by increased levels of the transcription factor NF-κB found in HIV-1-infected monocyte-derived macrophages and promonocytic

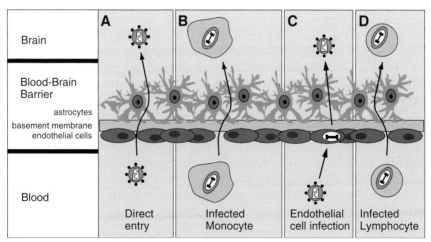

FIGURE 2   Several mechanisms have been proposed to explain HIV-1 entry into the CNS. HIV-1 may cross the blood–brain barrier (BBB) as cell-free virions (A), within infected monocytes (B), through infected brain microvascular endothelial cells (BMEC) or infected choroid plexus cells (C), or within infected lymphocytes (D).

cells (Bachelerie *et al.*, 1991). A recent study supports the role of TNFα in creating an environment within the brain that allows the transport of HIV-1-infected macrophages across the BBB (Fiala *et al.*, 1997). In a BBB model constructed using human endothelial cells and astrocytes grown on a collagen I or fibronectin matrix, the model BBB completely blocked the transmission of HIV-1 in the absence of TNFα. As TNFα was added, paracellular virus penetration through the model BBB increased in a dose-dependent manner. These results suggest that the BBB functions early in infection to protect the brain from HIV-1, but the production of immune mediators such as TNFα opens a paracellular route for HIV-1 entry into the brain.

HIV-1-infected macrophages also induce upregulation of cell adhesion molecules on vascular endothelial cells, which facilitates binding of the infected macrophages to the BBB. Unlike uninfected macrophages, HIV-1-infected macrophages induce expression of E-selectin on primary human brain microvascular endothelial cells (Nottet *et al.*, 1996). Furthermore, activated, HIV-1-infected macrophages induce greater expression of E-selectin and VCAM-1 on brain endothelial cells than do unactivated, infected cells (Nottet *et al.*, 1996), indicating that immune activation of monocytes and macrophages subsequent to viral infection of the CNS is a key component to facilitating transendothelial migration (Persidsky *et al.*, 1997). *In vivo* expression of E-selectin (and to a lesser extent VCAM-1) was higher in AIDS patients experiencing encephalitis than in the brains of AIDS patients suffering from other neurological diseases or patients without CNS disease. These *in vivo* observations support the hypothesis that adhesion molecule expression is relevant to BBB penetration by HIV-1-infected macrophages (Nottet *et al.*, 1996). Other investigators have determined that blood-derived macrophages that infiltrate the brain express high amounts of the pro-inflammatory cytokines TNFα and interleukin-1β (IL-1β) (Tyor *et al.*, 1992; Wesselingh *et al.*, 1993), which also induce adhesion molecule expression on the vascular endothelium. Furthermore, Tat induces E-selectin expression on microvascular endothelial cells (Hofman *et al.*, 1993) and may facilitate macrophage invasion through the BBB (Rappaport *et al.*, 1999).

HIV-1-infected macrophages also secrete signaling molecules that increase BBB permeability. For instance, *in vitro* activation of HIV-1-infected monocyte-derived macrophages (MDM) results in overexpression of TNFα, leukotrienes, and PAF, which alter the permeability of the endothelial cell layer (Bjork *et al.*, 1983; Black and Hoff, 1985). Once through the endothelial cell layer, invading macrophages must penetrate the basement membrane surrounding the abluminal side of the BBB. HIV-1 infection imparts upon macrophages the ability to digest and invade the basement membrane to a significantly greater extent than uninfected control cells (Dhawan *et al.*, 1992). This ability is the result of gelatinase B upregulation in HIV-1-infected macrophages (Dhawan *et al.*, 1995), which could be mimicked by treatment of uninfected macrophages with the HIV-1 Tat protein (Lafrenie *et al.*,

1996). In summary, HIV-1 infection of macrophages leads to the production of a cascade of cellular molecules that ultimately allow HIV-1 transmigration through the BBB.

A number of investigators have suggested an alternative hypothesis in which HIV-1 enters the CNS by infecting cells that serve to compartmentalize the brain. Direct infection of vascular endothelial cells may lead to release of budding virions on the parenchymal endothelial membrane surface, thereby seeding HIV-1 directly into the CNS (Fig. 2C). This theory is supported by demonstration of *in vitro* susceptibility of cultured endothelial cells to HIV-1 infection, *in vivo* detection of virus in endothelial cells by *in situ* hybridization (Bagasra *et al.*, 1996), and *in vivo* detection of HIV-1-infected cerebral vascular endothelium 60 weeks after intravenous inoculation of pigtailed macaques with the T-tropic HIV-1 strain LAI (Frumkin *et al.*, 1995). Other studies that used a neuroinvasive strain of SIV indicate that replication in brain microvascular endothelial cells (BMEC) may predict neuroinvasiveness *in vivo* and that virus passage in the brain generates neuroinvasive strains which infect BMECs *in vitro* with increased efficiency (Strelow *et al.*, 1998). Epithelial cells of the choroid plexus, which are also susceptible to HIV-1 infection (Bagasra *et al.*, 1996; Falangola *et al.*, 1995; Harouse *et al.*, 1989b), may also serve as an entry portal for HIV-1, as has been observed with visna virus (Johnson, 1982).

Other investigators suggest that lymphocytes may play a role in the Trojan horse mechanism (Fig. 2D). HIV-1 infection of CD4-positive T cells results in changes in cell adhesion molecule expression, including increased expression of integrins LFA-1 and VLA-4, which can interact with endothelial cell ligands such as ICAM-1 and VCAM-1 (Sloan *et al.*, 1992; Weidenheim *et al.*, 1993). This has the effect of enhancing lymphocyte binding to the endothelial cell surface. In addition, lymphocytes can secrete enzymes that degrade the basement membrane of the endothelial cells, which may allow migration of the HIV-1-infected T cells through the BBB (Sloan *et al.*, 1992). However, given the low frequency of infiltrating lymphocytes within the HIV-1-infected brain, their contribution to establishment and maintenance of viral load remains to be ascertained.

B. Chemokine Receptor Expression within the CNS and Involvement in HIV-1 Neuropathogenesis

Since their identification as coreceptors for HIV-1 entry in susceptible peripheral blood cells, chemokine receptors and their corresponding ligands have become an integral part of the HIV-1 pathogenic landscape (Feng *et al.*, 1996). They are no less important in studies of HIV-1-associated CNS disease. In fact, while the list of chemokine receptors expressed within the brain continues to expand (Fig. 3), our current understanding suggests that only a select few receptors play major roles in HIV-1 CNS infection. A

cell type	chemokine receptors expressed
Central nervous system	
Microglial cell	CCR3, CCR5, CXCR2, CXCR4
Astrocyte	CCR1, CCR5, CXCR2, CXCR4
Endothelial cell	CCR5, CXCR4, DARC
Neuron	CCR1, CCR3, CCR5, CXCR2, CXCR4, DARC
Oligodendrocyte	?
Immune system	
Lymphocyte	CCR1, CCR2, CCR3, CCR4, CCR5, CXCR4
Monocyte/ Macrophage	CCR1, CCR2, CCR3, CCR4, CCR5, CXCR4

FIGURE 3   Cells of the CNS express an array of chemokine receptors. Neuroglial cell types found in the CNS express numerous CC and CXC chemokine receptors (Hesselgesser and Horuk, 1999; Lavi *et al.*, 1998). Chemokine receptor expression on T lymphocytes and cells of monocyte/macrophage origin is shown for comparison. Receptors are listed without regard for expression level on each cell type; some receptors are expressed at higher levels than others.

number of reviews are available that address this topic in greater detail (Hesselgesser and Horuk, 1999; Lavi *et al.*, 1998).

The identification of chemokine receptors as coreceptors for HIV-1 entry and their differential usage during viral entry has resulted in the reclassification of HIV-1 strains based on their chemokine receptor usage (Berger *et al.*, 1998). Strains of HIV-1 generally use one of two coreceptor molecules in addition to the cell surface receptor CD4 expressed on cells susceptible to infection. Macrophage tropic strains of HIV-1, which are predominant within the HIV-1-infected brain, use primarily the CC chemokine (or β-chemokine) receptor CCR5. Strains that use CCR5 for entry are referred to as R5 strains. T-lymphocyte tropic strains of HIV-1 generally use the CXC chemokine (or α-chemokine) receptor CXCR4 and are classified as X4 strains of HIV-1.

Chemokine receptors expressed by microglial cells include CXCR4 (Lavi *et al.*, 1997) as well as CCR5 and CCR3 (He *et al.*, 1997; Shieh *et al.*, 1998). CXCR4- and CCR5-positive microglia have been identified in pediatric HIV-1-infected brains (Vallat *et al.*, 1998). Chemokine receptor-mediated HIV-1 entry into microglial cells is generally similar to utilization of chemokine receptors on their peripheral monocytic counterparts. The predominant coreceptor used by HIV-1 to enter microglial cells is CCR5, especially by viral isolates from patients with HIVD (Albright *et al.*, 1999). Heterozygous expression of the delta 32 variant of CCR5 causes delayed progression of HIV-1-associated disease and lower immune system viral loads (Meyer *et al.*, 1997). A recent observation has suggested that heterozygous CCR5 delta 32 expression in the brain may also be protective against development of HIVD (van Rij *et al.*, 1999), presumably by impairing infection of macrophages and microglia by R5 HIV-1 strains. However, some viral strains can also use CCR3 and CXCR4, as evidenced by viral studies using brain isolates and molecular clones (Shieh *et al.*, 1998). Another study demonstrated that both CCR3 and CCR5 can support efficient microglial cell infection by isolates that utilize these coreceptors and that those infections can be blocked using eotaxin or MIP-1β, ligands for CCR3 and CCR5, respectively (He *et al.*, 1997). However, another group demonstrated that CCR5- and CCR3-specific antibodies failed to block microglial cell infection, suggesting the possibility that other coreceptors may also facilitate HIV-1 entry (Ghorpade *et al.*, 1998).

As chemokine receptor expression is closely related to HIV-1 susceptibility, expression of these molecules is relevant to HIV-1 infection of astrocytes during the course of HIV-1-associated CNS disease. Murine astrocytes express functional CCR1 as well as CXCR4 (Tanabe *et al.*, 1997a,b). Human fetal astrocytes have been shown to express low levels of CXCR4 and CXCR2 (Hesselgesser *et al.*, 1997; Hesselgesser and Horuk, 1999). Regional *in vivo* astrocytic expression of CCR5 has been demonstrated in the human hippocampal and cerebellar regions of the brain (Rottman *et al.*, 1997).

Neurons also express an array of chemokine receptors. Numerous chemokine receptor messages are expressed in rat hippocampal neurons, including CCR1, CCR4, CCR5, CCR 9/10, CXCR2, CXCR4, and CX3CR1 (Meucci *et al.*, 1998). These studies also demonstrated that SDF-1α and other chemokines were able to block gp120-mediated neurotoxicity. *In vivo*, neuronal CXCR4 expression is localized to the hippocampus (Lavi *et al.*, 1997), while CCR5 expression is evident in the hippocampus and cerebellum (Rottman *et al.*, 1997). Neuronal cells produced by differentiation of the human NTera2 cell line express CXCR2, CXCR4, CCR1, and CCR5, which are fully functional in their capacity to bind their respective chemokines and elicit chemotactic responses (Hesselgesser *et al.*, 1997). This same study demonstrated that HIV-1 was antagonistic to SDF-1 binding to CXCR4, suggesting that gp120–CXCR4 binding may play a role in neuronal neuropathogenesis. Although chemokine receptor expression may also facilitate HIV-1 entry and infection of neurons, the very limited number of neurons identified in postmortem HIV-1-infected brains would suggest that such a mechanism is of little relevance to HIV-1-associated CNS disease.

As mentioned above, brain capillary endothelial cells are an integral part of the blood–brain barrier. Their susceptibility to HIV-1 infection indicates that they provide another portal for HIV-1 entry into the CNS. As cells susceptible to HIV-1 infection, endothelial cells must also express cell surface receptor molecules suitable for HIV-1 use. Chemokine receptors expressed on brain endothelial cells include DARC, CXCR4, and CCR5

(Hesselgesser and Horuk, 1999). Furthermore, the presentation of CXCR4 on the endothelial cell surface can be downregulated by interferon-γ (IFN-γ) and modulated transiently by IL-1β, TNFα, or LPS (Gupta *et al.*, 1998). Cytokine-mediated changes in endothelial cell CXCR4 expression may be relevant to endothelial cell HIV-1 infection in light of the cytokine deregulation associated with HIV-1 CNS infection.

Expression of chemokine receptors on oligodendrocytes has not been studied extensively as expression on other cell types. However, the CXC chemokine melanoma growth stimulatory activity (MGSA) can function synergistically with PDGF to promote oligodendrocyte growth, suggesting the presence of at least one class of chemokine receptor on oligodendrocytes (Hesselgesser and Horuk, 1999). Since oligodendrocytes do not support consistently detectable levels of HIV-1 infection, the primary function of oligodendrocyte chemokine receptors in HIVD may be to mediate cellular toxicity associated with soluble factors including CNS chemokines and gp120.

Aside from their secondary role as coreceptors for HIV-1, chemokine receptors function primarily as transducers of intracellular signals mediated by chemokine ligands. Chemokines, normally expressed in the CNS, are upregulated during CNS inflammation (Lavi *et al.*, 1998). In HIV-1-infected brains, expression of macrophage inflammatory protein-1α (MIP-1α) and MIP-1β (ligands of CCR5) was localized to cerebral cortex astrocytes and microglia and was increased with the appearance of dementia (Schmidtmay-erova *et al.*, 1996). In SIV-infected macaques, CNS infection resulted in upregulation of the CXC chemokine interferon inducible protein-10 (IP-10) and the CC chemokines MIP-1α, MIP-1β, MCP-3, IP-10, and RANTES (Sasseville *et al.*, 1996). Within the brain, chemokines are produced primarily by astrocytes and microglial cells (Hesselgesser and Horuk, 1999; Lavi *et al.*, 1998). Along the blood–brain barrier, endothelial cells are also capable of releasing an array of chemokines in response to various proinflammatory cytokines (Hesselgesser and Horuk, 1999). Similarly, during macaque CNS infection by SIV, cells identified as brain endothelial cells produced MIP-1α, MIP-1β, MCP-3, and RANTES (Sasseville *et al.*, 1996). Within the context of HIV-1-associated CNS disease, deregulation of chemokine expression may be involved in inflammatory processes associated with HIV-1 CNS disease, aberrant chemotactic recruitment of infected and uninfected leukocytes to the BBB and the CNS, and events leading to neurotoxicity and CNS dysfunction (Hesselgesser and Horuk, 1999; Lavi *et al.*, 1998).

C. Cellular Hosts to HIV-1 Infection within the CNS

Replication of HIV-1 within the CNS is a necessary component of the progression of HIV-1 CNS disease and HIVD. As numerous studies have shown, cells of monocyte/macrophage lineage and brain microglial cells are the primary host cells for productive HIV-1 CNS infection. However, it is apparent that other cell types within the CNS also play active roles in the development of HIVD by supporting viral replication. However, these populations of cells, which include astrocytes, neurons, brain microvascular endothelial cells, and oligodendrocytes, do not support the levels of HIV-1 replication found in cells of monocyte/macrophage/microglial origin. Nevertheless, their dysfunction as a consequence of HIV-1 replication within the CNS may be an integral part of HIV-1-associated CNS disease and the appearance of HIVD.

I. Microglial Cells

Microglial cells, which are generally accepted as the primary resident CNS cells that support productive HIV-1 replication within the brain, function as immune effector cells within the CNS. Like their monocyte and macrophage counterparts within the peripheral immune system, microglial cells are recruited to sites of injury or infection where they become phagocytic and release inflammatory cytokines as well as neurotoxic soluble mediators such as TNFα. Microglial cells express many of the same cell surface markers as monocytes and macrophages, but can be distinguished by quantitative differences in expression as well as morphology. Microglia are generally classified by their location within the brain and proximity to blood vessels that perfuse the brain. Parenchymal microglia are long-lived cells that enter the CNS as monocytes during development. Perivascular microglia have a higher rate of turnover and are renewed by monocytes that migrate into the CNS from the peripheral blood (Gonzalez-Scarano and Baltuch, 1999).

Despite the level of neuronal degeneration in HIVD suggested by the large body of clinical findings, HIV-1 infection within the CNS is supported not by neurons, but by microglial cells. Numerous studies have demonstrated that infected microglial cells, macrophages, and MGCs (formed by the fusion of infected and uninfected microglia and macrophages) were present in HIV-1-infected brains at numbers that greatly exceeded the populations of any other infected cell type (Epstein *et al.*, 1984; Gabuzda *et al.*, 1986; Stoler *et al.*, 1986; Wiley *et al.*, 1986). Even when PCR was employed to detect low copy numbers of HIV-1 nucleic acids (Bagasra *et al.*, 1996; Nuovo *et al.*, 1994), cells identified as microglia and macrophages were shown to be the predominant HIV-1-infected cell type.

Like macrophages, microglial cells are readily infected by macrophage-tropic strains of HIV-1 (Watkins *et al.*, 1990) that use the CCR5 coreceptor (Albright *et al.*, 1999; Shieh *et al.*, 1998) and may be less able to support infection by lymphotropic strains (McCarthy *et al.*, 1998). The observation that isolation of infectious virus from the brain yields predominantly macrophage-tropic strains of HIV-1 is consistent with the abundance of infected microglia and macrophages within the brain. Preferential infection of microglial cells may constitute a subset of macrophage tropism, since some primary brain isolates of HIV-1 replicate better in microglial cells than in monocyte-derived macrophage cells (Strizki *et al.*, 1996).

The prevalence of HIV-1-infected microglial cells suggests that the appearance of HIVD may be related to replication within microglial cell populations. However, such correlations have not yet been definitively established. Although several postmortem studies documented levels of microglial cell and macrophage HIV-1 expression consistent with the diagnosis of HIVD, they also found similar levels of expression in a number of nondemented patients and little or no HIV-1 expression in some severely demented patients (Brew *et al.*, 1995b; Glass *et al.*, 1995; Wiley and Achim, 1994). These seemingly incongruous observations might be explained by a temporal relationship between HIVD and viral load in which HIV-1 replication in microglial cells and macrophages and HIV-1 encephalitis precedes the onset of HIVD (Wiley and Achim, 1994). Markers for microglial cell activation have also been correlated with the severity and progression of HIVD (Adamson *et al.*, 1999). Other studies demonstrated that elevated levels of unintegrated HIV-1 DNA were associated with HIV-1 encephalitis (Pang *et al.*, 1990) as well as increased HIV-1 replication and the appearance of MGCs and dementia (Teo *et al.*, 1997). Unintegrated HIV-1 DNA in infected microglial cells may serve as a template for production of low levels of progeny virus (Cara *et al.*, 1996) or neurotoxic viral proteins such as gp120 (Stevenson *et al.*, 1990).

Infection of microglial cells by HIV-1 also elicits the release of soluble viral and cellular factors, which are thought to produce neuronal degeneration ultimately leading to dementia. HIV-1 infection of microglial cells may contribute indirectly to neurotoxicity through interactions between gp120 and microglial cell-expressed CXCR4, which lead to macrophage/microglial cell activation and release of soluble mediators that induce neuronal apoptosis (Kaul and Lipton, 1999). This hypothesis is corroborated by detection of neuronal apoptosis coincident with the regional appearance of multinucleated giant cells and p24 expression in HIV-1-infected brains (Adle-Biassette *et al.*, 1999). Glycoprotein-120 shed by infected microglia may also indirectly impact neuronal function by binding to oligodendrocytes and causing demyelination (Kimura-Kuroda *et al.*, 1994). HIV-1-infected microglial cells may release other viral proteins, including Tat and Nef, which are also potential neurotoxins.

Cellular products commonly implicated as neurotoxins released from microglial cells include cytokines (including TNFα), chemokines, arachidonic acid, and nitric oxide. Primary macrophages and the U-937 monocytic cell line both release increased levels of the CC chemokine monocyte chemotactic protein 1 (MCP-1) in response to HIV-1 infection, suggesting that the same may occur in microglial cells (Mengozzi *et al.*, 1999). Elevated levels of MCP-1 in the CSF, which correlate with the appearance of HIV-1 encephalitis (Cinque *et al.*, 1998), may result in recruitment of greater numbers of HIV-1-susceptible mononuclear immune cells to the CNS. HIV-1 infection of microglial cells has also been correlated with increased expression of inducible nitric oxide synthase (iNOS) in demented brains (Rostasy *et al.*, 1999).

2. Astrocytes

The contribution of HIV-1 replication in astrocytes to CNS viral burden and neuropathogenesis, the recent subject of an extensive review (Brack-Werner, 1999), is still a topic of active debate. Early studies of patients with HIV-1-associated encephalopathy indicated that HIV-1 was located predominantly in MGCs, macrophages, and microglia, but could also be identified to a lesser extent in astrocytes (Epstein *et al.*, 1984; Stoler *et al.*, 1986; Wiley *et al.*, 1986). Investigators in subsequent studies have demonstrated the presence of HIV-1-specific nucleic acids and antigens in cells identified biochemically and morphologically as astrocytes within postmortem tissue samples from AIDS patients with encephalopathy or HIVD (Bagasra *et al.*, 1996; Nuovo *et al.*, 1994; Ranki *et al.*, 1995; Saito *et al.*, 1994). Supporting *in vitro* investigations have shown that human astrocytes in both primary cultures and continuous cell lines are susceptible to HIV-1 infection (Cheng-Mayer *et al.*, 1987; Chiodi *et al.*, 1987; Dewhurst *et al.*, 1987; Harouse *et al.*, 1989a; Tornatore *et al.*, 1991). Infection of these cells results in low levels of HIV-1 replication, characterized by limited expression of viral antigens and nucleic acids, no changes in cell mortality or morphology, and very low to undetectable levels of progeny virus production (Cheng-Mayer *et al.*, 1987; Chiodi *et al.*, 1987; Dewhurst *et al.*, 1987, 1988). Studies have proposed that the restrictive level of replication in astrocytes may be due not only to the lack of CD4 expression, but also to cell type-specific blocks in the replicative cycle (Neumann *et al.*, 1995; Niikura *et al.*, 1996). Astrocytes supporting restricted HIV-1 replication may act as viral reservoirs in the CNS, producing elevated levels of HIV-1 following reactivation by cytokines associated with CNS infection. Astrocyte infection may also be enhanced by the presence of co-infecting pathogens like cytomegalovirus (McCarthy *et al.*, 1998).

Evidence supporting infection of astrocytic cells also comes from investigations that have demonstrated selective expression of the HIV-1 genome within infected astrocytes. Studies of viral expression in the brain have shown that *nef*-specific messenger RNA, transcribed from *nef* gene quasispecies found within the brain (Blumberg *et al.*, 1992), can be detected in astrocytes *in vivo* (Epstein and Gendelman, 1993). Nef overexpression in astrocytes is also a feature of pediatric CNS infection (Saito *et al.*, 1994; Tornatore *et al.*, 1994). Additional evidence concerning the involvement of Nef in HIV-1 glial cell infection comes from *in vitro* studies in which infection of continuous astrocytoma and glioma cell lines resulted in a restricted infection with increased levels of Nef production (Brack-Werner *et al.*, 1992).

Nef expression also resulted in enhanced viral expression and maintenance of replication in primary human astrocytes (Bencheikh et al., 1999). Although Nef expression in immune cell populations results in downregulation of viral expression (Ahmad and Venkatesan, 1988; Bandres and Ratner, 1994) as well as endocytosis of the cell surface CD4 receptor molecule (Aiken et al., 1994), the role of Nef in astrocyte infection is not fully understood.

There is also other *in vitro* evidence indicating that HIV-1 replication can be supported in astrocytes. Primary human fetal astrocytes sustained very low levels of replication after infection with cell-free virus, but produced higher, viral clone-specific levels of p24 expression following transfection with full-length molecular clones (Bencheikh et al., 1999). Although the levels of virus were not as high as those produced by productively infected T lymphocytes, the latter results indicate that astrocytes are able to support HIV-1 transcription as directed from the LTR. In human astrocytic glioblastoma cells, HIV-1 transcription is modulated by TAR-independent Tat transactivation (Taylor et al., 1992). This alternative pathway of Tat activation is mediated by the HIV-1 NF-κB region (Taylor et al., 1994) and involves a factor specific to neuroglial cells (Taylor et al., 1995). *In vitro* transcription assays in astrocytic cells demonstrated that Tat transactivation was sufficient to allow replication of TAR-deletion mutants, indicating an independence from traditional mechanisms of Tat transactivation. Other research groups have also described a TAR-independent Tat activity in HeLa cells (Kim and Panganiban, 1993) and unstimulated human hepatoma cells (Zhu et al., 1996), indicating that Tat can be functional in cells other than the traditional CD4-positive T-lymphocyte, monocyte, and macrophage target cells.

3. Neurons

The extent of neuronal HIV-1 infection and the contribution to HIV-1-associated neuropathogenesis is not yet clear. Early studies using less sensitive detection techniques suggested that neurons were not infected during HIV-1 infection of the CNS. However, recent investigations using the more sensitive *in situ* PCR technique have identified small populations of HIV-1-infected neurons (Bagasra et al., 1996; Nuovo et al., 1994). In one study, infected neurons were identified in 17 of the 22 brains examined and the number of infected neurons correlated loosely with the severity of HIVD (Bagasra et al., 1996). Furthermore, few neurons expressed unspliced HIV-1 mRNA, which is indicative of productive viral replication (Bagasra et al., 1996). The relevance of these small populations of HIV-1-infected neurons to the development of HIVD has yet to be determined. However, their contribution to the pathogenesis of HIV-1 CNS infection is almost certainly considerably less than the impact of neurotoxicity and neuronal death caused indirectly by production of viral and cellular neurotoxins by other neuroglial cell populations.

As is the case with astrocytes, the limited infection of neuronal populations may be the consequence of cell-dependent restrictions in HIV-1 replication. For example, replicative studies in several neuronal cell lines have suggested that neuronal differentiation may be a factor in restricting HIV-1 replication (Truckenmiller *et al.*, 1993; Vesanen *et al.*, 1994). Additionally, HIV-1 LTR-directed reporter gene expression in transgenic mice resulted in neuronal expression in specific regions of the brain when an LTR from a CNS-derived HIV-1 strain was used, suggesting that LTR function may be limited by LTR sequence as well as by the physiology of specific neuronal subpopulations (Corboy *et al.*, 1992). In similar experiments in which the HIV-1 genome was expressed in transgenic mice, transgene expression was anatomically localized and accompanied by a distinct neurological syndrome which involved both the CNS and peripheral nervous system (Thomas *et al.*, 1994).

4. Oligodendrocytes

Although oligodendrocyte infection has been demonstrated *in vitro* (Albright *et al.*, 1996), substantial evidence supporting *in vivo* infection has yet to be reported. Although oligodendrocyte abnormalities have been described in examinations of HIV-1-infected postmortem brains (Esiri *et al.*, 1991), the cellular damage was likely caused by indirect mechanisms of cytotoxicity and not by direct infection. Using *in situ* PCR, Bagasra and colleagues demonstrated infection of oligodendrocytes in numbers less than either astrocytes or neurons (Bagasra *et al.*, 1996). The relevance of these findings has yet to be fully understood.

5. Cells at the CNS Periphery: Endothelial and Choroid Plexus Cells

Endothelial cells, which form the first barrier between the peripheral circulation and the CNS, may provide an entry portal through which HIV-1 may enter the CNS. HIV-1 replication in brain capillary endothelial cells may lead to virus budding into the interior of the brain and infection of adjacent neuroglial cells within the brain by direct cell-to-cell contact or the spread of cell-free viral progeny. *In vitro* and *in vivo* studies suggest that BMECs can support HIV-1 replication. *In vitro*, human BMECs are susceptible to HIV-1 infection (Moses *et al.*, 1993) facilitated by CD4-independent viral entry. *In vivo*, HIV-1 has been detected using *in situ* PCR in microvascular endothelial cells at levels exceeding 65% in postmortem CNS tissue of AIDS patients (Bagasra *et al.*, 1996). Endothelial cells are also infected in animal models of HIV-1 infection. HIV-1 infection of the cerebral vascular endothelium was demonstrated 60 weeks after intravenous inoculation of pigtailed macaques with the T-tropic strain LAI (Frumkin *et al.*, 1995).

HIV-1 may also enter through replication in epithelial cells of the choroid plexus, as has been observed with visna virus (Johnson, 1982). Choroid plexus cells have been shown to be susceptible to HIV-1 infection (Bagasra *et al.;* 1996; Falangola *et al.,* 1995; Harouse *et al.,* 1989b). This port of entry may contribute to the regional distribution of HIV-1 observed within the brain, particularly within deep white matter and diencephalic and mesencephalic structures, but not within the cortex (Dickson *et al.,* 1991; Kure *et al.,* 1990a; Price, 1994).

D. Infection and Control of HIV-I Gene Expression in Cells of the Monocyte/Macrophage Lineage

The presence of multinucleated giant cells in HIV-1-infected brains, one of the hallmarks of HIV-1 CNS infection, is an indicator of ongoing viral replication within cells of monocyte/macrophage and microglial origin. As in other cells susceptible to HIV-1 infection, viral replication in monocytoid cells is subject to controls imposed by the host cell. For example, coreceptor-mediated viral entry is dependent on the host cell phenotype, since differentiation from monocyte to macrophage is accompanied by upregulation of CCR5 expression, which renders the cell more permissive to infection by R5 viruses. Transcription of HIV-1 from the provival genome also offers numerous opportunities to modulate viral replication in macrophages and microglia. Regulated HIV-1 gene expression requires interactions between host cell transcription factors, the RNA polymerase II complex, and the HIV-1 long terminal repeat (LTR), that section of the proviral DNA which serves as the HIV-1 promoter. Although numerous transcription factors interact with the LTR and modulate viral transcription, members of the Sp and C/EBP transcription factor families have particularly distinct roles in monocyte/macrophage HIV-1 transcription and are discussed in detail below.

The retroviral replication cycle consists of virus binding and entry, reverse transcription, viral DNA integration into the host cell genome, transcription and translation of virus-specific nucleic acids and proteins, and progeny virus assembly and release. During the process of reverse transcription, the HIV-1 RNA genome is transcribed by the HIV-1-encoded reverse transcriptase enzyme into a double-stranded, linear DNA molecule that is integrated into the host genome. The integrated proviral DNA molecule consists of the coding regions for all of the structural, enzymatic, and regulatory proteins flanked by two identical LTRs. The LTRs perform multiple functions during integration and viral expression. In its 5' position, the LTR serves as the promoter for the entire complement of HIV-1 genes.

The complete LTR is comprised of approximately 650 nucleotides and is divided into the U3, R, and U5 regions, named according to their origins within the single-stranded RNA viral genome (Fig. 4). Of the three, the U3 and R regions are most intimately involved in regulating HIV-1 transcription. The R region, the 5' end of which marks the initiation site of transcription, contains binding sites for several transcription factors. The R region also

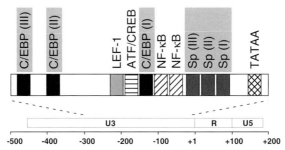

FIGURE 4   The 5' HIV-1 LTR regulates transcription of the viral genome through interactions with numerous transcription factors. The illustration depicts the HIV-1 5' LTR and the relative positions of important transcription binding sites. Binding sites for the Sp and C/EBP transcription factor families play important roles in HIV-1 regulation in cells of monocyte/macrophage/microglial origin.

contains sequences that encode the TAR element, the RNA structure through which the HIV-1 transactivator protein Tat greatly increases HIV-1 expression by promoting elongation of nascent RNA transcripts. The U3 region contains numerous binding sites for cellular and viral transcription factors (Kingsman and Kingsman, 1996). Included within this region are elements found in other eukaryotic promoters, including a TATAA box and binding sites for transcription factors Sp1, NF-κB, C/EBP, and LEF-1 (Fig. 4).

LTR regulation within monocytic cells and other cell types susceptible to HIV-1 infection is dependent on both availability of transcription factors that bind the LTR and cell type-dependent posttranslational modification of those factors. The array of transcription factors available to the LTR within macrophages and microglial cells is, in turn, closely related to the differentiation state of the cell. Macrophage differentiation begins with monocyte precursor cells that arise from pluripotent stem cells in the bone marrow. Monocyte precursors spend approximately 6 days in the marrow before moving into circulation where they differentiate into mature monocytes (Hoffbrand and Pettit, 1993). Monocytes remain in circulation 8 to 72 h before entering a given tissue and differentiating further into macrophages (Hoffbrand and Pettit, 1993). Those that take up long-term residency are known as tissue macrophages and may become even more differentiated than macrophages recruited in response to an immunologic challenge. In the brain, resident macrophages are known as microglial cells. During this progressive differentiation, HIV-1 LTR activity and viral expression may be altered as a consequence of cellular differentiation and the changing milieu of transcription factors within the nucleus. The following sections describe two transcription factor families that are tied to both monocytic differentiation and LTR regulation of HIV-1 expression and are, therefore, relevant to HIV-1 replication within monocytic cells within the peripheral blood as well as brain.

I. Sp1 Transcription Factor Family

During the process of monocyte/macrophage differentiation, alterations in the milieu of active transcription factors prompt the expression of genes specific to the differentiated monocyte (Clarke and Gordon, 1998). Recent studies have implicated cis-acting promoter elements that bind members of the Sp transcription factor family in this process. Deletion analysis of the lysosomal acid lipase promoter illustrated that the Sp and AP-2 binding sites played an important role in the differentiation-specific expression of the lipase gene product (Ries *et al.*, 1998). We have also performed studies that suggest that Sp factors may also have important roles in HIV-1 replication in cells of monocyte/macrophage/microglial lineage (Millhouse, 1998).

Sp1 was originally isolated and cloned by way of its interaction with the G/C rich region of the SV40 early promoter (Briggs *et al.*, 1986; Kadonaga *et al.*, 1987). Like most transcription factors, Sp1 acts to enhance transcriptional initiation by stabilizing the basal promoter complex and recruiting additional factors via protein–protein interactions. Initially, the primary role of Sp1 was believed to be in basal transcription of housekeeping genes. However, recent work has revealed a greater level of complexity in its role in cellular transcription. Of particular interest was the discovery of additional Sp factors with high levels of homology to Sp1. The Sp family of transcription factors now includes Sp1, Sp2, Sp3, Sp4, and a number of more distantly related factors such as the BTEB factors (Hagen *et al.*, 1992; Imataka *et al.*, 1992; Kingsley and Winoto, 1992; Sogawa *et al.*, 1993). Both Sp1 and Sp3 are ubiquitously expressed in tissues throughout the body. In contrast, Sp4 is expressed in a brain-restricted manner (Hagen *et al.*, 1992).

Recognition of cis-acting binding sites by members of the Sp factor family is closely tied to the impact of these factors on gene expression. Due to the strong homology of their zinc finger DNA binding domains, Sp1, Sp3, and Sp4 share a similar affinity for binding G/C-rich sequences (Hagen *et al.*, 1992, 1994; Kingsley and Winoto, 1992). Unlike the other Sp family members, Sp2 favors binding to an alternative GT-rich nucleotide sequence found in the promoter of a T cell receptor gene (Kingsley and Winoto, 1992). Due to the near equivalent nucleotide recognition preferences of the Sp factors, competition between various Sp factors present in a given cell type may influence the identify of the factor that predominates within the basal promoter initiation complex. Such competition is functionally interesting, since Sp1 and Sp3 exhibit different abilities to activate transcription. Sp1 has been shown to be a strong activator of transcription (Kadonaga *et al.*, 1987). Interestingly, transcription from the Sp3 promoter produces transcripts with two translational initiation sites, resulting in the generation of both full-length and truncated Sp3 factors (Kennett *et al.*, 1997). Although the full-length protein may retain some ability to facilitate transcription, it is clearly a weaker transcriptional activator than Sp1, and in several viral and eukaryotic promoters it facilitates repression of Sp1-activated transcription (Birnbaum *et al.*, 1995; Hagen *et al.*, 1994). Hence, the use of Sp3 in systems primarily driven by Sp1-mediated transcriptional activation results in repression of transcription (De Luca *et al.*, 1996). Additional studies have indicated that the Sp3 protein also contains a portable domain which represses transcription independent of its ability to compete with Sp1 for factor binding sites (De Luca *et al.*, 1996; Majello *et al.*, 1994, 1997). The truncated factor acts as a competitive inhibitor (Kennett *et al.*, 1997). In this case, inhibition is generated by competition for factor binding sites among other Sp factors and truncated Sp3, which lacks a functional transcriptional activation domain while maintaining a functional DNA binding domain. The expression of Sp4 within cells in the brain adds an additional level of complexity to transcriptional regulation by Sp family members. Like Sp1, Sp4 also acts to enhance transcription (Hagen *et al.*, 1994, 1995).

Our understanding of the regulation of Sp factors within the nucleus is still incomplete. Although Sp1 can be phosphorylated by a DNA-dependent protein kinase and modified with O-linked oligosaccharides (Jackson *et al.*, 1990; Jackson and Tjian, 1988), potential regulatory roles for such modifications have yet to be fully established. Although glycosylation of Sp factors has not been examined extensively, studies have been performed to examine the impact of phosphorylation on Sp factor activity. Sp1 binding to cis-acting G/C elements has been shown to induce phosphorylation of transacting Sp factors in a manner that correlates with function (Jackson *et al.*, 1990). Several studies have shown that Sp1 exists in a phosphorylated state in the liver, brain, kidney, and spleen and that this phosphorylated state reduces Sp1 DNA binding activity (Armstrong *et al.*, 1997; Leggett *et al.*, 1995). Both studies imply a regulatory role associated with phosphorylation of Sp1. Phosphorylation may reduce the ability of Sp1 to activate transcription by reducing its binding affinity or increase its ability to activate transcription after DNA binding is established. These two potential regulatory roles for Sp factor phosphorylation are not exclusive and may represent segments of a larger regulatory cascade.

Regulation of Sp factors at a cellular level would clearly impact the activity of cellular and viral promoter structures within a given cellular environment. Sp factor activity may also be regulated by the quantity of the various Sp family members present in the nucleus of a given cell type. Under such circumstances the nuclear levels of the strongly activating Sp1 relative to levels of the weaker full-length Sp3 and the repressive truncated Sp3 proteins would be of particular interest. Cell types expressing high levels of Sp family activators relative to repressor forms would be expected to impart higher activity on promoter structures containing cis-acting Sp elements, while those expressing high levels of Sp family repressors would be expected to repress these same promoters. Cell type- or differentiation-dependent regulation of Sp family expression may thus represent an additional level of complexity in the regulation of Sp factor-mediated transcriptional activation.

In light of these findings it may be necessary to reexamine the role of the Sp family of transcriptional factors in HIV-1 gene expression and viral replication. Clearly, the G/C box array is a central element within the LTR structure (Fig. 4). The positioning of the three G/C-rich elements within the LTR (nucleotides -45 to -77 relative to the site of RNA transcription initiation) allows for recruitment of multiple Sp factors directly adjacent to the TATA box and subsequent upregulation of transcription (Jones *et al.*, 1986; Zeichner *et al.*, 1991). The Sp sites are also important in Tat-induced transcription (Berkhout and Jeang, 1992). Mutations which compromise all three cis-acting Sp elements were shown to have profound effects on both basal and Tat-mediated transcriptional activation in HeLa cells (Harrich *et al.*, 1989). Interestingly, compromise of a single Sp cis-acting element in this system resulted in only a minimal decrease in basal and Tat-induced activity. In addition, synthetic promoters have been generated which are Tat-inducible in a Sp1-dependent manner (Kamine *et al.*, 1991). Consistent with this observation is the finding that Tat and Sp1 interact directly, as demonstrated via affinity purification (Jeang *et al.*, 1993). This interaction may influence Sp factor activity by directly modulating phosphorylation of the Sp1 protein (Chun *et al.*, 1998). The Sp/Tat interaction is also position-dependent since insertion of additional sequences between the G/C array and the TATA box resulted in strong reductions in Tat inducibility (Huang and Jeang, 1993). Clearly, these studies, as well as studies performed by others, have shown that the G/C box array and the Sp family members it recruits are important to both basal and Tat-induced LTR activity. The cis-acting Sp sites may also play an important role in NF-κB-mediated LTR activation. The induction of LTR enhancer activity by mitogens has been shown to be dependent on a protein–protein interaction between NF-κB and Sp1 (Perkins *et al.*, 1993). This interaction is thought to be dependent both on position and orientation of the cis-acting Sp element. The biologic relevance of the various transcriptional effects mediated by the Sp family of transcription factors is underscored by studies which showed significant alterations in viral replication stemming from alterations in the G/C box array (Chang *et al.*, 1993; Ross *et al.*, 1991).

Majello and coworkers have attempted to dissect the role of the individual Sp family members on HIV-1 gene transcription (Fig. 5). Their studies have indicated that Sp1 acts as a strong activator of both basal and Tat-induced transcription. It can also facilitate NF-κB-mediated LTR enhancement. Like Sp1, Sp4 was shown to activate basal and Tat-induced transcription from the LTR. However, Sp4 lacks the ability to successfully facilitate NF-κB-mediated LTR enhancement. Conversely, Sp3 was unable to activate either basal or Tat-induced LTR activity. Moreover, when coexpressed with Sp1, Sp3 acted to repress the ability of Sp1 to activate transcription. Because

A

Factor	Protein	Expression	Binding Site	Function	
				General	HIV-1 LTR
Sp1	90 KDa	Ubiquitous	GC box	Activator	Activator
Sp2	90 KDa	Ubiquitous	GT box	Activator	N/A
Sp3	97 KDa	Ubiquitous	GC box	Activator/Repressor	Repressor
	60 KDa	Ubiquitous	GC box	Repressor	Repressor
Sp4	90 KDa	Brain	GC box	Activator	Activator

B

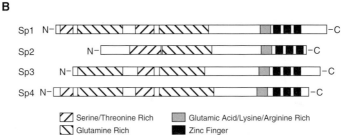

FIGURE 5   Sp factors and their function. The size, cellular expression, and binding site preference of Sp factors Sp1, Sp2, Sp3, and Sp4 are listed. Their general function as well as HIV-1 LTR-specific activity are also indicated.

of the recent observation that Sp3 transcripts generate both full-length and truncated proteins (Kennett et al., 1997), these coexpression results must be reevaluated. Whether repression was generated by full-length Sp3, truncated Sp3, or some combination of the two is not immediately apparent. One possible interpretation is that competitive inhibition by truncated Sp3, which lacks an activation domain, was responsible for the observed repression. This concern may be placated by the observation that the DNA binding and transcriptional repression domains of Sp3 can be successfully separated in a GAL4–Sp3 fusion protein. Hence, Sp3 acts as a repressor of HIV-1 LTR-mediated transcription (Majello et al., 1994). Although Majello did not address the effect of truncated Sp3 protein, one can safely infer, based on the physical structure of Sp3, that it would act as a competitive inhibitor of Sp1- or Sp4-mediated transcriptional activation. This would indicate that Sp3 and its truncated form would oppose Sp1- and or Sp4-mediated LTR activation, which again illustrates the importance of understanding the cell- and differentiation-dependent mechanisms that regulate the predominance of Sp activators or Sp repressors and the modification of Sp factors.

We have shown that the amounts of activating Sp1 factor relative to the levels of the repressive truncated Sp3 factor increase in cell lines representing a greater state of monocytic differentiation (McAllister and Wigdahl, unpublished data). These findings mirror studies performed using differentiating keratinocytes, which demonstrated that Sp3 is in greater abundance than Sp1 before differentiation and that this relationship inverts upon *in vitro* keratinocyte differentiation (Apt *et al.*, 1996). We have also shown that LTR sequence variation occurring during the course of HIV-1 infection may impact the binding and ultimate function of Sp factors during viral replication. The brain-derived YU-2 strain of HIV-1 contains a single base pair change in the promoter distal Sp binding site. Our studies have shown that this naturally occurring binding site is severely limited in its capacity to bind Sp factors. Furthermore, in the context of a functional LTR, the presence of this site results in a significant reduction in transient LTR activity in lymphocyte (Millhouse *et al.*, 1998) and monocyte cell lines (McAllister and Wigdahl, unpublished data). These observations suggest a role for cis-acting Sp elements in regulating HIV-1 replication within cells of monocytic origin located in either the peripheral blood or brain.

The long-standing importance of the G/C box array in the HIV-1 promoter and the newly described importance of cis-acting Sp elements in the regulation of monocytic specific promoters indicates a need for further investigations into the role of the Sp factors during HIV-1 replication in cells of the monocytic lineage. A better understanding of the interactions which drive replication in this system will assist clarification of our view of HIV-1 replication in the nervous system and the pathogenesis of HIVD.

2. C/EBP Transcription Factor Family

In 1988, Landschulz and colleagues identified and cloned a protein that interacted with the CCAAT box of several promoters as well as a core homology sequence (TGTGGA/TA/TA/TG) found in several viral enhancers (Landschulz *et al.*, 1988). This protein was subsequently designated the CAAT/enhancer binding protein (C/EBP). At least eight proteins have been identified which constitute the C/EBP protein family. These proteins appear to be important in the transcriptional regulation of HIV-1 as well as several different families of cellular genes. As regulators of transcription, C/EBP factors function independently and via numerous protein–protein interactions with other transcription factor family members. Their cell type-specific expression patterns and activities also suggest that they may have important roles in HIV-1 infection and associated pathogenesis within CNS.

The C/EBP family of proteins is a subset of the b-ZIP family of transcription factors (which also includes members of the ATF/CREB transcription factor family). C/EBP proteins exhibit strong homology in the basic and leucine zipper regions at their C termini. Over eight C/EBP proteins have been cloned and characterized (Wedel and Ziegler-Heitbrock, 1995). In many cases, the same proteins were cloned by different investigators and in different animal systems, giving many of these proteins multiple synonyms. Some members activate transcription, while others serve to repress it. C/EBP family members C/EBPα, NF-IL6 (C/EBPβ), C/EBPδ, C/EBPε, and CRPI are transcriptional activators. Truncated forms of C/EBPα (a 30-kDa C/EBPα protein) and C/EBPβ (a 21-kDa C/EBPβ protein, referred to as LIP) are negative regulators of transcription and lack some of the N-terminal protein sequences, where the transactivation domains are located in the full-length proteins. Since these proteins bind DNA but do not trans-activate (Descombes and Schibler, 1991), it is thought that they dimerize with full-length activator proteins and thereby block transactivation (Sears and Sealy, 1994). It was previously thought that CHOP-10 was a trans-dominant repressor protein, due to its inability to bind DNA (Ron and Habener, 1992). However, recent evidence suggests that CHOP–C/EBP het-erodimers can bind a select and relatively specific subset of genes that contain special C/EBP sites (Ubeda et al., 1996). C/EBPγ also functions as a trans-dominant negative repressor (Cooper et al., 1995). Interestingly, C/EBPγ is ubiquitously expressed in all adult tissues and cell lines investigated (Roman et al., 1990), indicating that this specific protein may act as a general buffer for C/EBP activator proteins in a cell-type-independent manner (Cooper et al., 1995).

As is the case with other b-ZIP protein families, C/EBP family members form homo- and heterodimers with one another. C/EBP family members are apparently not restricted in their ability to bind with other members of the C/EBP family (Cao et al., 1991; Roman et al., 1990; Williams et al., 1991). However, since the pattern of certain amino acids within the leucine zippers of b-ZIP proteins differs between family members, specific dimeriza-tion pairs may be preferred (O'Shea et al., 1991; Vinson et al., 1993).

While C/EBP proteins are expressed in fibroblasts, placenta, pituitary gland, glioblastoma, endothelial cells (Wedel and Ziegler-Heitbrock, 1995), and intestinal cells (Chandrasekaran and Gordon, 1993), high levels of C/EBP mRNA and protein expression are limited to several cell types, includ-ing hepatocytes, adipocytes, neurons, myeloid cells, and B cells. Most rele-vant to studies of HIV-1 CNS infection is the observation that C/EBP proteins appear to play a prominent role in the regulation of gene expression in myelomonocytic cells. Regulatory regions of many genes expressed in mono-cytes and macrophages contain C/EBP binding sites, including MIP-1α, TNFα, interleukin-6 (IL-6) (Bretz et al., 1994; Matsusaka et al., 1993; Tanaka et al., 1995), and IL-8 (Matsusaka et al., 1993; Stein and Baldwin, 1993). Furthermore, C/EBP proteins have been shown to regulate the tran-scription of these monocyte/macrophage-expressed genes, with specific cell-signaling events playing a significant role in the regulation of C/EBP fam-ily members.

C/EBP factors are also linked to cellular activation. LPS has been shown to increase the amounts of C/EBPβ and -δ in several tissues and may contrib-ute, in part, to LPS induction of various genes (Kinoshita et al., 1992; Natsuka et al., 1992). Studies have also shown that C/EBPβ is translocated to the nucleus of LPS-stimulated peripheral blood monocytes and contributes to the activation of IL-6, MCP-1, and IL-1β (Bretz *et al.*, 1994). However, another study has shown that all three proteins (C/EBPα, C/EBPβ, and C/EBPδ) are involved in mediating LPS-inducible expression of IL-6 and MCP-1 within lymphoblasts. Lymphoblasts normally lack C/EBP proteins and do not display LPS induction of proinflammatory cytokines, indicating some of the activities of these proteins can be redundant with regard to IL-6 and MCP-1 expression (Hu *et al.*, 1998). Additionally, when U-937 cells, a human promonocytic cell line, were treated with PMA (phorbol 12-myristate 13-acetate) to induce differentiation along the monocyte/macrophage pathway, enhanced binding of C/EBPβ but not -α or -δ was encountered (Combates *et al.*, 1997).

C/EBPε may also play a role in normal myeloid development (Antonson *et al.*, 1996; Chumakov *et al.*, 1997). Expression studies using RT–PCR demonstrated expression of C/EBPε mRNA in myeloblastic, myelomonoblastic, and T-cell lymphoblastic leukemic cell lines, which appeared to be upregulated during differentiation toward a granulocyte phenotype and downregulated when cells were induced to differentiate toward a macrophage phenotype (Morosetti *et al.*, 1997; Yamanaka *et al.*, 1997). Additional studies using antisense oligonucleotides determined that inhibition of C/EBPε expression in HL-60 and NB4 myloblasts and promyelocytes decreased their proliferative capacity. More recent studies indicated that C/EBPε may be particularly important in the expression of different chemokines and chemoattractants as well as an important regulator of myeloid cell type differentiation and cytokine gene expression in differentiated cells (Williams *et al.*, 1998).

C/EBPβ has also been shown to be a neuronal transcriptional regulator in the mouse (Sterneck and Johnson, 1998). Utilizing *in situ* hybridization analysis, these studies demonstrated that C/EBPβ is widely expressed in the adult mouse CNS in cells of the hippocampus and dentate gyrus and in cerebellar Purkinje and granule cells. Using PC12 cells (a pheochromocytoma cell line that differentiates into neuronlike cells in response to nerve growth factor), it was shown that C/EBPβ mRNA levels increased and protein levels correspondingly decreased with differentiation of the PC12 cells. Thus, C/EBPβ may also participate in neurotrophin signaling within the brain.

A number of studies have demonstrated important interactions between C/EBP family members and other transcription factors in the context of a wide array of promoters (Fig. 6). Some of the major protein families which have been implicated in these interactions include Sp1, ATF/CREB, and NF-κB (Fig. 6A). These interactions also provide insight into the functions of C/EBP family members. Interactions with Sp, ATF/CREB, and NF-κB factors are particularly relevant to transcriptional regulation of HIV-1 in cells of monocyte/macrophage/microglial origin, since all of these factors interact functionally with the LTR during transcriptional regulation of HIV-1 within this cell lineage. Furthermore, the close proximity of these

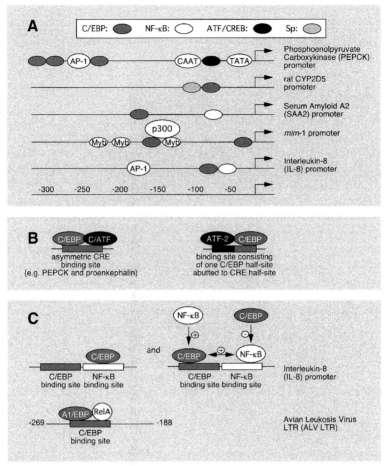

FIGURE 6   C/EBP proteins interact with numerous proteins from other transcription factor families to synergistically increase transcription. (A) C/EBP proteins have been shown to interact with ATF/CREB, Sp, and NF-κB proteins in a number of promoter systems. Relative positions of binding sites are depicted to scale. (B) C/EBP proteins can also heterodimerize with other protein family members, such as different ATF/CREB proteins. (C) C/EBP proteins have been shown to bind NF-κB elements, heterodimerize with κB proteins, and exhibit enhanced binding due to interactions with NF-κB family members. This figure was adapted from data in previously published reports (Bowers *et al.*, 1996; Lee *et al.*, 1997, 1994; Mink *et al.*, 1997; Ray *et al.*, 1995; Roesler *et al.*, 1996; Shuman *et al.*, 1997; Stein and Baldwin, 1993; Vallejo *et al.*, 1993).

factor binding sites to important C/EBP binding sites in the LTR may facilitate functional interactions between C/EBP and the above transcription factors (Fig. 4).

C/EBP proteins have been shown to interact with Sp1 proteins to regulate expression of CD11c integrin, another protein produced in myelomonocytic cells (Lopez-Rodriguez *et al.*, 1997). In this case, C/EBPα, Sp1, and AP-1 recognized a composite regulatory element. Additionally, studies by Lee and coworkers demonstrated that Sp1, bound to a nearby Sp1 binding site, can facilitate C/EBPβ binding to a C/EBP binding site which is not normally recognized by C/EBP factors (Lee et al., 1994, 1997). These studies also demonstrated a differential interaction between C/EBPα and Sp1, such that C/EBPβ and not C/EBPα could interact with Sp1 to activate a rat liver gene (CYP2D5 P-450). Domain switch studies indicated that the interaction between Sp1 and C/EBPβ was dependent on the leucine zipper and activation domain of C/EBPβ. Therefore, two members of the same transcription factor family could achieve differences in target binding sequence specificity due to differential interactions with cooperating Sp family members.

Binding site sequence specificity may also be affected due to interactions with ATF/CREB family members (Fig. 6B). ATF-2 can dimerize with C/EBPα to bind asymmetric binding sites composed of half-site consensus sequences for each monomer (Shuman et al., 1997). While this interaction increases activation from these asymmetric sites, it decreases activation from consensus C/EBP binding sites. Vallejo and colleagues reported the interaction of C/EBPβ and C/ATF (a member of the ATF/CREB family) (Vallejo et al., 1993). These dimers bind to a subclass of asymmetric CRE sites rather than C/EBP sites. Both of these studies indicate that cross talk between these two families may be important and may serve to integrate different hormonal and developmental stimuli. We have performed studies which have suggested that C/EBP binding may be related to occupancy of an adjacent ATF/CREB binding site within the HIV-1 LTR (Krebs et al., 1997) (Fig. 4). Further studies have indicated that ATF/CREB binding may augment C/EBP binding to an otherwise low-affinity C/EBP binding site sequence that occurs naturally in LTRs from a large number of HIV-1 strains (Ross and Wigdahl, unpublished data).

ATF/CREB family members also control C/EBP activation in a different manner. CREB protein binds to two cognate sequences within the promoter of the NF-IL6 gene (Niehof et al., 1997). Thus, expression of C/EBP family members appears to be intimately related to ATF/CREB signaling pathways. This further strengthens the importance of cross talk between these two families of transcription factors. Furthermore, C/EBPα recruited to the cAMP response element has been implicated as an important respondent to liver-specific cAMP responsiveness of the PEPCK (phosphoenolpyruvate carboxykinase) promoter (Roesler et al., 1996). These studies indicate that the complex interactions between the ATF/CREB and C/EBP factors need to be further investigated to determine the extent of their interactions in the context of regulating HIV-1 gene expression.

Extensive studies indicate that members of the NF-κB transcription factor family also interact with C/EBP proteins to regulate gene expression (Fig. 6C). The p50 subunit of NF-κB associates with NF-IL6 (LeClair et al., 1992) through the Rel homology domain and leucine-zipper motif, respectively. This discovery was soon followed by the demonstration of a general interaction between these two families of proteins, since p65, p50, and Rel could functionally synergize with C/EBPα, C/EBPβ, and C/EBPδ. These interactions actually lead to the inhibition of promoters containing κB elements and the synergistic stimulation of promoters containing C/EBP binding elements. Interactions between Rel A and C/EBPβ at a composite site created by adjacent NF-κB and C/EBP sites was also recently reported in the intercellular adhesion molecule-1 (ICAM-1) promoter in several endothelial cell lines (Catron *et al.*, 1998). Studies in the regulation of the serum amyloid A gene indicated that not only did members of these two transcription factor families interact, but heteromeric complexes were much more potent transactivators of gene expression than either family alone (Ray *et al.*, 1995). These heteromeric complexes were actually capable of efficiently promoting transcription from binding sites for both κB and C/EBP. This is indicative of yet another mechanism by which cross talk between different transcription factor family members may impact the regulation of a large number of genes.

In 1993, it was discovered that C/EBPβ could transactivate the HIV-1 LTR in transient transfection assays and that the LTR contained three binding sites for this protein (Tesmer *et al.*, 1993). Since then, evidence for the importance of C/EBP family members in HIV-1 replication has grown. It was demonstrated that C/EBP proteins transactivated the HIV-1 LTR in the U-937 promonocytic cell line (Henderson *et al.*, 1995). Utilizing site-directed mutagenesis, it was shown that LTR-directed transcription in these cells required one functional C/EBP site and that the high-affinity C/EBP sites were functionally equivalent. Further studies indicated that these two C/EBP binding sites were required for replication of an infectious HIV-1 molecular clone in U-937 promonocytic cell lines as well as in primary monocyte/macrophage populations. However, they were dispensable for replication of an infectious molecular clone in various T-cell lines and primary T-cell populations (Henderson and Calame, 1997; Henderson *et al.*, 1996).

A recent study by Honda and coworkers illustrated an interesting relationship between HIV-1 infection of monocytic cells and C/EBP. HIV-1 replication in THP-1 derived macrophages (THP-1 promonocytic cell line treated with LPS or IFN-β) was inhibited after *Mycobacterium tuberculosis* infection, even though HIV-1 replication is normally increased during tuberculosis. The reason for this replication suppression was that M. *tuberculosis*, LPS, and IFN-β all induced the inhibitory C/EBPβ protein LIP. *In vivo*, alveolar macrophages from normal lung also strongly expressed this inhibitory protein, but proinflammatory stimulation due to pulmonary tuberculosis abolished LIP expression and induced a novel C/EBP DNA binding protein. Thus, the THP-1-derived macrophages are similar to alveolar macrophages *in vivo*, and a normal cellular immune response leads to a switching of C/EBP family members, which allows for high levels of HIV-1 replication (Honda *et al.*, 1998).

Binding of HIV-1 to a potential host cell has recently been shown to activate cell signaling pathways which increase C/EBP protein expression (Fig. 7). Popik and colleagues demonstrated that binding of HIV-1 glycoproteins (from both macrophage-tropic and T-cell-tropic strains) to a functional CD4 receptor rapidly activated the ERK/mitogen-activated protein (MAP) kinase pathway. This resulted in phosphorylation and increased promoter binding of C/EBP (as well as AP-1 and NF-κB), which stimulated the expression of cytokine and chemokine genes including IL-6 (Popik *et al.*, 1998). These results imply that interactions between infected cells and viral particles or soluble envelope glycoproteins may stimulate HIV-1 replication through C/EBP transcription factors. Such a mechanism may serve to increase levels of virus produced by HIV-1-infected macrophages and microglial cells in the brain.

HIV-1 Tat protein also induces IL-6 (Ambrosino *et al.*, 1997). This observation is consistent with previous results that have reported an increased level of IL-6 in serum and CSF of HIV-1-infected patients (Gallo *et al.*, 1989). As previously indicated, IL-6 signaling leads to an increase in expression of C/EBP family members, which in turn transactivate the LTR. Therefore, Tat expression effectively increases the levels of C/EBP available

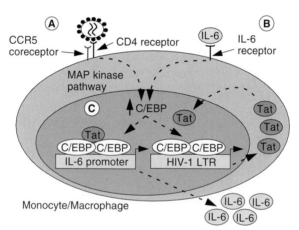

FIGURE 7   HIV-1 infection of monocyte/macrophage populations is intimately connected to expression of C/EBP family members. (A) Binding of the HIV-1 glycoproteins to the CD4 receptor and (B) binding of interleukin-6 (IL-6) to its cognate receptor results in (C) increased levels of activated nuclear C/EBPβ and increased binding of C/EBP factors. HIV-1 Tat protein has also been shown to increase C/EBPβ binding to the IL-6 promoter, resulting in increased levels of secreted IL-6. In summary, HIV-1 infection increases C/EBP nuclear protein expression by at least two different pathways, creating a C/EBP activation loop. This figure was adapted from data in previously published reports (Akira *et al.*, 1990; Ambrosino *et al.*, 1997; Popik *et al.*, 1998).

for further transactivation of the LTR. Additionally, Ambrosino and co-workers discovered that Tat can function by enhancing the C/EBP DNA binding activity and can actually complex with *in vitro* translated C/EBPβ. The investigators also observed an *in vivo* interaction between Tat and C/EBPβ utilizing the yeast two-hybrid system.

Several groups also reported interactions with a family of general transcription factors that may be important to C/EBP-directed transcription. NF-κB and C/EBP interactions both have functional roles in HIV-1 regulation. Complexes were observed between p50 and C/EBPβ and -δ in cells transfected with a p50 expression plasmid (Ruocco *et al.*, 1996), which strongly activated the HIV-1 LTR through the κB binding sites. Mutational analysis indicated that this transactivation relied on the DNA-binding domain of p50 and the transactivation domain of C/EBPβ. To complicate the picture, studies conducted by Mondal and coworkers indicated that an NF-κB-proximal C/EBP binding site within the HIV-1 LTR negatively regulated the HIV-1 promoter in brain-derived cells due to interactions with NF-κB (Mondal *et al.*, 1994). The studies indicated that HIV-1 LTR activity was downregulated due to increased DNA–protein complex formation between endogenous proteins and an NF-κB DNA probe with no detectable association with C/EBP in the complex. The differential use of transcription factor families and individual family members within specific cell populations adds an additional level of complexity to the regulation of HIV-1 gene expression.

E. Cellular Interactions Mediating HIV-1-Associated CNS Damage

The progression from initial CNS infection to development of HIV-1-associated CNS dysfunction and HIVD is the consequence of complex interactions that take place between cells within the CNS. During the spread of HIV-1 within the brain, infected and uninfected cells interact with each other through numerous cellular and viral soluble factors, resulting in changes in the collective function of these resident CNS cells and physical alterations to the structure of the brain. To better illustrate the multitude of cellular interactions, increasingly complex models of neuroglial cell damage have been proposed.

Most of the observations made in brains of patients suffering from HIVD demonstrate that the primary cell types infected within the CNS are CD4-positive, CCR5-positive cells of macrophage and microglial origin. The levels of HIV-1 replication in these cell populations support not only direct cellular damage as a consequence of intracellular viral replication, but also indirect cellular damage caused by the release of inflammatory or toxic soluble mediators. While some evidence does exist to support HIV-1 infection of neuroglial cells, the apparent level of infection would seem to be insufficient to account for more than a small part of the clinical and pathological effects of HIV-1 CNS infection. Hypothetical models of HIVD must include not only the contributions made by HIV-1-infected macrophages and microglial cells, but also their complex interactions with largely uninfected neuroglial cell populations. Accurate models of HIVD must strive to include the following: (1) HIV-1 infection of macrophage and microglial cells that results in disruption of normal cellular functions and propagation of HIV-1 within the CNS; (2) limited levels of HIV-1 replication in astrocytes and, perhaps, in neurons; and (3) release of abnormal levels of soluble mediators as a consequence of HIV-1 infection or as a secondary response by uninfected cells to extracellular factors or cell-to-cell interactions. All of these issues have been addressed in increasingly complex models described in a previous review of HIVD (Price, 1994).

1. Direct Effects of Neuroglial Cell Damage

Viral infection of a neuroglial cell within the CNS can result in perturbation of cellular processes and, in the extreme, compromise the integrity of the cell to the point of cell death. Changes that accompany viral replication can directly affect the role of that cell within the context of normal CNS function. Poliovirus, a virus which infects neurons and causes neuronal cytolytic damage as a direct consequence of infection, is an example of a pathogen that directly affects its host cell, and, by doing so, causes disease. Like poliovirus, HIV-1 can infect host cells within the CNS and cause direct cellular damage through alterations of cellular metabolism and function, intracellular toxicity of viral proteins, or insertional mutagenesis and transactivation. However, unlike poliovirus, the principal cell type that supports viral replication (cells of macrophage/microglial cell lineage) is not the same cell type that underlies the pathogenesis of HIVD (neurons). Although the direct damage to macrophages and microglia caused by HIV-1 infection is the foundation for development of HIVD, its importance lies in the indirect impact on other cells within the CNS, including neurons. Levels of astrocytic HIV-1 infection documented in infected brains may also directly affect astrocytic function and indirectly affect neuronal function. Although neurons are largely uninfected, the contribution of low-level neuronal infection to CNS damage may still warrant consideration.

2. Indirect Effects of Neuroglial Cell Infection

The pathological and clinical findings associated with HIV-1 CNS infection and HIVD are not consistent with a model in which infection of a single cell type is the sole cause of neurological disease. Any hypothetical reconstruction of the mechanisms and events that lead to HIVD must also take into account the impact of HIV-1 replication on uninfected "bystander" cells and the contribution of their dysfunction to the onset of disease. An accurate model of HIVD must include cells that play at least four roles in the events that precipitate the onset of dementia (Fig. 8) (Price, 1994). In interactions between two cell types, an initiator cell releases soluble mediators as a consequence of HIV-1 replication and perturbation of the initiator cell's normal function. These molecules, which include viral proteins gp120 and Tat or cell-encoded molecules such as neopterin, quinolinic acid, or β_2-microglobulin (β_2M), diffuse extracellularly and adversely affect the function of secondary, uninfected target cells. Addition of a third cell type, called an amplifier cell, accounts for secondary responses of uninfected cells to infected cells in close proximity. Amplifier cells serve as intermediaries, responding to signaling molecules produced by the initiator cell and producing their own set of mediators (cytokines or chemokines) which, in turn, disrupt target cell function. The relevance of amplifier cell types to HIV-1 CNS infection is that the damage capacity of a limited number of HIV-1-infected cells can be greatly magnified, accounting for the disproportionate relationship between the limited number of infected cells observed in the CNS and the appearance of widespread pathological effects. Addition of a fourth cell type accounts for the element of time and the extended course of HIV-1-associated CNS disease. The modulator cell acts to regulate HIV-1 infection of the initiator cell. As immune function diminishes during the progression of AIDS, the modulator cell, as a representative of the immune system, would gradually lose its capacity to keep viral replication in check. Adding additional complexity to this model is the observation that cellular interac-

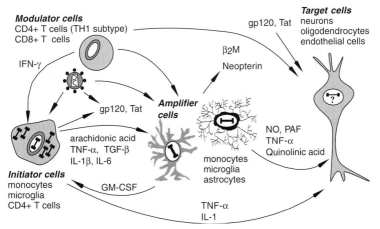

FIGURE 8   Events underlying HIV-1-associated CNS disease and HIVD involve complex interactions between uninfected and HIV-1-infected cells within the CNS. HIV-1 infection of initiator cells causes a cascade of cellular effects mediated by soluble signaling molecules and temporally regulated by the action of modulator cells. References include Price (1994) and Lipton (1998). Abbreviations: β_2M, β_2-microglobulin; IFN-γ, interferon-γ; IL-1β, interleukin-1β; IL-6, interleukin 6; NO, nitric oxide; PAF, platelet activating factor; TGF-β, transforming growth factor-β; TNFα, tumor necrosis factor-α.

tions are often not unidirectional. Feedback loops between cell types may serve to enhance cellular dysfunction and toxicity associated with HIV-1 CNS infection.

The roles of initiator, amplifier, modulator, and target cells may be played by multiple cell types, dependent on the susceptibility to HIV-1 infection, the presence of a productive infection, the level of cellular differentiation, and the state of cell activation. During the early stages of infection, the initiator cells are generally those cells that have crossed the BBB carrying infectious virus. Whether those cells are CD4-positive monocytes or activated T lymphocytes depends on the accuracy of models that predict the identity of infected cells that traffic through the BBB. During the progression of HIVD and the decline of CD4-positive T lymphocytes in the peripheral circulation, infected macrophages and microglia may take on the role of maintaining macrophage-tropic viral strains that predominate in the brain. These cells produce and release soluble gp120, Tat, and Nef (Lipton, 1992b; Sabatier *et al.*, 1991; Werner *et al.*, 1991), which serve as both signaling molecules and as antigens that elicit an immune response. Astrocytes and cells of monocyte/macrophage/microglial lineage may serve as amplifier cells since both cell types can respond to and produce cytokines. These cells may function particularly well in this role since the numbers of both cell types are increased during the course of disease. Candidate soluble mediators of the amplifier cells include quinolinic acid, TNFα, nitric oxide, interleukin-1, and arachidonic acid metabolites. Modulator cells are generally cells of immune system origin that have the capacity to initiate virus clearance or neutralization or modulate viral replication in the initiator cells through the release of cytokines. The role of modulator cell may be played by helper CD4-positive T lymphocytes, CD8-positive T lymphocytes, or B cells that produce HIV-1-specific antibodies. Target cells are generally those cell types that may be killed or altered functionally by soluble mediators released by the amplifier or initiator cell types. Although all cell types within the CNS may be subject to the effects of changes in the extracellular milieu, target cells are identified primarily as neurons and oligodendrocytes. There is evidence to support neuronal death and oligodendrocyte dysfunction during the course of HIVD (Everall *et al.*, 1993), as well as changes in cortical architecture (Masliah *et al.*, 1992a). Vascular endothelial cells and astrocytes may also be considered target cells, since their roles in supporting the integrity and function of the CNS are also compromised.

3. Mediators of Neuroglial Cell Damage

The central instruments of damage in these models are soluble mediators released in response to HIV-1 CNS infection. The HIV-1-mediated deregulation of normal cellular metabolic processes causes both infected and uninfected cells to release cellular and viral products that are toxic to resident CNS cells. For example, HIV-1-infected macrophages secrete soluble products shown to be toxic to neuroglial cells (Pulliam *et al.,* 1991). Results from other investigations have shown that uninfected astrocytes, in concert with cocultured HIV-1-infected monocytes, produce factors toxic to human neurons (Genis *et al.,* 1992). The body of evidence regarding HIV-1-associated CNS damage indicates that the release of soluble mediators is the consequence of complex interactions between infected and uninfected cells in the CNS (Fig. 8).

The list of candidate neurotoxins involved in inducing neurologic dysfunction is lengthy and includes several classes of cellular and viral soluble mediators. Studies have demonstrated elevated levels of cytokines TNFα (Grimaldi *et al.,* 1991; Perrella *et al.,* 1992), IL-6 (Gallo *et al.,* 1989; Perrella *et al.,* 1992), IL-1β (Gallo *et al.,* 1989), and IL-1α (Perrella *et al.,* 1992) in the cerebral spinal fluid of HIV-1-infected patients. Elevated levels of TNFα, IL-1β, and transforming growth factor β (TGFβ) mRNA expression in HIV-1-infected brains have also been reported (Wahl *et al.,* 1991; Wesselingh *et al.,* 1993). Increased expression of TNFα has also been associated with the diagnosis of HIVD (Wesselingh *et al.,* 1993). Increases in β_2M, neopterin, and quinolinic acid, all considered markers of immune activation, have been observed in the CSF of HIV-1-infected patients with HIVD (Brew *et al.,* 1990, 1992; Heyes *et al.,* 1991a). β_2M, the light chain of the major histocompatibility class I (MHC-I) cell surface molecule, is upregulated by a number of cytokines, including IFN-γ (Price, 1994). PAF, secreted by HIV-1-infected monocytes (Genis *et al.,* 1992), is also found at elevated levels in HIVD patients (Perry *et al.,* 1998).

Elevated levels of cytokine expression during HIV-1 CNS infection have numerous effects on the normal function and architecture of the brain. TNFα can cause damage to myelin and myelin-producing oligodendrocytes (Selmaj and Raine, 1988; Wilt *et al.,* 1995) as well as influence HIV-1 gene expression in monocytes (Vitkovic *et al.,* 1990) and glial cells (Tornatore *et al.,* 1991). TNFα, produced in high levels by HIV-1-infected monocytes cocultured with astrocytes (Genis *et al.,* 1992), is also toxic to primary human neurons (Gelbard *et al.,* 1993). PAF, which is also found in the CSF during HIV-1 infection, is also toxic to primary neurons (Gelbard *et al.,* 1994) as well as the NTera 2 neuronal cell line (Westmoreland *et al.,* 1996). Exposure to PAF can result in neuronal apoptosis (Perry *et al.,* 1998), which may be mediated by signal transduction pathways that differ from those used by TNFα (Pulliam *et al.,* 1998). Although TGFβ, IL-1β, and IL-6 are all elevated during HIVD or HIV-1 CNS infection, their impacts on neuroglial function/survival and their participation in HIVD are not as clearly defined (Kolson *et al.,* 1998).

Viral proteins proposed to be neurotoxic candidates include Tat, Nef, and the envelope protein gp120. Given that a biologically active form of the HIV-1 transactivator protein Tat is released from HIV-1-infected cells (Ensoli *et al.,* 1990, 1993) and can be taken up by a variety of cell types,

Tat is a rather attractive neurotoxin candidate. Once inside other cells, Tat can be transported to the nucleus where it is capable of transactivating both the HIV-1 LTR and cellular promoters (Frankel and Pabo, 1988; Kolson *et al.*, 1994; Rappaport *et al.*, 1999). Tat has been shown to transactivate promoters for cytokines such as TGF-β1 (Cupp *et al.*, 1993; Zauli *et al.*, 1992), TNFα, and IL-2 (Westendorp *et al.*, 1994) and to repress both MHC class promoter activity (Howcroft *et al.*, 1993) and antigen-specific T-cell responsiveness (Subramanyam *et al.*, 1993). Recent studies also indicate that Tat induces IL-6 mRNA expression in human brain endothelial cells, which may play a role in altering the blood–brain barrier (Zidovetzki *et al.*, 1998) and augment HIV-1 replication through C/EBP transcription factors (Tesmer *et al.*, 1993). Tat gene sequence heterogeneity is higher in brains from HIVD patients, indicating a relationship between selective pressures in the CNS and the development of dementia (Bratanich *et al.*, 1998).

Various studies have reported recombinant Tat to be neurotoxic. Micromolar concentrations of Tat have been reported to kill murine glioma and neuroblastoma cell lines and strongly depolarize excitable cells (Sabatier *et al.*, 1991). *In vivo* studies demonstrated that intracerebral injections of recombinant Tat into mice resulted in paralysis and death. Tat is also toxic to human fetal neurons and causes depolarization of adult and human fetal neurons after single or repeated application at femtomolar concentrations (Cheng *et al.*, 1998; Magnuson *et al.*, 1995). Furthermore, Nath and coworkers demonstrated that Tat was mildly toxic to human fetal neurons (Nath *et al.*, 1996). In contrast, other investigators have found no direct neurotoxic effects in 3-day-old rat brain cultures, but did document changes in neuronal and glial architecture in the presence of Tat (Kolson *et al.*, 1993). Recently, the combination of Tat and TNFα was shown to induce neuronal apoptosis to a greater extent than either treatment alone (Shi *et al.*, 1998). The induction of apoptosis appeared to be due to an increase in cellular oxidative stress. However, antioxidants did not completely inhibit apoptosis, suggesting several mechanisms may be involved in apoptotic induction. In a similar study, exposure to Tat induced human neuronal apoptotic death through TNFα and activation of non-N-methyl-D-aspartate (NMDA) receptors using an NF-κB-independent mechanism (New *et al.*, 1998).

The viral regulatory protein Nef has been implicated as a neurotoxic modulator due to the levels of Nef expression in the CNS of AIDS patients (Ranki *et al.*, 1995; Saito *et al.*, 1994; Tornatore *et al.*, 1994). In particular, subcortical astrocytes from pediatric brain tissue infected with HIV-1 were found to express high levels of the Nef protein (Saito *et al.*, 1994; Tornatore *et al.*, 1994). In another study, 7 of 14 autopsied brains from adult AIDS patients showed expression of Nef in astrocytes (Ranki *et al.*, 1995). Six of the seven patients who exhibited Nef expression suffered from moderate to severe dementia, but did not possess the neuropathologic markers typically associated with HIVD. The implication of this study is that the impact of

Nef on dementia may be independent of other factors responsible for causing HIVD. Although the functional basis for Nef neurotoxicity has yet to be determined, Nef protein has some sequence similarities to scorpion toxin (Werner *et al.*, 1991).

Another viral protein implicated in mediating neurotoxicity is the envelope glycoprotein gp120. One of the earliest studies concerning the neurotoxicity of gp120 demonstrated that recombinant gp120 killed embryonic rat hippocampal neurons at picomolar concentrations (Brenneman *et al.*, 1988). Cultured mouse neurons are also sensitive to the neurotoxic effect of gp120 (Dreyer *et al.*, 1990). In cultures of human neuroglial cells, gp120 induced apoptosis in neurons and microglial cells, but not in astrocytes (Lannuzel *et al.*, 1997). Neuronal apoptosis may be mediated by interactions between gp120 and the chemokine receptor CXCR4, as demonstrated using a human neuronal cell line (Hesselgesser *et al.*, 1998). In addition, the ability of gp120 to induce neuroglial apoptosis may be strain-dependent and related to the appearance of HIV-1 strains in the late stages of HIV-1-associated disease (Ohagen *et al.*, 1999). Additional studies have suggested that the neurotoxicity of gp120 is the result of neuronal stimulation through the NMDA receptor. In studies along this line, gp120 acted synergistically with endogenous glutamate to induce neurotoxicity (Lipton *et al.*, 1991), which could be blocked by the NMDA antagonist memantine (Lipton, 1992a). HIV-1 gp120 may act indirectly through the NMDA receptor, since neuronal exposure to gp120 results in changes in calcium homeostasis and modified NMDA-dependent NMDA receptor activity (Lannuzel *et al.*, 1995). Nitric oxide (NO) has also been implicated in gp120 neurotoxicity (Dawson and Dawson, 1994). Other studies have demonstrated that natural and synthetic glucocorticoids exacerbate the neuronal toxicity of gp120 (Brooke *et al.*, 1998; Iyer *et al.*, 1998). These data suggest that synthetic glucocorticoids used to treat severe cases of *Pneumocystis carinii* pneumonia may have an adverse effect on neuronal viability within the HIV-1-infected nervous system. Regardless of the mechanism of neurotoxicity, it appears clear that gp120 is an important mediator of neuronal degradation. This point is underscored by studies in which HIVD-like neuropathology (including astrocytosis and neuronal loss) was demonstrated in transgenic mice expressing gp120 under the direction of an astrocytic promoter (Toggas *et al.*, 1994). *In vivo* and *in vitro* models that include astrocytes, gp120 may act on neurons indirectly through astrocytes, which undergo alterations in GFAP expression (Pulliam *et al.*, 1993) and ion transport (Benos *et al.*, 1994) after exposure to gp120.

Genetic diversity and the appearance of HIV-1 quasispecies during the course of extended replication may also be a determinant in gp120 neurotoxicity. The limited fidelity of reverse transcriptase, selective pressures of the host immune system, and selection imposed by antiretroviral drug therapies result in the production of virus particles which, as a population, have a great deal of genetic variability throughout the viral genome, including the envelope gene. The resulting population of viruses is referred to as a quasispecies (Wain-Hobson, 1989). The envelope genes of viral quasispecies isolated from the brain have reduced genetic variability when compared to the variability of viruses from other organs (Epstein *et al.*, 1991), suggesting that replicating viruses, compartmentalized within the brain, are under selective pressure. Brain-derived V1 and V2 gp120 sequences from patients with HIVD cluster phylogenetically and are distinct from similar sequences derived from nondemented patients (Power *et al.*, 1998). Furthermore, macrophage tropic recombinant viruses which include the V1-V3 sequences from HIVD patients produce soluble factors which induce neuronal death to a greater extent than viruses with V1-V3 from nondemented patients (Power *et al.*, 1998). These studies suggest that viral sequence diversity may be a determinant in the appearance of HIVD.

Other factors believed to be possible mediators of neurotoxicity include numerous cellular metabolities such as arachidonic acid, nitric oxide, and quinolinic acid. Arachidonic acid, released by macrophages, astrocytes, and neurons, potentiates the activity of the NMDA receptor (Miller *et al.*, 1992) and has implications for neurotoxicity associated with HIV-1 infection and gp120 release. Prostaglandins, products of arachidonic acid catabolism, are elevated in the CSF of patients with HIVD (Griffin *et al.*, 1994) and may also be neurotoxic.

Nitric oxide (NO) was proposed as being neurotoxic based on studies which demonstrated that murine microglial cells stimulated with lipopolysaccharide (LPS) or γ-interferon released the NO metabolite nitrite (Zielasek *et al.*, 1992), which was shown to be neurotoxic (Chao *et al.*, 1992). NO production specific to HIV-1-infected monocyte-derived macrophages (MDMs) has been observed, albeit at modest levels (Bukrinsky *et al.*, 1995). NO may also be involved in HIV-1 gp120 neurotoxicity (Dawson and Dawson, 1994). Bukrinsky and coworkers also demonstrated that inducible NO synthase (iNOS) transcripts were present in a brain with advanced HIV-1 CNS disease, but not in brains with lesser apparent neuropathogenesis (Bukrinsky *et al.*, 1995). However, a more recent study of postmortem HIV-1-infected brains did not demonstrate elevated NOS levels (Bagasra *et al.*, 1997). Therefore, the importance of NO as a neurotoxic mediator in the development of HIVD is still debatable.

Quinolinic acid is a product of tryptophan metabolism and is produced by human macrophages in response to various stimuli, including HIV-1 infection (Brew *et al.*, 1995a). Chronic exposure of rat striatum cultures to levels of quinolinic acid found in the CSF of HIVD patients induced neurotoxicity (Whetsell and Schwarcz, 1989). Furthermore, elevated quinolinic acid levels in the CSF of patients with HIVD can be correlated with severity of neurologic dysfunction (Achim *et al.*, 1993; Heyes *et al.*, 1991b). Elevated levels of quinolinic acid are also associated with SIV CNS infection in macaques and rhesus monkeys (Heyes *et al.*, 1990; Jordan and Heyes,

1993). Since quinolinic acid may also serve as an NMDA receptor agonist, the neurotoxicity of quinolinic acid may be a consequence of interactions with neuronal NMDA receptors. These findings directly implicate quinolinic acid in CNS pathogenesis associated with HIV-1 infection.

IV. Chemotherapeutic Treatment of HIVD

Efforts mounted in response to the AIDS epidemic have been concentrated in three directions: prevention of transmission, development of an HIV-1 vaccine, and treatment of individuals already infected with HIV-1. The overriding goal of each of these approaches is the reduction in incidence and/or severity of disease processes associated with HIV-1 infection, including HIVD. With respect to HIV-1 CNS infection and development of HIVD, the benefit of blocking transmission is obvious: individuals not infected with HIV-1 do not develop HIVD. An HIV-1 vaccine will provide immunologic protection from HIV-1 infection and would also presumably decrease or eliminate the appearance of HIVD. However, an effective vaccine has yet to be developed. Finally, chemotherapeutic treatments of the immunologic and neurologic consequences of HIV-1 infection, the most successful of which to date have been combination drug regimens, may reverse or prevent the progression of HIV-1-associated neurologic disease.

A. Treatment of HIV-1 CNS Infection Using Single- and Multi-drug Strategies

Initial pharmaceutical treatments of HIV-1 infection concentrated on checking systemic viral replication by inhibiting the activity of reverse transcriptase (RT), an enzyme unique to HIV-1 as well as other retroviruses. One of the first RT inhibitors, approved for clinical use and employed successfully as an antiretroviral agent in 1987, was the nucleoside analog 3'-azido-3'-deoxythymidine (AZT or zidovudine). Following its introduction, AZT was shown to be an effective drug in the treatment of HIV-1 infection and the development of AIDS (Fischl et al., 1987). Early studies demonstrated that AZT treatment lessened cognitive deficits related to HIV-1 infection in both adults (Schmitt et al., 1988) and children (Pizzo et al., 1988). AZT crosses the BBB and accumulates in the CSF at levels that give it the highest CSF-to-plasma ratio among available nucleoside analogs (Acosta et al., 1996). In studies which included patients diagnosed with HIVD, administration of AZT monotherapy regimens resulted in significant, dose-dependent improvements in neurologic function (Sidtis et al., 1993) and reductions in CSF p24 (Royal et al., 1994), the appearance of MGCs, concomitant neuropathologic damage, and the incidence of severe dementia at the time of death (Vago et al., 1993). AZT also reversed increases in cerebral blood volume that accompany CNS HIV-1 infection (Tracey *et al.*, 1998). In another study, AZT administration resulted in reductions in CSF HIV-1 RNA (Gisslen *et al.*, 1997). Although AZT may reduce the risk of HIVD (Baldeweg *et al.*, 1998), treatment may not always result in broad improvements in neurologic function (Gorman *et al.*, 1993) and may only result in transient gains in neurologic function (Tozzi *et al.*, 1993). Unfortunately, the efficacy of AZT in the treatment of peripheral and CNS HIV-1 infection is substantially diminished by the appearance of AZT-resistant HIV-1 variants after therapy extended beyond 6 months (Larder *et al.*, 1989).

Following the introduction of AZT, other RT inhibitors were introduced for the treatment of AIDS. Nucleoside analog RT inhibitors (NRTIs) include didanosine (ddI), zalcitabine (ddC), stavudine (d4T), lamivudine (3TC), and, recently, abacavir (1592U89). Nonnucleoside RT inhibitor (NNRTI) drugs include nevirapine, delavirdine, and efavirenz. NRTI and NNRTI drugs continue to be introduced as products of a greater understanding of drug–RT interactions. As more RT inhibitors were approved for clinical use, numerous treatment regimens using each drug singly or in combination were developed. Used singly, ddI was as effective as AZT as measured by neuropsychological parameters (Sidtis *et al.*, 1997), but less effective than AZT when assessed by CSF HIV-1 RNA levels (Gisslen *et al.*, 1997). Treatment with ddI also resulted in cognitive improvements in HIV-1-infected children (Butler *et al.*, 1991). In a study that combined 3TC with either AZT or d4T, administration of either combination resulted in consistent penetration of each drug into the CSF and undetectable levels of CSF HIV-1 RNA (Foudraine *et al.*, 1998). Similarly, treatment of CNS-compromised patients with a combination of AZT and ddI resulted in measurable improvements in neuropsychological functions associated with HIVD (Brouwers *et al.*, 1997). In HIV-1 infected children, treatment combining AZT and ddI was more effective against HIV-1-associated pediatric CNS disease than monotherapy with either AZT or ddI (Raskino *et al.*, 1999). RT inhibitors may be limited by their differential abilities to cross the BBB (Glynn and Yazdanian, 1998) and, like AZT, the evolution of drug-resistant HIV-1 strains within the CNS and the peripheral circulation.

B. Combination Therapies and Highly Active Antiretroviral Therapy (HAART): Treatments That Include Protease Inhibitors and Affect HIV-1 CNS Infection

The development and introduction of drugs in 1995 that inhibit the activity of the HIV-1 protease enzyme broadened the arsenal of agents available to treat HIV-1 infections. Available protease inhibitors (PIs) include indinavir, saquinavir, nelfinavir, ritonavir, and amprenavir. Combination therapy using protease inhibitors in combination with NRTIs and NNRTIs resulted in dramatic reductions in circulating virus (Perelson *et al.*, 1997), decreases in viral load in the blood below the limits of detection, and reductions in disease progression and mortality. Despite early expectations to the contrary, this treatment regimen, referred to as highly active antiretroviral therapy (HAART), does not result in complete eradication of HIV-1, which finds sanctuary in regions of the body where drug penetration is low or in cells that express little or no virus (Schrager and D'Souza, 1998). However, since the introduction of HAART, the incidence of HIVD has dropped dramatically (Brodt *et al.*, 1997), suggesting that the course of HIV-1-associated disease in the CNS can be altered despite the role of the brain as a potential reservoir for HIV-1 during HAART.

Clinical studies have demonstrated that HAART is effective against HIV-1 CNS disease. In a patient diagnosed with HIVD, 12 weeks of HAART (AZT combined with indinavir and lamivudine) and nonsteroidal anti-inflammatory agents resulted in measurable improvements in neurologic function as well as reductions in CSF neurotoxin levels and decreases in plasma viremia to undetectable levels (Gendelman *et al.*, 1998). In a recent clinical study, administration of combination therapies that included a PI resulted in improvements in neuropsychological parameters compared to patients receiving monotherapy or no antiretroviral therapy (Sacktor *et al.*, 1999). In another study of 34 HIV-1-infected patients, significant cognitive improvement, sustained to 1 year after initiation of HAART, was observed in over two-thirds of the subjects (Sacktor *et al.*, 1998). HAART has also been used to successfully treat pediatric HIV-1 CNS infections (Tepper *et al.*, 1998). HAART treatment during the early stages of HIVD may reverse the progression of disease (Chang *et al.*, 1999).

PIs used alone or combined with RT inhibitors have demonstrable activity against HIV-1 within the CNS, but may be limited by bioavailability in the brain. In a study which combined saquinavir and ritonavir, measurable levels of both PIs correlated with decreases in CSF and plasma HIV-1 RNA (Kravcik *et al.*, 1999). PI treatment was also associated with improvements in cognitive abilities and neuropathological MRI findings (Filippi *et al.*, 1998). However, in combination therapies that included nelfinavir, levels of nelfinavir found in the CSF were negligible (Aweeka *et al.*, 1999). The inability of a particular PI to penetrate the CSF may be related to active drug transport within the brain. In mice with a homozygous knockout of the drug transporter P-glycoprotein, levels of indinavir, saquinavir, and nelfinavir in the brain were elevated 7- to 36-fold over levels in wild-type mice (Kim *et al.*, 1998), suggesting that PI penetration may be limited by active drug efflux and that therapeutic inhibition of P-glycoprotein activity may result in improved PI penetration into the CSF and brain.

Current findings indicate that administration of combination drug therapies (including HAART) can result in measurable improvements in neurological function and reductions in the incidence or severity of HIVD. However, like treatments with AZT, combination therapies may not provide indefinite benefits. The limited penetration of some drugs into the CNS and the potential of the CNS to act as a reservoir for HIV-1 during antiretroviral treatments may restrict the effectiveness of drug therapies against HIV-1-associated CNS disease.

V. Summary

Despite more than 15 years of extensive investigative efforts, a complete understanding of the neurological consequences of HIV-1 CNS infection remains elusive. Although the resources of numerous investigators have been focused on studies of HIV-1-associated CNS disease, the complex nature of the disease processes that underlie the clinical, pathological, and cellular manifestations of HIV-1 CNS infection have required a larger volume of studies than was initially envisioned. Several major areas remain as the focus of current research efforts.

One of the more pressing issues facing researchers and clinicians alike is the search for correlates to the development of HIV-1-associated CNS neuropathology and the onset of HIVD. Although numerous parameters have been studied, none have been shown to be absolute predictors or markers of HIV-1-related CNS dysfunction. The identification of solid correlates of HIVD is an important goal that would permit clinical identification of individuals at risk for developing potentially crippling, life-threatening CNS abnormalities and would facilitate early treatment of nascent neurological problems.

A more complete comprehension of the cellular foundations of CNS dysfunction and HIVD is also a fundamental part of strategies designed to treat or prevent HIV-1-associated CNS disease. Future investigations will strive to expand the body of knowledge concerning the complex interactions between infected and uninfected neuroglial cells and the roles of numerous cytokines, chemokines, and other soluble agents that are deregulated during HIV-1 CNS infection. In particular, a thorough understanding of the mechanisms of neurotoxicity may facilitate the development of new therapies that alleviate or eliminate the clinical consequences of CNS infection.

Finally, investigators will continue to study HIVD within the context of single and combination drug therapies used in the treatment of HIV-1 infection and AIDS. As newer and more effective systemic treatments for HIV-1 infection and AIDS are introduced, the effects of these treatments on the onset, incidence, and severity of HIVD will also require intensive study. The impact of drug therapies on the ability of the CNS to act as an HIV-1 reservoir will also need to be addressed. Introduction of each new drug or drug combination will necessitate studies of drug penetration into the CNS and efficacy against the development of CNS abnormalities. Furthermore, as more effective treatments prolong the lifespan of individuals infected with HIV-1, the impact of extended survival on the occurrence and severity of HIVD will also require further investigations.

The quest for answers to these and other questions will be complicated by the diversity of experimental systems used to study different aspects of HIV-1 CNS infection and HIVD. Each system has its own unique strengths and weaknesses. Clinical observations provide a continuous spectrum of symptomatic findings but reveal little about the underlying mechanisms of disease. *In vivo* imaging techniques, such as CT and MRI, also provide a continuum of observations, but the images are limited in their resolution. Neuropathological examinations of postmortem HIV-1-infected brains offer gross, cellular, and molecular views (including phenotypic and genotypic analyses of CNS viral isolates) of the diseased brain, but only provide a snapshot of the end-stage neurologic dysfunction. Studies that rely on animal surrogates for HIV-1, including SIV, simian-HIV (SHIV), feline immunodeficiency virus (FIV), visna virus, and HIV-1 SCID-hu models, permit experimental protocols that cannot be carried out in humans, but are limited by the fidelity with which each virus and animal model emulates the conditions and events observed in the human host. Finally, *in vitro* techniques, which include the use of primary cells and cell lines, adult or fetal human cell cultures, and BBB barrier model systems, are also convenient means by which aspects of HIVD can be studied. However, experiments conducted using primary cell populations can be affected by donor variability and cellular changes that arise during *in vitro* growth of primary cells.

The multitude of investigational approaches also poses the problem of integration. Each new finding must be compared to results obtained using other systems and evaluated for its relevance to *in vivo* conditions. For example, numerous *in vitro* studies using animal cell cultures and whole-animal models have shown that gp120 can act as a potent neurotoxin. However, these results must be integrated with observations obtained using human *in vitro* systems and assessed within the context of neuropathogenesis associated with HIV-1 CNS infection *in vivo*. Clearly, a complete understanding of the molecular mechanisms of HIV-1 CNS infection and HIVD will require melding numerous observations made using a variety of experimental systems.

HIV-1-associated CNS disease and HIV-1 dementia are complex disease states whose clinical and pathological hallmarks have their basis in events that take place at the cellular and molecular levels during HIV-1 replication within the CNS. Continued study of these events will certainly help further our understanding of the consequences of HIV-1 infection of the CNS and advance the search for effective treatments of HIVD.

Acknowledgements

Preparation of this review was supported by Public Health Service Grant R01 NS32092-09.

References

Achim, C. L., Heyes, M. P., and Wiley, C. A. (1993). Quantitation of human immunodeficiency virus, immune activation factors, and quinolinic acid in AIDS brains. *J. Clin. Invest.* **91,** 2769–2775.

Acosta, E. P., Page, L. M., and Fletcher, C. V. (1996). Clinical pharmacokinetics of zidovudine: An update. *Clin. Pharmacokinet.* **30,** 251–262.

Adamson, D. C., McArthur, J. C., Dawson, T. M., and Dawson, V. L. (1999). Rate and severity of HIV-associated dementia (HAD): Correlations with gp41 and iNOS. *Mol. Med.* **5,** 98–109.

Adle-Biassette, H., Chretien, F., Wingertsmann, L., Hery, C., Ereau, T., Scaravilli, F., Tardieu, M., and Gray, F. (1999). Neuronal apoptosis does not correlate with dementia in HIV infection but is related to microglial activation and axonal damage. *Neuropathol. Appl. Neurobiol.* **25,** 123–133.

Adle-Biassette, H., Levy, Y., Colombel, M., Poron, F., Natchev, S., Keohane, C., and Gray, F. (1995). Neuronal apoptosis in HIV infection in adults. *Neuropathol. Appl. Neurobiol.* **21,** 218–227.

Ahmad, N., and Venkatesan, S. (1988). Nef protein of HIV-1 is a transcriptional repressor of HIV-1 LTR. *Science* **241,** 1481–1485.

Aiken, C., Konner, J., Landau, N. R., Lenburg, M. E., and Trono, D. (1994). Nef induces CD4 endocytosis: Requirement for a critical dileucine motif in the membrane-proximal CD4 cytoplasmic domain. *Cell* **76,** 853–864.

Akira, S., Isshiki, H., Sugita, T., Tanabe, O., Kinoshita, S., Nishio, Y., Nakajima, T., Hirano, T., and Kishimoto, T. (1990). A nuclear factor for IL-6 expression (NF-IL6) is a member of a C/EBP family. *EMBO J.* **9,** 1897–1906.

Albright, A. V., Shieh, J. T., Itoh, T., Lee, B., Pleasure, D., O'Connor, M. J., Doms, R. W., and Gonzalez-Scarano, F. (1999). Microglia express CCR5, CXCR4, and CCR3, but of these, CCR5 is the principal coreceptor for human immunodeficiency virus type 1 dementia isolates. *J. Virol.* **73,** 205–213.

Albright, A. V., Strizki, J., Harouse, J. M., Lavi, E., O'Connor, M., and Gonzalez-Scarano, F. (1996). HIV-1 infection of cultured human adult oligodendrocytes. *Virology* **217,** 211–219.

Ambrosino, C., Ruocco, M. R., Chen, X., Mallardo, M., Baudi, F., Trematerra, S., Quinto, I., Venuta, S., and Scala, G. (1997). HIV-1 Tat induces the expression of the interleukin-6 (IL6) gene by binding to the IL6 leader RNA and by interacting with CAAT enhancer-binding protein beta (NF-IL6) transcription factors. *J. Biol. Chem.* **272,** 14883–14892.

Anders, K. H., Guerra, W. F., Tomiyasu, U., Verity, M. A., and Vinters, H. V. (1986). The neuropathology of AIDS. UCLA experience and review. *Am. J. Pathol.* **124,** 537–558.

Antonson, P., Stellan, B., Yamanaka, R., and Xanthopoulos, K. G. (1996). A novel human CCAAT/enhancer binding protein gene, C/EBP epsilon, is expressed in cells of lymphoid and myeloid lineages and is localized on chromosome 14q11.2 close to the T-cell receptor alpha/delta locus. *Genomics* **35,** 30–38.

Apt, D., Watts, R. M., Suske, G., and Bernard, H. U. (1996). High Sp1/Sp3 ratios in epithelial cells during epithelial differentiation and cellular transformation correlate with the activation of the HPV-16 promoter. *Virology* **224,** 281–291.

Armstrong, S. A., Barry, D. A., Leggett, R. W., and Mueller, C. R. (1997). Casein kinase II-mediated phosphorylation of the C terminus of Sp1 decreases its DNA binding activity. *J. Biol. Chem.* **272,** 13489–13495.

Aweeka, F., Jayewardene, A., Staprans, S., Bellibas, S. E., Kearney, B., Lizak, P., Novakovic-Agopian, T., and Price, R. W. (1999). Failure to detect nelfinavir in the cerebrospinal fluid of HIV-1–infected patients with and without AIDS dementia complex. *J. Acq. Immune Defic. Syndr. Hum. Retrovirol.* **20,** 39–43.

Aylward, E. H., Brettschneider, P. D., McArthur, J. C., Harris, G. J., Schlaepfer, T. E., Henderer, J. D., Barta, P. E., Tien, A. Y., and Pearlson, G. D. (1995). Magnetic resonance imaging measurement of gray matter volume reductions in HIV dementia. *Am. J. Psychiat.* **152,** 987–994.

Bachelerie, F., Alcami, J., Arezana-Seisdedos, F., and Virelizier, J. L. (1991). HIV enhancer activity perpetuated by NF-kappa B induction on infection of monocytes. *Nature* **350,** 709–712.

Bagasra, O., Bobroski, L., Sarker, A., Bagasra, A., Saikumari, P., and Pomerantz, R. J. (1997). Absence of the inducible form of nitric oxide synthase in the brains of patients with the acquired immunodeficiency syndrome. *J. Neurovirol.* **3,** 153–167.

Bagasra, O., Lavi, E., Bobroski, L., Khalili, K., Pestaner, J. P., Tawadros, R., and Pomerantz, R. J. (1996). Cellular reservoirs of HIV-1 in the central nervous system of infected individual: Identification by the combination of *in situ* polymerase chain reaction and immunohistochemistry. *AIDS* **10,** 573–585.

Baldeweg, T., Catalan, J., and Gazzard, B. G. (1998). Risk of HIV dementia and opportunistic brain disease in AIDS and zidovudine therapy. *J. Neurol. Neurosurg. Psychiat.* **65,** 34–41.

Bandres, J. C., and Ratner, L. (1994). Human immunodeficiency virus type 1 Nef protein down-regulates transcription factors NF-kappa B and AP-1 in human T cells *in vitro* after T-cell receptor stimulation. *J. Virol.* **68,** 3243–3249.

Belman, A. L. (1994). HIV-1-associated CNS disease in infants and children. In "HIV, AIDS, and the Brain" (R. W. Price and S. W. Perry III, Eds.), pp. 289–310. Raven Press, New York.

Belman, A. L., Lantos, G., Horoupian, D., Novick, B. E., Ultmann, M. H., Dickson, D. W., and Rubinstein, A. (1986). AIDS: Calcification of the basal ganglia in infants and children. *Neurology* **36,** 1192–1199.

Bencheikh, M., Bentsman, G., Sarkissian, N., Canki, M., and Volsky, D. J. (1999). Replication of different clones of human immunodeficiency virus type 1 in primary fetal human astrocytes: Enhancement of viral gene expression by Nef. *J. Neurovirol.* **5,** 115–124.

Bencherif, B., and Rottenberg, D. A. (1998). Neuroimaging of the AIDS dementia complex. *AIDS* **12,** 233–244.

Benos, D. J., Hahn, B. H., Bubien, J. K., Ghosh, S. K., Mashburn, N. A., Chaikin, M. A., Shaw, G. M., and Benveniste, E. N. (1994). Envelope glycoprotein gp120 of human immunodeficiency virus type 1 alters ion transport in astrocytes: Implications for AIDS dementia complex. *Proc. Natl. Acad. Sci. USA* **91,** 494–498.

Berger, E. A., Doms, R. W., Fenyo, E. M., Korber, B. T., Littman, D. R., Moore, J. P., Sattentau, Q. J., Schuitemaker, H., Sodroski, J., and Weiss, R. A. (1998). A new classification for HIV-1 [letter]. *Nature* **391,** 240.

Berkhout, B., and Jeang, K. T. (1992). Functional roles for the TATA promoter and enhancers in basal and Tat-induced expression of the human immunodeficiency virus type 1 long terminal repeat. *J. Virol.* **66,** 139–149.

Birnbaum, M. J., van Wijnen, A. J., Odgren, P. R., Last, T. J., Suske, G., Stein, G. S., and Stein, J. L. (1995). Sp1 trans-activation of cell cycle regulated promoters is selectively repressed by Sp3. *Biochemistry* **34,** 16503–16508.

Bjork, J., Lindbom, L., Gerdin, B., Smedegard, G., Arfors, K. E., and Benveniste, J. (1983). Paf-acether (platelet-activating factor) increases microvascular permeability and affects endothelium-granulocyte interaction in microvascular beds. *Acta Physiol. Scand.* **119,** 305–308.

Black, K. L., and Hoff, J. T. (1985). Leukotrienes increase blood–brain barrier permeability following intraparenchymal injections in rats. *Ann. Neurol.* **18,** 349–351.

Blanche, S., Tardieu, M., Duliege, A., Rouzioux, C., Le Deist, F., Fukunaga, K., Caniglia, M., Jacomet, C., Messiah, A., and Griscelli, C. (1990). Longitudinal study of 94 symptomatic infants with perinatally acquired human immunodeficiency virus infection: Evidence for a bimodal expression of clinical and biological symptoms. *Am. J. Dis. Child* **144**, 1210–1215.

Blumberg, B. M., Epstein, L. G., Saito, Y., Chen, D., Sharer, L. R., and Anand, R. (1992). Human immunodeficiency virus type 1 nef quasispecies in pathological tissue. *J. Virol.* **66**, 5256–5264.

Boni, J., Emmerich, B. S., Leib, S. L., Wiestler, O. D., Schupbach, J., and Kleihues, P. (1993). PCR identification of HIV-1 DNA sequences in brain tissue of patients with AIDS encephalopathy. *Neurology* **43**, 1813–1817.

Bouwman, F. H., Skolasky, R. L., Hes, D., Selnes, O. A., Glass, J. D., Nance-Sproson, T. E., Royal, W., Dal Pan, G. J., and McArthur, J. C. (1998). Variable progression of HIV-associated dementia. *Neurology* **50**, 1814–1820.

Bowers, W. J., Baglia, L. A., and Ruddel, A. (1996). Regulation of avian leukosis virus long terminal repeat-enhanced transcription by C/EBP-Rel interactions. *J. Virol.* **70**, 3051–3059.

Brack-Werner, R. (1999). Astrocytes: HIV cellular reservoirs and important participants in neuropathogenesis. *AIDS* **13**, 1–22.

Brack-Werner, R., Kleinschmidt, A., Ludvigsen, A., Mellert, W., Neumann, M., Herrmann, R., Khim, M. C., Burny, A., Muller-Lantzsch, N., Stavrou, D., and Erfle, V. (1992). Infection of human brain cells by HIV-1: Restricted virus production in chronically infected human glial cell lines. *AIDS* **6**, 273–285.

Bratanich, A. C., Liu, C., McArthur, J. C., Fudyk, T., Glass, J. D., Mittoo, S., Klassen, G. A., and Power, C. (1998). Brain-derived HIV-1 tat sequences from AIDS patients with dementia show increased molecular heterogeneity. *J. Neurovirol.* **4**, 387–393.

Brenneman, D. E., Westbrook, G. L., Fitzgerald, S. P., Ennist, D. L., Elkins, K. L., Ruff, M. R., and Pert, C. B. (1988). Neuronal cell killing by the envelope protein of HIV and its prevention by vasoactive intestinal peptide. *Nature* **335**, 639–642.

Bretz, J. D., Williams, S. C., Baer, M., Johnson, P. F., and Schwartz, R. C. (1994). C/EBP-related protein 2 confers lipopolysaccharide-inducible expression of interleukin 6 and monocyte chemoattractant protein 1 to a lymphoblastic cell line. *Proc. Natl. Acad. Sci. USA* **91**, 7306–7310.

Brew, B. J., Bhalla, R. B., Paul, M., Gallardo, H., McArthur, J. C., Schwartz, M. K., and Price, R. W. (1990). Cerebrospinal fluid neopterin in human immunodeficiency virus type 1 infection. *Ann. Neurol.* **28**, 556–560.

Brew, B. J., Bhalla, R. B., Paul, M., Sidtis, J. J., Keilp, J. J., Sadler, A. E., Gallardo, H., McArthur, J. C., Schwartz, M. K., and Price, R. W. (1992). Cerebrospinal fluid beta 2-microglobulin in patients with AIDS dementia complex: An expanded series including response to zidovudine treatment. *AIDS* **6**, 461–465.

Brew, B. J., Corbeil, J., Pemberton, L., Evans, L., Saito, K., Penny, R., Cooper, D. A., and Heyes, M. P. (1995a). Quinolinic acid production is related to macrophage tropic isolates of HIV-1. *J. Neurovirol.* **1**, 369–374.

Brew, B. J., Pemberton, L., Cunningham, P., and Law, M. G. (1997). Levels of human immunodeficiency virus type 1 RNA in cerebrospinal fluid correlate with AIDS dementia stage. *J. Infect. Dis.* **175**, 963–966.

Brew, B. J., Rosenblum, M., Cronin, K., and Price, R. W. (1995b). AIDS dementia complex and HIV-1 brain infection: Clinical–virological correlations. *Ann. Neurol.* **38**, 563–570.

Briggs, M. R., Kadonaga, J. T., Bell, S. P., and Tjian, R. (1986). Purification and biochemical characterization of the promoter-specific transcription factor, Spl. *Science* **234**, 47–52.

Britton, C. B., and Miller, J. R. (1984). Neurologic complications in acquired immunodeficiency syndrome (AIDS). *Neurol. Clin.* **2**, 315–339.

Brodt, H. R., Kamps, B. S., Gute, P., Knupp, B., Staszewski, S., and Helm, E. B. (1997). Changing incidence of AIDS-defining illnesses in the era of antiretroviral combination therapy. *AIDS* **11**, 1731–1738.

Brooke, S. M., Howard, S. A., and Sapolsky, R. M. (1998). Energy dependency of glucocorticoid exacerbation of gp120 neurotoxicity. *J. Neurochem.* **71**, 1187–1193.

Brouwers, P., Hendricks, M., Lietzau, J. A., Pluda, J. M., Mitsuya, H., Broder, S., and Yarchoan, R. (1997). Effect of combination therapy with zidovudine and didanosine on neuropsychological functioning in patients with symptomatic HIV disease: A comparison of simultaneous and alternating regimens. *AIDS* **11**, 59–66.

Budka, H. (1990). Human immunodeficiency virus (HIV) envelope and core proteins in CNS tissues of patients with the acquired immune deficiency syndrome (AIDS). *Acta Neuropathol.* **79**, 611–619.

Budka, H. (1991). Neuropathology of human immunodeficiency virus infection. *Brain Pathol.* **1**, 163–175.

Budka, H., Costanzi, G., Cristina, S., Lechi, A., Parravicini, C., Trabattoni, R., and Vago, L. (1987). Brain pathology induced by infection with the human immunodeficiency virus (HIV): A histological, immunocytochemical, and electron microscopical study of 100 autopsy cases. *Acta Neuropathol.* **75**, 185–198.

Buffet, R., Agut, H., Chieze, F., Katlama, C., Bolgert, F., Devillechabrolle, A., Diquet, B., Schuller, E., Pierrot-Deseilligny, C., Gentilini, M. *et al.* (1991). Virological markers in the cerebrospinal fluid from HIV-1-infected individuals. *AIDS* **5**, 1419–1424.

Bukrinsky, M. I., Nottet, H. S., Schmidtmayerova, H., Dubrovsky, L., Flanagan, C. R., Mullins, M. E., Lipton, S. A., and Gendelman, H. E. (1995). Regulation of nitric oxide synthase activity in human immunodeficiency virus type 1 (HIV-1)-infected monocytes: Implications for HIV-associated neurological disease. *J. Exp. Med.* **181**, 735–745.

Butler, K. M., Husson, R. N., Balis, F. M., Brouwers, P., Eddy, J., el-Amin, D., Gress, J., Hawkins, M., Jarosinski, P., Moss, H. *et al.* (1991). Dideoxyinosine in children with symptomatic human immunodeficiency virus infection. *N. Engl. J. Med.* **324**, 137–144.

Cao Z., Umek, R. M., and McKnight, S. L. (1991). Regulated expression of three C/EBP isoforms during adipose conversion of 3T3-L1 cells. *Genes Dev.* **5**, 1538–1552.

Cara, A., Cereseto, A., Lori, F., and Reitz, M. S., Jr. (1996). HIV-1 protein expression from synthetic circles of DNA mimicking the extrachromosomal forms of viral DNA. *J. Biol. Chem.* **271**, 5393–5397.

Catron, K. M., Brickwood, J. R., Shang, C., Li, Y., Shannon, M. F., and Parks, T. P. (1998). Cooperative binding and synergistic activation by RelA and C/EBP beta on the intercellular adhesion molecule-1 promoter. *Cell Growth Differ.* **9**, 949–959.

Centers for Disease Control (1981). Pneumocystis pneumonia—Los Angeles. *Mortal. Morbid. Weekly Rep.* **30**, 250–252.

Chandrasekaran, C., and Gordon, J. I. (1993). Cell lineage-specific and differentiation-dependent patterns of CCAAT/enhancer binding protein alpha expression in the gut epithelium of normal and transgenic mice. *Proc. Natl. Acad. Sci. USA* **90**, 8871–8875.

Chang, L., Ernst, T., Leonido-Yee, M., Witt, M., Speck, O., Walot, I., and Miller, E. N. (1999). Highly active antiretoviral therapy reverses brain metabolite abnormalities in mild HIV dementia. *Neurology* **53**, 782–789.

Chang, L. J., McNulty, E., and Martin, M. (1993). Human immunodeficiency viruses containing heterologous enhancer/promoters are replication competent and exhibit different lymphocyte tropisms. *J. Virol.* **67**, 743–752.

Chao, C. C., Hu, S., Molitor, T. W., Shaskan, E. G., and Peterson, P. K. (1992). Activated microglia mediate neuronal cell injury via a nitric oxide mechanism. *J. Immunol.* **149**, 2736–2741.

Cheng, J., Nath, A., Knudsen, B., Hochman, S., Geiger, J. D., Ma, M., and Magnuson, D. S. (1998). Neuronal excitatory properties of human immunodeficiency virus type 1 Tat protein. *Neuroscience* **82**, 97–106.

Cheng-Mayer, C., Rutka, J. T., Rosenblum, M. L., McHugh, T., Stites, D. P., and Levy, J. A. (1987). Human immunodeficiency virus can productively infect cultured human glial cells. *Proc. Natl. Acad. Sci. USA* **84**, 3526–3530.

Chiodi, F., Fuerstenberg, S., Gidlund, M., Asjo, B., and Fenyo, E. M. (1987). Infection of brain-derived cells with the human immunodeficiency virus. *J. Virol.* **61,** 1244–1247.

Chiodi, F., Norkrans, G., Hagberg, L., Sonnerborg, A., Gaines, H., Froland, S., Fenyo, E. M., Norrby, E., and Vandvik, B. (1988). Human immunodeficiency virus infection of the brain. II. Detection of intrathecally synthesized antibodies by enzyme linked immunosorbent assay and imprint immunofixation. *J. Neurol. Sci.* **87,** 37–48.

Chrysikopoulos, H. S., Press, G. A., Grafe, M. R., Hesselink, J. R., and Wiley, C. A. (1990). Encephalitis caused by human immunodeficiency virus: CT and MR imaging manifestations with clinical and pathologic correlation. *Radiology* **175,** 185–191.

Chumakov, A. M., Grillier, I., Chumakova, E., Chih, D., Slater, J., and Koeffler, H. P. (1997). Cloning of the novel human myeloid-cell-specific C/EBP-epsilon transcription factor. *Mol. Cell. Biol.* **17,** 1375–1386.

Chun, R. F., Semmes, O. J., Neuveut, C., and Jeang, K. T. (1998). Modulation of Sp1 phosphorylation by human immunodeficiency virus type 1 Tat. *J. Virol.* **72,** 2615–2629.

Cinque, P., Vago, L., Mengozzi, M., Torri, V., Ceresa, D., Vicenzi, E., Transidico, P., Vagani, A., Sozzani, S., Mantovani, A., Lazzarin, A., and Poli, G. (1998). Elevated cerebrospinal fluid levels of monocyte chemotactic protein-1 correlate with HIV-1 encephalitis and local viral replication. *AIDS* **12,** 1327–1332.

Clarke, S., and Gordon, S. (1998). Myeloid-specific gene expression. *J. Leukoc. Biol.* **63,** 153–168.

Combates, N. J., Kwon, P. O., Rzepka, R. W., and Cohen, D. (1997). Involvement of the transcription factor NF-IL6 in phorbol ester induction of P-glycoprotein in U937 cells. *Cell Growth Differ.* **8,** 213–219.

Cooper, C., Henderson, A., Artandi, S., Avitahl, N., and Calame, K. (1995). Ig/EBP (C/EBP gamma) is a transdominant negative inhibitor of C/EBP family transcriptional activators. *Nucleic Acids Res.* **23,** 4371–4377.

Corboy, J. R., Buzy, J. M., Zink, M. C., and Clements, J. E. (1992). Expression directed from HIV long terminal repeats in the central nervous system of transgenic mice. *Science* **258,** 1804–1808.

Cortey, A., Jarvik, J. G., Lenkinski, R. E., Grossman, R. I., Frank, I., and Delivoria-Papadopoulos, M. (1994). Proton MR spectroscopy of brain abnormalities in neonates born to HIV-positive mothers. *Am. J. Neuroradiol.* **15,** 1853–1859.

Cupp, C., Taylor, J. P., Khalili, K., and Amini, S. (1993). Evidence for stimulation of the transforming growth factor beta 1 promoter by HIV-1 Tat in cells dervied from CNS. *Oncogene* **8,** 2231–2236.

D'Addario, M., Roulston, A., Wainberg, M. A., and Hiscott, J. (1990). Coordinate enhancement of cytokine gene expression in human immunodeficiency virus type 1-infected promonocytic cells. *J. Virol.* **64,** 6080–6089.

D'Addario, M., Wainberg, M. A., and Hiscott, J. (1992). Activation of cytokine genes in HIV-1 infected myelomonoblastic cells by phorbol ester and tumor necrosis factor. *J. Immunol.* **148,** 1222–1229.

Dal Pan, G. J., McArthur, J. H., Aylward, E., Selnes, O. A., Nance-Sproson, T. E., Kumar, A. J., Mellits, E. D., and McArthur, J. C. (1992). Patterns of cerebral atrophy in HIV-1-infected individuals: Results of a quantitative MRI analysis. *Neurology* **42,** 2125–2130.

Davis, L. E., Hjelle, B. L., Miller, V. E., Palmer, D. L., Llewellyn, A. L., Merlin, T. L., Young, S. A., Mills, R. G., Wachsman, W., and Wiley, C. A. (1992). Early viral brain invasion in iatrogenic human immunodeficiency virus infection. *Neurology* **42,** 1736–1739.

Dawson, T. M., and Dawson, V. L. (1994). gp120 neurotoxicity in primary cortical cultures. *Adv. Neuroimmunol.* **4,** 167–173.

De Luca, P., Majello, B., and Lania, L. (1996). Sp3 represses transcription when tethered to promoter DNA or targeted to promoter proximal RNA. *J. Biol. Chem.* **271,** 8533–8536.

DeCarli, C., Fugate, L., Falloon, J., Eddy, J., Katz, D. A., Friedland, R. P., Rapoport, S. I., Brouwers, P., and Pizzo, P. A. (1991). Brain growth and cognitive improvement in children with human immunodeficiency virus-induced encephalopathy after 6 months of continuous infusion zidovudine therapy. *J. Acq. Immune Defic. Syndr.* **4**, 585–592.

Descombes, P., and Schibler, U. (1991). A liver-enriched transcriptional activator protein, LAP, and a transcriptional inhibitory protein, LIP, are translated from the same mRNA. *Cell* **67**, 569–579.

Dewhurst, S., Sakai, K., Bresser, J., Stevenson, M., Evinger-Hodges, M. J., and Volsky, D. J. (1987). Persistent productive infection of human glial cells by human immunodeficiency virus (HIV) and by infectious molecular clones of HIV. *J. Virol.* **61**, 3774–3782.

Dewhurst, S., Sakai, K. Zhang, X. H., Wasiak, A., and Volsky, D. J. (1988). Establishment of human glial cell lines chronically infected with the human immunodeficiency virus. *Virology* **162**, 151–159.

Dhawan, S., Toro, L. A., Jones, B. E., and Meltzer, M. S. (1992). Interactions between HIV-infected monocytes and the extracellular matrix: HIV-infected monocytes secrete neutral metalloproteases that degrade basement membrane protein matrices. *J. Leukoc. Biol.* **52**, 244–248.

Dhawan, S., Weeks, B. S., Soderland, C., Schnaper, H. W., Toro, L. A., Asthana, S. P., Hewlett, I. K., Stetler-Stevenson, W. G., Yamada, S. S., Yamada, K. M., and Meltzer, M. S. (1995). HIV-1 infection alters monocyte interactions with human microvascular endothelial cells. *J. Immunol.* **154**, 422–432.

Dickson, D. W., Mattiace, L. A., Kure, K., Hutchins, K., Lyman, W. D., and Brosnan, C. F. (1991). Microglia in human disease, with an emphasis on acquired immune deficiency syndrome. *Lab. Invest.* **64**, 135–156.

Dreyer, E. B., Kaiser, P. K., Offermann, J. T., and Lipton, S. A. (1990). HIV-1 coat protein neurotoxicity prevented by calcium channel antagonists. *Science* **248**, 364–367.

Elder, G. A., and Sever, J. L. (1998). Neurologic disorders associated with AIDS retroviral infection. *Rev. Infect. Dis.* **10**, 286–302.

Ensoli, B., Barillari, G., Salahuddin, S. Z., Gallo, R. C., and Wong-Staal, F. (1990). Tat protein of HIV-1 stimulates growth of cells derived from Kaposi's sarcoma lesions of AIDS patients. *Nature* **345**, 84–86.

Ensoli, B., Buonaguro, L., Barillari, G., Fiorelli, V., Gendelman, R., Morgan, R. A., Wingfield, P., and Gallo, R. C. (1993). Release, uptake, and effects of extracellular human immunodeficiency virus type 1 Tat protein on cell growth and viral transactivation. *J. Virol.* **67**, 277–287.

Epstein, L. G., and Gendelman, H. E. (1993). Human immunodeficiency virus type 1 infection of the nervous system: Pathogenetic mechanisms. *Ann. Neurol.* **33**, 429–436.

Epstein, L. G., Kuiken, C., Blumberg, B. M., Hartman, S., Sharer, L. R., Clement, M., and Goudsmit, J. (1991). HIV-1 V3 domain variation in brain and spleen of children with AIDS: Tissue-specific evolution within host-determined quasispecies. *Virology* **180**, 583–590.

Epstein, L. G., Sharer, L. R., Cho, E. S., Myenhofer, M., Navia, B., and Price, R. W. (1984). HTLV-III/LAV-like retrovirus particles in the brains of patients with AIDS encephalopathy. *AIDS Res.* **1**, 447–454.

Epstein, L. G., Sharer, L. R., and Goudsmit, J. (1988). Neurological and neuropathological features of human immunodeficiency virus infection in children. *Ann. Neurol.* **23**, S19–S23.

Epstein, L. G., Sharer, L. R., Joshi, V. V., Fojas, M. M., Koenigsberger, M. R., and Oleske, J. M. (1985). Progressive encephalopathy in children with acquired immune deficiency syndrome. *Ann. Neurol.* **17**, 488–496.

Epstein, L. G., Sharer, L. R., Oleske, J. M., Connor, E. M., Goudsmit, J., Bagdon, L., Robert-Guroff, M., and Koenigsberger, M. R. (1986). Neurologic manifestations of human immunodeficiency virus infection in children. *Pediatrics* **78**, 678–687.

Esiri, M. M., Morris, C. S., and Millard, P. R. (1991). Fate of oligodendrocytes in HIV-1 infection. *AIDS* **5**, 1081–1088.

Everall, I. Luthert, P., and Lantos, P. (1993). A review of neuronal damage in human immunodeficiency virus infection: Its assessment, possible mechanism and relationship to dementia. *J. Neuropathol. Exp. Neurol.* **52**, 561–566.

Everall, I. P., Glass, J. D., McArthur, J., Spargo, E., and Lantos, P. (1994). Neuronal density in the superior frontal and temporal gyri does not correlate with the degree of human immunodeficiency virus-associated dementia. *Acta Neuropathol.* **88**, 538–544.

Falangola, M. F., Hanly, A., Galvao-Castro, B., and Petito, C. K. (1995). HIV infection of human choroid plexus: A possible mechanism of viral entry into the CNS. *J. Neuropathol. Exp. Neurol.* **54**, 497–503.

Feng, Y., Broder, C. C., Kennedy. P. E., and Berger, E. A. (1996). HIV-1 entry cofactor: Functional cDNA cloning of a seven-transmembrane, G protein-coupled receptor. *Science* **272**, 872–877.

Fiala, M., Looney, D. J., Stins, M., Way, D. D., Zhang, L., Gan, X., Chiappelli, F., Schweitzer, E. S., Shapshak, P., Weinand, M., Graves, M. C., Witte, M., and Kim, K. S. (1997). TNF-alpha opens a paracellular route for HIV-1 invasion across the blood–brain barrier. *Mol. Med.* **3**, 553–564.

Filippi, C. G., Sze, G., Farber, S. J., Shahmanesh, M., and Selwyn, P. A. (1998). Regression of HIV encephalopathy and basal ganglia signal intensity abnormality at MR imaging in patients with AIDS after the initiation of protease inhibitor therapy. *Radiology* **206**, 491–498.

Fischl, M. A., Richman, D. D., Grieco, M. H., Gottlieb, M. S., Volberding, P. A., Laskin, O. L., Leedom, J. M., Groopman, J. E., Mildvan, D., Schooley, R. T. *et al.* (1987). The efficacy of azidothymidine (AZT) in the treatment of patients with AIDS and AIDS-related complex: A double-blind, placebo-controlled trial. *N. Engl. J. Med.* **317**, 185–191.

Foudraine, N. A., Hoetelmans, R. M., Lange, J. M., de Wolf, F., van Benthem, B. H., Maas, J. J., Keet, I. P., and Portegies, P. (1998). Cerebrospinal-fluid HIV-1 RNA and drug concentrations after treatment with lamivudine plus zidovudine or stavudine. *Lancet* **351**, 1547–1551.

Frankel, A. D., and Pabo, C. O. (1988). Cellular uptake of the tat protein from human immunodeficiency virus. *Cell* **55**, 1189–1193.

Frumkin, L. R., Patterson, B. K., Leverenz, J. B., Agy, M. B., Wolinsky, S. M., Morton, W. R., and Corey, L. (1995). Infection of *Macaca nemestrina* brain with human immunodeficiency virus type 1. *J. Gen. Virol.* **76**, 2467–2476.

Gabuzda, D. H., Ho, D. D., de la Monte, S. M., Hirsch, M. S., Rota, T. R., and Sobel, R. A. (1986). Immunohistochemical identification of HTLV-III antigen in brains of patients with AIDS. *Ann. Neurol.* **20**, 289–295.

Gallo, P., Frei, K., Rordorf, C., Lazdins, J., Tavolato, B., and Fontana, A. (1989). Human immunodeficiency virus type 1 (HIV-1) infection of the central nervous system: an evaluation of cytokines in cerebrospinal fluid. *J. Neuroimmunol.* **23**, 109–116.

Gelbard, H. A., Dzenko, K. A., DiLoreto, D., del Cerro, C., del Cerro, M., and Epstein, L. G. (1993). Neurotoxic effects of tumor necrosis factor alpha in primary human neuronal cultures are mediated by activation of the glutamate AMPA receptor subtype: Implications for AIDS neuropathogenesis. *Dev. Neurosci.* **15**, 417–422.

Gelbard, H. A., James, H. J., Sharer, L. R., Perry, S. W., Saito, Y., Kazee, A. M., Blumberg, B. M., and Epstein, L. G. (1995). Apoptotic neurons in brains from paediatric patients with HIV-1 encephalitis and progressive encephalopathy. *Neuropathol. Appl. Neurobiol.* **21**, 208–217.

Gelbard, H. A., Nottet, H. S., Swindells, S., Jett, M., Dzenko, K. A., Genis, P., White, R., Wang, L., Choi, Y. B., Zhang, D., Lipton, S. A., Tourtellotte, W. W., Epstein, L. G., and Gendelman, H. E. (1994). Platelet-activating factor: a candidate human immunodeficiency virus type 1-induced neurotoxin. *J. Virol.* **68**, 4628–4635.

Geleziunas, R., Schipper, H. M., and Wainberg, M. A. (1992). Pathogenesis and therapy of HIV-1 infection of the central nervous system. *AIDS* **6**, 1411–1426.

Gelman, B. B., and Guinto, F. C., Jr. (1992). Morphometry, histopathology, and tomography of cerebral atrophy in the acquired immunodeficiency syndrome. *Ann. Neurol.* **32**, 31–40.

Gendelman, H. E., Zheng, J., Coulter, C. L., Ghorpade, A., Che, M., Thylin, M., Rubocki, R., Persidsky, Y., Hahn, F., Reinhard, J., Jr., and Swindells, S. (1998). Suppression of inflammatory neurotoxins by highly active antiretroviral therapy in human immunodeficiency virus-associated dementia. *J. Infect. Dis.* **178**, 1000–1007.

Genis, P., Jett, M., Bernton, E. W., Boyle, T., Gelbard, H. A., Dzenko, K., Keane, R. W., Resnick, L., Mizrachi, Y., Volsky, D. J., Epstein, L. G., and Gendelman, H. E. (1992). Cytokines and arachidonic metabolites produced during human immunodeficiency virus (HIV)-infected macrophage-astroglia interactions: Implications for the neuropathogenesis of HIV disease. *J. Exp. Med.* **176**, 1703–1718.

Ghorpade, A., Xia, M. Q., Hyman, B. T., Persidsky, Y., Nukuna, A., Bock, P., Che, M., Limoges, J., Gendelman, H. E., and Mackay, C. R. (1998). Role of the beta-chemokine receptors CCR3 and CCR5 in human immunodeficiency virus type 1 infection of monocytes and microglia. *J. Virol.* **72**, 3351–3361.

Gisslen, M., Norkrans, G., Svennerholm, B., and Hagberg, L. (1997). The effect on human immunodeficiency virus type 1 RNA levels in cerebrospinal fluid after initiation of zidovudine or didanosine. *J. Infect. Dis.* **175**, 434–437.

Glass, J. D., Fedor, H., Wesselingh, S. L., and McArthur, J. C. (1995). Immunocytochemical quantitation of human immunodeficiency virus in the brain: Correlations with dementia. *Ann. Neurol.* **38**, 755–762.

Glynn, S. L., and Yazdanian, M. (1998). *In vitro* blood–brain barrier permeability of nevirapine compared to other HIV antiretroviral agents. *J. Pharm. Sci.* **87**, 306–310.

Gonzalez-Scarano, F., and Baltuch, G. (1999). Microglia as mediators of inflammatory and degenerative diseases. *Annu. Rev. Neurosci.* **22**, 219–240.

Gonzalez-Scarano, F., Nathanson, N., and Wong, P. K. Y. (1995). Retroviruses and the nervous system. *In* "The Retroviridae" (J. A. Levy, Ed.), Vol. 4, pp. 409–490. Plenum, New York.

Gorman, J. M., Mayeux, R., Stern, Y., Williams, J. B., Rabkin, J., Goetz, R. R., and Ehrhardt, A. A. (1993). The effect of zidovudine on neuropsychiatric measures in HIV-infected men. *Am. J. Psychiatry.* **150**, 505–507.

Goswami, K. K., Miller, R. F., Harrison, M. J., Hamel, D. J., Daniels, R. S., and Tedder, R. S. (1991). Expression of HIV-1 in the cerebrospinal fluid detected by the polymerase chain reaction and its correlation with central nervous system disease. *AIDS* **5**, 797–803.

Grafe, M. R., Press, G. A., Berthoty, D. P., Hesselink, J. R., and Wiley, C. A. (1990). Abnormalities of the brain in AIDS patients: Correlation of postmortem MR findings with neuropathology. *Am. J. Neuroradiol.* **11**, 905–911; discussion, 912–903.

Griffin, D. E., Wesselingh, S. L., and McArthur, J. C. (1994). Elevated central nervous system prostaglandins in human immunodeficiency virus-associated dementia. *Ann. Neurol.* **35**, 592–597.

Grimaldi, L. M., Martino, G. V., Franciotta, D. M., Brustia, R., Castagna, A., Pristera, R., and Lazzarin, A. (1991). Elevated alpha-tumor necrosis factor levels in spinal fluid from HIV-1-infected patients with central nervous system involvement. *Ann. Neurol.* **29**, 21–25.

Gupta, S.K., Lysko, P. G., Pillarisetti, K., Ohlstein, E., and Stadel, J. M. (1998). Chemokine receptors in human endothelial cells: Functional expression of CXCR4 and its transcriptional regulation by inflammatory cytokines. *J. Biol. Chem.* **273**, 4282–4287.

Haase, A. T. (1986). Pathogenesis of lentivirus infections. *Nature* **322**, 130–136.

Hagen, G., Dennig, J., Preiss, A., Beato, M., and Suske, G. (1995). Functional analyses of the transcription factor Sp4 reveal properties distinct from Sp1 and Sp3. *J. Biol. Chem.* **270**, 24989–24994.

Hagen, G., Muller, S., Beato, M., and Suske, G. (1992). Cloning by recognition site screening of two novel GT box binding proteins: A family of Sp1 related genes. *Nucleic Acids Res.* **20**, 5519–5525.

Hagen, G., Muller, S., Beato, M., and Suske, G. (1994). Sp1-mediated transcriptional activation is repressed by Sp3. *EMBO J.* **13**, 3843–3851.

Harouse, J. M., Kunsch, C., Hartle, H. T., Laughlin, M. A., Hoxie, J. A., Wigdahl, B., and Gonzalez-Scarano, F. (1989a). CD4-independent infection of human neural cells by human immunodeficiency virus type 1. *J. Virol.* **63**, 2527–2533.

Harouse, J. M., Wroblewska, Z., Laughlin, M. A., Hickey, W. F., Schonwetter, B. S., and Gonzalez-Scarano, F. (1989b). Human choroid plexus cells can be latently infected with human immunodeficiency virus. *Ann. Neurol.* **25**, 406–411.

Harrich, D., Garcia, J., Wu, F., Mitsuyasu, R., Gonazalez, J., and Gaynor, R. (1989). Role of SP1-binding domains in *in vivo* transcriptional regulation of the human immunodeficiency virus type 1 long terminal repeat. *J. Virol.* **63**, 2585–2591.

He, J., Chen, Y., Farzan, M., Choe, H., Ohagen, A., Gartner, S., Busciglio, J., Yang, X., Hofmann, W., Newman, W., Mackay, C. R., Sodroski, J., and Gabuzda, D. (1997). CCR3 and CCR5 are co-receptors for HIV-1 infection of microglia. *Nature* **385**, 645–649.

Henderson, A. J., and Calame, K. L. (1997). CCAAT/enhancer binding protein (C/EBP) sites are required for HIV-1 replication in primary macrophages but not CD4(+) T cells. *Proc. Natl. Acad. Sci. USA* **94**, 8714–8719.

Henderson, A. J., Connor, R. I., and Calame, K. L. (1996). C/EBP activators are required for HIV-1 replication and proviral induction in monocytic cell lines. *Immunity* **5**, 91–101.

Henderson, A. J., Zou, X., and Calame, K. L. (1995). C/EBP proteins activate transcription from the human immunodeficiency virus type 1 long terminal repeat in macrophages/monocytes. *J. Virol.* **69**, 5337–5344.

Hesselgesser, J., Halks-Miller, M., DelVecchio, V., Peiper, S. C., Hoxie, J., Kolson, D. L., Taub, D., and Horuk, R. (1997). CD4-independent association between HIV-1 gp120 and CXCR4: Functional chemokine receptors are expressed in human neurons. *Curr. Biol.* **7**, 112–121.

Hesselgesser, J., and Horuk, R. (1999). Chemokine and chemokine receptor expression in the central nervous system. *J. Neurovirol.* **5**, 13–26.

Hesselgesser, J., Taub. D., Baskar, P., Greenberg, M., Hoxie, J., Kolson, D. L., and Horuk, R. (1998). Neuronal apoptosis induced by HIV-1 gp120 and the chemokine SDF-1 alpha is mediated by the chemokine receptor CXCR4. *Curr. Biol.* **8**, 595–598.

Heyes, M. P., Brew, B., Martin, A., Markey, S. P., Price, R. W., Bhalla, R. B., and Salazar, A. (1991a). Cerebrospinal fluid quinolinic acid concentrations are increased in acquired immune deficiency syndrome. *Adv. Exp. Med. Biol.* **294**, 687–690.

Heyes, M. P., Brew, B. J., Martin, A., Price, R. W., Salazar, A. M., Sidtis, J. J., Yergey, J. A., Mouradian, M. M., Sadler, A. E., Keilp, J., Rubinow, D., and Markey, S. P. (1991b). Quinolinic acid in cerebrospinal fluid and serum in HIV-1 infection: Relationship to clinical and neurological status. *Ann. Neurol.* **29**, 202–209.

Heyes, M. P., Gravell, M., London, W. T., Eckhaus, M., Vickers, J. H., Yergey, J. A., April, M., Blackmore, D., and Markey, S. P. (1990). Sustained increases in cerebrospinal fluid quinolinic acid concentrations in rhesus macaques (*Macaca mulatta*) naturally infected with simian retrovirus type-D. *Brain Res.* **531**, 148–158.

Hickey, W. F. (1999). Leukocyte traffic in the central nervous system: The participants and their roles. *Semin. Immunol.* **11**, 125–137.

Ho, D. D., Bredesen, D. E., Vinters, H. V., and Daar, E. S. (1989). The acquired immunodeficiency syndrome (AIDS) dementia complex. *Ann. Intern. Med.* **111**, 400–410.

Ho, D. D., Rota, T. R., Schooley, R. T., Kaplan, J. C., Allan, J. D., Groopman, J. E., Resnick, L., Felsenstein, D., Andrews, C. A., and Hirsch, M. S. (1985). Isolation of HTLV-III from cerebrospinal fluid and neural tissues of patients with neurologic syndromes related to the acquired immunodeficiency syndrome. *N. Engl. J. Med.* **313**, 1493–1497.

Hoffbrand, A. V., and Pettit, J. E. (1993). "Essential Haematology," Blackwell Scientific, Oxford, UK.

Hofman, F. M., Wright, A. D., Dohadwala, M. M., Wong-Staal, F., and Walker, S. M. (1993). Exogenous tat protein activates human endothelial cells. *Blood* 82, 2774–2780.

Hollander, H. (1991). Neurologic and psychiatric manifestations of HIV disease. *J. Gen. Intern. Med.* 6, S24–S31.

Honda, Y., Rogers, L., Nakata, K., Zhao, B. Y., Pine, R., Nakai, Y., Kurosu, K., Rom, W. N., and Weiden, M. (1998). Type I interferon induces inhibitory 16-kD CCAAT/ enhancer binding protein (C/EBP) beta, repressing the HIV-1 long terminal repeat in macrophages: Pulmonary tuberculosis alters C/EBP expression, enhancing HIV-1 replication. *J. Exp. Med.* 188, 1255–1265.

Howcroft, T. K., Strebel, K., Martin, M. A., and Singer, D. S. (1993). Repression of MHC class I gene promoter activity by two-exon Tat of HIV. *Science* 260, 1320–1322.

Hu, H. M., Baer, M., Williams, S. C., Johnson, P. F., and Schwartz, R. C. (1998). Redundancy of C/EBP alpha, beta and delta in supporting the lipopolysaccharide-induced transcription of IL-6 and monocyte chemoattractant protein-1. *J. Immunol.* 160, 2334–2342.

Huang, L. M., and Jeang, K. T. (1993). Increased spacing between Sp1 and TATAA renders human immunodeficiency virus type 1 replication defective: Implication for Tat function. *J. Virol.* 67, 6937–6944.

Husstedt, I. W., Evers, S., Reichelt, D., Sprinz, A., Riedasch, M., and Grotemeyer, K. H. (1998). Progressive peripheral and central sensory tract lesion in HIV-infected patients evidenced by evoked potentials (a three-year follow-up study). *J. Neurol. Sci.* 159, 54–59.

Imataka, H., Sogawa, K., Yasumoto, K., Kikuchi, Y., Sasano, K., Kobayashi, A., Hayami, M., and Fujii-Kuriyama, Y. (1992). Two regulatory proteins that bind to the basic transcription element (BTE), a GC box sequence in the promoter region of the rat P-4501A1 gene. *EMBO J.* 11, 3663–3671.

Iyer, A. M., Brooke, S. M., and Sapolsky, R. M. (1998). Glucocorticoids interact with gp120 in causing neurotoxicity in striatal cultures. *Brain Res.* 808, 305–309.

Jackson, S. P., MacDonald, J. J., Lees-Miller, S., and Tjian, R. (1990). GC box binding induces phosphorylation of Sp1 by a DNA-dependent protein kinase. *Cell* 63, 155–165.

Jackson, S. P., and Tjian, R. (1988). O-glycosylation of eukaryotic transcription factors: Implications for mechanisms of transcriptional regulation. *Cell* 55, 125–133.

Jakobsen, J., Gyldensted, C., Brun, B., Bruhn, P., Helweg-Larsen, S., and Arlien-Soborg, P. (1989). Cerebral ventricular enlargement relates to neuropsychological measures in unselected AIDS patients. *Acta Neurol. Scand.* 79, 59–62.

Janssen, R. S., Cornblatz, D. R., Epstein, L. G., Foa, R. P., McArthur, J. C., and Price, R. W. (1991). Nomenclature and research case definitions for neurologic manifestations of human immunodeficiency virus-type 1 (HIV-1) infection: Report of a Working Group of the American Academy of Neurology AIDS Task Force. *Neurology* 41, 778–785.

Janssen, R. S., Nwanyanwu, O. C., Selik, R. M., and Stehr-Green, J. K. (1992). Epidemiology of human immunodeficiency virus encephalopathy in the United States. *Neurology* 42, 1472–1476.

Jarvik, J. G., Hesselink, J. R., Kennedy, C., Teschke, R., Wiley, C., Spector, S., Richman, D., and McCutchan, J. A. (1988). Acquired immunodeficiency syndrome. Magnetic resonance patterns of brain involvement with pathologic correlation. *Arch. Neurol.* 45, 731–736.

Jeang, K. T., Chun, R., Lin, N. H., Gatignol, A., Glabe, C. G., and Fan, H. (1993). *In vitro* and *in vivo* binding of human immunodeficiency virus type 1 Tat protein and Sp1 transcription factor. *J. Virol.* 67, 6224–6233.

Johann-Liang, R., Cervia, J. S., and Noel, G. J. (1997). Characteristics of human immunodeficiency virus-infected children at the time of death: An experience in the 1990s. *Pediatr. Infect. Dis. J.* 16, 1145–1150.

Johnson, R. T. (1982). "Viral Infections of the Nervous System." Raven Press, New York.

Johnson, R. T., Glass, J. D., McArthur, J. C., and Chesebro, B. W. (1996). Quantitation of human immunodeficiency virus in brains of demented and nondemented patients with acquired immunodeficiency syndrome. *Ann. Neurol.* 39, 392–395.

Jones, K. A., Kadonaga, J. T., Luciw, P. A., and Tjian, R. (1986). Activation of the AIDS retrovirus promoter by the cellular transcription factor, Sp1. *Science* **232**, 755–759.

Jordan, E. K., and Heyes, M. P. (1993). Virus isolation and quinolinic acid in primary and chronic simian immunodeficiency virus infection. *AIDS* **7**, 1173–1179.

Kadonaga, J. T., Carner, K. R., Masiarz, F. R., and Tjian, R. (1987). Isolation of cDNA encoding transcription factor Sp1 and functional analysis of the DNA binding domain. *Cell* **51**, 1079–1090.

Kamine, J., Subramanian, T., and Chinnadurai, G. (1991). Sp1-dependent activation of a synthetic promoter by human immunodeficiency virus type 1 Tat protein. *Proc. Natl. Acad. Sci. USA* **88**, 8510–8514.

Kato, T., Hirano, A., Llena, J. F., and Dembitzer, H. M. (1987). Neuropathology of acquired immune deficiency syndrome (AIDS) in 53 autopsy cases with particular emphasis on microglial nodules and multinucleated giant cells. *Acta Neuropathol.* **73**, 287–294.

Kaul, M., and Lipton, S. A. (1999). Chemokines and activated macrophages in HIV gp120-induced neuronal apoptosis. *Proc. Natl. Acad. Sci. USA* **96**, 8212–8216.

Kennett, S. B., Udvadia, A. J., and Horowitz, J. M. (1997). Sp3 encodes multiple proteins that differ in their capacity to stimulate or repress transcription. *Nucleic Acids Res.* **25**, 3110–3117.

Ketzler, S., Weis, S., Haug, H., and Budka, H. (1990). Loss of neurons in the frontal cortex in AIDS brains. *Acta Neuropathol.* **80**, 92–94.

Kim, R. B., Fromm, M. F., Wandel, C., Leake, B., Wood, A. J., Roden, D. M., and Wilkinson, G. R. (1998). The drug transporter P-glycoprotein limits oral absorption and brain entry of HIV-1 protease inhibitors. *J. Clin. Invest.* **101**, 289–294.

Kim, Y. S., and Panganiban, A. T. (1993). The full-length Tat protein is required for TAR-independent, posttranscriptional trans activation of human immunodeficiency virus type 1 env gene expression. *J. Virol.* **67**, 3739–3747.

Kimura-Kuroda, J., Nagashima, K., and Yasui, K. (1994). Inhibition of myelin formation by HIV-1 gp120 in rat cerebral cortex culture. *Arch. Virol.* **137**, 81–99.

Kingsley, C., and Winoto, A. (1992). Cloning of GT box-binding proteins: A novel Sp1 multigene family regulating T-cell receptor gene expression. *Mol. Cell. Biol.* **12**, 4251–4261.

Kingsman, S. M., and Kingsman, A. J. (1996). The regulation of human immunodeficiency virus type-1 gene expression. *Eur. J. Biochem.* **240**, 491–507.

Kinoshita, S., Akira, S., and Kishimoto, T. (1992). A member of the C/EBP family, NF-IL6 beta, forms a heterodimer and transcriptionally synergizes with NF-IL6. *Proc. Natl. Acad. Sci. USA* **89**, 1473–1476.

Koenig, S., Gendelman, H. E., Orenstein, J. M., Dal Canto, M. C., Pezeshkpour, G. H., Yungbluth, M., Janotta, F., Aksamit, A., Martin, M. A., and Fauci, A. S. (1986). Detection of AIDS virus in macrophages in brain tissue from AIDS patients with encephalopathy. *Science* **233**, 1089–1093.

Kolson, D. L., Buchhalter, J., Collman, R., Hellmig, B., Farrell, C. F., Debouck, C., and Gonzalez-Scarano, F. (1993). HIV-1 Tat alters normal organization of neurons and astrocytes in primary rodent brain cell cultures: RGD sequence dependence. *AIDS Res. Hum. Retroviruses* **9**, 677–685.

Kolson, D. L., Collman, R., Hrin, R., Balliet, J. W., Laughlin, M., McGann, K. A., Debouck, C., and Gonzalez-Scarano, F. (1994). Human immunodeficiency virus type 1 Tat activity in human neuronal cells: Uptake and trans-activation. *J. Gen. Virol.* **75**, 1927–1934.

Kolson, D. L., Lavi, E., and Gonzalez-Scarano, F. (1998). The effects of human immunodeficiency virus in the central nervous system. *Adv. Virus Res.* **50**, 1–47.

Kozlowski, P. B., Brudkowska, J., Kraszpulski, M., Sersen, E. A., Wrzolek, M. A., Anzil, A. P., Rao, C., and Wisniewski, H. M. (1997). Microencephaly in children congenitally infected with human immunodeficiency virus—A gross-anatomical morphometric study. *Acta Neuropathol.* **93**, 136–145.

Kravcik, S., Gallicano, K., Roth, V., Cassol, S., Hawley-Foss, N., Badley, A., and Cameron, D. W. (1999). Cerebrospinal fluid HIV RNA and drug levels with combination ritonavir and saquinavir. *J. Acq. Immune Defic. Syndr.* **21**, 371–375.

Krebs, F. C., Goodenow, M. M., and Wigdahl, B. (1997). Neuroglial ATF/CREB factors interact with the human immunodeficiency virus type 1 long terminal repeat. *J. Neurovirol.* 3(Suppl. 1), S28–S32.

Kure, K., Lyman, W. D., Weidenheim, K. M., and Dickson, D. W. (1990a). Cellular localization of an HIV-1 antigen in subacute AIDS encephalitis using an improved double-labeling immunohistochemical method. *Am. J. Pathol.* **136**, 1085–1092.

Kure, K., Weidenheim, K. M., Lyman, W. D., and Dickson, D. W. (1990b). Morphology and distribution of HIV-1 gp41-positive microglia in subacute AIDS encephalitis: Pattern of involvement resembling a multisystem degeneration. *Acta Neuropathol.* **80**, 393–400.

Lafrenie, R. M., Wahl, L. M., Epstein, J. S., Hewlett, I. K., Yamada, K. M., and Dhawan, S. (1996). HIV-1-Tat modulates the function of monocytes and alters their interactions with microvessel endothelial cells: A mechanism of HIV pathogenesis. *J. Immunol.* **156**, 1638–1645.

Landschulz, W. H., Johnson, P. F., Adashi, E. Y., Graves, B. J., and McKnight, S. L. (1988). Isolation of a recombinant copy of the gene encoding C/EBP. *Genes Dev.* **2**, 786–800.

Lannuzel, A., Barnier, J. V., Hery, C., Huynh, V. T., Guibert, B., Gray, F., Vincent, J. D., and Tardieu, M. (1997). Human immunodeficiency virus type 1 and its coat protein gp120 induce apoptosis and activate JNK and ERK mitogen-activated protein kinases in human neurons. *Ann. Neurol.* **42**, 847–856.

Lannuzel, A., Lledo, P. M., Lamghitnia, H. O., Vincent, J. D., and Tardieu, M. (1995). HIV-1 envelope proteins gp120 and gp160 potentiate NMDA-induced $[Ca^{2+}]_i$ increase, alter $[Ca^{2+}]_i$ homeostasis and induce neurotoxicity in human embryonic neurons. *Eur. J. Neurosci.* **7**, 2285–2293.

Larder, B. A., Darby, G., and Richman, D. D. (1989). HIV with reduced sensitivity to zidovudine (AZT) isolated during prolonged therapy. *Science* **243**, 1731–1734.

Lavi, E., Kolson, D. L., Ulrich, A. M., Fu, L., and Gonzalez-Scarano, F. (1998). Chemokine receptors in the human brain and their relationship to HIV infection. *J. Neurovirol.* **4**, 301–311.

Lavi, E., Strizki, J. M., Ulrich, A. M., Zhang, W., Fu, L., Wang, Q., O'Connor, M., Hoxie, J. A., and Gonzalez-Scarano, F. (1997). CXCR-4 (Fusin), a co-receptor for the type 1 human immunodeficiency virus (HIV-1), is expressed in the human brain in a variety of cell types, including microglia and neurons. *Am. J. Pathol.* **151**, 1035–1042.

LeClair, K. P., Blanar, M. A., and Sharp, P. A. (1992). The p50 subunit of NF-kappa B associates with the NF-IL6 transcription factor. *Proc. Natl. Acad. Sci. USA* **89**, 8145–8149.

Lee, Y. H., Sauer, B., Johnson, P. F., and Gonzalez, F. J. (1997). Disruption of the C/EBP alpha gene in adult mouse liver. *Mol. Cell. Biol.* **17**, 6014–6022.

Lee, Y. H., Yano, M., Liu, S. Y., Matsunaga, E., Johnson, P. F., and Gonzalez, F. J. (1994). A novel cis-acting element controlling the rat CYP2D5 gene and requiring cooperativity between C/EBP beta and an Sp1 factor. *Mol. Cell. Biol.* **14**, 1383–1394.

Leggett, R. W., Armstrong, S. A., Barry, D., and Mueller, C. R. (1995). Sp1 is phosphorylated and its DNA binding activity down-regulated upon terminal differentiation of the liver. *J. Biol. Chem.* **270**, 25879–25884.

Levin, H. S., Williams, D. H., Borucki, M. J., Hillman, G. R., Williams, J. B., Guinto, F. C., Jr., Amparo, E. G., Crow, W. N., and Pollard, R. B. (1990). Magnetic resonance imaging and neuropsychological findings in human immunodeficiency virus infection. *J. Acq. Immune Defic. Syndr.* **3**, 757–762.

Levy, J. A., Shimabukuro, J., Hollander, H., Mills, J., and Kaminsky, L. (1985). Isolation of AIDS-associated retroviruses from cerebrospinal fluid and brain of patients with neurological symptoms. *Lancet* **2**, 586–588.

Lipton, S. A. (1992a). Memantine prevents HIV coat protein-induced neuronal injury *in vitro*. *Neurology* **42**, 1403–1405.

Lipton, S. A. (1992b). Requirement for macrophages in neuronal injury induced by HIV envelope protein gp120. *NeuroReport* **3**, 913–915.

Lipton, S. A., Sucher, N. J., Kaiser, P. K., and Dreyer, E. B. (1991). Synergistic effects of HIV coat protein and NMDA receptor-mediated neurotoxicity. *Neuron* **7**, 111–118.

Lopez-Rodriguez, C., Botella, L., and Corbi, A. L. (1997). CCAAT-enhancer-binding proteins (C/EBP) regulate the tissue specific activity of the CD11c integrin gene promoter through functional interactions with Sp1 proteins. *J. Biol. Chem.* **272**, 29120–29126.

Magnuson, D. S., Knudsen, B. E., Geiger, J. D., Brownstone, R. M., and Nath, A. (1995). Human immunodeficiency virus type 1 tat activates non-N-methyl-D-aspartate excitatory amino acid receptors and causes neurotoxicity. *Ann. Neurol.* **37**, 373–380.

Majello, B., De Luca, P., Hagen, G., Suske, G., and Lania, L. (1994). Different members of the Sp1 multigene family exert opposite transcriptional regulation of the long terminal repeat of HIV-1. *Nucleic Acids Res.* **22**, 4914–4921.

Majello, B., De Luca, P., and Lania, L. (1997). Sp3 is a bifunctional transcription regulator with modular independent activation and repression domains. *J. Biol. Chem.* **272**, 4021–4026.

Masliah, E., Achim, C. L., Ge, N., DeTeresa, R., Terry, R. D., and Wiley, C. A. (1992a). Spectrum of human immunodeficiency virus-associated neocortical damage. *Ann. Neurol.* **32**, 321–329.

Masliah, E., Ge, N., Achim, C. L., Hansen, L. A., and Wiley, C. A. (1992b). Selective neuronal vulnerability in HIV encephalitis. *J. Neuropathol. Exp. Neurol.* **51**, 585–593.

Matsusaka, T., Fujikawa, K., Nishio, Y., Mukaida, N., Matsushima, K., Kishimoto, T., and Akira, S. (1993). Transcription factors NF-IL6 and NF-kappa B synergistically activate transcription of the inflammatory cytokines, interleukin 6 and interleukin 8. *Proc. Natl. Acad. Sci. USA* **90**, 10193–10197.

Mayeux, R., Stern, Y., Tang, M. X., Todak, G., Marder, K., Sano, M., Richards, M., Stein, Z., Ehrhardt, A. A., and Gorman, J. M. (1993). Mortality risks in gay men with human immunodeficiency virus infection and cognitive impairment. *Neurology* **43**, 176–182.

McArthur, J. C. (1987). Neurologic manifestations of AIDS. *Medicine (Baltimore)* **66**, 407–437.

McArthur, J. C., Hoover, D. R., Bacellar, H., Miller, E. N., Cohen, B. A., Becker, J. T., Graham, N. M., McArthur, J. H., Selnes, O. A., Jacobson, L. P., Visscher, B. R., Concha, M., and Saah, A. (1993). Dementia in AIDS patients: Incidence and risk factors: Multicenter AIDS Cohort Study. *Neurology* **43**, 2245–2252.

McArthur, J. C., McClernon, D. R., Cronin, M. F., Nance-Sproson, T. E., Saah, A. J., St. Clair, M., and Lanier, E. R. (1997). Relationship between human immunodeficiency virus-associated dementia and viral load in cerebrospinal fluid and brain. *Ann. Neurol.* **42**, 689–698.

McArthur, J. C., and Selnes, O. A. (1997). Human immunodeficiency virus dementia. *In* "AIDS and the Nervous system" (J. R. Berger and R. M. Levy, Eds.), pp. 527–567. Lippincott–Raven, Philadelphia.

McArthur, J. C., Selnes, O. A., Glass, J. D., Hoover, D. R., and Bacellar, H. (1994). HIV Dementia: Incidence and risk factors. *In* "HIV, AIDS, and the Brain" (R. W. Price and S. W. Perry, Eds.), Vol. 72, pp. 251–272. Raven Press, New York.

McCarthy, M., He, J., and Wood, C. (1998). HIV-1 strain-associated variability in infection of primary neuroglia. *J. Neurovirol.* **4**, 80–89.

Mengozzi, M., De Filippi, C., Transidico, P., Biswas, P., Cota, M., Ghezzi, S., Vicenzi, E., Mantovani, A., Sozzani, S., and Poli, G. (1999). Human immunodeficiency virus replication induces monocyte chemotactic protein-1 in human macrophages and U937 promonocytic cells. *Blood* **93**, 1851–1857.

Meucci, O., Fatatis, A., Simen, A. A., Bushell, T. J., Gray, P. W., and Miller, R. J. (1998). Chemokines regulate hippocampal neuronal signaling and gp120 neurotoxicity. *Proc. Natl. Acad. Sci. USA* **95**, 14500–14505.

Meyer, L., Magierowska, M., Hubert, J. B., Rouzioux, C., Deveau, C., Sanson, F., Debre, P., Delfraissy, J. F., and Theodorou, I. (1997). Early protective effect of CCR-5 delta 32 heterozygosity on HIV-1 disease progression: Relationship with viral load: The SEROCO Study Group. *AIDS* 11, F73–78.

Michaels, J., Price, R. W., and Rosenblum, M. K. (1988). Microglia in the giant cell encephalitis of acquired immune deficiency syndrome: Proliferation, infection and fusion. *Acta Neuropathol.* 76, 373–379.

Miller, B., Sarantis, M., Traynelis, S. F., and Attwell, D. (1992). Potentiation of NMDA receptor currents by arachidonic acid. *Nature* 355, 722–725.

Millhouse, S., Krebs, F. C., Yao, J., McAllister, J. J., Conner, J., Ross, H., and Wigdahl, B. (1998). Sp1 and related factors fail to interact with the NF-kappaB-proximal G/C box in the LTR of a replication competent, brain-derived strain of HIV-1 (YU-2). *J. Neurovirol.* 4, 312–323.

Mink, S., Haenig, B., and Klempnauer, K. H. (1997). Interaction and functional collaboration of p300 and C/EBPbeta. *Mol. Cell. Biol.* 17, 6609–6617.

Mondal, D., Alam, J., and Prakash, O. (1994). NF-kappa B site-mediated negative regulation of the HIV-1 promoter by CCAAT/enhancer binding proteins in brain-derived cells. *J. Mol. Neurosci.* 5, 241–258.

Morosetti, R., Park, D. J., Chumakov, A. M., Grillier, I., Shiohara, M., Gombart, A. F., Nakamaki, T., Weinberg, K., and Koeffler, H. P. (1997). A novel, myeloid transcription factor, C/EBP epsilon, is upregulated during granulocytic, but not monocytic, differentiation. *Blood* 90, 2591–2600.

Moses, A. V., Bloom, F. E., Pauza, C. D., and Nelson, J. A. (1993). Human immunodeficiency virus infection of human brain capillary endothelial cells occurs via a CD4/galactosylceramide-independent mechanism. *Proc. Natl. Acad. Sci. USA* 90, 10474–10478.

Nath, A., Psooy, K., Martin, C., Knudsen, B., Magnuson, D. S., Haughey, N., and Geiger, J. D. (1996). Identification of a human immunodeficiency virus type 1 Tat epitope that is neuroexcitatory and neurotoxic. *J. Virol.* 70, 1475–1480.

Natsuka, S., Akira, S., Nishio, Y., Hashimoto, S., Sugita, T., Isshiki, H., and Kishimoto, T. (1992). Macrophage differentiation-specific expression of NF-IL6, a transcription factor for interleukin-6. *Blood* 79, 460–466.

Navia, B. A., Cho, E. S., Petito, C. K., and Price, R. W.(1986a). The AIDS dementia complex. II. Neuropathology. *Ann. Neurol.* 19, 525–535.

Navia, B. A., Jordan, B. D., and Price, R. W. (1986b). The AIDS dementia complex. I. Clinical features. *Ann. Neurol.* 19, 517–524.

Navia, B. A., Petito, C. K., Gold, J. W., Cho, E. S., Jordan, B. D., and Price, R. W. (1986c). Cerebral toxoplasmosis complicating the acquired immune deficiency syndrome: Clinical and neuropathological findings in 27 patients. *Ann. Neurol.* 19, 224–238.

Navia, B. A., and Price, R. W. (1987). The acquired immunodeficiency syndrome dementia complex as the presenting or sole manifestation of human immunodeficiency virus infection. *Arch. Neurol.* 44, 65–69.

Neumann, M., Felber, B. K., Kleinschmidt, A., Froese, B., Erfle, V., Pavlakis, G. N., and Brack-Werner, R. (1995). Restriction of human immunodeficiency virus type 1 production in a human astrocytoma cell line is associated with a cellular block in Rev function. *J. Virol.* 69, 2159–2167.

New, D. R., Maggirwar, S. B., Epstein, L. G., Dewhurst, S., and Gelbard, H. A. (1998). HIV-1 Tat induces neuronal death via tumor necrosis factor-alpha and activation of non-N-methyl-D-aspartate receptors by a NF-kappa B-independent mechanism. *J. Biol. Chem.* 273, 17852–17858.

Nicholson, W. J., Shepherd, A. J., and Aw, D. W. (1996). Detection of unintegrated HIV type 1 DNA in cell culture and clinical peripheral blood mononuclear cell samples: Correlation to disease stage. *AIDS Res. Hum. Retroviruses* 12, 315–323.

Niehof, M., Manns, M. P., and Trautwein, C. (1997). CREB controls LAP/C/EBP beta transcription. *Mol. Cell. Biol.* **17**, 3600–3613.

Niikura, M., Dornadula, G., Zhang, H., Mukhtar, M., Lingxun, D., Khalili, K., Bagasra, O., and Pomerantz, R. J. (1996). Mechanisms of transcriptional transactivation and restriction of human immunodeficiency virus type 1 replication in an astrocytic glial cell. *Oncogene* **13**, 313–322.

Nottet, H. S., and Gendelman, H. E. (1995). Unraveling the neuroimmune mechanisms for the HIV-1-associated cognitive/motor complex. *Immunol. Today* **16**, 441–448.

Nottet, H. S., Jett, M., Flanagan, C. R., Zhai, Q. H., Persidsky, Y., Rizzino, A., Bernton, E. W., Genis, P., Baldwin, T., Schwartz, J., LaBenz, C. J., and Gendelman, H. E. (1995). A regulatory role for astrocytes in HIV-1 encephalitis: An overexpression of eicosanoids, platelet-activating factor, and tumor necrosis factor-alpha by activated HIV-1-infected monocytes is attenuated by primary human astrocytes. *J. Immunol.* **154**, 3567–3581.

Nottet, H. S., Persidsky, Y., Sasseville, V. G., Nukuna, A. N., Bock, P., Zhai, Q. H., Sharer, L. R., McComb, R. D., Swindells, S., Soderland, C., and Gendelman, H. E. (1996). Mechanisms for the transendothelial migration of HIV-1-infected monocytes into brain. *J. Immunol.* **156**, 1284–1295.

Nuovo, G. J., Gallery, F., MacConnell, P., and Braun, A. (1994). In situ detection of polymerase chain reaction-amplified HIV-1 nucleic acids and tumor necrosis factor-alpha RNA in the central nervous system. *Am. J. Pathol.* **144**, 659–666.

O'Shea, E. K., Klemm, J. D., Kim, P. S., and Alber, T. (1991). X-ray structure of the GCN4 leucine zipper, a two-stranded, parallel coiled coil. *Science* **254**, 539–544.

Ohagen, A., Ghosh, S., He, J., Huang, K., Chen, Y., Yuan, M., Osathanondh, R., Gartner, S., Shi, B., Shaw, G., and Gabuzda, D. (1999). Apoptosis induced by infection of primary brain cultures with diverse human immunodeficiency virus type 1 isolates: Evidence for a role of the envelope. *J. Virol.* **73**, 897–906.

Oleske, J., Minnefor, A., Cooper, R., Jr., Thomas, K., dela Cruz, A., Ahdieh, H., Guerrero, I., Joshi, V. V., and Desposito, F. (1983). Immune deficiency syndrome in children. *J. Am. Med. Assoc.* **249**, 2345–2349.

Pang, S., Koyanagi, Y., Miles, S., Wiley, C., Vinters, H. V., and Chen, I. S. (1990). High levels of unintegrated HIV-1 DNA in brain tissue of AIDS dementia patients. *Nature* **343**, 85–89.

Perelson, A. S., Essunger, P., Cao, Y., Vesanen, M., Hurley, A., Saksela, K., Markowitz, M., and Ho, D. D. (1997). Decay characteristics of HIV-1-infected compartments during combination therapy [see comments]. *Nature* **387**, 188–191.

Perkins, N. D., Edwards, N. L., Duckett, C. S., Agranoff, A. B., Schmid, R. M., and Nabel, G. J. (1993). A cooperative interaction between NF-kappa B and Sp1 is required for HIV-1 enhancer activation. *EMBO J.* **12**, 3551–3558.

Perrella, O., Carrieri, P. B., Guarnaccia, D., and Soscia, M. (1992). Cerebrospinal fluid cytokines in AIDS dementia complex. *J. Neurol.* **239**, 387–388.

Perry, S. W., Hamilton, J. A., Tjoelker, L. W., Dbaibo, G., Dzenko, K. A., Epstein, L. G., Hannun, Y., Whittaker, J. S., Dewhurst, S., and Gelbard, H. A. (1998). Platelet-activating factor receptor activation: An initiator step in HIV-1 neuropathogenesis. *J. Biol. Chem.* **273**, 17660–17664.

Perry, V. H., and Gordon, S. (1988). Macrophages and microglia in the nervous system. *Trends Neurosci.* **11**, 273–277.

Persidsky, Y., Stins, M., Way, D., Witte, M. H., Weinand, M., Kim, K. S., Bock, P., Gendelman, H. E., and Fiala, M. (1997). A model for monocyte migration through the blood–brain barrier during HIV-1 encephalitis. *J. Immunol.* **158**, 3499–3510.

Petito, C. K., Navia, B. A., Cho, E. S., Jordan, B. D., George, D. C., and Price, R. W. (1985). Vacuolar myelopathy pathologically resembling subacute combined degeneration in patients with the acquired immunodeficiency syndrome. *N. Engl. J. Med.* **312**, 874–879.

Peudenier, S., Hery, C., Montagnier, L., and Tardieu, M. (1991). Human microglial cells: Characterization in cerebral tissue and in primary culture, and study of their susceptibility to HIV-1 infection. *Ann. Neurol.* **29**, 152–161.

Pizzo, P. A., Eddy, J., Falloon, J., Balis, F. M., Murphy, R. F., Moss, H., Wolters, P., Brouwers, P., Jarosinski, P., Rubin, M. *et al.* (1988). Effect of continuous intravenous infusion of zidovudine (AZT) in children with symptomatic HIV infection [see comments]. *N. Engl. J. Med.* 319, 889–896.

Popik, W., Hesselgesser, J. E., and Pitha, P. M. (1998). Binding of human immunodeficiency virus type 1 to CD4 and CXCR4 receptors differentially regulates expression of inflammatory genes and activates the MEK/ERK signaling pathway. *J. Virol.* 72, 6406–6413.

Portegies, P. (1994). AIDS dementia complex: A review. *J. Acq. Immune Defic. Syndr.* 7, S38–S48; discussion, S48–S39.

Power, C., McArthur, J. C., Nath, A., Wehrly, K., Mayne, M., Nishio, J., Langelier, T., Johnson, R. T., and Chesebro, B. (1998). Neuronal death induced by brain-derived human immunodeficiency virus type 1 envelope genes differs between demented and nondemented AIDS patients. *J. Virol.* 72, 9045–9053.

Price, R. W. (1994). Understanding the AIDS Dementia Complex (ADC): the challenge of HIV and its effects on the central nervous system. *In* "HIV, AIDS, and the Brain" (R. W. Price and S. W. Perry III, Eds.), pp. 1–45. Raven Press, New York.

Price, R. W., Brew, B., Sidtis, J., Rosenblum, M., Scheck, A. C., and Cleary, P. (1988). The brain in AIDS: Central nervous system HIV-1 infection and AIDS dementia complex. *Science* 239, 586–592.

Price, R. W., and Brew, B. J. (1988). The AIDS dementia complex. *J. Infect. Dis.* 158, 1079–1083.

Price, R. W., and Sidtis, J. J. (1992). The AIDS dementia complex. *In* "AIDS and Other Manifestations of HIV Infection" (G. P. Wormser, Ed.), pp. 373–382. Raven Press, New York.

Pulliam, L., Herndier, B. G., Tang, N. M., and McGrath, M. S. (1991). Human immunodeficiency virus-infected macrophages produce soluble factors that cause histological and neurochemical alterations in cultured human brains. *J. Clin. Invest.* 87, 503–512.

Pulliam, L., West, D., Haigwood, N., and Swanson, R. A. (1993). HIV-1 envelope gp120 alters astrocytes in human brain cultures. *AIDS Res. Hum. Retroviruses* 9, 439–444.

Pulliam, L., Zhou, M., Stubblebine, M., and Bitler, C. M. (1998). Differential modulation of cell death proteins in human brain cells by tumor necrosis factor alpha and platelet activating factor. *J. Neurosci. Res.* 54, 530–538.

Pumarola-Sune, T., Navia, B. A., Cordon-Cardo, C., Cho, E. S., and Price, R. W. (1987). HIV antigen in the brains of patients with the AIDS dementia complex. *Ann. Neurol.* 21, 490–496.

Ranki, A., Nyberg, M., Ovod, V., Haltia, M., Elovaara, I., Raininko, R., Haapasalo, H., and Krohn, K. (1995). Abundant expression of HIV Nef and Rev proteins in brain astrocytes in vivo is associated with dementia. *AIDS* 9, 1001–1008.

Rappaport, J., Joseph, J., Croul, S., Alexander, G., Del Valle, L., Amini, S., and Khalili, K. (1999). Molecular pathway involved in HIV-1-induced CNS pathology: Role of viral regulatory protein, Tat. *J. Leukoc. Biol.* 65, 458–465.

Raskino, C., Pearson, D. A., Baker, C. J., Lifschitz, M. H., O'Donnell, K., Mintz, M., Nozyce, M., Brouwers, P., McKinney, R. E., Jimenez, E., and Englund, J. A. (1999). Neurologic, neurocognitive, and brain growth outcomes in human immunodeficiency virus-infected children receiving different nucleoside antiretroviral regimens. *Pediatrics* 104, e32.

Ray, A., Hannink, M., and Ray, B. K. (1995). Concerted participation of NF-kappa B and C/EBP heteromer in lipopolysaccharide induction of serum amyloid A gene expression in liver. *J. Biol. Chem.* 270, 7365–7374.

Ries, S., Buchler, C., Langmann, T., Fehringer, P., Aslanidis, C., and Schmitz, G. (1998). Transcriptional regulation of lysosomal acid lipase in differentiating monocytes is mediated by transcription factors Sp1 and AP-2. *J. Lipid Res.* 39, 2125–2134.

Roesler, W. J., Crosson, S. M., Vinson, C., and McFie, P. J. (1996). The alpha-isoform of the CCAAT/enhancer-binding protein is required for mediating cAMP responsiveness of the phosphoenolpyruvate carboxykinase promoter in hepatoma cells. *J. Biol. Chem.* **271**, 8068–8074.

Roman, C., Platero, J. S., Shuman, J., and Calame, K. (1990). Ig/EBP-1: A ubiquitously expressed immunoglobulin enhancer binding protein that is similar to C/EBP and heterodimerizes with C/EBP. *Genes Dev.* **4**, 1404–1415.

Ron, D., and Habener, J. F. (1992). CHOP, a novel developmentally regulated nuclear protein that dimerizes with transcription factors C/EBP and LAP and functions as a dominant-negative inhibitor of gene transcription. *Genes Dev.* **6**, 439–453.

Rosenblum, M. K. (1990). Infection of the central nervous system by the human immunodeficiency virus type 1: Morphology and relation to syndromes of progressive encephalopathy and myelopathy in patients with AIDS. *Pathol. Annu.* **25**, 117–169.

Ross, E. K., Buckler-White, A. J., Rabson, A. B., Englund, G., and Martin, M. A. (1991). Contribution of NF-kappa B and Sp1 binding motifs to the replicative capacity of human immunodeficiency virus type 1: Distinct patterns of viral growth are determined by T-cell types. *J. Virol.* **65**, 4350–4358.

Rostasy, K., Monti, L., Yiannoutsos, C., Kneissl, M., Bell, J., Kemper, T. L., Hedreen, J. C., and Navia, B. A. (1999). Human immunodeficiency virus infection, inducible nitric oxide synthase expression, and microglial activation: Pathogenetic relationship to the acquired immunodeficiency syndrome dementia complex. *Ann. Neurol.* **46**, 207–216.

Rothenberg, R., Woelfel, M., Stoneburner, R., Milberg, J., Parker, R., and Truman, B. (1987). Survival with the acquired immunodeficiency syndrome: Experience with 5833 cases in New York City. *N. Engl. J. Med.* **317**, 1297–1302.

Rottenberg, D. A., Moeller, J. R., Strother, S. C., Sidtis, J. J., Navia, B. A., Dhawan, V., Ginos, J. Z., and Price, R. W. (1987). The metabolic pathology of the AIDS dementia complex. *Ann. Neurol.* **22**, 700–706.

Rottman, J. B., Ganley, K. P., Williams, K., Wu, L., Mackay, C. R., and Ringler, D. J. (1997). Cellular localization of the chemokine receptor CCR5. Correlation to cellular targets of HIV-1 infection. *Am. J. Pathol.* **151**, 1341–1351.

Roy, S., Geoffroy, G., Lapointe, N., and Michaud, J. (1992). Neurological findings in HIV-infected children: A review of 49 cases. *Can. J. Neurol. Sci.* **19**, 453–457.

Royal, W., III, Selnes, O. A., Concha, M., Nance-Sproson, T. E., and McArthur, J. C. (1994). Cerebrospinal fluid human immunodeficiency virus type 1 (HIV-1) p24 antigen levels in HIV-1-related dementia. *Ann. Neurol.* **36**, 32–39.

Ruocco, M. R., Chen, X., Ambrosino, C., Dragonetti, E., Liu, W., Mallardo, M., De Falco, G., Palmieri, C., Franzoso, G., Quinto, I., Venuta, S., and Scala, G. (1996). Regulation of HIV-1 long terminal repeats by interaction of C/EBP (NF- IL6) and NF-kappa B/Rel transcription factors. *J. Biol. Chem.* **271**, 22479–22486.

Sabatier, J. M., Vives, E., Mabrouk, K., Benjouad, A., Rochat, H., Duval, A., Hue, B., and Bahraoui, E. (1991). Evidence for neurotoxic activity of tat from human immunodeficiency virus type 1. *J. Virol.* **65**, 961–967.

Sacktor, N. C., Lyles, R. H., Skolasky, R. L., Anderson, D. E., McArthur, J. C., McFarlane, G., Selnes, O. A., Becker, J. T., Cohen, B., Wesch, J., and Miller, E. N. (1999). Combination antiretroviral therapy improves psychomotor speed performance in HIV-seropositive homosexual men: Multicenter AIDS Cohort Study (MACS). *Neurology* **52**, 1640–1647.

Sacktor, N. C., Skolasky, R. L., Esposito, D., et al., (1998). Combination therapy including protease inhibitors improves psychomotor speed performance in HIV infection. *Neurology* **50**, A248–A249.

Saito, Y., Sharer, L. R., Epstein, L. G., Michaels, J., Mintz, M., Louder, M., Golding, K., Cvetkovich, T. A., and Blumberg, B. M. (1994). Overexpression of nef as a marker for restricted HIV-1 infection of astrocytes in post-mortem pediatric central nervous tissues. *Neurology* **44**, 474–481.

Sasseville, V. G., Smith, M. M., Mackay, C. R., Pauley, D. R., Mansfield, K. G., Ringler, D. J., and Lackner, A. A. (1996). Chemokine expression in simian immunodeficiency virus-induced AIDS encephalitis. *Am. J. Pathol.* **149**, 1459–1467.

Scarmato, V., Frank, Y., Rozenstein, A., Lu, D., Hyman, R., Bakshi, S., Pahwa, S., and Pavlakis, S. (1996). Central brain atrophy in childhood AIDS encephalopathy. *AIDS* **10**, 1227–1231.

Schmidtmayerova, H., Nottet, H. S., Nuovo, G., Raabe, T., Flanagan, C. R., Dubrovsky, L., Gendelman, H. E., Cerami, A., Bukrinsky, M., and Sherry, B. (1996). Human immunodeficiency virus type 1 infection alters chemokine beta peptide expression in human monocytes: implications for recruitment of leukocytes into brain and lymph nodes. *Proc. Natl. Acad. Sci. USA* **93**, 700–704.

Schmitt, F. A., Bigley, J. W., McKinnis, R., Logue, P. E., Evans, R. W., and Drucker, J. L. (1988). Neuropsychological outcome of zidovudine (AZT) treatment of patients with AIDS and AIDS-related complex. *N. Engl. J. Med.* **319**, 1573–1578.

Schrager, L. K., and D'Souza, M. P. (1998). Cellular and anatomical reservoirs of HIV-1 in patients receiving potent antiretroviral combination therapy. *J. Am. Med. Assoc.* **280**, 67–71.

Sears, R. C., and Sealy, L. (1994). Multiple forms of C/EBP beta bind the EFII enhancer sequence in the Rous sarcoma virus long terminal repeat. *Mol. Cell. Biol.* **14**, 4855–4871.

Sei, S., Stewart, S. K., Farley, M., Mueller, B. U., Lane, J. R., Robb, M. L., Brouwers, P., and Pizzo, P. A. (1996). Evaluation of human immunodeficiency virus (HIV) type 1 RNA levels in cerebrospinal fluid and viral resistance to zidovudine in children with HIV encephalopathy. *J. Infect. Dis.* **174**, 1200–1206.

Selmaj, K. W., and Raine, C. S. (1988). Tumor necrosis factor mediates myelin and oligodendrocyte damage *in vitro*. *Ann. Neurol.* **23**, 339–346.

Selnes, O. A., Miller, E., McArthur, J., Gordon, B., Munoz, A., Sheridan, K., Fox, R., and Saah, A. J. (1990). HIV-1 infection: No evidence of cognitive decline during the asymptomatic stages: The Multicenter AIDS Cohort Study. *Neurology* **40**, 204–208.

Sharer, L. R. (1992). Pathology of HIV-1 infection of the central nervous system: A review. *J. Neuropathol. Exp. Neurol.* **51**, 3–11.

Sharer, L. R., Michaels, J., Murphey-Corb, M., Hu, F. S., Kuebler, D. J., Martin, L. N., and Baskin, G. B. (1991). Serial pathogenesis study of SIV brain infection. *J. Med. Primatol.* **20**, 211–217.

Shaw, G. M., Harper, M. E., Hahn, B. H., Epstein, L. G., Gajdusek, D. C., Price, R. W., Navia, B. A., Petito, C. K., O'Hara, C. J., Groopman, J. E., Cho, E.-S., Oleske, J. M., Wong-Staal, F., and Gallo, R. C. (1985). HTLV-III infection in brains of children and adults with AIDS encephalopathy. *Science* **227**, 177–182.

Shi, B., De Girolami, U., He, J., Wang, S., Lorenzo, A., Busciglio, J., and Gabuzda, D. (1996). Apoptosis induced by HIV-1 infection of the central nervous system. *J. Clin. Invest.* **98**, 1979–1990.

Shi, B., Raina, J., Lorenzo, A., Busciglio, J., and Gabuzda, D. (1998). Neuronal apoptosis induced by HIV-1 Tat protein and TNF-alpha: Potentiation of neurotoxicity mediated by oxidative stress and implications for HIV-1 dementia. *J. Neurovirol.* **4**, 281–290.

Shieh, J. T., Albright, A. V., Sharron, M., Gartner, S., Strizki, J., Doms, R. W., and Gonzalez-Scarano, F. (1998). Chemokine receptor utilization by human immunodeficiency virus type 1 isolates that replicate in microglia. *J. Virol.* **72**, 4243–4249.

Shuman, J. D., Cheong, J., and Coligan, J. E. (1997). ATF-2 and C/EBPalpha can form a heterodimeric DNA binding complex *in vitro*: Functional implications for transcriptional regulation. *J. Biol. Chem.* **272**, 12793–12800.

Sidtis, J. J., Dafni, U., Slasor, P., Hall, C., Price, R. W., Kieburtz, K., Tucker, T., and Clifford, D. B. (1997). Stable neurological function in subjects treated with 2'3'-dideoxyinosine. *J. Neurovirol.* **3**, 233–240.

Sidtis, J. J., Gatsonis, C., Price, R. W., Singer, E. J., Collier, A. C., Richman, D. D., Hirsch, M. S., Schaerf, F. W., Fischl, M. A., Kieburtz, K. *et al.* (1993). Zidovudine treatment of the AIDS dementia complex: results of a placebo-controlled trial: AIDS Clinical Trials Group. *Ann. Neurol.* **33**, 343–349.

Sloan, D. J., Wood, M. J., and Charlton, H. M. (1992). Leucocyte recruitment and inflammation in the CNS. *Trends Neurosci.* **15**, 276–278.

Snider, W. D., Simpson, D. M., Nielsen, S., Gold, J. W., Metroka, C. E., and Posner, J. B. (1983). Neurological complications of acquired immune deficiency syndrome: Analysis of 50 patients. *Ann. Neurol.* **14**, 403–418.

Sogawa, K., Imataka, H., Yamasaki, Y., Kusume, H., Abe, H., and Fujii-Kuriyama, Y. (1993). cDNA cloning and transcriptional properties of a novel GC box-binding protein, BTEB2. *Nucleic Acids Res.* **21**, 1527–1532.

Sonza, S., Kiernan, R. E., Maerz, A. L., Deacon, N. J., McPhee, D. A., and Crowe, S. M. (1994). Accumulation of unintegrated circular viral DNA in monocytes and growth-arrested T cells following infection with HIV-1. *J. Leukoc. Biol.* **56**, 289–293.

Spencer, D. C., and Price, R. W. (1992). Human immunodeficiency virus and the central nervous system. *Annu. Rev. Microbiol.* **46**, 655–693.

Stein, B., and Baldwin, A. S., Jr. (1993). Distinct mechanisms for regulation of the interleukin-8 gene involve synergism and cooperativity between C/EBP and NF-kappa B. *Mol. Cell. Biol.* **13**, 7191–7198.

Sterneck, E., and Johnson, P. F. (1998). CCAAT/enhancer binding protein beta is a neuronal transcriptional regulator activated by nerve growth factor receptor signaling. *J. Neurochem.* **70**, 2424–2433.

Stevenson, M., Haggerty, S., Lamonica, C. A., Meier, C. M., Welch, S. K., and Wasaik, A. J. (1990). Integration is not necessary for expression of human immunodeficiency virus type 1 protein products. *J. Virol.* **64**, 2421–2425.

Stoler, M. H., Eskin, T. A., Benn, S., Angerer, R. C., and Angerer, L. M. (1986). Human T-cell lymphotropic virus type III infection of the central nervous system: A preliminary *in situ* analysis. *J. Am. Med. Assoc.* **256**, 2360–2364.

Strelow, L. I., Watry, D. D., Fox, H. S., and Nelson, J. A. (1998). Efficient infection of brain microvascular endothelial cells by an *in vivo*-selected neuroinvasive SIVmac variant. *J. Neurovirol.* **4**, 269–280.

Strizki, J. M., Albright, A. V., Sheng, H., O'Connor, M., Perrin, L., and Gonzalez-Scarano, F. (1996). Infection of primary human microglia and monocyte-derived macrophages with human immunodeficiency virus type 1 isolates: Evidence of differential tropism. *J. Virol.* **70**, 7654–7662.

Subramanyam, M., Gutheil, W. G., Bachovchin, W. W., and Huber, B. T. (1993). Mechanism of HIV-1 Tat induced inhibition of antigen-specific T cell responsiveness. *J. Immunol.* **150**, 2544–2553.

Tanabe, S., Heesen, M., Berman, M. A., Fischer, M. B., Yoshizawa, I., Luo, Y., and Dorf, M. E. (1997a). Murine astrocytes express a functional chemokine receptor. *J. Neurosci.* **17**, 6522–6528.

Tanabe, S., Heesen, M., Yoshizawa, I., Berman, M. A., Luo, Y., Bleul, C. C., Springer, T. A., Okuda, K., Gerard, N., and Dorf, M. E. (1997b). Functional expression of the CXC-chemokine receptor-4/fusin on mouse microglial cells and astrocytes. *J. Immunol.* **159**, 905–911.

Tanaka, T., Akira, S., Yoshida, K., Umemoto, M., Yoneda, Y., Shirafuji, N., Fujiwara, H., Suematsu, S., Yoshida, N., and Kishimoto, T. (1995). Targeted disruption of the NF-IL6 gene discloses its essential role in bacteria killing and tumor cytotoxicity by macrophages. *Cell* **80**, 353–361.

Tardieu, M. (1998). HIV-1 and the developing central nervous system. *Dev. Med. Child Neurol.* **40**, 843–846.

Taylor, J. P., Pomerantz, R., Bagasra, O., Chowdhury, M., Rappaport, J., Khalili, K., and Amini, S. (1992). TAR-independent transactivation by Tat in cells derived from the CNS: A novel mechanism of HIV-1 gene regulation. *EMBO J.* **11**, 3395–3403.

Taylor, J. P., Pomerantz, R. J., Oakes, J. W., Khalili, K., and Amini, S. (1995). A CNS-enriched factor that binds to NF-kappa B and is required for interaction with HIV-1 tat. *Oncogene* **10**, 395–400.

Taylor, J. P., Pomerantz, R. J., Raj, G. V., Kashanchi, F., Brady, J. N., Amini, S., and Khalili, K. (1994). Central nervous system-derived cells express a kappa B-binding activity that enhances human immunodeficiency virus type 1 transcription *in vitro* and facilitates TAR-independent transactivation by Tat. *J. Virol.* **68**, 3971–3981.

Teo, I., Veryard, C., Barnes, H., An, S. F., Jones, M., Lantos, P. L., Luthert, P., and Shaunak, S. (1997). Circular forms of unintegrated human immunodeficiency virus type 1 DNA and high levels of viral protein expression: Association with dementia and multinucleated giant cells in the brains of patients with AIDS. *J. Virol.* **71**, 2928–2933.

Tepper, V. J., Farley, J. J., Rothman, M. I., Houck, D. L., Davis, K. F., Collins-Jones, T. L., and Wachtel, R. C. (1998). Neurodevelopmental/neuroradiologic recovery of a child infected with HIV after treatment with combination antiretroviral therapy using the HIV-specific protease inhibitor ritonavir. *Pediatrics* **101**, E7.

Tesmer, V. M., Rajadhyaksha, A., Babin, J., and Bina, M. (1993). NF-IL6-mediated transcriptional activation of the long terminal repeat of the human immunodeficiency virus type 1. *Proc. Natl. Acad. Sci. USA* **90**, 7298–7302.

Thomas, F. P., Chalk, C., Lalonde, R., Robitaille, Y., and Jolicoeur, P. (1994). Expression of human immunodeficiency virus type 1 in the nervous system of transgenic mice leads to neurological disease. *J. Virol.* **68**, 7099–7107.

Toggas, S. M., Masliah, E., Rockenstein, E. M., Rall, G. F., Abraham, C. R., and Mucke, L. (1994). Central nervous system damage produced by expression of the HIV-1 coat protein gp120 in transgenic mice. *Nature* **367**, 188–193.

Tornatore, C., Chandra, R., Berger, J. R., and Major, E. O. (1994). HIV-1 infection of subcortical astrocytes in the pediatric central nervous system. *Neurology* **44**, 481–487.

Tornatore, C., Nath, A., Amemiya, K., and Major, E. O. (1991). Persistent human immunodeficiency virus type 1 infection in human fetal glial cells reactivated by T-cell factor(s) or by the cytokines tumor necrosis factor alpha and interleukin-1 beta. *J. Virol.* **65**, 6094–6100.

Tozzi, V., Narciso, P., Galgani, S., Sette, P., Balestra, P., Gerace, C., Pau, F. M., Pigorini, F., Volpini, V., Camporiondo, M. P. *et al.* (1993). Effects of zidovudine in 30 patients with mild to end-stage AIDS dementia complex. *AIDS* **7**, 683–692.

Tracey, I., Hamberg, L. M., Guimaraes, A. R., Hunter, G., Chang, I., Navia, B. A., and Gonzalez, R. G. (1998). Increased cerebral blood volume in HIV-positive patients detected by functional MRI. *Neurology* **50**, 1821–1826.

Truckenmiller, M. E., Kulaga, H., Coggiano, M., Wyatt, R., Snyder, S. H., and Sweetnam, P. M. (1993). Human cortical neuronal cell line: A model for HIV-1 infection in an immature neuronal system. *AIDS Res. Hum. Retroviruses* **9**, 445–453.

Tyor, W. R., Glass, J. D., Griffin, J. W., Becker, P. S., McArthur, J. C., Bezman, L., and Griffin, D. E. (1992). Cytokine expression in the brain during the acquired immunodeficiency syndrome. *Ann. Neurol.* **31**, 349–360.

Ubeda, M., Wang, X. Z., Zinszner, H., Wu, I., Habener, J. F., and Ron, D. (1996). Stress-induced binding of the transcriptional factor CHOP to a novel DNA control element. *Mol. Cell. Biol.* **16**, 1479–1489.

Vago, L., Castagna, A., Lazzarin, A., Trabattoni, G., Cinque, P., and Costanzi, G. (1993). Reduced frequency of HIV-induced brain lesions in AIDS patients treated with zidovudine. *J. Acq. Immune Defic. Syndr.* **6**, 42–45.

Vallat, A. V., De Girolami, U., He, J., Mhashilkar, A., Marasco, W., Shi, B., Gray, F., Bell, J., Keohane, C., Smith, T. W., and Gabuzda, D. (1998). Localization of HIV-1 co-receptors CCR5 and CXCR4 in the brain of children with AIDS. *Am. J. Pathol.* **152**, 167–178.

Vallejo, M., Ron, D., Miller, C. P., and Habener, J. F. (1993). C/ATF, a member of the activating transcription factor family of DNA-binding proteins, dimerizes with CAAT/enhancer-binding proteins and directs their binding to cAMP response elements. *Proc. Natl. Acad. Sci. USA* **90**, 4679–4683.

van Rij, R. P., Portegies, P., Hallaby, T., Lange, J. M., Visser, J., Husman, A. M., van't Wout, A. B., and Schuitemaker, H. (1999). Reduced prevalence of the CCR5 delta32 heterozygous genotype in human immunodeficiency virus-infected individuals with AIDS dementia complex. *J. Infect. Dis.* **180**, 854–857.

Vazeux, R., Brousse, N., Jarry, A., Henin, D., Marche, C., Vedrenne, C., Mikol, J., Wolff, M., Michon, C., Rozenbaum, W., Bureau, J.-F., Montagnier, L., and Brahic, M. (1987). AIDS subacute encephalitis: Identification of HIV-infected cells. *Am. J. Pathol.* **126**, 403–410.

Vazeux, R., Lacroix-Ciaudo, C., Blanche, S., Cumont, M. C., Henin, D., Gray, F., Boccon-Gibod, L., and Tardieu, M. (1992). Low levels of human immunodeficiency virus replication in the brain tissue of children with severe acquired immunodeficiency syndrome encephalopathy. *Am. J. Pathol.* **140**, 137–144.

Vesanen, M., Salminen, M., Wessman, M., Lankinen, H., Sistonen, P., and Vaheri, A. (1994). Morphological differentiation of human SH-SY5Y neuroblastoma cells inhibits human immunodeficiency virus type 1 infection. *J. Gen. Virol.* **75**, 201–206.

Vinson, C. R., Hai, T., and Boyd, S. M. (1993). Dimerization specificity of the leucine zipper-containing bZIP motif on DNA binding: Prediction and rational design. *Genes Dev.* **7**, 1047–1058.

Vitkovic, L., Kalebic, T., de Cunha, A., and Fauci, A. S. (1990). Astrocyte-conditioned medium stimulates HIV-1 expression in a chronically infected promonocyte clone. *J. Neuroimmunol.* **30**, 153–160.

Wahl, S. M., Allen, J. B., McCartney-Francis, N., Morganti-Kossmann, M. C., Kossmann, T., Ellingsworth, L., Mai, U. E., Mergenhagen, S. E., and Orenstein, J. M. (1991). Macrophage- and astrocyte-derived transforming growth factor beta as a mediator of central nervous system dysfunction in acquired immune deficiency syndrome. *J. Exp. Med.* **173**, 981–991.

Wain-Hobson, S. (1989). HIV genome variability *in vivo*. *AIDS* **3**, S13–S18.

Watkins, B. A., Dorn, H. H., Kelly, W. B., Armstrong, R. C., Potts, B. J., Michaels, F., Kufta, C. V., and Dubois-Dalcq, M. (1990). Specific tropism of HIV-1 for microglial cells in primary human brain cultures. *Science* **249**, 549–553.

Wedel, A., and Ziegler-Heitbrock, H. W. (1995). The C/EBP family of transcription factors. *Immunobiology* **193**, 171–185.

Weidenheim, K. M., Epshteyn, I., and Lyman, W. D. (1993). Immunocytochemical identification of T-cells in HIV-1 encephalitis: Implications for pathogenesis of CNS disease. *Mod. Pathol.* **6**, 167–174.

Wekerle, H., Unington, C., Lassmann, H., and Meyermann, R. (1986). Cellular immune reactivity within the CNS. *Trends Neurosci.* **9**, 271–277.

Werner, T., Ferroni, S., Saermark, T., Brack-Werner, R., Banati, R. B., Mager, R., Steinaa, L., Kreutzberg, G. W., and Erfle, V. (1991). HIV-1 Nef protein exhibits structural and functional similarity to scorpion peptides interacting with K+ channels. *AIDS* **5**, 1301–1308.

Wesselingh, S. L., Power, C., Glass, J. D., Tyor, W. R., McArthur, J. C., Farber, J. M., Griffin, J. W., and Griffin, D. E. (1993). Intracerebral cytokine messenger RNA expression in acquired immunodeficiency syndrome dementia. *Ann. Neurol.* **33**, 576–582.

Westendorp, M. O., Li-Weber, M., Frank, R. W., and Krammer, P. H. (1994). Human immunodeficiency virus type 1 Tat upregulates interleukin-2 secretion in activated T cells. *J. Virol.* **68**, 4177–4185.

Westmoreland, S. V., Kolson, D., and Gonzalez-Scarano, F. (1996). Toxicity of TNF alpha and platelet activating factor for human NT2N neurons: A tissue culture model for human immunodeficiency virus dementia. *J. Neurovirol.* **2**, 118–126.

Whetsell, W. O., Jr., and Schwarcz, R. (1989). Prolonged exposure to submicromolar concentrations of quinolinic acid causes excitotoxic damage in organotypic cultures of rat corticostriatal system. *Neurosci. Lett.* **97**, 271–275.

Wiley, C. A., and Achim, C. (1994). Human immunodeficiency virus encephalitis is the pathological correlate of dementia in acquired immunodeficiency syndrome. *Ann. Neurol.* **36**, 673–676.

Wiley, C. A., Schrier, R. D., Nelson, J. A., Lampert, P. W., and Oldstone, M. B. (1986). Cellular localization of human immunodeficiency virus infection within the brains of acquired immune deficiency syndrome patients. *Proc. Natl. Acad. Sci. USA* **83**, 7089–7093.

Wiley, C. A., Soontornniyomkij, V., Radhakrishnan, L., Masliah, E., Mellors, J., Hermann, S. A., Dailey, P., and Achim, C. L. (1998). Distribution of brain HIV load in AIDS. *Brain Pathol.* **8**, 277–284.

Williams, S. C., Cantwell, C. A., and Johnson, P. F. (1991). A family of C/EBP-related proteins capable of forming covalently linked leucine zipper dimers *in vitro*. *Genes Dev.* **5**, 1553–1567.

Williams, S. C., Du, Y., Schwartz, R. C., Weiler, S. R., Ortiz, M., Keller, J. R., and Johnson, P. F. (1998). C/EBP epsilon is a myeloid-specific activator of cytokine, chemokine, and macrophage-colony-stimulating factor receptor genes. *J. Biol. Chem.* **273**, 13493–13501.

Wilt, S. G., Milward, E., Zhou, J. M., Nagasato, K., Patton, H., Rusten, R., Griffin, D. E., O'Connor, M., and Dubois-Dalcq, M. (1995). *In vitro* evidence for a dual role of tumor necrosis factor-alpha in human immunodeficiency virus type 1 encephalopathy. *Ann. Neurol.* **37**, 381–394.

Worley, J., and Price, R. W. (1992). Management of neurologic complications of HIV-1 infection and AIDS. *In* "The Medical Management of AIDS" (M. A. Sande and P. A. Volberding, Eds.), pp. 193–217. W. B. Saunders, Philadelphia.

Yamanaka, R., Kim, G. D., Radomska, H. S., Lekstrom-Himes, J., Smith, L. T., Antonson, P., Tenen, D. G., and Xanthopoulos, K. G. (1997). CCAAT/enhancer binding protein epsilon is preferentially up-regulated during granulocytic differentiation and its functional versatility is determined by alternative use of promoters and differential splicing. *Proc. Natl. Acad. Sci. USA* **94**, 6462–6467.

Zauli, G., Davis, B. R., Re, M. C., Visani, G., Furlini, G., and La Placa, M. (1992). Tat protein stimulates production of transforming growth factor-beta 1 by marrow macrophages: A potential mechanism for human immunodeficiency virus type 1-induced hematopoietic suppression. *Blood* **80**, 3036–3043.

Zeichner, S. L., Kim, J. Y., and Alwine, J. C. (1991). Linker-scanning mutational analysis of the transcriptional activity of the human immunodeficiency virus type 1 long terminal repeat. *J. Virol.* **65**, 2436–2444.

Zhu, M., Duan, L., and Pomerantz, R. J. (1996). TAR- and Tat-independent replication of human immunodeficiency virus type 1 in human hepatoma cells. *AIDS Res. Hum. Retroviruses* **12**, 1093–1101.

Zidovetzki, R., Wang, J. L., Chen, P., Jeyaseelan, R., and Hofman, F. (1998). Human immunodeficiency virus Tat protein induces interleukin 6 mRNA expression in human brain endothelial cells via protein kinase C- and cAMP-dependent protein kinase pathways. *AIDS Res. Hum. Retroviruses* **14**, 825–833.

Zielasek, J., Tausch, M., Toyka, K. V., and Hartung, H. P. (1992). Production of nitrite by neonatal rat microglial cells/brain macrophages. *Cell. Immunol.* **141**, 111–120.

Nafees Ahmad

Department of Microbiology and Immunology
College of Medicine
The University of Arizona Health Sciences Center
Tucson, Arizona 85724

Molecular Mechanisms of Human Immunodeficiency Virus Type 1 Mother–Infant Transmission

I. Introduction

AIDS in children is one of the fastest growing aspects of the AIDS pandemic, as a greater number of women in the child-bearing age group are infected with human immunodeficiency virus type 1 (HIV-1). The World Health Organization (WHO) estimates that by the year 2000, 10 million children will have been born HIV infected (Chin, 1990; WHO, 1992). Each year, an estimated 300,000 infants worldwide are born with HIV-1 infection. The United Nations Program on AIDS (UNAIDS) and the United States Agency for International Development (USAID) project a higher number of children will be infected with HIV-1 by the year 2010, with 90% of them in the developing countries. In less developed countries, a significant proportion of all HIV-infected people are children (Chin, 1990), and it is estimated that one-third of all childhood deaths will be due to HIV-1 infection (Bennett and Rogers, 1991). Maternal–fetal transmission accounts for approximately

Advances in Pharmacology, Volume 49

387

80% of all HIV-1 infections in children (Ahmad, 1994, 1996; Blanche *et al.*, 1989; European Collaboration Study, 1998; Italian Multiculture Study, 1988; Mok *et al.*, 1987; Ryder *et al.*, 1988; Sprecher *et al.*, 1986). Infants born to mothers infected with HIV-1 are at risk of acquiring HIV-1 and of subsequently developing AIDS (Anderson and Medley, 1989; Scott *et al.*, 1985; Hoff *et al.*, 1988). Mother-to-infant transmission of HIV-1 predominately occurs perinatally at an estimated rate of more than 30%. However, the rate of transmission also depends on the symptoms of the disease and the frequency of delivery and at estimated rates of 24% in symptom-free mothers and 65% in mothers with the disease or who have had a previous child with AIDS (Scott *et al.*, 1985; European Collaborative Study, 1988; Italian Multiculture Study, 1988). In contrast, the rate of HIV-2 mother–infant transmission is much lower compared to HIV-1 (Adjoriolo-Johnson *et al.*, 1994; Morgan *et al.*, 1990) and dually infected mothers could transmit both viruses, but transmission of HIV-1 has generally been observed (Adjoriolo-Johnson *et al.*, 1994)

While the actual mechanisms of perinatal transmission are not known, the timing of HIV-1 transmission from mother to infant can occur mainly at three stages: prepartum (transplacental passage), intrapartum (exposure of infant's skin and mucus membrane to maternal blood and vaginal secretions), and postpartum (breast milk). To date, there are no clearly defined factors, viral or host, associated with maternal transmission of HIV-1. However, maternal parameters including advanced clinical stages of the mother, low CD4+ lymphocyte counts, maternal immune response to HIV-1, recent infection, high level of circulating HIV-1, and maternal disease progression have been implicated in an increased risk of mother-to-infant transmission of HIV-1 (Ahmad, 1996; Mok *et al.*, 1987; Scott *et al.*, 1985; Blanche *et al.*, 1989; Hira *et al.*, 1989; Ryder *et al.*, 1988). Several other factors such as acute infection during pregnancy, the presence of other sexually transmitted diseases or other chronic infections, disruption of placental integrity secondary to chorioamnionitis, and tobacco smoking have been shown to be associated with mother-to-infant transmission of HIV-1 (Report of a Consensus Workshop, 1992). Several studies have demonstrated a direct association between the presence of maternal antibody against the V3 domain of the envelope protein and a lower rate of transmission of HIV-1 (Rossi *et al.*, 1989; Devash *et al.*, 1990), whereas others have showed lack of correlation (Hasley *et al.*, 1992; Parekh *et al.*, 1991). The ability of maternal antibody to neutralize its own isolate (autologous neutralization) may be particularly important because it has been suggested that the virus mutants that are selected under immune pressure and that cannot be neutralized may play a role in transmission (Scarlatti *et al.*, 1991; Bryson *et al.*, 1993).

Thus, prevention of mother-to-infant transmission of HIV-1 must be an urgent global health priority. At present, the molecular mechanisms of perinatal transmissions are not known, which makes it very difficult to define strategies for effective treatment and prevention of HIV-1 infection in children. We (Ahmad *et al.*, 1995) and others (Mulder-Kapinga, 1993; Scarlatti *et al.*, 1993; Wolinsky *et al.*, 1992) have been involved in understanding the molecular mechanisms of HIV-1 maternal–fetal transmission, especially in the characterization of HIV-1 isolates that are involved in mother-to-infant transmission. Selective transmission of HIV-1 from mothers to infants has been proposed. Our hypothesis is that there are specific molecular and biological properties of HIV-1 that are critical determinants of perinatal transmission. We have shown transmission of a minor variant with a macrophage-tropic and non-syncytium-inducing phenotype from mothers to infants. We should target our preventive strategies on the properties of the transmitted viruses.

II. Timing of HIV-1 Mother–Infant Transmission _____

The actual mechanisms of perinatal transmission are not known; however, the timing of HIV-1 transmission from mother-to-infant can occur mainly at three stages: prepartum (transplacental passage), intrapartum (exposure of infant's skin and mucus membrane to maternal blood and vaginal secretions), and postpartum (breast milk). There is evidence that HIV-1 can infect the placenta at all stages of pregnancy. The intact amniotic sac can be infected through placental tears, with transfusion of infected blood into the fetal circulation. Based on examination of placental and fetal tissue following termination of pregnancies, there is substantial evidence that intrauterine infection of HIV-1 occurs (Douglas and King, 1992; Mulder-Kampinga *et al.*, 1993). Several studies have demonstrated the infection of placentas or fetuses by histologic methods, polymerase chain reaction (PCR), or *in situ* hybridization (Brossard *et al.*, 1993; Lyman *et al.*, 1988), including the ability of certain placenta-derived cells to support HIV-1 replication *in vitro* (David *et al.*, 1993; Siegal *et al.*, 1990). In addition, the ability of HIV-1 to pass through an intact placental barrier maintained *ex vivo* has been demonstrated (Bawdon *et al.*, 1993; Schwartz and Nahmias, 1991). Moreover, HIV-1 antigens have been found in the amniotic fluid (Clavelli *et al.*, 1991; Viscarello *et al.*, 1992) and may be related to the time of transmission.

In HIV-1-infected mothers, the zygote can be infected as early as the time of conception, probably by the virus present in vaginal secretions (Laimore *et al.*, 1993). The zygote then travels down the oviduct to the uterus. The oviduct, which is lined with macrophages, can also infect the dividing embryo on its passage to the uterus. The embryo may also be exposed to virus by uteral macrophages or by residual seminal fluid (Laimore *et al.*, 1993). These hypotheses are further supported by the presence of HIV-1 sequences and antigen in an 8-week fetus (Laimore *et al.*, 1993).

The formation of the placenta takes place shortly after the attachment of fetus to the uterine wall and most maternal–fetal exchanges are mediated by the placenta (Scott, 1994). In addition, HIV-1 infection has been demonstrated in placental tissues (Muary et al., 1989; Sprecher et al., 1986). Since placental tissue is rich in monocytes and macrophages, the macrophage-tropic viral isolate can infect placental tissue and be transmitted to the fetus. The exact route by which HIV-1 crosses the placenta is not known; however, the outermost layers of the placenta, the terminal villi, and the trophoblast do posses the CD4 receptor (Chandwani et al., 1991), which can be infected by HIV-1. The terminal villi and the trophoblasts are bathed in maternal blood and yet all fetuses are not infected by HIV-1. It is possible that HIV-1 replication is not supported in these cells (Chandwani et al., 1991) or a particular genotype or phenotype is needed to replicate in these cells. However, disruption of these cells could allow migration of the virus to the underlying cytotrophoblasts, which may support HIV-1 replication and spread the virus via CD4-positive Hofbauer or endothelial cells to the fetal cells. HIV-1 can also be transmitted from mother to infant through disruption of the placental membrane as a result of viral, bacterial, and/or fungal infections; sexually transmitted diseases; and smoking (Nair et al., 1993; Scott, 1994). In addition, disruption of the placental membrane can occur as a result of chorioamnionitis (Scott, 1994). Chorioamnionitis has been shown to occur in approximately 20% of normal pregnancies (Scott, 1994) and that percentage increases greatly in HIV-1-infected women (Nair et al., 1993). Moreover, it is likely that this would expose the underlying cells to maternal blood and increase the chance of transmission.

At least 50% of infection occurs during or shortly before birth (intrapartum) (Bryson et al., 1992). In addition to exposure to maternal blood at the time of labor and in the birth canal during delivery, HIV-1 has been found in cervical and vaginal secretions of infected women (Henin et al., 1993). This result provides evidence that there could be an additional source of HIV-1 exposure for vaginally delivered infants. It has been suggested that if HIV-1 can be detected by virus culture or PCR in peripheral blood within 48 h of birth the infection should be termed as intrauterine (Bryson et al., 1992; Henin et al., 1993). Some indirect evidence suggests that transmission may occur around the time of delivery. Moreover, the risk of transmission of HIV-1 to a firstborn twin has been found to be twofold higher than the secondborn twin (Duliege et al., 1992; Goedert et al., 1991; Goedert 1992) because the first child is exposed for a longer time to HIV-infected material in the birth canal than the second child. Further analysis of the European Collaborative Study (1994) has shown that the infection rate in children born by cesarean section was lower than those born via vaginal delivery.

The isolation and detection of HIV-1 by PCR in infants just after birth supports the view that some infants acquire infection at or very near the time of birth. Several studies have demonstrated a lack of detectable virus in some infants at the time of birth, which turned positive after 3 to 6 months (Bryson *et al.*, 1993; Ehrnst *et al.*, 1991; Krivine *et al.*, 1992; Rouzioux *et al.*, 1993). This could be attributed to the relative insensitivity of testing at birth, a very small virus inoculum, or sequestration of the virus. Furthermore, studies of an animal model of perinatal transmission using simian immunodeficiency virus (SIV) have shown that fetal monkeys can be infected during pregnancy (Davison-Fairburn *et al.*, 1992; Fazley *et al.*, 1993). In one study, placental disruption appeared to be an additional requirement for infection of fetus in monkeys (Davison-Fairburn *et al.*, 1992).

Postpartum transmission due to breast milk feeding has been documented in several cases (Lepage *et al.*, 1987; Weinbreck *et al.*, 1988; Ziegler *et al.*, 1985). HIV-1 transmission by breast milk may be related to the duration of exposure to breast milk, infectivity of the milk, specific susceptibility of the infant, or the timing of exposure (van't Wout *et al.*, 1994). Most of the infections through breast feeding occur within 3–6 months after birth, probably from colostrum or early milk.

III. Factors Influencing HIV-1 Mother–Infant Transmission

While the risk of transmission of HIV-1 from mother to infant is high, certain factors, viral or host, protect 70% of children born to HIV-1-infected mothers Connor *et al.*, 1993; Hira *et al.*, 1989; Oleske *et al.*, 1983; Thomas *et al.*, 1989). These factors are an important area of research and must be identified and characterized. To date, there are no clearly defined factors, viral or host, associated with maternal transmission of HIV-1. However, maternal parameters including advanced clinical stages of the mother, low CD4+ lymphocyte counts, maternal immune response to HIV-1, recent infection, high level of circulating HIV-1, and maternal disease progression have been implicated in an increased risk of mother-to-infant transmission of HIV-1 (Anderson and Medley 1988; Blanche *et al.*, 1989; European Collaborative Study 1994; Hira *et al.*, 1989; Report on a Consensus Workshop 1992; Ryder *et al.*, 1988). In a French Cohort study involving a 7-year follow-up, two factors were identified as being associated with an increased risk of maternofetal transmission: p24 antigenemia and elevated maternal age (Mayaux *et al.*, 1995). Furthermore, the risk of transmission increased gradually from 15% at counts of >600 CD4+ cells to 43% at counts of <200 and was also related to the percentage of CD8+ cells with the lowest risk (12%) when the CD4+ cell count was >500 and the highest risk (50%) for <200 (Mayaux *et al.*, 1995). Several studies indicate that elevated maternal viral load, plasma HIV-1 RNA levels, may play an important role in perinatal transmission. Two groups observed different threshold effects for transmission, with 80% of the women transmitting who had HIV-1 RNA level of over 100,000 copies/μl (Fang *et al.*, 1995) and 75% of the women transmitting with HIV-1 RNA levels over 50,000 copies/ μl (Dickover *et al.*, 1996). However, larger studies have been unable to significantly correlate a high viral load with increased risk of vertical transmission. Cao *et al.* (1997) and the investigators of the Ariel Project in the United States reported that the risk of transmission increased slightly with a higher viral load, but no threshold value of virus load was identified which discriminated between transmitters and nontransmitters. They further concluded that a high maternal viral load is insufficient to fully explain vertical transmission of HIV-1 (Cao *et al.*, 1997). Several other studies reported similar results, i.e., no predictive threshold for maternal HIV-1 RNA was observed for vertical transmission (Mayaux *et al.*, 1997).

Several other factors such as acute infection during pregnancy, the presence of other sexually transmitted diseases or other chronic infections, disruption of placental integrity secondary to chorioamnionitis, or smoking have been shown to be associated with mother-to-infant transmission of HIV-1 (Report of a Consensus Workshop, 1992). Several studies have demonstrated a direct association between the presence of maternal antibody against the V3 domain of the envelope protein and a lower rate of transmission of HIV-1 (Devash *et al.*, 1990; Rossi *et al.*, 1989), whereas others have shown lack of a correlation (Hasley *et al.*, 1992; Parekh *et al.*, 1991). The ability of maternal antibody to neutralize its own isolate (autologous neutralization) may be particularly important because it has been suggested that the virus mutants that are selected under immune pressure and cannot be neutralized may play a role in transmission (Bryson *et al.*, 1993; Scarlatti *et al.*, 1991). Obstetrical factors such as mode of delivery, invasive monitoring, or duration of ruptured membranes may alter the risk of intrapartum transmission (Douglas *et al.*, 1992; Konduri *et al.*, 1993; Newell *et al.*, 1993; Report of a Consensus Workshop, 1992). The type of breast milk (colostrum vs. later milk); duration of breast-feeding; and maternal factors such as viral load, antibody content of milk, and duration of mother's infection may also influence transmission (Report of a Consensus Workshop, 1992). In developed countries such as the United States and others, breast-feeding is not recommended to HIV-1-infected mothers. In addition, the possibility of viral factors affecting mother-to-infant transmission cannot be ruled out, since 70% of the children born to HIV-1-infected mothers are uninfected.

IV. Effect of Smoking on HIV-1 Mother–Infant Transmission

Smoking has been linked with an increased risk of acquiring HIV-1 infection in women (Hasley *et al.*, 1992) and rapid progression to AIDS in

HIV-1-infected asymptomatic individuals (Burns *et al.*, 1991; Nieman *et al.*, 1993; Royce *et al.*, 1990) and it could potentially affect mother-to-infant transmission of HIV-1. Moreover, the development of interventions to prevent or reduce maternal transmission of HIV-1 requires an understanding of the multiple factors involved and the mechanisms by which they interact. One such factor, a low CD4+ level in the mother, has been shown to be associated with an increased risk of transmission by a number of perinatal studies (European Collaborative Study, 1988; Ryder *et al.*, 1988; St. Louis *et al.*, 1993; Tibaldi *et al.*, 1993). Furthermore, substances that can reduce CD4+ levels, like cigarette smoking, can influence maternal transmission of HIV-1.

In a New York City cohort, HIV-1-seropositive women who smoked cigarettes after the first trimester and who had a prenatal CD4+ level less than 20%, had a greater than threefold increased risk of transmitting their infections to their infants (Burns *et al.*, 1994). Turner *et al.* (1997) have also shown an increase in the rate of maternal–fetal transmission of HIV-1 as a result of smoking. There are several ways in which cigarette smoking during pregnancy might interact with a low CD4+ level to increase the risk of maternal transmission. Smoking has been associated with multiple alterations in the immune system, including decreased B-lymphocyte capping and *in vitro* antibody production and a decrease in the number, proportion, and function of natural killer cells (Savage *et al.*, 1991). Constituents of cigarette smoke have been identified in the lining of the cervix of smokers (Schiffman *et al.*, 1987), and smoking has been associated with changes in numerical density of cervical Langerthans cells, macrophage, and CD4+ lymphocytes (Barton *et al.*, 1988). Accordingly, smoking may have an important impact on the immune response, including local immune function in female genital tract (Smoking and Immunity, 1990). In addition, cigarette smoking during pregnancy has been associated with alterations in placental morphology and function (Arnholdt *et al.*, 1990: Jauniax *et al.*, 1992) and since HIV-1 infection has been demonstrated in placenta (Muary *et al.*, 1989; Sprecher *et al.*, 1986), tobacco smoking could have a direct effect on prepartum (transplacental passage) transmission of HIV-1. Several obstetric complications, including antepartum vaginal bleeding and placental abruption (Myer and Tonascia, 1977) and premature rupture of membranes (PROM) and preterm PROM (Ekwo *et al.*, 1993; Henin *et al.*, 1993), have been associated with cigarette smoking during pregnancy. While the presence of HIV-1 in cervicovaginal secretions has been demonstrated to be increased during pregnancy (Clemetson *et al.*, 1993; Henin *et al.*, 1993) and several cases of mother-to-child transmission occur at or near delivery (Ehrnst *et al.*, 1991; European Collaborative Study, 1988), any process (such as smoking) that increases the amount of blood or secretions in the birth canal would be expected to increase the risk of intrapartum transmission.

V. Diagnosis of HIV-1 Infection in Perinatally Infected Infants

The maternal IgG HIV-1 antibodies that cross the placenta to the fetus can persist for up to 18 months in the infant (European Collaborative Study, 1988; Mok *et al.*, 1987). Thus, using conventional ELISA for the detection of HIV-1 antibodies in infants is not immediately useful for serodiagnosis of vertically transmitted infection because the infant's IgG cannot be distinguished from those acquired from the mother. The infants born to HIV-1-infected mothers are evaluated in regular follow-ups up to 3 years before they are declared uninfected based on Center for Disease Control guidelines (Center for Disease Control, 1992). A specific immune response of the infant that would indicate infection is the presence of IgA or IgM antibodies, which do not cross the placenta (Nicholas *et al.*, 1989) and can be used to diagnose HIV-1 infection in children born to seropositive mothers (Weiblen *et al.*, 1990). HIV-1-infected children may be identified by measuring antibody production in the newborn's PBMC cultures using a B-cell mitogen (Amadori *et al.*, 1988; Laure *et al.*, 1988). However, the specific and reliable methods to detect HIV-1 in infants are PCR, antigen detection, virus culture, and *in vitro* antibody production. In a cohort study, the sensitivities of these tests were estimated and compared among each other and were found to be 81.5% (PCR), 70.3% (virus culture), 92.5% (*in vitro* antibody production), and 44.4% (antigen) (De Rossi *et al.*, 1991). In general, a positive PCR and virus culture in a newborn is considered to be indicative of HIV-1 infection. Moreover, PCR should be done on HIV-1 proviral DNA in order to detect the presence of HIV-1 in infected infants, as shown previously (Rogers *et al.*, 1989).

VI. Immunologic Abnormalities in HIV-1-Infected Infants

HIV-1-infected infants' progress more rapidly from asymptomatic to symptomatic infection and from onset of symptomatic infection to death. The basis of more rapid progression to AIDS in infants is unknown. It is possible that HIV-1 interacts with the neonate's immune system in a different way than that observed in adults. In neonates, HIV-1 replication is probably supported in the thymus. Thymocytes appear to differ from T cells in that they may support the replication of HIV-1 in the absence of any stimulatory effects (Schnittman *et al.*, 1990). Thus, thymic injury from HIV-1 infection may have a profound impact on development of the immune system in HIV-1-infected fetuses and infants. Moreover, the thymus is critically important in infants and neonates in populating the immune system with T cells. Pediatric HIV-1 infection differs in immune dysfunction from that seen in adults. The immunologic abnormalities observed in HIV-1-infected infants and children include a decreased percentage of thymic CD4+ cells, drastic reduction in cortical CD4/CD8 double-positive cells (Rubenstein 1993; Stanley *et al.*, 1993), and an increased percentage of CD8+ cells (Koup *et al.*, 1993). In addition, the stromal cells, which support thymocyte development, are damaged. In the periphery, an inverted ratio of CD4/CD8, increased quantitative Ig, decreased *in vitro* response to mitogens/antigens, decreased cytotoxic T-lymphocyte (CTL) response, and decreased phagocytosis have been reported (Koup *et al.*, 1993). In addition, poor antibody response to vaccination with T-dependent and T-independent antigens; increased production of IL-1β, IL-2, IL-6, and interferon-γ in lymph nodes; and decreased production of IL-2, IL-4, and interferon-γ by CD4 cells have been observed in HIV-1-infected children (Koup *et al.*, 1993).

VII. Clinical Manifestations in HIV-1-Infected Infants

In HIV-1-infected mother–infant pairs, the clinical manifestations in infected infants differ from those seen in their mothers (Clavelli *et al.*, 1990). Recurrent bacterial infections, lymphocyte interstitial pneumonitis, encephalopathy, and neurological and physical growth deficits are commonly observed in children with HIV-1 infection (Claveli *et al.*, 1990). HIV-1-infected infants tend to have more CNS involvement than infected adults. This is further supported by the presence of HIV-1 in fetal CNS tissue (Lyman *et al.*, 1990). The common opportunistic infection in both children and adults is *Pneumocystis carnii* pneumonia (PCP) (Center for Disease Control 1992). Several other common opportunistic infections in children include cytomegalovirus (CMV) infections, Epstein-Bar virus (EBV) infections, herpes zoster, mycobacterium avium, crytosporidium enteritis, chronic and recurrent mucosal and esophageal candadiasis, and mucocutaneous herpes simplex virus infection (Van Dyke, 1993). A higher incidence of common childhood infections such as otitis media, sinusitis, viral respiratory infections, bacterial pneumonia, bacteremia, and meningitis have been observed in HIV-1-infected children (Van Dyke *et al.*, 1993).

VIII. Prevention of HIV-1 Mother–Infant Transmission by Antiretroviral Therapy

The AIDS Clinical Trials Group (ACTG) protocol 076 suggests that women with greater than 200 CD4 counts who initiate treatment with zidovudine (ZDV) during pregnancy can prevent transmission of HIV-1 to their infants in about two-thirds of the cases (Cooper *et al.*, 1996). Oral

ZDV was given to HIV-1-infected pregnant women between 14 to 34 weeks gestation and continued throughout pregnancy to target *in utero* transmission. ZDV was not administered during the first trimester because of the potential toxicity during the period of maximal organ development. ZDV was intravenously administered during labor because it rapidly crosses the placenta and provides active levels of drug to the fetus during passage through the birth canal. Finally, oral ZDV was given to newborns for 6 weeks to inhibit viral replication if virus or infected maternal cells passed into the infant's circulation during uterine contraction. Furthermore, an ACTG 185 perinatal trial that examined women with more advanced disease and lower CD4 counts than ACTG 076 (<200) showed a similar reduction in the transmission rate. The efficacy of ZDV in clinical practice has been evaluated and the results are consistent with the ACTG 076 and 185 trials. This treatment has become standard where the status of HIV-1 infection of pregnant women is known and has resulted in the reduction of HIV-1 transmission from mother to infant. Recently, a short course of zidovudine (300 mg twice daily from 36 weeks gestation and every 3 h from onset of labor until delivery) reduced perinatal transmission in Thailand (Shaffer *et al.*, 1999) and Cote d'Ivoire (Wiktor *et al.*, 1999).

One of the critical aspects of ZDV use is the long-term toxicity of the treatment. There is a particular concern regarding intrauterine ZDV exposure, especially for those infants who will not be infected. The lethal effect of ZDV on embryonic development has been demonstrated in experimental animals (McGaughey *et al.*, 1992; Wattier *et al.*, 1993). However, the long-term toxicity is being evaluated, in ACTG 219 protocols, in infants up until age 21 and in ACTG 288 protocols, which follow women from ACTG 076 for 3 years postpartum. The other concern is the development of ZDV resistance in women receiving ZDV during pregnancy and the possibility that the ZDV-resistant mutant can be transmitted to their infants. If we can differentiate transmitters from nontransmitters early in pregnancy by analyzing their viral genotypes in conjunction with maternal parameters (CD4 counts, etc.), it will be helpful in making the decision as to which infected pregnant women should be placed on antiretroviral agents. This would then avoid adverse effects of the antivirals, if any, used during pregnancy on the growth and development of uninfected children born to infected mothers.

IX. Pathogenesis of HIV-1 Infection and Disease Progression

HIV-1 encodes three structural (*gag, pol,* and *env*) and six regulatory/accessory (*tat, rev, nef, vif, vpu,* and *vpr*) genes. The potential pathogenic region of HIV-1 probably resides within the *env* gene (Shioda *et al.*, 1991).

The hypervariable regions, designated V1 to V5, are interspersed among conserved regions along the *env* gene. The V3 loop is functionally important in virus infectivity (Wiley *et al.*, 1989), virus neutralization (Lasky *et al.*, 1986; Mathews *et al.*, 1986), and host cellular tropism (Hwang *et al.*, 1991; Shioda *et al.*, 1991), whereas the V1–V2 regions influence replication efficiency in macrophages by affecting virus spread (Toohey *et al.*, 1995). Genetic variability in HIV-1, especially in the variable region 3 (V3) of the envelope gene, has been observed within infected individuals (Ahmad *et al.*, 1995; Mulder-Kampinga *et al.*, 1993, 1995; Myers *et al.*, 1995; Scarlatti *et al.*, 1993; Wolinsky *et al.*, 1992, 1995). These variants arise during retroviral replication by errors in reverse transcription (Dougherty and Temin, 1988; Peterson *et al.*, 1988; Roberts, 1988). Several reasons for the existence of different genetic variants within an infected individual could be postulated such as immunologic pressure for change, alteration in cell tropism, and replication efficiency (Hwang *et al.*, 1991; Shioda *et al.*, 1991; Siliciano *et al.*, 1988). Although the CD4+ lymphocyte is the major target for HIV-1 replication, cells of monocyte-macrophage lineage represent the predominant HIV-1-infected cell type in most tissues, including the central nervous system (Gartner *et al.*, 1986; Gendelman *et al.*, 1989; Koenig *et al.*, 1986; Potts *et al.*, 1990). Moreover, the presence of a heterogeneous population of HIV-1 within infected individuals poses a major problem in the development of strategies for prevention and treatment of HIV-1 infection. It is very difficult to inhibit or neutralize a heterogeneous population of HIV-1.

People infected with HIV-1 exhibit variable rates of disease progression. The basis of this variability is unknown, but it has been suggested to be due to interplay between the strength of the host immune system and the phenotypic characteristics of the initially infecting or evolving virus subtype (Lu *et al.*, 1997; Sheppard *et al.*, 1993; Tersmette *et al.*, 1989). Furthermore, disease progression is reflected in decreased CD4+ counts, and the rate of decline can be used to divide HIV-1-infected individuals into three general groups: (i) rapid progressors (10%), (ii) progressors (80%), and (iii) nonprogressors or long-term survivors (10%) (Sheppard *et al.*, 1992). *In vitro* studies have demonstrated that HIV-1 isolates from asymptomatic individuals tend to replicate more slowly than HIV-1 isolates from 50% of the patients with AIDS (Cheng-Myer *et al.*, 1990). In addition, T-lymphotropic-syncytium-inducing (SI) HIV-1 variants appear during the asymptomatic phase of infection in about 50% of infected individuals and are associated with CD4 depletion and more rapid progression to AIDS (Koot *et al.*, 1993). In general, macrophage-tropic-nonsyncytium-inducing (NSI) variants of HIV-1 have been observed in the early asymptomatic phase (Schuitmaker *et al.*, 1991). Furthermore, these macrophage-tropic NSI variants have been found to exist during all stages of infection (Schuitmaker *et al.*, 1991, 1992). The molecular basis of some of these phenotypes has been partially elucidated and mapped to the *env* gene. In addition, a differential pattern of unspliced and spliced RNAs has been observed in rapid progressors and nonprogressors, with a higher level of unspliced RNA in rapid progressors and relatively stable levels of unspliced and spliced viral mRNAs in slow progressors (Furtado *et al.*, 1995). Moreover, the extent of viral transcription and replication correlates with the rate of CD4+ T-cell loss (Furtado *et al.*, 1995), suggesting that the HIV-1 population in rapid progressors replicates faster than the HIV-1 population in slow progressors (Cheng-Myer *et al.*, 1990; Tersmette *et al.*, 1988). However, the mutations in the regulatory and accessory genes that might have any impact on the course of HIV-1 infection is currently unknown. Recently, persistence of attenuated or defective regulatory and accessory genes have been shown in HIV-1-infected long-term survivors that contributed to their lack of disease progression (Iverson *et al.*, 1995). In HIV-1 mother–infant transmission, the genotypic and phenotypic diversity and the variability in viral genes may be important because of a direct correlation between increased risk of transmission and maternal disease progression.

X. Molecular and Biological Characterization of HIV-1 Involved in Mother-to-Infant Transmission _____

We hypothesize that the molecular and biological properties of HIV-1 involved in mother-to-infant transmission are critical determinants of perinatal transmission. Wolinsky *et al.* (177) were the first group that analyzed and compared HIV-1 DNA sequences in the V3 and V4–V5 regions of the *env* gene from three mother–infant pairs and suggested that a minor subtype of maternal virus from a genetically heterogeneous virus population could be transmitted to the infant. Studies from my laboratory compared the HIV-1 DNA sequences in the V3 region of the envelope gene from seven mother–infant pairs following perinatal transmission (Ahmad *et al.*, 1995). These results suggest that a minor subtype of maternal virus from the genetically heterogeneous virus population was transmitted to the infant (Ahmad *et al.*, 1995). The minor HIV-1 genotype predominates initially as a homogeneous population in the infant and then becomes diverse as the infant grows older (Ahmad *et al.*, 1995). Several other groups have also reported transmission of minor (Contag *et al.*, 1998; Mulder-Kampinga *et al.*, 1993, 1995; Pasquier *et al.*, 1998; Sato *et al.*, 1999; Scarlatti *et al.*, 1993), major (Scarlatti *et al.*, 1993), and multiple (Lamers *et al.*, 1994; Pasquier *et al.*, 1998) HIV-1 genotypes from mother to infant. In addition, sequence analysis of five maternal–fetal macaque pairs has revealed the selective transmission of a single SIV genotype from mother to infant (Amedee *et al.*, 1995). In this transplacental transmission, the mothers harbored a heterogeneous virus population compared to their infants (Amedee *et al.*, 1996).

Nucleotide sequencing directly derived from HIV-1 DNA isolated from infected mother–infant pairs' peripheral blood mononuclear cells (PBMC), alignment of deduced amino acid sequences, and phylogenetic analysis (Felsenstein, 1989) have provided a powerful tool to analyze and identify HIV-1 genotypes transmitted from mother to infant. HIV-1-infected mothers harbored a heterogeneous virus population compared to the infants' virus population (Ahmad *et al.*, 1995; Mulder-Kampinga *et al.*, 1993, 1995; Scarlatti *et al.*, 1993; Wolinsky *et al.*, 1992). Ahmad *et al.* (1995) have shown that the HIV-1 sequences in younger infants are more homogeneous than the sequences of older infants. All the infants' sequences were different but even the older infants' sequences displayed patterns similar to those seen in their mothers (Ahmad *et al.*, 1995). Multiple alignments of deduced amino acid sequences and phylogenetic analysis of a mother–infant pair have allowed us and others to identify HIV-1 genotypes that could be transmitted from mother to infant. Multiple alignment shows that the infants' sequences have a pattern or signature region seen in their mothers (Ahmad *et al.*, 1995; Mulder-Kampinga *et al.*, 1993, 1995; Scarlatti *et al.*, 1993; Wolinsky *et al.*, 1992). Similarly, phylogenetic analysis performed revealed several subtypes in mothers and mainly one subtype in infants. In these studies, the mothers' minor sequence was found to be closer to the infant's sequence. These studies are providing new directions in the understanding HIV-1 heterogeneity, transmission, and disease progression.

Similar observations of selective transmission of HIV-1 have also been found in transmitter–recipient partners involving sexual transmission including a homogeneous sequence population present in the recipients (Cichutek *et al.*, 1992; McNearny *et al.*, 1992; Pang *et al.*, 1992; Wolfs *et al.*, 1992; Zhang *et al.*, 1993; Zhu *et al.*, 1993). Three models have been proposed to explain this feature: (i) the random dilution effect, in which a low inoculum of the virus is transmitted from the transmitter to the recipient; (ii) selective amplification, in which multiple HIV-1 variants may enter the recipient but only one is selectively amplified; and (iii) selective transmission, in which one viral variant has a selective advantage in penetrating the mucosal barrier of the new host (Zhu *et al.*, 1993). In mother–infant transmission, the selective transmission model looks the most favorable, as evidenced by the selective transmission of HIV-1 variants from mother to infant (Ahmad *et al.*, 1995; Mulder-Kampinga *et al.*, 1993, 1995; Scarlatti *et al.*, 1993; Wolinsky *et al.*, 1992). These findings are based only on the analysis of few small regions in the *env* gene. Further studies involving analysis of several other important regions in the HIV-1 genome are needed in order to molecularly characterize the HIV-1 transmitted from mother to infant.

Little is known about the biological properties of HIV-1 transmitted from mother to infant. However, the viral phenotype involved in sexual transmission has been elucidated. Zhu *et al.* (1993) have shown the viral phenotype to be uniformly macrophage-tropic (MT) and non-syncytium-

inducing (NSI) in five HIV-1 seroconvertors, including a homogeneous sequence population in the recipients. Several other studies have suggested
the viral phenotype to be macrophage-tropic-NSI viruses, based on the
sequence analysis of the V3 region (MnNearny *et al.*, 1991). In addition,
van't Wout *et al.* (1994) have shown that macrophage-tropic HIV-1 variants
initiate infection after sexual, parenteral, and vertical transmission. In contrast, Kliks *et al.* (1994), in a small-cohort study, have demonstrated the
transmission of a rapid or high-titered replicating, T-cell tropic, and neutral-
ization-resistant HIV-1 variant from mother to child. Moreover, the viral
phenotype in SIV transmission from mother to infant transplacentally was
found to be macrophage-tropic (Amedee *et al.*, 1995). However, more research is needed to better understand the viral phenotype and other biological
properties of the HIV-1 transmitted from mother to infant.

   To further characterize the biological properties of HIV-1 associated
with mother-to-infant transmission, we have evaluated the functional role
of the V3 region from mother–infant isolates because of it's being a major
determinant of replication efficiency, cell tropism, and cytopathic effects.
The V3 region of mother–infant isolates characterized before (Ahmad *et
al.*, 1995) were reciprocally transferred into an HIV-1-infectious molecular
clone and the biological properties, including replication efficiency, cellular
tropism, and cytopathic effects, were evaluated (Ahmad *et al.*, 1999, Matala
et al., 1999). The V3-region chimeras were transfected into HeLa cells by
electroporation and virus production was measured by RT assay (Ahmad
et al., 1999; Matala *et al.*, 1999). We found that all our V3 chimeras
replicated in HeLa cells, as evidenced by RT activity in the culture media.
The V3-region chimeras replicated at the same level of the parent clone
(pNL4-3), suggesting that the V3-region substitution did not alter the *env*
open reading frame. We then determined the replication of mother–infant
V3 chimeras in a T-lymphocyte cell line (A3.01) and found these chimeras
were unable to replicate in T-lymohocyte cell lines. These data suggest
that the V3 region from mother–infant isolates changed the tropism of the
lymphotropic parent clone NL4-3. We next examined the replication of
these chimeras in the HOS CD4-CCR5 cell line, which contains the receptor
(CD4) and coreceptor (CCR5) for the macrophage-tropic clone. Figure 1
shows the replication of the V3-region chimeras in the HOS CD4-CCR5
cell line. Interestingly, the mother–infant V3-region chimeras infected and
replicated in the HOS CD4-CCR5 cell line, including the macrophage-tropic
Ada-M, whereas NL4-3 was unable to replicate in this cell line. The data
suggest that the V3 region from mothers and infants conferred macrophage
tropism to the virus (Matal *et al.*, 1999). Furthermore, the syncytium-
inducing ability of these chimeras was examined on MT-2 cells and found
to be non-syncytium-inducing (NSI) (Matala, *et al.*, 1999) and are now
referred to as R5 virus. The characterization of the molecular and biological
properties of HIV-1 variants transmitted from mother to infant will allow

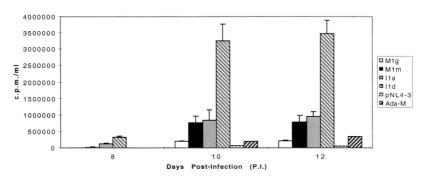

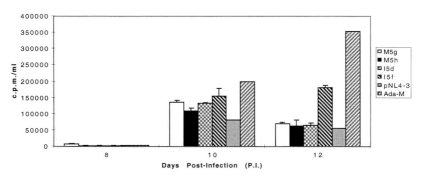

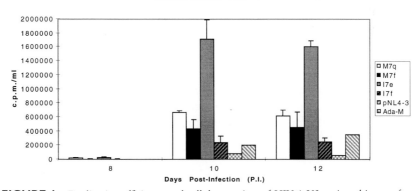

FIGURE I Replication efficiency and cellular tropism of HIV-1 V3 region chimeras from mother–infant isolates following perinatal transmission. The V3 regions from three mother (M)–infant (1) pairs (1, 5, and 7; Ahmad *et al.*, 1995) were reciprocally inserted into a lymphotropic molecular clone, NL4-3. Viruses generated from the V3 region chimeras, NL4-3 and Ada-M, were used to infect Hos-CD4-CCR5 cell line. Equal amounts of viruses were used to infect the cell line and virus production was measured by reverse transcriptase assay in the culture media. The V3 regions from mother–infant pairs conferred macrophage-tropism to the virus.

us to understand the molecular mechanisms of maternal transmission of HIV-1. These studies may be helpful in the further development of methods for the prevention and treatment of HIV-1 infection in children.

The characterization of HIV-1 genotypes associated with maternal transmission has been performed using only the blood samples of mother–infant pairs (Ahmad *et al.*, 1995; Mulder-Kampinga *et al.*, 1993, 1995; Scarlatti *et al.*, 1993; Wolinsky *et al.*, 1992). The minor genotypes of the heterogeneous population of HIV-1-infected mothers have been shown to predominate in their infants following perinatal transmission. The possibility remains, however, that the minor genotypes of HIV-1 found in maternal blood could be either the same or different from those present in other sources of transmission such as vaginal secretions and placenta. Substantial evidence exists supporting *in utero* transmission of HIV-1 from infected mothers to the developing fetus (Muary *et al.*, 1989; Sprecher *et al.*, 1988). Since HIV-1 has been detected in placenta (Courgnaud *et al.*, 1991; Lyman *et al.*, 1990), the HIV-1 genotypes found in the placenta or vaginal secretions are likely to be transmitted to the infants. Therefore, it will be important to perform a comparative sequence analysis on the maternal blood samples along with vaginal secretions and placental tissue.

XI. Chemokine Receptors and HIV-I Mother–Infant Transmission

Two distinct coreceptors, CXCR4 and CCR5, have been identified for the entry of T-lymphotropic and macrophage-tropic HIV-1, respectively (Alkhatib *et al.*, 1996; Feng *et al.*, 1996). The region responsible for determining coreceptor utilization was examined by Choe *et al.* (1996) and showed that the V3 region was responsible for interacting with this coreceptor. The role of the V3 region becomes very important in determining the tropism that may play an important role in transmission, infection, and disease progression. M-tropic viruses are the more commonly transmitted viruses in sexual (Zhu *et al.*, 1993) and vertical (Matala *et al.*, 1999) transmissions. While M-tropic (R5) viruses predominate initially in most infected individuals, T-tropic viruses (X4 viruses) are more virulent and associated with a faster rate of $CD4^+$ T-cell loss following several years of infection (Connor and Ho, 1994; Richmann and Bozzette, 1994). In addition, individuals homozygous for a 32-bp deletion in their CCR5 genes were substantially protected from HIV-1 infection and heterozygous for a 32-bp deletion in CCR5 genes had a slow disease progression (Dean *et al.*, 1996). Several other coreceptors that interact with HIV-1 have also been identified, including CCR2b, CCR3, CCR8, BOB, BONZO, and CX3CR1. Thus, the study of the viral genotypes and phenotypes controlled by the *env* gp120 and its interaction with the coreceptors (CXCR4, CCR5, etc.) may have significance for understanding viral transmission, pathogenesis, and disease progression. We and others have shown that R5 viruses are involved in maternal–fetal transmission.

Several studies have examined the role of CCR5 on maternal–fetal transmission and have found that infants who have two copies of a 32-bp deletion in CCR5 were infectable by X4 viruses following perinatal transmission (Koup, 1997; Salvatori *et al.*, 1998). They further suggested that that CCR5 deletion had a minimal to nonexistent effect on maternal–fetal transmission.

XII. Genetic Analysis of HIV-1 Regions Following Mother-to-Infant Transmission

Genetic analysis of HIV-1 sequences in other regions of the genome, in addition to the variable regions of *env*, from mother–infant pairs following perinatal transmission has been very limited (Myers *et al.*, 1995). The possibility exists, however, that several other regions or motifs in the HIV-1 genome may be involved in mother-to-infant transmission and could be critical determinants of perinatal transmission. My laboratory has been actively involved in the analyses of various HIV-1 regions following perinatal transmission, with the idea that a complete molecular and biological characterization of HIV-1 associated with maternal–fetal transmission may provide relevant information for the strategies of prevention and treatment. We should target the properties of HIV-1 that are involved in transmission for preventive strategies.

HIV-1 encodes *gag* p17 matrix protein, which plays a pivotal role in the virus life cycle, including virus entry, localization to the nucleus, and virus assembly and release, and may have a role in transmission. In addition, the accessory genes *vif* and *vpr* are found to be highly conserved and functional during natural infection (Sova *et al.*, 1995; Goh *et al.*, 1997), suggesting that *vif* and *vpr* are important for HIV-1 pathogenesis in maternal–fetal isolates. Since the transmitted viruses from mothers to infants are macrophage-tropic and NSI (R5 viruses), the role of HIV-1 accessory genes *vif* and *vpr* and p17 matrix becomes important because these proteins are necessary for HIV-1 replication in macrophages and quiescent T cells (Connor *et al.*, 1995; Gabuzda *et al.*, 1994; Von Schwedler *et al.*, 1994). Therefore, we performed a complete analysis of HIV-1 *vif* and *vpr* sequences from mother–infant pairs following perinatal transmission. Better characterization of HIV-1 transmitted from mothers to infants may provide relevant information toward the development of strategies for prevention and treatment because the strategies involved should be targeted at the properties of the transmitted viruses.

To determine the coding potential of the *env, gag* p17, *vif,* and *vpr* genes, we analyzed these sequences from seven infected mother–infant pairs following perinatal transmission. The frequencies of the coding potential of the Gag p17, Vif, Vpr, Env (V3 region), and Env V1–V5 open reading frames were 86.2, 89.8, 97.12, 98.4, and 92%, respectively (Fig. 2). These data suggest that these open reading frames are highly conserved following mother-to-infant transmission of HIV-1 and may have a role in perinatal transmission. The degree of variability of HIV-1 sequences, including *env* V3-region, *gag* p17 *vif,* and *vpr* sequences, is shown in Figs. 3 and 4. There was a low degree of sequence variability of HIV-1 sequences in the regions

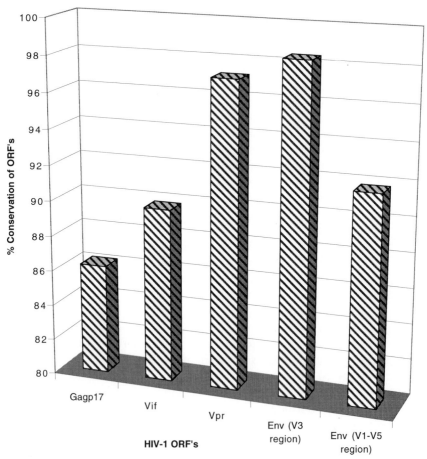

FIGURE 2   Conservation of intact HIV-1 Gag p17, Vif, Vpr, Env (V3 region), and Env (V1–V5 regions) open reading frames in mother–infant isolates following perinatal transmission. The frequency of conservation is expressed as a percentages of intact open reading frames in mother–infant isolates.

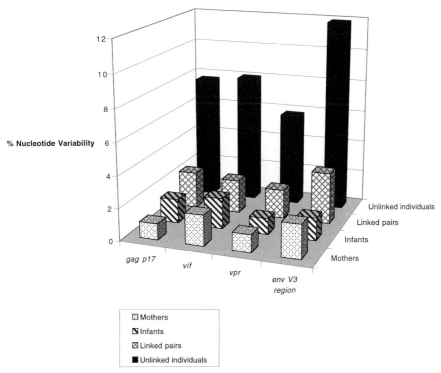

12
10
8
% Nucleotide Variability 6
4
2
0

Unlinked individuals
Linked pairs
Infants
Mothers

gag p17
vif
vpr
env V3
region

☒ Mothers
◩ Infants
☒ Linked pairs
■ Unlinked individuals

FIGURE 3 Genetic variability of *gag* p17, *vif*, *vpr*, and *env* (V3 region) nucleotide sequences. The percentages of mismatches were calculated between nucleotide sequences within the same mother's set, within the same infant's set, between epidemiologically linked mother–infant pairs, and between epidemiologically unlinked individuals. The distance percentages were rounded off to the nearest decimal.

of *gag* p17, *vif*, and *vpr* compared to *env* V3-region sequences. In addition, the mothers' V3-region sequences were more heterogeneous compared to infants' sequences, suggesting selective transmission. The genetic variability of HIV-1 nucleotide sequences in the regions of *gag* p17, *vif*, *vpr*, and *env* V3 were determined within mothers, within infants, and between epidemiologically linked and unlinked individuals, as shown in Fig. 3. The data suggested that HIV-1 sequences from epidemiologically linked mother–infant pairs were closer than those from epidemiologically unlinked individuals. Interestingly, HIV-1 sequences in the conserved and less variable regions, including *gag* p17, *vif*, and *vpr*, were distinguishable from epidemiologically linked and unlinked individuals. Furthermore, the HIV-1 amino acid variabilities in the regions of Gag p17, Vif, Vpr, and Env (V3 region within mothers, within infants, and between linked mother–infant pairs) are shown in Fig. 4. The phylogenetic analysis of HIV-1 *env* V3-region, *gag* p17, *vif*, and *vpr* sequences from seven mother–infant pair isolates following perinatal

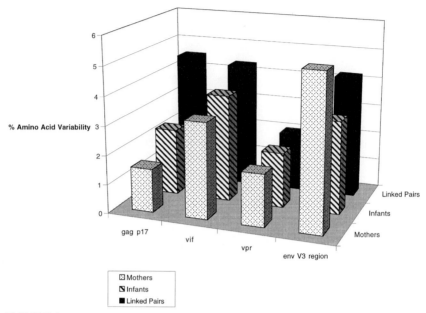

FIGURE 4   Distribution of Gag p17, Vif, Vpr, and Env (V3) amino acid distances within mothers, within infants, and between epidemiologically linked mother–infant pairs. The percentages of mismatches were calculated between amino acid sequences within the same mother's set, within the same infant's set, and between epidemiologically linked mother–infant pairs. The distance percentages were rounded off to the nearest decimal.

transmission were performed. The phylogenetic tree for 263 sequences for V3 region (Ahmad *et al.*, 1995), 166 sequences for *gag* p17 (Hahn *et al.*, 1999), 137 sequences for *vif* (Yedavalli *et al.*, 1998a), and 166 sequences for *vpr* (Yedavalli *et al.*, 1998b) from seven mother–infant pairs revealed that all the mother–infant pairs were well discriminated, separated, and confined within subtrees (not shown), indicating that the epidemiologically linked mother–infant pairs were closer to each other and that there was no PCR product. We also performed a global phylogenetic tree for all our mother–infant pairs' *gag* p17, *vif,* and *vpr* sequences and other available HIV-1 sequences in these regions in the HIV databases and found that the mother–infant sequences were separated from other HIV-1 sequences. In addition, our mother–infant sequences grouped with the subtype, or clade, B. The functional domains required for Gag p17 matrix, Vif and Vpr activities, were examined in the deduced amino acid sequences of mother–infant isolates and found to be highly conserved. Several motifs in Gag p17, including glutamic acid (E) or aspartic acid (D) at position 55, tyrosine (Y) or phenylalanine (F) at position 79, an aspartic acid (D) or glutamic acid (E) at positions 93 and 102, and dipeptide alanine–glutamic acids (AD) at positions 122–123 were present in most of the mother–infant pairs' sequences (Hahn *et al.*, 1999). Furthermore, the p17 motifs that were previously shown to be significantly associated with transmission (Narwa *et al.*, 1996), including a glutamic acid (E) at position 93, was found in one of seven mother–infant pairs and KIEEEQN at 103–109 was found in six of the seven mother–infant pairs. Taken together, these findings suggest that an intact an functional *gag* p17 open reading frame is essential for HIV-1 replication in mothers and infants and contains several motifs that may be associated with perinatal transmission.

While perinatal transmission may be multifactorial in nature, characterization of HIV-1 transmission from mother to infant may provide relevant information for the development of strategies for prevention and treatment of HIV-1 infection in children. In this context, we (Ahmad *et al.*, 1995) and others (Mulder-Kapinga, 1993; Scarlatti *et al.*, 1993; Wolinsky *et al.*, 1992) have shown a selective transmission of HIV-1 from mothers to their infants. Further molecular and biological characterization may help AIDS researchers to target the specific variant of HIV-1 involved in transmission. We also show that the minor genotype transmitted from mother to infant is macrophage-tropic and NSI (R5 virus) (Matala *et al.*, 1999). In addition, intact and functional *gag* p17, *vif*, and *vpr* genes were conserved following mother–infant transmission. The conservation of intact and functional *vif* and *vpr* genes, which are essential for HIV-1 replication in macrophages, further supports our findings of macrophage-tropic and NSI phenotypes of the transmitted viruses (R5 viruses). These results might be helpful in understanding the pathogenesis of HIV-1 infection in mothers and infants, including the molecular mechanisms involved in perinatal transmission, which may aid in the development of better strategies for prevention and treatment. In conclusion, we should target our preventive strategies on these molecular and biological properties of the virus.

Acknowledgments

The author thanks Dr. Colombe Chappey, National Center for Biotechnology Information, National Library of Medicine, for assistance in nucleotide sequence analysis; and Dr. Raymond C. Baker, Department of General Pediatrics, Children's Hospital Medical Center, Cincinnati, Ohio; and Dr. Ziad M. Shehab, Department of Pedatrics, College of Medicine, University of Arizona, for providing HIV-1-infected mother–infant pairs' samples. The author also thanks Dr. John J. Marchalonis, Chairman, Department of Microbiology & Immunology, University of Arizona, for support and encouragement; his coworkers Erik Matala, Venkat Sita R. K. Yedavalli, Tobias Hahn, Mohammed Husain, David Brooks, Scott Martin, Ulrike Philippar, Alison Holzer, and Andrew Harodner for their contribution toward the ongoing research in the laboratory; and AIDS Research and Reference Reagent Program for providing HOS-CD4-CCR5 (contributed by N. Landau) and Ada-M (contributed by H. Gendelman). The National Institute of Allergy and infectious Disease (AI 40378) and the Arizona Disease Control Research Commission (9601) are acknowledged for funding the research proposals.

References

Adjoriolo-Johnson, G., De Cock, K. M., Ekpini, E., Vetter, K. M., Sibailly, T. *et al.* (1994). Prospective comparison of mother-to-child transmission of HIV-1 and HIV-2 in Abidjan, Ivory Coast. *J. Am. Med. Assoc.* **272,** 462–466.

Ahmad, N. (1994). Mother-to-infant transmission of AIDS. In "HIV," pp. 121–116. *Society for AIDS Prevention and Education Publication,* Kathmandu, Nepal.

Ahmad, N. (1996). Maternal–fetal transmission of human immunodeficiency virus. *J. Biomed. Sci.* **2,** 238–250.

Ahmad, N., Baroudy, B. M., Baker, R. C., and Chappey, C. (1995). Genetic analysis of human immunodeficiency virus type 1 envelope V3 region isolates from mothers and infants after perinatal transmission. *J. Virol.* **69,** 1001–1012.

Ahmad, N., Matala, E., Yedavalli, V. R. K., Hahn. T., and Husain, M. (1999). "Characterization of Human Immunodeficiency Virus Type 1 Involved in Maternal–Fetal Transmission." International Center for Genetic Engineering and Biotechnology Virology Symposium Proceedings, Oxford Publications, (in press).

Alkhatib, G., Combadiere, C., Broder, C., Feng, Y., Kennedy, P., Murphy, P., and Berger, E. (1996). CC CKR5: A RANTES, MIP-1 alpha, MIP-1beta receptor as a fusion cofactor for macrophage-tropic HIV-1. *Science* **272,** 1955–1958.

Amadori, A., de Rossi, A., Giaquinto, C., Faulkner-Valle, G., Zacchello, F., and Chieco-Bianchi, L. (1988). *In vitro* production of HIV-specific antibody in children at risk of AIDS. *Lancet* **ii,** 852–854.

Amedee, A. M., Lacour, N., Gierman, J. L., Martin, L. N., Clements, J. E., Bohm, R., Harrison, R. M., and Murphey-Corb, M. (1995). Genotypic selection of simian immunodeficiency virus in macaque infants infected transplacentally. *J. Virol.* **69,** 7982–7990.

Arnholdt, H., Meisel, F., Fandrey, K. *et al.* (1990). Proliferation of villous trophoblast of the human placenta in normal and abnormal pregnancies. *Virchows Arch. Cell. Pathol. B* **60,** 365–372.

Barton, S. E., Maddox, P. H., Jenkins, D. *et al.* (1988). Effect of cigarette smoking on cervical epithelial immunity: A mechanism for neoplastic changes? *Lancet* **2,** 652–654.

Bawdon, R., Gravell, M., and Sever, J. (1993). Studies on the transmission of cell free human immunodeficiency virus in the *ex vivo* human placental model. *Pediatr. Res.* **33,** 288.

Bennett, J. V., and Rogers, M. F. (1991). Child survival and perinatal infections with human immunodeficiency virus. *Am. J. Dis. Child* **145,** 1242–1247.

Blanche, S., Rouzious, C., Moscato, M.-I. G. *et al.* (1989). A prospective study of infants born to women seropositive for human and immunological virus type 1. *N. Engl. J. Med.* **320,** 1643–1648.

Bryson, Y., Lehman, D., Garratty, E. *et al.* (1993). The role of maternal autologous neutralizing antibody in preventing maternal fetal HIV-1 transmission. *J. Cell Biochem.,* 9517E. [abstract 951:7E]

Bryson, Y. J., Luzuriaga, K., Sullivan, J. L., and Wara, D. W. (1992). Proposed definitions for *in utero* versus intrapartum transmission of HIV-1. *N. Engl. J. Med.* **327,** 1246–1247.

Burns, D. N., Kramer, A., Yellin, F., Fuchs, D. *et al.* (1991). Cigarette smoking: A modifier of human immunodeficiency virus type 1 infection. *J. Acq. Immuno. Syndr.* **4,** 76–83.

Burns, D. N., Landesman, S., Munez, L. R. *et al.* (1994). Cigarette smoking, premature rupture of membranes, and vertical transmission of HIV-1 among women with low CD4+ levels. *J. Acq. Immune Defic. Syndr.* **7,** 718–726.

Cao, Y., Krogstad, P., Korber, B., Koup, R., Muldoon, M., Macken, C., Song, J.-L. Jin, Z., Zhao, J.-Q., Clapp, S., Chen, I., Ho, D., Ammann, A. and the Ariel Project Investigators (1997). Maternal HIV-1 viral load and vertical transmission of infection: The Ariel Project for the prevention of HIV transmission from mother to infant. *Nat. Med.* **3,** 549–552.

Chandwani, S., Greco, M. A., Mittal, K., Antoine, C., Krasinski, K., and Borkowsky, W. (1991). Pathology and HIV expression in placenta of seropositive women. *J. Infect. Dis.* **163**, 1134–1138.

Center for Disease Control (1992). HIV/AIDS Surveillance Report, January 16.

Cheng-Myer, C., Seto, D., Tateno, M., and Levy, J. A. (1990). Biologic features of HIV-1 that correlate with virulence in the host. *Science* **240**, 80–82.

Chin, J. (1990). Current and future dimensions of HIV/AIDS pandemic in women and children. *Lancet* **336**, 221–224.

Choe, H., Farzan, M., Sun, Y., Sullivan, N., Rollins, B., Ponath, P. D., Wu, L., Mackay, C. R., LaRosa, G., Newman, W., Gerard, N., Gerard, C., and Sodroski, J. (1996). The beta chemokine receptors CCR3 and CCR5 facilitateinfection by primary HIV-1 isolates. *Cell* **85**, 1135–1815.

Cichutek, K., Merget, H., Norley, S., Linde, R., Kreuz, W., Gahr, M., and Kurth, R. (1992). Development of quasispecies of human immunodeficiency virus type 1 *in vivo. Proc. Natl. Acad. Sci. USA* **89**, 7365–7369.

Clavelli, T. A., and Rubenstein, A. (1990). Pediatric HIV-1 infection: A review. *Immunodefic. Rev.* **2**, 83–127.

Clemetson, D. B. A., Moss, G. B., Willerford, D. M. *et al.* (1993). Detection of HIV DNA in cervical and vaginal secretions. *J. Am. Med. Assoc.* **269**, 2860–2864.

Connor, R. I., and Ho, D. D. (1994). Human immunodeficiency virus type 1 varaints with increased replication capacity develop during asymptomatic stage before disease progression. *J. Virol.* **68**, 4400–4408.

Connor, R. L., Chen, B. K., Chen, S., and Landau, N. R. (1995). Vpr is required for efficient replication of human immunodeficiency virus type 1 in mononuclear phagocytes. *Virology* **206**, 935–944.

Contag, C. H., Ehrnst, A., Duda, J., Bohlin, A.-B., Lindgren, S., Learn, G. H., and Mullins, J. J. (1998). Mother-to-infant transmission of human immunodeficiency virus type 1 involving five envelope sequence subtypes. *J. Virol.* **71**, 1292–1300.

Cooper, R. R., Nugnet, P. R., Diaz, C. *et al.* (1996). After AIDS Clinical Trial 076: The changing pattern of zidovudine use during pregnancy, and the subsequent reduction in vertical transmission of human immunodeficiency virus in a cohort of infected women and their infants. *J. Infect. Dis.* **174**, 1207–1211.

Courgnaud, V., Laure, F., and Brossard, A. (1991). Frequent and early *in utero* HIV-1 infection. *AIDS Res. Human Retroviruses* **7**, 337–341.

David, F. J., Tran, H. C., and Autran, B. (1993). Maternal–fetal HIV transmission: Placental cells CD4 expression and permisivity to infection with HIV. *In* "Program and Abstracts of the VII International Conference on AIDS." Florence, Itlay. [Abstract M.A. 79]

Davison-Fairburn, B., Baskin, G., and Murphey-Corb, M. (1992). Maternal–fetal transmission of SIV in 2 species of monkeys. *Symp. Nonhum. Primate Models AIDS* **10**, 29.

De Rossi, A., Ades, A. E., Mammano, F., Mistro, A. D., Amadori, A., Giaquinto, C., and Chieco-Bianchi, L. (1991). Antigen detection, virus culture, polymerase chain reaction, and *in vitro* antibody production in the diagnosis of vertically transmitted HIV-1 infection. *AIDS* **5**, 15–20.

De Rossi, A., Ometto, L., Mammano, F., Zanotto, C. *et al.* (1992). Vertical transmission of HIV-1: Detection of virus in peripheral blood cells of children at birth. *AIDS* **6**, 117–1120.

Dean, M. M., Carrington, C., Winkler, G. A., Huttley, M. W. *et al.* (1996). Genetic restriction of HIV-1 infection and progression to AIDS by a deletion allele of the CKR5 structural gene. *Science* **273**, 1856–1862.

Devash, Y., Calvelli, T. A., Wood, D., Regan, K. J., and Rubinstein, A. (1990). Vertical transmission of human immunodeficiency virus is correlated with the absence of high affinity/acidity maternal antibodies to gp120 neutralizing domain. *Proc. Natl. Acad. Sci. USA* **87**, 3444–3449.

Dickover, R. E., Garraty, E. M., Herman, S. et al. (1996). Identification of levels of maternal HIV-1 RNA assocaited with risl of perinatal transmission. J. Am. Med. Assoc. **275**, 599–605.

Dougherty, J., and Temin, H. (1988). Determination of the rate of base-pair substitution and insertion mutations in retrovirus replication. J. Virol. **62**, 2817–2822.

Douglas, G. C., and King, B. F. (1992). Maternal–fetal transmission of HIV: A review of possible routes and cellular mechanisms of infection. Clin. Infect. Dis. **15**, 678–691.

Duliege, A. M., Felton S., and Gerdert, J. J. (1992). High risk of HIV-1 infection for first born twin: The role of intrapartum transmission. In "Advances in Vaccine Development Groups Fifth Annual Meeting of the National Cooperative Vaccine Development Groups for AIDS." Chaintly, VA, August–September. [Abstract]

Ehrnst, A., Lindergren, S., Dictor, M., Johanon, B., Sonnerborg, A., Czajkowski, J., Sundin, G., and Bohlin, A.-E. (1991). HIV in pregnant women and their offsprings: Evidence for late transmission Lancet **338**, 203–207.

Ekwo, E. E., Gosselwink, C. A., Woolson, R. et al. (1993). Risks for preterm rupture of amniotic membranes. Int. J. Epidemiol. **22**, 495–503.

European Collaboration Study (1988). Mother to child transmission of HIV infection. Lancet. **ii**, 1043–1046.

European Collaborative Study (1994). Cesarean section and risk of vertical transmission of HIV-1. Lancet **342**, 1464–1467.

Fang, G. et al. (1995). Maternal plasmahuman immunodeficiency virus type 1 RNA level: A determinant and projected threshold for mother-to-child transmission. Proc. Natl. Acad. Sci USA **92**, 12100–12104.

Fazely, F., Sharma, P. L., Fratazzi, C. et al. (1993). Simian immunodeficiency virus infection via amniotic fluid: a model to study fetal immunopathogenesis and prophylaxis. J. Acq. Immune Defic. Syndr. **6**, 107–114.

Felsenstein, J. (1989). Phylip phylogenetic inference package (version 3.2). Cladiatice **5**, 164–166.

Feng, Y., Broder, C., Kennedy, P., and Berger, E. (1996). HIV-1 entry cofactor: Functional cDNA cloning of a seven transmembrane, G protein-coupled receptor. Science **272**, 872–877.

Furtado, M. R., Kingsley, L. A., and Wolinsky, S. M. (1995). Changes in viral mRNA expression pattern correlate with a rapid rate of CD4+ T-cell number decline in human immunodeficiency virus type 1 infected individuals. J. Virol. **69**, 2092–2100.

Gartner, S., Markovits, P., Markovitz, D. M., Kaplan, M. H., Gallo, R. C., and Popovich, M. (1996). The role of mononuclear phagocytes in HTLV-III/LAV infection. Science **233**, 215–219.

Gendelman, H. E., Orenstein, J. M., Baca, L. M., Weiser, B., Burger, H., Kaltzer, D. C., and Meltzer, M. S. (1989). The macrophage in the persistence and pathogenesis of HIV infection. AIDS 3, 475–495.

Goedert, J. J. (1992). HIV exposed twins. Lancet **339**, 628.

Goedert, J. J., Duliege, A.-M., Amos, C. I., Felton, S., and Biggar, R. J. (1991). High risk of HIV-1 infection for first-born twins. Lancet **338**, 1471–1475.

Goedert, J. J., Mendez, H., Drummond, J. E. et al. (1989). Mother to infant transmission of human immunodeficiency virus type 1 associated with prematurity or tow anti gp120. Lancet. **2**, 1351–1354.

Gabuzda, D. H., Li, H., Lawrence, K., Vasir, B. S., Crawford, K., and Langhoff, E. (1994). Essential role of vif in establishing productive HIV-1 infection in peripheral blood T lymphocytes and monocytes/macrophages. J. Acq. Immune Defic. Syndr. **7**, 908–915.

Ganeshan, S., Dickover, R. E., Korber, B. T. M., Bryson, Y. J., and Wolinsky, S. M. (1997). Human immunodeficiency virus type 1 genetic evolution in children with divergent rates of development of disease. J. Virol. **71**, 663–677.

Goh, W. C., Rogel, M. E., Kinsey, C. M., Michael, S. C., Fultz, P. N., Nowak, M. A., Hahn, B. H., and Emerman, M. (1998). HIV-1 Vpr increases viral expression by manipulation of the cell cycle: A mechanism for selection of Vpr *in vivo*. *Nat. Med.* **4**, 65–71.

Hahn, T., Matala, E., Chappey, C., and Ahmad, N. (1999). Characterization of mother–infant HIV-1 gag p17 sequences associated with perinatal transmission. *AIDS Res. Hum. Retroviruses* **15**, 875–888.

Hasley, N. A., Coberly, J. S., Holt, E. *et al.* (1992). Sexual behavior, smoking, and HIV infection in Haitian women. *J. Am. Med. Assoc.* **267**, 2062–2066.

Halsey, N. A., Markham, R., Wahren, B., Boulos, R., Rossi, P., and Wigzell, H. (1992). Lack of association between antibodies to V3 loop peptides and maternal–infant HIV-1 transmission. *J. Acq. Immune Defic. Syndr.* **5**, 153–157.

Henin, Y., Mandelbrot, L., Henrion, R. *et al.* (1993). Virus excretion in the cervicovaginal secretions of pregnant women and non-pregnant HIV-infected HIV infected women. *J. Acq. Immune Defic. Syndr Hum. Retroviruses* **6**, 72–72.

Hira, S., Kamanga, J., Bhat, G. J. *et al.* (1989). Perinatal transmission of HIV-1 in Lusaka Zambia. *Br. Med. J.* **299**, 1250–1252.

Hoff, R., Berardi, V., Weiblan, B. *et al.* (1988). Seroprevalence of HIV among childbearing women: estimation by testing samples by blood from newborns. *N. Engl. J. Med.* **318**, 525.

Hwang, S. S., Boyle, T. J., Lyerly, H. K., and Cullen, B. R. (1991). Identification of the envelope V3 loop as the primary determinant of cell tropism. *Science* **253**, 71–74.

Italian Multiculture Study (1988). Epidemiology, clinical features and prognostic factors of pediatric HIV infection. *Lancet* ii, 1043–1046.

Iversen, A. K., Shpaer, E. G., Rodgiro, A. G. *et al.* (1995). Persistence of attenuated rev genes in HIV-1 infected asymptomatic individuals. *J. Virol.* **69**, 5743–5753.

Jauniax, E., and Burton, G. J. (1992). The effect of smoking in pregnancy on early placental morphology. *Obstet. Gynecol.* **79**, 645–648.

Kliks, S. C., Wara, D. W., Landers, D. V., and Levy, J. A. (1994). Features of HIV-1 that could influence maternal–child transmission. *J. Am. Med. Assoc.* **272**, 467–474.

Koenig, S., Gendelman, H. E., Orenstein, J. M., Dal Canto, M. C., Pezeshkpour, G. H., Yungbluth, M., Janotta, F., Aksamit, A., Martin, M. A., and Fauci, A. S. (1986). Detection of AIDS virus in macrophages in brain tissue from AIDS patients with encephalopathy. *Science* **233**, 1089–1093.

Konduri, K., Jones, T., Moore, E. *et al.* (1993). Peripartum blood secretion contact linked to vertical transmission of HIV infection. *Am. J. Obstet. Gynecol.* **1443**, 420.

Koot, M., Keet, I. P. M., Vos, A. H. V. *et al.* (1993). Prognostic value of human immunodeficiency virus type 1 biological phenotype for rate od CD4 cell depletion and progression to AIDS. *Ann. Intem. Med.* **118**, 681–688.

Koup, R. A., and Wilson, C. B. (1993). Clinical immunology of HIV-infected children. *In* "Pediatric AIDS: The Challenge of HIV Infection in Infants, Children, and Adolescents" (P. A. Pizzo and C. M. Wilfert, Eds.), pp. 158–165. Williams and Wilkins, New York.

Koup, R. A. (1997). Genetics of host resistance to HIV transmission: U.S. studies. In "Conference on Global Strategies for the Prevention of HIV Transmission from Mothers to Infants," pp. 91–93. Washington, DC.

Krivine, A., Firtion, G., Cao, L. *et al.* (1992). HIV replication during the first weeks of life. *Lancet* **339**, 1187–1189.

Laimore, M. D., Cuthbert, P. S., and Utley, L. I. (1993). Cellular localization of CD4 in human placenta: Implications for maternal-to-fetal transmission of human immunodeficiency virus *J. Immunol.* **151**, 1673–1681.

Lamers, S. L., Sleasman, J. W., She, J. X., Barrie, K. A., Pomeroy, S. M., Barrett, D. J., and Goodenow, M. (1994). Persistence of multiple maternal genotypes of human immunodeficiency virus type 1 in infants by vertical transmission. *J. Clin. Invest.* **93**, 380–390.

Lasky, L. A., Groopman, J. E., Fennie, C. W., Benz, P. M., Capon, D. J., Dowbenko, D. J., Nakamura, G. R., Nunes, W. M., Renz, M. E., and Berman, P. W. (1986). Neutralization of the AIDS retrovirus by antibodies to a recombinant envelope glycoprotein. *Science* **233**, 209–212.

Laure, F., Courgnand, V., Rouzioux, C. *et al.* (1988). Detection of HIV-1 DNA in infants and children by means of polymerase chain reaction. *Lancet* ii, 538–541.

Lepage, P., Vande Perre, P., Carad, M. *et al.* (1987). Postnatal transmission of HIV from mother to child. *Lancet* **2**, 400.

Lyman, W. D., Kress, Y., Kure, K., Rashbaum, W. K., Rubinstein, A., Soeiro, R. (1990). Detection of HIV in fetal central nervous system tissue. *AIDS* **4**, 917–920.

Lyman, W. D., Kress, Y., and Rashbaum, W. K. (1988). An AIDS-virus associated antigen localized in fetal tissues. *Ann. N. Y. Acad. Sci.* **549**, 258–259.

Lu, S.-L., Schacker, T., Musey, L., Shriner, D., Mcelrath, M. J., Corey, L., and Mullins, J. I. (1997). Divergent patterns of progression to AIDS after infection from the same source: Human immunodeficiency virus type 1 evolution and antiviral responses. *J. Virol.* **71**, 4284–4295.

Mangione, E. J. (1994). Zidovudine reduces maternal transmission of HIV. *Colorado Med.* **91**, 191.

Matala, E., Yedavalli, V. R. K., and Ahmad, N. (1999). Biological characterization of HIV-1 envelope V3 region isolates assocaited with maternal-fetal transmission (manuscript in preparation).

Mathews, T. J., Langlois, A. J., Robey, W. G., Chang, N. T., Gallo, R. C., Fischinger, P. J., and Bolognesi, D. P. (1986). Restricted neutralization of divergent human T-lymphotrophic virus type III isolates by antibodies to the major envelope glycoprotein. *Proc. Natl. Acad. Sci. USA* **83**, 9709–9713.

Mayaux, M.-J., Blanche, S., Rouzioux, C. *et al.* (1995). Maternal factors associated with perinatal HIV-1 transmission: The French cohort study: 7 years of follow up observation. *J. AIDS Hum. Retroviruses* **8**, 188–194.

McGaughey, R. W., Usaha, V., Gallicano, G. I., Capco, D. G., and Jacobs, B. L. (1992). "Azidothymidine (AZT), A Viral Reverse Transcriptase Inhibitor, Disrupts Early Development in Mice." Special Poster Session, American Society for Cell Biology.

McNearny, T., Westervelt, P., Thielan, B. J., Trowbridge, D. B., Garcia, J., Whittler, R., and Ratner, L. (1990). Limited sequence heterogeneity among biologically distinct human immunodeficiency virus type 1 isolates from Individuals involved in a clustered infectious outbreak. *Proc. Natl. Acad. Sci. USA* **87**, 1917–1922.

Mok, J. Q., Giaquinto, C., DeRossi, A. *et al.* (1987). Infants born to mothers seropositive for human immunodeficiency virus-preliminary findings from a multiculture European study. *Lancet* i, 1164–1168.

Morgan, G., Wilkins, H. A., Pepin, J., Jobe, O., Brewster, D., and Whittle, H. (1990). AIDS following mother-to-child transmission of HIV-2. *AIDS* **4**, 879–882.

Muary, W., Potts, B. J., and Rabson, A. B. (1989). HIV-1 infection of first trimester and term human placental tissue, a possible mode of maternal-fetal transmission. *J. Infect. Dis.* **460**, 583–588.

Mulder-Kampinga, G. A., Kuiken, C., Dekker, J., Scherpbier, H. J., Boer, K., and Goudsmit, J. (1993). Genomic human immunodeficiency virus type 1 RNA variation in mother and child following intrauterine virus transmission. *J. Gen. Virol.* **74**, 1747–1756.

Mulder-Kampinga, G. A., Simonon, A., Kuiken, C. L., Dekker, J., Scherpbier, H. J., van de Perre, P., Boer, K., and Goudsmit, J. (1995). Similarity in env and gag genes between genomic RNAs of human immunodeficiency virus type 1 (HIV-1) from mother and infant is unrelated to time of HIV-1 RNA positively in the child. *J. Virol.* **69**, 2285–2296.

Myers, G., Korber, B., Hahn, B. H., Jeang, K-T., Mellors, J. W., McCutchan, F. E., Henderson, L. E., Pavlakis, G. N. (1995). "Human Retroviruses and AIDS Database." Theoretical Biology, Los Alamos National Laboratory, Los Alamos, NM.

Nair, P., Alger, L., Hines, S., Seiden, S., Hebel, R., and Johnson, J. P., (1993). Maternal and neonatal characteristics associated with HIV infection in infants of seropositive women *J. Acq. Immune Defic. Syndr.* **6**, 298–302.

Narwa, R., Roques, P., Courpotin, C., Parnrt-Mathieu, F., Boussin, F., Roane, A., Marce, D., Lasfargues, G., and Dormont, D. (1996). Characterization of human immunodeficiency virus type 1 p17 matrix protein motifs associated with mother-to-child transmission. *J. Virol.* **70,** 4474–4483.

Nei, M., and Gojobori, T. (1986). Simple methods for estimating the number of synonymous and nonsynonymous nucleotide substitutions. *Mol. Biol. Evol.* **3,** 418.

Newell, M.-L., and Peckham, C. S. (1993). Risk factors for vertical transmission of HIV-1 and early markers of HIV-1 infection in children. *AIDS* **1,** S91–S97.

Nicholas, S. W., Sondhemier, D. L., Willoughby, A. D., Yaffe, S. J., and Katz, S. L. (1989). Human immunodeficiency virus infection in childhood, adolesence, and pregnancy: A status report and national research agenda. *Pediatrics* **83,** 293–308.

Nieman, R. B., Fleming, J., Coker, R. J. *et al.* (1993). The effect of cigarette smoking on the development of AIDS in HIV-1-seropositive individuals. *AIDS* **7,** 705–710.

Oleske, J., Minnefor, A., Cooper, R, Jr. *et al.* (1993). Immune deficiency syndrome in children. *J. Am. Med. Assoc.* **249,** 2350.

Pang, S., Shlesinger, Y., Darr, E. S., Moudgh, T., and Ho, D. D. (1992). Rapid generation of sequence variation during primary HIV-1 infection. *AIDS* **6,** 453–460.

Parekh, B. S., Shaffer, N., Pau, C.-P., Abrams, E., Thomas, P., Pollack, H., Bamji, M., Kaul, A., Schochetman, G., Rogers, M., George J. R. and the NYC Collaborative Group (1991). Lack of correlation between maternal antibodies to V3 loop peptides of gp120 and perinatal HIV-1 transmission. *AIDS* **5,** 1179–1184.

Pasquier, C., Cayrou, C., Blancher, A., Tourne-Petheil, C., Berrebi, A., Tricoire, J., Puel, J., and Izopet, J. (1998). Molecular evidence for mother-to-child transmission of multiple variants by analysis of RNA and DNA sequences of human immunodeficiency virus type 1. *J. Virol.* **72,** 8943–8948.

Peterson, B. D., Poresz, B. J., and Loeb J. A. (1988). Fidelity of HIV-1 reverse transcriptase. *Science* **242,** 1168–1167.

Potts, B. J., Maury, W., and Martin, M. A. (1990). Replication of HIV-1 in primary monocyte cultures. *Virology* **175,** 465–476.

Report of a Consensus Workshop, Siena (Italy), January 17–18 (1992). Maternal factors involved in mother-to-child transmission of HIV-1. *J. Acq. Immune Defic. Syndr.* **992**(5), 1019–1029.

Richman, D. D., and Bozzette, S. A. (1994). The impact of syncytium-inducing phenotype of human immuodeficiency virus on disease progression. *J. Infect. Dis.* **169,** 968–974.

Roberts, J. D. (1988). Fidelity of two retroviral reverse transcriptase during DNA-dependent synthesis *in vitro. Mol. Cell. Biol.* **9,** 469–476.

Rogers, M. F., Ou, C.-Y., Rayfield, M. *et al.* (1989). Use of the polymerase chain reaction for early detection of the proviral sequences of HIV in infants born to seropositive mothers. *N. Engl. J. Med.* **320,** 1649–1654.

Rossi, P., Moschese, V., Broliden, P. A. *et al.* (1989). Presence of maternal antibodies to human immunodeficiency virus type 1 envelope glycoprotein gp120 epitopes correlates with uninfected status of children born to seropositive mothers. *Proc. Natl. Acad. Sci. USA* **86,** 8055–8058.

Rouzioux, C., Costagliola, D., and Burgard, M. (1993). Timing of mother-to-child transmission depends on maternal AIDS. *AIDS* **7,** S49–S52.

Royce, R. A., and Winkelstein, W., Jr. (1990). HIV infection, cigarette smoking and CD4+ T-lymphocyte counts: Preliminary results from San Francisco Men's health study. *AIDS* **4,** 327–333.

Rubenstein, A., Sicklick, M., Gupta A. *et al.* (1993). Acquired immunodeficiency with reserved T4/T8 ratios in infants born to promiscuous and drug-addicted mothers. *J. Am. Med. Assoc.* **249,** 2350.

Ryder, R. W., Nsa, W., Hassig, S. E., Behets, F., Rayfi Áeld, M., Kungola, E., Nelson, A., Mulenda, U., Francis, H., Mwandagalirwa, K., Davachi, F., Rogers, M., Nzilambi, N.,

Greenberg, A., Mann, J., Quinn, T. C., Piot, P., and Curran, J. W. (1988). Perinatal transmission of human immunodeficiency virus type 1 to infants of seropositive women in Zaire. *N. Engl. J. Med.* **320**, 1637–1642.

Salvatori, F., Romiti, M. L., Colognesi, C., Orlandi, P., Tresoldi, E., Amoroso, A., Plebani, A., Tovo, P. A., Rossi, P., and Scaralatti, G. (1998). HIV-1 coreceptor usage and CCR5 defective allele in mother-to-child transmission. In "The Immunologists, 10th International Congress of Immunology." New Delhi.

Sato, H., Shiino, T., Kodaka, N., Taniguchi, K., Tomita, Y., kato, K., Miyakuni, T., and Takebe, Y (1999). Evolution and biological characterization of human immunodeficiency virus type 1 subtype E gp120 V3 sequences following horizontal and vertical transmission in a single family. *J. Virol.* **73**, 3551–3559.

Savage, S. M., Donaldson, L. A., Cherian, S. *et al.* (1991). Effects of cigarette smoke on the immune response: Chronic exposure to cigarette smoke inhibits surface immunoglobulins mediated responses in B-cells. *Toxicol. Appl. Pharmacol.* **111**, 523–539.

Scarlatti, G., Albert, J., Rossi, P. *et al.* (1993). Mother-to-child transmission of HIV-1: Correlation with neutralizing antibodies against primary isolates. *J. Infect. Dis.* **168**, 207–210.

Scarlatti, G., Leitner, T., Hapi, E., Wahlberg, J. *et al.* (1993). Comparison of variable region 3 sequences of human immunodeficiency virus type 1 from infected children with the RNA DNA sequences of the virus population of their mothers. *Proc. Natl. Acad. Sci. USA* **90**, 1721–1725.

Scarlatti, G., Valtez, L., Plebani, A. *et al.* (1991). Polymerase chain reaction, virus isolation and antigen assay of HIV-1 to antibody positive mothers and their children. *AIDS* **5**, 1173–1178.

Schiffman, M. H., Haley, N. J., Felton, J. S. *et al.* (1987). Biochemical epidemiology of cervical neoplasia: Measuring cigarette smoke constituents in cervix. *Cancer Res.* **47**, 3886–3888.

Schnittman, S. M., Denning, S. M., Greenhouse, J. J. *et al.* (1990). Evidence for suceptibility of intrathymic T-cell precursors and their progeny carrying T-cell antigen receptor phenotypes TCR+ and TCR+ to human immunodeficiency virus infection: A mechanism for CD4+ (T4) lymphocyte depletion. *Proc. Natl. Acad. Sci. USA* **87**, 7727–7731.

Schwartz, D. A., and Nahmias, A. J. (1991). Human immunodeficiency virus and placenta: Current concepts of vertical transmission in relation to other viral agents. *Ann. Clin. Lab. Sci.* **21**, 264–274.

Scott, G. M., Buck, B. E., Leterman, J. G. *et al.* (1985). Acquired immunodeficiency syndrome in infants. *N. Engl. J. Med.* **310**, 76.

Scott, R. J. (1994). "Danforths Obsterics and Gynecology," 7th ed. Lippincott, Philadelphia.

Schuitmaker, H. N., Kootstra, N. A., De Goede, R. E. Y. *et al.* (1991). Monocytotropic human immunodeficiency virus 1 variants detectable in all stages of HIV infection lack T cell line tropism and syncytium inducing ability in primary T cell culture. *J. Virol.* **65**, 356–363.

Schuitmaker, H. N., Koot, H. M., Kootstra, N. A. *et al.* (1992). Biological phenotype of HIV-1 clones at different stages of infection: Progression of disease isassociated with a shift from macrophage-tropic to T-cell tropic virus population. *J. Virol.* **66**, 1354–1360.

Shaffer, N., Chauachoowong, R., Mock, P. A. *et al.* and Bangkok Collaborative Perinatal Transmission Study Group (1999). Short-course zidovudine for perinatal HIV-1 transmission in Bangkok, Thailand: A randomised controlled trial. *Lancet* **353**, 773–780.

Sheppard, H. W., Lang, W., Ascher, M. S., Vittinghoff, E., and Winkelstein, W. (1993). The characterization of nonprogressors long term HIV-1 infection with stable CD4+ T-cell levels. *AIDS* **7**, 1159–1166.

Shioda, T., Levy, J. A., and Cheng-Mayer, C. (1991). Macrophage and T cell line tropism of HIV-1 are determined by specific regions of the envelope gp120 gene. *Nature* **349**, 167–169.

Siegel, G., Schafer, A., and Ungar, M. (1990). HIV-1 in fetal organs and in embryonic placental HIV-1 positive mothers. In "Program and Abstracts of the VIth International Conference on AIDS." San Francisco. [Abstract FB 445]

Siena Consensus Workshop II (1995). Strategies for prevention of perinatal transmission of HIV infetion. *J. Acq. Immune Defic. Syndr Hum. Retroviruses* **8**, 161–175.

Siliciano, R. F., Lawton, T., Knall, R. W., Karr, R. W., Berman, P., Gregory, T., and Reinherz, E. L. (1988). Analysis of host–virus Interactions in AIDS with anti-gp120 T cell clones: Effect of HIV sequence variation and a mechanism for CD4+ cell depletion. *Cell* **54**, 561–575.

Sott, G. B., Fischl, M. A., Khmas, N. *et al.* (1985). Mothers of infants with acquired immunodeficiency syndrome: evidence for both symptomatic and asymptomatic carriers. *J. Am. Med. Assoc.* **253**, 363–366.

Sprecher, S., Soumenkoff, G., Puissant, F. *et al.* (1986). Vertical transmission of HIV in 15-week fetus. *Lancet* **2**, 288–289.

Stanley, S. K., McCune, J. M., Kaneshima, H., Justement, J. S., *et al.* (1993). Human immunodeficiency virus infection of human thymus and disruption of thymic microenvironment in the SCID-hu mouse. *J. Exp. Med.* **178**, 1151–1163.

St. Louis, M. E., Kamenga, M., Brown, C. *et al.* (1993). Risk for perinatal HIV transmission according to maternal immunologic, virologic, and placental factors. *J. Am. Med. Assoc.* **269**, 2853–2859.

Sova, P., van Rannst, M., Gupta, P., Balachandran, R., Chao, W., Itescu, S., Mckinley, G., and Volsky, D. J. (1995). Conservation of an intact human immunodeficiency virus type 1 vif gene *in vitro* and *in vivo*. *J. Virol.* **69**, 2557–2564.

Smoking and Immunity (1990). *Lancet* **335**, 1561–1562.

Sprecher, S., Soumenkoff, G., Puissant, F., and Degueldre, M. (1986). Vertical transmission of HIV in 15 week fetus. *Lancet* **2**, 288–289.

Tersmette, M., Gruters, R. A., deWolf, F. *et al.* (1989). Evidence for a role of virulent HIV-1 variants in the pathogenesis of AIDS: Studies on sequential HIV isolates. *J. Virol.* **63**, 2118–2125.

Tersmette, M., Lange, J. M. A., deGoeda, R. E. Y. *et al.* (1989). Association between biological properties of human immunodeficiency virus variants and risk for AIDS and AIDS mortality. *Lancet* **1**, 983–985.

Tersmette, M., Rudd, E. Y., deGoeda, B. J. M. *et al.* (1988). Differential syncytium-inducing capacity of HIV isolates: Frequent detection of syncytium-inducing isolates in patients with AIDS and AIDS related complex. *J. Virol.* **62**, 2026–2032.

Tibaldi, C., Tovo, P. A., Ziarati, N. *et al.* (1993). Asymptomatic women at high risk of vertical HIV-1 transmission to their fetuses. *Br. J. Obstet. Gynaecol.* **100**, 334–337.

Toohey, K., Wehrly, K., Nisho, J., Perryman, S., and Chesebro, B. (1995). HIV env V1 and V2 regions influence replication efficiency in macrophages by affecting virus spread. *Virology* **213**, 70–79.

Turner, B. J., Hauck, W. W., Fanning, T. R., and Markson, L. E. (1997). Cigarette smoking and matenal-child transmission. *J. Acq. Immune Defic. Syndr. Hum. Retroviruses* **14**, 327–337.

Van Dyke, R. B. (1993). Opportunistic infections in pediatric HIV disease. *In* "Pediatric AIDS: Clinical, Pathologic, and Basic Science Perspectives" (W. D. Lyman and A. Rubinstein. Eds.), pp. 129–157. New York Academy of Sciences.

van't Wout, A. B., Kootstra, N. A., Mulder-kampinga, G. A., Albrecht-van Lent, N., Scherpbier, H. J., Veenstra, J., Boer, K., Coutinho, R. A., Miedema, F., and Schuitmaker, H. (1994). Macrophage-tropic variants initiate human immunodeficiency virus type 1 infection after sexual, parenteral, and vertical transmission. *J. Clin. Invest.* **94**, 2060–2067.

von Schwedler, U., Kornbluth, R. S., and Trono, D. (1994). The nuclear localization signal of the matrix protein of human immunodeficiency virus type 1 allows the establishment of infection in macrophages and quiescent T lymphocytes. *Proc. Natl. Acad. Sci. USA* **91**, 6992–6996.

Viscarello, R. R., Cullen, M. T., DeGennaro, N. J., and Hobbins, J. C. (1992). Fetal blood sampling in human immunodeficiency virus-seropositive women before elective midtrimester termination of pregnancy. *Am. J. Obstet. Gynecol.* **1677**, 1075–1079.

Wattler, R. D., Usaha, V., Gallicano, G. I., Shors, S. T., Capco, D. G., Jacobs, B. L., and McGaughey, R. W. (1993). Preimplantation development in mice is enhanced and lethal onset of AZT decreased by mouse-alpha interferon induction of an interferon-specific protein. Annual meeting of American Society of Cell Biology.

Weiblen, B. J., Lee, F. K., Cooper, E. R., Landesman, S. H. *et al.* (1990). Early diagnosis of HIV infection in infants by detection of IgA HIV antibodies. *Lancet* **335**, 988–990.

Weinbreck, P., Loustand, V., Denis, F. *et al.* (1988). Postnatal transmission of HIV infection. *Lancet* **1**, 482.

Wiktor, S. Z., Ekpini, E., Karon, J. M., Nkengasong, J. *et al.* (1999). Short-course oral zidovudine for prevention of mother-to-child transmission of HIV-1 in Abidjan, Cote of Ivoire: A randomized trial. *Lancet* **353**, 781–785.

Wiley, R. L., Ross, E. K., Buckler-White, A. J., Theodore, T. S., Earl, P. L., and Martin, M. A. (1989). Functional interactions of constant and variable domains of HIV that is critical for infectivity. *J. Virol.* **62**, 139–147.

Wolfs, T. F.W., Zwart, G., Bakker, M., and Goudsmit, J. (1992). HIV-1 genomic RNA diversification following sexual and parental virus transmission. *Virology* **189**, 103–110.

Wolinsky, S. M., Korber, B. T. M., Neumann, A. U., Daniels, M., Kunstman, K. J., Whetsell, A. J., Furtado, M. R., Cao, Y., Ho, D. D., Safrit, J. T., and Koup, R. A. (1996). Adaptive evolution of human immunodeficiency virus-type 1 during the natural course of infection. *Science* **272**, 537–542.

Wolinsky, S. M., Wike, C. M., Korber, B. T. M., Hutto, C., Parks, W. P., Rosenblum, L. L., Kuntsman, K. J., Furtado, M. R., and Munoz, I. L. (1992). Selective transmission of human immunodeficiency virus type-1 variants from mother to infants. *Science* **255**, 1134–1137.

World Health Organization Global Program on AIDS (1992). Current and future dimensions of HIV-1 AIDS pandemic: A capsule summary. WHO-GPA January WHO/GPA/RES/SFI/92.1.

Yedavalli, V. R. K., Chappey, C., Matala, E., and Ahmad, N. (1998a). Conservation of an intact *vif* gene of human immunodeficiency virus type 1 during maternal-fetal transmission. *J. Virol.* **72**, 1092–1102.

Yedavalli, V. R. K., Chappey, C., and Ahmad, N. (1998b). Maintenance of an intact human immunodeficiency virus type 1 *vpr* gene following mother-to-infant transmission. *J. Virol.* **72**, 6937–6943.

Zhang. J., MacKenzie, L. Q., Cleland, A., Hoims, E. C., Leigh Brown, A. J., and Simmonds, P. (1993). Selection for specific sequences in the external envelope protein of human immunodeficiency virus type 1 upon primary infection. *J. Virol.* **67**, 3345–3356.

Zhu, T., Mo, H., Wang, N., Nam, D. S., Cao, Y., Koup, R. A., and Ho D. D. (1993). Genotypic and Phenotypic characterization of HIV-1 in patients with primary infection. *Science* **261**, 1179–1181.

Ziegler, J. B., Cooper, D. A., Johnson, R. *et al.* (1985). Postnatal transmission of AIDS associated retrovirus from mother to infant. *Lancet* **1**, 896.

Mao-Yuan Chen* and Chun-Nan Lee†

*Department of Internal Medicine
National Taiwan University Hospital
Taipei, Taiwan 10016
†School and Graduate Institute of Medical Technology
College of Medicine
National Taiwan University
Taipei, Taiwan 10016

Molecular Epidemiology of HIV-1: An Example of Asia

I. Introduction

The human immunodeficiency virus type 1 (HIV-1) epidemic continues to grow in countries where high prevalence rates already exist and has spread to areas where extremely low levels of infection were found before 1990. It was estimated that in the world over 30 million people were infected with HIV-1 by the beginning of 1998 (UNAIDS, 1998). Over two-thirds of all the people infected with HIV-1 live in sub-Saharan Africa. The HIV-1 epidemic in Africa, which was initially most severe in areas stretching from West Africa across to the Indian Ocean, in the 1990s marched gradually to the southern countries of Africa. The HIV-1 seroprevalence surveys among pregnant women in South Africa showed seroprevalence that was below 5% before 1992 but had reached 41.2% in late December 1998 (Wilkinson *et al.*, 1999). HIV-1 in sub-Saharan Africa has mostly spread through heterosexual contact. But the HIV-1 epidemic varies between African countries as

Advances in Pharmacology, Volume 49

the result of different sexual behaviors and sociocultural and economic factors. The way HIV-1 moves through the countries can also influence the magnitude of the epidemic and the predominance of certain HIV-1 subtypes.

HIV-1 infection rates appear to be dropping in Western Europe and North America. Transmission through homosexual contact has reduced greatly. However, new infections among intravenous drug users (IDUs) and through heterosexual contact are still increasing in some disadvantaged sections of society. In Latin America and the Caribbean, rising infection rates in women and in poorer and less educated members of the population are an ominous sign for a future epidemic. But the most surprising recent HIV-1 epidemic has occurred in Eastern Europe. An explosive increase in the number of HIV-1-infected individuals has been documented in Ukraine, Russia, and Belarus since 1995. The overlap of IDUs and sex workers and the dramatic increase in sexual transmitted diseases (STDs) are warning signs of a widespread HIV-1 epidemic entering the general population in this area.

Asia, where the two countries, China and India, with the largest populations in the world are located had low HIV-1 infection rates in 1980s. By the late 1980s, Thailand was the first Asian country to experience an explosive spreading of HIV-1 among IDUs and female commercial sex workers (CSWs) (Weniger *et al.*, 1991). Shortly afterward, the HIV epidemic among IDUs expanded to neighboring areas, including Myanmar, the Yunnan province in China, and the northeastern Indian states (Weniger, 1996). In the early 1990s, a sharp rise in HIV-1 infections among CSWs and STD patients not related to Thailand epidemic was detected in western and southeastern India (Bollinger *et al.*, 1995). By the late 1990s, successive transmission of HIV-1 had entered the low-risk general population through heterosexual contact in both Thailand and India, with the estimated number of HIV-1 infections reaching 0.78 and 4 million respectively. The epidemic now marches to China and countries bordering Thailand. If this trend continues, Asia, with 5 times the population, will soon surpass sub-Saharan Africa in the number of people living with HIV-1.

HIV-1 can be divided into two groups: M (major) and O (outlier). Group M can be further classified into at least 10 subtypes, designated by letters A through J. An HIV-1 subtype is usually determined by the similarities and differences of nucleotide sequences in the *env* or *gag* gene. Phylogenetic analysis of circulating HIV-1 subtypes is of great value in tracing when and how the virus is introduced into a specific area (Weniger *et al.*, 1994). Rapid heteroduplex mobility assays (HMA) can also be applied to classify subtypes and estimate genetic diversity without expensive and labor-intensive DNA sequencing (Delwart *et al.*, 1993). In Asia, V3-loop peptide-enzyme immunoassay (PEIA) is widely used for subtype identification (Cheinsong-Popov *et al.*, 1994). This technique can be used to screen samples on large scale, but is limited by cross-reactivity when several HIV-1 subtypes cocirculate in the population studied (Nkengasong *et al.*, 1998). Further-

TABLE I Seroprevalence Rates (in Percentages) among Different Risk Groups in Asian Countries in the Advanced Stage of the HIV-1 Epidemic as Compared with Myanmar

Risk groups	Thailand	India	Cambodia	Myanmar
IDUs	33.1	55.7	—	72.2
CSWs	13	27.3	39.3	21
STDs	6.8	33	—	7
Pregnant women	1.3[a]	4.3	0.8	0.8
	1.7[b]	3.4	3.5	1.0

Abbreviations: IDUs, intravenous drug users; CSWs, commercial sex workers; STDs, sexually transmitted diseases.
Source: UNAIDS: Report on the global HIV epidemic (1998).
[a] Major urban areas.
[b] Outside major urban areas.

more, intrasubtype genetic divergence cannot be determined by this method. Nevertheless, PEIA is suitable for countries where one or two subtypes are found and the HIV-1 epidemic is explosive and of recent onset. However, the recent discovery of recombinant HIV-1 strains in China (Shao *et al.,* 1998b) can be an obstacle to the study of molecular epidemiology. Simple PEIA is misleading under this circumstance. Genotyping by simple nested polymerase chain reaction (PCR) using subtype-specific primers encompassing both *gag* and *vpu* or *env* may be a good solution to this problem (Chen, 1998; Kondo *et al.,* 1998; Lee *et al.,* 2000).

The geographic distribution of HIV-1 subtypes is greatly affected by political, ethnic, social, and economic interactions between neighboring countries. Through various interactions, the virus may enter different susceptible populations by chance. Molecular epidemiology is important in understanding how the virus was introduced into a country and then spread among risk groups. The nature of the HIV-1 epidemic in many Asian countries had changed significantly since the last review in 1994 (Weniger *et al.,* 1994). In this chapter, we review the published reports concerning the molecular epidemiology of HIV-1 infection in several Asian countries, grouped according to the latest data concerning HIV prevalence rates. As shown in Table I, the categorizations of the HIV-1 epidemic are based mainly on the seroprevalence rates of pregnant women, which can best reflect the extent of widespread HIV-1 infection in general population.

II. Countries in the Advanced Stage of the HIV-1 Epidemic

A. Thailand

The epidemiology of HIV-1 infection in Thailand had been well characterized since its beginning. Similar to most other Asian countries, the first case, detected in 1985, was a homosexual man. However, Thailand was the first Asian country to experience an explosive increase in HIV-1 incidence, which climbed from about 1% among IDUs at the start of 1988 to 32–43% by August–September 1988 in Bangkok (Weniger *et al.*, 1991). National surveys conducted 5 years later showed that the prevalence rate among IDUs stabilized at 35–40% (Brown *et al.*, 1994), perhaps the result of needles not being shared as frequently. Unfortunately, the explosive spread of HIV-1 among IDUs was followed by a second wave of infection among lower class brothel-based prostitutes in northern Thailand (Chiang Mai). The first national serosurvey detected a seroprevalence rate of 44% among female prostitutes in this area in June 1989. The HIV-1 epidemic quickly spread to CSWs in other areas. As anticipated, HIV-1 was transmitted successively to male clients and then to the girlfriends and wives of these men. The high mobility of the labor and sex-worker populations and the frequent premarital and extramarital sexual contact with the relative small population of CSWs in Thailand provide a possible explanation for the speed with which HIV-1 seroprevalence has grown. By mid-1993, HIV-1 prevalence reached 35% among IDUs, 29% among female sex workers in brothels, 8% among male patients with sexually transmitted disease, 4% among military conscripts, and 1.4% among pregnant woman. These data show that the HIV-1 epidemic had firmly established itself in the general population.

The viral strains circulating in Thailand prior to 1988 were genetically similar to the subtype B viruses found in the Americas and in Europe (Kalish *et al.*, 1994). However, two distinct genotypes were found to segregate by mode of transmission after the outbreak of the HIV-1 epidemic in Thailand (Ou *et al.*, 1993). Subtype B′ infected approximately 75% of IDUs while 86% of heterosexually infected patients had subtype E. Although subtype B′ is clustered phylogenetically with subtype B, it is genetically distinct from the other subtype B viruses found in the Americas and in Europe. It is not known how the two Thai strains were introduced into separate dynamic, high-risk subgroups that spread the virus more rapidly. Both genotypes show similar interperson nucleotide divergence rates, implying that they may have been introduced into Thailand at much the same time. The source of subtype E infection among CSWs in central Thailand (including Bangkok) might have come from northern Thailand. Studies on a limited number of IDUs in Chiang Mai and Chiang Rai showed that subtype E infection prevailed during both 1991 and 1994–1995, in contrast to central Thailand where subtype B′ prevailed among IDUs (Ou *et al.*, 1993; Subbarao *et al.*, 1998). It is possible that HIV-1 subtype E infected northern Thai IDUs initially and then spread to CSWs. This theory is compatible with the fact that high HIV-1 prevalence rates were first identified among both IDUs and female CSWs in northern Thailand and nearly two-thirds of the HIV-1-infected CSWs were born in this area (Subbarao *et al.*, 1998).

Low genetic variations were found in nucleotide sequences of Thai subtypes B' and E in 1991. However, genetic diversity increased over time. The sequence divergence that occurred in Thailand had been used as the standard to estimate roughly the duration of the rapidly spreading HIV-1 epidemics. Monitoring the molecular epidemiology of HIV-1 infection in Thailand also showed that new infections with subtype E viruses among IDUs in Bangkok increased significantly from 2.6% in 1988–1989 to 43.8% in 1992–1993 (Wasi et al., 1995; Kalish et al., 1995). The chronological shift was assumed to result from increasing sexually acquired infections and the reduction in needle-sharing among IDUs. Among heterosexuals, the proportion of subtype E infection increased from 86% in 1991 to 98% in 1994–1995 (Subbarao et al., 1998). At present, most of the circulating HIV-1 viruses in Thailand are subtype E because of the much larger heterosexual population. In this way, heterosexual prostitution is responsible for the Thai HIV-1 epidemic.

B. India

India is now considered the country with the largest number (an estimated 4 million people infected with HIV) of people infected with HIV-1 in the world. The HIV-1 epidemic started in the mid-1980. By 1988–1989, seroprevalence rates showed that HIV-1 had already entered several risk groups including CSWs, STD clinic patients, IDUs, and paid blood donors (Bollinger et al., 1995). Commercial blood donors were probably infected through sexual contact with female CSWs or by contaminated equipment used in plasma extraction (Navarro et al., 1988). From 1990 to 1995, the prevalence rate had risen from less than 10% to between 40–50% among female sex workers in Bombay, India (Lalvani et al., 1996). As of December 1997, heterosexual contact was the major mode of transmission (74.1%) among 5145 reported AIDS cases (Mishra et al., 1998). At present, HIV-1 has spread beyond high-risk groups into the general population (Gangakhedkar et al., 1997). HIV-2 was found to have been extensively transmitted along with HIV-1 in Bombay (Grez et al., 1994). However, its present status in India remains unclear.

The early isolates from western India were closely related to subtype C, as determined by phylogenetic analysis (Grez et al., 1994). The specific subtype could have been introduced from South Africa, where subtype C prevails, because of a close historic relationship between the two countries. A later analysis also found that the prevailing subtype in western India and New Delhi was subtype C (Tripathy et al., 1996). In a previous review (Weniger et al., 1994), the intrasubtype divergence rates suggested that HIV-1 had circulated longer in India than in Thailand. Subtype C is also transmitted rapidly through heterosexual contact in South Africa (van Harmelen et al., 1999). But high intrasubtype diversity in South Africa suggests that multiple introductions of subtype C occurred rather than through the clonal epidemic developed in India. Other subtypes had been detected in India, such as subtype A in Bombay and the American/European subtype B in southcentral India (Baskar *et al.*, 1994), but they were not known to represent major circulating subtypes. Among IDUs living in Manipur, a state bordering Myanmar, both North American subtype B and Thai subtype B' were discovered (Panda *et al.*, 1996). The absence of subtype C is puzzling since subtype C was supposed to have been introduced into the Yunnan province in China and possibly into neighboring Northern Myanmar from India (Fig. 1). However, molecular epidemiology results in India could be biased by small sample size.

C. Cambodia

HIV-1 infection was not detected in Cambodia until 1991. However, the HIV-1 epidemic in Cambodia seems to have spread through heterosexual contact, with a speed similar to that in Thailand. The median HIV-1 rates in seroprevalence surveys were 37.9% in commercial sex workers, 8.15% in soldiers, and 2.6% in pregnant women (Phalla *et al.*, 1999). In contrast to neighboring countries, there is little evidence for transmission by IDUs.

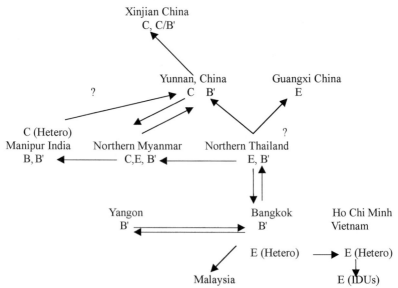

FIGURE I   The spread of HIV-1 infection along drug-trafficking routes and the distribution of HIV-1 subtypes among IDUs in Asia.

The phylogenetic analysis of *env* C2/V3 sequences obtained from nine HIV-1-infected patients showed that all Cambodian HIV-1 strains fell into the subtype E cluster (Kusagawa *et al.*, 1999). The absence of monophylogenetic clustering of sequences and the comparable level of contemporaneous divergence relative to that of Thailand did not support the possibility of a single founder strain. The most likely reason for the prevalence of HIV-1 subtype E in Cambodia is the cross-border population migration with neighboring countries.

III. Countries with a High Potential for Rapid Heterosexual Transmission

Countries in this group have high levels of HIV infection among IDUs. Different modes of interaction between IDUs and heterosexuals may influence the transmission of HIV-1 infection.

A. Myanmar

HIV-1 was detected in 23.7% of 228 IDUs in 1989 while none was found among 46 IDUs in 1988. This was just 1 year after an HIV-1 outbreak among IDUs in Thailand. HIV serosurveillance conducted in 1992–1993 revealed high HIV-1 prevalence rates ranging from 27 to 95% among IDUs in different areas of Myanmar (Htoon *et al.*, 1994). Seroprevalence rate also rose from 1.8% in September 1992 to 6.8% in September 1993 among CSWs in Yangon. The higher positive rate (2 to 12%) among antenatal women in areas close to the border with Thailand indicates cross-border transmission. Myanmar has now approached the advanced level of the HIV-1 epidemic (Table I).

In 1991, six subtype B' viruses and one subtype E virus were identified by phylogenetic analysis of C2/V3 regions in seven specimens collected from Myanmar (Cassol *et al.*, 1996). All four IDUs were infected by subtype B'. Both subtypes E and B' were found among heterosexual patients. A nationwide study using V3 PEIA in 1995 showed that only 1 among 47 IDUs and 2 among 18 heterosexual patients (including CSWs) were infected with subtype E in Yangon (Kusagawa *et al.*, 1998). Phylogenetic analysis of HIV subtypes also showed consistent results. The extent of interpersonal nucleotide sequence divergence by year in Yangon is close to that of Thailand, suggesting a similar duration of the HIV-1 epidemic. However, the HIV-1 epidemic in Yangon differed from that in Bangkok in speed of transmission and in the predominant HIV subtypes among CSWs. Subtype B' continued to prevail among CSWs despite introduction of subtype E into the heterosexual population. In other areas of Myanmar, the molecular epidemiology of HIV-1 infection showed a close relationship with those of neighboring countries. Therefore, subtype E prevailed among heterosexuals in areas close to Thailand and subtype C could be almost certain to exist in areas close to Yunnan province, China, in addition to subtypes B' and E (Kusagawa *et al.*, 1998).

B. China

As of December 1994, a cumulative number of 1445 native residents were detected as HIV-1 positive (Yu *et al.*, 1996). Of these reported cases, 88% were from Yunnan, a southwestern province bordering Myanmar, Laos, and Vietnam. Heroin traffic usually enters Guangxi through Yunnan or Vietnam en route to Hong Kong. Not surprisingly, 90.7% of the 1291 patients with known risk factors were IDUs. HIV-1 was also found among female sex workers returning from Thailand (Cheng *et al.*, 1994). The cumulative reported cases increased more than threefold from 1994 to the end of 1996 (Neild *et al.*, 1997). Surveys conducted from 1992 to 1997 in Yunnan showed an explosive increase in HIV-1 prevalence rates among IDUs at several sites in just 1 year (Zhang, 1998). For example, the prevalence rates in Dali were 0% in 1994, 4.7% in 1995, and 47.5% in 1996. The latest prevalence rates were higher than 45% in 7 of 11 sites. In Guangxi province, which neighbors Yunnan, 40% of 227 IDUs were found to be HIV-1 positive in 1996–1997 (Yu *et al.*, 1998). Furthermore, HIV-1 had spread to Xinjian in west China via drug traffic. The prevalence rate was 20.8% among IDUs in 1997 (Qu *et al.*, 1998). Another warning bell is the increase of prevalence rate among prostitutes from below 0.5% before 1997 to 1–2% in 1997. Migration, a resurgence of prostitution, and high rates of STDs among prostitutes (Gil *et al.*, 1996) after China's economic transformation created a huge HIV-1 susceptible population in a country with 1.2 billion people. Therefore, one of the urgent studies will be the prevailing subtype and seroprevalence rate among prostitutes in densely populated areas, since heterosexual contact is the fastest growing method of transmission.

The early isolates from Yunnan were genetically characterized and classified as subtype B (Weniger *et al.*, 1994). During 1993–1994, subtype E infections, determined by PEIA, were detected among five HIV-1-seropositive prostitutes returning from Thailand (Cheng *et al.*, 1994). Phylogenetic analysis of 11 C2V3 sequences collected during 1992–1993 from Yunnan showed that besides four subtype B' infections, seven subtype C infections were discovered for the first time (Luo *et al.*, 1995). Using the same method, nine subtype E and five subtype C viruses were identified among 14 C2V3 sequences obtained from Guangxi province. Interestingly, all subtype E viruses were detected in Pingxiang City, which borders Vietnam, and all subtype C in Baise City, which borders Yunnan (Yu *et al.*, 1998). The geographic relationship reflects the two drug traffic routes through Vietnam and Yunnan into Guangxi. Another study conducted in Guangxi had similar findings among IDUs (Chen *et al.*, 1999). They also found HIV-1 among a low-risk group, commercial blood donors, similarly to that in India. They were all infected with subtype B', but one was infected with subtype D. Sequence divergence of these subtype B' isolates suggested that these infections were recently acquired and probably had the same soure, which could be the contaminated equipment used in blood collection. The distribution of HIV-1 subtypes in China during 1996–1997 was found to have geographic variations. While subtype E was still limited to the border and coastal regions, subtype B' had spread to central and eastern China and subtype C to far western China (Shao *et al.*, 1998a). In addition, subtype B had been found among homosexuals and subtypes A and D among visitors to Africa. As has been done in Thailand, studies of molecular epidemiology must be conducted in different geographic areas and among different risk groups in the presence of at least three predominant HIV subtypes circulating in China. Recently, recombinant HIV-1 strains between subtypes B' and C had been identified in two separate areas: the Sichuan province of southwestern China and the Xinjiang province of far west China (Shao *et al.*, 1998b). These two areas can be linked by a drug-trafficking route. Phylogenetic analysis of both *env* and *gag* gene will be needed to monitor molecular epidemiology in these areas.

C. Vietnam

The first case of HIV-1 infection was identified in southern Vietnam in 1990. A significant rise in the seroprevalence rate has been noted among IDUs and CSWs since 1993 (Lindan *et al.*, 1997). By the end of December 1996, a total of 4961 HIV-1 infections had been reported in Vietnam. About half of the reported HIV-1 cases lived in Ho Chi Minh City and most of them were IDUs. The HIV-1 prevalence rate rose dramatically from 1% in 1992 to 42% in 1995 among IDUs in Ho Chi Minh City. During the same period, the rates of increase in HIV-1 prevalence were much slower among CSWs (from 0 to 1.2%) and STD clinic patients (from 0.2 to 1.0%). The use of the same paraphernalia for multiple customers could explain quick transmission of HIV-1 infection among IDUs and might result in a founder effect.

According to the analysis of 50 specimens by HMA, HIV-1 subtype E prevails in southern Vietnam (Menu *et al.*, 1996). Phylogenetic analyses of C2/V3 sequences obtained from two sex workers and three IDUs also showed that HIV-1 strains circulating in southern Vietnam are clustered to subtype E and form a monophylogenetic group (Nerurkar *et al.*, 1996). It was assumed that HIV-1 infection spread from Thailand to Cambodia then to southern Vietnam. The higher HIV-1 prevalence rate detected among CSWs who live in provinces bordering Cambodia is consistent with the hypothesis.

However, the extremely different mode of HIV-1 transmission in Cambodia from that in Vietnam still remains an enigma.

No information is available about the HIV-1 subtypes circulating in the northernmost province. Almost all known HIV-1 infections living in this area are IDUs. The geographical relationship to the drug traffic route suggests that the same subtype may prevail in nearby China and northern Vietnam.

D. Malaysia

The earliest documented, after the first case reported in 1986, HIV-1-infected patients in Malaysia were hemophiliacs, homosexual men, and IDUs. Prevalence of HIV-1 infection among IDUs increased rapidly to 30% in 1992 (Singh *et al.*, 1996). As of February 1996, IDUs comprised about 77% of the 15,100 HIV-1-infected cases detected by the Malaysian National AIDS Reference Laboratory. Phylogenetic tree analysis among IDUs showed that the predominant subtype was Thai B′ (11 of 13) between 1992 and 1993 (Brown *et al.*, 1996). In another report, 89 individuals were studied in 1992–1996. HIV-1 subtypes determined by PEIA showed that 81% of heterosexual patients were infected with subtype E. Although subtype B′ was still the predominant subtype (55%) among IDUs, 36% of IDUs were infected with subtype E (Beyrer *et al.*, 1998). Overall, more patients were infected with subtype E (54%) than subtype B′ (38%). Furthermore, 88% of commercial sex workers in the study were found to be Thais. Therefore, the HIV-1 epidemic in Malaysia can be regarded as an extension of the Thai epidemic. The future trend of the HIV-1 epidemic in Malaysia can only be predicted from the results of surveyed CSWs and STD clinic patients.

IV. Countries with Low HIV Prevalence Rates

The HIV-1 epidemics in these countries are characterized by low prevalence rates among IDUs and CSWs. However, incidence of HIV-1 is still increasing in most countries through heterosexual contact.

A. Taiwan

As of March 1999, the number of HIV-1 infections reported among native residents in Taiwan reached 2022 (Bureau of Communicable Disease Control, 1999). Of the 1573 infected persons with known risk factors, 1461 (92.9%) were infected through sexual contact, including 699 (44.4%) heterosexuals, 443 (28.2%) homosexuals, and 319 (20.3%) bisexuals. In 1984–1987, HIV-1 infections were found only among homosexual men and hemophiliacs. The risk factors diversified to include heterosexuals and IDUs in the following years. The number of annual reported cases has increased by an average rate of 24% since 1991 (Chen, *et al.*, 1994; Chang, 1998). During this period, HIV-1 infection through heterosexual contact increased rapidly and gradually became predominant. However, the prevalence rate among homosexuals also increased from 4.7% in 1988–1991 to 9.5% in 1995–1996 (Ko *et al.*, 1992, 1996). In Taiwan, only 47 HIV-1 infections were documented among IDUs. Survey of prisoners who were IDUs showed that the prevalence rates were between 0 and 0.7%. The low incidence of HIV-1 infection among IDUs is possibly due to not sharing needles. Most of the female sex workers are illegal in Taiwan and are difficult to approach. In one study, a low prevalence rate (2 of 1036) among female sex workers was found in 1993–1996 (Chen *et al.*, 1998a). In contrast, the prevalence rate of HIV-1 infection was 2.7% among male STD patients in 1995. Among low-risk groups such as blood donors and military conscripts, the prevalence rates were 0.0023 and 0.0072% in 1996.

HIV-1 subtypes determined mainly by PEIA from 1993 to 1996 among 288 HIV/AIDS patients showed that 63.2% were subtype B and 30.6% were subtype E (Chen *et al.*, 1998b). Most of the homosexuals (75.8%) and bisexuals (84.4%) were infected with subtype B, as expected. Men with subtype E infection were more often heterosexuals (56.2%), whereas all 21 female patients were infected most often with subtype E (71.4%). This result was incompatible with the demographic data, which revealed that 51% of heterosexual male patients were infected with subtype B. One possible explanation is that homosexuality is discriminated against by the society, therefore the self-reported risk factors may not be correct. Travel histories showed that many infected heterosexual men might have become infected through sexual contact with prostitutes in Southeast Asian countries. Subtype E may become the prevailing subtype in the future because an increasing proportion of heterosexual and homosexual men were recently infected with subtype E.

The intrasubtype divergence of the C2/V3 region of subtypes B and E was 13.6 and 7.4% respectively. The latter is close to the degree of sequence divergence of subtype E detected in Thailand in 1994–1995 (Subbarao *et al.*, 1998). Multiple introductions of different strains into Taiwan seemed more likely according to the extent of genetic diversity. The Thailand variant subtype B′ was not found in Taiwan. Other HIV-1 subtypes including A, C, F, and G had been reported in Taiwan (Guo *et al.*, 1993; Chang *et al.*, 1997; Lee *et al.*, 1998). Some of these patients were infected abroad. For example, the first subtype G in Taiwan was detected in a woman who was the spouse of a bisexual German. Three patients infected with subtype A were detected in 1989, 1991, and 1995 respectively. The first case was a sailor who might have been infected abroad, but two heterosexual women, whose infection was discovered after 1991, were probably infected domestically. However, three subtype A isolates were phylogenetically unrelated, as shown by the genetic variations, which ranged between 12 and 15% in the *env* V3 gene. The sporadic detection of non-B, non-E HIV-1 infections and the wide intrasubtype genetic diversity showed that these rare HIV-1 subtypes did not enter highly dynamic risk groups. Recently, four genetically clustered subtype G isolates (nucleotide differences ranged from 0.3 to 2.0% in the *gag* gene) were found. They were genetically distinct from the first subtype G reported in Taiwan. Epidemiological data showed that a juvenile prostitute was probably first infected through sexual contact with a white male. Later, she transmitted the virus to one customer and to her boyfriend. Her boyfriend then infected another female sex partner (Lee *et al.*, 1999). Our findings suggest that the introduction of foreign subtypes more often than not entered a population with a very low dynamic of transmission in a country with low HIV-1 prevalence.

B. Japan

In the 1980s, HIV-1 infection was found mainly among hemophiliacs and homosexuals. The incidence of HIV-1 infection is still increasing in Japan (Kihara *et al.*, 1998). In 1997, the annual reported HIV-1 infections were 397 in Japan as compared to 353 in Taiwan in the same year. As is in Taiwan, HIV-1 infection through heterosexual contact increased significantly. Through heterosexual contact with prostitutes from Southeastern Asia, subtype E has become the predominant subtype among heterosexuals since 1994 (Kondo *et al.*, 1998). The trend of the HIV-1 epidemic in Japan is very similar to that in Taiwan.

C. Philippines

As of August 1997, 933 cumulative cases of HIV-1 infection had been reported in Philippines by the National AIDS registry. Most of the reported cases (71%) were infected through heterosexual contact. Only 0.6% of the reported cases are IDUs (Paladin *et al.*, 1998) The distribution of HIV-1 subtypes in the Philippines studied in one report might not provide answer to the prevailing HIV-1 subtypes among native risk groups because travel histories show that 65% of the enrolled subjects were presumably infected outside of the Philippines. In addition, another 6% were wives of infected spouses who resided in the United States. There were five subtypes found in that study: A, B, C, D, and E. Subtypes E and B were predominant among a small number of patients infected indigenously. Interestingly, three female sex workers were infected with African E strains. The trend of the HIV-1 epidemic in the Philippines remains unclear. Sentinel surveys of CSWs and STD clinic patients as well as continuous monitoring of molecular epidemiology are obviously required.

D. South Korea

The rate of HIV-1 infection is relatively low in South Korea as compared with other Asian countries. As of March 1997, the cumulative number of people known to be infected was 645. Among them, 66% were infected through heterosexual contact and 19% through homosexual contact. The low percentage of female carries (12%) could be explained partly by the high proportion of infections through heterosexual contact abroad (32 and 37.7% in two reports; Kang et al., 1997; Kim et al., 1999).

Sequence variations of the *nef* gene among 46 HIV-1-infected patients were studied between 1993 and 1997 (Kang et al., 1997). Of the 46 sequences studied, 41 were classified to subtype B, 3 to subtype A, 1 was classified to subtype D, and 1 was unclassifiable. Nucleotide variations of 32 closely clustered subtype B sequences ranged from 1.9 to 8.8%. Of these 32 patients, the majority were infected through homosexual contact (17 patients) or through transfusion of blood or blood products (11 patients). Closely related sequences found in the majority of Korean homosexual carriers indicated the introduction of a single strain or few HIV-1 strains into the risk group. In another report (Kim et al., 1999), a phylogenetic tree was constructed by standard analysis of the C2/V3 region. The predominant HIV-1 strain is still subtype B (51 of 58 isolates). But two subtype Cs, one subtype H, and four subtype As were also found.

Infections through homosexual contact and the transfusion of blood-related products were overpresented in both reports. Therefore, the number of non-B HIV-1 subtypes detected among patients who were infected abroad through heterosexual contact might actually be greater than reported. However, genetic diversity of non-B subtypes in Korea did not show evidence of domestic transmission. The HIV-1 epidemic in Korea is distinguished by persistent prevailing subtype B infection among heterosexuals.

E. Singapore

In Singapore, there has been a similarly rapid increase in the incidence of HIV-1 infection since 1991, as in Taiwan. Infection through heterosexual contact is the dominant route of transmission (Se-Thoe et al., 1998). The heterosexual risk group is infected almost equally with subtypes B and E while all homosexual/bisexual patients are infected with subtype B. Subtype B' was found among IDUs. Other documented subtypes, A and C, were few and related to foreign origin. It is not known if the chronological shift in molecular epidemiology of HIV-1 infection from subtype B to E will happen in Singapore.

V. Conclusions _____

The most severe HIV-1 epidemic is occuring in sub-Saharan Africa. Widespread HIV-1 infection cannot occur unless the infection moves into the heterosexual group, which is the largest susceptible population. The high HIV-1 prevalence rates in Africa were the result of the migration of populations from rural to urban centers between 1960 and 1980 (Quinn, 1994). Urbanization lead to demographic, economic, and social changes as well as to an increasing number of prostitutes due to the growing sex industry. The accompanying high prevalence rates of STDs were then responsible for the explosive HIV epidemic when HIV-1-infected individuals migrated from low-endemic rural areas to uninfected urban areas. There are no simple explanations for the diverse extent and speed in the transmission of HIV through heterosexual contact. The seroprevalence of local CSWs, the frequency of visiting CSWs, the rate of condom use, the incidence of STDs, and the number of casual sex partners can all attribute to the HIV-1 epidemic. In Asia, both African and American/European patterns of HIV-1 infection are found. Obviously, Asian countries with either a high or low level of HIV-1 infection can be distinguished by their economic status. The repeated African experience in Thailand and India shows that the similar socioeconomic upswing provided the fertile ground for the explosive HIV-1 epidemic. In China, the revival of widescale female prostitution following economic success also resulted in high STDs among CSWs (Gil *et al.*, 1996). Recently, the opening of many former socialist countries enabled social and economic transformations. The incidence of syphilis increased sharply in Eastern Europe after economic liberation (UNAIDS WHO, 1998). Although similar risk behaviors are noted in China and Eastern Europe, predicting the future level of the HIV-1 epidemic is problematic as compared to India and Thailand because of quite different social and cultural structures.

Asian countries where economic success has continued in recent decades have low HIV-1 prevalence. The seroprevalence rates among IDUs are low in these countries possibly because of easy access to needles, as in Taiwan. The trend of HIV infection among homosexuals is similar to those North America and Western Europe. Various HIV-1 subtypes have been introduced but the phylogenetic data do not show widespread transmission. However, there has been a significant increase in HIV-1 infection through heterosexual contact in these countries since 1991. For example, the annual number of newly diagnosed cases of HIV-1 infection has increased, with a mean rate of 25%, since 1991 in Taiwan. In some countries, subtype E infection accounts for a significant part of the increase in HIV-1 infections and thought to be of Southeastern Asian origin. Nevertheless, the HIV-1 epidemic among heterosexuals is still on a very small scale in these countries, as reflected by the number of vertical transmissions.

The Golden Triangle, where the borders of Myanmar, Laos, and Thailand meet, is a major heroin-producing, -refining, and -trading area. The Golden Triangle is unique to the Asian HIV pandemic. IDUs gather along the drug traffic route from this heroin-producing area to the West. There is a common feature of the explosive HIV-1 epidemic among IDUs—the founder effect (Lukashov *et al.*, 1998). Recombinant strains can be the predominant subtypes among IDUs under this special situation (Liitsola *et al.*, 1998; Shao *et al.*, 1998b). The circulating HIV-1 subtypes in areas bordering the Golden Triangle are theoretically related to each other (Fig. 1). The first wave of HIV infection occurred in northern Thailand and Bangkok. From here, subtype E, prevalent in Northern Thailand, spread east (to Guangxi) via northern Vietnam and subtype B', prevalent in Bangkok, spread north (to Yunnan), south (to Malaysia), and west (to Yangon and India). However, the discovery of the American subtype B and subtype C among IDUs in northeastern India and Yunnan respectively was puzzling. Subtype C prevailed among heterosexuals in India, but was not detected among IDUs in northeastern India (Panda *et al.*, 1996). It is possible that these viruses may have entered the IDU group by chance through heterosexual contact. In China, subtype C was transmitted further to far-western Xinjiang and subtype B' spread to central and eastern China along the drug-trafficking route. Surveys of IDUs, CSWs, and STD clinic patients conducted in densely populated cities such as Peking, Shanghai, and Guangzhou are therefore urgently needed to keep the HIV-1 epidemic in check. The monitoring of molecular epidemiology is important in China in order to understand the future trend of the HIV-1 epidemic.

There are two patterns of HIV-1 transmission: among IDUs by sharing needles and among non-IDUs through heterosexual contact. In Bangkok HIV-1 was transmitted quickly among both CSWs and IDUs simultaneously. In Yangon transmission began with a rapid increase in HIV-1 infection among IDUs which was followed by a slower rate of transmission among CSWs. As to the latest Asian HIV-1 epidemic, in Vietnam the pattern was similar to that in the early stage of the Yangon epidemic, while in Malaysia it seemed more similar to that in Bangkok. In Yangon, it took 3 years for HIV-1 infection to spill over from IDUs to CSWs. It seems unlikely that HIV-1 interchange between IDUs and CSWs occurred on a large scale. Other sources of introducing HIV-1 infection, such as migrant sex workers, may therefore become predominant if sufficient numbers are present.

The HIV-1 epidemic may not spread at the same rate among heterosexuals in different areas. For example, the spread of HIV-1 in Cambodia and Vietnam is closely linked, but the prevalence rates among CSWs varies significantly between the two countries. Several factors, including frequency of visiting prostitutes, population size of prostitutes, migration of populations, rate of condom use, and incidence of genital ulcers might have contribution to the differences. Obviously, prevalence of risky sexual behaviors and incidence of STDs can be use to levels of HIV-1 infection in a particular country. The recent decline in HIV infection among young men in Thailand shows that changes in sexual behavior was a successful intervention in preventing HIV-1 infection (Nelson *et al.*, 1996).

Three major HIV-1 subtypes are now circulating in Asia. Since most Asian countries cannot afford expensive highly active antiretroviral therapy, the development of an effective vaccine seems to be cost-effective. However, cross-subtype immunity may not be produced by the candidate vaccines derived from subtype B strains. Vaccines against the more prevalent subtypes C, E, and B' should be developed for evaluation in Asian countries (van der Groen *et al.*, 1998). More importantly, each Asian government should learn from the past and respond to the most recent HIV-1 epidemic appropriately.

References

Baskar, P. V., Ray, S. C., Rao, R., Quinn, T. C., Hildreth, J. E. K., and Bollinger, R. C. (1994). Presence in India of HIV type 1 similar to North American strains. *AIDS Res. Hum. Retroviruses* **10**, 1039–1041.

Beyrer, C., Vancott, T. C., Peng, N. K., Artenstein, A., Duriasamy, G., Nagaratnam, M., Saw, T. L., Hegerich, P. A., Loomis-Price, L. D., Hallberg, P. L., Ettore, C. A., and Nelson, K. E. (1998). HIV type 1 subtypes in Malaysia, determined with serologic assays: 1992–1996. *AIDS Res. Hum. Retroviruses* **14**, 1687–1691.

Bollinger, R. C., Tripathy, S. P., and Quinn, T. C. (1995). The human immunodeficiency virus epidemic in India: Current magnitude and future projection. *Medicine* **74**, 97–106.

Brown, T., Sittitrai, W., Vanichseni, S., and Thisyakorn, U. (1994). The recent epidemiology of HIV and AIDS in Thailand. *AIDS* 8(Suppl. 2), S131–S141.

Brown, T. S., Robbins, K. E., Sinniah, M., Saraswathy, T. S., Lee, V., Hooi, L. S., Viayamalar, B., Luo, C. C., Ou, C. Y., Rapier, J., Schochetman, G., and Karlish, M. (1996). HIV type 1 subtypes in Malaysia include B, C and E. *AIDS Res. Hum. Retroviruses* **12**, 1655–1657.

Bureau of Communicable Disease Control (1998). "*AIDS Statistics Report*, Taiwan." Department of Health, Taiwan.

Cassol, S., Weniger, B. G., Babu, G., Salminen, M. O., Zheng, X., Htoon, M. T., Delaney, A., O'Shaughnessy, M., and Ou, C. Y. (1996). Detection of HIV type 1 *env* subtypes A, B, C and E in Asia using dried blood spots: A new surveillance tool for molecular epidemiology. *AIDS Res. Hum. Retroviruses* **12**, 1435–1441.

Chang, H. J. (1998). The current status of HIV-AIDS in Taiwan. *J. Publ. Health Med.* **20**: 11–15.

Chang, K. S. S., Lin, C. I., Chen, J. H., Shih, C. H., Lin, H. C., Lin, R. Y., Twu, S. C., and Saliminen, M. O. (1997). HIV type I *env* gene diversity detected in Taiwan. *AIDS Res. Hum. Retroviruses* **13**: 201–204.

Cheingsong-Popov, R., Lister, S., Callow, D., Kaleebu, P., Beddows, S., Weber, J. and the WHO Network for HIV Isolation and Characterization (1994). Serotyping HIV type 1 by antibody binding to the V3 loop: Relation to viral genotype. *AIDS Res. Hum. Retroviruses* **11**, 1379–1386.

Chen, J., Young, N. L., Subbarao, S., Warachit, P., Saguanwongse, S., Wongsheree, S., Jayavasu, C., Luo, C. C., and Mastro, T. D. (1999). HIV type 1 subtypes in Guangxi province, China, 1996. *AIDS Res. Hum. Retroviruses* **15**, 81–84.

Chen, M. Y., Wang, G. R., Chuang, C. Y., and Shih, Y. T. (1994). Human immunodeficiency virus infection in Taiwan, 1984 to 1994. *J. Formos. Med. Assoc.* **93**, 901–905.

Chen, M. Y. (1998). "Rapid Detection of HIV-1 Subtype E Infection by Polymerase Chain Reaction (PCR) in Taiwan." Twelfth international Conference on AIDS, Geneva, Switzerland. [Abstract 13146]

Chen, Y.-M. A., Yu, P. S., Lin, C. C., and Jen, I. (1998a). Surveys of HIV-1, HTLV-1, and other sexually transmitted diseases among female sex workers in Taipei city and prefectures, Taiwan from 1993 to 1996. *J. Acq. Immune Defic. Syndr. Hum. Retrovirol.* **18**, 299–303.

Chen, Y.-M. A., Lee, C. M., Lin, R. Y., and Chang, H. J. (1998b). Molecular epidemiology and trends of HIV-1 subtypes in Taiwan. *J. Acq. Immune Defic. Syndr. Hum. Retrovirol* **19**, 393–402.

Cheng, H., Zhang, J., Capizzi, J., Young, N. L., and Mastro, T. D. (1994). HIV-1 subtype E in Yunnan, China. *Lancet* **344**, 953–954.

Delwart, E. L., Shpaer, E. G., and Louwagie, J. (1993). Genetic relationships determined by a DNA heteroduplex mobility assay: Analysis of HIV-1 env genes. *Science* **262**, 1257–1261.

Gangakhedkar, R. R., Bentley, M. E., Divekar, A. D., Gadkari, D., Mehendale, S. M., Shepherd, M. E., Bollinger, R. C., and Quinn, T. C. (1997). Spread of HIV infection in married monogamous woman in India. *J. Am. Med. Assoc.* **278**, 2090–2092.

Gil, V. E., Wang, M. S., Anderson, A. F., Lin, G. M., and Wu, Z. O. (1996). Prostitutes, prostitution and STD/HIV transmission in mainland China. *Soc. Sci. Med.* **42**, 141–152.

Grez, M., Dietrich, U., Balfe, P., von Briesen, H., Maniar, J. K., Manhambre, G., Delwart, E. L., Mullins, J. I., and Rubsamen-Waigmann, H. (1994). Genetic analysis of human immunodeficiency virus type 1 and 2 (HIV-1 and HIV-2) mixed infections in India reveals a rescent spread of HIV-1 and HIV-2 from a single ancestor for each of these viruses. *J. Virol.* **68**, 2161–2168.

Guo, H. G., Reitz, M. S., Gallo, R. C., Ko, Y. C., and Chang, K. S. S. (1993). A new subtype of HIV-1 gag sequence detected in Taiwan. *AIDS Res. Hum. Retroviruses* **9**, 925–927.

Htoon, M. T., Lwin, H. H., San, K. O., Zan, E., and Thwe, M. (1994). HIV/AIDS in Myanmar. *AIDS* **8**(Suppl. 2), S105–S109.

Kalish, M. L., Baldwin, A., Raktham, S., Wasi, C., Luo, C. C., Schochetman, G., Mastro, T. D., Young, N., Vanicheseni, S., Rubsamen-Waigmann, H., Briesen, H., Mullins, J. I., Delwart, E., Herring, B., Esparza, J., Heyward, W. L., and Osmanov, S. (1995). The evolving molecular epidemiology of HIV-1 envelope subtypes in injecting drug users in Bangkok, Thailand: Implications for HIV vaccine trials. *AIDS* **9**, 851–857.

Kalish, M. L., Luo, C. C., Weniger, B. G., Limpakarnjanarat, K., Young, N., Ou, C. Y., and Schochetman, G. (1994). Early HIV type 1 strains in Thailand were not responsible for the current epidemic. *AIDS Res. Hum. Retroviruses* **10**, 1573–1575.

Kan, M. R., Cho, Y. K., Chun, J., Kim, Y. B., Lee, I. S., Lee, H. J., Kim, S. H., Kim, Y. K., Yoon, K., Yang, J. M., Kim, J. M., Shin, Y. O., Kang, C., Lee, J. S., Choi, K. W., Kim, D. G., Fitch, W. M., and Kim, S. (1998). Phylogenetic analysis of the *nef* gene reveals a distinctive monophyletic clade in Korean HIV-1 cases. *J. Acq. Immune Defic. Syndr. Hum. Retrovirol.* **17**, 58–68.

Kihara, M., Yamazaki, S., Kihara, M., and Nakatami, H. (1998). "Revised Surveillance Report of HIV/AIDS in Japan, 1997." Twelfth international Conference on AIDS, Geneva, Switzerland. [Abstract 13141]

Kim, Y. B., Cho, Y. K., Lee, H. J., Kim, C. K., Kim, Y. K., and Yang, J. M. (1999). Molecular phylogenetic analysis of human immunodeficiency virus type 1 strains obtained from Korean patients: *env* gene sequences. *AIDS Res. Hum. Retroviruses* **15**, 303–307.

Ko, N. Y., Chung, H. H., and Chang, S. J. (1996). The relationship between self-efficacy, perceived AIDS threat and sexual behaviors: Analysis of 108 male homosexuals in southern Taiwan. *Nursing Care* **4**, 285–297.

Ko, Y. C., and Chang, S. J. (1992). Sex patterns and human immunodeficiency virus infection among homosexuals in Taiwan. *Sex Transm. Dis.* **19**, 335–338.

Konodo, M., Saito, T. S., Ito, A. I., Kawata, K. K., Sagara, H. S., Kihara, M. K., Sato, H. S., Takebe, Y. T., Nishioka, K. N., and Imai, M. I. (1998). "Simple Method for Typing HIV-1 (B, E) by PCR—Chronological Shift of Subtype Distribution of HIV-1 (B to E) in Japan." Twelfth international Conference on AIDS, Geneva, Switzerland. [Abstract 13138]

Kusagawa, S., Sato, H., Kato, K., Nohtomi, K., Shino, T., Samrith, C., Leng, H. B., Phalla, T., Heng, M. B., and Takebe, Y. (1999). HIV type 1 *env* subtype E in Cambodia. *AIDS Res. Hum. Retroviruses* **15**, 91–94.

Kusagawa, S., Sato, H., Watanabe, S., Nohtomi, K., Kato, K., Shino, T., Thwe, M., Oo, K. Y., Lwin, S., Mra, R., Kywe, B., Yamazaki, S., and Takebe, Y. (1998). Genetic and serologic characterization of HIV type 1 prevailing in Myanmar. *AIDS Res. Hum. Retroviruses* **14**, 1379–1385.

Lalvani, A., and Shastri, J. S. (1996). HIV epidemic in India. *Lancet* **347**, 1349–1350.

Lee, C. N., Wang, W. K., Fan, W. S., Twu, S. J., Chen, S. C., Sheng, M. C., and Chen, M. Y. (2000). Determination of human immunodeficiency virus type 1 subtypes in Taiwan by vpu gene analysis. *J. Clin. Microbiol.* **38**, in press.

Lee, C. N., Chen, M. Y., Fan, W. S., Twu, S. J., Lin, R. Y. (1999). Domestic transmission of HIV type 1 subtype G strains in Taiwan. *AIDS Res. Hum. Retroviruses* **15**, 1137–1140.

Lee, C. N., Chen, M. Y., Lin, H. S., Lee, M. C., Luo, C. C., Twu, S. J., Lin, R. Y., and Chuang, C. Y. (1998). HIV type 1 *env* subtype A variants in Taiwan. *AIDS Res. Hum. Retroviruses* **14**, 807–809.

Liitsola, K., Tashkinova, I., Laukkanen, T., Korovina, G., Smolskaja, T., Momot, O., Mashkil-leyson, N., Chaplinskas, S., Brummer-Korvenkontio, H., Vanhatalo, J., Leinikki, P., and Salminen, M. O. (1998). HIV-1 genetic subtype A/B recombinant strains causing an explosive epidemic in injecting drug users in Kalinigrad. *AIDS* **12**, 1907–1919.

Lindan, C. P., Lieu, T. X., Giang, L. T., Lap, V. D., Thuc, N. V., Thinh, T., Lurie, P., and Mandel, J. S. (1997). Rising HIV infection rates in Ho Chi Minh City herald emerging AIDS epidemic in Vietnam. *AIDS* **11**(Suppl. 1), S5–S13.

Lukashov, V. V., Karamov, E. V., Eremin, V. F., Titov, L. P., and Goudsmit, J. (1998). Extreme founder effect in an HIV type 1 subtype A epidemic among drug users in Svetlogorsk, Belarus. *AIDS Res. Hum. Retroviruses* **14**, 1299–1303.

Luo, C. C., Tian, C., Hu, D. J., Kai, M., Dondero, T. J., and Zheng, X. (1995). HIV subtype C in China. *Lancet* **345**, 1051–1052.

Menu, E., Lien, T. T. X., Lafon, M. E., Lan, N. T. H., Muller-Trutwin, M. C., Thuy, N. T., Deslandres, A., Chaouat, G., Trung, D. Q., Khiem, H. B., Fleury, H. J. A., and Barre-Sinoussi, S. F. (1996). HIV-1 type 1 Thai subtype E is predominant in South Vietnam. *AIDS Res. Hum. Retroviruses* **12**, 629–633.

Mishra, S. N., Bhargava, N. C., Satpathy, S. K., and Prasada Rao, J. V. R. (1998). "HIV/AIDS in India—Need for a National STD Surveillance System." Twelfth international Conference on AIDS, Geneva, Switzerland. [Abstract 13292]

Navarro, V., Roig, P., Nieto, A., Jimenez, J., Tuset, C., Tuset, L., Navarro, R., and Juan, G. (1988). A small outbreak of HIV infection among commercial plasma donors. *Lancet* **ii**, 42.

Neild, P. J., and Gazzard, B. G. (1997). HIV infection in China. *Lancet* **350**, 963.

Nelson, K. E., Celentano, D. D., Eiumtrakol, S., Hoover, D. R., Beyrer, C., Suprasert, S., Kuntolbutra, S., and Khamboonruang, C. (1996). Changes in sexual behavior and a decline in HIV infection among young men in Thailand. *N. Engl. J. Med.* **355**, 297–303.

Nerurkar, V. R., Nguyen, H. T., Dashwood, W. M., Hoffmann, P. R., Yin, C., Morens, D. M., Kaplan, A. H., Detels, R., and Yanagihara, R. (1996). HIV type 1 subtype E in commercial sex workers and injection drug users in Southern Vietnam. *AIDS Res. Hum. Retroviruses* **12**, 841–843.

Nkengasong, J. N., Willems, B., Janssens, W., Cheingsong-Popov, R., Heyndrickx, L., Barin, F., Ondoa, P., Fransen, K., Goudsmit, J., and van der Groen, G. (1998). Lack of correlation between V3-loop peptide enzyme immunoassay serologic subtyping and genetic sequencing. *AIDS* **12**, 1405–1412.

Ou, C. Y., Takebe, Y., Weniger, B. G., Luo, C. C., Kalish, M. L., Auwanit, W., Yamazaki, S., Gayle, H. D., Young, N. L., and Schochetman, G. (1993). Independent introduction of two major HIV-1 genotypes into distinct high-risk populations in Thailand. *Lancet* **341**, 1171–1174.

Paladin, F. J. E., Monzon, O. T., Tsuchie, H., Aplasca, M. R. A., Learn, G. H., and Kurimura, T. (1998). Genetic subtypes of HIV-1 in the Philippines. *AIDS* **12**, 291–300.

Panda, S., Wang, G., Sarkar, S., Perez, C. M., Chakraborty, S., Agarwal, A., Dorman, K., Sarkar, K., Detels, R., and Kaplan, A. H. (1996) Characterization of V3 loop of HIV type 1 spreading rapidly among injection drug users of Manipur, India: A molecular epidemiological perspective. *AIDS Res. Hum. Retroviruses* **12**, 1571–1573.

Phalla, T., Leng, H. B., Mills, S., Bennet, A., Wienrawee, P., Gorbach, P., and Chin, J. (1998). HIV and STD epidemiology, risk behaviors, and prevention and care response in Cambodia. *AIDS* **12**(Suppl. B), S11–S18.

Qu, S., Sun, X. H., Zheng, X., and Shen, J. (1998). "HIV Sentinel Surveillance in China in 1997." Twelfth international Conference on AIDS, Geneva, Switzerland. [Abstract 13145]

Quinn, T. C. (1994). Population migration and the spread of type 1 and 2 human immunodeficiency viruses. *Proc. Natl. Acad. Sci. USA* **91**, 2407–2414.

Se-Thoe, S. Y., Foley, B. T., Chan, S. Y., Lin, R. V. T. P., Oh, H. M. L., Ling, A. E., Chew, S. K., Snodgrass, I., and Sng, J. E. H. (1998). Analysis of sequence diversity in the C2-V3 regions of the external glycoproteins of HIV type 1 in Singapore. *AIDS Res. Hum. Retroviruses* **14**, 1601–1604.

Shao, Y., Su, L., Sun, X. H., Xing, H., Pan, P. L., Wolf, H., and Shen, J. (1998a). "Molecular Epidemiology of HIV Infection in China." Twelfth international Conference on AIDS, Geneva, Switzerland. [Abstract 13132]

Shao, Y., Su, L., Zhao, F., Xing, H., Zhang, Y. Z., Wolf, H., and Zhang, L. L. (1998b). "Genetic Recombination of HIV-1 Strains Identified in China." Twelfth international Conference on AIDS, Geneva, Switzerland. [Abstract 11179]

Singh, S., and Crofts, N. (1993). HIV infection among injecting drug users in north-east Malaysia, 1992. *AIDS Care* **5** (3), 273–281.

Subbarao, S., Limpakarnjanarat, K., Mastro, T. D., Bhumisawasdi, J., Warachit, P., Jayavasu, C., Young, N. L., Luo, C. C., Shaffer, N., Kalish, M. L., and Schochetman, G. (1998). HIV type 1 in Thailand, 1994–1995: Persistence of two subtypes with low genetic diversity. *AIDS Res. Hum. Retroviruses* **14**, 319–327.

Tripathy, S., Renjifo, B., Wang, W. K., McLane, M. F., Bollinger, R., Rodrigues, J., Osterman, J., Tripathy, S., and Essex, M. (1996). Envelope glycoprotein 120 sequences of primary HIV type 1 isolates from Pune and New Delhi, India. *AIDS Res. Hum. Retroviruses* **12**, 1199–1202.

UNAIDS and WHO (1998). Report on the global HIV/AIDS epidemic—June 1998.

van der Groen, G., Nyambi, P. N., Beirnaert, E., Davis, D., Fransen, K., Heyndrickx, L., Ondoa, P., van der Auwera, G., and Janssens, W. (1998). Genetic variation of HIV type 1: Relevance of interclade variation to vaccine development. *AIDS Res. Hum. Retroviruses* **14**, S211–S221.

van Harmelen, J. H., van der Ryst, E., Loubser, A. S., York, D., Madurai, S., Lyons, S., Wood, R., and Williamson, C. (1999). A predominantly HIV type 1 subtype C-restricted epidemic in South African urban populations. *AIDS Res. Hum. Retroviruses* **15**, 395–398.

Wasi, C., Herring, B., Raktham, S., Vanichseni, S., Mastro, T. D., Yang, N. L., Rubsamen-Waigmann, H., Briesen, H., Kalish, M. L., Luo, C. C., Pau, C. P., Baldwin, A., Mullins, J. I., Delwart, E. L., Esparza, J., Heyward, W. L., and Osmanov, S. (1995). Determination of HIV-1 subtypes in injecting drug users in Bangkok, Thailand, using peptide-binding enzyme immunoassay and heteroduplex mobility assay: Evidence of increasing infection with HIV-1 subtype E. *AIDS* **9**, 843–849.

Weniger, B. G., Limpakarnjanarat, K., Ungchusak, K., Thanprasertsuk, S., Choopanya, K., Vanichseni, S., Uneklabh, T., Thongcharoen, P., and Wasi, C. (1991). The epidemiology of HIV infection and AIDS in Thailand. *AIDS* **5**(Suppl. 2), S71–S85.

Weniger, B. G., Takebe, Y., Ou, C. Y., and Yamazaki, S. (1994). The molecular epidemiology of HIV in Asia. *AIDS* **8**(Suppl. 2), S13–S18.

Wilkinson, D., Connolly, C., and Rotchford, K. (1999). Continued explosive rise in HIV prevalence among pregnant women in rural South Africa. *AIDS* **13**, 740.

Yu, E. S. H., Xie, Q., Zhang, K., Lu, P., and Chan, L. L. (1996). HIV infection and AIDS in China, 1985 through 1994. *Am. J. Public Health* **86,** 1116–1122.

Yu, X. F., Chen, J., Shao, Y., Beyrer, C., and Lai, S. (1998). Two subtypes of HIV-1 among injection-drug users in southern China. *Lancet* **351,** 1250.

Zhang, J. (1998). "HIV Prevalence Trends among Intravenous Drug Users in Yunnan Province between 1992–1997". Twelfth international Conference on AIDS, Geneva, Switzerland. [Abstract 43459]

Vanessa M. Hirsch
Jeffrey D. Lifson

Laboratory of Molecular Microbiology
National Institute of Allergy and Infectious Diseases (NIAID)
National Institutes of Health
Twinbrook II Facility
Rockville, Maryland 20852 and

Laboratory of Retroviral Pathogenesis
AIDS Vaccine Program
SAIC Frederick
NCI/FCRDC
Frederick, Maryland 21702

Simian Immunodeficiency Virus Infection of Monkeys as a Model System for the Study of AIDS Pathogenesis, Treatment, and Prevention

I. Introduction

Simian immunodeficiency viruses (SIV) are a large family of primate lentiviruses that naturally infect a wide range of African primates. These viruses are highly relevant models for the study of human AIDS since upon experimental infection of macaques, they induce an immunodeficiency that is remarkably similar to AIDS in humans. This has lead to extensive characterization of a number of isolates of SIV which are presently used in the study of AIDS pathogenesis, the development of vaccines, and the assessment of antiviral therapies. An essential component of these animal studies has been the use of plasma viral RNA assays for assessing viral replication. This chapter reviews the relative pathogenicity of different isolates of SIV and discusses the use of plasma viremia as an early readout for the study of pathogenesis, therapy, and vaccine development.

At the present time, these primate lentiviruses can be classified into five lineages based upon sequence and functional genetic organization. These

Advances in Pharmacology, Volume 49

five lineages are represented by (1) SIVcpz from chimpanzees (*Pan troglodytes*), (2) SIVsm from sooty mangabeys (*Cercocebus torquatus atys*), (3) SIVagm from four species of African green monkeys (members of the *Chlorocebus aethiops* superspecies), (4) SIVsyk from Sykes' monkeys (*Cercopithecus mitis albogularis*), and (5) SIVmnd from a mandrill (*Mandrillus sphinx*) together with SIVlhoest from l'hoest monkeys (*Cercopithecus l'hoesti lhoesti*) (Hirsch *et al.*, 1999) and SIVsun from sun-tailed monkeys (*Cercopithecus l'hoesti solatus*) (Beer *et al.*, 1999). More detailed information on the phylogenetic relationships between these viruses have been reviewed previously (Franchini and Retiz, 1994; Hirsch and Johnson 1993; Johnson and Hirsch, 1993; Sharp *et al.*, 1995). Recently SIVrcm from redcapped mangabeys (SIVrcm; *Cercocebus torquatus torquatus*) was partially characterized by Georges Courbot *et al.* (1998) and SIVdrl from drills (SIVdrl; *Mandrillus leucophaeus*) was partially characterized by Clewley *et al.* (1998). Complete analysis of their genomes will be required to determine whether these are representative of new lineages. The various SIV strains are listed in Table I and the phylogenetic relationship between fully characterized

TABLE I Major Lineages of Simian Immunodeficiency Virus

SIV strain	Species of origin	Scientific name
SIVsm	Sooty mangabey	*Cercocebus atys*
SIVmac	Macaque	*Macaca sp.*
SIVstm	Stumptailed macaque	*Macaca arctoides*
SIVmne	Pigtailed macaque	*Macaca nemestrina*
HIV-2	Human	*Homo sapiens*
SIVagm	African green monkey	*Chlorocebus aethiops sp.*
SIVagm/ver	Vervet monkey	*Chlorocebus aethiops pygerythrus*
SIVagm/gri	Grivet monkey	*Chlorocebus aethiops aethiops*
SIVagm/tan	Tantalus monkey	*Chlorocebus aethiops tantalus*
SIVagm/sab	Sabaeus monkey	*Chlorocabus aethiops sabaeus*
SIVsyk	Sykes' monkey	*Ceropithecus mitis albogularis*
SIVl'hoest	L'hoest monkey	*Cercopithecus l'hoesti*
SIVsun	Suntailed monkey	*Cercopithecus solatus*
SIVmnd	Mandrill	*Mandrillus sphinx*
SIVrcm	Redcapped mangabey	*Cercocebus torquatus torquatus*
SIVdrl	Drill	*Mandrillus leucophaeus*
SIVcpz	Chimpanzee	*Pan troglodytes*
HIV-1	Human	*Homo sapiens*

lineages of SIV and HIV is depicted in Fig. 1. Each of the lineages share approximately 50% identity between the most highly conserved *gag* and *pol* genes. The identification of unique but related SIV isolates within the members of the African green monkey lineage implies that these viruses are ancient, since speciation is estimated to have occurred many thousands of years ago. It is therefore believed that the SIVs coevolved with their host species. A similar situation has been observed recently in members of the l'hoesti superspecies (l'hoest monkeys and sun-tailed monkeys) (Beer *et al.*, 1999).

In addition to this evidence of long-term evolution within African primates, there are situations that can only be explained by recent cross-species transmission (reviewed in Sharp *et al.*, 1995). For example, a remarkable phylogenetic relationship exists between SIV isolated from sooty mangabey monkeys (SIVsm; *Cercocebus torquatus atys*) and HIV-2 in West African humans (Hirsch *et al.*, 1989; Marx *et al.*, 1991; Gao *et al.*, 1992; reviewed in Sharp *et al.*, 1995). Indeed, some of these viruses cannot be distinguished phylogenetically, implying that HIV-2 arose by cross-species transmission

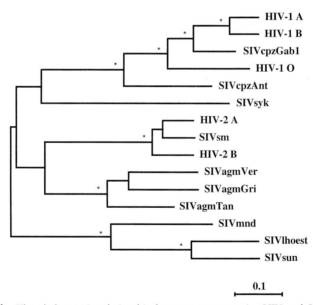

0.1

FIGURE I   The phylogenetic relationship between representative HIV and SIV strains is shown in this maximum-likelihood analysis of concatenated Gag-Pol-Vif-Env-Nef proteins. Five SIV lineages are represented respectively by SIVcpz, SIVsyk, SIVsm, SIVagm, and SIVl'-hoest. SIVrcm and SIVdrl are not shown since sequence analyses of their complete genomes were not available. Horizontal branch lengths indicate the degree of divergence as compared to the scale at the bottom of the figure (0.1 amino acid replacement per site). Asterisks indicate that the clade to the right was found in 100% of the bootstrap values of the neighbor-joing analysis.

from sooty mangabeys to humans (Gao *et al.*, 1991). More recently, a similar relationship has been described between SIV isolates from chimpanzees (SIVcpz; *Pan troglodytes*) and HIV-1, consistent with the origins of the HIV-1 epidemic in chimpanzees (Huet *et al.*, 1990; Janssens *et al.*, 1994; Gao *et al.*, 1999).

Given the genetic relatedness of the immunodeficiency viruses of nonhuman primates with the etiologic agents of the human acquired immunodeficiency syndrome (AIDS), the human immunodeficiency viruses (HIV-1 and HIV-2), it is perhaps not surprising that these viruses share many biological properties. Indeed, much of the interest in these viruses stems from their similarities to HIV-1 and HIV-2 in genetic structure, gene regulation, tropism, and cellular receptor usage. SIV and HIV share tropism for CD4+ T lymphocytes and macrophages, and utilize CD4 as well as the chemokine receptor molecule, CCR5, for viral entry. The vast majority of SIV isolates do not use the CXCR4 chemokine receptor molecule for entry which is one characteristic that distiguishes them from HIV-1 (Unutmaz *et al.*, 1998). As shown in Fig. 2, the genetic organization of SIV and HIV are similar. The basic genome structure of the majority of the primate lentiviruses represented in SIVagm, SIVmnd, SIVsyk, SIVsun, and SIVlhoest is *gag-pol-vif-vpr-tat-rev-env-nef*. SIVsm and HIV-2 share a common novel gene, *vpx*, in the central region of their genomes, and SIVcpz and HIV-1 share the *vpu* gene. The major utility of SIV as an animal model for AIDS arises from the observations that many SIV isolates can infect Asian macaques (*Macaca sp.*) and induce an AIDS-like syndrome similar to HIV infection of humans,

A. **SIVagm, SIVlhoest, SIVsun, SIVmnd, SIVsyk**

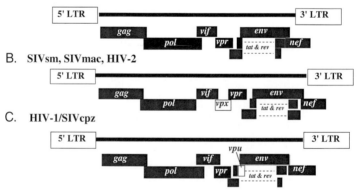

B. **SIVsm, SIVmac, HIV-2**

C. **HIV-1/SIVcpz**

FIGURE 2   Genomic organization of SIV. A schematic representation of the genome structure of various SIV and HIV strains is shown. (A) The majority of SIV strains have a structure as depicted (*gag-pol-vif-vpr-tat-rev-env-nef*) where each gene is represented by a black rectangle. (B) SIVsm and HIV-2 have an additional gene, Vpx, shown by the white rectangle and (C) SIVcpz and HIV-1 have an additional gene, Vpu, shown by the white rectangle.

as reviewed in Allan (1991), Hirsch and Johnson (1994), and Letvin and King (1990).

II. SIV as a Model for Human AIDS

A. Natural Infection

Although SIV infection appears to be highly prevalent among free-living African primates, there is no evidence that infection is associated with any adverse consequences. The best evidence for the apathogenic nature of SIV infection in African primates comes from studies of sooty mangabeys housed in North American primate centers. Although up to 90% seropositivity has been reported in these colonies, there is no evidence of AIDS in observation over the entire life span of these animals. The lack of disease association provides a model to study successful host mechanisms in dealing with lentiviral infection. Unfortunately, there are few of such animals in captivity, their immunology is poorly characterized, and the viruses infecting such animals are genetically diverse. Therefore experimental models of natural infection can provide a system for examining the host mechanisms responsible for protecting against development of AIDS.

One such model is experimental SIVagm infection of African green monkeys (AGM), which also does not result in disease development in AGM. This model becomes more interesting when one realizes that experimental transfer of SIVagm from a naturally infected AGM to one of the Asian macaque species frequently results in an AIDS-like syndrome with remarkable similarities to human AIDS (Hirsch et al., 1995). Thus SIV strains are not attenuated per se; rather it is the unique virus–host interaction in African monkeys that results in lack of disease. At the present time the host mechanism(s) responsible for the lack of virulence of these viruses in their natural host species have not been delineated.

B. Pathogenic Experimental Infection

The observation that SIV induces AIDS in macaques actually came about by serendipity. In the 1980s, an unusual clustering of lymphomas and immunodeficiency-associated disorders was noted in a colony of captive macaques at the New England Regional Primate Research Center. These observations eventually led to the isolation of simian immunodeficiency virus (SIV), which was designated SIVmac to indicate its apparent origin in macaques (Daniel et al., 1985; Letvin et al., 1985). Additional related SIV isolates were identified in stumptailed macaques (SIVstm) at the California Regional Primate Center and a pigtailed macaque from the Washington Regional Primate Research Center (SIVmne; Benveniste et al., 1986). In parallel, investigators at the Tulane primate center, conducting leprosy studies in sooty mangabeys, observed that transplantation of tissues from a sooty mangabey to a rhesus macaque resulted in AIDS (Murphey-Corb *et al.*, 1986). Investigators at the Yerkes primate Center also identified SIV in their colony of sooty mangabey monkeys (Fultz *et al.*, 1986) and the viruses from these two centers were designated SIVsm. After molecular characterization of SIVmac and SIVsm, it became apparent that these were highly related viruses (Hirsch *et al.*, 1989). Based upon the presence of SIVsm in feral populations of sooty mangabeys in West Africa (Marx *et al.*, 1991) and sequence analysis of these and North American isolates of SIVsm/SIVmac (Chakrabarti *et al.*, 1987; Hirsch *et al.*, 1989), researchers have concluded that SIVmac, SIVstm, and SIVmne are actually inadvertant transmissions of SIVsm into macaque populations through housing with sooty mangabeys in captivity (Hirsch and Johnson, 1994; Sharp *et al.*, 1995).

Many of the SIVsm and SIVmac isolates that have been studied for pathogenesis in primates are described in Table II. Other SIV isolates from each of the primate lentivirus lineages have been characterized for their pathogenic effects in their natural host species or macaques (reviewed in Allan *et al.*, 1991; Johnson and Hirsch, 1994), as summarized in Table III. Both SIVsm and SIVagm (Hirsch *et al.*, 1995) can induce AIDS in experimentally inoculated macaques. Some of these viruses do not appear

TABLE II   Genetic and Pathogenic Diversity of SIV Strains

Subtype	Strain	Isolate form	Disease potential
SIVmac	SIVmac251	SIVmac/251, uncloned	High, AIDS
		SIVmac/32H, uncloned	Moderate, AIDS
		SIVmac/J5, molecular clone	Low, AIDS
		SIVmac/BK28, molecular clone	Low, AIDS
		SIVmac/1A11, molecular clone	Attenuated
	SIVmac239	SIVmac/239, molecular clone	High, AIDS
	SIVmne	SIVmne, uncloned	Moderate, AIDS
		SIVmne/E11S, biological clone	Low, AIDS
		SIVmne/c18, molecular clone	Low, AIDS
SIVsm	SIVsmB670	SIVsmB670, uncloned	High, AIDS
	SIVsmF236	SIVsmF236, uncloned	Moderate
		SIVsmH-4, molecular clone	Low, AIDS
		SIVsmH-3, molecular clone	Low, AIDS
	SIVsmE660	SIVsmE660, uncloned	High, AIDS
	SIVsmE543	SIVsmE543, uncloned	High, AIDS
		SIVsmE543-3, molecular clone	High, AIDS
	SIVsmm9	SIVsmm9, uncloned	Moderate, AIDS
	SIVsmPBj	SIVsmPBj14, biologically cloned	High, Acute disease
		SIVsmPBj6.6, molecular clone	High, Acute Disease

TABLE III Pathogenesis of Other SIV Lineages

Lineage	Isolate	Pathogenesis
SIVagm	SIVagm/ver-3	No disease (Cyn, Pt, AGM)
	SIVagm/verTyo1	No disease (Cyn, AGM)
	SIVagm/ver155	No disease (Pt, Rh, AGM)
	SIVagm/ver90	AIDS (PT); No disease (Rh)
	SIVagm/ver9063	AIDS (PT); No disease (Rh, AGM)
	SIVagm/gri-1	n.t.
	SIVagm/sab	No disease (Rh)
	SIVagm/tan	No disease (Cyn, AGM)
SIVsyk	SIVsyk173	No disease (Rh, Pt, Cyn)
SIVmnd	SIVmnd/GB-1	n.t.
SIVlhoest	SIVlhoest-7	AIDS (Pt)

Abbreviations: Cyn, cynomolugus macaque; Pt, pigtailed macaque; AGM, African green monkey; Rh, rhesus macaque; n.t., not tested.

to be pathogenic. For example, SIVsyk infects various macaque species but does not appear to result in AIDS in these animals. In contrast, SIVagm can produce AIDS in pigtailed macaques (but not rhesus macaques or the natural host, AGM; Hirsch *et al.*, 1995). In addition, SIVlhoest from a l'hoest monkey induces characteristic CD4 depletion in pigtailed macaques, also consistent with virulence in this species (Hirsch *et al.*, 1999).

The majority of studies have focused on the SIVsm and SIVmac viruses (reviewed in Letvin and King, 1990) since these viruses were the first to be demonstrated to induce an immunodeficiency syndrome in macaques. Both SIVsm and SIVmac cause a fatal immunodeficiency in a variety of species of macaque monkeys with an accompanying depletion of circulating CD4 lymphocytes and the onset of opportunistic infections and virally induced meningoencephalitis. The resulting disease is remarkably similar in pathology and apparent pathogenesis to human AIDS. However, in contrast to human HIV infection, where progression from initial infection to AIDS may take more than a decade, in many SIV infected macaque models these events are compressed into a 1- to 2-year period and thus into the realm of experimental feasibility.

I. Phases of SIV Infection

Similar to the human disease, experimental infection of macaques with SIVsm/SIVmac can be divided into three distinct phases, the primary infection, an asymptomatic phase, and a late phase, termed AIDS. Primary infection occurs within the first 3 weeks after intravenous or mucosal inoculation and is characterized by massive viremia; a transient leukopenia; and clinical signs such as fever, lymphadenopathy, diarrhea, rash, anorexia, and general malaise (Letvin and King 1990). SIV-specific antibodies and cytotoxic T lymphocytes (CTL) develop (Kuroda *et al.*, 1998) and the plasma viremia resolves as the animal enters a clinically asymptomatic phase. The animals remain in an apparently healthy state although many exhibit significant lymphadenopathy and continual but gradual decline in the absolute circulating CD4 lymphocytes, along with evidence of immune activation. The sequential progression of the lymph node pathology ranges from early lymphoid hyperplasia, to dissolution of germinal centers, eventually leading to severe follicular and paracortical lymphoid depletion during the late stages of the disease (Hirsch *et al.*, 1991). The final phase, AIDS, is characterized primarily by severe depletion of CD4 lymphocytes and the onset of opportunistic infections such as cytomegalovirus (CMV), *Pneumocystis carnii* pneumonia, and mycobacterial infections (Baskin *et al.*, 1988; Hirsch and Johnson, 1994; McClure *et al.*, 1989; Zhang *et al.*, 1988). In HIV-1 infected humans, a switch in the major coreceptor (from CCR5 to CXCR4) used by HIV-1 has been observed in about 50% of patients as they progress to AIDS. Coreceptor use of SIV and HIV is reviewed in Unutmaz *et al.* (1998). The coincidence of this switch has lead some to postulate that the increased cytopathic effects of these viruses (due to the corecptor switch) is associated with an increase in virulence of late-stage isolates (reviewed by Fenyo *et al.*, 1989). These theories are impossible to address with HIV in humans. Recent studies with the SIVmne/macaque model demonstrate that viruses that evolve in an infected macaque become more virulent as assessed by increased rapidity in AIDS induction in naive macaques (Edmondson *et al.*, 1998; Kimata *et al.*, 1999). Interestingly, the increase in virulence of SIVmne is not associated with a change in coreceptor use (Kimata *et al.*, 1999).

2. AIDS-Inducing Strains of SIV

As is evident from Table II, the pathogenicity of SIVsm/mac strains varies significantly from attenuated to highly pathogenic. A wide range of isolates are available as uncloned virus stocks along with numerous molecularly or biologically cloned viruses. The clinical course of infected animals varies significantly based on the virus strain and the individual animals' response to infection. Some of these factors that can influence the pathogenicity of SIV isolates are the source of the virus, strain of virus, whether it is molecularly cloned, the species of animal inoculated, and the tissue culture passage history of the virus. Some strains are relatively nonpathogenic, others result in AIDS after a long latency, and some induce AIDS more rapidly.

This situation affords a spectrum of experimental options that is broad enough to be confusing. However, it also means that virtually regardless of the specific aspect of HIV pathogenesis of interest, there is an SIV infection model that nicely recapitulates the essential aspects of the process. Thus, uncloned SIVmac251 (Letvin *et al.*, 1985), SIVsmB670 (Zhang *et al.*, 1988), SIVsmE660, and SIVsmE543 (Hirsch and Johnson, 1994) are all highly pathogenic isolates. Characteristically, 10–30% of animals inoculated with such strains fail to develop SIV-specific immune responses and die rapidly within 6 months of inoculation, as depicted in a Kaplan–Meier plot of survival of SIV-infected macaques in Fig. 3. It is fairly rare to observe long-term survivors (long-term nonprogressors) of infection with such highly pathogenic strains. In contrast, there are other AIDS-inducing strains that are slightly less pathogenic which do not induce rapidly progressive disease. Such strains include SIVmne (Benveniste *et al.*, 1988), SIVsmF236 (Zhang *et al.*, 1988; Hirsch and Johnson, 1994), and SIVsmm9 (McClure *et al.*, 1989). In general, all of the uncloned SIVsm/SIVmac isolates exhibit some virulence. There are fewer examples of pathogenic molecularly cloned (or biologically cloned) SIV isolates. Most of the molecularly cloned viruses are minimally, if at all, pathogenic. These viruses include SIVmac1A11 (Luciw *et al.*, 1992; Marthas *et al.*, 1993), SIVsmH-4 (Hirsch *et al.*, 1989), SIVmacBK-28 (Edmondson *et al.*, 1998), and SIVsm62d (Hirsch *et al.*, 1998a). There are only a handful of AIDS-inducing molecularly cloned SIVs, including SIVmac239 (Kestler *et al.*, 1988) and SIVsmE543-3 (Hirsch *et al.*, 1998).

3. Strains with Variant Pathogenesis

Only one strain of SIV (SIVsm/PBj) appears to be acutely lethal. Experimental infection of pigtail macaques results in a highly reproducible syn-

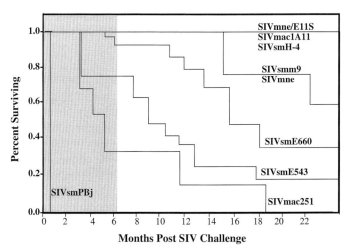

FIGURE 3 Variability in survival of macaques infected with various strains of SIVsm and SIVmac as depicted in a Kaplan–Meier plot. Some strains such as SIVmac251, SIVsmE660, and SIVsmE543 are highly pathogenic, whereas other strains are minimally pathogenic. Characteristically, 10 to 30% of macaques inoculated with highly pathogenic strains develop AIDS within 6 months of inoculation. These plots were constructed from data on survival of macaques inoculated with different SIV strains (Baskin *et al.*, 1988; McClure *et al.*, 1989; Hirsch and Johnson, 1994; Letvin *et al.*, 1985; Letvin and King, 1990).

drome of severe diarrhea and death by 7 to 14 days postinoculation (Fultz *et al.*, 1989; Fultz and Zack, 1990; Dewhurst *et al.*, 1990; Novembre *et al.*, 1993). Originally isolated from a pigtailed macaques inoculated with the AIDS-inducing SIVsmm9 strain, SIVsmPBj is an interesting virus characterized by distinct *in vitro* properties and *in vivo* pathogenesis. *In vitro*, in contrast to other SIV and HIV isolates, the virus replicates efficiently in cultures of resting T lymphocytes. At least part of this phenotype appears to be related to a characteristic mutation in the *nef* gene that introduces an ITAM motif associated with the ability to activate T cells. Introduction of this mutation into the *nef* gene of the AIDS-inducing SIVmac239 strain partially recapitulates the PBj phenotype (Du *et al.*, 1995). *In vivo* infection with the PBj virus is characterized by early high-level viral replication in the lymphoid tissues of the gastrointestinal tract. This is accompanied by a massive infiltration of lymphocytes and inflammatory changes and production of IL-6 and other cytokines, leading to diarrhea, dehydration, erosion of the mucosa, and death within 2 weeks following inoculation (Fultz and Zack, 1994).

In addition to the SIV viruses described above which cause a gradual depletion of CD4 cells that mimics the pattern seen in human infection with HIV-1, experimental infection systems have been described that result in dramatic, rapid, and virtually complete loss of CD4 cells. One such system involves infection of pigtailed macaques with the HIV-2 isolate HIV-2/287 (Hu *et al.*, 1993; Watson *et al.*, 1997a). This particular isolate was derived from the HIV-2/EHO isolate by serial passage in pigtailed macaques through which the virus acquired virulence. Following inoculation, high levels of viral replication are observed, while circulating CD4 cells decline to virtually unmeasurable levels over a period of weeks. Histopathologic analyses demonstrate extensive T-cell depletion of lymphoid tissues and confirm that the loss of measurable circulating cells is due to true loss of the cells.

A similar pattern of pathogenesis is seen with certain engineered viruses, designated SHIVs, for simian/human immunodeficiency viruses. Engineered by recombinant techniques for use in vaccine experiments in which investigators wished to study the envelope glycoprotein of HIV-1 in an *in vivo* nonhuman primate model, the SHIVs are chimeric viruses that essentially consist of viral cores composed of SIV internal structural proteins surrounded by HIV-1 envelopes. The subtleties of the exact construction of different SHIVs, including the source of accessory genes and regulatory sequences (HIV or SIV), are reviewed in Lu *et al.* (1996). The initial SHIVs that were inoculated into animals proved to replicate only transiently at low levels and were apathogenic (Li *et al.*, 1992; Shibata *et al.*, 1991). It was only after varying degrees of *in vivo* passage that pathogenic SHIVs were successfully isolated (Joag *et al.*, 1997; Lu *et al.*, 1998; Shibata *et al.*, 1997). Thus there are a number of independently isolated pathogenic SHIVs. SHIV/89.6P was derived from SHIV89.6, which expressed the primary isolate 89.6 envelope. SHIV/KU-1 (Joag *et al.*, 1997) was derived from a SHIV expressing the HIV-1/IIIB envelope (Li *et al.*, 1992), and SHIV/DH12R was derived from SHIV expressing the envelope of the primary HIV-1 isolate, DH12 (Shibata *et al.*, 1997). These viruses are associated with sustained, high-level viral replication (similar to levels seen with pathogenic SIV isolates) and rapid, virtually complete depletion of CD4 cells, strongly reminiscent of the pattern seen in HIV-2/287 infection of pigtail macaques. Each of the SHIV strains (SHIV89.6P, SHIV/KU-1 and SHIV/KU-2, and SHIV/DH12R) that acquired virulence express a CXCR4-utilizing envelope. Interestingly a number of mutations in the envelope gene acquired through the course of *in vivo* passage appear to be conserved between different SHIV isolates that became pathogenic through *in vivo* passage. Strains which express a CCR5-utilizing envelope do not appear to induce the rapid peripheral CD4 lymphocyte depletion (Harouse *et al.*, 1999). These viruses provide a system for the assessment of candidate vaccines that incorporate HIV-1 envelope glycoprotein as a part of the immunogen with a rigorous pathogenic challenge, allowing evaluation of both laboratory and clinical endpoints (Lu *et al.*, 1996).

4. Undefined Host Factors Influence Variable Disease Outcome

For a given virus, there is generally a characteristic associated range of pathogenicity in a given macaque species that is broadly consistent from experiment to experiment and correlates with the extent of viral replication. Interestingly, even a given virus can vary in replicative capacity and pathogenicity *in vivo* in different macaque species. The biologically cloned virus SIVmne/E11S (Benveniste *et al.*, 1994) and the isolate from which it was derived (SIVMne; Benveniste *et al.*, 1986) exhibit among the clearer examples of species-dependent pathogenicity upon experimental infection. *Macaca mulatta* can be infected, but appears to be relatively resistant to pathogenic consequences of infection. In contrast, infection is associated with depletion of CD4+ T cells and development of AIDS over 1–2 years in the majority of inoculated *Macaca nemestrina*, while experimental inoculation of *Macaca fasicularis* results in AIDS at a slightly lower frequency and typically after a longer duration. The underlying basis for this phenomenon is not well understood. These differences in pathogenicity in different species correlate broadly with the extent of replication by the different viruses in different species. The species of macaque used for experimental infection also has a major impact upon pathogenesis. In general, pigtailed macaques (*M. nemestrina*) appear to be the most susceptible to the majority of SIVs and can even be infected with strains of HIV-1 (albeit with very low viral replication levels) (Agy *et al.*, 1992). Thus SIVagm and SIVsmPBj are uniformly pathogenic in this species but not in rhesus macaques (Hirsch *et al.*, 1995; Lewis *et al.*, 1992). However, there are some exceptions to this rule.

For example, SIVmac239 is highly adapted to rhesus macaques and is significantly less pathogenic in pigtailed and cynomologus macaques.

Regardless of the relative virulence of a particular SIV isolate, considerable biologic variation can occur between different identically inoculated animals. Thus, the disease course in an individual animal can vary from rapid to intermediate to slow. There is a spectrum of potential responses to SIV infection with specific strains and the spectrum varies for each individual isolate. A small number of animals inoculated with a highly pathogenic strain of SIV may not develop SIV-specific antibody and will die rapidly, whereas others mount a more effective immune response and survive for longer periods of time (Zhang *et al.*, 1988). The specific host factors responsible for this great variation in response to infection are not known. The disease course in rhesus macaques can be predicted to some degree by assessment of *in vitro* susceptibility to SIV of PBMC from individual macaques (Lifson *et al.*, 1997).

III. Viral Load Measurements as a Prognostic Indicator ———

A. Measurement of Viral Load

In the course of characterizing various systems of different species of monkeys infected with different strains of SIV it has become clear that the extent of viral replication, or "viral load," is one of the most important determinants of pathogenesis (Hirsch *et al.*, 1996; Lifson *et al.*, 1997; Watson *et al.*, 1997; Staprans *et al.*, 1999). Viral load is most conveniently assessed by measurement of the level of virion-associated SIV RNA in plasma. This observation, which parallels similar observations in HIV-infected humans (Mellors *et al.*, 1996; O'Brien *et al.*, 1996), has been enabled by the development of laboratory methods that allow the sensitive, accurate, and precise quantitation of viral load in specimens from infected animals. Prior to considering in detail the role of viral load measurements in understanding SIV pathogenesis and in the evaluation of experimental vaccines and therapies, we briefly review the approaches used to perform such measurements.

1. Classic Methods of Measuring Viral Load

Initial approaches for measurement of viral load in SIV-infected animals used classic methods, such as limiting dilution infectivity cultures with indicator cells for plasma or PBMC, capture immunoassays for detection of viral proteins, or immunohistochemical/*in situ* hybridization analysis of tissues. However, infectivity cultures are expensive and time and labor intensive and suffer from limitations in assay reproducibility and dynamic range. Capture immunoassays for viral proteins are simple and convenient, but are of limited sensitivity and may be seriously confounded by interference from endogenous antibodies present in specimens from infected or vaccinated animals. Analysis of tissues by immunohistochemistry for SIV antigens or *in situ* hybridization for SIV RNA are critical methods for localizing infected cells in tissues but are difficult to standardize for quantitative purposes. In addition, there is considerable variation in the expression of virus in different tissues that could be affected by sampling. Finally, obtaining the samples is relatively invasive as compared to blood sampling, even for those tissues that are most readily analyzed (tonsils and peripheral lymph nodes). This limits longitudinal assessments of viral load. As was the case for HIV, it became apparent that nucleic-acid-based approaches to assay plasma viral RNA would provide the best combination of feasibility, cost-effectiveness, and assay performance characteristics for measurements of SIV viral load.

2. Branched DNA Methods

As for HIV, the available nucleic acid methods for quantifying viral load can be classified as either target-amplification or signal-amplification approaches. In each instance, an amplification step is introduced to allow indirect measurement of the very small (in absolute terms) amounts of viral RNA present in test samples, amounts too small to be measured without amplification. The primary signal amplification method in use for quantification of SIV uses an approach designated "branched DNA" or bDNA detection. In this method, solid-phase bound oligonucleotides are used for sequence-specific capture of SIV RNA, which is then subsequently detected by oligonucleotides containing complementary sequences for binding to the captured SIV RNA probes. Extensive "branched DNA" arms, to which are conjugated numerous alkaline phosphatase moieties, allow sensitive and quantitative chemiluminescent detection (Marx *et al.*, 1996; Staprans *et al.*, 1999). The method is robust, with good precision, although it requires a relatively large sample volume and has not been as sensitive as other methods (see below). A newer version of the basic assay has improved sensitivity.

3. Quantitative Competitive PCR Methods for Assaying Viral RNA

Other available methods depend on amplification of the target template itself to achieve quantitation in a measurable range. The most widely used of these methods are based on variations of the polymerase chain reaction. However, as reviewed in detail elsewhere, there are serious intrinsic problems in attempting to use PCR for quantitative applications (Piatak *et al.*, 1996; Piatak and Lifson, 1997). To overcome these problems, two main approaches have been employed. In the first approach, designated internally controlled PCR, or competitive PCR, or sometimes quantitative competitive PCR (QC-PCR), a synthetic internal control template is spiked into the test sample. This internal control template is designed to use the same primers as the test target template and to be reverse transcribed and PCR amplified with efficiency comparable to the test template, but to be independently quantifiable. The basic approach thus is based on the premise that by testing a fixed but unknown amount of test template against a limited bracketing range of spiked internal control template, the amounts of amplified product for unknown and test template at the end of PCR can be measured. Since both templates are reverse transcribed and amplified with comparable efficiency, the ratio of the measurable amounts of products following amplification should reflect the ratio of target templates prior to amplification. Based on a regression of the ratio of measured postamplification products for the two templates as a function of the input copy number of the control temples, the titration equivalence point can be determined by interpolation. The key feature of this approach is that it is based on relative quantitation of target and internal control templates, not absolute endpoint quantitation of target template. Thus it avoids many of the intrinsic problems associated with absolute quantitation of endpoint PCR amplifications (Piatak *et al.*, 1993, 1996; Piatak and Lifson, 1997).

4. Real-Time Methods for Assaying Viral RNA

Internally controlled PCR/RT–PCR approaches have proved to be sensitive, robust, and reliable and have been used extensively in pathogenesis studies and in the evaluation of experimental vaccines and therapies in various SIV model systems (Hirsch *et al.*, 1996, 1998; Nowak *et al.*, 1997; Lifson *et al.*, 1997; Tsai *et al.*, 1998, Van Rompay *et al.*, 1998). However, these approaches are time and labor intensive, which limits throughput. Recently, new approaches have been developed based on kinetic PCR or "real-time" PCR (Heid *et al.*, 1996; Gibson *et al.*, 1996; Livak *et al.*, 1995; Suryanarayana *et al.*, 1998). The key feature of these methods is that the measurement is based on kinetic or real-time measurements of accumulating product during ongoing PCR amplification rather than endpoint measurements of accumulated amplified PCR product at the conclusion of the reaction. This confers several advantages. First, it allows measurements to be performed during the earliest stages of the exponential phase of PCR amplification. This period exhibits the most consistent relationship between input template copy number and PCR product. Second, the kinetic nature of this approach provides an extremely broad linear dynamic range. Finally, since measurements are obtained during the PCR amplification, there is no need for separate analysis of amplified products at the conclusion of amplification. This factor increases throughput and minimizes potential for PCR back-contamination associated with manipulation of amplified material.

To realize these advantages requires the ability to sequentially and non-invasively monitor accumulation of specific amplicons during ongoing PCR reactions. This technical challenge has been elegantly solved with instrumentation that provides for light excitation of ongoing PCR reactions and quantitative collection of the resulting emitted fluorescence signal. The fluorescence signal is derived via a variety of reagents and schemes that depend on release from fluorescence resonance energy transfer-mediated quenching of fluorescence from fluorochrome-labeled hybridizing oligonucleotide primers or probes. Release from quenching is obligately and quantitatively linked in a proportional manner to specific amplification of the target sequence. While a detailed description of these methods is beyond the scope of this chapter, they provide excellent sensitivity, precision, and dynamic range, with excellent specimen throughput.

Real-time PCR methods represent a significant advance in these regards, although it is important to note that there can be pitfalls with this emerging technology. The potentially severe errors in quantitation can be introduced by sequence mismatches between probes and cognate target sequences to which they are intended to hybridize (Suryanarayana et al., 1998). In addition to PCR-based methods, there are a number of other technologies based on cyclical enzymatic target-amplification strategies. Discussion of these less widely used methods is beyond the scope of this chapter.

B. Viral Replication in Experimental Lentiviral Infection of Primates

Measurements of viral load through quantitation of virion associated viral RNA in plasma has been invaluable in defining the relationship between viral replication patterns and pathogenesis in different SIV infection models. As noted above, there are a number of different systems used in experimental infection studies, varying in both the species of macaque employed and the strains of virus used. Indeed, one of the principal advantages of the SIV-infected macaque as an animal model system for AIDS is the ability to define the amount, route, dose, and timing of inoculation and the identity of the inoculating virus.

I. SIV Infection of Adapted Natural Host Species

As described above for the families of SIVs, there appear to be natural host species in which the virus does not appear to be pathogenic. Given the well-established relationship between level of viral replication and pathogenicity, one obvious hypothesis is that the adapted host species have simply developed mechanisms to limit viral replication, thereby preventing pathogenicity. However, available studies of the natural host for SIVsm, sooty mangabeys, suggest that the levels of viral replication are comparable to the levels seen when the same viruses are used to infect new host species, such as rhesus or pigtail macaques (Rey-Cuille et al., 1997; Kaur et al., 1998, Villinger et al., 1996, 1999). The high viral load in such animals is confirmed and supported by the ease with which virus can be isolated from plasma of such animals as well as by in situ hybridization of lymphoid tissues of sooty mangabeys. Additional studies aimed at determining the basis of nonpathogenicity of these viruses in the adapted host will hopefully identify the responsible host factors. Progress in this area may provide important insights into understanding, and ultimately preventing, the pathogenesis of AIDS.

In a similar vein, analysis of African green monkeys infected with SI-Vagm strains is also of interest. The technical issues involved in plasma viral load measurement in African green monkeys are more complicated due to the greater genetic diversity among SIVs infecting these species. As a result, there is not a good consensus of understanding within the field concerning quantitation of plasma viral load levels in various different African green monkey species naturally or experimentally infected with different SIVagm isolates. Studies of tissues from naturally infected African green monkeys revealed that the majority have very low expression of virus in tissues (Beer *et al.*, 1996; V. Hirsch, unpublished observations). Real-time assays for viral RNA levels in plasma will be necessary to address whether the viral load in such animals is lower than that observed in sooty mangabeys. This will require the development of species-specific primers and probes to reliably detect the four specific SIVagm subtypes within African green monkeys (SIVagm/tan, SIVagm/ver, SIVagm/gri, and SIVagm/sab). However, once these problems are resolved, additional studies to characterize the rate and extet of viral replication, and the basis of nonpathogenicity in this adapted host species, will also be of great interest.

2. Experimental Pathogenic Infection of Macaques

There are a number of other experimental systems characterized by much greater pathogenicity, including infection of rhesus or pigtail macaques with SIV viruses such as SIVsmE660 (a biological swarm), SIVsmE543-3, SIVmac251, SIVmac239, or SIVsmB670. Infection with these viruses generally follows a consistent pattern, illustrated for SIVsmE660 in pigtail macaques in Fig. 4. Following intravenous inoculation, virus is first detectable in plasma within 3 to 7 days postinoculation. Plasma SIV RNA levels increase exponentially, reaching peak values from 10 to 20 days postinoculation. Over the next couple of weeks, there is typically a down-modulation of circulating virus levels, of varying degrees, leading in most animals to a relative stabilization of plasma viral load at what has been termed the postacute viral load "set point" or "inflection point," approximately 6–8 weeks postinoculation. Plasma viral load at this "set point" is broadly predictive of the subsequent clinical course, with animals that show higher viral loads at this time showing persistently elevated plasma virus levels and on average a more rapid progression to AIDS and death (Hirsch *et al.*, 1996; Watson *et al.*, 1997). Conversely, the small percentage of animals that show lower levels of plasma SIV over this time interval show persistently restricted viral load and much slower disease progression. In some instances, these animals exhibit a nonprogressive clinical course, with low or unmeasurable levels of plasma virus, implying host control of readily demonstrable persistent infection. This situation mimics the rare human patients with long-term

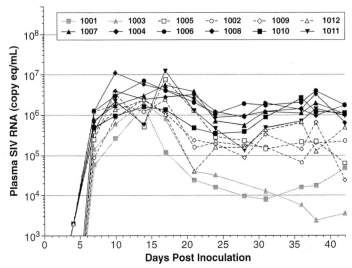

FIGURE 4 Variability in plasma viremia within a strain in a cohort of pigtailed macaques is shown over the first 40 days after intravenous inoculation of SIVsmE660 (Lifson *et al.,* 1997). Macaques with rapidly progressive infection that failed to seroconvert are shown in black symbols with solid lines. Those macaques with partial control of viremia are shown with open symbols and dashed lines and those which controlled viremia to a significant extent ($<10^4$/ml) are shown with shaded symbols.

nonprogressive HIV-1 infection (Panteleo *et al.,* 1995; Cao *et al.,* 1995). A significant percentage of animals show little or no evidence of control of viral replication, as measured by plasma viremia. These animals typically fail to seroconvert and experience a rapidly fatal clinical course, generally dying less than 6 months postinoculation with massive levels of plasma viremia (up to 10^9 copies/ml). As shown in Fig. 5, there is a remarkable correlation between relative viral load assessed by in situ hybridization and that assessed by plasma viral RNA assays. Thus plasma viral load measurements appear to reflect the ongoing virus expression in SIV-infected macaques (Lifson *et al.,* 1997).

Of course, the patterns described above are generalizations, and the behavior of individual animals may vary, at least in part due to the ongoing dynamics of viral evolution within the host and the influence of the host immune response. Thus, the inoculated virus can evolve through *in vivo* replication (Edmondson *et al.,* 1998; Kimata *et al.,* 1999), in some instances increasing in pathogenicity such that virus recovered from late in infection may show greater and more rapid pathogenicity on inoculation into new naïve hosts (Kimata *et al.,* 1999).

3. Minimally Pathogenic SIV Isolates

a. Spontaneous Attenuated SIV Variants The SIV strains described above are particularly useful for studies of comparatively rapid pathogenesis or

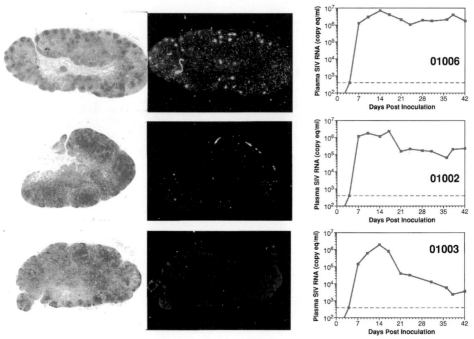

FIGURE 5   Correlation between the pattern of viral replication in three macaques inoculated with SIVsmE660 is shown on the right with the detection limit of the assay shown by a dotted line. The middle panel shows SIV-specific *in situ* hybridization of lymph node biopsies obtained at 4 weeks postinoculation and the left panel shows the corresponding H&E-stained histopathologic sections. The macaque at the top (01006) demonstrated uncontrolled viremia and progressed rapidly to AIDS with high lymph node expression of virus. The macaque in the middle panel (01002) seroconverted and decreased viremia to 100,000 copies/ml. The virus expression in the lymph node is moderate and there is evidence of trapping of virus–immune complexes on follicular dendritic cells. The macaque at the bottom (01003) controlled viremia to a greater degree to approximately 1000 copies/ml and few SIV-expressing cells were observed by *in situ* hybridization.

rigorous testing of vaccines or treatment approaches. However, there are other viruses that are essentially apathogenic, even when experimentally inoculated into host species readily susceptible to highly pathogenic infection with closely related viruses. Examples include the 1A11 molecular clone, derived from the highly pathogenic SIVmac251 swarm (Marthas *et al.*, 1989, 1993). This virus represents an interesting model in that it frequently mediates an "abortive" infection, especially after vaginal inoculation, characterized by a transient low-level viremia. The virus replicates to low levels and then becomes undetectable, while the animals remain clinically well, with no evidence of progressive SIV disease. This pattern has some similarities to another phenomenon, in which the SIVmac251 swarm, which generally replicates to high levels and is strongly pathogenic following intravenous inoculation, can be associated with a different profile following low-dose vaginal inoculation (McChesney *et al.*, 1998). In this latter instance, the majority of animals show typical productive infection and disease course; however, a small percentage of vaginally inoculated animals show a pattern of transient low-level viremia and limited, atypical immune responses (often with variable low-level cellular responses in the absence of seroconversion). Exhaustive efforts at necropsy several years postinoculation demonstrated the persistence of virus, including replication competent virus in many instances, in most animals. However, in the vast majority of macaques that show this transient viremia pattern the infection appears to be largely latent, both virologically and clinically (McChesney *et al.*, 1998). Other examples of minimally pathogenic but ultimately AIDS-inducing strains include molecularly cloned SIVsmH-4 (Johnson *et al.*, 1990), SIVmac/BK28 (Edmondson *et al.*, 1998), and SIVsm62d (Hirsch *et al.*, 1998). Characteristically, these attenuated and minimally pathogenic SIV strains exhibit low levels of both primary and chronic plasma viremia as demonstrated by the comparison of plasma viremia in macaques inoculated with SIVsmE660 and SIVsm62d in Figs. 6A and 6B. As evident in this figure, range in both primary and postacute plasma viremia is much lower in macaques infected with the less pathogenic strain, SIVsm62d, as compared to SIVsm62d-infected macaques, which maintain lower viremia and do not show disease progression, whereas macaques at the top of the spectrum of viremia levels develop AIDS.

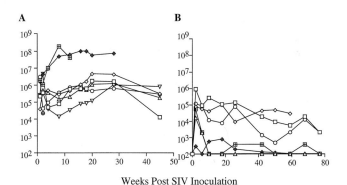

Weeks Post SIV Inoculation

FIGURE 6 Differences in the range of plasma viremia observed in macaques inoculated with a highly pathogenic strain (SIVsmE660) and a minimally pathogenic AIDS-inducing strain (SIVsm62d) are shown. (A) The range of viremia in SIVsmE660-inoculated rhesus macaques (Hirsch *et al.*, 1996). The two macaques with the highest viremia in A died with AIDS by 12 and 32 weeks respectively, whereas the others survived for approximately 1 year. (B) Lower plasma viremia in macaques inoculated with the less pathogenic SIVsm62d molecularly cloned virus. The two animals with higher relative viremia (open symbols) died with AIDS at 58 and 77 weeks postinoculation, whereas the other macaques remained healthy for over 3 years. These animals are described in Hirsch *et al.* (1998a).

b. Genetically Modified Attenuated SIV Variants   An additional example of the relationship between viral replication levels and pathogenicity can be drawn from studies of attenuated strains of SIV generated by mutational deletion of accessory genes in an effort to develop strains suitable for evaluation as candidate live attenuated vaccine strains. A series of deletion mutants have been developed and evaluated. The wild-type virus from which the deleted mutants were constructed, SIVmac239, establishes a high-level persistent viremia and pathogenic, progressive infection in the vast majority of inoculated rhesus macaques (Kestler *et al.*, 1988). In striking contrast, in the majority of inoculated animals the deleted mutants show a blunted peak viremia that resolves, with plasma SIV RNA levels generally decreasing to below the level of detection in most assays (Desrosiers *et al.*, 1998). The degree of blunting of the peak in *in vivo* viremia correlates with the extent of attenuation through mutation (Desrosiers *et al.*, 1998; Johnson *et al.*, 1999).

4. Viral Replication Patterns of Viruses with Variant Pathogenesis

In addition to the experimental SIV infection systems described above, there are some other systems involving experimental infection of macaques that result in variant patterns of pathogenesis rather than the typical progressive infection leading to AIDS. These systems have been effectively used for specific experimental purposes.

a. SIVsmPBj   The kinetics of viremia in SIVsmPBj-inoculated macaques is more rapid than observed with AIDS-inducing strains of SIV, frequently peaking by 7 days after intravenous inoculation (Hirsch *et al.*, 1998; O'Neil *et al.*, 1999) versus 11 to 14 days for other SIVmac and SIVsm strains (Lifson *et al.*, 1997). However, the actual peak levels are not significantly higher than that observed with AIDS-inducing strains. Plasma viremia is accompanied by a decline in all lymphocyte subsets, as illustrated in Fig. 7A. The kinetics of viremia in intrarectally inoculated macaques is delayed by 3 to 4 days as compared to those inoculated intravenously (Fig. 7B). In keeping with the relationship between viral replication levels and pathogenesis seen in other systems, site-directed mutants of SIVsmPBj (nef, vpr, or vpx mutants) that show blunted *in vivo* viral replication also show markedly blunted pathogenesis, with animals surviving infection (Novembre *et al.*, 1997; Hirsch *et al.*, 1998). While this pattern of pathogenesis is sufficiently different from the typical course of progressive SIV or HIV infection leading to AIDS to not be an optimal model, it has been usefully employed to study specific questions in AIDS pathogenesis (Hirsch *et al.*, 1998).

b. Acutely CD4 Depleting Viruses   Overall, the kinetics of viral replication following infection with the acutely CD4-depleting SHIV isolates, as reflected by plasma SIV RNA levels, parallel the pattern seen for pathogenic

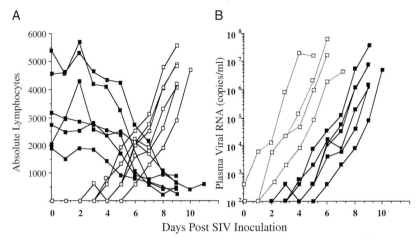

FIGURE 7   (A) The kinetics of viremia (open symbols) and accompaning lymphopenia (black symbols) in macaques inoculated with the acutely lethal, molecularly cloned, SIVsmPBj6.6 extracted from data presented in Hirsch *et al.* (1998b). (B) The delay in viremia observed in macaques inoculated intrarectally (black symbols, solid lines) as compared to those inoculated intravenously (open symbols, dotted lines).

SIV isolates (Reimann *et al.*, 1999). Infection with pathogenic SHIV isolates leads to the rapid development of high levels of plasma viremia, typically as high and as rapidly or slightly more rapidly than is observed for pathogenic SIV isolates (Lu *et al.*, 1998; Joag *et al.*, 1997, 1998; Reimann *et al.*, 1996; Shibata *et al.*, 1997). After viremia reaches a peak there can be some modest down-modulation of virus levels, but, in general, levels remain high through the period of CD4 depletion. With extensive depletion of CD4+ T cells there may be some modest decrease in levels of circulating virus, although it is interesting that moderate levels of plasma viremia are maintained even after depletion of CD4+ T cells from the circulation and lymph nodes is virtually complete.

In striking contrast, infection with the nonpathogenic SHIVs, from which the pathogenic, acutely CD4-depleting SHIVs are derived by *in vivo* passage, results in only transient viremia, reinforcing the relationship between levels of viral replication and pathogenesis *in vivo* (Reimann *et al.*, 1999; Li *et al.*, 1992; Shibata *et al.*, 1991). For the pathogenic SHIVs, a greater inherent cytopathicity for CD4+ T cells may also contribute to the dramatic CD4-depleting phenotype and overall pathogenesis *in vivo*. All of the SHIVs that induce the rapid CD4 depletion utilize CXCR4 as their coreceptor. Few SIVs that utilize CCR5 have been constructed (Luciw *et al.*, 1992; Harouse *et al.*, 1999) and these viruses do not appear to cause the acute peripheral CD4 depletion. A recent study has demonstrated, however, that a CCR5-using SHIV causes depletion of CD4+ intraepithelial lymphocytes in the gastrointestinal tract (Harouse *et al.*, 1999) similar to that seen in pathogenic SIV infection (Veazy *et al.*, 1998).

IV. Modulation of Viral Replication by
Partially Protective Vaccines _____

Vaccination includes the prophylactic immunization as well as the immunization to modify disease during the chronic stage of infection. Therapeutic vaccination is beyond the scope of this chapter and is not discussed. Quantitative assays of viral load provide an extremely useful tool in the experimental evaluation of vaccines for AIDS. Nucleic-acid-based viral-load studies provide among the most sensitive means of evaluating protection from challenge in vaccine studies. The absence of detectable viral RNA in plasma, and viral RNA or DNA in PBMC and lymph node cells, can be used to confirm complete protection from infection ("sterilizing immunity") in macaques in which there is a failure to isolate infectious virus and lack of an anamnestic antibody response. Even in many instances where complete protection from infection was not achieved, measurement of viral load has proved extremely valuable in the evaluation of experimental vaccines in the SIV system. The use of the SIV/macaque model to evaluate AIDS vaccines has been extensively reviewed (Almond and Heeney, 1998; Hu *et al.*, 1993; Letvin, 1998; Nathanson *et al.*, 1999; Schultz and Hu, 1993; Schultz and Stott, 1994). Thus the discussion in this chapter deals primarily with studies in which protection from infection was not achieved but where measurement of viral load in the postchallenge period has demonstrated vaccination-associated reductions in viral load in vaccinated animals that did become infected relative to unvaccinated controls. In many instances, long-term follow-up has shown that substantial reduction of viral load in the immediate postchallenge period can be associated with sustained modulation of viral replication and improved clinical course relative to control animals (Hirsch *et al.*, 1996).

At the present time, a number of vaccine strategies have demonstrated complete to partial protection in primate models. To those unfamiliar with the various strains of SIV (and SHIV) used in vaccine experiments, the results of challenge experiments can be difficult to decipher and vaccine modalities almost impossible to compare. It is critical to remember that the level of protection observed is impacted not only by the efficacy of the vaccine but also by the genetic relatedness of the vaccine virus and the challenge virus and by the virulence of the challenge virus. Assessment of vaccine studies should also include evaluation of the neutralization phenotype since some viruses may appear to be quite similar genetically but antibodies generated to one virus may not cross-neutralize the other strain. Thus a vaccine can appear to mediate complete protection from infection if the animals are challenged with a virus identical to the vaccine virus that has low virulence (low AIDS-inducing potential and low virus loads). However, the same vaccination regimen may afford little or no protection from a more robust challenge. Therefore, in evaluating vaccine studies it is critical to understand the typical viral replication profile and clinical course associated with the individual viruses used as challenge strains in vaccine trials, as described in Section III.

A. Attenuated Live SIV Vaccines

The most effective vaccine modality still appears to be attenuated live SIV. One of the first of such attenuated live SIV mutants to be used in such a fashion is the SIV mac/1A11 virus, which is a spontaneously generated attenuated molecular clone. Prior infection of macaques with the 1A11 virus resulted in protection from AIDS when the animals were challenged with a pathogenic SIV strain (Marthas *et al.*, 1990). Later studies with genetically modified strains of SIV with deletions in the accessory genes (*nef* and *vpr*) and the LTR revealed that prior infection can provide complete protection if challenge is delayed for 6 months to a year after vaccination (Connor *et al.*, 1998). Prior infection with these viruses has resulted in some of the more impressive vaccination related protection observed to date in the SIV system (Daniel *et al.*, 1992; Desrosiers *et al.*, 1998; Johnson *et al.*, 1999). However, even with this approach, broad protection against heterologous challenges has proven difficult (Lewis *et al.*, 1999; Desrosiers *et al.*, 1999). Even in such situations, a reproducible reduction in viremia has been observed (Desrosiers *et al.*, 1998; Johnson *et al.*, 1999; Lewis *et al.*, 1999). Since the level at which plasma viremia plateaus after the primary phase of infection is an excellent prognostic indicator, significant reductions in plasma viremia in such vaccinated monkeys are associated with a long-term clinical benefit in these animals (Hirsch *et al.*, 1996; Watson *et al.*, 1997b).

The potential human use of this approach is precluded for the foreseeable future by observations that in some animals inoculated with these deleted attenuated viruses, both neonates and juveniles, pathogenic infections have been observed (Baba *et al.*, 1998, 1999; Alexander *et al.*, 1999). It is nonetheless interesting that the pathogenicity was associated with much higher levels of viral replication than is typically observed in the majority of animals receiving these viruses and with "compensatory" mutations that might be expected to increase viral replication levels (Desrosiers *et al.*, 1998). Thus even the exceptions to the rule of the behavior of these mutant attenuated viruses reinforces the rule of the relationship between the extent of viral replication and pathogenesis.

B. Live Viral Vectors

The best evidence for reduction in viremia as a consequence of prior vaccination has been observed in macaques immunized with live viral vectors that express SIV antigens. Thus, as with attenuated live vaccines, priming with vaccinia virus SIV envelope recombinants followed by a recombinant envelope antigen boost can under ideal circumstances prevent infection. Complete protection has thus been observed in one study that used as the challenge virus a biologically cloned SIV isolate (SIVmne/E11S) that is minimally pathogenic in the species used for the trial (Hu *et al.*, 1992). In this situation, protection appears to be mediated by type-specific neutralizing antibodies that were fortunately matched to the challenge strain. However, when similarly immunized macaques were challenged with a slightly more pathogenic, and more heterogenous, isolate, uncloned SIVmne, only partial protection was observed spanning the spectrum from complete protection and transient infection to reduction in viral replication (Polacino *et al.*, 1999). The investigators have also observed a significant reduction in viremia in macaques immunized with vaccinia virus core antigen recombinants, suggesting that genes other than envelope can contribute in vaccine-mediated protective effects (Hu *et al.*, 1993). In a similar vein, immunization of macaques with vaccinia virus envelope recombinant virus and boosting with recombinant envelope did not prevent infection following challenge with the highly pathogenic SIVmac251, but modulation of viremia was observed (Ahmad *et al.*, 1994).

There are a number of poxviruses with unique properties that are available for use as vaccine vectors (reviewed in Paoletti *et al.*, 1996; Tartaglia *et al.*, 1998). This includes the conventional vaccinia viruses used for the smallpox eradication campaign [such as New York Board of Health (Wyeth) and Copenhagen strains]. The use of these viruses in populations where a fraction of the vaccinees might be immunosuppressed due to HIV-1 infection is problematic due to the risk of disseminated fatal vaccinia virus infection in such individuals. Therefore attenuated poxviruses have been developed for use as vaccine vectors (Meyer *et al.*, 1991; Paoletti *et al.*, 1996; Moss *et al.*, 1996). NYVAC and MVA (modified vaccinia virus Ankara) are attenuated vaccinia viruses (Blanchard *et al.*, 1998) that have been explored as potential AIDS vaccine vectors in primate models. NYVAC is genetically modified version of the New York Board of Health strain, whereas MVA was spontaneously generated through passage in chicken embry fibrobalsts. Both have severe host range restrictions in mammalian cells and are safe in immunosuppressed animal models. In addition, the avipoxviruses, ALVAC, and fowlpox are also attractive candidates since they would be immunogenic in vaccinia-immunized individuals (Andersson *et al.*, 1996). Protection from infection with HIV-2, which is apathogenic in macaques, has been observed in macaques immunized with poxvirus recombinants of HIV-2 (Abimiku *et al.*, 1995; Myagkikh *et al.*, 1995). Modulation of plasma viremia has been observed in macaques immunized with both NYVAC and MVA–SIV recombinant vaccines (Benson *et al.*, 1998; Hirsch *et al.*, 1997). Modulation of viral load was more pronounced in macaques challenged by the intrarectal route (Benson *et al.*, 1998). A small proportion of the vaccinees in both of these studies controlled virus replication to extremely low levels and have become macaque equivalents of long-term nonprogressors (LTNP) (Abimiku *et al.*, 1997; Hirsch *et al.*, 1997). The effect on viremia in macaques immunized with MVA–SIV recombinants as compared to those immunized with a Wyeth–SIV recombinant or nonrecombinant vaccinia virus is illustrated in Fig. 8. Fowlpox recombinants expressing HIV-1 antigens have been used in combination with DNA priming. This approach significantly boosts CTL responses and has been shown to protect macaques against HIV-1 infection (Kent *et al.*, 1998). However, the relevance of this protection is unclear, since HIV-1 infection of macaques is highly transient in nature and therefore this constitutes the weakest of vaccine challenges.

C. Other Viral Vectors

A number of other nonpathogenic viruses under consideration and investigation as potential viral vectors for an AIDS vaccine are poliovirus replicons (Morrow *et al.*, 1999); adeno-associated virus (AAV) in the very early stages of development (Clark *et al.*, 1995); adenovirus (Robert-Guroff *et al.*, 1998); alphaviruses, including Semliki forest virus (SFV; Berglund *et al.*, 1997; Mossman *et al.*, 1996); and venezuelan equine encephalitits virus (VEE; Caley *et al.*, 1997). Adenovirus recombinants of HIV provided protection when used in a prime boost strategy in chimpanzees when the challenge strain was matched genetically to the vaccine virus and are capable of preventing infection (Robert-Guroff *et al.*, 1998). When a more rigorous

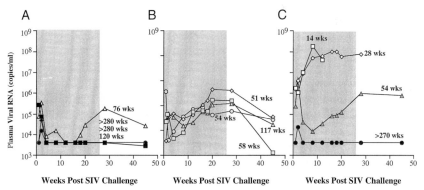

FIGURE 8   The effect of vaccination with a partially protective vaccine, MVA-expressing SIV env and gag-pol (Hirsch *et al.*, 1996). Plasma viremia during the first 50 weeks postintravenous inoculation with SIVsmE660 of macaques vaccinated with (A) MVA-expressing SIV gag-pol and env, (B) Wyeth-expressing SIV gag-pol and env, and (C) nonrecombinant vaccina virus. The survival of the macaques in weeks is indicated by the numbers near the plot of plasma viremia. The shaded area indicates the first 6 months after challenge, the period in which rapid progressor macaques develop AIDS. Rapid progressors are plotted with open symbols, slower progressors with shaded symbols, and nonprogressors with black symbols.

challenge is used such as in the SIV/macaque model, macaques immunized with a adenovirus–SIVenv recombinant and boosted with SIVmac gp120 were not protected from infection following intravaginal challenge with pathogenic SIVmac251. However, significant reduction in plasma viremia was observed in these vaccinated macaques (Buge *et al.*, 1997). Macaques immunized with SFV-expressing envelope were protected from acutely lethal disease when challenged with SIVsmPBj, although all animals became infected. Although viremia was not characterized in these animals, it is likely that the protection observed with SFV recombinant viruses was due to blunting of acute viremia. There are no published reports on challenge studies in macaques immunized with either VEE– or AAV–SIV recombinants although these studies are in progress.

D. DNA Immunization

Another exciting strategy for generating both cellular and humoral immunity is the use of naked DNA as an immunogen as reviewed by Robinson (1997). DNA can be adminstered either intramuscularly or coated on gold particles by gene gun. Although this method is extremely immunogenic in mice, there have been difficulties in generating similar responses in primates. Nevertheless, there are some preliminary trials in which protection from infection was achieved in chimpanzees immunized with HIV-1 envelope and challenged with HIV-1/SF-2 (Boyer *et al.*, 1997). The SF-2 strain of HIV-1 is considered to be a less rigorous challenge than other HIV-1 strains such as IIIB, since it shows restricted replication in nonimmunized chimps. Others have investigated the use of HIV-1 envelope DNA in rhesus macaques. Macaques immunized with env and boosted with recombinant env were protected from infection after an intravenous challenge with SHIV/IIIB (Letvin *et al.*, 1997). As discussed above (Section II,B,3), the original parental SHIVs such as SHIV/IIIB are not pathogenic and do not replicate efficiently in macaques. Therefore, as with the experiments with chimps, this is not a rigorous challenge and will require further validation with a more robust challenge. When DNA immunization has been evaluated in the SIV/macaque model (Fuller *et al.*, 1997; Haigwood *et al.*, 1999; Lu *et al.*, 1996) reduction in viremia has been observed, consistent with a partially protective effect of this vaccine regimen. As discussed briefly above under poxvirus vectors, there is considerable promise in the approach of combining DNA immunization with a viral vector such as the attenuated poxviruses. Preliminary studies suggest that such an approach significantly boosts CTL responses (Kent *et al.*, 1998; Robinson *et al.*, 1999).

In summary, a pattern begins to emerge from evaluation of challenge results with various immunization protocols. With the exception of live attenuated SIV vaccines, the degree of protection observed in many of the trials is less than ideal. This translates into complete protection if an attenuated virus is used for challenge and significant reduction in viremia if a more robust pathogenic challenge virus is employed. However, it should also be noted that the challenges typically employed in vaccine studies in macaques are, for practical reasons, typically much more rigorous than estimates of the type of exposures involved in human infection with HIV-1. Thus, vaccine studies generally are designed using a challenge inoculum that will result in productive infection of all nonvaccinated control animals. This contrasts with estimates of infection rates for sexual transmission of HIV-1 that are generally less than 1% per exposure episode. This factor, in combination with the observation that, in general, there has been greater success in protecting against mucosal challenge than against intravenous challenge, suggests that even vaccine approaches that are less than completely protective in macaque/SIV models may still show some degree of efficacy in people. Experience with clinical studies of candidate vaccines will be required to further clarify this issue and perhaps help refine challenge models to further optimize the evaluation of vaccines in macaques.

V. Antiviral Therapy

Experimental models of SIV infection have also proven valuable in the evaluation of antiviral therapies. However, since many anti-HIV drugs are targeted to specific enzymes such as the viral reverse transcriptase and protease, and there are subtle differences in the corresponding enzymes in SIV, some anti-HIV compounds may not work as potently against SIV as they do against HIV. Nevertheless, SIV-infected macaques remain a useful model for the evaluation of compounds having good potency against SIV and HIV. In addition, they provide an extremely important model to explore questions related to pathogenesis and treatment that cannot be readily approached in HIV-1-infected human subjects, due to logistical constraints, ethical considerations, or other issues.

A. Treatment of Macaques with Antiretroviral Drugs

One compound with broad activity against many retroviruses, including SIV, is the reverse transcriptase inhibitor 9-[2-(R)-(phosphonomethoxy)propyl]adenine (PMPA). This compound is particularly convenient to use in SIV-infected macaques and due to its potency and pharamcokinetic profile, effective drug levels can be maintained with a single daily dose, given by subcutaneous injection, without any need to anesthetize the animals. This avoids the practical difficulties encountered in trying to achieve controlled administration of other drugs, the pharmacological or pharmacokinetic properties of which often require multiple daily doses, oral administration, or parenteral administration via routes less convenient than subcutaneous injection. Due in part to this profile, PMPA has been used extensively in SIV studies to address a number of different, important questions.

B. Treatment of Chronic SIV Infection

PMPA has been shown to potently suppress viral replication in chronically SIV-infected macaques, although drug levels returned to essentially pretreatment baseline levels upon drug discontinutation (Nowak *et al.*, 1997; Tsai *et al.*, 1997). An example of a 14-day treatment with PMPA of three chronically SIV-infected macaques is shown in Fig. 9. Careful measurements of plasma viral load prior to, during, and upon discontinuation of drug treatment have allowed the estimation of viral dynamics parameters in SIV-infected macaques (Nowak *et al.*, 1997). These parameters are broadly comparable with the same parameters measured in HIV-infected patients treated with antiviral drugs (Ho *et al.*, 1995; Wei *et al.*, 1995), underscoring the similarity of HIV and SIV infection and reinforcing the relevance of studies in SIV-infected macaques for understanding HIV infection in humans. In addition to PMPA, other compounds that have shown activity in SIV-infected macaques include d4T, ddI, and hydroxyurea, among others.

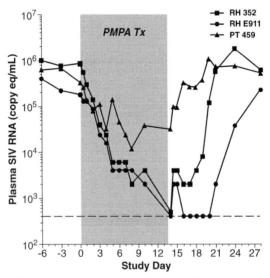

FIGURE 9   The effect of therapy with the antiviral drug PMPA on plasma viral RNA levels during chronic SIV infection for three macaques. Macaque RH 352 and RH 911 were inoculated with SIVsmE543-3 and macaque PT 459 was inoculated with SIVsmE660 (Nowak *et al.*, 1997). The shaded area indicates the period of drug treatment. Note the rapid decline in plasma viral RNA levels during the 14 days of treatment and the rapid rebound to pretreatment values after withdrawal of drug treatment.

C. Treatment of Acute SIV Infection

A variety of different treatment approaches have been evaluated for their ability to impact acute SIV infection. Both immunological and pharmacological approaches have been tried, often in studies designed to test the ability of a given regimen to mediate postexposure prophylaxis, i.e., to prevent the establishment of persistent pathogenic infection or modify the subsequent course of infection by treatment begun after exposure to infectious virus. Passive transfer of SIV immune globulin has been shown to modulate viral replication during primary infection (Haigwood *et al.*, 1997). Interestingly, the apparent effects of the immune globulin infusion persist long after circulating levels of the infused antibody had declined to below the threshold of detection. Treated animals showed lower levels of circulating virus and prolonged survival relative to controls, suggesting the possibility of long-lasting effects as a consequence of modulation of primary infection.

I. Studies with SIVMne

In postinoculation treatment models, PMPA treatment begun shortly after inoculation was able to prevent establishment of persistent infection with SIVMne (Tsai *et al.*, 1995, 1998; Van Rompay *et al.*, 1998, 1999). Both the interval between inoculation and initiation of treatment and the duration of treatment affected effectiveness in preventing persistent infection (Tsai *et al.*, 1998). Even in instances where the establishment of persistent infection was not prevented by postinoculation PMPA treatment, viral replication levels and clinical course were impacted, including in studies involving postinoculation treatment of neonatal macaques (Van Rompay *et al.*, 1996).

The treatment of neonatal macaques also underscores a unique SIV/ macaque system of great value in validating the feasibility for impacting a preventable form of infection, i.e., perinatal infection. In an SIV/neonatal macaque model of perinatal infection, PMPA has been shown to be capable of preventing the establishment of persistent infection, even using as few as two doses, bracketing the period of virus exposure (Van Rompay *et al.*, 1999).

These studies also underscore the potential of treatments impacting viral replication during primary infection to fundamentally modulate the subsequent pattern of viral replication and pathogenesis, including long-lasting effects manifested well after discontinuation of the treatment. This suggests that such treatment may induce a basic change in the dynamics of the relationship between the virus and the host, perhaps with regard to facilitating the development of immune responses capable of achieving long-term suppression of the virus. Studies in these types of postinoculation treatment models may usefully inform vaccine development efforts.

2. Studies with HIV-2

Transient postinoculation antiretroviral treatment has been shown to produce sustained, long-term impact on viral replication patterns, pathogenesis, clinical course, and survival in a different experimental model. As described above, infection of pigtail macaques with HIV-2/287 typically results in rapid, virtually complete destruction of the CD4+ T-cell population, with development of AIDS and death. Postinoculation treatment of HIV-2$_{287}$-infected animals with the reverse transcriptase inhibitor d4T produced lowered circulating viral loads, with persistent modulation of viral load even after drug treatment was discontinued (Watson *et al.*, 1997a). This effect was associated with prolonged survival in the treated animals compared to identically inoculated, untreated controls.

Like HIV-2$_{287}$ of pigtail macaques, SHIV KU2 infection of rhesus macaques results in rapid, virtually complete destruction of the CD4+ T-cell population (Joag *et al.*, 1998). PMPA treatment, begun 1 week postinoculation, at a time when plasma SIV levels were in excess of 10^7 copy Eq/ml, with widely disseminated infection, was nevertheless capable of fundamentally altering the virus/host relationship. Although treatment did not result in clearance of the infection, following discontinuation of drug treatment plasma virus levels did not rebound to levels seen in the postacute phase of infection in untreated animals. Rather, circulating virus levels fluctuated at greatly reduced values, ranging from undetectable to peak values that were still orders of magnitude lower than those seen in untreated animals. This reduced level of viral replication, reflected by reduced levels of viremia, was not associated with the depletion of CD4+ T cells that is hallmark of pathogenic SHIV KU2 infection, through more than a year of follow-up after drug discontinuation. The two studies described above underscore the power of macaque infection models for studies of pathogenesis questions directly relevant to critical issues in human HIV infection, in this instance, achievement of host control of infection in the absence of continual lifelong antiretroviral therapy (Watson *et al.*, 1997a).

VI. Summary

As presented in this review, there are a number of different models of both natural and experimental infection of monkeys with primate lentiviruses. There are numerous different viruses and multiple different monkey species, making for a potentially large number of different combinations. The fact that each different combination of virus isolate and host macaque species may show different behavior underscores the need to understand the different models and their key features. On the one hand, this diversity of systems underscores the need to provide some standardization of the systems used for certain kinds of studies, such as vaccine evaluations, in order to facilitate the comparison of results obtained in different experiments, but in essentially the same experimental system. On the other hand, the rich diversity of different systems, with different features and behaviors, represents a tremendous resource, among other things allowing the investigator to select the system that best recapitulates particular aspects of human HIV infection for study in a relevant nonhuman primate model. Such studies have provided, and may be expected to continue to provide, important insights to guide HIV treatment and vaccine development in the future.

References

Abimiku, A. G., Franchini, G., Tartaglia, J., Aldrich, K., Myagkikh, M,. Markham, P. D., Chong, P., Klein, M., Kieny, M. P., Paoletti, E. *et al.* (1995). HIV-1 recombinant poxvirus vaccine induces cross-protection against HIV-2 challenge in rhesus macaques. *Nat. Med.* **1**, 321–329.

Abimiku, A. G., Robert-Guroff M., Benson J., Tartaglia J., Paoletti E., Gallo R. C., Markham P. D., and Franchini G. (1997). Long-term survival of SIV mac251-infected macaques previously immunized with NYVAC-SIV vaccines. *J. Acq. Immune Defic. Syndr.* **15**, S78–S85.

Agy, M. B., Frumkin, L. R., Corey, L., Coombs, R. W., Wolinsky, S. M., Koehler, J., Morton, W. R., and Katze, M. G. (1992). Infection of *Macaca nemestrina* by human immunodeficiency virus type-1. *Science* **257**, 103–106.

Ahmad, S., Lohman, B., Marthas, M., Giavedoni, L., el-Amad, Z., Haigwood, N. L., Scandella, C. J., Gardner, M. B., Luciw, P. A., and Yilma, T. (1994). Reduced virus load in rhesus macaques immunized with recombinant gp160 and challenged with simian immunodeficiency virus. *AIDS Res. Hum. Retroviruses* **10**, 195–204.

Alexander, L., Du, Z., Howe, A. Y., Czajak, S., and Desrosiers, R. C. (1999). Induction of AIDS in rhesus monkeys by a recombinant simian immunodeficiency virus expressing nef of human immunodeficiency virus type 1. *J. Virol.* **73**, 5814–5825.

Allan, J. S. (1991). Pathogenic properties of simian immunodeficiency viruses in nonhuman primates. *In* "Annual Review of AIDS Research" (W. Koff, Ed.), Vol. 1, pp. 191–206. Dekker, New York.

Almond, N., Kent, K., Crange, M., Rud, E., Clarke, B., and Stott, E. J. (1995). Protection by attenuated simian immunodeficiency virus in macaques against challenge with virus-infected cells. *Lancet* **345**, 1342–1344.

Almond, N. M., and Heeney, J. L. (1998). AIDS vaccine development in primate models. *AIDS* **12**, S133–S140.

Andersson, S., Makitalo, B., Thorstensson, R., Franchini, G., Tartaglia, J., Limbach, K., Paoletti, E., Putkonen, P., and Biberfeld, G. (1996). Immunogenicity and protective efficacy of a human immunodeficiency virus type 2 recombinant canarypox (ALVAC) vaccine candidate in cynomolgus monkeys. *J. Infect. Dis.* **174**, 977–985.

Arthur, L. O., Bess, J. W., Jr, Chertova, E. N., Rossio, J. L., Esser, M. T., Benveniste, R. E., Henderson, L. E., and Lifson, J. D. (1998). Chemical inactivation of retroviral infectivity by targeting nucleocapsid protein zinc fingers: A candidate SIV vaccine. *AIDS Res. Hum. Retroviruses*, (Suppl. 3), S311–S319.

Baba, T. W., Jeong, Y. S., Penninck, D., Bronson, R., Greene, M. F., and Ruprecht, R. M. (1995). Pathogenicity of live, attenuated SIV after mucosal infection of neonatal macaques. *Science* **267**, 1820–1825.

Baba, T. W., Liska, V., Khimani, A. H., Ray, N. B., Dailey, P. J., Penninck, D., Bronson, R., Greene, M. F., McClure, H. M., Martin, L. N., and Ruprecht, R. M. (1999). Live attenuated, multiply deleted simian immunodeficiency virus causes AIDS in infant and adult. *Nat. Med.* **5**, 194–203.

Baskin, G. B., Murphey-Corb, M., Watson, E. A., and Martin, L. N. (1988). Necropsy findings in rhesus monkeys experimentally infected with cultured simian immunodeficiency virus (SIV/Delta). *Vet. Path.* **25**, 456–467.

Beer, B., Bailes, E., Goeken, R., Dapolito, G., Coulibaly, C., Norley, S. G., Kurth, R., Gautier, J.-P., Gautier-Hion, A., Vallet, D., Sharp, P. M., and Hirsch, V. M. (1999). Simian immunodeficiency virus (SIV) from sun-tailed monkeys (*Cercopithecus solatus*): Evidence for host dependent evolution of SIV within the C. l'hoesti superspecies. *J. Virol.* **73**(9), 7734–7744.

Beer, B., Scherer, J., zur Megede, J., Norley, S., Baier, M., and Kurth, R. (1996). Lack of dichotomy between virus load of peripheral blood and lymph nodes during long-term simian immunodeficiency virus infection of African green monkeys. *Virology* **219**, 367–375.

Benson, J., Chougnet, C., Robert-Guroff, M., Montefiori, D., Markham, P., Shearer, G., Gallo, R. C., Cranage, M., Paoletti, E., Limbach, K., Venzon, D., Tartaglia, J., and Franchini, G. (1998). Recombinant vaccine-induced protection against the highly pathogenic simian immunodeficiency virus SIV mac 251: Dependence on route of challenge exposure. *J. Virol.* **72**, 4170–4182.

Benveniste, R. E., Arthur, L. O., Tsai, C.-C., Sowder, R., Copeland, T. D., Henderson, L. E., and Oroszlan, S. (1986). Isolation of a lentivirus from a macaque with lymphoma: Comparison with HTLV-III/LAV and other lentiviruses. *J. Virol.* **60**, 483–490.

Benvensite, R. E., Hill, R. W., Eron, L. J., Csaikl, U. M., Knott, W. B., Henderson, L. E., Sowder, R. C., Nagashima, K., and Gonda, M. A. (1990). Characterization of clones of HIV-1 HuT 78 cells defective in gag gene processing and of SIV clones producing large amounts of envelope glycoprotein. *J. Med. Primatol.* **19**, 351–366.

Benveniste, R. E., Roodman, S. T., Hill, R. W., Knott, W. B., Ribas, J. L., Lewis, M. G., and Eddy, G. A. (1994). Infectivity of titered doses of simian immunodeficiency virus clone E11S inoculated intravenously into rhesus macaques. *J. Med. Primatol.* **23**, 83–88.

Berglund, P., Quesada-Rolander, M., Putkonen, P., Biberfeld, G., Thorstensson, R., and Liljes-trom, P. (1997). Outcome of immunization of cynomolgus monkeys with recombinant Semliki Forest virus encoding human immunodeficiency virus type 1 envelope protein and challenge with a high dose of SHIV-4 virus. *AIDS Res. Hum. Retroviruses* **13**, 1487–1495.

Bess, J. W., Jr., Gorelick, R. J., Bosche, W. J., Henderson, L. E., and Arthur, L. O. (1997). Microvesicles are a source of contaminating cellular proteins found in purified HIV-1 preparations. *Virology* **230**, 134–144.

Blanchard, T. J., Alcami, A., Panayotam A., and Smith G. L. (1998). Modified vaccinia virus Ankara undergoes limited replication in human cells and lack several immunomodulatory proteins: Implications for use as a human vaccine. *J. Gen. Virol.* **79**, 1159–1167.

Boyer, J. D., Ugen, K. E., Wamg, B., Agadjanyan, M., Gilbert, L., Bagarazzi, M. L., Chattergoon, M., Frost, P., Javadian, A., Williams, W. V., Refaeli, Y., Ciccarelli, R. B., McCallus, D., Coney, L., and Weiner, D. B. (1997). Protection of chimpanzees from high-dose heterologous challenge by DNA vaccination. *Nat. Med.* **3**, 526–532.

Buge, S. L., Richardson, E., Alipanah, S., Markham, P., Cheng, S., Kalyan, N., Miller, C. J., Lubeck, M., Udem, S., Eldridge, J., and Robert-Guroff, M. (1997). An adenovirus simian immunodeficiency virus env vaccine elicits humoral, cellular, and mucosal immune responses in rhesus macaques and decreases viral burden following vaginal challenge. *J. Virol.* **71**, 8531–8541.

Caley, I. J., Betts, M. R., Irlbeck, D. M., Davis, N. L., Swanstrom, R., Frelinger, J. A., and Johnston, R. E. (1997). Humoral, mucosal, and cellular immunity in response to a human immunodeficiency virus type 1 immunogen expressed by a Venezuelan equine encephalitis virus vaccine vector. *J. Virol.* **71**, 3031–3038.

Cao, Y., Qin, L., Zhang, L., Safrit, J., and Ho, D. (1995). Virologic and immunologic characterization of long-term survivors of human immunodeficiency virus type 1 infection. *N. Engl. J. Med.* **332**, 201–208.

Chakrabarti, L., Guyader, M., Alizon, M., Daniel, M. D., Desroisiers, R., Tiollais, P., and Sonigo, P. (1987). Sequence of simian immunodeficiency virus from macaque and its relationship to other human and simian retroviruses. *Nature* **328**, 543–545.

Clark, K. R., Voulgaropoulou, F., Fraley, D. M., and Johnson, P. R. (1995). Cell lines for the production of recombinant adeno-associated virus. *Hum. Gene Ther.* **6**, 1329–1341.

Clements, J. E., Montelaro, R. C., Zink, M. C. *et al.* (1995). Cross-protective immune responses induced in rhesus macaques by immunization with attenuated macrophage-tropic simian immunodeficiency virus. *J. Virol.* **69**, 2737–2744.

Clewley, J. P., Lewsi, J. C. M., Brown, D. W. G., and Gadsby, E. L. (1998). A novel simian immunodefieincy virus (SIVdrl) *pol* sequence from the drill monkey, *Mandrillus leucophaeus. J. Virol.* **72**, 10305–10309.

Connor, R. I., Montefiori, D. C., Binley, J. M. *et al.* (1998). Temporal analyses of virus replication, immune responses, and efficacy in rhesus macaques immunized with a live, attenuated simian immunodeficiency virus vaccine. *J. Virol.* **72**, 7501–7509.

Daniel, M. D., Kirchhoff, F., Czajak, S. C., Sehgal, P. K., and Desrosiers, R. C. (1992). Protective effects of a live attenuated SIV vaccine with a deletion in the nef gene. *Science* **258**, 1938–1941.

Daniel, M. D., Kirchhoff, F., Czajak, S. C., Sehgal, P. K., and Desrosiers, R. C., (1992). Protective effects of a live attenuated SIV vaccine with a deletion in the nef gene. *Science* **258**, 1938–1941.

Daniel, M. D., Letvin, N. L., King, N. W., Kannagi, M., Seghal, P., Hunt, R. D., Kanki, P., Essex, M., and Desroisiers, R. C. (1985). Isolation of a T-cell tropic HTLV-III-like retrovirus from macaques. *Science* **228**, 1201–1204.

Deeks, S. G., Barditch-Crovo, P., Lietman, P. S., Hwang, F., Cundy, K. C., Rooney, J. F., Hellman, N. S., Safrin, S., and Kahn, J. O. (1998). Safety, pharmacokinetics, and antiretroviral activity of intravenous 9-[2-(*R*)-(phosphonomethoxy)propyl]adenine a novel anti-human immunodeficiency virus (HIV) therapy in HIV infected adults. *Antimicrob. Agents Chemother.* **42**, 2380–2384.

Desrosiers, R. C., Lifson, J. D., Gibbs, J. S., Czajak, S. C., Howe, A. Y., Arthur, L. O., and Johnson, R. P. (1998). Identification of highly attenuated mutants of simian immunodeficiency virus. *J. Virol.* **72**, 1431–1437.

Dewhurst, S., Embertson, J. E., Anderson, D. C., Mullins, J. I., and Fultz, P. N. (1990). Sequence analysis and pathogenicity of molecularly cloned SIV smm-PBj14. *Nature* **345**, 636–640.

Dittmer, U., Brooks, D. M., and Hasenkrug, J. K. J. (1999). Requirement for multiple lymphocyte subsets in protection by a live attenuated vaccine against retroviral infection. *Nat. Med.* **4**, 1890–1893.

Du, Z., Lang, S. M., Sasseville, V. G., Lackner, A. A., Ilyinskii, P. O., Daniel, M. D., Jung, J. U., and Desrosiers, R. C. (1995). Identification of a nef allele that causes lymphocyte activation and acute disease in macaque monkeys. *Cell* **82**, 665–674.

Edmonson, P., Murphey-Corb, M., Martin, L. N., Delahunty, C., Heeney, J., Kornfeld, H., Donahue, P. R., Learn, G. H., Hood, L., and Mullins, J. I. (1998). Evolution of a simian immunodeficiency virus pathogen. *J. Virol.* **72**, 405–414.

Embertson, J., Zupancic, J. L., Ribas, J., and Haase, H. (1994). Massive covert infection of helper T lymphocytes by HIV during the incubation period of AIDS. *Science* **259**, 359–362.

Fenyo, E. M., Albert, J., and Asjo, B. (1989). Replicative capacity, cytopathic effect and cell tropism of HIV. *AIDS* **3**(Suppl. 1), S5–S12.

Finzi, D., Blankson, J., Siliciano, J. D., Margolick, J. B., Chadwick, K., Pierson, T, Smith, K., Lisziewicz, J., Lori, F., Flexner, C., Quinn, T. C., Chaisson, R. E., Rosenberg, E., Walker, B., Gange, S., Gallant, J., and Siliciano, R. F. (1999). Latent infection of CD4+ T cells provides a mechanism for lifelong persistence of HIV-1 in patients on effective combination therapy. *Nat. Med.* **5**, 512–517.

Fomsgaard, A., Hirsch, V. M., Allan, J. S., and Johnson, P. R. (1991). A highly divergent proviral DNA clone of SIV from a distinct species of African green monkey. *Virology* **182**, 397–402.

Franchini, G., and Reitz, M. S. (1994). Phylogenesis and genetic complexity of the nonhuman primate retroviridae. *AIDS Res. Hum. Retroviruses* **10**(9), 1047–1060.

Franchini, G., Robert-Guroff, M., Tartaglia, J., Aggarwal, A., Abimiku, A., Benson, J., Markham, P., Limbach, K., Hurteau, G., Fullen, J. *et al.* (1995). Highly attenuated HIV type 2 recombinant poxviruses, but not HIV-2 recombinant Salmonella vaccines, induce long-lasting protection in rhesus macaques. *AIDS Res. Hum. Retroviruses* **11**, 909–920.

Fuller, D. H., Simpson, L., Cole, K. S. *et al.* (1997). Gene gun-based nucleic acid immunization alone or in combination with recombinant vaccinia vectors suppresses virus burden in rhesus macaques challenged with a heterologous SIV. *Immunol. Cell Biol.* **75**, 389–396.

Fultz, P. N. (1991). Human immunodeficiency virus infection of chimpanzees: An animal model for asymptomatic HIV carriers and vaccine efficacy. *In* "Annual Review of AIDS Research" (W. Koff, Ed.), Vol. 1, pp. 207–218. Dekker, New York.

Fultz, P. N., and Zack, P. M. (1994). Unique lentivirus–host interactions: SIVsmmPBj14 infection of macaques. *Virus Res.* **32**, 205–225.

Fultz, P. N., McClure, H. M., Anderson, D. C., Swenson, R. B., Ananad, R., and Srinvasan, A. (1986). Isolation of a T-lymphotropic retrovirus from naturally-infected sooty mangabey monkeys (*Cercocebus atys*). *Proc. Natl. Acad. Sci. USA* **83**, 5286–5290.

Fultz, P. N., McClure, H. M., Anderson, D. C., and Switzer, W. W. (1989). Identification and biologic characterization of an acutely lethal variant of simian immunodeficiency virus from sooty mangabeys (SIV/SMM). *AIDS Res. Hum. Retroviruses* **5**, 397–409.

Georges-Courbot, M. C., Lu, C. Y., Makuwa, M., Tefler, P., Onanga, R., Dubreuil, G., Chen, Z., Smith, S. M., Georges, A., Gao, F., Hahn, B. H., and Marx, P. A. (1998). Natural infection of a household per red-capped mangabey (*Cercocebus torquatus torquatus*) with a new simian immunodefieincy virus. *J. Virol.* **72**, 600–608.

Gao, F. L., Yue, L., White, A. T., Pappas, P. G., Barchue, J., Hanson, A. P., Greene, B. M., Sharp, P. M., Shaw, G. M., and Hahn, B. H. (1992). Human infection by genetically diverse SIVsm-related HIV-2 in West Africa. *Nature* **358**, 495–499.

Gibson, U. E. M., Heid, C. A., and Williams, P. M. (1996). A novel method for real time quantitative RT PCR. *Gen. Res.* **6**, 995–1001.

Haigwood, N. L., Pierce, C. C., Robertson, M. N., Watson, A. J., Montefiori, D. C., Rabin, M., Lynch, J. B., Kuller, L., Thompson, J., Morton, W. R., Benveniste, R. E., Hu, S. L., Greenberg, P., and Mossman, S. P. (1999). Protection from pathogenic challenge using multigenic DNA vaccines. *Immunol. Lett.* **66**, 183–188.

Haigwood, N. L., Watson, A., Sutton, W. F. *et al.* (1996) Passive immune globulin therapy in SIV/macaque model: early intervention can alter disease profile. *Immunol. Lett.* **51**, 107–114.

Harouse, J. M., Gettie, A., Tan, R. C., Blanchard, J., and Cheng-Mayer, C. (1999). Distinct pathogenic sequelae in rhesus macaques infected with CCR5 or CXCR4 utilizing SHIVs. *Science* **284**, 816–819.

Hasenkrug, K. J., Brooks, D. M., and Dittmer, U. (1998). Critical role for CD4+ T cells in controlling retrovirus replication and spread in persistently infected mice. *J. Virol.* **72**, 6559–6564.

Heid, C. A., Stevens, J., Livak, K. J., and Williams, P. M. (1996). Real-time quantiative RT–PCR. *Gen. Res.* **6**, 986–994.

Hirsch, V. M., Olmsted, R. A., Murphey-Corb, M., Purcell, R. H., and Johnson, P. R. (1989). An African primate lentivirus (SIVsmm) closely related to HIV-2. *Nature* **339**, 389–391.

Hirsch, V. M., and Johnson, P. R. (1993). Genetic diversity and phylogeny of primate lentiviruses. *In* "HIV Molecular Organization Pathogenicity and Treatment" (J. Morrow and N. Haigwood, Eds.), pp. 221–240. Elsevier, The Netherlands.

Hirsch, V. M., and Johnson, P. R. (1994). Pathogenic diversity of simian immunodeficiency viruses. *Virus Res.* **32**, 183–203.

Hirsch, V. M., Adger-Johnson, D., Campbell, B., Goldstein, S., Brown, C., Elkins, W. R., and Montefiori, D. C. (1997). A molecularly cloned pathogenic neutralization resitant simian immunodeficiency virus, SIV smE543-3. *J. Virol.* **71**, 1608–1615.

Hirsch, V. M., Dapolito, G., Goeken, R., and Capbell, B. J. (1995a). Phylogeny and natural history of the primate lentiviruses, SIV and HIV. *Curr. Opin. Genet. Dev.* **5**, 798–806.

Hirsch, V. M., Dapolito, G., Johnson, P. R., Elkins, W. R., London, W. T., Montali, R. J., Goldstein, S., and Brown, C. (1995b). Induction of AIDS by simian immunodeficiency virus from an African green monkey: Species-specific variation in pathogenicity correlates with the extent of *in vivo* replication. *J. Virol.* **69**, 955–967.

Hirsch, V. M., Dapolito, G., Hahn, A., Lifson, J., Montefiori, D., Brown, C. R., and Goeken, R. (1998a). Viral genetic evolution in macaques infected with molecularly cloned simian immunodeficiency virus correlates with the extent of persistent viremia. *J. Virol.* **8**, 6482–6489.

Hirsch, V. M., Sharkey, M. E., Brown, C. R., Brichacek, B., Goldstein, S., Wakefield, J., Byrum, R., Elkins, W. R., Hahn, B. H., Lifson, J. D., and Stevenson, M. (1998b). Vpx is required for dissemination and pathogenesis of SIVsmPBj: Evidence for macrophage-dependent viral amplification. *Nat. Med.* **4**, 1401–1408.

Hirsch, V. M., Fuerst, T. R., Sutter, G., Carroll, M. W., Yang, L. C., Goldstein, S., Piatak, M., Elkins, W. R., Alvord, W. G., Montefiori, D. C., Moss, B., and Lifson, J. D. (1996). Patterns of viral replication correlate with outcome in simian immunodeficiency virus (SIV)-infected macaques: Effect of prior immunization with a trivalent SIV vaccine in modified vaccinia virus Ankara *J. Virol.* **70**, 3741–3752.

Ho, D., Neumann, A. U., Perelson, A. S., Chen, W., Leonard, J. M., and Markowitz, M. (1995). Rapid turnover of plasma virions and CD4 lymphocytes in HIV-1 infection. *Nature* **373**, 123–126.

Hu, S.-L., Abrams, K., Barber, G. N. *et al.* (1992). Protection of macaques against SIV infection by recombinant subunit vaccines using SIV env glycoprotein gp160. *Science* **255**, 456–459.

Hu, S.-L., Haigwood, N. L., and Morton, W. R. (1993). Non-human primate models for AIDS research. *In* "HIV Molecular Organization Pathogenicity and Treatment" (J. Morrow and N. Haigwood, Eds.), pp. 294–327. Elsevier, The Netherlands.

Huet, T., Cheynier, R., Meyrhans, A., Roelants, G., and Wain-Hobson, S. (1990). Genetic organization of a chimpanzee lentivirus related to HIV-1. *Nature* **345**, 356–359.

Janssens, W., Fransen, K., Peeters, M., Hendrickx, L., Motte, J., Bedjabaga, L., Delaporte, E., Piot, P., and Vandergroen, G. (1994). Phylogenetic analysis of a new chimpanzee lentivirus SIV cpz-gab-2 from a wild-captured chimpnazee from Gabon. *AIDS Res. Hum. Retroviruses* **10**, 1191–1192.

Joag, S. V., Adany, I., Li, Z. *et al.* (1997). Animal model of mucosally transmitted human immunodeficiency virus type 1 disease: Intravaginal and oral deposition of simian/human immunodeficiency virus in macaques results in systemic infection, elimination of CD4+T cells and AIDS. *J. Virol.* **71**, 4016–4023.

Joag, S. V., Li, Z., Wang, C., Jia, F., Foresman, L., Adany, I., Pinson, D. M., Stephens, E. B., and Narayan, O. (1998). Chimeric SHIV that causes CD4+T cell loss and AIDS in rhesus macaques. *J. Med. Primatol.* **27**, 59–64.

Johnson, R. P. *et al.* (1998). Protective immunity induced by live attenuated simian immunodeficiency virus. *Curr. Opin. Immunol.* **10**, 436–443.

Johnson, P. R., Goldstein, S., London, W. T., Fomsgaard, A., and Hirsch, V. M. (1990). Molecular clones of SIVsm and SIVagm: experimental infection of macaques and African green monkeys. *J. Med. Primatol.* **19**, 279–286.

Johnson, R. P., Lifson, J. D., Czajak, S. C., Cole, K. S., Manson, K. H., Glickman, R., Yang, J., Montefiori, D. C., Montelaro, R., Wyand, M. S., and Desrosiers, R. C. (1999). Highly attenuated vaccine strains of simian immunodeficiency virus protect against vaginal challenge: Inverse relation of degree of protection with level of attenuation. *J. Virol.* **73**, 4952–4961.

Johnson, P. R., Myers, G., and Hirsch, V. M. (1991). Genetic diversity and phylogeny of non-human primate lentiviruses. *In* "Annual Review of AIDS Research" (W. Koff *et al.*, Eds.), Vol. 1, pp. 47–62. Dekker, New York.

Kaur, A., Grant, R. M., Means, R. E., McClure, H., Feinberg, M., and Johnson, R. P. (1998). Diverse host responses and outcomes following simian immunodeficiency virus SIVmac239 infection in sooty mangabeys and rhesus macaques. *J. Virol.* **72**, 9597–9611.

Kent, S. J., Zhao, A., Best, S. J., Chandler, J. D., Boyle, D. B., and Ramshaw, I. A. (1998). Enhanced T-cell immunogenicity and protective efficacy of a human immunodeficiency virus type 1 vaccine regimen consisting of consecutive priming with DNA and boosting with recombinant fowlpox virus. *J. Virol.* **72**, 10180–10188.

Kestler, H. W., Kodama, T., Ringler, D., Marthas, M., Pederson, N., Lackner, A., Rieger, D., Seghal, P., Daniel, M., King, N., and Desrosiers, R. (1988). Induction of AIDS in rhesus monkeys by molecularly cloned simian immunodeficiency virus. *Nature* **248**, 1109–1112.

Kimata, J. T., Kuller, L., Anderson, D. B., Dailey, P., and Overbaugh, J. (1999). Emerging cytopathic and antigenic simian immunodeficiency virus variants influence AIDS progression. *Nat. Med.* **5**, 535–541.

Kuroda, M. J., Schmitz, J. E., Charini, W. A., Nickerson, C. E., Lifton, M. A., Lord, C. I., Forman, M. A., and Letvin, N. L. (1999). Emergence of CTL coincides with clearance of virus during primary simian immunodeficiency virus infection in rhesus monkeys. *J. Immunol.* **162**, 5127–5133.

Letvin, N. L. (1998). Progress in the development of an HIV-1 vaccine. *Science* **280**, 1875–1880.

Letvin, N. L., and King, N. (1990). Immunopathologic and pathologic manifestations of the infection of rhesus monkeys with simian immunodeficiency virus of macaques. *J. Acq. Immune Defic. Sydr.* **3**, 1023–1040.

Letvin, N. L., Daniel, M. D., Seghal, P. K., Desrosiers, R. C., Hunt, R. D., Waldron, L. M., MacKey, J. J., Schmidt, D. K., Chalifoux, L. V., and King, N. W. (1985). Induction of AIDS-like disease in macaque monkeys with T-cell tropic retrovirus STLV-III. *Science* **230**, 71–73.

Letvin, N. L., Montefiori, D., Yasutami, Y., Perry, H. C., Davies, M. E., Lekutis, C., Alroy, M., Freed, D. C., Lord, C. I., Handt, L. K., Liu, M. A., and Shiver, J. W. (1997). Potent, protective anti-HIV immune response generated by bimodal HIV envelope DNA plus protein vaccination. *Proc. Natl. Acad. Sci. USA* **94**, 9378–9383.

Lewis, M. G., Yalley-Ogunro, J., Greenhouse, J. J., Brennan, T. P., Jiang, J. B., VanCott, T. C., Lu, Y., Eddy, G. A., and Birx, D. L. (1999). Limited protection from a pathogenic chimeric simian-human immunodeficiency virus challenge following immunization with attenuated simian immunodeficiency virus. *J. Virol.* **73**, 1262–1270.

Lewis, M. G., Zack, P. M., Elkins, W. R., and Jahrling, P. B. (1992). Infection of rhesus and cynomolgus macaques with a rapidly fatal SIV (SIVSMM/PBj) isolate from sooty mangabeys. *AIDS Res. Hum. Retroviruses* **8**, 1631–1639.

Li, J., Lord, C., Haseltine, W., Letvin, N. L., and Sodroski, J. (1992). Infection of cynomolgus monkeys with a chimeric HIV-1SIV mac virus that expresses the HIV-1 envelope glycoproteins. *J. Acq. Immune Defic. Syndr.* **5**, 639–646.

Lifson, J. D., and Piatak, M. (1997). Internally controlled PCR approaches for quantification of retroviral replication and other applications. *In* "Quantitative PCR Techniques," pp. 43–58. Eaton, Natick, MA.

Lifson, J. D., Nowak, M., Goldstein, S., Rossio, J., Kinter, A., Vasquez, G., Wiltrout, T. A., Brown, C., Schneider, D., Wahl, L., Lloyd, A., Elkins, W. R., Fauci, A. S., and Hirsch, V. M. (1997). The extent of early virus replication is a critical determinant of the natural history of AIDS virus infection. *J. Virol.* **71**, 9508–9514.

Lisziewicz, J., Rosenberg, E., Lieberman, J., Jessen, H., Lopalco, L., Siliciano, R., Walker, B., and Lori, F. (1999). *N. Engl. J. Med.* **340**, 1683–1684. [letter]

Livak, K. J., Flood, S. J., Marmarao, J., Giusti, W., and Deetz, K. (1995). Oligonucleotides with fluorescent dyes at opposite ends provide a quenched probe system useful for detecting PCR product and nucleic acid hybridization. *PCR Methods Appl.* **4**, 357–362.

Lu, S., Arthos, J., Montefiori, D. C., Yasutomi, Y., Manson, K., Mustafa, F., Johnson, E., Santoro, J. C., Wissink, J., Mullins, J. I., Haynes, J. R., Letvin, N. L., Wyand, M., and

Robinson, H. L. (1996). Simian immunodeficiency virus DNA vaccine trial in macaques. *J. Virol.* **70**, 3978–3991.

Lu, Y. (1997). HIV-1 vaccine candidate evaluation in non-human primates. *Crit. Rev. Oncog.* **8**, 273–291.

Lu, Y., Pauza, C. D., Montefiori, D. C., and Miller, C. J. (1998). Rhesus macaques that become systemically infected with pathogenic SHIV 89.6-PD after intravenous, rectal, or vaginal inoculation and fail to make an antiviral antibody response rapidly develop AIDS. *J. Acq. Immune Defic. Syndr.* **19**, 6–18.

Lu, Y., Salvato, M., Pauza, C. D. *et al.* (1996). Utility of SHIV for testing HIV-1 vaccine candidates in macaques. *J. Acq. Immune Defic. Syndr.* **12**, 99–106.

Luciw, P. A., Shaw, K. E., Unger, R. E., Planelles, V., Stout, M. W., Lackner, J. E., Pratt-Lowe, E., Leung, N. J., Banapour, B., and Marthas, M. L. (1992). Genetic and biological comparisons of pathogenic and nonpathogenic molecular clones of simian immunodeficiency virus (SIVmac). *AIDS Res. Hum. Retroviruses* **8**, 395–402.

Marthas, M. L., Ramos, R. A., Lohman, B. L., Van Rompay, K. K., Unger, R. E., Miller, C. J., Banapour, B., Pedersen, N. C., and Luciw, P. A. (1993). Viral determinants of simian immunodeficiency virus (SIV) virulence in rhesus macaques assessed by using attenuated and pathogenic clones of SIVmac. *J. Virol.* **67**, 6047–6055.

Marthas, M. L., Sutjipto, S., Higgins, J., Lohman, B., Torten, J., Luciw, P. A., Marx, P. A., and Pedersen, N. C. (1990). Immunization with a live, attenuated simian immunodeficiency virus (SIV) prevents early disease but not infection in rhesus macaques challenged with pathogenic SIV. *J. Virol.* **64**, 3694–3700.

Marthas, M. L., Sutjipto, S., Higgins, J. *et al.* (1990). Immunization with a live, attenuated simian immunodeficiency virus (SIV) prevents early disease but not infection in rhesus macaques challenged with pathogenic SIV. *J. Virol.* **64**, 3694–3700.

Marx, P., Li, Y., Lerche, N. W., Sutjipto, S., Gettie, A., Yee, J. A., Brotman, B. H., Prince, A. M., Hanson, A., Webster, R. G., and Desrosiers, R. C. (1991). Isolation of a simian immunodeficiency virus related human immunodeficiency virus type 2 from a West African pet sooty mangabey. *J. Virol.* **65**, 4480–4485.

Marx, P. A., Spira, A. I., Gettie, A., Dailey, P. J., Veazey, R. S., Lackner, A. A., Miller, C. J., Claypool, L. E., Ho, D. D., and Alexander, N. J. (1996). Progesterone implants enhance SIV vaginal transmission and early virus load. *Nat. Med.* **2**, 1084–1089.

McChesney, M. B., Collins, J. R., Lu, D., Lu, X., Torten, J., Ashley, R. L., Cloyd, M., and Miller, C. J. (1998). Occult systemic infection and persistent simian immunodeficiency virus (SIV)-specific CD4+-T-cell proliferative responses in rhesus macaques that were transiently viremic after intravaginal inoculation of SIV. *J. Virol.* **72**, 10029–10035.

McClure, H. M., Anderson, D. C., Fultz, P. N., Ansari, A. A., Lockwood, E., and Brodie, A. (1989). Spectrum of disease in macaques monkeys chronically infected with SIVsmm. *Vet. Immunol. Immunopathol.* **21**, 13–24.

Mellors, J. W., Rinaldo, C. R., Jr., Gupta, P., White, R. M., Todd, J. A., and Kingsley, L. A. (1996). Prognosis in HIV-1 infection predicted by the quantity of virus in plasma. *Science* **272**, 1167–1170.

Meyer, H., Sutter, G., and Mayer, A. (1991). Mapping of deletions in the genome of the highly attenuated vaccinia virus MVA and their influence on virulence. *J. Gen. Virol.* **72**, 1031–1038.

Morrow, C. D., Novak, M. J., Ansardi, D. C., Porter, D. C., and Moldoveanu, Z. (1999). Recombinant viruses as vectors for mucosal immunity. *Curr. Top. Microbiol. Immunol.* **236**, 255–273.

Moss, B., Carroll, M. W., Wyatt, L. S., Bennink J. R., Hirsch V. M., Goldstein, S., Elkins, W. R., Lifson, J. D., Piatak, M., Restifo, N. P., Owerwijk, W., Chamberlain, R., Rosenberg, S. A., Sutter, G. (1996). Host range restricted, non-replicating vaccinia virus vectors as vaccine candidates. *Adv. Exp. Med. Biol.* **397**, 7–13.

Mossman, S. P., Bex, F., Berglund, P., Arthos, J., O'Neil, S. P., Riley, D., Maul, D. H., Bruck, C., Momin, P., Burny, A., Fultz, P. N., Mullins, J. I., Liljestrom, P., and Hoover, E. A. (1996). Protection against lethal simian immunodeficiency virus SIVsmmPBj14 disease by a recombinant semliki forest virus gp160 vaccine and by a gp120 subunit vaccine. *J. Virol.* **70**, 1953–1960.

Murphey-Corb, M., Martin, L. N., Rangan, S. R. S., Baskin, G. B., Gormus, B. J., Wolf, R. H., Andes, W. A., West, M., and Montelaro, R. C. (1986). Isolation of an HTLV-III-related retrovirus from macaques with simian AIDS and its possible origin in asymptomatic mangabeys. *Nature* **321**, 435–437.

Myagkikh, M., Alipanah, S., Markham, P. D., Tartaglia, J., Paoletti, E., Gallo, R. C., Franchini, G., and Robert-Guroff, M. (1996). Multiple immunizations with attenuated poxvirus HIV type 2 recombinants and subunit boosts required for protection of rhesus macaques. *AIDS Res. Hum. Retroviruses* **12**, 985–992.

Nathanson, N., Hirsch, V. M., and Matheson, B. J. (2000). The role of nonhuman primate models in the development of an AIDS vaccine. *AIDS* (in press).

Novembre, F. J., Johnson, P. R., Lewis, M. G., Anderson, D. C., Klumpp, S., McClure, H. M., and Hirsch, V. M. (1993). Multiple viral determinants contribute to pathogenicity of the acutely lethal simian immunodeficiency virus SIVsmmPBjvariant. *J. Virol.* **67**, 2466–2474.

Novembre, F. J., Lewis, M. G., Saucier, M. M., Yalley-Ogunro, J., Brennan, T., McKinnon, K., Bellah, S., and McClure, H. M. (1996). Deletion of the nef gene abrogates the ability of SIVsmmPBj to induce acutely lethal disease in pigtail macaques. *AIDS Res. Hum. Retroviruses* **12**, 727–736.

Nowak, M. A., Lloyd, A. L., Vasquez, G. M., Wiltrout, T. A., Wahl, L. M., Bischofberger, N., Williams, J., Kinter, A., Fauci, A. S., Hirsch, V. M., and Lifson, J. D. (1997). Viral dynamics of primary viremia and antiretroviral therapy in simian immunodeficiency virus infection. *J. Virol.* **71**, 7518–7525.

O'Brien, T. R., Blattner, W. A., Waters, D., Eyster, E., Hilgartner, M. W., Cohen, A. R., Luban, N., Hatzakis, A., Aledort, L. M., Rosenberg, P. S., Miley, W. J., Kroner, B. L., and Goedert, J. J. (1996). Serum HIV-1 RNA levels and time to development of AIDS in the Multicenter Hemophilia Cohort Study. *J. Am. Med. Assoc.* **276**, 105–110.

O'Neil, S. P., Mossman, S. P., Maul, D. H., and Hoover, E. A. (1999). Virus threshold determines disease in SIVsmmPBj14-infected macaques. *AIDS Res. Hum. Retroviruses* **15**, 183–194.

Polacino, P., Stallard, V., Montefiori, D. C., Brown, C. R., Richardson, B. A., Morton, W. R., Benveniste, R. E., and Hu, S. L. (1999). Protection of macaques against intrarectal infection by a combination immunization regimen with recombinant simian immunodeficiency virus SIVmne gp160 vaccines. *J. Virol.* **73**, 3134–3146.

Panteleo, G., Menzo, S., Vaccarezza, M., Graziosi, C., Cohen, O. J., Demarest, J. F., Montefiori, D., Orenstein, J. M., Fox, C., Schrager, L. K., Margolick, J. B., Buchbinder, S., Giorgi, J. V., and Fauci, A. S. (1995). Studies in subjects with long-term nonprogressive human immunodeficiency virus infection. *N. Engl. J. Med.* **332**, 209–216.

Paoletti, E. (1996). Applications of pox virus vectors to vaccination: An update. *Proc. Natl. Acad. Sci. USA* **93**, 11349–11353.

Peeters, M., Honore, C., Huet, T., Bedjaba, L., Ossari, S., Bussi, P., Cooper, R. W., and Delaporte, E. (1989). Isolation and partial characterization of an HIV-related virus occurring naturally in chimpanzees in Gabon. *AIDS* **3**, 625–630.

Piatak, M., Saag, M. S., Yang, L. C., Clark, S. J., Kappes, K. C., Luk, K. C., Hahn, B. H., Shaw, G. M., and Lifson, J. D. (1993). High levels of HIV-1 in plasma during all stages of infection determined by competitive PCR. *Science* **259**, 1749–1754.

Piatak, M., Wages, J., Luk, K.-C., and Lifson, J. D. (1996). Competitive RT PCR for quantification of RNA: Theoretical considerations and practical advice. *In* "A Laboratory Guide to RNA: Isolation, Analysis, and Synthesis" (P. A. Krieg, Ed.), pp. 191–221. Wiley–Liss, New York.

Reimann, K. A., Li, J. T., Voss, G., Lekutis, C., Tenner-Racz, K., Racz, P., Lin, W., Montefiori, D. C., Lee-Parritz, D. E., Lu, Y., Collman, R. G., Sodroski, J., and Letvin, N. L. (1996). A chimeric simian/human immunodeficiency virus expressing a primary patient human immunodeficiency virus type 1 isolate env causes an AIDS-like disease after *in vivo* passage in rhesus monkeys. *J. Virol.* **70**, 6922–6928.

Reimann, K. A., Watson, A., Dailey, P. J., Lin, W., Lord, C. I., Steenbeke, T. D., Parker, R. A., Axthelm, M. K., and Karlsson, G. B. (1999). Viral burden and disease progression in rhesus monkeys infected with chimeric simian-human immunodeficiency viruses. *Virology* **256**, 15–21.

Rey-Cuille, M. A., Berthier, J. L., Bomsel-Demontoy, M. C., Chaduc Y., Montagnier, L., Hovanessian, A. G., and Chakrabarti, L. A. (1998). Simian immunodeficiency virus replicates to high levels in sooty mangabeys without inducing disease. *J. Virol.* **72**, 3872–3886.

Robert-Guroff, M., Kaur, H., Patterson, L. J., Leno, M., Conley, A. J., McKenna, P. M., Markham, P. D., Richardson, E., Aldrich, K., Arora, K., Murty, L., Carter, L., Zolla-Pazner, S., and Sinangil, F. (1998). Vaccine protection against a heterologous, non-syncytium-inducing, primary human immunodeficiency virus. *J. Virol.* **72**, 10275–10280.

Robinson, H. (1997). DNA vaccines for immunodeficiency viruses. *AIDS* **11**(Suppl. A), S109–S119.

Robinson, H. L., Montefiori, D. C., Johnson, R. P., Manson, K. H., Kalish, M. L., Lifson, J. D., Rizvi, T. A., Lu, S., Hu, S.-L., Mazzara, G. P., Pamicali, D. L., Herndon, J. G., Glickman, R., Candido, M. A., Lydy, S. L., Wyand, M. S., and McClure, H. M. (1999). Neutralizing antibody-independent containment of immunodeficiency virus challenges by DNA priming and recombinant pox virus booster immunizations. *Nat. Med.* **5**, 526–534.

Ruprecht, R. M., Baba, T. W., Rasmussen, R., Hu, Y., and Sharma, P. L. (1996). Murine and simian retrovirus models: The threshold hypothesis. *AIDS* **10**(Suppl. A), S33–S40.

Schultz, A. M., and Hu, S. L. (1993). Primate models for HIV vaccines. *AIDS* **7**, S161–S170.

Schultz, A. M., and Stott, E. J. (1994). Primate models for AIDS vaccines. *AIDS* **8**, S203–S212.

Sharp, P., and Li, W.-H. (1988). Understanding the origins of the AIDS viruses. *Nature* **336**, 315.

Sharp, P. M., Robertson, D. L., and Hahn, B. H. (1995). Cross-species transmission and recombination of AIDS viruses. *Philos. Trans. R. Soc. London Ser. B* **349**, 41–47.

Shibata, R., Kawamura, M., Sakai, H., Hayami, M., Ishimoto, A., and Adachi, A. (1991). Generation of chimeric human and simian immunodeficiency virus infections to monkey peripheral blood mononuclear cells. *J. Virol.* **65**, 3514–3520.

Shibata, R., Maldarelli, F., Siemon, C. *et al.* (1997). Infection and pathogenicity of chimeric simian–human immunodeficiency viruses in macaques: Determinants of high virus loads and CD4 cell killing. *J. Infect. Dis.* **176**, 362–373.

Shibata, R., Siemon, C., Czajak, S. C., Desrosiers, R. C., and Martin, M. A. (1997). Live, attenuated simian immunodeficiency virus vaccines elicit potent resistance against a challenge with a human immunodeficiency virus type 1 chimeric virus. *J. Virol.* **71**, 8141–8148.

Shinohara, K., Sakai, K., Ando, S., Ami, Y., Yoshino, N., Takahashi, E., Someya, K., Suzaki, Y., Nakasone, T., Sasaki, Y., Kaizu, M., Lu, Y., and Honda, M. (1999). A highly pathogenic simian/human immunodeficiency virus with genetic changes in cynomolgus monkey. *J. Gen. Virol.* **80**, 1231–1240.

Staprans, S. I., Dailey, P. J., Rosenthal, A., Horton, C., Grant, R. M., Lerche, N., and Feinberg, M. B. (1999). Simian immunodeficiency virus disease course is predicted by the extent of virus replication during primary infection. *J. Virol.* **73**, 4829–4839.

Suryanarayana, K., Wiltrout, T. A., Vasquez, G. M., Hirsch, V. M., and Lifson, J. D. (1998). Plasma SIV RNA viral load by real time quantification of product generation in RT PCR. *AIDS Res. Hum. Retroviruses* **14**, 183–189.

Tartaglia, J., Benson, J., Cornet, B. *et al.* (1998). Potential improvements for poxvirus-based immunization vehicles. *In* "Retroviruses of Human AIDS and Related Animal Diseases" (M. Girard and B. Dodet, Eds.), pp. 187–197. Elsevier, Paris.

Tsai, C. C., Emau, P., Follis, K. E., Beck, T. W., Benveniste, R. E., Bischofberger, N., Lifson J. D., and Morton, W. R. (1998). Effectiveness of postinoculation 9-[2-(R)-(phosphono-methoxy)propyl]adenine treatment for prevention of persistent simian immunodeficiency virus SIVmne infection depends critically on timing of initiation and duration of treatment. *J. Virol.* **75**, 4265–4273.

Tsai, C. C., Follis, K. E., Beck, T. W., Sabo, A., Bischofberger, N., and Dailey, P. J. (1997). Effects of 9-[2-(R)-(phosphonomethoxy)propyl]adenine monotherapy on chronic SIV infection in macaques. *AIDS Res. Hum. Retroviruses* **13**, 707–712.

Tsai, C. C., Follis, K. E., Sabo, A., Beck, T. W., Grant, R. F., Bischofberger, N., Benveniste, R. E., and Black, R. (1995). Prevention of SIV infection in macaques by 9-[2-(R)-(phosphono-methoxy)propyl]adenine. *Science* **270**, 1121–1122.

Unutmaz, D., KewalRamani, V. N., and Littman, D. R. (1998). G-protein coupled receptors in HIV and SIV: New perspectives on lentivirus–host interactions and on the utility of animal models. *Semin. Immunol.* **10**, 225–236.

Van Rompay, K. K., Berardi, C. J., Aguirre, N. L., Bischofberger, N., Lietman, P. S., Pedersen, N. C., and Marthas, N. L. (1998). Two doses of PMP protect newborn macaques against oral simian immunodeficiency virus infection. *AIDS* **12**, F79–F83.

Van Rompay, K. K., Cherrington, J. M., Marthas, M. L., Berardi, C. J., Mulato, A. S., Spinner, A., Tarara, R. P., Canfield, D. R., Telm, S., Bischofberger, N., and Pedersen, N. C. (1996). 9-[2-(R)-(phosphonomethoxy)propyl]adenine therapy of established simian immunodefi-ciency virus infection in infant rhesus macaques. *Antimicrob. Agents Chemother.* **40**, 2586–2591.

Van Rompay, K. K., Dailey, P. J., Tarara, R. P., Canrield, D. R., Aguirre, N. L., Cherrington, J. M., Lamy, P. D., Bischofberger, N., Pederson, N. C., and Marthas, M. L. (1999). Early short term 9-[2-(R)-(phosphonomethoxy) propyl] adenine treatment favorably alters the subsequent disease course in simian immunodeficiency virus infected newborn rhesus macaques. *J. Virol.* **73**, 2947–2955.

Van Rompay, K. K., Marthas, M. L., Lifson, J. D., Berardi, C. J., Vasquez, G. M., Agatep, E., Dehqanzada, Z. A., Cundy, K. C., Bischofberger, N., and Pedersen, N. C. (1998). Administration of 9-[2-(R)-(phosphonomethoxy)propyl]adenine (PMPA) for prevention of perinatal simian immunodeficiency virus infection in rhesus macaques. *AIDS Res. Hum. Retroviruses* **14**, 761–763.

Veazey, R. S., DeMaria, M., Chalifoux, L. V., Shvetz, D. E., Pauley, D. R., Knight, H. L., Rosenzweig, M., Johnson, R. P., Desrosiers, R. C., and Lackner, A. A. (1998). Gastrointes-tinal tract as a major site of CD4 depletion and viral replication in SIV infection. *Science* **280**, 427–431.

Villinger, F., Brice, G. T., Mayne, A., Bostik, P., and Ansari, A. A. (1999). Control mechanisms of virus replication in naturally SIVsmm infected mangabeys and experimentally infected macaques. *Immunol. Lett.* **66**, 37–46.

Villinger, F., Folks, T. M., Lauro, S., Powell, J. D., Sundstrom, J. B., Mayne, A., and Ansari, A. A. (1996). Immunological and virological studies of natural SIV infection of disease-resistant nonhuman primates. *Immunol Lett.* **51**, 59–68.

Watson, A., McClure, J., Ranchalis, J., Scheibel, M., Schmidt, A., Kennedy, B., Morton, W. R., Haigwood, N. L., and Hu, S. L. (1997a). Early postinfection antiviral treatment reduces viral load and prevents CD4+ cell decline in HIV type 2-infected macaques. *AIDS Res. Hum. Retroviruses* **13**, 1375–1381.

Watson, A., Ranchalis, J., Travis, B., McClure, J., Sutton, W., Johnson, P. R., Hu, S. L., and Haigwood, N. L. (1997b). Plasma viremia in macaques infected with simian immunodefi-ciency virus: Plasma viral load early in infection predicts survival. *J. Virol.* **71**, 2884–2890.

Wei, X., Ghosh, S. K., Taylor, M. E., Johnson, V. A., Emini, E. A., Deutsch, P., Lifson, J. D., Bonhoeffer, S., Nowak, M. A., Hahn, B. H., Saag, M. S., and Shaw, G. W. (1995). Viral dynamics in human immunodeficiency virus type 1 infection. *Nature* **373**, 117–122.

Wyand, M. S., Manson, K. H., Garcia-Moll, M., Montefiori, D., and Desrosiers, R. C. (1996). Vaccine protection by a triple deletion mutant of simian immunodeficiency virus. *J. Virol.* **70,** 3724–3733.

Zhang, J., Martin, L. N., Warson, E. A., Montelaro, R. C., West, M., Epstein, L., and Murphey-Corb, M. (1988). Simian immunodeficiency virus/Delta-induced immunodeficiency disease in rhesus monkeys: Relation of antibody response and antigenemia. *J. Infect. Dis.* **158,** 1277–1286.

John J. Trimble*,
Janelle R. Salkowitz†,
Harry W. Kestler‡

*Biology Department
Saint Francis College
Loretto, Pennsylvania 15940

†Case Western Reserve University
Department of Infectious Diseases
Cleveland, Ohio 44106

‡Department Science and Mathematics
Lorain County Community College
Elyria, Ohio 44035

Animal Models for AIDS Pathogenesis

I. Introduction

Two hundred years ago Edward Jenner described the cost of domestication of animals as an increased risk of the transmission of infectious diseases (Jenner, 1802). He also noticed that zoonotic infectious agents could attenuate when first passed in an alternate species. His third and most important observation was that the attenuated variant microbe could be used as a vaccine against pathogenic versions of the organism. The success of his vaccine was not fully realized until 150 years (1979) after his death. Today smallpox is confined to a couple of research laboratories in the United States and Russia.

HIV is a lentivirus that has been recently introduced into humans most likely as a result of a cross-species transmission of the virus from another primate. Recent evidence strongly suggests that HIV was transmitted from chimpanzees to humans (Gao *et al.*, 1999) (see Section VIII). A close associa-

Advances in Pharmacology, Volume 49

tion of humans with chimps from the wild appears to be responsible for a new disease in humans. Jenner had predicted that human intrusion into the habitats of wild animals could result in the transmission of infectious agents.

Like smallpox viruses, AIDS viruses also have a great differences in pathogenic potential. Attenuation of primate lentiviruses has also been described. The first case of attenuation was not the result of passage of virus *in vivo* or by tissue culture passage, but *in vitro* using recombinant techniques (Kestler *et al.*, 1991). The elimination of the *nef* reading-frame rendered the molecular clone SIVmac239 infectious, but unable to induce disease in rhesus macaques. This suggested that *nef* attenuated SIV may represent a model for a "modern day cowpox virus" and the use of this weakened SIV as a potential vaccine has been described (Daniel *et al.*, 1992).

It is unlikely that Jenner's prevention strategy will ever be applied to the disease caused by HIV. The association of the attenuated vaccine with disease in immunocompromised rhesus monkeys (Baba *et al.*, 1995) diminishes the enthusiasm for such an approach. Furthermore, even in immuno-competent animals new evidence suggests that there is no escape from disease, only a delay in disease onset (Baba *et al.*, 1999). The results for rhesus monkeys with attenuated SIV is duplicated in humans infected with *nef*-deleted HIV. Individuals who harbor naturally attenuated HIV do not escape disease, but, like the rhesus macaques, experience a delay in its onset (Altman, 1999; Greenough *et al.*, 1999; Learmont *et al.*, 1999). These results virtually eliminate any live vaccine candidate where proviral DNA persists from further consideration. A critical difference in the disease that Jenner eliminated, smallpox, and AIDS is the persistence of the lentivirus while the poxvirus is cleared.

In order for a live attenuated vaccine to succeed, additional control or elimination of persistent proviral copies is necessary. A vaccine in which a conditional lethal genetic element has been added is a viable alternative to the live *nef*-attenuated vaccine and should be pursued (Kestler and Jeang, 1995; Chakrabarti *et al.*, 1996; Smith *et al.*, 1996; Kestler and Chakrabarti, 1997). In this model an additional target for antiretroviral therapy is added; however, therapy using conventional highly active antiretroviral therapy (HAART) could also be applied to *nef*-attenuated viral infection. Combination therapy could succeed in eliminating an attenuated virus from the infected individual and this treatment regimen should be tested in an animal model.

The primate lentiviruses have differential pathogenesis; not all combinations of virus and hosts produce disease. Pathogenesis is a function of both viral and host determinants. Furthermore, even within a species susceptible to disease caused by an AIDS virus, variability in manifestations of the disease exists. As an example, when Rhesus monkeys are inoculated with isogenic molecularly cloned virus, two different disease patterns are observed: one in which monkeys succumb to AIDS in a short time course (6 months) and one in which monkeys succumb to AIDS over a more protracted time course (2 to 3 years) (Kestler *et al.*, 1990). This can only be due to individual variations in the monkeys since the virus used in the inoculum is genetically homogenous. We know that the monkeys who respond differently have a difference in antibody response to the virus.

Recent evidence has suggested that even though few animals are infected with SIVcpz in the wild, *Pan troglodytes* may have been the reservoir for HIV transmission into humans. SIVcpz was found in so few animals that many have speculated that HIV transmission from humans to monkeys was more likely than SIVcpz transmission to humans. The answer to this "which came first question" may be found in some new data about chimpanzee infections (Gao *et al.*, 1999). Hann and coworkers have found that three subspecies of chimpanzees are infected with SIVcpz. These animals are geographically isolated from each other and the viruses of the subspecies are distinct (Golberg and Ruvolo, 1997). Therefore Gao and colleagues speculated that infection of the chimpanzees by the virus that was the ancestor of the modern SIVcpz occurred prior to subspeciation. If this is true, one can determine the minimal length of time chimps have been infected by calculating the time since the subspecies arose. This can be done by comparing the sequence of mitochondrial DNA that diverges at a relatively constant rate (Morin *et al.*, 1994). Thus by calculating homology, one can surmise the minimal length of time that SIV has been in the chimp population. Using this technique they have determined that chimpanzees have been infected with the virus for at least 10,000 years (Gao *et al.*, 1999).

If chimpanzees have been infected this long with the virus it is likely that monkeys who are resistant to infection or disease have been selected. Thus by mounting a "genetic defense" against the virus, chimpanzees have nearly eliminated SIV from its population. Thus the epidemic may be nearly over in the chimpanzee population and just beginning in the human population. A number of mutations exist in the human population that affect the ability of HIV to infect or produce disease, providing evidence that we are mounting a genetic defense against HIV. It appears that the mutations in humans, mapping to second receptor genes, will not be the same as those that were selected for in the chimp population since the second receptors in the chimpanzee are nearly identical to the wild-type human second receptors (see Section VIII). The question of why there is differential pathogenesis for HIV in the two primate species remains unanswered.

II. Primate Lentiviruses Cause AIDS

Infectious disease science has been guided for over 200 years by a set of postulates described by Robert Koch to establish microbial etiology for a disease (Koch and Cheyne, 1880). The postulates state that:

1. The same pathogen must be present in every case of the disease and must be isolated from the diseased host.
2. The pathogen must be grown in pure culture and characterized.
3. The pathogen from the pure culture must produce the same disease in an animal model.
4. The pathogen can be reisolated from the diseased tissue from the animal and must be identical to the original organism.

Koch's postulates have served for the identification of a host of bacterial, fungal, parasitic, and viral diseases. Koch's postulates have even been applied, albeit with some modification, to suspected prion diseases where the infectious agent may be void of nucleic acid. With the successes of these postulates, there have been failures where a microbe is responsible for the disease but Koch's postulates fail to identify a microbial etiology. These are best exemplified by syphilis and *Treponema palladum*. While it is widely accepted that that there is a bacterial cause for syphilis, the organism *Treponema palladum* has never been cultured on artificial media. The disease tuberculosis was shown to be due to *Mycobacterium tuberculosis* by Koch himself, but even this organism is known to produce pathology in tissues other than the lung.

Koch's postulates are being updated to reflect modern molecular technology (Shaw and Falkow, 1988). Recently Koch's postulates were used to show viral etiology for Kaposi's Sarcoma (KS). The order in which Koch's postulates were applied in this case was not the typical order. The investigators characterized the microorganism before they had isolated it. Chang and Moore made an assumption that the cause of KS was microbial and that there would be residual nucleic acid in the diseased tissue. The characterization they made was genetic. They used a technique called representational difference analysis to ask what was genetically different about diseased tissue from normal tissue in the same patient. They found a DNA sequence and it was used to search Genbank for homology. This search produced several "hits." The candidate sequences were several distinct herpesviruses (Chang *et al.*, 1994). They returned to the first of Koch's postulates and focused their isolation attempts on herpesvirues and thus discovered HHV-8 (Moore *et al.*, 1996).

There has been much spoken in the popular media about applying Koch's postulates to human AIDS. Establishing a microbial cause for AIDS using Koch's postulates has been problematic. However, when examining the total of all the information available, the evidence is quite good for a lentiviral etiology for acquired immunodeficiency syndrome. The central problem has been the lack of an animal model for the disease. While chimpanzees can be infected with HIV, they do not in general experience the severity of the disease that humans do. This may be due to the fact that now we are beginning to understand that the chimpanzee population has been infected with the virus for a long period of time and the chimpanzees of today may have been selected for resistance to lentiviral disease (Gao *et al.*, 1999). There have been recent reports of pathology in this animal model (Fultz, 1997; Mwaengo and Novembre, 1998; Wei and Fultz, 1998) (See Section VIII for more discussion). Other animal models supply evidence for a lentiviral etiology for AIDS. SIV infection of rhesus monkeys shows a disease that is exactly like human AIDS.

The isolation of molecular clones of SIV have enabled the investigation of the molecular genetics of primate lentiviruses and have provided the best evidence for a viral cause for AIDS. A molecular cloned virus contains the total genetic information for the virus on a proviral DNA copy. That copy can, by itself, program the production of the virus; in essence the virus life cycle can be started in the absence of complete virus. Thus, pure DNA encoding viral information can be used to test whether the virus causes AIDS or whether some contaminating substance copurifying in the virus preparation is necessary to produce the disease. The molecular clone SIV-mac239 was the first molecular cloned AIDS virus shown to be pathogenic (Kestler *et al.*, 1990). The hypothesis was extended further by the discovery that pathogenesis could be attenuated by the deletion of a gene called *nef* (Kestler *et al.*, 1991). The discovery of long-term HIV-infected AIDS non-progressors who harbor *nef*-deleted virus suggests that HIV is also the cause of AIDS in humans (Deacon *et al.*, 1995; Kirchhoff *et al.*, 1995).

The best evidence for a lentiviral cause of AIDS comes from an unfortunate laboratory accident. A well-characterized strain of HIV was accidentally introduced into a laboratory worker who eventually developed AIDS (Weiss *et al.*, 1988). The virus that was recovered from the individual was identical to the virus the individual was working with (Kong and Shaw, 1989).

III. SIV Infection of Rhesus Monkeys

SIV was used to study pathogenic effects of primate lentiviruses and to understand the molecular genetics of AIDS viruses. Additionally, SIV was instrumental in evaluating candidate vaccines and has itself uncovered a potential AIDS vaccine strategy. There are many animal models for AIDS and they differ in the extent to which disease is represented and the degree of genetic relatedness of the infectious agent or component of the agent used in the model. The first animal model for AIDS was the result of a discovery in the Primate Center System in the United States. A disease that was remarkably similar to human AIDS spontaneously arouse in several colonies around the country. Scientists who were at the time studying a retrovirus that produced a different disease that only loosely resembled human AIDS discovered the disease. These virologists were adept at recognizing the biochemical and physical properties of retroviruses and one of them,

M. D. Daniel, isolated the first simian immunodeficiency virus (SIVmac) from a captive rhesus macaque (*Macaca mulatta*) at the New England Regional Primate Research Center (Daniel *et al.*, 1985). Rhesus macaques infected with SIVmac suffered from a disease that was very similar to HIV disease in humans. With the first isolation and subsequent inoculation of naive animals, the first animal model for AIDS was developed (Letvin *et al.*, 1985). Since then, SIV has been obtained from sooty mangabeys (*Cercocebus atys*; Fultz *et al.*, 1986a), African green monkeys (*Cercopithecus aethiops*; Ohta *et al.*, 1988), Mandrill (*Papio sphinx*; Tsujimoto *et al.*, 1988), and chimpanzees (*Pan troglodies*; Peeters *et al.*, 1989, 1992; Huet *et al.*, 1990; Vanden Haesevelde *et al.*, 1996). The SIV model was accelerated by the isolation of molecular provrial clones of the virus (Kestler *et al.*, 1990). One clone was able to produce AIDS in the model (Kestler *et al.*, 1991), leading to the many studies on the molecular genetics of AIDS pathogenesis. The use of clones had a number of other applications, such as a detailed analysis of neutralization and the generation of escape mutations (Burns and Desrosiers, 1990) and the analysis of determinants needed for macrophage tropism as well as neurotropism (Kodama *et al.*, 1990; Mori *et al.*, 1992). The use of molecular clones has been questioned since in an infected individual the virus exist as a quasispecies, but data from HIV-infected individuals suggest that only one clone of a quasispecies is transmitted in a natural infection (Zhu *et al.*, 1993). An acutely pathogenic molecular clone has been isolated from an infected sooty mangabey called pBJ14 (Fultz *et al.*, 1989). The determinants of this virus needed for acute pathogenesis were in this virus.

IV. HIV Infection of Macaques

It came as a surprise to most primate lentivirologists that pigtailed macaques could be used as a host for HIV. Researchers at the University of Washington's Regional Primate Center succeeded in infecting pigtailed macaques (*Macaca nemestrina*) with HIV-1 (Agy *et al.*, 1992) despite conventional wisdom based on the findings that other closely related monkeys such as rhesus (*M. mulatta*), longtail macaques (*Macaca fascicularis*), and cynomogus monkeys resisted infection (Lewis *et al.*, 1992). Several logistical advantages over the only other nonhuman primates to be infected with HIV-1, chimpanzees and gibbon apes (Fultz *et al.*, 1986c; Lusso *et al.*, 1988), are that the pigtailed macaques are not as scarce, expensive, or an endangered species, yet do share hematological markers similar to humans. The early excitement over the potential of an animal model that may help in understanding the pathogenesis of early HIV-1 infection and evaluating candidate vaccines and antiviral therapies has since been tempered.

The possibility that pigtailed macaques might be infectable with HIV-1 arose from observations that the macaques and pigtail PBMCs were more susceptible to infection with HIV-2 and SIV (Agy *et al.*, 1992; Lewis *et al.*, 1992). Acute and persistent infection of eight pigtailed macaques by cell-free virus or HIV-infected cells was determined by cocultivation and the presence of HIV-1 DNA in PBMC, detectable by the polymerase chain reaction and seroconversion to produce antibodies to HIV gag and envelope antigens (Agy *et al.*, 1992). The animals developed moderate lymphadenopathy and two developed rash over the abdominal and inguinal regions. In contrast to experience with HIV-1-infected chimpanzees, one of the eight developed viremia, as evidenced by isolation of virus from cell-free plasma. The HIV_{LAI} and the molecularly cloned HIV_{NL4-3} strains were used to infect these animals.

The confirmatory report by Frumkin *et al.* (1993) extended the model to show infection of pigtailed macaque PBMCs by five laboratory-adapted and one clinical isolate of HIV-1 and *in vivo* infection of six macaques with cell-free and cell-associated virus. Coculture, PCR amplification of gag DNA, and seroconversion demonstrated acute and persistent infection. The infected macaques again had moderate lymphadenopathy and two had rash. In contrast to acute HIV infection of humans, none of the six macaques developed viremia and only one developed serum HIV-1 antigenemia. Typically, virus was not routinely recoverable after 2 months. Infections were not associated with any significant disturbances in CD4+ T-cell levels nor were there signs of immunodeficiency disease (Frumkin *et al.*, 1993; Bosch *et al.*, 1997). Evidence of HIV DNA in nonneuronal brain cells and mild brain pathology (Frumkin *et al.*, 1995; Anderson *et al.*, 1994) were not accompanied by viral encephalitis or behavioral changes often seen in advanced human brain disease.

The limitations of the pigtailed macaque/HIV-1 model began to be defined by studies that showed not all strains of HIV replicated in pigtailed macaque PBMCs. Twenty-four primary isolates of HIV-1 failed to replicate in pigtailed macaque PBMCs (Otten *et al.*, 1994). Also, intravenous inoculations with field isolates of HIV-1 and a high-titered laboratory strain failed to establish a productive infection using conditions that were successful using HIV-2. Only HIV_{IIIB} and related strains such as HIV_{CH69}, which was obtained from an HIV_{IIIB} persistently infected chimpanzee, replicated in pigtailed macaque PBMCs and even then at a delayed, 10 to 100 times decreased level compared with human cells (Frumkin *et al.*, 1993; Gartner *et al.*, 1994a; Kimball and Bosch, 1998). Individual pigtailed macaques did not seem to be responsible for this limitation since T cells from more than 40 macaques were infectable with HIV_{CH69} (Gartner *et al.*, 1994a). The observation that HIV-1 replication in pigtailed macaques cell appeared to be suboptimal yet could establish persistent infections with low-dose inoculations, allowing recovery of infectious virus more than 1 year later, was of some interest (Gartner *et al.*, 1994b). It showed that persistent infection, however reduced compared to human HIV infection, as demonstrated by virus recovery was not due to lingering virus present in cells following large inoculations. Also, the delay in seroconversion beyond that expected from an immune response to heterologous host cells and a large virus inoculum suggested the specific anti-HIV-1 immune response followed replication of virus *in vivo*.

The necessity of adaptation of HIV to the pigtailed macaque host cells was considered a possible limitation to replication *in vivo* and establishing persistent infection or developing pathogenesis. Passage on autologous PBMCs to prepare HIV-1 inoculum of pigtailed macaques did not increase viral replication *in vivo* or produce disease (Gartner *et al.*, 1994a). Serial transfer of HIV-1 to successive animals failed in one report, apparently due to a premature transfer before the peak of viral replication and adaptation was reached (Gartner *et al.*, 1994b). Transfer of HIV-1 in pigtailed macaques was successful between groups of pigtailed macaques, with HIV recoverable from macaque PBMCs up to 10 weeks after transfusion (Agy *et al.*, 1997). Plasma viremia detected as HIV-1 RNA was found in 11 of 12 animals in the study. Seroconversion following transfers between groups was also established, although none developed any clinical signs of immunodeficiency disease or drops in CD4+ T cells. Animals in the second group did show higher levels of HIV RNA than the original pair of inoculated animals, but that increase was not reproduced in the third group. A fourth group transfused with 10 ml of blood 8 weeks postinoculation from group three animals did not yield culturable virus or produce seroconversion (Agy *et al.*, 1997). The primary objective of the study, to recover an adapted, pathogenic variant of HIV-1, failed. Sequence analysis of env DNA from serial passage in pigtailed macaques showed little evidence of adaptation, presumably as a function of limited viral replication.

The block in viral replication of HIV-1 in pigtailed macaques *in vitro* and *in vivo* may be potentially overcome with HIV-1/-2 chimeric viruses (Otten *et al.*, 1994) or SIV/HIV-1 chimeras, the so-called SHIV constructs (Kimball and Bosch, 1998). Identification of a post-reverse transcription (RT) block on efficient HIV-1 replication in macaque cells *in vitro* suggested that a little as one SIV gene might be replaced in HIV to create a chimera capable of inducing CD4+ T-cell depletion and disease in pigtailed macaques (Kimball and Bosch, 1998).

The value of the pigtailed macaque/HIV-1 model began to be demonstrated by studies addressing specific aspects of HIV infection and pathogenesis. The earliest reports showed inhibition of HIV-1 replication *in vitro* by agents such as human recombinant soluble CD4+ (Agy *et al.*, 1992) and AZT (Frumkin *et al.*, 1993). The potential of pigtailed macaques as a nonhuman primate model for pharmacokinetic analysis of anti-HIV drugs has included effects of ddI (Pereira *et al.*, 1994) and d4T (Keller *et al.*, 1995) on neonatal pigtailed macaques. Prenatal, postnatal, and maternal toxicity testing and pharmacokinetic analyses of ddI and d4T (Odinecs *et al.*, 1996) and AZT (Tuntland *et al.*, 1998; Ha *et al.*, 1998) helped determine guide treatment regimens that may need to vary with age. The pigtailed macaque/HIV-1 model allowed study of vertical transmission between fetal and maternal circulation by intra-amniotic inoculation. This would allow a reverse route of infection of the mother by the fetus via the placenta (Ochs *et al.*, 1993). Other HIV transmission studies include inoculation of pigtailed macaque neonates via oral, anal, and intravenous routes (Bosch *et al.*, 1997). This report found that oral transmission was not more efficient that nontraumatic anal inoculation in HIV-1 transmission.

Independent evidence, other than coculture, PCR amplification of viral genes, and seroconversion, that HIV-1 productively and persistently infects pigtailed macaques includes studies detailing the anti-HIV-1 cellular immune response. The poor HIV-1 replication to levels below that permitting repeated reisolation after 20 weeks postinoculation might be due to the relatively poor replication relative to human cells. It may also be caused by an effective and sustained proliferative and cytotoxic T-lymphoctye response (CTL). Kent *et al.* (1995), investigating the immunological containment of HIV-1 in pigtailed macaques, found an early and persistent CD4+ T-cell proliferative response and CTL response starting 4–8 weeks after infection and persisting to 140 weeks. These CTL responses temporally correlate with reduction and continued control of HIV in macaques (Kent *et al.*, 1997). Comparisons with humans infected with HIV suppressing HIV similar to that seen in long-term nonprogressors must be tempered with the result that viremia in the pigtailed macaques is 2–3 logs lower than in acute HIV-1 infection of humans. Nevertheless, the intriguing result that HIV levels fall with the rise of an HIV-1-specific cellular immune response and continues to be contained in macaques lead to even more promising work. Vaccination with HIV DNA and boosting with a recombinant fowlpox vaccine encoding HIV antigens protected four macaques from intravenous challenge (Kent *et al.*, 1998). No signs of acute infection, plasma viremia, or culturable virus were obtained. Four unvaccinated animals became infected using these same criteria. This otherwise dramatic result of protective immunity may reflect the suboptimal replication and nonpathogenic infection of HIV-1 in pigtailed macaques.

V. SHIV Infection of Macaques

In the long search for an adequate animal model for HIV, a number of novel approaches have been taken. Infection of chimpanzees with HIV employs the virus that we would like to have more information about, but it does not produce significant disease. Thus, while useful, this model does not have the same properties of SIV infection of rhesus monkeys or HIV infection of humans. These properties include depletion of cells bearing the

CD4 antigen and production of high virus loads. SIV infection of rhesus monkeys employs a virus that is only a relative of the clinically significant virus HIV. To evaluate the behavior of HIV genes in the SIV animal model, a novel approach has been taken (Li *et al.*, 1992). Various investigators have developed SHIV strains of virus, that is, SIV HIV hybrid virus. Recently, it appears that some combinations of HIV and SIV genes can be made pathogenic for Rhesus macaques. SHIVs have great utility for a number of genetic applications. Perhaps the best use is to map genetic differences between HIV and SIV. They have also been used to evaluate host genetic differences as well as immune responses. Doms and coworkers have used a SHIV construct to find host determinants for species specificity (Hoffman *et al.*, 1998). Their analysis will help reconstruct molecular events that transpired when primate lentiviruses were crossing species barriers.

The most exciting use of SHIVs has been to determine the nature of AIDS virus pathogenesis. This goal was not immediately realized because of the limited pathogenesis of early SHIV constructs. The blockade was eliminated when Narayan applied an old technique used on another lentivirus to the problem. The technique, *in vivo* passage of virus, selects for virus that has fast replication kinetics. The technique was initially applied to visna virus (Clements *et al.*, 1982) and is performed by serially infecting animals with the virus. The virus that comes out is genetically distinct from the virus that was used in the initial infection. The mechanism by which it works is unknown, but probably involves selection of viruses that are adapted for replication in the tissue that is sampled when the passage is performed. Since lentiviruses have a great genetic plasticity, better replicating strains rapidly emerge. In the case of SHIV, the technique can be used to adapt the virus to its "new" host. Early replication kinetics are probably essential to lentiviral pathogenesis. Early replication kinetics are essential for establishment of infection. An HIV infection can be aborted if AZT is employed (Sperling *et al.*, 1996; Wade *et al.*, 1998). Naryan has used SHIVs to map pathogenic determinants by genotyping serial isolates from the *in vivo* passaging of the virus (McCormick-Davis *et al.*, 1998). The determinants of pathogenesis mapped to env, *nef*, and vpu.

VI. HSIV

A number of studies have been devoted to the investigation of chimeric SIV and HIV, called SHIV, in which HIV genes are inserted into an SIV backbone (see Section V). The purpose of such a construct is to address questions of HIV pathogenicity or immunogenicity in the macaque model. The opposite approach has not been explored; that is, to take SIV genes and insert them into an HIV backbone. Recently Yoon *et al.* (1998) have taken the *nef* gene from the pathogenic molecular clone SIVmac and cloned it into HIV-NL4-3. These workers found that the SIV allele of *nef* could complement the HIV-1 *nef* in an *in vitro* growth assay. They have provided further evidence for a lentiviral cause of AIDS. Since *nef* is a pathogenic determinant of SIV, and previous studies have shown that HIV *nef* can substitute for the SIV *nef* in a SHIV, this study demonstrated that the SIV *nef* can also substitute for the HIV *nef*. It is apparent that although genotypically distinct, the phenotype of either *nef* gene product is identical.

VII. HIV Infection of Chimpanzees

The HIV-1/chimpanzee (*P. troglodytes*) model is arguably the most relevant of several primate animals models due to the greater than 98% genetic identity between our species (King and Wilson, 1975) and the recent identification of the chimpanzee as the most likely origin of HIV (Gao *et al.*, 1999). The considerable expense and concrete limitation on the number of these endangered species available for research in HIV pathogenesis and drug or vaccine testing have been and are continuing problems. As viable candidate HIV vaccines reach status for testing in humans and animal models, ethical and practical debates about the use of the HIV-1/chimpanzee model have intensified.

HIV infection of the chimpanzee has been established by cell-associated and cell-free HIV-1 or by transfusion of blood or plasma (Alter *et al.*, 1984; Francis *et al.*, 1984; Gajdusek *et al.*, 1985; Fultz *et al.*, 1986b). These animals developed anti-HIV IgG antibody responses and HIV was recoverable from peripheral blood mononuclear cells (PBMC). While these reports and others suggest primary HIV infection would cause mild symptoms including lymphadenopathy or weight loss, no clinical disease is apparent in most instances. Laboratory results have characterized the immunological response, including neutralizing antibodies and antibody-dependent cellular cytotoxicities (Nara *et al.*, 1987) and cell-mediated immune responses (Eichberg *et al.*, 1987).

These early reports of a valuable animal model for AIDS began immediately to support other evidence concerning the transmission of HIV. For instance, infection of chimps with lymphocyte-poor plasma suggested the risk of exposure to HIV might be present in blood products such as clotting factors (Alter *et al.*, 1984). Also the inability of HIV to be transmitted to cagemates in close physical contact for long periods of time (Fultz *et al.*, 1986c; Fultz *et al.*, 1987) helped define the limits of HIV transmission methods, that is, excluding casual contact.

The value of the chimpanzee model was held up as representative of the early natural history of HIV infection, as that stage was rarely available for study in humans. The closest genetic companion to humans, *Homo sapiens*, among the great apes, may well have been expected to be infectable with HIV. However, the mere 1.5% genetic difference between chimpanzees and humans overall includes central major histocompatibility complex (MHC) genes and variation at MHC accounts for a large proportion of that difference (Leelayuwat *et al.*, 1993). Furthermore, this difference in MHC may influence the course of HIV infection (Malkovsky, 1996). Differences in clinical and immunological responses to HIV are more apparent. The acute primary stage of HIV disease in chimps was milder than that in humans, suggesting biological differences in either the virus stocks used to infect the animals or differences in the immunological responses between chimpanzees and humans. In chimpanzees, the appearance and waning of plasma viremia has been detected, often sporadically. The continued presence of virus in PBMCs subject to virus reisolation has been more reliable (Alter *et al.*, 1984; Fultz *et al.*, 1986c; Gajdusek *et al.*, 1985; Nara *et al.*, 1987; Nara *et al.*, 1989). The humoral response also mimics that of humans, with the appearance of neutralizing antibody titers typically produced after about 6 weeks and plateauing at about 6 months after infection (Fultz, 1993).

As in humans, chimpanzees after acute HIV-1 infection follow a long course of asymptomatic infection. No plasma viremia is detectable with coculture methods but virus may be infrequently recovered from PBMC cocultivation. The ability of the virus to persist in the presence of neutralizing antibody suggested that that humoral response played a lesser role. Antibody to virus that was initially type specific was found to broaden with time of infection to neutralize other more diverse isolates (Morrow *et al.*, 1989; Nara *et al.*, 1987). With the advent of PCR methods, reanalysis of HIV-1 infection of chimpanzees became more approachable. Largely due to the limited numbers of chimpanzees available, chimps were often "recycled" in various studies. The diversification and recombination among HIV strains following superinfection (Fultz *et al.*, 1987; Fultz *et al.*, 1997; Wei and Fultz, 1998) or challenge with heterologous HIV strains highlighted the ability of HIV to change considerably during the asymptomatic, yet active viral replication period of infection in this animal (Saksela *et al.*, 1993). Wooley and colleagues have shown recombination between different strains of SIV in a dually infected rhesus monkey (Wooley *et al.*, 1997). The extraordinary genetic diversity of HIV remains a hurdle for vaccine design, testing, and analysis in animal models.

With analysis of difference in HIV strains being more testable than biological differences between *Pan* and humans, several reports began to distinguish differences among HIV isolates and strains in their ability to infect chimpanzees. Most macrophage-tropic isolates replicated poorly in chimpanzee cells *in vitro* (Gendelman *et al.*, 1991; Schuitemaker *et al.*, 1993), mirroring the relatively poor replication seen in acutely infected animals. Lymphotropic HIV strains can replicate in chimpanzee cells *in vitro*. The extensive passage of HIV-1 in human T-cell lines or direct isolates from patients may select against well-replicating strains in chimpanzees. The poorly replicating strains, HIV$_{LAI}$, has been used for most vaccine trials where challenges have been used. Other *in vitro* or *in vivo* poorly replicating strains include HIV_{SF2} (Berman *et al.*, 1990; Berman *et al.*, 1996). The better replicating strains, HIV_{IIIb}, HIV_{RF}, and HIV_{MN} (Girard *et al.*, 1995) are found to have been passaged *in vivo* (HIV LAI-1B; Gendelman *et al.*, 1991; Watanabe *et al.*, 1991) or *in vitro* (HIV_{DH12} and $HIV_{E/90CR402}$; Shibata *et al.*, 1995; Girard *et al.*, 1996; Barre-Sinoussi *et al.*, 1997). The improvement in replication in chimpanzees or their cells following passage in the same is not unexpected due to mutation and selection of the better replicating variants within viral quasispecies. Other considerations such as the presence of heterologous antigens in the lipid envelope may have contributed to the immune response of chimpanzees infected with cell free-virus or virally infected cells from human cell lines or patients.

Examination of the mild acute HIV infection of chimpanzees has lead to the understanding that the differences may be due to immunological differences. The CD4+ depletion seen in humans has been investigated in *Pan*. The largely apathogenic strains of HIV used in most studies did not cause the dramatic depletion of CD4+ cells in chimpanzees. When CD4+ depletion was noted, increased apotosis in this cell population was an important factor and a correlate of disease (Davis *et al.*, 1998). Other studies found the lack of apoptosis correlated with absence of immune dysfunction and disease (Heeney *et al.*, 1993; Estaquier *et al.*, 1994; Gougeon *et al.*, 1997). The effects of xenostimulation of the immune system was found to increase viral replication to levels that had not previously been detectable (Shibata *et al.*, 1997). The effect of imunosuppression, such as via steroid therapy, demonstrated that the increase in HIV replication responds to immune system function (Morrow *et al.*, 1987; Shibata *et al.*, 1997).

A few studies have established that infection via the mucosal route is possible (Fultz *et al.*, 1986b; Girard *et al.*, 1998; Davis *et al.*, 1998). The scarcity of chimpanzees has largely dictated that vaccination challenge studies use the intravenous route, which is more quantifiable and better characterized. If more animals were available a mucosal vaccine challenge, which would be more predictive of the usefulness of the vaccine for humans, could be employed in the animal model. Even the coreceptors of HIV infection between human and chimpanzees, CXCR4 and CCR5, are identical or show differences as little as two substitutions. This suggests that the poor replication of some strains of HIV is not due to differences in CD4 receptors or coreceptors (Pretet *et al.*, 1997)

The potential for animal models, including HIV-1/chimpanzees, for not only defining aspects about HIV infectiousness and pathogenesis but also serving as venues for testing antiviral or immunosupportive therapies and eventually vaccines was noted early. The scarcity, expense, and largely apathogenic HIV infections have limited the usefulness of HIV-1/chimpanzees in preclinical drug evaluations. Rather, mechanisms for drug testing in clinical trials already existed and AIDS patients clamored to enter them. No pressing need for an animal model beyond toxicity testing existed. Now, with combination therapy available, HAART (highly active antiretroviral therapy) undermines the potential for evaluation of vaccines in humans and increases the need for a suitable animal model for vaccine testing. That model may have just appeared on the horizon.

A chimpanzee infected with three different isolates of HIV-1 over 10 years progressed to AIDS (Novembre *et al.*, 1997). A decline in CD4+ cells over 3 years, an increase in cell-free virus in plasma, and opportunistic infections characterize the disease. Of particular interest was the presence of a new strain, HIV_{JC}, that is cytopathic for chimpanzee peripheral blood cells, significantly divergent from all inoculating viruses, and suggestive of a large quasispecies. This suggests that not only may some HIV-1 strains produce AIDS in chimpanzees but also that passage may result in a more pathogenic virus. Transfusion of blood into other chimpanzees resulted in a rapid and progressive loss of CD4+ T cells, high viral burdens, immune hperactivation, and increased levels of CD4+ T-cell apopotosis (Novembre *et al.*, 1997; Davis *et al.*, 1998). The conclusion that some HIV-1 strains may be pathogenic for chimpanzees and that HIV-1-induced disease is possible invigorated researchers' interest in using this model. Objections to using a highly pathogenic strain of HIV in chimps arose (Leigh, 1998; Prince and Andrus, 1998; Ryder, 1998; Prince *et al.*, 1999), largely arguing that our nearest genetic relative deserves additional ethical consideration. The scientific advantage of a highly pathogenic HIV strain that may be used as a challenge for vaccine trials was always limited by the number and expense of chimpanzees and especially time if disease only developed on a decade-long scale. The advantages of using the highly pathogenic strain of HIV in chimpanzee vaccine efficacy studies, rather than a nonpathogenic strain, HIV_{HAN-2}, permit a fair test of vaccination, with an endpoint as close as possible to that of human HIV-1 disease, that is, AIDS (Letvin, 1998).

The first attempts to vaccinate chimpanzees met with mixed success. Claims of protection or lack of protection were difficult to evaluate due to the small sample size, below that necessary for statistical significance; additionally, differences in methods, including various challenge HIV strains or doses, made drawing general conclusions about the data difficult. For example, with no known correlates of protective immunity, reports on the lack of neutralization of virus following HIV envelope, gp120 vaccination (Arthur *et al.*, 1989; Nara *et al.*, 1990) appeared to be a setback in vaccine-development efforts. A test of clade B HIV-1-vaccinated chimpanzees with challenge with heterologous HIV-1 (clade E) also did not prevent infection (Girard *et al.*, 1996). Unlike other established antiviral vaccines, it became apparent that neutralizing antibody was not critical to limiting viral replication or disease progression. The relatively rare and late appearance of antibody-dependent cytotoxicity in chimpanzees differs significantly from that seen in human HIV infection (Ferrari *et al.*, 1994). Virus-specific cytotoxic T-cell response that coincides with the drop in HIV during acute infection indicates its importance (Eichberg *et al.*, 1987). Other antiviral vaccines did not necessarily induce cellular T-cell responses. Successes in vaccination of chimpanzees have included whole inactivated virus, HIV–poxvirus recombinants, or purified recombinant proteins such as surface glycoprotein gp160 (Hu *et al.*, 1987; Berman *et al.*, 1990; Girard *et al.*, 1991; Fultz *et al.*, 1992; Girard, 1995). However, these positive reports of protection often used strains of HIV that replicated poorly in chimpanzees for challenge or had very extensive immunization and booster schedules unlikely to be practical. The strains used were not representative of clinical isolates of HIV-1 and were delivered in dosages or routes unlike typical human transmission of HIV.

The description of HIV-1 infection of chimpanzees as basically apathogenic with some minor variations such as detectable lymphopenia or thrombocytopenia began to change a disease model as the length of time of infection for some of the chimpanzees stretched into a decade (Fultz *et al.*, 1991). The search for strains of HIV with a highly pathogenic potential for chimpanzees began with surveying patient isolates for their ability to grow in chimpanzee cells *in vitro* (Shibata *et al.*, 1995). Only 3 of 23 isolates were infectious in cell culture and one strain, HIV_{DH12}, had extremely rapid replication kinetics, profound pathogenicity, and dual tropism for macrophages and T cells. Unlike other strains of HIV which do not cause viremia in chimpanzees, HIV_{DH12} caused viremia, lymphadenopathy, and disseminated rash. The peak load of HIV_{DH12} in one infected chimpanzee during primary infection and 6 months postinfection rose to levels comparable to that of HIV-infected humans. Without question, the infectability of chimpanzees with HIV contributed to our early understandings about transmission, primary HIV infection, and the immune response. What will the HIV/chimp model offer future vaccine and anti-HIV therapeutic evaluations? The answer, however positive, is preordained by the limited numbers of this endangered animal. A role in vaccine, rather than drug, studies seems likely. The current debates about the choice of HIV vaccine challenge strains for chimpanzees demonstrates our experience and advancement in using this HIV animal model. It may well determine our ability to develop and fairly test a safe and effective vaccine against HIV.

VIII. HIV Infection of SCID Mice

Mice genetically bred to have immunodeficiency called severe combined immunodeficiency (SCID) can be implanted with human tissue. The implanted tissue will develop a partial humanlike immune system, and HIV can infect the engrafted animal. The model has been useful in confirming viral pathogenicity predicted from the rhesus monkey model and has also been useful in testing potential vaccines. However, the model, like primate models, are not widely available to investigators because of cost and the expertise necessary to maintain facilities for immunologically crippled animals.

Most animal models for HIV are approximations of the human HIV situation, with numerous limitations. They are either animal lentiviruses infecting nonhuman animal hosts such as SIV/macaques or HIV is forced into a context beyond its natural tropism such as nonhuman primates, rabbits, or transgenic mice.

The need for a small-animal model for HIV infection was set back by the failures to infect various animal cells *in vitro,* even following the xenogenic expression of the principle HIV receptor, the human CD4+ protein. CD4+ rodents cell were refractory to HIV infection and since the second receptors for HIV, the CCR5 and CXCR4 chemokine receptors, were not identified until 1996 (Dragic *et al.,* 1996; Deng *et al.,* 1996; Feng *et al.,* 1996), direct infection of rodent cells has not been achieved. Nevertheless small-animal models have played important roles in studies of viral replication, viral pathogenesis, preclinical evaluation of antiviral drugs, and vaccines, resulting in a reduction in cost and more rapid development. Specifically, without a small-animal model for HIV infection, candidate antiviral compounds would leap from *in vitro* efficacy against the virus, with some animal toxicity testing, directly to clinical studies in humans. The assessment of dose–response relationships *in vivo,* serum binding, tissue penetration, and pharmacokinetics, while not equal in animals and humans, would, in the least, give an important first approximation. Additionally, a postexposure efficacy trial, such as postexposure prophylaxis with AZT following HIV exposure in the workplace (Henderson and Gerberding, 1989), would be difficult to pursue in a clinical trial, but approachable with a small-animal model.

The description in 1983 of a severe combined immunodeficiency mouse (Bosma *et al.,* 1988) resulted from a defect in rearrangement of receptors for B and T cells due to recombinase deficiency. The animals have normal erythrogenic, myelomonocytic, and megakaryocytic lineages, but lack mature B and T cells. The lack of humoral or cell-mediated immunity is therefore called severe combined immunodeficiency. They cannot produce antibodies and cannot reject allografts or xenogratfts. Normal innate immunity as well as functional macrophages, NK cells, and elevated hemolytic complement activity remain. Mice homozygous for the *scid* mutation, such as C.B-17-*scid/scid* mice, have shown some degree of "leakiness." Some mice that produce peripheral populations of mature lymphocytes and antibodies, perhaps due to somatic reversion events, are routinely excluded from experiments. While detecting murine Ig in serum by ELISA is a straightforward procedure, no standard for such exclusion has been established and has ranged from 0.4 μg/ml to 50 μg/ml by different investigators (Bosma *et al.,* 1988; Nonoyama *et al.,* 1993; Mosier *et al.,* 1993).

The xenotransplantation of SCID mice with components of the human immune system infectable with HIV began in 1988. Several variations have since been tested in order to improve reproducibility and the length or constitution of the human immune cell transplantation. These variations have served to help define aspects of HIV pathogenesis, antiviral safety and efficacy, passive and active vaccination effectiveness, model mucosal transmission, pediatric AIDS, and HIV encephalitis, among others. A review of the different SCID mouse models reconstituted with human cells or tissues and the usefulness in studying HIV processes follows.

A. SCID Hu thy/liv

Human reconstituted SCID mice have been developed using fragments of human fetal thymus and liver organ transplants under the renal capsule (McCune et al., 1988, 1990; Namikawa et al., 1990). A local chimeric lymphoid compartment is fused to create a conjoined organ capable of long-term multilinear human hematopoesis. Fetal human liver provides a source of human hematopoietic stem cells, allowing self-renewal. The thymus provides an appropriate environment and architecture for immune-cell maturation. In addition, the use of fetal organs assures that no mature human T cells are present that would mount a graft versus host (GVH) response to the SCID mouse. In fact, developing human immune cells become tolerant to the murine tissue. The thy/liv SCID-hu mice, as they are called, produce B and T cells of human origin in the peripheral circulation and secrete human IgG (Krowka et al., 1991; Namikawa et al., 1990).

The ability of the transplanted tissue to become infected with HIV (Namikawa et al., 1988) was immediately found to be effected by the strains of HIV used. The SCID-hu mouse model served as testing for antiviral medication such as AZT (McCune et al., 1990). It was found that tissue culture-adapted HIV were not infectious, whereas primary patient isolates were (Namikawa et al., 1988). The kinetics of replication and CD4+ cell depletion in the SCID-hu mouse varied according to phenotypic cell tropism of the particular HIV isolate (Jamieson et al., 1995; Uittenbogaart et al., 1996). HIV that is T-cell tropic with an SI (syncytium-inducing) phenotype is more pathogenic, replicates to higher viral titers, and results in more severe thymic and CD4+ depletion compared to non-SI, macrophage-tropic isolates (Jamieson et al., 1995; Kaneshima et al., 1994; Kollmann et al., 1996). The high viral burden and rapid CD+ depletion suggested to some investigators that direct killing, and not apoptosis, was responsible during the initial phase of infection in SCID-hu mice (Jamieson et al., 1997).

The report by Kitchen and Zack (1997) identified the greater expression of CXCR4 cytokine receptor, an HIV coreceptor with CD4, in immature thymocytes. This suggested that the rapid disease progression in some children may be due to efficient infection of the thymus by certain SI strains of HIV.

A dramatic example of the analysis possible in SCID-hu mice not possible in humans includes identification of HIV-1 determinants necessary for replication *in vivo* as opposed to *in vitro* (Su *et al.*, 1997). Following the accidental infection of a lab worker with the HIV Lai/IIIB strain that may be attenuated by passage in T-cell line, HIV reisolation occurred only with primary PBMC and not T-cell lines. To determine whether the replication of HXB2, the molecular clone of HIV Lai/IIIB, contained rare members able to replicate *in vivo* or reversion events were required, the SCID-hu model was used. HXB2 didn't grow out efficiently in the infected lab worker or in SCIDhu mice. The apparent attenuation in the envelope region was supported by making recombinant viruses between HXB2 and the envelope gene isolated from the lab worker (Kong and Shaw, 1989). In addition to envelope, studies identified other gene regions of HIV responsible for attenuation, including deletions in *nef* (Jamieson *et al.*, 1994; Aldrovandi and Zack, 1996), vif, and vpu (Aldrovandi and Zack, 1996).

Several major or minor variations in thy/liv SCID-hu mouse models for HIV pathogenesis have also been pursued. Increasing the amount of human thymus and liver tissue under both kidney capsules increased the number of human T cells in the peripheral blood. This resulted in a more disseminated infection following HIV injection (Kollmann *et al.*, 1994, 1995). In addition to thy/liv tissue in the kidney capsule, fragments of autologous lungs were coimplanted into the peritoneal cavity in a study of thymocyte abnormalities occurring in the first month of HIV infection. The lung tissue served as a peripheral immune tissue rich in macrophages, allowing a systemic and multifocal infection (Autran *et al.*, 1996). An interesting model for mucosal transmission of HIV included implantation of human intestine in thy/liv-SCID hu mice (Gibbons *et al.*, 1997). The thy/liv implants apparently provide a continuing supply of circulating T cells that populate the human intestinal implant. Inoculation into the lumen allows infection to occur at a mucosal surface, although whether human T cells, monocytes, or dendritic cells are first infected is not yet clear. The potential remains for this model system to yield valuable insight into mucosal transmission of HIV and the viral tropism for certain cell types or coreceptors.

B. SCID Hu PBL

The other major model system for reconstituting SCID mice uses human peripheral blood lymphocytes (PBL) from HIV-negative, EBV-negative donors (Mosier *et al.*, 1988). The Hu PBL-SCID mice are infectable with HIV (Mosier *et al.*, 1991), resulting in rapid CD4+ cell decline using noncytopathic HIV isolates (Mosier *et al.*, 1993b). No prior activation of the human cells is necessary, presumably because mature cells become activated in the murine environment, including inducing a graft versus host response and disease. Results consistent with the SCID-hu model of HIV infection include envelope's role in cell tropism, SI phenotype, and CD4+ T-cell depletion (Mosier et al., 1993b; Gulizia et al., 1996; Markham et al., 1996; Picchio et al., 1998).

The Hu-PBL SCID mouse has shown that resistance of donors vaccinated with Vaccinia gp160 and boosted with recombinant HIV gp160 can be protective against homologous challenge (Mosier et al., 1993a). Of three donors, all yielded some resistance to hu-PBL-SCID mice even though only one made neutralizing antibody. The T-cell response did correlate with the level of protection. This demonstrates the usefulness of the Hu-SCID mice model for evaluating future vaccine candidates and for defining the nature of protective immunity.

The most effective therapy for HIV infection will likely combine antiviral therapy with immunomodulators. The potential for pharmacological intervention to act as immunomodulators has also been tested in SCID mice. Linomide, a synthetic derivative of quinoline, has strong immunomodulatory effects in humans by limiting inflammatory cytokine overproduction and preventing apoptosis. Since the ability of HIV to replicate depends in part on the activation state, reduction of inflammatory cytokines and maintenance of CD4+ cell demonstrated by linomide may help reconstitute the immune system and boost the anti-HIV-1 response (del Real et al., 1998). The effect of IL-10, but not IL-12, on a modified SCID-hu mouse model with disseminated HIV infection and plasma viremia showed potential as therapy for acute HIV infection (Kollmann et al., 1996). The virtue of the model for further evaluation of exogenous cytokine therapy in a preclinical tests was established.

The need for a model of pediatric AIDS and the attendant problems in conducting human clinical trails in newborns may be solved in part by the description of a neonatal C.B-17 SCID mouse (nSCID) model using human cord blood lymphocytes (Hu-CBLs; Reinhardt et al., 1994). The engraftment of human cells in peripheral lymphoid organs and blood was much greater than that using adult human blood donors and adult SCID mice. Both laboratory and clinical HIV isolates were infectious and pathogenic in the hu-CBL-nSCID mice. The hu-CBL-nSCID mice had greater engraftment and contained higher numbers of CD4+ cells and few memory T cells and functionally immature T and B cells. Therefore this system mimics certain aspects of pediatric infection and should be useful for testing antivirals and strategies for blocking HIV infection of neonates.

Improving the SCID mouse model specifically for testing antivirals led Alder et al. (1995) to use multiple reconstitutions with human lymphocytes and a large inoculum. The advantage gained was splenic HIV p24 antigen levels detectable by ELISA. Using AZT as an example, potency and tissue concentration correlated with treatment efficacy. The high viral burden and p24 measurements were reported to be more quantitative than with other methods using PCR. Difficulties reported were the rarely detectable p24 levels in plasma, necessitating a terminal assay for assessing drug efficacy; biohazard concerns of injecting live mice with HIV; and variability in achieving splenic P24 levels in all trials. Despite the problems, the authors maintain that SCID mice models for pre clinical testing of antiviral would save time and money compared to human trails designed to evaluate potential HIV therapeutics.

Rather than use normal human cells in SCID mice, human tumor cells U937 were injected subcutaneously with cell-free HIV-1 or HIV-1-infected cells (Lapenta *et al.*, 1997). Pretreatment with antibody to either mouse-IFN or granulocytes resulted in tumor take and higher levels of HIV p24 antigenemia. The long-lasting antigenemia, over 3 months, was inhibited by AZT, as were the levels of virus expression and number of infected cells at the tumor site. The features of easy establishment of U937 tumor cells in SCID mice, high reproducibility, well-defined viral kinetics, and high levels of persistent viremia offer advantages for testing antivirals. A novelty of the model is the ability to serially reimplant tumor cells from HIV-infected mice treated with antivirals to investigate long-term antiviral therapy for the selection of HIV-1 drug-resistant strains.

Two models of the SCID mouse reconstituted with human cells have addressed the "leakiness" of the murine immunodeficiency and the remaining murine immune function other than B- and T-lymphocytic responses. The SCID-Beige mouse, closely related to C.B-17 SCID mice, also has defects in natural killer cells in addition to a lack of mature B and T cells (MacDougall *et al.*, 1990). This additional defect may allow for greater acceptance of xenografts, help establish viral infections, and remove the murine NK function from troubling the interpretation of antiviral protection studies. The *beige* defect also makes the mice less "leaky" compared to *scid/scid* mice (Mosier *et al.*, 1993b). Reconstitution efficiencies approaching 100% in SCID-Beige mice, functional human immune responses to KLH antigen, and intaperitoneal infection by HIV were reported (McBride *et al.*, 1995). Backcrossing of the *scid/scid* mice with nonobese diabetic NOD/Lt stain increases the percentage of engraftment of human PBLs in mouse spleens by 5- to 10-fold over that in inbred C.B-17 *scid/scid* mice (Hesselton *et al.* 1995). The success in engraftment and HIV infection of back-crossed NOD/LtSz-*scid/scid* mice may be due to loss of aspects of innate immunity still present in C.B-17 SCID mice, including macrophages, NK cells, and hemolytic complement. The resulting increase in human cell engraftment may make the NOD/LtSz-*scid/scid* mice a better choice for antiviral drug testing.

The blood–brain barrier presents another problem for the discovery and evaluation of anti-HIV therapies. HIV encephalitis is characterized by the infiltration of virus-infected macrophages, the presence of multinucleated giant cells, and a striking gliosis. The clinical manifestation of AIDS dementia and its relationship to HIV encephalitis is still not clear (Persidsky *et al.*, 1997). Modeling HIV encephalitis by injecting HIV and human PBLs intracerebrally into SCID mice yielded similar pathologies (Tyor *et al.*, 1993; Persidsky *et al.*, 1996).

IX. HIV Infection of Rabbits

The development of animal models for the limited host range of HIV has hampered AIDS. The restriction is apparently at the level of both the receptor CD4 and coreceptor CCR5. Dunn *et al.* (1995) made a transgenic rabbit model that specifically and stably expresses human CD4 on the surface of the lymphocytes. These lymphocytes were more susceptible to HIV-1 IIIB infection than normal rabbits. They have been shown to be capable of infecting these rabbits with IIIB and retrieving viral DNA from these rabbits. Another group (Speck *et al.*, 1998) has shown that when human CCR5 is added along with human CD4 in a rabbit cell line (SIRC), it renders the cells permissive to infection with HIV-1 macrophage-tropic strains. A rabbit transgenic for both CD4 and CCR5 may develop disease and would be very useful to access candidate therapies and vaccines.

X. Mice Transgenic for Primate Lentivirus Genes

Transgenic mice encoding AIDS viral genes have been used to analyze the functions of the gene products. A transgenic animal has a foreign gene integrated into its genome as an embryo. Thus, as the mouse develops, all cells of the animal encode the foreign gene. Expression of the foreign gene is dependent on regulatory signals introduced with the transgene. Tissue specific and developmentally specific promoter sequences can be used to limit gene expression to a particular time or organ. Alternatively, the use of a pan-active promoter will enable the assessment of phenotypes in a variety of contexts and may be more useful in uncovering fundamental mechanisms of the transgene. The model has been useful in determining the several properties of HIV and SIV proteins. Studies on *tat* show that the protein may facilitate the development of Kaposi's sarcoma (Vogel *et al.*, 1988). The product of the *env* gene may be required for the development of neurological lesions (Thomas *et al.*, 1994). The *nef* gene product appears to interact with the cellular machinery to establish conditions favorable to replication of HIV and SIV (Brady *et al.*, 1993; Skowronski *et al.*, 1993; Lindemann *et al.*, 1994; Hanna *et al.*, 1998a; Larsen *et al.*, 1998).

Primate lentiviruses contain at least six other genes in addition to *gag*, *pol*, and *env*. The genes *tat, rev, nef, vif, vpr,* and *vpu* are important in regulating the virus life cycle. AIDS-defining malignancies and other, less clear AIDS-associated pathologies have been reproduced in HIV-1 and SIVmac239 transgenic mice (Vogel *et al.*, 1988; Larsen *et al.*, 1998). Mice are not natural hosts to HIV-1, HIV-2, or SIV. This is probably due to the lack of appropriate primary and secondary receptors, since human CD4 transgenic mice do not support HIV infection (Lores *et al.*, 1992). A human CD4/human second receptor (CKR5R or fusin) doubly transgenic mouse has not been developed. The use of a transgenic mouse model has circumvented the problem of species specificity since genetic transformation of the mouse embryo is independent of virus infectivity (Gordon and Ruddle, 1981; Hogan, 1986; Taketo *et al.*, 1991). Mice transgenic for primate lentivirus genes have allowed us to study the following effects of a given HIV-1, HIV-2, or SIV gene on a developing animal system: development, regulation, expression, and function. These transgenic models have contributed significantly to our understanding of HIV pathogenesis and disease.

A. Proviral Transgenic Mice

One of the first HIV-1 transgenic mouse lines was deficient in *gag* and *pol*, as described by Dickie *et al.* (1991). Deletion of the *gag* and *pol* genes was performed as a safety measure to be sure that the virus would be defective in its ability to replicate. As safety concerns abated, researchers chose to use full-length virus under the control of murine active promoters. Jolicoeur expressed full-length HIV under the control of the mouse mammary tumor virus promoter (MMTV LTR) (Jolicoeur *et al.*, 1992). They did not find an abnormal phenotype, only expression of the proteins as directed by the promoter. Dickie *et al.* (1993) showed, using a MLV/HIV construct, that the development of cataracts was the predominating phenotype. They later showed that this phenotype was linked to the *nef* gene (Dickie, 1996). Iwakura *et al.* (1992) also found cataracts in transgenic mice containing proviral DNA. Interestingly enough, all of these proviral constructs have an intact *nef* gene. From the later work by Hanna, one can infer that the reason for disease is the presence of the *nef* gene (Hanna *et al.*, 1998b).

The value of the model was realized recently when investigators found pathogenesis very similar to that observed in human AIDS using a proviral transgenic mouse model (Hanna *et al.*, 1998b). Using a large number of animals and a second promoter (CD4) in addition to the LTR to drive gene expression of the provirus, they showed depletion of CD4 a hallmark of AIDS. Furthermore, the severity of disease was positively correlated with the level of expression of the *nef* gene product and mice with proviral transgenes that lacked *nef* did not develop any signs of AIDS. This model will be instrumental in the development of antiretrovirals that target Nef.

B. Tat Transgenic Mice

HIV-1 produces a small nuclear protein, tat, which is required for viral replication. The product of the tat gene is a potent transactivator that is capable of upregulating gene expression in cultured cells by both transcriptional and posttranscriptional mechanisms. Tat has been found to function as an extracellular protein during an acute HIV-1 infection or a transfection of tat (Ensoli *et al.*, 1990). This extracellular protein can stimulate the growth of spindlelike cells taken from KS lesion of an HIV-1-infected individual (Ensoli *et al.*, 1990). Tat was also implicated in the upregulation of numerous cytokines, which are found to be elevated in HIV-1-infected patients. TAT-transgenic mice that have been developed give strong evidence to the pathogenic potential of the tat gene in HIV-1 disease.

When tat is under the control of the proviral LTR in transgenic mice, only male mice developed KS-like lesions on the skin at 12–18 months of age (Vogel *et al.*, 1988). Upon careful examination of the skin, the most prominent findings were hypercellularity of spindle-shaped cells in the dermis and subsequent development of malignant tumors. After long-term follow-up, a significant percentage of these mice developed liver tumor past the age of 18 months (Vogel *et al.*, 1991). This evidence has shown that tat may be a cofactor for the development of KS (HIV-1 co-infection with HHV8) and hepatocellular carcinoma in HIV-1-infected humans (Vogel *et al.*, 1991).

Brady *et al.* (1995) made transgenic mice which expressed tat in T cells under the transcriptional control of the human CD2 promoter. They looked at cytokine mRNA and protein expression in activated T cells and were able to show elevated levels of TNF-β in primary activated T cells CD2-Tat transgenic mice, which correlated with an *in vitro* tat-transfected T-cell line (Buonaguro *et al.*, 1992). Their aging mice expressed high levels of Tat, but did not develop skin lesions.

Evidence exists which shows that Tat (Ensoli *et al.*, 1990). TNF (Brady *et al.*, 1995), and other cytokines (Buonaguro *et al.*, 1992) promote the growth and proliferation of KS-derived cells from an AIDS patient. The dermal lesions seen by Vogel (Vogel *et al.*, 1988) may be expression of Tat outside the lymphoid system.

C. Env Transgenic Mice

Neurological complications such as dementia and paralysis have been attributed not only to opportunistic infections, but also to HIV-1 infection of the CNS (Cinque *et al.*, 1998; Robertson *et al.*, 1998). To better understand the pathogenic role that *env*, specifically gp 120, plays in the CNS, three different models that describe transgenic mice expression of the *env* protein are presented. Thomas made transgenic mice expressing gp 120 under the control of the human neurofilament light (NFL) promoter (Thomas *et al.*, 1994). Their observations were that NFL-HIV-transgenic mice had expression of HIV in neurons that triggered degenerative changes in the nervous system. Berrada describes a transgenic mouse model similar to that of Thomas that looked at the effect of full-length gp 160 under the control of the human NFL gene promoter (Berrada *et al.*, 1995). The protein was expressed in the brain stem and spinal chord of the transgenic mice. The phenotype that is seen mostly in these mice was neuronal toxicity, although there was no difference in motor function between the transgenic mice and the controls.

Toggas placed gp 120 under the control of the murine glial fibrillary acidic protein (GFAP) in transgenic mice (Toggas *et al.*, 1994). This protein was expressed and induced pathological effects in astrocytes, neurons, and microglia in the CNS. These data provide strong *in vivo* evidence that *env* expression is detrimental to CNS tissue.

D. Nef Transgenic Mice

Nef has been shown to be essential for the development of disease and in maintaining high viral titers in SIVmac239 infection of rhesus monkeys (Kestler *et al.*, 1991) and in HIV-1-infected SCID-hu mice (Jamieson *et al.*, 1994). Monkeys infected with virus that has a deletion in the *nef* gene have nonprogressive disease characterized by normal CD4 cell counts and low viral loads. There are some humans infected with HIV-1 that have deletions in *nef*. These individuals also appear to have nonprogressive disease and are classified as long-term nonprogressors (LTNP) (Deacon *et al.*, 1995; Kirchhoff *et al.*, 1995). For 15–20 years after infection, they too have normal CD4 cell counts and low viral loads. It appears that *nef* is extremely important in the determination of pathogenicity in primate lentiviruses *in vivo*, yet is not required for viral replication *in vitro* (Terwilliger *et al.*, 1986; Niederman *et al.*, 1989).

Mice transgenic for the *nef* gene of HIV-1 have been studied extensively. Several labs have shown that the NEF protein expressed by human T-cell promoters/enhancers in these mice can down-modulate surface CD4 and deplete CD4+ T cells in the thymus (Brady *et al.*, 1993; Skowronski *et al.*, 1993; Lindemann *et al.* 1994). Oddly, Dickie *et al.*, (1996) have shown that mice transgenic for HIV-1 *nef* develop cataracts. The most convincing evidence that HIV-1 *nef* is essential for pathogenicity in the transgenic mouse model was from Hanna *et al.* (1998b). They described a mutational analysis of an HIV-1 proviral DNA under the control of the human CD4 promoter. Transgenic mice exhibited down-modulation of CD4 in the thymus and periphery, sickness such as diarrhea and wasting, and early death with any of the constructs that contained *nef*. In fact, the amount of protein that was expressed correlated with the severity of disease.

We have described a transgenic animal system to examine the function of SIVmac239 *nef,* the allele from the pathogenic molecular clone that has been demonstrated to be required for pathogenesis in rhesus monkeys (Salkowitz, 1998; Larsen *et al.,* 1998). In this construct, *nef* from the pathogenic molecular clone SIVmac239 is under the control of the potent human CMV major immediate-early promoter (MIEP). The promoter of the transgene plays the primary role of a single-gene or multiple-gene construct. A broad-acting promoter such as CMV will allow expression of the transgene in many tissues, unlike using a promoter specifically for a lymphocyte subtype.

Using MIEP to control the expression of the SIVmac239 *nef* gene, the transgenic mice express NEF in the broadest variety of tissue types. This is unlike using a T-cell-specific promoter where expression will be limited to the T-cell lineage. This transgenic mouse line would express Nef in other mouse cell types and induce some of the manifestations of the monkey disease.

Three independent SIV *nef* transgenic mouse lines were derived. There was no enhanced lethality due to the presence of the *nef* transgene on the fetal development of the mice as evidenced by litter size (our average litter size was 9 ± 3 for both transgenic and wild-type mice). Additionally, the inheritance pattern in the test cross of a heterozygote and a wild-type mouse was also at the expected ratio (50% transgene). *Nef* transgenic mice were born healthy and expressed low levels of NEF throughout the first year of life (Salkowitz, 1998).

As early as 4 months of age, tumors began to develop in all transgenic lines, while wild-type FVB/N littermates remained tumor free. Tumors were found in 5, 10, and 28% of the three transgenic lines. A significant number of these tumors were either adenocarcinoma (42%), primarily of the lung, or fibrosarcoma (33%). Interestingly, the observed growth of tumors shows many parallels with tissues that are likely to be infected by human CMV (spleen, lung, mucous membrane, kidneys, and genital tract). Baskar *et al.* (1996) found that MIEP-driven expression of β-galactosidase in transgenic mice overlaps with the sites of natural human CMV infection, suggesting that viral transcriptional factors that are needed for virus replication are present in these cell types.

Fluorescent staining of peripheral blood mononuclear cells (PBMC) showed that altered levels of CD4+, CD8+, and CD3+ T cells accompany disease in *nef*-transgenic mice. PBMC from diseased mice at the time of sacrifice showed a significant decrease in the level of CD4+ T cells which was far below those in *nef*+ healthy mice and wild-type control groups. This down-modulation was statistically significant. There was also, on average, a decreased number of CD8+ and CD3+ T cells in diseased mice (Salkowitz, 1998), but these differences were not statistically significant. Interestingly, the CD8+ T cells in healthy *nef*+ animals were elevated to 21%, a value that was significantly elevated above the wild-type and sick animals, demonstrating that the SIV *nef*-transgenic mice, like the HIV-1 *nef*-transgenic mice, show a similar loss of CD4+ T cells (Brady, Pennington *et al.*, 1993; Skowronski *et al.*, 1993; Lindemann *et al.*, 1994; Hanna *et al.*, 1998a,b).

Seven immortalized tumor cell lines were established from the primary tumors in the animals. The types of tumors seen in these mice are adenocarcinoma, mostly of the lung, and fibrosarcoma, mostly of the skin or of the reproductive organs. Tumor cell lines from the mice reached the point of countinuous growth, or crisis, at varying times. Most tumor cell lines reached crisis within 3 weeks. The tumor cell lines in culture showed decreased dependence on growth factors and no inhibition of growth due to contact.

Three of seven tumor cell lines were tumorgenic *in vivo*. Intraperitoneal injection of these three tumors in recipient *nef*+ or wild-type animals produced tumors at the site of injection in approximately 2 weeks. By 1 month postinjection, mice were overcome by the tumor and were sacrificed. Mice had normal levels of peripheral CD4+, CD8+, and CD3+ T cells at the time of sacrifice. There was no difference in the course of tumor growth or the incidence of death between *nef*-transgenic and wild-type recipient mice. All naïve mice died when they were transplanted with the three tumor cell lines. Tumors were still able to develop with as little as 2500 cells from these three tumor cell lines when injected intraperitoneally. All three tumor cell lines were fibrosarcomas. None of the other tumor cell lines derived from transgenic animals or *nef*-transformed cell lines (NIH3T3/*nef* and NIH3T3/*nef*-stop), or NIH3T3, were able to produce tumors when transplanted into recipient mice. Subcutaneous injection of the tumor cell lines on the back left side above the hind leg of recipient mice slowed the progression of three tumor-producing cell lines. Mice were still overcome by the uninhibited progression of the tumor by 1 month after transplantation. These tumors are unique in their ability to cause disease in a normal immunocompetent host (Salkowitz, 1998).

p53 is a tumor suppressor gene (Lodish, 1999). The absence of this gene leads to uncontrolled cell proliferation and contributes to tumor (Lodish, 1999). *p53* as a tumor suppressor has been demonstrated in transgenic mice that lack a functional *p53* gene. These mice are viable at birth, but they are highly susceptible to the development of tumors, especially malignant lymphoma (Marin *et al.*, 1994). Lack of a functional *p53* gene is also associated with a higher rate of metastasis and the ready establishment of the tumor in tissue culture (Baxter *et al.*, 1996), both of which we have seen in our tumor cell lines. Mutations in *p53* are also common in human tumors of various cell types.

All *nef* tumor cell lines lacked functional *p53* activity. NIH3T3 showed an upregulation of expression of *p53* in the same experiments. Mutations in the *p53* gene resulting in tumors may indicate a defect in the cell-mediated immunity branch of the immune system due to loss of immune surveillance (Lodish, 1999).

The immune system of tg10198 transgenic mouse line described here has also been studied by Larsen *et al.* (1998). They found that this line exhibited a significantly increased mortality rate when challenged with HSV-1. These mice also showed a tremendous antibody response after virus challenge; however, their HSV-1-neutralizing antibodies were severely diminished as compared to wild type. These results indicate that there may be a defect in B-cell responses. They also showed that the spleen cells from *nef*-transgenic mice have a decreased response to phytohemagglutinin (PHA), which showed that there was a compromise in T-cell function. Taken together their results suggest that the presence of *nef* is sufficient to induce immune dysfunction in SIVmac239-transgenic mice. The work described here and Hanna *et al.* (1998b) show that *nef* is the major determinant of pathogenicity and it alone is sufficient to induce AIDS in transgenic mice. The role of the CMV MIEP may be to direct expression of *nef* outside the lymphoid tissues with the resulting tumor development that was not seen by other groups.

References

Agy, M. B., Frumkin, L. R. *et al.* (1992). Infection of *Macaca nemestrina* by human immunodeficiency virus type-1. *Science* 257(5066), 103–106.

Agy, M. B., Schmidt, A. *et al.* (1997). Serial *in vivo* passage of HIV-1 infection in *Macaca nemestrina*. *Virology* 238(2), 336–343.

Alder, J., Hui, Y. H. *et al.* (1995). Efficacy of AZT therapy in reducing p24 antigen burden in a modified SCID mouse model of HIV infection. *Antiviral Res* 27(1–2), 85–97.

Aldrovandi, G. M., and Zack, J. A. (1996). Replication and pathogenicity of human immunodeficiency virus type 1 accessory gene mutants in SCID-hu mice. *J. Virol.* 70(3), 1505–1511.

Alter, H. J., Eichberg, J. W. *et al.* (1984). Transmission of HTLV-III infection from human plasma to chimpanzees: An animal model for AIDS. *Science* 226(4674), 549–552.

Altman, L. (1999). After 17 Healthy years, hope of safe H.I.V. dies. *New York Times.*

Anderson, D. M., Agy, M. B. *et al.* (1994). HIV infection in non-human primates: The *Macaca nemestrina* model. *Virus Res.* 32(2), 269–282.

Arthur, L. O., Bess, J. W., Jr., *et al.* (1989). Challenge of chimpanzees (*Pan troglodytes*) immunized with human immunodeficiency virus envelope glycoprotein gp 120. *J. Virol.* 63(12), 5046–5053.

Autran, B., Guiet, P. *et al.* (1996). Thymocyte and thymic microenvironment alterations during a systemic HIV infection in a severe combined immunodeficient mouse model. *AIDS* 10(7), 717–727.

Baba, T. W., Jeong, Y. S. *et al.* (1995). Pathogenicity of live, attenuated SIV after mucosal infection of neonatal macaques. *Science* 267(5205), 1820–1825.

Baba, T. W., Liska, V. *et al.* (1999). Live attenuated, multiply deleted simian immunodeficiency virus causes AIDS in infant and adult macaques. *Nat. Med.* 5(2), 194–203.

Barre-Sinoussi, F., Georges-Courbot, M. C. *et al.* (1997). Characterization and titration of an HIV type 1 subtype E chimpanzee challenge stock. *AIDS Res. Hum. Retroviruses* 13(7), 583–591.

Baskar, J. F., Smith, P. P. *et al.* (1996). Developmental analysis of the cytomegalovirus enhancer in transgenic animals. *J. Virol.* **70**(5), 3215–3226.

Baxter, E. W., Blyth, K. *et al.* (1996). Moloney murine leukemia virus-induced lymphomas in p53-deficient mice: overlapping pathways in tumor development? *J. Virol.* **70**(4), 2095–2100.

Berman, P. W., Gregory, T. J. *et al.* (1990). Protection of chimpanzees from infection by HIV-1 after vaccination with recombinant glycoprotein gp 120 but not gp 160. *Nature* **345**(6276), 622–625.

Berman, P. W., Murthy, K. K. *et al.* (1996). Protection of MN-rgp 120-immunized chimpanzees from heterologous infection with a primary isolate of human immunodeficiency virus type 1. *J. Infect. Dis.* **173**(1), 52–59.

Berrada, F., Ma, D. *et al.* (1995). Neuronal expression of human immunodeficiency virus type 1 env proteins in transgenic mice: Distribution in the central nervous system and pathological alterations. *J. Virol.* **69**(11), 6770–6778.

Bosch, M. L., Schmidt, A. *et al.* (1997). Infection of Macaca nemestrina neonates with HIV-1 via different routes of inoculation. *AIDS* **11**(13), 1555–1563.

Bosma, G. C., Fried, M. *et al.* (1988). Evidence of functional lymphocytes in some (leaky) scid mice. *J. Exp. Med.* **167**(3), 1016–1033.

Brady, H. J., Abraham, D. J. *et al.* (1995). Altered cytokine expression in T lymphocytes from human immunodeficiency virus Tat transgenic mice. *J. Virol.* **69**(12), 7622–7629.

Brady, H. J., Pennington, D. J. *et al.* (1993). CD4 cell surface downregulation in HIV-1 Nef transgenic mice is a consequence of intracellular sequestration. *EMBO J.* **12**(13), 4923–4932.

Buonaguro, L., Barillari, G. *et al.* (1992). Effects of the human immunodeficiency virus type 1 Tat protein on the expression of inflammatory cytokines. *J. Virol.* **66**(12), 7159–7167.

Burns, D. P., and Desrosiers, R. C. (1990). Sequence variability of simian immunodeficiency virus in a persistently infected rhesus monkey. *J. Med. Primatol.* **19**(3–4), 317–326.

Chakrabarti, B. K., Maitra, R. K. *et al.* (1996). A candidate live inactivatable attenuated vaccine for AIDS. *Proc. Natl. Acad. Sci. USA* **93**(18), 9810–9815.

Chang, Y., Cesarman, E. *et al.* (1994). Identification of herpesvirus-like DNA sequences in AIDS-associated Kaposi's sarcoma. *Science* **266**(5192), 1865–1869.

Cinque, P., Vago, L. *et al.* (1998). Cerebrospinal fluid HIV-1 RNA levels: Correlation with HIV encephalitis. *AIDS* **12**(4), 389–394.

Clements, J. E., D'Antonio, N. *et al.* (1982). Genomic changes associated with antigenic variation of visna virus. II. Common nucleotide sequence changes detected in variants from independent isolations. *J. Mol. Biol.* **158**(3), 415–434.

Daniel, M. D., Kirchhoff, F. *et al.* (1992). Protective effects of a live attenuated SIV vaccine with a deletion in the nef gene. *Science* **258**(5090), 1938–1941.

Daniel, M. D., Letvin, N. L. *et al.* (1985). Isolation of T-cell tropic HTL V-III-like retrovirus from macaques. *Science* **228**(4704), 1201–1204.

Davis, I. C., Girard, M. *et al.* (1998). Loss of CD4+ T cells in human immunodeficiency virus type 1-infected chimpanzees is associated with increased lymphocyte apoptosis. *J. Virol.* **72**(6), 4623–4632.

Deacon, N. J., Tsykin, A. *et al.* (1995). Genomic structure of an attenuated quasi species of HIV-1 from a blood transfusion donor and recipients. *Science* **270**(5238), 988–991.

del Real, G., Llorente, M. *et al.* (1998). Suppression of HIV-1 infection in linomide-treated SCID-hu-PBL mice. *AIDS* **12**(8), 865–872.

Deng, H., Liu, R. *et al.* (1996). Identification of a major co-receptor for primary isolates of HIV-1. *Nature* **381**(6584), 661–666.

Dickie, P. (1996). HIV type 1 Nef perturbs eye lens development in transgenic mice. *AIDS Res. Hum. Retroviruses* **12**(3), 177–189.

Dickie, P., Felser, J. *et al.* (1991). HIV-associated nephropathy in transgenic mice expressing HIV-1 genes. *Virology* **185**(1), 109–119.

Dickie, P., Ramsdell, F. *et al.* (1993). Spontaneous and inducible epidermal hyperplasia in transgenic mice expressing HIV-1 Nef. *Virology* 197(1), 431–438.

Dragic, T., Litwin, V. *et al.* (1996). HIV-1 entry into CD4+ cells is mediated by the chemokine receptor CC-CKR-5. *Nature* 381(6584), 667–673.

Dunn, C. S., Mehtali, M. *et al.* (1995). Human immunodeficiency virus type 1 infection of human CD4-transgenic rabbits. *J. Gen. Virol.* 76(6), 1327–1336.

Eichberg, J. W., Zarling, J. M. *et al.* (1987). T-cell responses to human immunodeficiency virus (HIV) and its recombinant antigens in HIV-infected chimpanzees. *J. Virol.* 61(12), 3804–3808.

Ensoli, B., Barillari, G. *et al.* (1990). Tat protein of HIV-1 stimulates growth of cells derived from Kaposi's sarcoma lesions of AIDS patients. *Nature* 345(6270), 84–86.

Estaquier, J., Idziorek, T. *et al.* (1994). Programmed cell death and AIDS: Significance of T-cell apoptosis in pathogenic and nonpathogenic primate lentiviral infections. *Proc. Natl. Acad. Sci. USA* 91(20), 9431–9435.

Feng, Y., Broder, C. C. *et al.* (1996). HIV-1 entry cofactor: Functional cDNA cloning of a seven-transmembrane, G protein-coupled receptor. *Science* 272(5263), 872–877.

Ferrari, G., Place, C. A. *et al.* (1994). Comparison of anti-HIV-1 ADCC reactivities in infected humans and chimpanzees. *J. Acq. Immune Defic. Syndr.* 7(4), 325–331.

Francis, D. P., Feorino, P. M. *et al.* (1984). Infection of chimpanzees with lymphadenopathy-associated virus. *Lancet* 2(8414), 1276–1277.

Frumkin, L. R., Agy, M. B. *et al.* (1993). Acute infection of *Macaca nemestrina* by human immunodeficiency virus type 1. *Virology* 195(2), 422–431.

Frumkin, L. R., Patterson, B. K. *et al.* (1995). Infection of Macaca nemestrina brain with human immunodeficiency virus type 1. *J. Gen. Virol.* 76(10), 2467–2476.

Fultz, P. N. (1993). Nonhuman primate models for AIDS. *Clin. Infect. Dis.* 17(Suppl. 1), S230–S235.

Fultz, P. N. (1997). HIV type 1 strains pathogenic for chimpanzees. *AIDS Res. Hum Retroviruses* 13(15), 1261.

Fultz, P. N., McClure, H. M. *et al.* (1986a). Isolation of a T-lymphotropic retrovirus from naturally infected sooty mangabey monkeys (*Cercocebus atys*). *Proc. Natl. Acad. Sci. USA* 83(14), 5286–5290.

Fultz, P. N., McClure, H. M. *et al.* (1989). Identification and biologic characterization of an acutely lethal variant of simian immunodeficiency virus from sooty mangabeys (SIV/SMM). *AIDS Res. Hum. Retroviruses* 5(4), 397–409.

Fultz, P. N., McClure, H. M. *et al.* (1986b). Vaginal transmission of human immunodeficiency virus (HIV) to a chimpanzee. *J. Infect. Dis.* 154(5), 896–900.

Fultz, P. N., McClure, H. M. *et al.* (1986c). Persistent infection of chimpanzees with human T-lymphotropic virus type III/lymphadenopathy-associated virus: A potential model for acquired immunodeficiency syndrome. *J. Virol.* 58(1), 116–124.

Fultz, P. N., Nara, P. *et al.* (1992). Vaccine protection of chimpanzees against challenge with HIV-1-infected peripheral blood mononuclear cells. *Science* 256(5064), 1687–1690.

Fultz, P. N., Siegel, R. L. *et al.* (1991). Prolonged CD4+ lymphocytopenia and thrombocytopenia in a chimpanzee persistently infected with human immunodeficiency virus type 1. *J. Infect. Dis.* 163(3), 441–447.

Fultz, P. N., Srinivasan, A. *et al.* (1987). Superinfection of a chimpanzee with a second strain of human immunodeficiency virus. *J. Virol.* 61(12), 4026–4029.

Fultz, P. N., Yue, L. *et al.* (1997). Human immunodeficiency virus type 1 intersubtype (B/E) recombination in a superinfected chimpanzee. *J. Virol.* 71(10), 7990–7995.

Gajdusek, D. C., Amyx, H. L., *et al.* (1985). Infection of chimpanzees by human T-lymphotropic retroviruses in brain and other tissues from AIDS patients. *Lancet* 1(8419), 55–56.

Gao, F., Bailes, E. *et al.* (1999). Origin of HIV-1 in the chimpanzee Pan troglodytes troglodytes. *Nature* 397(6718), 436–441.

Gartner, S., Liu, Y. *et al.* (1994a). HIV-1 infection in pigtailed macaques. *AIDS Res. Hum. Retroviruses* **10**(Suppl. 2), S129–S133.

Gartner, S., Liu, Y. *et al.* (1994b). Adaptation of HIV-1 to pigtailed macaques. *J. Med. Primatol.* **23**(2–3), 155–163.

Gendelman, H. E., Ehrlich, G. D. *et al.* (1991). The inability of human immunodeficiency virus to infect chimpanzee monocytes can be overcome by serial viral passage *in vivo. J. Virol.* **65**(7), 3853–3863.

Gibbons, C., Kollmann, T. R. *et al.* (1997). Thy/Liv-SCID-Hu mice implanted with human intestine: an in vivo model for investigation of mucosal transmission of HIV. *AIDS Res. Hum. Retroviruses* **13**(17), 1453–1460.

Girard, M. (1995). Present status of vaccination against HIV-1 infection. *Int. J. Immunopharmacol.* **17**(2), 75–78.

Girard, M., Barre-Sinoussi, F. *et al.* (1996). Vaccination of chimpanzees against HIV-1. *Antibiot. Chemother.* **48**, 121–124.

Girard, M., Kieny, M. P. *et al.* (1991). Immunization of chimpanzees confers protection against challenge with human immunodeficiency virus. *Proc. Natl. Acad. Sci. USA* **88**(2), 542–546.

Girard, M., Mahoney, J. *et al.* (1998). Genital infection of female chimpanzees with human immunodeficiency virus type 1. *AIDS Res. Hum. Retroviruses* **14**(15), 1357–1367.

Girard, M., Meignier, B. *et al.* (1995). Vaccine-induced protection of chimpanzees against infection by a heterologous human immunodeficiency virus type 1. *J. Virol.* **69**(10), 6239–6248.

Girard, M., Yue, L. *et al.* (1996). Failure of a human immunodeficiency virus type 1 (HIV-1) subtype B-derived vaccine to prevent infection of chimpanzees by an HIV-1 subtype E strain. *J. Virol.* **70**(11), 8229–8233.

Goldberg, T. L., and Ruvolo, M. (1997). The geographic apportionment of mitochondrial genetic diversity in east African chimpanzees, *Pan troglodytes schweinfurthii. Mol. Biol. Evol.* **14**(9), 976–984.

Gordon, J. W., and Ruddle, F. H. (1981). Integration and stable germ line transmission of genes injected into mouse pronuclei. *Science* **214**(4526), 1244–1246.

Gougeon, M. L., Lecoeur, H. *et al.* (1997). Lack of chronic immune activation in HIV-infected chimpanzees correlates with the resistance of T cells to Fas/Apo-1 (CD95)-induced apoptosis and preservation of a T helper 1 phenotype. *J. Immunol.* **158**(6), 2964–2976.

Greenough, T. C., Sullivan, J. L. *et al.* (1999). Declining CD4 T-cell counts in a person infected with nef-deleted HIV-1. *N. Engl. J. Med.* **340**(3), 236–237.

Gulizia, R. J., Levy, J. A. *et al.* (1996). The envelope gp120 gene of human immunodeficiency virus type 1 determines the rate of CD4-positive T-cell depletion in SCID mice engrafted with human peripheral blood leukocytes. *J. Virol.* **70**(6), 4184–4187.

Ha, J. C., Nosbisch, C. *et al.* (1998). Fetal, infant, and maternal toxicity of zidovudine (azidothymidine) administered throughout pregnancy in Macaca nemestrina. *J. Acq. Immune Defic. Syndr. Hum. Retrovirol.* **18**(1), 27–38.

Hanna, Z., Kay, D. G. *et al.* (1998a). Transgenic mice expressing human immunodeficiency virus type 1 in immune cells develop a severe AIDS-like disease. *J. Virol.* **72**(1), 121–132.

Hanna, Z., Kay, D. G. *et al.* (1998b). Nef harbors a major determinant of pathogenicity for an AIDS-like disease induced by HIV-1 in transgenic mice. *Cell* **95**(2), 163–175.

Heeney, J., Jonker, R. *et al.* (1993). The resistance of HIV-infected chimpanzees to progression to AIDS correlates with absence of HIV-related T-cell dysfunction. *J. Med. Primatol.* **22**(2–3), 194–200.

Henderson, D. K., and Gerberding, J. L. (1989). Prophylactic zidovudine after occupational exposure to the human immunodeficiency virus: An interim analysis. *J. Infect. Dis.* **160**(2), 321–327.

Hesselton, R. M., Greiner, D. L. *et al.* (1995). High levels of human peripheral blood mononuclear cell engraftment and enhanced susceptibility to human immunodeficiency virus type 1 infection in NOD/LtSz-scid/scid mice. *J. Infect. Dis.* **172**(4), 974–982.

Hoffman, T. L., Stephens, E. B. *et al.* (1998). HIV type I envelope determinants for use of the CCR2b, CCR3, STRL33, and APJ coreceptors. *Proc. Natl. Acad. Sci. USA* 95(19), 11360–11365.

Hogan, B., Constantini, F., and Lacy, E. (1986). *In* "Manipulating the Mouse Embryo." Cold Spring Harbor Laboratory Press, Cold Spring Harbor, New York.

Hu, S. L., Fultz, P. N. *et al.* (1987). Effect of immunization with a vaccinia-HIV env recombinant on HIV infection of chimpanzees. *Nature* 328(6132), 721–723.

Huet, T., Cheynier, R. *et al.* (1990). Genetic organization of a chimpanzee lentivirus related to HIV-1. *Nature* 345(6273), 356–359.

Iwakura, Y., Shioda, T. *et al.* (1992). The induction of cataracts by HIV-1 in transgenic mice. *AIDS* 6(10), 1069–1075.

Jamieson, B. D., Aldrovandi, G. M. *et al.* (1994). Requirement of human immunodeficiency virus type 1 nef for *in vivo* replication and pathogenicity. *J. Virol.* 68(6), 3478–3485.

Jamieson, B. D., Pang, S. *et al.* (1995). *In vivo* pathogenic properties of two clonal human immunodeficiency virus type 1 isolates. *J. Virol.* 69(10), 6259–6264.

Jamieson, B. D., Uittenbogaart, C. H. *et al.* (1997). High viral burden and rapid CD4+ cell depletion in human immunodeficiency virus type 1-infected SCID-hu mice suggest direct viral killing of thymocytes *in vivo*. *J. Virol.* 71(11), 8245–8253.

Jenner, E. (1802). "An Inquiry into the Causes and Effects of the Variolae Vaccinae: A Disease Discovered in Some of the Western Counties of England, Particularly Gloucestershire, and Known by the Name of the Cow Pox," 2nd. ed. Ashley & Brewer, Springfield.

Jolicoeur, P., Laperriere, A. *et al.* (1992). Efficient production of human immunodeficiency virus proteins in transgenic mice. *J. Virol.* 66(6), 3904–3908.

Kaneshima, H., Su, L. *et al.* (1994). Rapid-high, syncytium-inducing isolates of human immunodeficiency virus type 1 induce cytopathicity in the human thymus of the SCID-hu mouse. *J. Virol.* 68(12), 8188–8192.

Keller, R. D., Nosbisch, C. *et al.* (1995). Pharmacokinetics of stavudine (2′,3′-didehydro-3′-deoxythymidine) in the neonatal macaque (*Macaca nemestrina*). *Antimicrob. Agents Chemother.* 39(12), 2829–2831.

Kent, K. A. (1995). Neutralising epitopes of simian immunodeficiency virus envelope glycoprotein. *J. Med. Primatol.* 24(3), 145–149.

Kent, S. J., Woodward, A. *et al.* (1997). Human immunodeficiency virus type 1 (HIV-1)-specific T cell responses correlate with control of acute HIV-1 infection in macaques. *J. Infect. Dis.* 176(5), 1188–1197.

Kent, S. J., Zhao, A. *et al.* (1998). Enhanced T-cell immunogenicity and protective efficacy of a human immunodeficiency virus type 1 vaccine regimen consisting of consecutive priming with DNA and boosting with recombinant fowlpox virus. *J. Virol.* 72(12), 10180–10188.

Kestler, H., Kodama, T. *et al.* (1990). Induction of AIDS in rhesus monkeys by molecularly cloned simian immunodeficiency virus. *Science* 248(4959), 1109–1112.

Kestler, H. W., and Chakrabarti, B. K. (1997). A live-virus "suicide" vaccine for human immunodeficiency virus. *Clin. J. Med.* 64(5), 269–274.

Kestler, H. W., and Jeang, K. T. (1995). Attenuated retrovirus vaccines and AIDS. *Science* 270(5239), 1219; discussion 1220–1222.

Kestler, H. W. d., Ringler, D. J. *et al.* (1991). Importance of the nef gene for maintenance of high virus loads and for development of AIDS. *Cell* 65(4), 651–662.

Kimball, L. E., and Bosch, M. L. (1998). *In vitro* HIV-1 infection in *Macaca nemestrina* PBMCs is blocked at a step beyond reverse transcription. *J. Med. Primatol.* 27(2–3), 99–103.

King, M. C., and Wilson, A. C. (1975). Evolution at two levels in humans and chimpanzees. *Science* 188(4184), 107–116.

Kirchhoff, F., Greenough, T. C. *et al.* (1995). Brief report: Absence of intact nef sequences in a long-term survivor with nonprogressive HIV-1 infection. *N. Engl. J. Med.* 332(4), 228–232.

Kitchen, S. G., and Zack, J. A. (1997). CXCR4 expression during lymphopoiesis: implications for human immunodeficiency virus type 1 infection of the thymus. *J. Virol.* **71**(9), 6928–6934.

Koch, R., and Cheyne, W. W. (1880). "Investigations into the Etiology of Traumatic Infective Diseases." The New Sydenham Society, London.

Kodama, T., Burns, D. P. *et al.* (1990). Molecular changes associated with replication of simian immunodeficiency virus in human cells. *J. Med. Primatol.* **19**(3–4), 431–437.

Kollmann, T. R., Kim, A. *et al.* (1995). Divergent effects of chronic HIV-1 infection on human thymocyte maturation in SCID-hu mice. *J. Immunol.* **154**(2), 907–921.

Kollmann, T. R., Pettoello-Mantovani, M. *et al.* (1996). Inhibition of acute *in vivo* human immunodeficiency virus infection by human interleukin 10 treatment of SCID mice implanted with human fetal thymus and liver. *Proc. Natl. Acad. Sci. USA* **93**(7), 3126–3131.

Kollmann, T. R., Pettoello-Mantovani, M. *et al.* (1994). Disseminated human immunodeficiency virus 1 (HIV-1) infection in SCID-hu mice after peripheral inoculation with HIV-1. *J. Exp. Med.* **179**(2), 513–522.

Kong, L., Taylor, M. E., Watwers, D., Blattner, W. A., Hahn, B. H., and Shaw, G. (1989). "Genetic Analysis of Sequential HIV-1 Isolates from an Infected Lab Worker." Fifth International Conference on AIDS. Montreal, Quebec International AIDS Society.

Krowka, J. F., Sarin, S. *et al.* (1991). Human T cells in the SCID-hu mouse are phenotypically normal and functionally competent. *J. Immunol.* **146**(11), 3751–3756.

Lapenta, C., Fais, S. *et al.* (1997). U937-SCID mouse xenografts: A new model for acute *in vivo* HIV-1 infection suitable to test antiviral strategies. *Antiviral Res.* **36**(2), 81–90.

Larsen, N. B., Kestler, H. W. *et al.* (1998). Mice transgenic for simian immunodeficiency virus nef are immunologically compromised. *J. Biomed. Sci.* **5**(4), 260–266.

Learmont, J. C., Geczy, A. F. *et al.* (1999). Immunologic and virologic status after 14 to 18 years of infection with an attenuated strain of HIV-1—A report from the Sydney Blood Bank Cohort. *N. Engl. J. Med.* **340**(22), 1715–1722.

Leelayuwat, C., Zhang, W. J. *et al.* (1993). Differences in the central major histocompatibility complex between humans and chimpanzees: Implications for development of autoimmunity and acquired immune deficiency syndrome. *Hum. Immunol.* **38**(1), 30–41.

Leigh, S. R. (1998). Chimp research. *Science* **282**(5386), 47.

Letvin, N. L. (1998). Progress in the development of an HIV-1 vaccine. *Science* **280**(5371), 1875–1880.

Letvin, N. L., Daniel, M. D. *et al.* (1985). Induction of AIDS-like disease in macaque monkeys with T-cell tropic retrovirus STLV-III. *Science* **230**(4721), 71–73.

Lewis, M. G., Zack, P. M. *et al.* (1992). Infection of rhesus and cynomolgus macaques with a rapidly fatal SIV (SIVSMM/PBj) isolate from sooty mangabeys. *AIDS Res. Hum. Retroviruses* **8**(9), 1631–1639.

Li, J., Lord, C. I. *et al.* (1992). Infection of cynomolgus monkeys with a chimeric HIV-1/SIVmac virus that expresses the HIV-1 envelope glycoproteins. *J. Acq. Immune Defic. Syndr.* **5**(7), 639–646.

Lindemann, D., Wilhelm, R. *et al.* (1994). Severe immunodeficiency associated with a human immunodeficiency virus 1 NEF/3'-long terminal repeat transgene. *J. Exp. Med.* **179**(3), 797–807.

Lodish, H. F. (1999). "Molecular Cell Biology." Scientific American Books, New York.

Lores, P., Boucher, V. *et al.* (1992). Expression of human CD4 in transgenic mice does not confer sensitivity to human immunodeficiency virus infection. *AIDS Res. Hum. Retroviruses* **8**(12), 2063–2071.

Lusso, P., Markham, P. D. *et al.* (1988). Cell-mediated immune response toward viral envelope and core antigens in gibbon apes (*Hylobates lar*) chronically infected with human immunodeficiency virus-1. *J. Immunol.* **141**(7), 2467–2473.

MacDougall, J. R., Croy, B. A. *et al.* (1990). Demonstration of a splenic cytotoxic effector cell in mice of genotype SCID/SCID.BG/BG. *Cell Immunol.* **130**(1), 106–117.

Malkovsky, M. (1996). HLA and natural history of HIV infection. *Lancet* 348(9021), 143–143.

Marin, M. C., Hsu, B. *et al.* (1994). Evidence that p53 and bcl-2 are regulators of a common cell death pathway important for *in vivo* lymphomagenesis. *Oncogene* 9(11), 3107–3112.

Markham, R. B., Schwartz, D. H. *et al.* (1996). Selective transmission of human immunodeficiency virus type 1 variants to SCID mice reconstituted with human peripheral blood monoclonal cells. *J. Virol.* 70(10), 6947–6954.

McBride, B. W., Easterbrook, L. M. *et al.* (1995). Human immunodeficiency virus infection of xenografted SCID-beige mice. *J. Med. Virol.* 47(2), 130–138.

McCormick-Davis, C., Zhao, L. J. *et al.* (1998). Chronology of genetic changes in the vpu, env, and Nef genes of chimeric simian-human immunodeficiency virus (strain HXB2) during acquisition of virulence for pig-tailed macaques. *Virology* 248(2), 275–283.

McCune, J. M., Namikawa, R. *et al.* (1988). The SCID-hu mouse: Murine model for the analysis of human hematolymphoid differentiation and function. *Science* 241(4873), 1632–1639.

McCune, J. M., Namikawa, R. *et al.* (1990). Suppression of HIV infection of AZT-treated SCID-hu mice. *Science* 247(4942), 564–566.

Moore, P. S., Gao, S. J. *et al.* (1996). Primary characterization of a herpesvirus agent associated with Kaposi's sarcomae. *J. Virol.* 70(1), 549–558.

Mori, K., Ringler, D. J. *et al.* (1992). Complex determinants of macrophage tropism in env of simian immunodeficiency virus. *J. Virol.* 66(4), 2067–2075.

Morin, P. A., Moore, J. J. *et al.* (1994). Kin selection, social structure, gene flow, and the evolution of chimpanzees. *Science* 265(5176), 1193–1201.

Morrow, W. J., Homsy, J. *et al.* (1989). Long-term observation of baboons, rhesus monkeys, and chimpanzees inoculated with HIV and given periodic immunosuppressive treatment. *AIDS Res. Hum. Retroviruses* 5(2), 233–245.

Morrow, W. J., Wharton, M. *et al.* (1987). Small animals are not susceptible to human immunodeficiency virus infection. *J. Gen. Virol.* 68(8), 2253–2257.

Mosier, D. E., Gulizia, R. J. *et al.* (1988). Transfer of a functional human immune system to mice with severe combined immunodeficiency. *Nature* 335(6187), 256–259.

Mosier, D. E., Gulizia, R. J. *et al.* (1991). Human immunodeficiency virus infection of human-PBL-SCID mice. *Science* 251(4995), 791–794.

Mosier, D. E., Gulizia, R. J. *et al.* (1993a). Resistance to human immunodeficiency virus 1 infection of SCID mice reconstituted with peripheral blood leukocytes from donors vaccinated with vaccinia gp160 and recombinant gp160. *Proc. Natl. Acad. Sci. USA* 90(6), 2443–7.

Mosier, D. E., Gulizia, R. J. *et al.* (1993b). Rapid loss of CD4+ T cells in human-PBL-SCID mice by noncytopathic HIV isolates. *Science* 260(5108), 689–692.

Mwaengo, D. M., and Novembre, F. J. (1998). Molecular cloning and characterization of viruses isolated from chimpanzees with pathogenic human immunodeficiency virus type 1 infections. *J. Virol.* 72(11), 8976–8987.

Namikawa, R., KAneshima, H. *et al.* (1988). Infection of the SCID-hu mouse by HIV-1. *Science* 242(4886), 1684–1686.

Namikawa, R., Weilbaecher, K. N. *et al.* (1990). Long-term human hematopoiesis in the SCID-hu mouse. *J. Exp. Med.* 172(4), 1055–1063.

Nara, P., Hatch, W. *et al.* (1989). The biology of human immunodeficiency virus-1 IIIB infection in the chimpanzee: *In vivo* and *in vitro* correlations. *J. Med. Primatol.* 18(3–4), 343–355.

Nara, P. L., Robey, W. G. *et al.* (1987). Persistent infection of chimpanzees with human immunodeficiency virus: Serological responses and properties of reisolated viruses. *J. Virol.* 61(10), 3173–3180.

Nara, P. L., Smit, L. *et al.* (1990). Emergence of viruses resistant to neutralization by V3-specific antibodies in experimental human immunodeficiency virus type 1 IIIB infection of chimpanzees. *J. Virol.* 64(8), 3779–3791.

Niederman, T. M., Thielan, B. J. *et al.* (1989). Human immunodeficiency virus type 1 negative factor is a transcriptional silencer. *Proc. Natl. Acad. Sci. USA* 86(4), 1128–1132.

Nonoyama, S., Smith, F. O. *et al.* (1993). Strain-dependent leakiness of mice with severe combined immune deficiency. *J. Immunol.* **150**(9), 3817–3824.

Novembre, F. J., Saucier, M. *et al.* (1997). Development of AIDS in a chimpanzee infected with human immunodeficiency virus type 1. *J. Virol.* **71**(5), 4086–4091.

Ochs, H. D., Morton, W. R. *et al.* (1993). Intra-amniotic inoculation of pigtailed macaque (*Macaca nemestrina*) fetuses with SIV and HIV-1. *J. Med. Primatol.* **22**(2–3), 162–168.

Odinecs, A., Pereira, C. *et al.* (1996). Prenatal and postpartum pharmacokinetics of stavudine (2',3'-didehydro-3'-deoxythymidine) and didanosine (dideoxyinosine) in pigtailed macaques (*Macaca nemestrina*). *Antimicrob. Agents Chemother.* **40**(10), 2423–2425.

Ohta, Y., Masuda, T. *et al.* (1988). Isolation of simian immunodeficiency virus from African green monkeys and seroepidemiologic survey of the virus in various non-human primates. *Int. J. Cancer* **41**(1), 115–122.

Otten, R. A., Brown, B. G. *et al.* (1994). Differential replication and pathogenic effects of HIV-1 and HIV-2 in *Macaca nemestrina*. *AIDS* **8**(3), 297–306.

Peeters, M., Fransen, K. *et al.* (1992). Isolation and characterization of a new chimpanzee lentivirus (simian immunodeficiency virus isolate cpz-ant) from a wild-captured chimpanzee. *AIDS* **6**(5), 447–451.

Peeters, M., Honore, C. *et al.* (1989). Isolation and partial characterization of an HIV-related virus occurring naturally in chimpanzees in Gabon. *AIDS* **3**(10), 625–630.

Pereira, C. M., Nosbisch, C. *et al.* (1994). Pharmacokinetics of dideoxyinosine in neonatal pigtailed macaques. *Antimicrob. Agents Chemother.* **38**(4), 787–789.

Persidsky, Y., Buttini, M. *et al.* (1997). An analysis of HIV-1-associated inflammatory products in brain tissue of humans and SCID mice with HIV-1 encephalitis. *J. Neurovirol.* **3**(6), 401–416.

Persidsky, Y., Limoges, J. *et al.* (1996). Human immunodeficiency virus encephalitis in SCID mice. *Am. J. Pathol.* **149**(3), 1027–1053.

Picchio, G. R., Gulizia, R. J. *et al.* (1998). The cell tropism of human immunodeficiency virus type 1 determines the kinetics of plasma viremia in SCID mice reconstituted with human peripheral blood leukocytes. *J. Virol.* **72**(3), 2002–2009.

Pretet, J. L., Zerbib, A. C. *et al.* (1997). Chimpanzees CXCR4 and CCR5 act as coreceptors for HIV type 1. *AIDS Res. Hum. Retroviruses* **13**(18), 1583–1587.

Prince, A. M., Allan, J. *et al.* (1999). Virulent HIV strains, chimpanzees, and trial vaccines. *Science* **283**(5405), 1117–1118.

Prince, A. M., and Andrus, L. (1998). AIDS vaccine trials in chimpanzees. *Science* **282**(5397), 2195–2196.

Reinhardt, B., Torbett, B. E. *et al.* (1994). Human immunodeficiency virus type 1 infection of neonatal severe combined immunodeficient mice xenografted with human cord blood cells. *AIDS Res. Hum. Retroviruses* **10**(2), 131–141.

Robertson, K., Fiscus, S. *et al.* (1998). CSF, plasma viral load and HIV associated dementia. *J. Neurovirol.* **4**(1), 90–94.

Ryder, O. A. (1998). Chimp research. *Science* **282**(5386), 47.

Saksela, K., Muchmore, E. *et al.* (1993). High viral load in lymph nodes and latent human immunodeficiency virus (HIV) in peripheral blood cells of HIV-1-infected chimpanzee. *J. Virol.* **67**(12), 7423–7427.

Salkowitz, J. (1998). The nef gene of SIVmac239 causes disease in transgenic mice. *In* "Pathology." Case Western Reserve University School of Medicine, Cleveland, OH.

Schuitemaker, H., Meyaard, L. *et al.* (1993). Lack of T cell dysfunction and programmed cell death in human immunodeficiency virus type 1-infected chimpanzees correlates with absence of monocytotropic variants. *J. Infect. Dis.* **168**(5), 1140–1147.

Shaw, J. H., and Falkow, S. (1988). Model for invasion of human tissue culture cells by Neisseria gonorrhoeae. *Infect. Immun.* **56**(6), 1625–1632.

Shibata, R., Hoggan, M. D. *et al.* (1995). Isolation and characterization of a syncytium-inducing, macrophage/T-cell line-tropic human immunodeficiency virus type 1 isolate that readily infects chimpanzee cells *in vitro* and *in vivo*. *J. Virol.* **69**(7), 4453–4462.

Shibata, R., Siemon, C. *et al.* (1997). Reactivation of HIV type 1 in chronically infected chimpanzees following xenostimulation with human cells or with pulses of corticosteroid. *AIDS Res. Hum. Retroviruses* **13**(5), 377–381.

Skowronski, J., Parks, D. *et al.* (1993). Altered T cell activation and development in transgenic mice expressing the HIV-1 nef gene. *EMBO J.* **12**(2), 703–713.

Smith, S. M., Markham, R. B. *et al.* (1996). Conditional reduction of human immunodeficiency virus type 1 replication by a gain-of-herpes simplex virus 1 thymidine kinase function. *Proc. Natl. Acad. Sci. USA* **93**(15), 7955–7960.

Speck, R. F., Penn, M. L. *et al.* (1998). Rabbit cells expressing human CD4 and human CCR5 are highly permissive for human immunodeficiency virus type 1 infection. *J. Virol.* **72**(7), 5728–5734.

Sperling, R. S., Shapiro, D. E. *et al.* (1996). Maternal viral load, zidovudine treatment, and the risk of transmission of human immunodeficiency virus type 1 from mother to infant: Pediatric AIDS Clinical Trials Group Protocol 076 Study Group. *N. Engl. J. Med.* **335**(22), 1621–1629.

Su, L., Kaneshima, H. *et al.* (1997). Identification of HIV-1 determinants for replication *in vivo*. *Virology* **227**(1), 45–52.

Taketo, M., Schroeder, A. C. *et al.* (1991). FVB/N: An inbred mouse strain preferable for transgenic analyses. *Proc. Natl. Acad. Sci. USA.* **88**(6), 2065–2069.

Terwilliger, E., Sodroski, J. G. *et al.* (1986). Effects of mutations within the 3' orf open reading frame region of human T-cell lymphotropic virus type III (HTLV-III/LAV) on replication and cytopathogenicity. *J. Virol.* **60**(2), 754–760.

Thomas, F. P., Chalk, C. *et al.* (1994). Expression of human immunodeficiency virus type 1 in the nervous system of transgenic mice leads to neurological disease. *J. Virol.* **68**(11), 7099–7107.

Toggas, S. M., Masliah, E. *et al.* (1994). Central nervous system damage produced by expression of the HIV-1 coat protein gp120 in transgenic mice. *Nature* **367**(6459), 188–193.

Tsujimoto, H., Cooper, R. W. *et al.* (1988). Isolation and characterization of simian immunodeficiency virus from mandrills in Africa and its relationship to other human and simian immunodeficiency viruses. *J. Virol.* **62**(11), 4044–4050.

Tuntland, T., Odinecs, A. *et al.* (1998). *In vivo* maternal-fetal-amniotic fluid pharmacokinetics of zidovudine in the pigtailed macaque: comparison of steady-state and single-dose regimens. *J. Pharmacol. Exp. Ther.* **285**(1), 54–62.

Tyor, W. R., Power, C. *et al.* (1993). A model of human immunodeficiency virus encephalitis in scid mice. *Proc. Natl. Acad. Sci. USA* **90**(18), 8658-8662.

Uittenbogaart, C. H., Anisman, D. J. *et al.* (1996). Differential tropism of HIV-1 isolates for distinct thymocyte subsets *in vitro*. *AIDS* **10**(7), F9–F16.

Vanden Haesevelde, M. M., Peeters, M. *et al.* (1996). Sequence analysis of a highly divergent HIV-1-related lentivirus isolated from a wild captured chimpanzee. *Virology* **221**(2), 346–350.

Vogel, J., Hinrichs, S. H. *et al.* (1991). Liver cancer in transgenic mice carrying the human immunodeficiency virus tat gene. *Cancer Res.* **51**(24), 6686–6690.

Vogel, J., Hinrichs, S. H. *et al.* (1988). The HIV tat gene induces dermal lesions resembling Kaposi's sarcoma in transgenic mice. *Nature* **335**(6191), 606–611.

Wade, N. A., Birkhead, G. S. *et al.* (1998). Abbreviated regimens of zidovudine prophylaxis and perinatal transmission of the human immunodeficiency virus. *N. Engl. J. Med.* **339**(20), 1409–1414.

Watanabe, M., Ringler, D. J. *et al.* (1991). A chimpanzee-passaged human immunodeficiency virus isolate is cytopathic for chimpanzee cells but does not induce disease. *J. Virol.* **65**(6), 3344–3348.

Wei, Q., and Fultz, P. N. (1998). Extensive diversification of human immunodeficiency virus type 1 subtype B strains during dual infection of a chimpanzee that progressed to AIDS. *J. Virol.* **72**(4), 3005–3017.

Weiss, S. H., Goedert, J. J. *et al.* (1988). Risk of human immunodeficiency virus (HIV-1) infection among laboratory workers. *Science* **239**(4835), 68–71.

Wooley, D. P., Smith, R. A. *et al.* (1997). Direct demonstrat of retroviral recombination in a rhesus monkey. *J. Virol.* **71**(12), 9650–9653.

Yoon, K., Kestler, H. W., *et al.* (1998). Growth properties of HSIVnef: HIV-1 containing the nef gene from pathogenic molecular clone SIVmac239. *Virus Res.* **57**(1), 27–34.

Zhu, T., Mo, H. *et al.* (1993). Genotypic and phenotypic characterization of HIV-1 patients with primary infection. *Science* **261**(5125), 1179–1181.

Index

Contents of Previous Volumes

Volume 41

Role of *p53* in Apoptosis
Christine E. Canman and Michael B. Kastan

Chemotherapy-Induced Apoptosis
Peter W. Mesner, Jr., I. Imawati Budihardjo, and Scott H. Kaufmann

Bcl-2 Family Proteins: Strategies for Overcoming Chemoresistance in Cancer
John C. Reed

Role of Bcr-Abl Kinase in Resistance to Apoptosis
Afshin Samali, Adrienne M. Gorman, and Thomas G. Cotter

Apoptosis in Hormone-Responsive Malignancies
Samuel R. Denmeade, Diane E. McCloskey, Ingrid B. J. K. Joseph, Hillary A. Hahm, John T. Isaacs, and Nancy E. Davidson

Volume 42

Catecholamine: Bridging Basic Science
Edited by David S. Goldstein, Graeme Eisenhofer, and Richard McCarty

Part A. Catecholamine Synthesis and Release

Part B. Catecholamine Reuptake and Storage

Part C. Catecholamine Metabolism

Part D. Catecholamine Receptors and Signal Transduction

Part E. Catecholamine in the Periphery

Part F. Catecholamine in the Central Nervous System

Part G. Novel Catecholaminergic Systems

Part H. Development and Plasticity

Part I. Drug Abuse and Alcoholism

Volume 43

Overview: Pharmacokinetic Drug–Drug Interactions
Albert P. Li and Malle Jurima-Romet